Rouault (Ed.)

**Iron-Sulfur Clusters in Chemistry and Biology**

# Also of interest

Handbook of Flavoproteins - Volume 1 Oxidases, Dehydrogenases and Related Systems
Russ Hille, Susan M. Miller, Bruce Palfey (Eds.), 2013
ISBN 978-3-11-026842-3, e-ISBN 978-3-11-026891-1,
Set-ISBN 978-3-11-026892-8

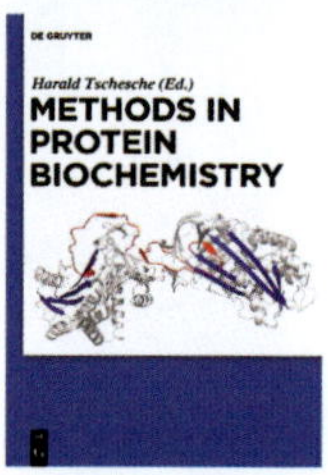


Methods in Protein Biochemistry
Harald Tschesche, 2012
ISBN 978-3-11-025233-0, e-ISBN 978-3-11-025236-1,
Set-ISBN 978-3-11-218934-4

Bioinorganic Chemistry: A Practical Course
Nils Metzler-Nolte, Ulrich Schatzschneider, 2009
ISBN 978-3-11-020954-9, e-ISBN 978-3-11-021136-8

---

BioInorganic Reaction Mechanisms
Ivana Ivanovic-Burmazovic (Editor-in-Chief)
ISSN 2191-2491

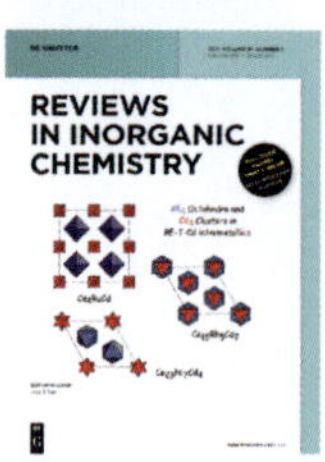


Reviews in Inorganic Chemistry
Axel Schulz (Editor-in-Chief)
ISSN 2191-0277

www.degruyter.com

# Iron-Sulfur Clusters in Chemistry and Biology

Edited by
Tracey A. Rouault

DE GRUYTER

**Editor**
Tracey A. Rouault
9000 Rockville Pike
Bethesda, MD 20892
USA

ISBN 978-3-11-030832-7
e-ISBN 978-3-11-030842-6
Set-ISBN 978-3-11-030843-3

*Library of Congress Cataloging-in-Publication data*
A CIP catalog record for this book has been applied for at the Library of Congress.

*Bibliographic information published by the Deutsche Nationalbibliothek*
The Deutsche Nationalbibliothek lists this publication in the Deutsche Nationalbibliografie; detailed bibliographic data are available in the Internet at http://dnb.dnb.de



Typesetting: Compuscript Limited, Shannon, Ireland
Printing and binding: CPI books GmbH, Leck
Cover image: Klemmer/Getty Images

∞ Printed on acid-free paper
Printed in Germany
www.degruyter.com

MIX
Papier aus verantwortungsvollen Quellen
FSC® C003147
FSC www.fsc.org

# Preface

Iron-sulfur (Fe-S) clusters are versatile prosthetic groups that enable their associated proteins to perform numerous functions, ranging from electron transport to substrate ligation, to structural support, and to DNA repair. Fe-S proteins did not become a focus of research until the late 1950s, when spectroscopy techniques evolved sufficiently to identify features that were specific for Fe-S clusters. Initially identified in mammalian succinate dehydrogenase, Fe-S clusters were subsequently found in numerous bacterial proteins that performed complex functions, including nitrogenase, which transforms atmospheric nitrogen into ammonia, generating an accessible source of nitrogen for synthesis of proteins and nucleic acids. Understanding how Fe-S clusters and proteins work has occupied many scientists for decades, and important breakthroughs regarding the mechanisms of nitrogenase and hydrogenase have occurred in just the last few years.

Not only is it a challenge to understand how Fe-S proteins work, but it is also a challenge to understand how Fe-S clusters are synthesized and inserted into Fe-S proteins in living organisms. Studies originally performed in bacterial model systems have revealed basic mechanisms of biogenesis that are conserved in all the kingdoms of life. Moreover, it has become apparent that flaws in the Fe-S assembly process cause several human diseases. As a result, biomedical researchers working on the pathophysiology of rare diseases such as Friedreich ataxia have begun attending conferences at which chemists and physicists discuss Fe-S research based on complex spectroscopic studies and computational analyses. Researchers from different ends of the spectrum have struggled to bridge the large gap between the physics and chemistry of Fe-S clusters and the important biological questions associated with their functions.

Despite a growing need for cross-disciplinary communication, there has been no single book devoted to Fe-S proteins that provided a basic and broad overview of the subject as it has evolved over the last several decades. This book was borne out of a desire to make the subject of Fe-S proteins more accessible by including a short history of Fe-S research, chapters that highlight the unique chemistry of Fe-S clusters and techniques important in analysis, and reviews from leading researchers on well-known Fe-S proteins such as nitrogenase and hydrogenase. In addition, numerous chapters focus on Fe-S synthesis and regulation in model organisms and in mammalian biogenesis, DNA metabolism, and human disease. Concluding with a discussion on the potential role of Fe-S clusters in capturing reducing power and contributing to the origin of life on earth, the final chapter touches on questions about how metabolic pathways initially developed.

I am indebted to my many outstanding and generous colleagues, who spent considerable time and effort in writing the chapters in this book. I hope that this book will be useful to those interested in the subject of Fe-S from many different perspectives and that researchers from related disciplines will gain a greater sense for the context of their own work.

I want to thank Stephanie Dawson, who perceived that there was an unmet intellectual need and initiated this project while she was an editor at De Gruyter. I also thank Julia Lauterbach of De Gruyter for her tireless support and guidance in turning this book into a reality. My family and friends graciously supported me when I needed time to work on the project known to them as "the book", and I am thankful for their help.

Tracey A. Rouault

# Tracey A. Rouault biography

Tracey A. Rouault is a leading researcher in the area of mammalian iron-sulfur proteins, an area she began to pursue after discovering an important role for an iron-sulfur protein in the regulation of mammalian iron metabolism. She received a degree in Biology from Yale College and an MD degree from Duke University Medical School, where she completed her training in internal medicine. She completed a medical fellowship at the National Institutes of Health in Bethesda, Maryland, and has since focused on the regulation of mammalian iron metabolism and its relationship to human diseases. Her main interests include elucidating mechanisms of mammalian iron-sulfur cluster biogenesis and exploring the pathophysiology of diseases related to ineffective iron-sulfur cluster biogenesis, several hematologic disorders, genetic cancer syndromes, and neurodegenerative diseases. Her early research in the role of iron-sulfur proteins in regulation led to a productive collaboration with Helmut Beinert, a researcher responsible for numerous ground-breaking advances related to iron-sulfur proteins. She has also collaborated with Richard Holm, whose pioneering work led to the inorganic synthesis of numerous iron-sulfur clusters and revealed that many properties of iron-sulfur proteins derive from intrinsic features of their iron-sulfur clusters. She is an active member of the rapidly growing iron-sulfur protein research community.

# Contents

# Contributing authors

**Francesco Bonomi**
DeFENS
University of Milan
Milan, Italy
e-mail: francesco.bonomi@unimi.it
chapter 1

**Toshiko Ichiye**
Department of Chemistry
Georgetown University
Washington, DC, USA
e-mail: ti9@georgetown.edu
chapter 2

**Michael Hendrich**
Department of Chemistry
Carnegie Mellon University
4400 5th Avenue
Pittsburgh, PA 15213-3890, USA
e-mail: hendrich@cmu.edu
chapter 3

**Doros T. Petasis**
Department of Physics
Allegheny College
520 N. Main Street
Meadville, PA 16335, USA
e-mail: dpetasis@allegheny.edu
chapter 3

**Mrinmoy Chakrabarti**
Department of Chemistry
Texas A&M University
College Station, TX 77843-3255, USA
e-mail: mrinmoy.chak@chem.tamu.edu
chapter 4

**Paul Lindahl**
Department of Chemistry
Texas A&M University
College Station, TX 77843-3255, USA
e-mail: lindahl@chem.tamu.edu
chapter 4

**Yilin Hu**
Department Molecular Biology and Biochemistry
University of California, Irvine
Irvine, CA 92697-3900, USA
e-mail: yilinh@uci.edu
chapter 5

**Markus Ribbe**
Department Molecular Biology and Biochemistry
University of California, Irvine
Irvine, CA 92697-3900, USA
e-mail: mribbe@uci.edu
chapter 5

**Susana L. A. Andrade**
Institut für organische Chemie und Biochemie
Universität Freiburg
Freiburg, Germany
chapter 6

**Oliver Einsle**
Institut für organische Chemie und Biochemie
Universität Freiburg
Freiburg, Germany
e-mail: einsle@bio.chemie.uni-freiburg.de
chapter 6

**Thomas Spatzal**
Institut für organische Chemie und Biochemie
Universität Freiburg
Freiburg, Germany
chapter 6

**Joe Jarrett**
Department of Chemistry
University of Hawai'i at Manoa
Honolulu, HI 96822, USA
e-mail: jtj@hawaii.edu
chapter 7

**Russ Hille**
Department of Biochemistry
University of California
Riverside, CA 92521, USA
e-mail: russ.hille@ucr.edu
chapter 8

**Squire J. Booker**
Department of Chemistry
Pennsylvania State University
302 Chemistry Building
University Park, PA 16802, USA
e-mail: sjb14@psu.edu
chapter 9

**Nicholas D. Lanz**
Departments of Biochemistry and Molecular Biology
The Pennsylvania State University
University Park, PA 16802, USA
e-mail: ndl111@psu.edu
chapter 9

**Juan C. Fontecilla-Camps**
Metalloproteins Unit
Institut de Biologie Structurale J.P. Ebel
Université Grenoble-Alpes, Commissariat à l'Energie Atomique and Centre National de la Recherche Scientifique
F-38027 Grenoble, France
e-mail: juan.fontecilla@ibs.fr
chapter 10

**Yvain Nicolet**
Metalloproteins Unit
Institut de Biologie Structurale J.P. Ebel
Université Grenoble-Alpes, Commissariat à l'Energie Atomique and Centre National de la Recherche Scientifique
F-38027 Grenoble, France
e-mail: yvain.nicolet@ibs.fr
chapter 10

**Dennis R. Dean**
Virginia Polytechnic Institute and State University
Blacksburg, VA, USA
e-mail: deandr@vt.edu
chapter 11

**Patricia Dos Santos**
Department of Chemistry
Wake Forest University
Winston-Salem, NC, USA
e-mail: dossanpc@wfu.edu
chapters 11 and 14

**F. Wayne Outten**
Department of Chemistry and Biochemistry
University of South Carolina
Columbia, SC, USA
e-mail: outtenf@mailbox.sc.edu
chapter 12

**Patricia Kiley**
Department of Biomolecular Chemistry
University of Wisconsin
Madison, WI, USA
e-mail: pjkiley@wisc.edu
chapter 13

**Erin L. Mettert**
Department of Biomolecular Chemistry
University of Wisconsin
Madison, WI, USA
e-mail: elmettert@wisc.edu
chapter 13

**Nicole T. Perna**
Department of Genetics
University of Wisconsin
Madison, WI, USA
e-mail: ntperna@wisc.edu
chapter 13

**Andrew Dancis**
Biomedical Graduate Studies
University of Pennsylvania
Philadelphia, PA, USA
e-mail: adancis@mail.med.upenn.edu
chapter 15

**Debkumar Pain**
Department of Pharmacology and Physiology
New Jersey Medical School
Rutgers University
Newark, NJ 07101, USA
e-mail: painde@njms.rutgers.edu
chapter 15

**Caryn E. Outten**
Department of Chemistry and Biochemistry
University of South Carolina
Columbia, SC 29208, USA
e-mail: outten@mailbox.sc.edu
chapter 16

**Tracey A. Rouault**
9000 Rockville Pike
Bethesda, MD 20892, USA
e-mail: traceyrouault@me.com
chapters 1 and 17

**Wing Hang Tong**
National Institutes of Health
Bethesda, MD 20892, USA
e-mail: tongw@mail.nih.gov
chapter 18

**Silke Leimkühler**
Molekulare Enzymologie
Universität Potsdam
Karl-Liebknecht-Strasse 24-25
14476 Potsdam, Germany
Potsdam, Germany
e-mail: sleim@uni-potsdam.de
chapter 19

**Kerstin Gari**
Institute of Molecular Cancer Research
University of Zurich
Winterthurerstrasse 190
8057 Zurich, Switzerland
e-mail: gari@imcr.uzh.ch
chapter 20

**Roland Lill**
Institut für Zytobiologie
Philipps-Universität Marburg
Robert-Koch Strasse 6
35032 Marburg, Germany
e-mail: Lill@staff.uni-marburg.de
chapter 21

**Marta Uzarska**
Institut für Zytobiologie
Philipps-Universität Marburg
Robert-Koch Strasse 6
35032 Marburg, Germany
chapter 21

**James Wohlschlegel**
Department of Biological Chemistry
David Geffen School of Medicine
University of California
Los Angeles, CA 90095, USA
e-mail: jwohl@mednet.ucla.edu
chapter 21

**Hong Ye**
Key Laboratory of Plant Resources
Conservation and Sustainable Utilization
South China Botanical Garden
Chinese Academy of Sciences
Guangzhou 510650, China
e-mail: hye@scib.ac.cn
chapter 22

**Michael W. W. Adams**
Department of Biochemistry and
Molecular Biology
University of Georgia
Athens, GA, USA
chapter 23

**Joan B. Broderick**
Department of Chemistry and
Biochemistry
Montana State University
Bozeman, MT, USA
chapter 23

**Eric S. Boyd**
Department of Microbiology
Montana State University
Bozeman, MT, USA
e-mail: eboyd@montana.edu
chapter 23

**John Peters**
Department of Chemistry and
Biochemistry
Montana State University
Bozeman, MT, USA
e-mail: john.peters.cab@gmail.com
chapter 23

**Gerrit J. Schut**
Department of Biochemistry and
Molecular Biology
University of Georgia
Athens, GA, USA
chapter 23

**Eric M. Shepard**
Department of Chemistry and
Biochemistry
Montana State University
chapter 23
Bozeman, MT, USA

# 1 Iron-sulfur proteins: a historical perspective

Francesco Bonomi and Tracey A. Rouault

## 1.1 Framing the scene

Although iron-sulfur proteins (Fe-S) are now recognized as being pervasive throughout all three kingdoms of life, they were not among the prosthetic groups that were recognized or studied during the first half of the twentieth century [1]. One reason for their relatively late appearance on the research scene was that they often lacked a distinctive visible color that commanded attention, unlike proteins that incorporate a heme cofactor or other metallic cofactors. Furthermore, Fe-S centers are often destabilized by exposure to oxygen, and working with Fe-S proteins requires special techniques for measuring iron and sulfur and equipment, such as anaerobic hoods, electron paramagnetic resonance (EPR) machinery and Mössbauer spectroscopy, in addition to the more commonly used ultraviolet and visible spectrophometric methods. Upon consideration of the importance of new instrumentation and techniques for the discovery and characterization of Fe-S centers, Helmut Beinert [2] concluded in a retrospective about Fe-S research that, "there was scarcely a way that these discoveries could have been made earlier."

## 1.2 The early days of "nonheme iron"

In 1951, researchers observed that a dark brown fraction from ammonium sulfate fractionation of leaf extracts was able to catalyze reduction of met-hemoglobin [3]. This report likely represents one of the earliest mentions of Fe-S activity, but no further insight into the nature of these reducing proteins was reported. In the ensuing years, from 1956 to 1958, tightly bound nonheme iron was reproducibly detected in animal tissues, particularly in lysates from their mitochondria [4].

A burst of knowledge was unleashed by the use of EPR imaging techniques, which were developed to assess materials that contained unpaired electrons during the 1940s and 1950s and became commercially available in 1956 [2]. EPR signals emanating from this nonheme iron were first detected in succinate dehydrogenase (SDH) [5, 6]. The development of sensitive microassays for iron and sulfide contents indicated that SDH also contained labile sulfide [7, 8]. Mitochondria and chloroplasts were the subject of intensive investigations by many gifted scientists, providing an ideal "testing ground" for the application of these novel techniques. This combination led to the first EPR spectra of FeS components in the respiratory chain and to a first coarse outline of the essential participation of these redox carriers to the electron flow within the system (Fig. 1.1) [9, 10].

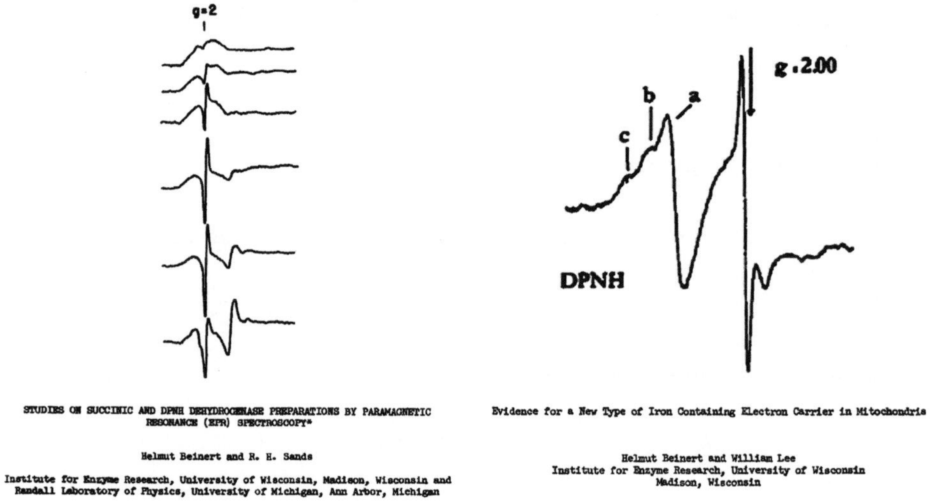


**Fig. 1.1:** The first spectra of Fe-S proteins obtained by EPR and published in 1960 and 1961. (Modified from Beinert H, Sands RH, *Biochem Biophys Res Commun*, 3, 41–46, 1960, and Beinert H, Lee W, *Biochem Biophys Res Commun*, 5, 40–45, 1961.)

Shortly after the pioneering work on mitochondrial nonheme iron proteins was performed, some of the most stable and abundant FeS proteins, namely clostridial (4Fe-4S) ferredoxins, were isolated and named by a group led by Len Mortenson, then at the DuPont Co. [11]. Len Mortenson was later instrumental in building the Chemistry Department at the University of Georgia (USA), laying the groundwork for the development of the Center for Metalloenzyme Studies, which grew during the ensuing years.

Multiple other types of FeS proteins were discovered, including proteins from the anaerobic photosynthetic purple sulfur bacterium *Chromatium vinosum* in which a nonstandard redox form of a [4Fe-4S] cluster was identified [12]. Others were found in which two histidine residues replaced half of the standard cysteines as iron ligands [13] and others in which a single iron atom was tetrahedrally coordinated by four cysteines in the absence of additional "inorganic sulfide," known as rubredoxin [14]. The relevance of all these contributions (and of many more that cannot be mentioned here for lack of space) found expression in the milestone book *Non Heme Iron Proteins*, which was edited by Anthony San Pietro and appeared in 1965.

## 1.3 Of proteins and analogues

No computers were involved when a structural model for a [2Fe-2S] cluster was proposed as early as 1966 based on the interpretation of the g = 1.94 EPR signal of plant-type ferredoxin [15]. Isotopic substitution and analysis of the hyperfine splitting pattern in EPR spectra later confirmed that the proposed structure was correct and that the two sulfur (or selenium) atoms were indeed indistinguishable [16]. In this regard,

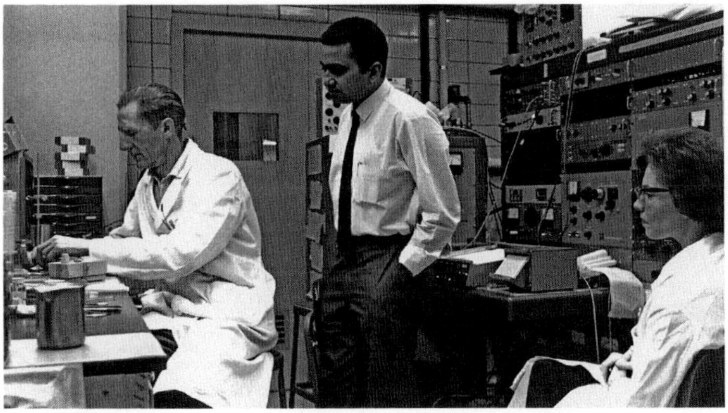

**Fig. 1.2:** A picture of Helmut Beinert at work (left) with two colleagues. (Courtesy of the University of Wisconsin and Dr. Elizabeth Craig.)

it is again worth remembering that the FeS proteins field has represented a very significant environment in which methodologies that work at the interface among physics, chemistry, and biochemistry have been deployed (Fig. 1.2). These methodologies cover the whole gamut of the electromagnetic spectrum, from microwaves to X-rays and beyond.

Confirming these hypothetical structures by X-ray crystallography required several years. Crystals of clostridial-type 2[4Fe-4S] ferredoxins had been obtained as early as 1966 [17], but it was not until the 1970s that crystallographic structures became available. The Lovenberg group presented a structure of rubredoxin [18], and this was quickly followed by reports on the structure of HiPIP [19] and of clostridial-type ferredoxin [20]. The structure of a plant-type ferredoxin [21] was solved and later supported by data from nuclear magnetic resonance (NMR) spectroscopy [22]. Applications of NMR to this particular field have been important because of the intrinsic difficulties associated with characterizing paramagnetic centers [23].

After those pioneering efforts, numerous structures were solved at high resolution. The complexity of the investigated systems also grew progressively from the 1960s to the present. Studies progressed from single-iron rubredoxins and two-iron ferredoxins to incredibly complex flavo-molybdo-iron proteins, often made up of several subunits, and included cases where metal clusters shared ligands from separate polypeptides or included non-amino acid ligands. The first proposed structure of nitrogenase [24] was an exciting milestone, and the intricacy of the chemistry and structural biochemistry of these complicated systems is still a subject of intense research interest.

In the 1970s, chemists were able to synthesize and characterize a number of structural analogues of FeS clusters at the atomic level [25]. The relative stability of these clusters as a function of their nuclearity and of the nature, size, and reactivity of terminal ligands was investigated by the Holm (Fig. 1.3) group and by many others (Fig. 1.3). These collective efforts led to the elucidation of the sequence of individual reactions resulting in the self-assembly of the clusters (for a comprehensive review of

**Fig. 1.3:** Richard Holm, with valued colleagues, synthesized most types of FeS clusters *in vitro* and thereby proved that FeS clusters could interconvert and assemble independently of protein structure. (Courtesy of Dr. Richard Holm.)

30 years of progress, see [26]). The original work was carried out in nonaqueous systems, but shortly afterward, it was shown that essentially the same chemistry worked in micellar systems and aqueous buffers as well as with other metals using enzymes to catalyze some individual steps of the overall chain of events (for example, [27]). The ability to reconstruct a replicate of various FeS centers found in proteins proved that the protein structure was unnecessary for the sites' existence [1] and supported the concept that FeS centers are modular centers that have an unusual ability to interconvert between species, allowing 2[2Fe-2S] clusters to readily form a single [4Fe-4S] complex [26] (Fig. 1.4). Moreover, FeS clusters proved to be more robust and cofactor-like than had been originally thought [1], and the chemical characteristics of sulfur were recognized for their unique contribution to the chemistry of FeS clusters [28].

In short, roughly one decade after the San Pietro book mentioned in Section 1.2, the knowledge in the field required had grown to the point that a two-volume book (properly titled *Iron-Sulfur Proteins* and edited by Walt Lovenberg) was not sufficient and was followed by a third volume several years later. The wealth of information within the book was great, and a wide variety of approaches and techniques originating from chemistry, physics, and biochemistry were put to synergistic use to clarify many puzzling issues. An important breakthrough occurred when researchers recognized that particular EPR signals emanated from two interacting iron ions, a ferrous ion and a ferric ion (reviewed in [2]), rather than from a single metal site (Fig. 1.3).

By the mid-1970s, there was enough information on the structural features of FeS proteins and on their distribution throughout the kingdoms of life to begin to consider when FeS proteins first appeared in life. Studies in molecular evolution led to increasing awareness that FeS structures and the proteins around them likely had been around since the earliest days of anaerobic life on this planet, and these structures may have been merged, reshuffled, and repurposed through fusion and duplication [29].

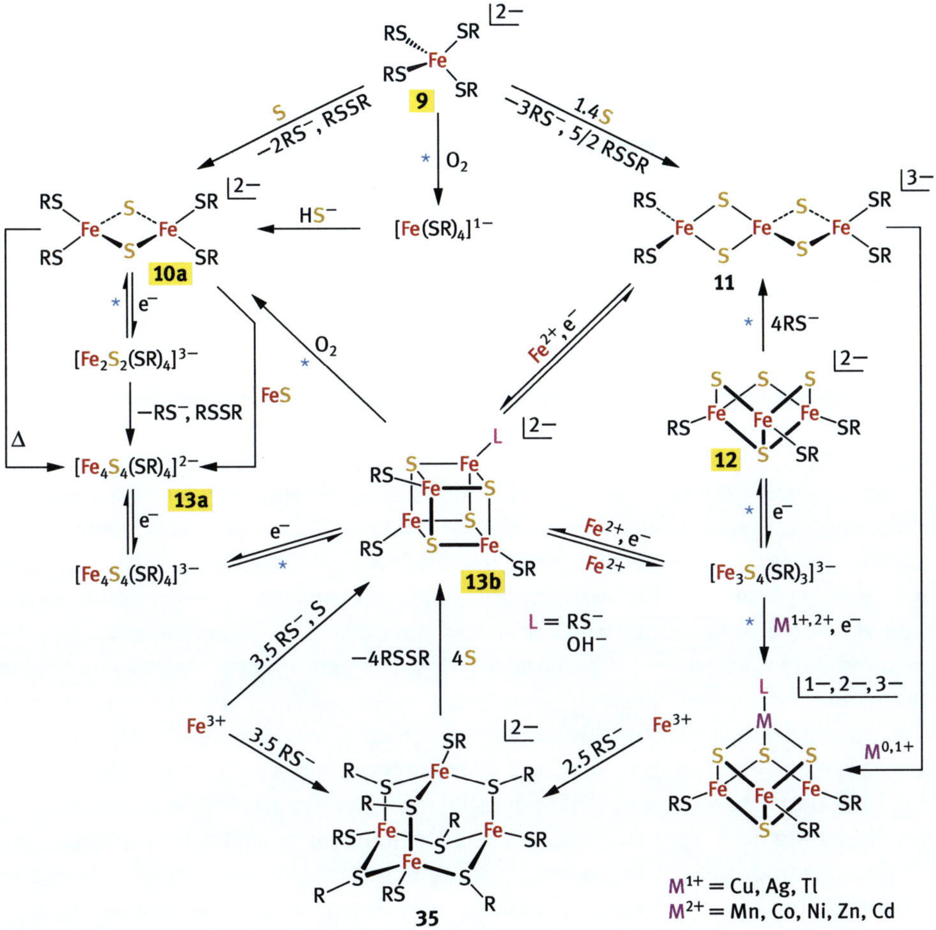


**Fig. 1.4:** Synthetic routes to assembly of FeS analogues. The structures most often encountered in proteins are highlighted. FeS clusters are highly interchangeable, and the integrity of FeS clusters does not depend on protein scaffolds. (Redrawn from Rao VP, Holm RH, *Chem Rev*, 104, 527–559, 2004.)

Thus, proteins might have evolved in the primordial environment around submarine volcanic vents and incorporated FeS centers into fundamental biochemical processes.

These concepts, along with the apparent ease of self-assembly of FeS structures and of their relative tolerance toward various types of ligands resulted in hypotheses based on the supposition that life arose in an "iron-sulfur" world. Biochemistry, as we know it, was hypothesized to have taken place first on the positively charged surface of pyrite crystals [30], and genuine FeS structures (not dissimilar from those "captured" by protein thiolates in a later stage of evolution) could have been responsible of providing the earliest catalysts in a nonprotein world, perhaps in separate compartments, which might be regarded as the earliest protocells [31]. Although still much debated [32, 33], these hypotheses continue to fascinate because they address critical questions about how various life forms may capture and store energy from the environment.

## 1.4 Beyond electron shuttles

The ability of FeS centers to accept and donate single electrons had led to the focus on their roles as electron shuttles. Helmut Beinert was once again among the first to recognize that a non-redox enzyme – namely, mammalian mitochondrial aconitase – was an iron-sulfur protein and to understand that transition from the non-active to the active form of the enzyme required conversion of a [3Fe-4S] into a [4Fe-4S] cluster [34]. Assessing this unequivocally took a rather unique combination of spectroscopic skills and analytical accuracy. Helmut Beinert had both, as testified by the back-to-back reports that appeared in 1983 [35, 36]. On a more personal note, it is worth remembering that, during the celebration of Helmut's ninety-second birthday in Madison in 2005, the distinguished spectroscopist Eckard Munck spoke about having introduced the concept of "millibeinerts" to score the reliability and accuracy of measurements that were performed in his own laboratory (FB, personal recollection). The conversion of a [3Fe-4S] cluster into a [4Fe-4S] cluster in mitochondrial aconitase apparently required only the addition of iron and a reducing agent, and the reverse conversion appeared to occur spontaneously when iron was not present. In fact, the fourth labile iron was involved in the direct ligation of the substrate, citrate or isocitrate [37]. Thus, a new role for FeS in ligating enzyme substrates was discovered.

In the early 1990s, yet another potential role of FeS proteins as sensors emerged when investigators were studying the regulation of intracellular iron metabolism. The mammalian protein responsible for regulating the translation of ferritin and stabilizing the transcript that encodes the transferrin receptor, known as the iron-responsive element binding protein (IRE-BP), was identified, and it unexpectedly had a high sequence similarity to mitochondrial aconitase [38], which had been crystallized and further characterized [39]. Mammalian cells had been known to possess a second aconitase, which was in the cytosol, and purification of the aconitase activity and peptide sequencing revealed that the IRE-BP, which was an apoprotein [40], and cytosolic aconitase with its [4Fe-4S] cluster were identical proteins [41]. To encompass the two activities of the proteins, it was renamed iron regulatory protein 1, and multiple studies revealed that the key to the transition from functioning as an active aconitase to an iron regulatory protein involved the loss of the [4Fe-4S] cluster (reviewed in [42, 43]).

These new concepts and the underlying evidence were discussed in a memorable meeting held in Konstanz in 1994, to celebrate Helmut's eightieth birthday and to present all these "novel" breakthroughs. Not long after, another example in which the FeS cluster served as a sensor was uncovered in bacteria in studies of fumarate nitrate reductase, where a labile Fe-S cluster was recognized as the key to sensing oxygen and remodeling transcription to direct a switch from aerobic to anaerobic metabolism (reviewed in [44]).

Scores of non-redox functions for Fe-S proteins accumulated over the years. In a review that appeared in 1997, the accumulated evidence was summarized by stating that, “Iron-sulfur clusters now rank with such biological prosthetic groups as hemes and flavins in pervasive occurrence and multiplicity of function” [1]. New roles continue to emerge, and Fe-S proteins are now recognized to play an important role in DNA metabolism and maintenance of DNA integrity [45, 46] and in human diseases [47].

## 1.5 How are FeS clusters synthesized in cells?

Despite the “self-assembling” nature of Fe-S clusters discussed earlier, it was not clear how cells could synthesize Fe-S clusters without encountering problems with the cytotoxicity of sulfide and with the insolubility of iron(III) sulfides. Protein-bound zero-valence sulfur had been found as a cysteine-bound persulfide at the active site of sulfurtransferases, and this form of “elemental” sulfur was demonstrated to undergo easy reduction to sulfide by addition of suitable thiols [48]. Bovine liver rhodanese was recognized as the epitome of this class of enzymes, and in the mid-1970s, it was found that liver rhodanese could rescue damaged Fe-S clusters in mitochondrial SDH by replenishing some of the missing cluster sulfide [49] or serve as a source of cluster sulfide in Fe-S proteins [27] and in their chemical analogues [27]. Nevertheless, rhodanese was known to be absent from scores of FeS-rich organisms, making it difficult to consider that rhodanese activity was of general relevance [50]. Indeed, evidence was accumulating that cysteine was a likely source of sulfide for the biogenesis of Fe-S structures in chloroplasts [51].

Advances in genetics, sequencing, biochemistry, and biophysics led to the discovery of the bacterial genes involved in nitrogen fixation (the *nif* gene cluster) in *Azotobacter vinelandii* [52] and to the identification of a cysteine desulfurase as the essential sulfide-generating component of the system (reviewed by [53]). Later, the isc (iron-sulfur cluster assembly) operon used for the general synthesis of Fe-S proteins was discovered in *A. vinelandii* [54], in other bacteria [55], in yeast model systems (see the review by [56]), and in mammals [57]. The role of scaffold proteins as intermediates in the assembly process was discovered [58], along with the importance of a chaperone-co-chaperone pair for cluster delivery [59, 60] and proposed roles for intermediate scaffolds [61]. Studies of Fe-S proteins and chemistry are ongoing, the field is vibrant, and unexpectedly, mutations in FeS assembly proteins have proven to be the cause of several important human diseases, including Friedreich ataxia, ISCU myopathy, a rare type of sideroblastic anemia, and lactic acidosis in infants (reviewed in [47]).

Indeed, how these FeS proteins work and are generated is the subject of many excellent ongoing research, which will be described in chapters that follow in this book. Topics range from nitrogen fixation, to hydrogenase function, to plant growth, and to the origin of life itself, with numerous implications for industrial processes, food production, and for human disease.

## Acknowledgment

We thank Jacques Meyer for generously providing his overview of the history of iron-sulfur research in his tribute to Helmut Beinert.

## References

[1] Beinert H, Holm RH, Munck E. Iron-sulfur clusters: nature's modular, multipurpose structures. Science 1997;277:653–9.
[2] Beinert H. Spectroscopy of succinate dehydrogenases, a historical perspective. Biochim Biophys Acta 2002;1553:7–22.
[3] Davenport HE, Hill R, Whatley FR. A natural factor catalyzing reduction of methaemoglobin by isolated chloroplasts. Proc R Soc Lond B Biol Sci 1952;139:346–58.
[4] Crane FL, Hatefi Y, Lester RL, Widmer C. Isolation of a quinone from beef heart mitochondria. Biochim. Biophys. Acta 1957;25:220–221.
[5] Beinert H, Sands RH. Studies on succinic and DPNH dehydrogenase preparations by paramagnetic resonance (EPR) spectroscopy. Biochem Biophys Res Commun 1960;3:41–6.
[6] Sands RH, Beinert H. Studies on mitochondria and submitochondrial particles by paramagnetic resonance (EPR) spectroscopy. Biochem Biophys Res Commun 1960;3:47–52.
[7] Massey V. Studies on succinic dehydrogenase. VII. Valency state of the iron in beef heart succinic dehydrogenase. J Biol Chem 1957;229:763–70.
[8] Brumby PE, Miller RW, Massey V. The content and possible catalytic significance of labile sulfide in some metalloflavoproteins. J Biol Chem 1965;240:2222–8.
[9] Beinert H, Lee W. Evidence for a new type of iron containing electron carrier in mitochondria. Biochem Biophys Res Commun 1961;5:40–5.
[10] Beinert H, Griffiths DE, Wharton DC, Sands RH. Properties of the copper associated with cytochrome oxidase as studied by paramagnetic resonance spectroscopy. J Biol Chem 1962;237:2337–46.
[11] Mortenson LE, Valentine RC, Carnahan JE. An electron transport factor from Clostridium pasteurianum. Biochem Biophys Res Commun 1962;7:448–52.
[12] Dus K, De Klerk H, Sletten K, Bartsch RG. Chemical characterization of high potential iron proteins from Chromatium and Rhodopseudomonas gelatinosa. Biochim Biophys Acta 1967;140:291–311.
[13] Rieske JS, Hansen RE, Zaugg WS. Studies on the electron transfer system. 58. Properties of a new oxidation-reduction component of the respiratory chain as studied by electron paramagnetic resonance spectroscopy. J Biol Chem 1964;239:3017–22.
[14] Lovenberg W, Sobel BE. Rubredoxin: a new electron transfer protein from Clostridium pasteurianum. Proc Natl Acad Sci USA 1965;54:193–9.
[15] Brintzinger H, Palmer G, Sands RH. On the ligand field of iron in ferredoxin from spinach chloroplasts and related nonheme iron enzymes. Proc Natl Acad Sci USA 1966;55:397–404.

[16] Orme-Johnson WH, Hansen RE, Beinert H, et al. On the sulfur components of iron-sulfur proteins. I. The number of acid-labile sulfur groups sharing an unpaired electron with iron. Proc Natl Acad Sci USA 1968;60:368–72.
[17] Lovenberg W, Buchanan BB, Rabinowitz JC. Studies on the chemical nature of clostridial ferredoxin. J Biol Chem 1963;238:3899–913.
[18] Herriott JR, Sieker LC, Jensen LH, Lovenberg W. Structure of rubredoxin: an x-ray study to 2.5 Å resolution. J Mol Biol 1970;50:391–406.
[19] Carter CWJ, Freer ST, Xuong NH, Alden RA, Kraut J. Structure of the iron-sulfur cluster in the Chromatium iron protein at 2.25 Angstrom resolution. Cold Spring Harb Symp Quant Biol 1972;36:381–5.
[20] Sieker LC, Adman E, Jensen LH. Structure of the Fe-S complex in a bacterial ferredoxin. Nature 1972;235:40–2.
[21] Tsukihara T, Homma K, Fukuyama K, et al. Preliminary x-ray diffraction studies on a [4Fe-4S] ferredoxin from Bacillus thermoproteolyticus. J Mol Biol 1981;152:821–3.
[22] Im SC, Liu G, Luchinat C, Sykes AG, Bertini I. The solution structure of parsley [2Fe-2S] ferredoxin. Eur J Biochem 1998;258:465–77.
[23] Bertini I, Luchinat C, Parigi G, Pierattelli R. NMR spectroscopy of paramagnetic metalloproteins. Chembiochem 2005;6:1536–49.
[24] Chan MK, Kim J, Rees DC. The nitrogenase FeMo-cofactor and P-cluster pair: 2.2 Å resolution structures. Science 1993;260:792–4.
[25] Orme-Johnson WH, Holm RH. Identification of iron-sulfur clusters in proteins. Methods Enzymol 1978;53:268–74.
[26] Rao VP, Holm RH. Synthetic analogues of the active sites of iron-sulfur proteins. Chem Rev 2004;104:527–59.
[27] Bonomi F, Pagani S, Kurtz DMJ. Enzymic synthesis of the 4Fe-4S clusters of Clostridium pasteurianum ferredoxin. Eur J Biochem 1985;148:67–73.
[28] Beinert H. A tribute to sulfur. Eur J Biochem 2000;267:5657–64.
[29] Meyer J. Iron-sulfur protein folds, iron-sulfur chemistry, and evolution. J Biol Inorg Chem 2008;13:157–70.
[30] Wachtershauser G. Before enzymes and templates: theory of surface metabolism. Microbiol Rev 1988;52:452–84.
[31] Kaschke M, Russell MJ, Cole WJ. [FeS/FeS2], a redox system for the origin of life (some experiments on the pyrite-hypothesis). Orig Life Evol Biosph 1994;24:43–56.
[32] De Duve C. The other revolution in the life sciences. Science 2013;339:1148.
[33] Russell MJ, Nitschke W, Branscomb E. The inevitable journey to being. Philos Trans R Soc Lond B Biol Sci 2013;368:20120254.
[34] Kent TA, Dreyer JL, Kennedy MC, et al. Mossbauer studies of beef heart aconitase: evidence for facile interconversions of iron-sulfur clusters. Proc Natl Acad Sci USA 1982;79:1096–100.
[35] Emptage MH, Dreyers JL, Kennedy MC, Beinert H. Optical and EPR characterization of different species of active and inactive aconitase. J Biol Chem 1983;258:11106–11.
[36] Kennedy MC, Emptage MH, Dreyer JL, Beinert H. The role of iron in the activation-inactivation of aconitase. J Biol Chem 1983;258:11098–105.
[37] Beinert H, Kennedy MC. 19th Sir Hans Krebs lecture. Engineering of protein bound iron-sulfur clusters. A tool for the study of protein and cluster chemistry and mechanism of iron-sulfur enzymes. Eur J Biochem 1989;186:5–15.
[38] Rouault TA, Stout CD, Kaptain S, Harford JB, Klausner RD. Structural relationship between an iron-regulated RNA-binding protein (IRE-BP) and aconitase: functional implications. Cell 1991;64:881–3.
[39] Robbins AH, Stout CD. Structure of activated aconitase: formation of the [4Fe-4S] cluster in the crystal. Proc Natl Acad Sci USA 1989;86:3639–43.

[40] Haile DJ, Rouault TA, Harford JB, et al. Cellular regulation of the iron-responsive element binding protein: disassembly of the cubane iron-sulfur cluster results in high-affinity RNA binding. Proc Natl Acad Sci USA 1992;89:11735–9.
[41] Kennedy MC, Mende-Mueller L, Blondin GA, Beinert H. Purification and characterization of cytosolic aconitase from beef liver and its relationship to the iron-responsive element binding protein. Proc Natl Acad Sci USA 1992;89:11730–4.
[42] Beinert H, Kennedy MC, Stout CD. Aconitase as iron-sulfur protein, enzyme, and iron-regulatory protein. Chem Rev 1996;96:2335–74.
[43] Rouault TA. The role of iron regulatory proteins in mammalian iron homeostasis and disease. Nat Chem Biol 2006;2:406–14.
[44] Kiley PJ, Beinert H. The role of Fe-S proteins in sensing and regulation in bacteria. Curr Opin Microbiol 2003;6:181–5.
[45] Stehling O, Vashisht AA, Mascarenhas J, et al. MMS19 assembles iron-sulfur proteins required for DNA metabolism and genomic integrity. Science 2012;337:195–9.
[46] Gari K, Leon Ortiz AM, Borel V, et al. MMS19 links cytoplasmic iron-sulfur cluster assembly to DNA metabolism. Science 2012;337:243–5.
[47] Rouault TA. Biogenesis of iron-sulfur clusters in mammalian cells: new insights and relevance to human disease. Dis Model Mech 2012;5:155–64.
[48] Pecci L, Pensa B, Costa M, Cignini PL, Cannella C. Reaction of rhodanese with dithiothreitol. Biochim Biophys Acta 1976;445:104–11.
[49] Bonomi F, Pagani S, Cerletti P, Cannella C. Rhodanese-mediated sulfur transfer to succinate dehydrogenase. Eur J Biochem 1977;72:17–24.
[50] Sandberg W, Graves MC, Rabinowitz JC. Role for rhodanese in Fe-S formation is doubtful. Trends Biochem Sci 1987;12:56.
[51] Takahashi Y, Mitsui A, Hase T, Matsubara H. Formation of the iron-sulfur cluster of ferredoxin in isolated chloroplasts. Proc Natl Acad Sci USA 1986;83:2434–7.
[52] Brigle KE, Newton WE, Dean DR. Complete nucleotide sequence of the Azotobacter vinelandii nitrogenase structural gene cluster. Gene 1985;37:37–44.
[53] Peters JW, Fisher K, Dean DR. Nitrogenase structure and function: a biochemical-genetic perspective. Annu Rev Microbiol 1995;49:335–66.
[54] Zheng L, Cash VL, Flint DH, Dean DR. Assembly of iron-sulfur clusters. Identification of an iscSUA-hscBA-fdx gene cluster from Azotobacter vinelandii. J Biol Chem 1998;273:13264–72.
[55] Takahashi Y, Nakamura M. Functional assignment of the ORF2-iscS-iscU-iscA-hscB-hscA-fdx-ORF3 gene cluster involved in the assembly of Fe-S clusters in Escherichia coli. J Biochem 1999;126:917–26.
[56] Lill R, Muhlenhoff U. Maturation of iron-sulfur proteins in eukaryotes: mechanisms, connected processes, and diseases. Annu Rev Biochem 2008;77:669–700.
[57] Ye H, Rouault TA. Human iron-sulfur cluster assembly, cellular iron homeostasis, and disease. Biochemistry 2010;49:4945–56.
[58] Johnson DC, Dean DR, Smith AD, Johnson MK. Structure, function, and formation of biological iron-sulfur clusters. Annu Rev Biochem 2005;74:247–81.
[59] Vickery LE, Cupp-Vickery JR. Molecular chaperones HscA/Ssq1 and HscB/Jac1 and their roles in iron-sulfur protein maturation. Crit Rev Biochem Mol Biol 2007;42:95–111.
[60] Kampinga HH, Craig EA. The HSP70 chaperone machinery: J proteins as drivers of functional specificity. Nat Rev Mol Cell Biol 2010;11:579–92.
[61] Shakamuri P, Zhang B, Johnson MK. Monothiol glutaredoxins function in storing and transporting [Fe2S2] clusters assembled on IscU scaffold proteins. J Am Chem Soc 2012;134:15213–6.

# 2 Chemistry of iron-sulfur clusters

Toshiko Ichiye

## 2.1 Introduction

Fe-S proteins are ubiquitous throughout all living organisms and participate in a wide variety of electron transfer and biosynthetic processes as well as important non-redox catalytic and regulatory functions [1–3]. The abundance of Fe-S redox sites, which characterize Fe-S proteins, may be in part due to the wide bioavailability of iron and sulfur and to their ability to form spontaneously [4, 5]. This has led to speculation that the simplest Fe-S proteins may have been the first electron transfer proteins [6]. In addition, these redox sites have many physical properties that make them unusually efficient for a variety of purposes.

The properties of Fe-S redox sites are governed by both iron and sulfur. Iron is a transition metal that usually occurs in the 2+, 3+, and sometimes 4+ oxidation states and can have different spin states. Sulfur can also occur in states from 2– to 6+ and makes and breaks bonds easily. The simplest Fe-S redox sites consist of 1 to 4 irons bound tetrahedrally by sulfur, which come from cysteinyl residues of the protein as well as inorganic sulfur (Fig. 2.1). Explicitly, these are the $[Fe(Cys)_4]^n$ site with a $[Fe]^{n+4}$ core, the $[Fe_2S_2(Cys)_4]^n$ site with a $[Fe_2S_2]^{n+4}$ core, and the $[Fe_4S_4(Cys)_4]^n$ site with a $[Fe_4S_4]^{n+4}$ core, where $n$ is the net charge of the complex. Because $n$ is of more direct relevance to electron transfer properties, we generally refer to the net charge of the entire redox site. The cores are also sometimes referred to using the notation [*i*Fe-*j*S], where *i* and *j* are the number of irons and inorganic sulfurs, respectively. Simple variations on these small clusters include $[Fe_3S_4(SR)_3]^n$, which has one iron and its cysteinyl ligand removed from the cubane core of $[Fe_4S_4(SR)_4]^n$ [7], and substitution of the cysteine ligands [4, 8]. Other variations include larger clusters, notably the P-cluster of nitrogenase, which contains a $Fe_8S_7$ core that is linked by six cysteines to the protein, and mixed clusters such as the FeMo cofactor also of nitrogenase, which contains a $MoFe_7S_9$ core linked to the protein by one cysteinyl ligand to a core Fe atom and one histidine ligand (to the Mo atom) [9].

Fe-S redox sites have been extensively studied by biomimetic chemistry [10]. Analogues for the Fe-S active sites of proteins mimic the structure and electronic structure of the protein sites to an extent not possible so far with other metal containing prosthetic groups. In addition, because they are stable outside of a protein environment, they can be characterized by numerous experimental techniques, usually much more accurately than within the protein. These studies have led to models for most of the Fe-S active sites found in Fe-S proteins, which have been invaluable in understanding their structure and function. However, although Fe-S clusters will form in certain apo-[Fe-S] proteins *in vitro* if $S^{2-}$ and $Fe^{2+/3+}$ are added, biological cluster assembly requires a complex biosynthetic machinery [5].

(a)

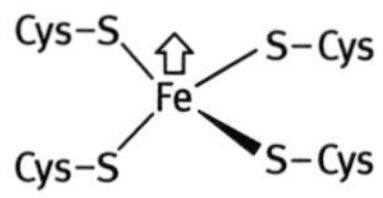


(b)

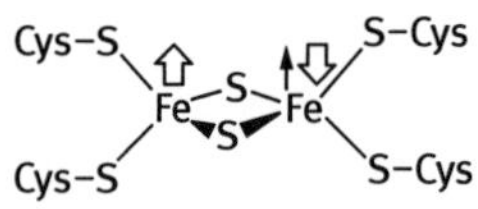


(c)

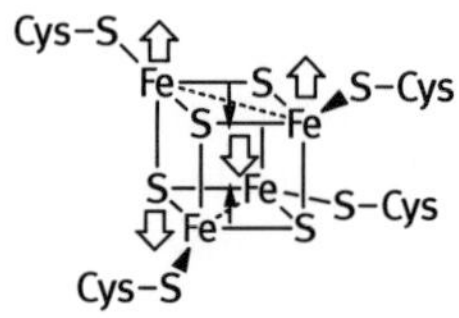


**Fig. 2.1:** (a) $[Fe(SCys)_4]^{2-}$, (b) $[Fe_2S_2(SCys)_4]^{2-}$, and (c) $[Fe_4S_4(SCys)_4]^{2-}$ redox sites. The five majority spins ($S_i$ = 5/2) are denoted by ⇧, and the minority spin ($s_i$ = ½) is denoted by ↑. In (b), the majority spins are antiferromagnetically coupled and the minority spin is localized on the right-hand Fe, whereas in (c), the top and bottom layers each have the majority spins ferromagnetically coupled and the minority spin is delocalized between the two Fe, with the two layers antiferromagnetically coupled.

## 2.2 Electronic structure of Fe-S complexes

The distinctive features of the electronic structure of the Fe-S complexes can be attributed to high-spin irons that are tetrahedrally coordinated by sulfurs. In particular, the properties of the 3d electrons with high-spin state found in transition metals are important. The features are illustrated here in the simple Fe-S complexes but also are relevant to the much larger complexes. Much of the general understanding of the spin-polarized electronic structure can be attributed to Noodleman and Case, which has been reviewed extensively by Noodleman *et al.* [11]. The unusual structural and redox properties of these clusters correlate with three characteristics: strong metal-ligand covalency, metal-metal spin-coupling, and metal-metal spin resonance delocalization; the latter two characteristics apply to clusters that contain more than one iron. Here, the electronic properties that give rise to these characteristics are described in this section, and the consequences of these properties to protein function will be described in Section 2.3.

### 2.2.1 Spin-polarization and strong metal-ligand bonds

The Fe-S complexes are characterized by strong metal-ligand bonds that arise from metal-ligand bonding interactions and strong spin-polarization effects. In addition, the strong spin-polarization effects lead to spin-coupling and spin-delocalization in clusters with more than one iron discussed subsequently. The spin-polarization arises because, in the weak ligand field of the sulfur, the metal $\alpha$ and $\beta$ spins show a strong splitting between the energy levels of majority ($\alpha$) and minority ($\beta$) spins. This strong splitting leads to molecular orbitals (MO) with an "inverted energy level scheme" [12, 13], which has been verified for most Fe-S complexes by optical and photoelectron spectroscopies [13–15]. In this scheme, the majority spin levels are lower in energy than the occupied thiolate ligand 3p levels and the empty minority spin levels lie above both (Fig. 2.2, right), in contrast to the normal-level scheme in which both the filled and empty metal levels lie above all of the filled ligand 3p levels (Fig. 2.2, left). Physically, the greater number (majority) of spin-up (or $\alpha$) spins than of spin-down (or $\beta$) spins means that the $\alpha$-spin electrons experience a much different environment than the $\beta$-spin electrons, and so the $\alpha$-spin levels are more energetically stabilized relative to those in the normal-level scheme. Consequently, the ligand orbitals can donate significantly to the metal levels in the inverted level scheme, thus forming the strong metal-ligand bonds.

The strongly polarized spins found in Fe-S complexes are not handled properly in standard spin-restricted density functional theory (DFT) calculations. The broken-symmetry (BS) method describes spin-polarization interactions by treating the $\alpha$- and $\beta$-spin electrons using different spatial Kohn-Sham orbitals [16]. Thus, the $\alpha$- and $\beta$-spin electrons can be treated separately in the different electron density functions. The BS-DFT approach has been shown to work well in a variety of Fe-S complexes especially when used with benchmarked hybrid density functionals [17].

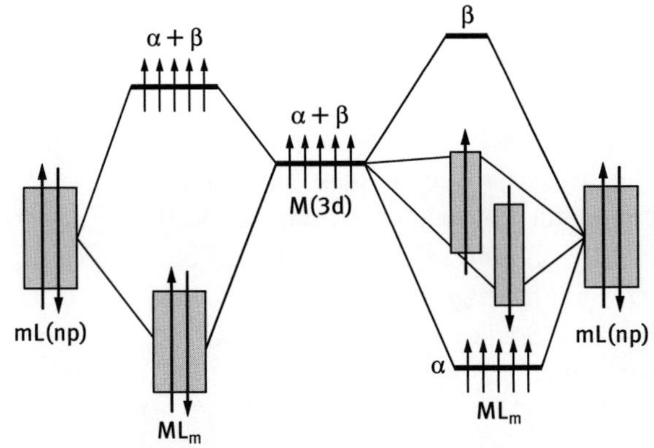


**Fig. 2.2:** Metal-ligand bonding interactions in the (left) normal- and (right) inverted-level schemes, in which the five metal 3d spins (center) interact with ligands (far left or far right, respectively). In the normal-level scheme, the filled and empty metal ($\alpha + \beta$) levels lie above the occupied ligand levels, whereas in the inverted-level scheme, the filled ($\alpha$) levels lie below the occupied ligand levels and the unfilled ($\beta$) levels lie above them.

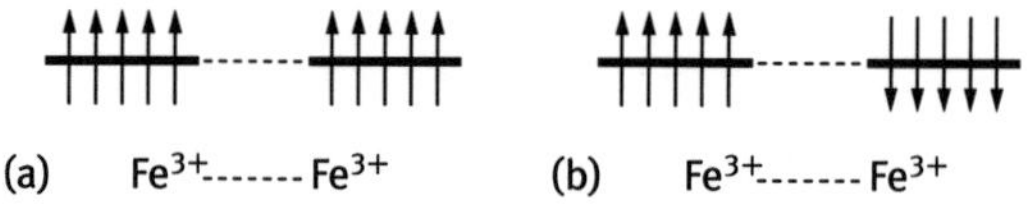


**Fig. 2.3:** Two high-spin ferric ions in which (a) the five spins are ferromagnetically coupled, such that all the spins are aligned, and (b) the five spins are antiferromagnetically coupled, such that the spins on a given iron are aligned but the spins of the two irons are anti-aligned.

### 2.2.2 Spin-coupling and metal-metal bonds

Another unusual feature of Fe-S complexes is weak metal-metal bonding interactions between the irons in complexes with two or more irons, which is attributed to spin-coupling between the spins on the individual iron sites [12]. These interactions are called Heisenberg exchange coupling, which typically favors opposite magnetic (i.e. antiferromagnetic) alignment of spins on neighboring irons over like (i.e. ferromagnetic) alignment (Fig. 2.3). In particular, this means, for proteins with a $[Fe_2S_2(Cys)_4]$ site, the alignment is typically antiferromagnetic, as shown in Fig. 2.1b. Furthermore, this leads to different Fe-S bond lengths in proteins with a $[Fe_4S_4(Cys)_4]$ site, which has a ferromagnetic alignment of spins on the top two irons, which form the "top" plane, and on the bottom two irons, which form the "bottom" plane, with antiferromagnetic alignment of spins on the top vs. bottom planes. Thus, the bond lengths within the top plane vs. the bottom plane will be the same, whereas the bonds between the top and the bottom planes will be different from the intra-plane bonds.

Because there are many allowed spin states of Fe-S complexes, they are difficult to represent accurately in a conventional quantum mechanical calculation. In a dinuclear system, the allowed coupled spins of two different metal sites of site spins in the uncoupled state of $S_1$ and $S_2$ (i.e. for $[Fe_2S_2]^{2+}$, $S_i = 1/2, 3/2, 5/2$) have a total spin $S_t$ that is an integer and between $|S_1 - S_2|$ and $|S_1 + S_2|$ (i.e. for $[Fe_2S_2]^{2+}$, $S_t = 0, 1, 2, 3, 4, 5$), forming a Heisenberg "spin ladder" of pure spin state multiplets. Because BS-DFT uses a single determinant to describe the ground state, the calculated ground state energy is not a pure spin state energy, but is instead a weighted average of pure spin states. Noodleman has developed spin projection methods based on the BS-DFT approach [16] for determining a correction for the ground state energy based on the calculated difference between the BS energy and the high-spin state energy. However, when hybrid density functionals are used, the BS energy of the Fe-S redox sites appears to correspond very well with the ground state energy because of the overestimation of spin-polarization interactions [18].

### 2.2.3 Spin resonance delocalization in mixed-valence iron pairs

A final unusual feature of Fe-S complexes is that ferromagnetically spin-coupled mixed-valence $Fe^{2+}$-$Fe^{3+}$ pairs, where each iron is internally high spin, are subject

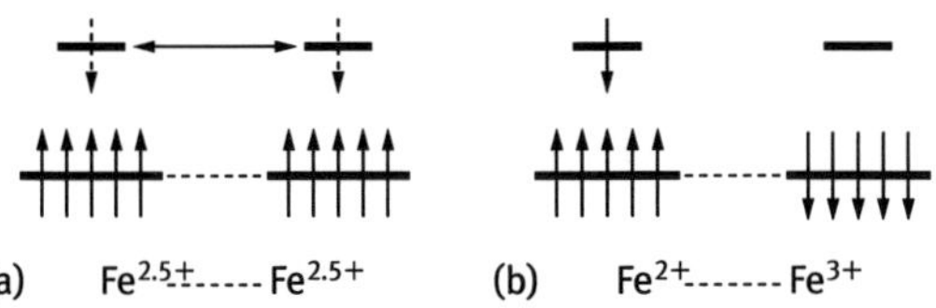


**Fig. 2.4:** A high-spin ferric-ferrous pair in which (a) the five majority spins are ferromagnetically coupled and the minority spin is delocalized between them and (b) the five majority spins are antiferromagnetically coupled and the minority spin is localized on the left-hand iron.

to resonance delocalization arising from double exchange interactions of the minority spin [4]. Because one electron becomes delocalized between the two irons, this effectively makes them into an $Fe^{2.5+}$-$Fe^{2.5+}$ pair, with the two irons being essentially equivalent (Fig. 2.4a). Moreover, resonance delocalization is most effective for tetrahedrally coordinated iron rather than in highly distorted tetrahedral coordination or trigonal or octahedral coordination. This phenomenon often occurs in the mixed-valence $[Fe_3S_4]$ and $[Fe_4S_4]$ cores but less so in mixed-valence $[Fe_2S_2]$ cores, although it has been observed in mutated forms of *Clostridium pasteurianum* $[Fe_2S_2]$ ferredoxins [19]. In particular, the $[Fe_4S_4]^{2+}$ core generally consists of two anti-ferromagnetically coupled planes (or layers), with each plane (or layer) consisting of a mixed-valence pair of irons that are ferromagnetically coupled and share a delocalized electron (Fig. 2.1c). However, vibronic and solvent effects as well as static asymmetries generally promote spin localization into discrete ion valences as shown by Mössbauer, ENDOR, and magnetic properties. In particular, ligand substitution has been shown to promote localization into a $Fe^{2+}$-$Fe^{3+}$ pair (Fig. 2.4b) in $[Fe_4S_4]$ cores [17].

Resonance delocalization can also be treated using a BS-DFT approach. The BS state is a weighted average of pure spin states that include those generated by the additional resonance delocalization. Although spin projection methods for treating the additional states have been developed [16], hybrid density functionals also give a BS state energy that is close to the true ground state energy, as in the case of simple spin-coupling alone [18].

## 2.3 Unique properties of Fe-S clusters

### 2.3.1 Stable rigid clusters mean low reorganization energy

The reorganization energy $\lambda$ in the Marcus theory for electron transfer is defined as the energy necessary to excite an electron from the reactant surface to the product surface without allowing geometric relaxation of the reactants. Thus, $\lambda$ is a measure of the energy difference due to the differences in the geometry of the reactants and products, and the closer the two are in geometry, the smaller the energy difference. Because small $\lambda$ generally leads to faster electron transfer rates, more efficient electron transfer proteins tend to have small geometric changes between the oxidized and reduced states.

For metal complexes either in a solution or in a protein, it is useful to divide $\lambda$ into an inner-sphere contribution from changes in the geometry of the metal complex upon addition or removal of an electron and an outer-sphere contribution from the reorganization of the environment, whether solvent or protein and solvent, in response to the change in charge of the metal complex. Thus, efficient metal clusters for electron transfer proteins should have the smallest inner-sphere reorganization energy.

Several properties of Fe-S clusters tend to reduce their reorganization energy. The strong Fe-S bonds increase in length slightly upon reduction from $Fe^{3+}$ to $Fe^{2+}$ but otherwise are not generally subject to bond breaking or other major changes. In addition, the strong preference for tetrahedral coordination in both $Fe^{3+}$ and $Fe^{2+}$ means there is little distortion upon reduction, whereas, for instance, four-coordinate complexes of $Cu^{2+}$ have a preference for square-planar and $Cu^{1+}$ has a preference for tetrahedral geometry [20]. The antiferromagnetic spin-coupling in the multi-Fe sites leads to further stabilization. For instance, the weak Fe-Fe bonding-type interactions in the $[Fe_2S_2]$ and $[Fe_4S_4]$ cores stabilize the interactions between irons, thus helping to make the complex very rigid. Also, the resonance delocalization in mixed-valence $[Fe_4S_4]$ cores leads to more continuous changes in geometry as an electron is added $Fe^{3+}$-$Fe^{3+}$ to $Fe^{2.5+}$-$Fe^{2.5+}$ or from $Fe^{2.5+}$-$Fe^{2.5+}$ to $Fe^{2+}$-$Fe^{2+}$.

### 2.3.2 Polynuclear clusters mean multiple valency

The spin-coupling interactions also promote the formation of multi-Fe complexes, which can form multiple net valence states. For instance, using $Fe^{3+}$ and $Fe^{2+}$ ions, the $[Fe_4S_4(SR)_4]^n$ site can apparently undergo at least four states, $n$ = 1-, 2-, 3-, 4-. Because different protein environments can make large contributions to the net reduction potential of a metalloprotein [21], the combination of different protein environments and different valencies leads to a large variation in reduction potentials found in $[Fe_4S_4]$ proteins.

### 2.3.3 Resonance delocalization and $[Fe_4S_4(Cys)_4]$ cluster conversion

The $[Fe_4S_4(Cys)_4]$ cluster can be formed from or cleaved into two $[Fe_2S_2]$ cores, processes known as cluster conversions. Both processes are important in biological functions such as assembly of $[Fe_4S_4]$ clusters from two $[Fe_2S_2]$ cores in $[Fe_4S_4]$-containing proteins and regulation of gene expression depending on whether the $[Fe_4S_4]$ is intact or cleaved to two $[Fe_2S_2]$ cores [5]. Because the $[Fe_4S_4]$ core generally consists of two antiferromagnetically coupled planes, with each plane consisting of a mixed-valence pair of ferromagnetically coupled irons that share a delocalized electron (Fig. 2.1c), cleavage might be expected to occur between the planes because the delocalized electron can generate high-spin products (Fig. 2.5, reaction to left product). Alternatively, cleavage perpendicular to the plane (Fig. 2.5, reaction to right product) would seem to imply cutting the delocalized electron on each plane in half between each iron, and cleaving all four bonds at once would give a very high reaction barrier. However, the cleavage products are observed to be low spin and thus antiferromagnetically coupled, implying that cleavage occurs perpendicular to the planes. Resolving this distinction between perpendicular and parallel cleavage is important in understanding the reaction barriers to cleavage, which is important for the kinetics of cluster conversion in proteins.

DFT studies combined with collision-induced dissociation and photoelectron spectroscopy experiments [22] have provided an explanation for how the cleavage can occur perpendicular to the plane. The barrier to fission perpendicular to the plane is substantially lowered by making a transition first from the spin-delocalized mixed-valence pair in the reactant state (i.e. $Fe^{2.5+}$-$Fe^{2.5+}$, left in Fig 2.6) to a spin-localized

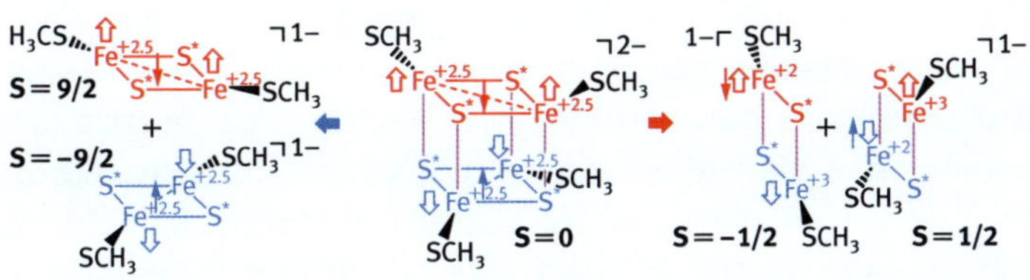


**Fig. 2.5:** Possible cleavage products of $[Fe_4S_4(L)_4]^{2-}$ (center). Cleavage between the top (red) and bottom (blue) redox planes results in two high spin products (left) while cleavage through the redox planes results in two low spin products (right). The five majority spins are donated by ⇧ and the minority spin is denoted by ↑.

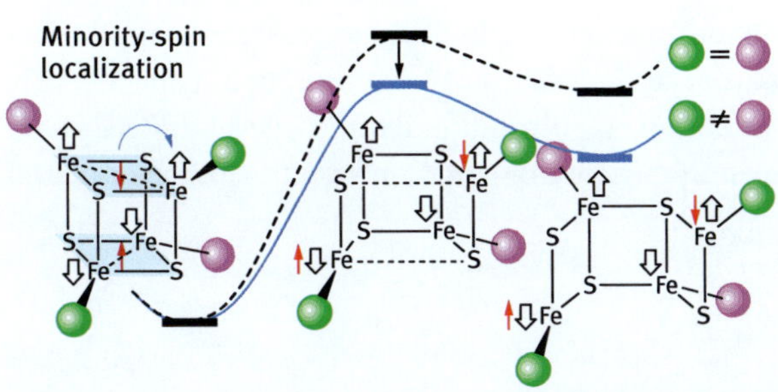


**Fig. 2.6:** Effects of ligands on cleavage of $[Fe_4S_4(L)_4]^{2-}$. Ligands are denoted by purple and green spheres. While the homoligand cluster (black) shows delocalization of the minority spins on each layer in the reactant state, heteroligands (blue) promote spin localization in the reactant state and thus enhance cleavage since the spins must be localized in the transition state. The five majority spins are donated by ⇧ and the minority spin is denoted by ↑. Lines represent the reaction pathways and bars represent from from left to right, the reactant, transition state, and cleavage intermediate. Blue shading on the initial cluster represents the planes with minority-spin delocalization.

mixed-valence pair (i.e. $Fe^{2+}$-$Fe^{3+}$, center in Fig. 2.6), thus creating a set of weak $Fe^{2+}$-S bonds and set of strong $Fe^{+3}$-S bonds. The barrier to this transition is small, so that the spin-localized pair can be considered isoenergetic with the reactant state on the scale in Fig. 2.6. Next, rather than the high barrier process of cleaving all four Fe-S bonds at once, the weak $Fe^{2+}$-S bonds are cleaved first via a transition state (center in Fig. 2.6) to a cleavage intermediate with two broken bonds (right in Fig. 2.6), followed by subsequent cleavage of the stronger $Fe^{3+}$-S bonds, resulting in cluster cleavage perpendicular to what were originally the two planes of delocalization with a lower overall reaction barrier than would be associated with attempting to break all four bonds at once. Furthermore, the propensity for cleavage can be enhanced by ligand substitution, which breaks the cuboidal symmetry and thus favors spin localization [17]. This reduces the barrier between the reactants and the transition state (blue line in Fig. 2.6) so the reaction becomes more favorable.

## 2.4 Summary

Iron-sulfur clusters have many properties that make them uniquely suited for a variety of functions in the proteins they are found in, including electron transfer proteins. These properties are a function of the high-spin irons that are tetrahedrally coordinated to sulfur, both inorganic and organic, which give rise to strong metal-ligand covalency, metal-metal bonding character, and electron delocalization.

## Acknowledgments

This work was supported by the National Institutes of Health under grant GM045303 and the William G. McGowan Foundation. The views and conclusions contained in this document are those of the author and should not be interpreted as necessarily representing the official policies or endorsements, either expressed or implied, of the U.S. Government. T.I. also thanks Shuqiang Niu for comments on the manuscript and for creating the figures.

## References

[1] Beinert H, Meyer J, Lill R. Iron-sulfur proteins. In: Lennarz WJ, Lane MD, eds. Encyclopedia of biological chemistry. Amsterdam: Elsevier; 2004;2:482–9.

[2] Johnson MK, Smith AD. Iron-sulfur proteins. In: Encyclopedia of inorganic and bioinorganic chemistry. Wiley; 2011. Online ISBN: 9781119951438 DOI: 10.1002/9781119951438.

[3] Cammack R. Iron-sulfur proteins. In: Lennarz WJ, Lane MD, eds. Encyclopedia of biological chemistry. Amsterdam: Elsevier; 2013;2:657–64.

[4] Beinert H, Holm RH, Münck E. Iron-sulfur clusters: nature's modular, multipurpose structures. Science 1997;277:653–9.
[5] Johnson DC, Dean DR, Smith AD, Johnson MK. Structure, function, and formation of biological iron-sulfur clusters. Ann Rev Biochem 2005;74:247–81.
[6] Hall DO, Cammack R, Rao KK. Role for ferredoxins in the origin of life and biological evolution. Nature 1971;233:136–8.
[7] Beinert H, Claire KM, Stout CD. Aconitase as iron-sulfur protein, enzyme, and iron-regulatory protein. Chem Rev 1996;96:2335–73.
[8] Frey PA, Hegeman AD, Ruzicka FJ. The radical SAM superfamily. Crit Rev Biochem Mol Biol 2008;43:63–88.
[9] Howard JB, Rees DC. How many metals does it take to fix $N_2$? A mechanistic overview of biological nitrogen fixation. Proc Natl Acad Sci USA 2006;103:17088–93.
[10] Rao PV, Holm RH. Synthetic analogues of the active sites of iron-sulfur proteins. Chem Rev 2004;104:527–59.
[11] Noodleman L, Lovell T, Liu T, Himo F, Torres RA. Insights into properties and energetics of iron-sulfur proteins from simple clusters to nitrogenase. Curr Opin Chem Biol 2002;6:259–73.
[12] Noodleman L, Peng CY, Case DA, Mouesca JM. Orbital interactions, electron delocalization and spin coupling in iron-sulfur clusters. Coord Chem Rev 1995;144:199–244.
[13] Niu S, Wang X-B, Nichols JA, Wang L-S, Ichiye T. Combined quantum chemistry and photoelectron spectroscopy study of the electronic structure and reduction potentials of rubredoxin redox site analogues. J Phys Chem A 2003;107:2898–907.
[14] Wang X-B, Niu S, Yang X, et al. Probing the intrinsic electronic structure of the cubane [4Fe-4S] cluster: nature's favorite cluster for electron transfer and storage. J Am Chem Soc 2003;125:14072–81.
[15] Solomon EI, Hedman B, Hodgson KO, Dey A, Szilagyi RK. Ligand K-edge X-ray absorption spectroscopy: covalency of ligand-metal bonds. Coord Chem Rev 2005;249:97–129.
[16] Noodleman L, Case DA. Density-functional theory of spin polarization and spin coupling in iron-sulfur clusters. In: Cammack R, ed. Adv Inorg Chem. San Diego (CA): Academic Press. 1992;38:423–70.
[17] Niu S, Ichiye T. Cleavage of [4Fe-4S]-type clusters: breaking the symmetry. J Phys Chem A 2009;113:5710–7.
[18] Niu S, Ichiye T. Density functional theory calculations of redox properties of iron-sulphur protein analogues. Mol Simul 2011;37:572–90.
[19] Johnson MK, Duin EC, Crouse BR, Golinelli M-P, Meyer J. Valence-delocalized $[Fe_2S_2]^+$ clusters. In: Solomon EI, Hodgson KO, eds. Spectroscopic methods in bioinorganic chemistry. Washington (DC): American Chemical Society; 1998:286–301.
[20] Lippard SJ, Berg JM. Principles of bioinorganic chemistry. Mill Valley (CA): University Science Books; 1994.
[21] Perrin BS Jr, Ichiye T. Fold versus sequence effects on the driving force for protein mediated electron transfer. Proteins 2010;78:2798–808.
[22] Niu S, Wang X-B, Yang X, Wang L-S, Ichiye T. Mechanistic insight into the symmetric fission of [4Fe-4S] analogue complexes and implications for cluster conversions in iron-sulfur proteins. J Phys Chem A 2004;108:6750–7.

# 3 Quantitative interpretation of EPR spectroscopy with applications for iron-sulfur proteins

Doros T. Petasis and Michael P. Hendrich

## 3.1 Introduction

Electron paramagnetic resonance (EPR) spectroscopy has long been a primary method for characterization of paramagnetic centers in materials and biological complexes [1]. EPR spectroscopy is only sensitive to paramagnetic centers, and because metal centers of biological complexes often have unpaired electrons, EPR spectroscopy provides a site-specific probe for the metal and its local environment. The common transition metals in biological complexes have valence d-orbitals that largely define the chemistry of the metal center. These d-orbitals have higher spin and significantly more unquenched angular momentum than p-orbitals, resulting in much wider shifts of EPR spectral features in comparison to organic radicals. Consequently, the spectra are distinctive for metal type, oxidation state, protein environment, substrates, and inhibitors. There are many books that cover the basic spin physics of magnetic resonance [2–6] and reviews addressing EPR spectroscopy of metalloproteins [7–10].

There are also a number of previous reviews especially addressing EPR spectroscopy of iron-sulfur complexes [11–13]. This review will therefore take a different and more practical approach. Over the years, the study of many metal centers in synthetic complexes and proteins in our and other laboratories has led to the development of a systematic methodology for quantitative interpretation of EPR spectra from a wide array of metal centers [14–31]. The methodology is now contained in the computer program SpinCount. SpinCount allows quantitative interpretation and simulation of EPR spectra from molecules containing transition metals. Although there are many programs available that allow simulation of EPR spectra for a specific spin system, SpinCount allows simulation of EPR spectra from any complex containing multiple sites composed of one or two metals in any spin state, and the calculations are rigorously quantitative. SpinCount can determine species concentrations from integration or simulation of spectra. This latter method is powerful because (1) it does not require a clean spectrum of a single species, (2) it is applicable to any spin system with any number of resonances, and (3) the spin standard can be any paramagnetic molecule with no relation to the unknown compound. SpinCount combines the ability to simulate EPR signals from general spin systems with the quantitative treatment of signal intensities. This review will focus on applications directed toward iron-sulfur centers and the use of this software for the interpretation of corresponding EPR spectra.

Although the software enables interpretation of EPR spectra from a much wider array of metal complexes than previously available, it is not "blackbox" in its utilization. The general quantitative treatment of a wide array of spin systems requires options for many electronic and magnetic parameters. However, most applications will use a relatively small subset, and the examples in this chapter will attempt to illustrate a more basic understanding of principles. The parameters in SpinCount have a physical rather than phenomenological basis for inclusion, which minimizes the number of parameters required to achieve an adequate simulation of the experimental spectrum and thereby contributes more significance to the result. SpinCount has a graphical interface that allows manipulation of general spectra. All routine procedures involving manipulation of spectra are present, in addition to many other tools that aid in spectral interpretation.

## 3.2 Basic EPR theory

The EPR technique is based on the Zeeman effect, which is the interaction of an external magnetic field **B** with the magnetic moments $\mu$ of unpaired electrons. The electronic magnetic moment is due to the spin angular momentum **S** of the electron. The interaction of the external magnetic field and the magnetic moment of an electron is described by the Zeeman Hamiltonian [2]

$$H_z = -\mu \cdot \mathbf{B} = g\beta \mathbf{S} \cdot \mathbf{B}, \tag{3.1}$$

where $g$ is the spectroscopic $g$ factor, $\beta$ is the Bohr magneton ($9.274 \times 10^{-24}$ J/T). The magnetic field **B** defines an axis of quantization (typically the $z$-axis), with $S_z$ as the projection of **S** onto **B**, which allows the dot product to be expressed as

$$\mathbf{S} \cdot \mathbf{B} = S_z B = m_S B, \tag{3.2}$$

where $m_S$ is the spin magnetic quantum number, with values of ±1/2. This leads to electronic energy states given by $E = \pm\frac{1}{2}g\beta B$ which produces a linear splitting of the degenerate $m_S$ spin energy levels as a function of magnetic field as shown in Fig. 3.1a. An electron on the lower energy level (spin-down) can absorb microwave radiation and make a transition to the upper state (spin-up), giving rise to an EPR signal.

The spin populations of the two states are given by the Boltzmann factor, which yields a ratio of

$$\frac{N_{-1/2}}{N_{+1/2}} = e^{-\frac{\Delta E}{k_B T}}, \tag{3.3}$$

with an excess of spins on the lower level. The quantitative interpretation of signal intensity requires signal measurements that do not saturate the signal to maintain this ratio in the spin populations. An electronic transition is possible only when the resonance condition is satisfied:

$$hv = \Delta E = g\beta B, \tag{3.4}$$

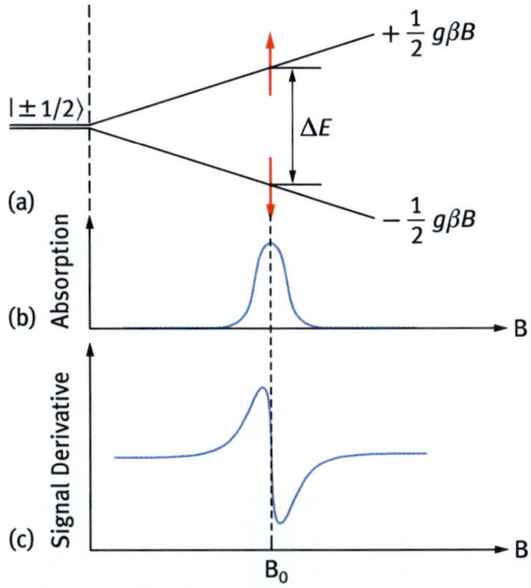


**Fig. 3.1:** (a) Zeeman splitting of the degenerate electronic spin states for an $S = 1/2$ system. (b) An electron can absorb energy to make a transition to the excited state resulting in an EPR absorption line. (c) Typical experimental EPR resonance line. The point where the line crosses the baseline determines the resonance magnetic field $B_0$ that allows the determination of the $g$ factor of the system.

where $h$ is Planck's constant ($6.626 \times 10^{-34}$ J s) and $\nu$ is the frequency of the electromagnetic radiation. The frequencies of the electromagnetic radiation are in the GHz region (microwave radiation) with magnetic fields between 0 and 1 T. Many transition series and rare earth ions are paramagnetic and give rise to EPR signals. In a typical EPR experiment, a microwave source produces radiation at a constant frequency while the magnetic field is swept through the desired range. Microwave energy is absorbed when the magnetic field goes through a value that satisfies the resonance condition. This absorption is called an "EPR resonance line" and appears as a Gaussian or Lorentzian curve in the microwave power spectrum (Fig. 3.1b). EPR spectrometers employ modulation of the magnetic field with phase-sensitive detection to significantly increase the signal-to-noise ratio, which results in the first derivative of this line, as shown in Fig. 3.1c. The area under the curve in Fig. 3.1b is proportional to the number of spins that contribute to the EPR signal. Integration of this line allows the determination of the species concentration of EPR samples. The resonance condition can be written in a form that makes it easy to convert magnetic fields into g values. With the frequency in gigahertz and the magnetic field in kilogauss, the constants $h$ and $\beta$ can be combined into one numerical constant that allows the resonance condition to be expressed in the form

$$g = 0.71449\frac{\nu(\text{GHz})}{B(\text{kG})}. \tag{3.5}$$

## 3.3 *g* Factor anisotropy

From the EPR spectrum and the resonance condition, the *g* value of the spin system can be determined. The *g* value is derived from a tensor with nine components that reflects the anisotropy of the molecule or crystal. It is represented by a 3 × 3 matrix, but with a proper choice of coordinate axes (principal axes), this 3 × 3 matrix can be expressed in a form where only the diagonal components $g_x$, $g_y$, and $g_z$ are nonzero. The experimental EPR line shown in Fig.3.1c is an example of an isotropic spectrum. This occurs when the system experiences cubic symmetry where all directions in space are equivalent and the *g* value is isotropic, i.e. $g_x = g_y = g_z = g$. If the symmetry is axial, then one axis of the molecule is unique (typically the *z*-axis), whereas the other two orientations are equivalent, producing two unique *g* factors: $g_z = g_{||}$ and $g_x = g_y = g_{\perp}$. If the symmetry is rhombic, there are no equivalent axes and three different *g* factors $g_x \neq g_y \neq g_z$ are obtained. In general, an EPR signal from a $S = 1/2$ system can show one of four different types of polycrystalline patterns (also referred to as a powder pattern) displayed in Fig. 3.2. Most commonly, a measurement is recorded on a solution, frozen solution, or powder sample. In such samples, all orientations of the molecule with respect to the magnetic field are possible. The polycrystalline pattern depends on the *g* values and shows characteristic features at magnetic fields corresponding to the principal *g* values, as shown in Fig. 3.2. For metal complexes in low symmetry, the orientation of the principal axis system of the **g** tensor with respect to the spatial arrangement of atoms will not be known but can be determined from measurement of single crystals.

## 3.4 Hyperfine structure

Many transition series ions have nuclei with a nonzero spin angular momentum (I) that gives rise to a nuclear magnetic moment that interacts with the magnetic moments of unpaired electrons. These electron-nuclear interactions produce

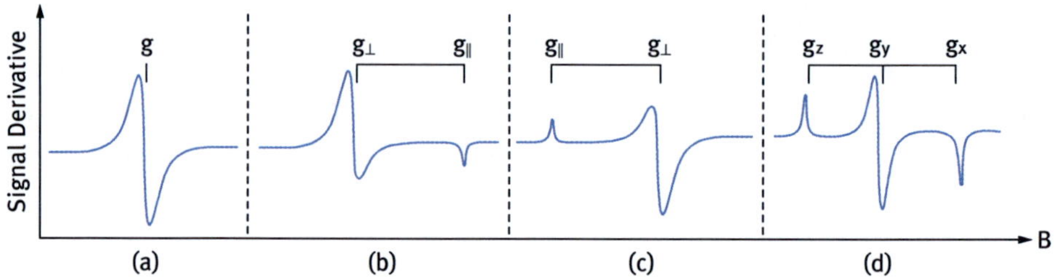


**Fig. 3.2:** Typical EPR powder spectra for various molecular symmetries: (a) isotropic, (b) axial with $g_x = g_y > g_z$, (c) axial with $g_x = g_y < g_z$, and (d) rhombic.

a splitting in the EPR spectra and give rise to hyperfine structure. The hyperfine contribution to the spin Hamiltonian has the general form

$$H_{\text{nuc}} = \mathbf{S} \cdot \mathbf{A} \cdot \mathbf{I}, \tag{3.6}$$

where **A** is the hyperfine tensor. In isotropic systems, the hyperfine tensor is replaced by the hyperfine constant $A$. This interaction of a spin $I$ nucleus with a spin $S$ electron results in the splitting of the $m_S$ electronic levels. The multiplicity of these levels is given by $(2I + 1)$ and each level is labeled by the quantum number $m_I$ ($m_I$ takes values from $-I$ to $+I$ in integer steps) as shown in Fig. 3.3a. Transitions are now allowed between the $m_I$ levels obeying the selection rules: $\Delta m_S = \pm 1$ and $\Delta m_I = 0$. These transitions give rise to $(2I + 1)$ equally spaced EPR lines of equal intensity. The magnetic field separation of the hyperfine lines is given by the hyperfine constant $A$, which is characteristic of the particular ion. Typical transition series ions with nuclear spin are $Cu^{2+}$ ($I = 3/2$), $Mn^{2+}$ ($I = 5/2$), and $Co^{2+}$ ($I = 7/2$).

Hyperfine interactions are also possible between a paramagnetic ion and nearby ligands possessing nuclear spin such as $^{14}N$ with $I = 1$. These so-called superhyperfine

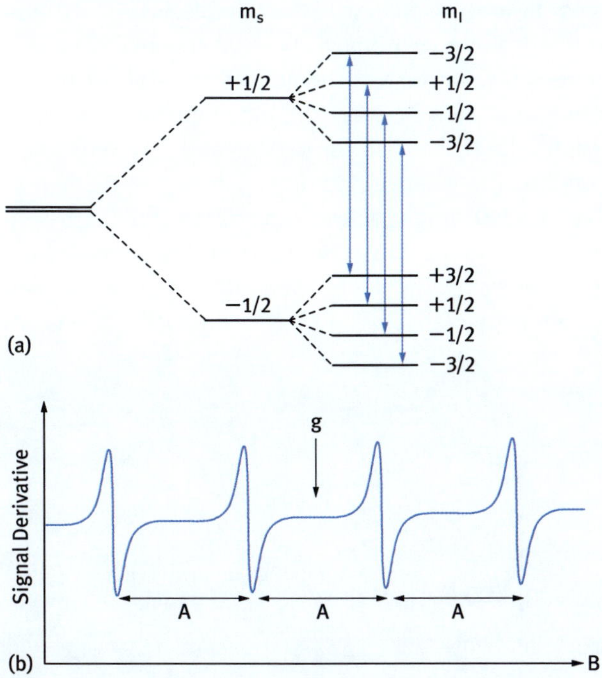


**Fig. 3.3:** (a) Hyperfine splittings in a system with $S = 1/2$, and $I = 3/2$. Allowed transitions between the nuclear levels obey the selection rule $\Delta m_S = \pm 1$ and $\Delta m_I = 0$ and are indicated by the arrows. (b) Ideal EPR hyperfine spectrum for a species with $I = 3/2$. The $g$ factor is determined from the field position of the center of the spectrum.

interactions produce groups of EPR lines of varying intensity. The interaction of the central ion with $n$ equivalent ligand nuclei produces $(2nI + 1)$ EPR lines. The value of the hyperfine constant $A$ is dependent on the spin density of the electron at the nucleus and therefore on bonding in the molecule. An electron in an orbital that is mainly metal based will have high spin density at the nucleus of the metal and a large hyperfine constant (roughly greater than 50 MHz). If the metal electron interacts with nearby ligands, the spin density and hyperfine constant are much less.

## 3.5 Ligand interactions

The paramagnetic ions encountered in biological systems are part of a molecular lattice and interact with the various ligands in the lattice. The interaction is due to the overlap of the d-orbitals of the transition series ions with the orbitals of the ligands. The ligand geometry is responsible for the extent of the overlap and the strength of the interactions. The most common symmetry is octahedral where the paramagnetic ion is at the origin of coordinates and six ligands are symmetrically arranged about it, forming pairs along the coordinate axes as shown in Fig. 3.4a. The octahedral ligands interact more strongly with the $e_g$ orbitals ($d_{z^2}$ and $d_{z^2-y^2}$) that have large electronic probability densities along the coordinate axes than with the $t_{2g}$ orbitals ($d_{xy}$, $d_{yz}$, and $d_{xz}$) resulting in a splitting of the five degenerate orbital d states, as shown in Fig. 3.4b.

The splitting leads to different spin states for a particular valence state of an ion, depending on the strength of the interactions between the transition series ion and the ligands. As an example, consider the $Fe^{III}$ ion: this is a $d^5$ ion and in a weak interaction, the splitting $\Delta$ is small, and the five valence electrons each occupy a different orbital according to Hund's rules. The resulting species has a spin $S = 5/2$ (high-spin $Fe^{III}$). In a strong ligand field, the splitting $\Delta$ is larger than the pairing energy of the electrons and now all five electrons must occupy the $t_{2g}$ states according to the Pauli principle. The result now is a system with total spin $S = 1/2$ (low-spin $Fe^{III}$), as shown in Fig. 3.5.

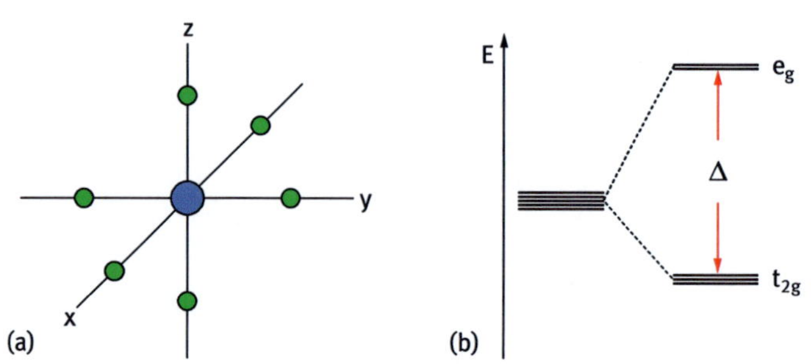


**Fig. 3.4:** (a) Octahedral arrangement of the ligands (small spheres) around a metal ion (large sphere). (b) The splitting of the five degenerate orbital d states into a triplet ($t_{2g}$) and doublet ($e_g$).

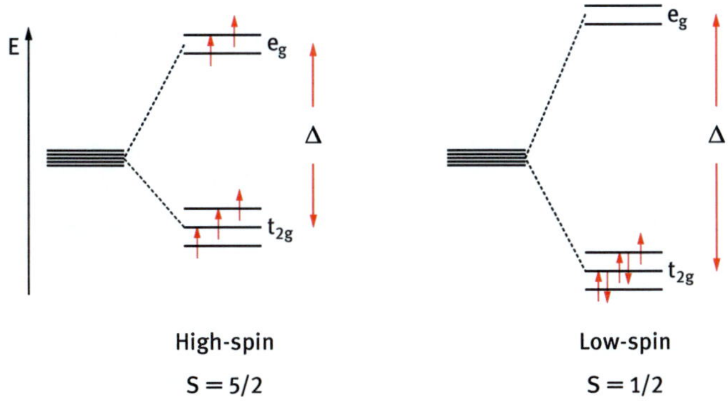


**Fig. 3.5:** High-spin $S = 5/2$ (left) and low-spin $S = 1/2$ (right) states of the $Fe^{III}$ ion in octahedral symmetry.

## 3.6 Spin Hamiltonian

A paramagnetic metal ion can have multiple unpaired electrons resulting in spin states greater than 1/2 occupying a number of discrete energy states. The spin Hamiltonian provides a compact way of representing the energy states of the metal ion with a relatively small number of terms that can be experimentally determined. The spin Hamiltonian used in SpinCount incorporates a wide array of interactions to be able to simulate EPR spectra from most any type of metal complex. Only a fraction of the possible terms in the spin Hamiltonian are required for a particular metal complex of interest. The more common terms are described next.

The spin Hamiltonian $H$ can be expressed in a summation of a number of terms as

$$H = H_1 + H_2 + H_{12} + H_{\text{nuc}}. \tag{3.7}$$

For the case of a system of two distinct metal ions (i = 1, 2), $H_i$ describes the interaction of the spin magnetic moment of each metal ion with the external magnetic field **B** and the ligand field in terms of its spin **S** and **g** tensor as [2, 4, 32].

$$H_i = \beta \mathbf{S_i} \cdot \mathbf{g_i} \cdot \mathbf{B} + D_i\left[S_{iz}^2 - \frac{S_i}{3}(S_i + 1) + \frac{E_i}{D_i}(S_{ix}^2 - S_{iy}^2)\right] \tag{3.8}$$

The first term is the Zeeman term and represents the interaction of each individual spin with the external magnetic field **B**. In the presence of spin-orbit coupling, the **g** tensor is anisotropic and can be expressed in terms of its principal values $g_x$, $g_y$, $g_z$. These intrinsic $g$ values of the metal center are typically close to 2 ($1.6 < g < 2.4$) and should not be confused with effective, fictitious, or observed $g$ values. The second term in Eq. 3.8 represents the zero-field splitting that removes the degeneracy of the $2S + 1$ manifold in zero magnetic field, which becomes relevant for ions with spin $S > 1/2$ in ligand symmetries lower than octahedral. $D$ and $E/D$ parameterize the axial and rhombic symmetry of the metal center. With an appropriate choice of coordinate system, the rhombic parameter is in the range $0 \leq E/D \leq 1/3$.

The hyperfine interaction is introduced in the spin Hamiltonian *via* the term $H_{nuc}$. The energy of interaction between a spin $\mathbf{S}_i$ interacting with a nuclear spin $\mathbf{I}_j$ is equal to $\mathbf{S}_i \cdot \mathbf{A}_{ij} \cdot \mathbf{I}_j$, where $\mathbf{A}_{ij}$ is the hyperfine tensor.

The term $H_{12}$ describes the interaction between two spin centers that are close:

$$H_{12} = \mathbf{S}_1 \cdot \mathbf{J} \cdot \mathbf{S}_2 + H_{dip}, \tag{3.9}$$

where the terms are the through-bond electronic exchange interaction, and the through-space magnetic dipole interaction. For Fe-S clusters, the magnitude of the exchange interactions between iron ions of the cluster is generally much larger in energy than the other terms of the spin Hamiltonian. Consequently, the individual spins of the iron ions couple to give total spin states [33], and usually one total spin state is low in energy (ground state) and well separated from all other total spin states. The energies of the ground total spin state, and the corresponding EPR spectra, can then be described with a single spin system using $H_1$.

## 3.7 Basic EPR instrumentation

Most commercially available EPR spectrometers operate at X-band frequencies (8.5–12 GHz) and employ a continuous-flow liquid helium cryostat to reach very low temperatures (typical temperature range 2–100 K). They utilize a microwave bypass arm to bias a diode detector for increased sensitivity, as shown in Fig. 3.6. The radiation from the microwave source (Gunn oscillator) travels to a resonant cavity where it is stored in the form of standing waves. When resonant absorption of radiation occurs in the sample, a small signal is reflected back from the resonant cavity to the diode detector, which acts as a microwave rectifier. The rectified signal from the detector is processed by an electronics console and sent to a computer for recording and analysis. The low temperatures are important to slow electronic relaxation between spin states and to improve the signal-to-noise ratio of the spectrum.

The sample is placed inside the resonant cavity at a position of maximum microwave magnetic field $\mathbf{B}_1$, as shown in Fig. 3.7. Bimodal rectangular cavities are available that allow the external magnetic field to be oriented in either a perpendicular or a parallel orientation relative to $\mathbf{B}_1$ for the detection of both half-integer and integer-spin systems, respectively. Alternatively, a cylindrical cavity can be employed with side-wall coupling as in the Q-band (34 GHz) spectrometer in our laboratory where the external magnetic field can be rotated between the perpendicular and parallel orientations [27]. In spectrometers operating at low temperatures, a continuous-flow cryostat is employed for cooling the sample to liquid helium temperatures. The cold helium gas flows through a nozzle equipped with a heater that allows control of the helium temperature. A thermocouple placed a few millimeters below the sample tube measures the temperature of the gas.

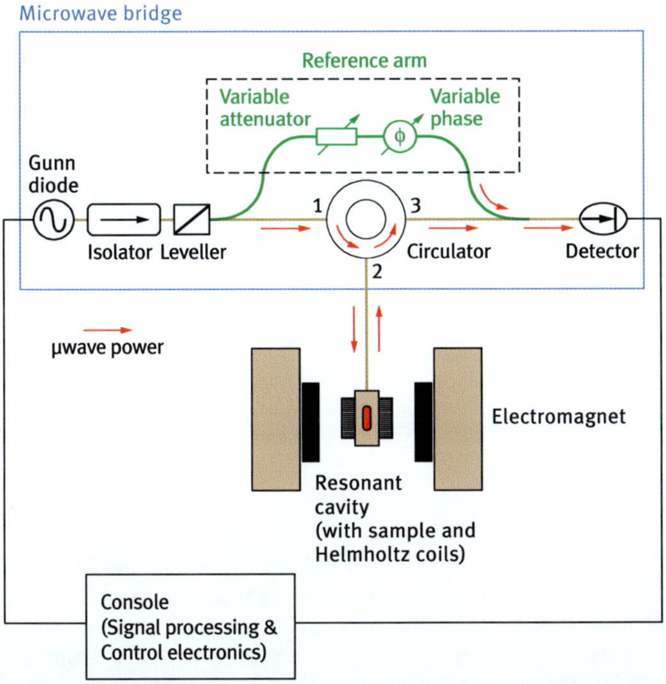


**Fig. 3.6:** Schematic diagram of the EPR spectrometer. The microwave radiation is produced by a Gunn diode (or klystron) with a small amount diverted through the bypass arm to the detector. The remaining radiation travels down a series of waveguides, an isolator (protects source from reflected microwave power), and a leveler (controls the microwave power) through a circulator to the sample in the resonant cavity (red arrows). When the instrument is properly tuned, no microwave power is reflected back from the cavity. When the magnetic field is swept through resonance, microwave power is absorbed by the sample in the cavity and a small signal is reflected back from the cavity that travels through the circulator and is mixed with the signal from the bypass arm. The combined signal then travels to the detector and the electronics console for processing.

## 3.8 Simulation of powder spectra

The majority of EPR data from metal centers are recorded from frozen solution samples because electronic relaxation rates near room temperature are fast, resulting in excessive line broadening, and loss of signal. Powder samples from synthetic complexes are often broadened by intermolecular dipolar interactions. Nevertheless, it should be standard practice to record such spectra because the occurrence of resonances in positions unchanged relative to a frozen solution sample indicates that the molecule does not undergo a structural change in a solvent system.

Whether measurements are from frozen solution or powder samples, the spectra of both are referred to as powder patterns. SpinCount will calculate spectra for single crystal or powder patterns. Simulations of powder spectra assume that all orientations of the molecule with respect to the magnetic field **B** are equally likely. A simulation is generated from the integral over a spherical surface of the magnetic field orientations.

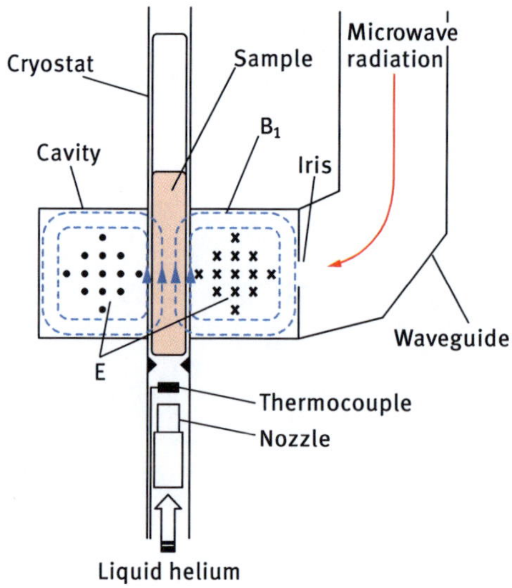


**Fig. 3.7:** Cross section of a rectangular $TE_{102}$ cavity with cryostat and sample. The microwave radiation is transmitted into the cavity through an adjustable circular iris and stored as standing waves. The magnetic field component of the microwave radiation (**$B_1$**: dotted blue lines) is 90° out of phase relative to the electric field component (**E:** circles and *x*'s). The arrows show the direction of the **$B_1$** vectors inside the cavity. The **E** vectors are at right angles relative to **$B_1$**, coming out of the page on the left half of the cavity (circles) and into the page on the right half of the page (*x*'s). The external magnetic field, provided by an electromagnet, is perpendicular to the plane of the page. The shown configuration allows the detection of species with half-integer spins. In a bimodal cavity, the microwave frequency can be adjusted to generate a different configuration of the standing wave pattern inside the cavity, that allows **$B_1$** to orient parallel to the external magnetic field for the detection of integer-spin systems.

A regular angular grid of points in cos $\theta$ and $\varphi$ is constructed over a specified range in $\theta$, $\varphi$. This range is usually one octant ($0 < \theta < 90$, $0 < \varphi < 90$), but lower symmetries with two metals will require calculation over additional octants. The number of points in the angular grid affects the smoothness of the simulation. More grid points will slow the calculation but generate less computation noise. Options for variable grids that will increase the grid density in angular regions with fast-changing magnetic field are present. At each angular grid point, three quantities are calculated using Eq. 3.7 for the two levels of interest in the spin system: the intensity, the linewidth, and the magnetic field satisfying the resonance condition $h\nu = \Delta E$, where $\Delta E$ is the energy spacing between the two levels. The intensity is the transition probability between the two states (from the time-dependent quantum interaction with the microwave magnetic field) times the Boltzmann population difference. A Gaussian or Lorentzian line is then added to the growing spectrum, centered at the resonance field with its calculated intensity and width. This represents the single crystal contribution to the powder spectrum for the specific grid angles.

The ability to generate quantitative simulations requires not only the correct position and intensity of resonances but also an accurate match to the spectral lineshape.

The sources of line broadening for metal centers generally fall into three categories: lifetime, unresolved hyperfine, and molecular disorder. In the absence of all other sources of broadening, the lineshape is Lorentzian with a width determined by the lifetime of the electronic state. Electronic lifetimes for metal centers at 4 K are roughly $10^{-6}$ s, giving linewidths of approximately 0.01 mT. The first and second coordination spheres of metal centers often contain multiple nuclei with spin $I \neq 0$ (e.g. $^{14}N$ or $^{1}H$), giving unresolved hyperfine broadening of the spectra, which is usually well modeled with a Gaussian lineshape. The width of this unresolved hyperfine is on the order of 1 mT and typically dominates the Lorentzian shape of lifetime broadening.

Numerous studies in our laboratory and others have shown that broadening from molecular disorder typically overwhelms the previous two sources at low temperatures. Upon freezing a sample, the variation in the metal-ligand bonds of each molecule in a sample, possibly from vibrations or random solvent interactions, are fixed into a static distribution. This bond variation results in a distribution of the orbital energy levels, and *via* the spin-orbit interaction, a distribution in various molecular parameters such as $g$, $D$, and $E/D$. This type of broadening in EPR spectra is commonly referred to as $g$ or $D$ strain. To model these distributions, SpinCount has corresponding distribution parameters (e.g. $\sigma_{gx}$, $\sigma_{E/D}$) and a minimum linewidth usually due to unresolved hyperfine splitting.

Many metal centers have spin states with $S > 1/2$, which show resonances from transitions between states other than $m_s = \pm 1/2$. In general, the number of possible transitions is $S(2S + 1)$, however, most of these will not contribute significant intensity. The most intense signals originate from transitions between doublets with $\Delta m_s = \pm 1$, but the observation of other resonances is often critical for determination of zero-field or exchange energies. Although common desktop computers are now available with fast numerical processing, as $S$ increases, it is important to optimize the calculation of the simulation to focus on those transitions that are observed experimentally, so that calculation time is not wasted on transitions that have negligible intensity. This is especially true for least-squares fitting of data. By default, SpinCount calculates all transitions, but options are available to restrict the calculations to those the user considers important.

## 3.9 Quantitative aspects

An important aspect of magnetic resonance spectroscopies is the quantitative intensity relation to the spectrum. The interaction of the microwave radiation with the sample can be modeled with time-dependent perturbation theory to give the transition rate for electrons in resonance and therefore the intensity of the corresponding signal. This calculation depends on the spin wavefunctions of the metal center. The spin wavefunctions can be accurately determined, and the intensity of the spectrum can thus be directly related to species concentration. EPR spectroscopy can provide a primary determination of concentrations of metal species for a sample. For simple spin $S = 1/2$ centers, or a single resonance from centers with $S > 1/2$, which can be effectively modeled as $S = 1/2$, the concentration of the species can be determined from double integration of the experimental spectrum. A double integration is necessary because the concentration depends on the area under the absorption curve, and the experimental spectrum is the first derivative with respect to the magnetic field. However, as is often the case for protein complexes, the signal of interest overlaps with signals from other spin centers or paramagnetic impurities rendering integration methods invalid. Integration methods are also invalid for spin systems with multiple states in resonance, or not reducible to an effective $S = 1/2$ center, or for integer-spin centers ($S = 1, 2, 3, \ldots$). For all cases in which integration methods are invalid, the concentration of species can be determined from simulation of the spectrum. A central feature of SpinCount is the ability to extract concentrations of species directly from spectra, either with double integration (when valid) or by reference to a simulation.

The experimental signal intensity is a function of the microwave components, receiver gain, digitization and dwell time, microwave mode, field modulation, sample volume, concentration, sample solvent and holder, temperature, microwave frequency and power, and all the theoretical considerations discussed earlier in the simulation of a spectrum. Many of these factors specifically depend on the properties of the particular EPR spectrometer that is recording the spectrum. SpinCount has a spectrometer-dependent component that links the required information from the spectrometer (referred to as an instrument) to a quantitative intensity scale within SpinCount. Each spectrometer requires its own instrument declaration, and the validity of the declaration must be initially checked with an intensity calibration procedure to verify that the intensity controls are correctly set. The intensity scale depends on a calibration constant specific to the spectrometer, which is easily determined from the spectrum of a spin standard. We routinely log a spin standard into SpinCount each day to account for any changes in instrumentation.

An accurate determination of the spin concentration is dependent on proper choice of experimental parameters and the correct sample temperature. A spin standard can be any compound for which the concentration of spin centers in the sample can be independently determined, but for simplicity, the standard should have $S = 1/2$. A common choice for low temperature studies is the ethylenediaminetetraacetic acid (EDTA) complex of $Cu^{II}$, at a concentration of 1 mM $Cu^{II}$ with 10-fold excess EDTA in a neutral buffer having 10% glycerol. Accurate standards of copper solutions are available from chemical suppliers. The accuracy of the intensity calibration will depend on several factors: careful analytical preparation of the standard sample, the effect on the microwaves by the sample in the cavity, sample geometry, measurement of the EPR spectrum under proper conditions, proper calculation of the second integral, an accurate value of the sample temperature. It is of critical importance to record the spin standard and unknowns under non-saturating microwave power conditions for quantitative information. The microwave power is nonsaturating if the signal intensity is proportional to the square root of the microwave power.

The introduction of a sample into the cavity distorts the microwaves and thus affects the magnitude of the microwave magnetic field incident on the sample (Fig. 3.7) [34]. The microwaves are affected both by the amount of quartz holding the sample and the solvent properties. The sample tubes should be of an approximately consistent quartz thickness, and the polarity of the solvent should be similar to the unknown. The effect of solvent can be checked by comparing integrated signal intensities of spin standard samples in various solvents. Also important is the difference in expansion/contraction of water *vs* organic solvents when freezing the samples. A water solution will expand by roughly 10% upon freezing, which reduces the spin concentration in the microwave cavity. Using a spin standard in water to quantify an unknown in an organic solvent would give an inaccurate spin concentration.

The sample geometry of the spin standard in the cavity should mimic that of the unknown. The sample tubes should be the same diameter and the length of the samples in the cavity should be the same. Typically, the length of the sample should span the entire length of the cavity (Fig. 3.7), typically 25 mm. Any part of the sample that is outside of the microwave cavity does not contribute to the signal intensity. The quartz stock used to make the sample tubes will often give small variations in the cross-sectional area of the tube. The cross-sectional area can be easily determined by measuring the length of a solution of known volume in the tube. SpinCount has corrections for the cross-sectional area and for short samples that do not span the length of the cavity.

A significant source of error in spin concentration determinations is incorrect sample temperature information. The widely used liquid helium flow cryostats have inherent temperature uncertainties because the metal temperature sensors cannot be positioned directly at the sample. The temperature probe is positioned approximately 5 mm from the bottom of the sample tube (Fig. 3.7). Consequently, the actual sample temperature is dependent on the flow and pressure of helium. The helium must be maintained at a sufficiently high flow to provide a uniform temperature along the sample. The temperature can be calibrated with a sensor mounted in an EPR tube, but then, it is important to maintain the whole helium system at the same pressures and flows for sample measurements. Ideally, the standard and unknown should be recorded at the same temperature. Provided the temperature information is good, or at least linearly correct, a $1/T$ correction can be applied for $S = 1/2$ complexes when determining spin concentrations. For samples with $S > 1/2$, other corrections will be necessary if double integration is used to determine the spin concentration. For spin concentration determinations from simulations, the appropriate factors are known and incorporated into the simulations.

## 3.10 Examples

### 3.10.1 $S = 1/2$ systems

Many iron-sulfur clusters show EPR signals from $S = 1/2$ spin states. Figure 3.8 shows a spectrum and a simulation for the $Fe_2S_2$ center of methane monooxygenase [35]. Under cellular conditions, the $Fe_2S_2$ cluster of proteins has two accessible oxidation

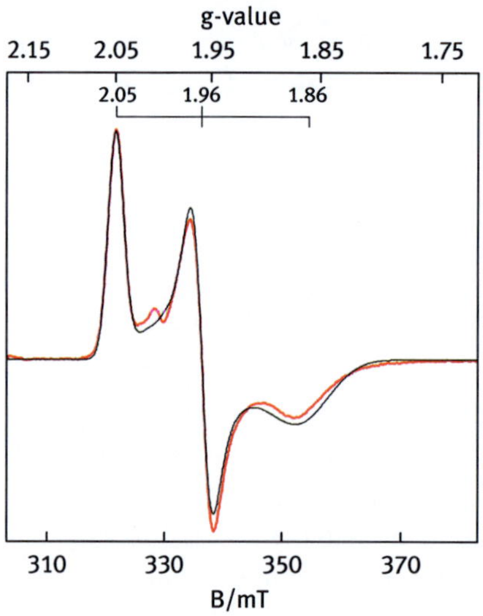


**Fig. 3.8:** EPR spectra and simulation (black line) of the reduced state of the $Fe_2S_2$ cluster of methane monooxygenase. Experimental conditions: temperature, 7.0 K; microwave frequency, 9.229 GHz; microwave power, 0.05 mW. Simulation parameters: $g$ = (1.860, 1.960, 2.050), $\sigma_g$ = (0.025, 0.009, 0.008). The weak signal near $g$ = 2.0 is an impurity.

states: oxidized ($Fe^{III}$ $Fe^{III}$) and reduced ($Fe^{III}$ $Fe^{II}$). The tetrahedral coordination of sulfur to each iron results in high-spin states for iron: $S$ = 5/2 for $Fe^{III}$, and $S$ = 2 for $Fe^{II}$. For both states of the cluster, the bridging inorganic sulfurs establish a strong electronic exchange interaction between the iron ions, causing coupling of the spins of each iron to give resultant total spin states. The ground total spin state is usually an antiparallel (antiferromagnetic) spin alignment of one iron relative to the other. Consequently, the oxidized state has spin $S$ = 0 (diamagnetic) and the reduced state has $S$ = 1/2.

The spectrum shown in Fig. 3.8 is a typical example. Spin $S$ = 1/2 species require a minimal set of parameters to characterize. Simulations are generated by specifying the principal components of a **g** tensor and a $g$ strain tensor $\boldsymbol{\sigma}_\mathbf{g}$ for a rhombic EPR spectrum, where the values for $\boldsymbol{\sigma}_\mathbf{g}$ are the width of the distribution of $g$ values. There are many potential sources for broadening of spectra, but the dominant source for $S$ = 1/2 species is often $g$ strain. As the value of $\Delta g_i = |g_i - 2.00|$ increases ($i = x, y, z$), the corresponding larger orbital contribution (and larger variation) result in broader lines, implying that the ratios $\sigma_{gi}/|\Delta g_i|$ will be approximately constant.

The **g** tensor of the total $S$ = 1/2 state is related to the **g** tensors of the ferric and ferrous irons by the spin-coupling relation [33]

$$\mathbf{g}_{1/2} = \frac{7}{3}\mathbf{g}_{5/2} - \frac{4}{3}\mathbf{g}_2. \tag{3.10}$$

High-spin $Fe^{III}$ has a ground orbital singlet state that is well separated from its excited orbital configurations, resulting in relatively small orbital contributions to the magnetic moment and principal values of $\mathbf{g}_{5/2}$ that are near 2.00. High-spin $Fe^{II}$ has significantly larger orbital contributions to the magnetic moment, which induces anisotropy into $\mathbf{g}_2$ and principal values deviating from 2.0 by up to roughly 0.2. The range of principal values of $\mathbf{g}_{1/2}$ is dominated by the **g** tensor of $Fe^{II}$ ($\mathbf{g}_2$), which results in $g$ values (like that of Fig. 3.8) significantly lower than 2.0.

Samples are often not of high purity without significant preparatory work directed at the isolation of a particular species. Initial sample preparations may show multiple species, or the isolation of a single species is not required, or not possible. For such cases, the determination of species concentrations from integration methods is not possible. SpinCount is particularly useful for simulating signals with multiple overlapping species. The program combines least-squares fitting and single-valued decomposition to allow deconvolution of overlapping signals. This allows identification of species and their respective concentrations. An example is taken from work on mitochondrial samples [22]. The EPR spectrum of whole mitochondria in the $g = 2$ region is shown in Fig. 3.9a (green line). This spectrum contains overlapping signals from many proteins. The software simulates multiple different spin centers and produces a sum of the simulations with an amount of each species that gives the best fit to the experimental spectrum. The spectrum of Fig. 3.9 was decomposed with this

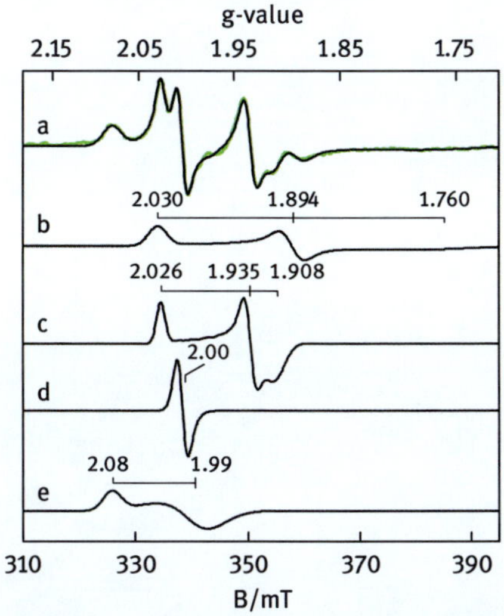


**Fig. 3.9:** (a) An EPR spectrum (green) and simulations (black) of isolated whole mitochondria [22]. Simulations (b–e) of the species contributing to the sum (black line) overlaid on the experimental spectrum. Instrumental parameters: microwave frequency, 9.487 GHz; microwave power, 0.2 mW; temperature 10 K. The **g** tensors of the simulations are listed on the figure. The concentrations determined from simulations: (b) Rieske center, 44 μM; (c) succinate dehydrogenase, 23 μM; (d) radical, 2 μM; (e) dehydrogenase 17 μM.

method to give the spectra of each species. The $S = 1/2$ signals are from the iron-sulfur clusters of the Rieske center (b), succinate dehydrogenase (c), an organic radical (d), and dehydrogenase (e). The sum of the four spectral simulations is overlaid on the experimental spectrum. The concentration of each species in the sample was determined from the simulations and listed in the figure caption.

When nuclear interactions are present, additional terms in the spin Hamiltonian (Eq. 3.7) must be added. Figure 3.10a shows an example that includes a hyperfine interaction from $^{14}NO$ bound to the iron of heme in the reduced form of the enzyme tryptophan dioxygenase [17]. The nuclear spin of $^{14}N$ ($I = 1$) and a corresponding hyperfine tensor $\boldsymbol{A}$ are required for characterization. The spectrum shows a triplet of lines ($2I + 1$) at each of the principal $g$ values (2.110, 2.031, 2.011). Although this appears to be a straightforward application, attempts to simulate the signal failed to match the spectrum, mainly because of additional signal intensity at $g = 1.985$. Additional seemingly spurious signals are often dismissed as impurities, as was the case for this enzyme, but careful consideration can uncover important properties.

The addition of another $S = 1/2$ species into the simulations provided an explanation of the extra signals. This is not surprising because more species allow more resonances and a better match, so we must look for additional evidence in support. The simulation of the two species are shown in Fig. 3.10b and c, and the sum is overlaid

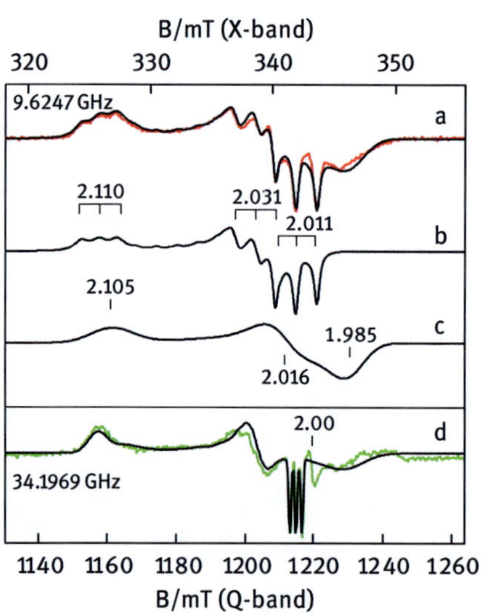


**Fig. 3.10:** (a) X- and (d) Q-band EPR spectra (colored lines) of the NO adduct of 0.3 mM tryptophan dioxygenase.[17] (b, c) The simulations (black lines) of the two heme species. (a, d) Simulation sums (black lines) of the two species at equal concentrations. Experimental conditions: microwave frequencies as listed; microwave power, 0.02 mW (X), 0.005 mW (Q); temperature, 20 K (X), 27 K (Q). The weak signal near $g = 1.96$ is due to a minor amount of free NO in solution. The **g** tensors of the simulations are listed on the figure. The **A** tensor for (b) is (48, 49, 39) MHz.

on the experimental spectrum of Fig. 3.10a. Multiple different enzyme preparations produced the same signals, and the ratio of the intensities was constant, providing the first clue that the additional intensity was not spurious. Next, the concentrations of the two species determined from the simulations was found to be equal, and each accounted for half of the total heme in the sample.

Further evidence was gathered from spectra at a second microwave frequency. The spectrum shown in Fig. 3.10d was recorded on a spectrometer operating at 34 GHz (referred to as Q-band). The spectra in the top and bottom parts of Fig. 3.10 are plotted for equal $g$ scale, meaning that the limits of the magnetic field range of the Q-band display are chosen to be 34.20/9.62 (ratio of frequencies) times the limits of the X-band display range. For such a plot, features that depend linearly on the magnetic field align vertically, whereas features that are not intrinsically magnetic field dependent do not align. The hyperfine interaction $H_{nuc}$ (Eq. 3.6) has no field dependence. The hyperfine splittings appear to be less in the bottom plot, but in fact, the splitting value is the same as measured from the correct scale. This also has experimental significance. The ability to resolve hyperfine splitting in spectra increases as the microwave frequency is decreased because the linewidth of resonances has a field-dependent component. Simultaneous least-square fits of the spectra produced the simulations shown in Fig. 3.10 for the two microwave frequencies, both with the same set of parameters. The simulation sums for equal species concentrations predicted both spectra. Species of Fig. 3.10b shows a three-line pattern with hyperfine tensor $\mathbf{A}$ = (48, 49, 39) MHz from the $^{14}N$ nuclei of NO. As with the **g** tensor, the hyperfine tensor can also be anisotropic. The isotropic part of **A**, $A_{iso} = (A_x + A_y + A_z)/3$, is a measure of the spin density on the nucleus. For the heme-NO complex of Fig. 3.10, $A_{iso}$ = 45 MHz, which, in comparison to that of free NO (roughly 70 MHz), indicates significant radical character on NO. The hyperfine parameters for the second species could not be determined due to its broad features because of larger $g$ strain in the second species. The analysis shown in Fig. 3.10 clearly indicated that the spectra contained two distinct $S$ = 1/2 rhombic species from heme-NO adducts. Tryptophan dioxygenase has four subunits per protein, with each subunit containing one heme. The speciation from the EPR spectra demonstrated that electronic configuration of the hemes was as a dimer of dimers.

### 3.10.2 Spin systems with $S$ = 3/2, 5/2, 7/2, etc.

For spin systems with $S > 1/2$, terms containing the axial ($D$) and rhombic ($E/D$) zero-field splitting parameters contribute to the spin Hamiltonian (Eq. 3.7). These terms originate from the spin-orbit interaction and cause the spin levels to split in energy in the absence of a magnetic field. The values of $D$ and $E/D$ depend on the metal, the ligands to the metal, and the geometry of the ligand coordination. For iron complexes, the magnitude of $D$ is typically in the range of 1–10 $cm^{-1}$. The magnitude of $D$ is often larger than the quantum of energy of the common X-band EPR spectrometers ($\nu$ = 9.5 GHz, $h\nu$ = 0.3 $cm^{-1}$).

This results in doublets of spin states (|±$m_s$>), which are split in energy, and often, the resonance from the |+1/2>, |−1/2> transition dominates the spectrum. The resonances occur at observed $g$ values far from $g = 2.0$ due to $S > 1/2$ and the zero-field splitting.

An energy level diagram as a function of magnetic field is shown in Fig. 3.11 for $S = 5/2$ with $D = 1\ \mathrm{cm}^{-1}$ and $E/D = 0$. The $z$-axis is usually chosen along a symmetry axis of the molecule, e.g. perpendicular to the plane of the porphyrin for a heme center. For the magnetic field aligned with the $z$-axis ($B \parallel z$), the splitting of the doublets due to the magnetic field is proportionally larger for higher $m_s$ values. For $B \parallel x$, only the |±1/2> doublet splits appreciably with magnetic field, and this splitting is greater than that for the $z$-axis. The black vertical lines in Fig. 3.11 mark the magnetic field values for which the resonance condition $\Delta E = h\nu$ is obeyed, where $\Delta E$ is the splitting in energy between two spin levels. For the |±1/2> doublet, the resonances occur at fields of 340 mT ($B \parallel z$) or 110 mT ($B \parallel x$ or $B \parallel y$), corresponding to $g$ values of 2 and 6, respectively. These transitions are allowed, whereas the transitions for the |±3/2> and |±5/2> doublets are not allowed for $E/D = 0$. Consequently, for metal centers in near axial symmetry, the spectra are dominated by the |±1/2> doublet.

The spin doublets |±$m_s$> of Fig. 3.11 are separated by an energy greater than $h\nu \approx 0.3\ \mathrm{cm}^{-1}$. These isolated doublets can each be represented by an effective $S' = 1/2$ system $S'$ whose principal $g$ values are determined from calculations based on Eq. 3.7. Figure 3.12 shows the principal $g$ values for each of the three doublets of an $S = 5/2$ spin system as a function of $E/D$. Figure 3.13 shows EPR spectra and simulations from three heme-containing proteins: aquometmyoglobin (Mb), horseradish peroxidase (HRP), and catalase. The $Fe^{III}$ ion of Mb is nearly axial ($E/D = 0$), and the principal $g$ values (6, 6, 2) are read off of the edge of Fig. 3.12 for the |±1/2> doublet (red lines). The spectrum of Mb at $g = 6$ is more intense than that at $g = 2$ because the

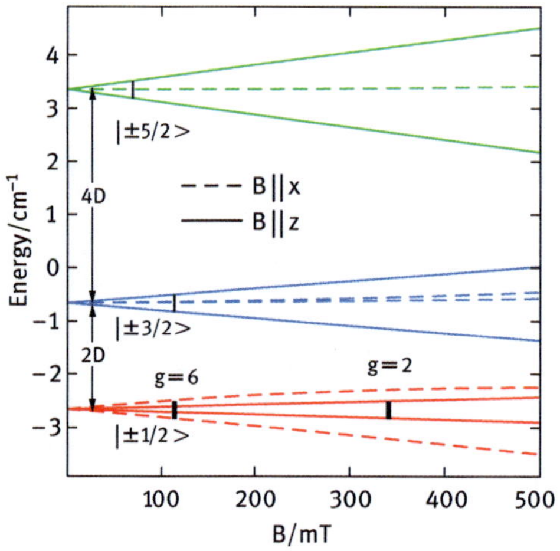


**Fig. 3.11:** The energies of the spin states for $S = 5/2$, $D = 1\ \mathrm{cm}^{-1}$, $E/D = 0$ as a function of the magnetic field along the $x$-axis (dashed lines) and $z$-axis (solid lines).

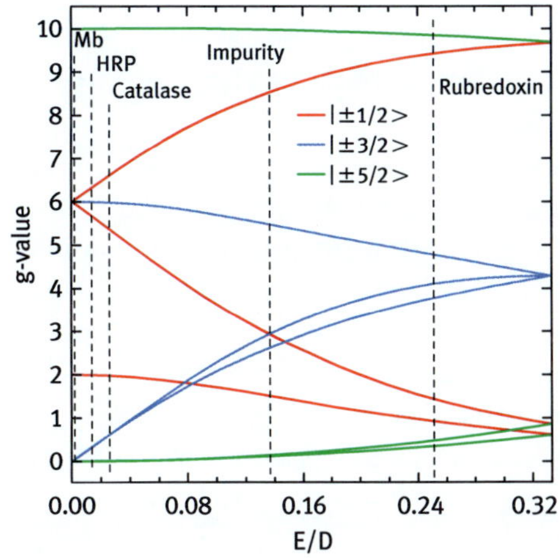


**Fig. 3.12:** The principal $g$ values for the three doublets of an $S = 5/2$ spin system as a function of $E/D$.

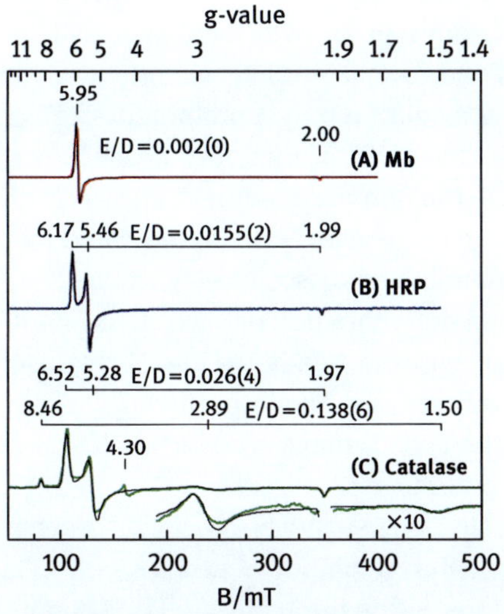


**Fig. 3.13:** EPR spectra (colored lines) and simulations (black lines) of the heme proteins listed in the figure. Instrumental parameters: microwave frequency, 9.662 GHz; temperature, 10 K. The simulation parameters are as listed. The parenthetical values are the width $\sigma_{E/D}$ of the distribution in $E/D$.

polycrystalline (frozen solution) sample has more molecules with **B** aligned near the $x$- and $y$-axes (porphyrin plane) than the $z$-axis of the molecule. As the symmetry of the metal center becomes less axial ($E/D$ greater, more rhombic), the principal $g$ values for the |±1/2> doublet shift from 2 and 6. The $g$ values in Fig. 3.12 can be used to identify the resonances of a single species. For HRP, the vertical dashed line at $E/D$ = 0.014 in

Fig. 3.12 crosses three *g* values (red lines), which match the principal *g* values of HRP determined from the spectrum of Fig. 3.13b. For catalase, the dashed line at $E/D$ = 0.026 matches the principal *g* values determined from Fig. 3.13c.

For EPR signals that can be described by an effective $S'$ = 1/2 system, the concentration of species can be determined from double integration of spectra with a correction for the use of an effective $S'$ = 1/2 system [36]. The correction factor is determined by the *g* values and incorporated into SpinCount. Figure 3.12 is useful for a qualitative understanding, but simulations provide verification and concentration determination if multiple species are present. SpinCount calculates spectra from the spin Hamiltonian of Eq. 3.7 and therefore incorporates the information displayed in Fig. 3.12 in addition to accurate spectral intensities. The simulations of the Mb and HRP spectra shown in Fig. 3.13 match the experimental spectra well for the parameters listed in the figure caption. The simulations quantitatively account for the polycrystalline sum, in addition to changes in the transition probability (intensities) as a function of field. The catalase spectrum provides a good example of the use of simulations for more complicated spectra that are likely to be observed in typical laboratory situations. This catalase sample was prepared from a protein stock solution purchased from Sigma. The catalase spectrum had an additional species with *g* values of 8.46, 2.89, and 1.50, which correspond to $E/D$ = 0.138, as indicated for the "impurity" species in Fig. 3.12. The impurity displayed only a prominent peak at $g$ = 8.46 and one might be tempted to dismiss this signal or incorrectly associate it with the small peak at $g$ = 4.3. The simulation overlaid on the experimental spectrum is composed of two species with parameters given in the figure, which indicated that the $g$ = 8.46 and weaker $g$ = 2.89 and 1.50 resonances are associated with the same species. The simulations allowed a concentration determination of each species. Although the resonance features of the impurity appear weak relative to that of catalase, the impurity accounted for 30% of the iron in this sample. The impurity species has weaker intensity because it spans a larger magnetic field range. The large $E/D$ value of the impurity suggests degradation of the porphyrin.

For metal centers not coordinated to a macrocycle, such as iron-sulfur centers, the symmetry is typically more rhombic, closer to the maximum value $E/D$ = 0.33. As the rhombicity increases from axial, the transitions within the |±3/2> and |±5/2> doublets become allowed. For $E/D$ near to or greater than 0.2, the signal from the |±3/2> doublet dominates the spectrum near $g$ = 4.3. Figure 3.12 shows that as the value of $E/D$ approaches 0.33, the three principal *g* values of the |±3/2> doublet coalesce at 4.3, whereas the *g* values of the other doublets are widely spread from 10 to <1. The $g$ = 4.3 signal is intrinsically more intense because the resonances all occur near $g$ = 4.3, independent of the orientation of the molecule with respect to the magnetic field. Figure 3.14 shows an EPR spectrum and simulation of the oxidized ($Fe^{III}$, $S$ = 5/2) state of the $FeS_4$ center of rubredoxin [37, 38]. The spectrum has a strong resonance at $g$ = 4.3 from the |±3/2> doublet and a weak resonance at $g$ = 9.0 from the |±1/2>

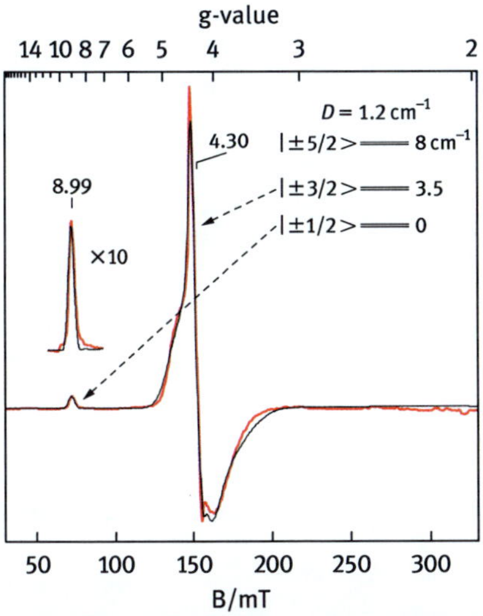


**Fig. 3.14:** An EPR spectrum (colored line, 9.15 GHz) and simulation of the oxidized state of rubredoxin at a temperature of 10 K. The simulation (black line, $D$ = 1.2 cm$^{-1}$ and $E/D$ = 0.258, $\sigma_{E/D}$ = 0.04) correctly predicts the relative intensities of the $g$ = 9 and 4.3 peaks. The energies and assignments of the $S$ = 5/2 spin state are shown.

doublet. The value of $E/D$ must be determined from simulation because the $g$ values of the spectrum are relatively insensitive to $E/D$ near 0.33. For rubredoxin, $E/D$ = 0.25, as indicated in Fig. 3.12. The axial zero-field parameter ($D$ = 1.2 cm$^{-1}$) can be determined from simulation or the temperature dependence of the signal. The population of the spin states within the $S$ = 5/2 manifold obeys the Boltzmann distribution. For rubredoxin, the |±3/2> doublet is 3.5 cm$^{-1}$ above the |±1/2> doublet as shown in the figure inset. As the sample temperature is raised higher than roughly that of liquid helium (4 K), the population of the |±3/2> doublet increases, causing the intensity of the $g$ = 4.3 resonance to increase relative to the $g$ = 9.0 resonance.

The reader is strongly cautioned regarding assignments of $g$ = 4.3 signals. Oxidized adventitious iron or iron impurities in low amounts often show surprisingly intense signals at $g$ = 4.3. Too often, such signals in the literature are incorrectly assigned as the majority species of samples. The determination of species concentrations from spectra should be standard practice. A $g$ = 4.3 signal is often sufficiently sharp that double integration methods, as an effective $S'$ = 1/2 species, will give an approximate concentration. For this determination, the signal should be recorded for higher sample temperature (>50 K) to assure that each of the three doublets has approximately equal population. The double integration of the region encompassing most of the $g$ = 4.3 signal (but not any other signals) is then multiplied by 3 to account the spin population in the other doublets that are not included in the integration. Alternatively, simulations also give the concentration of the species.

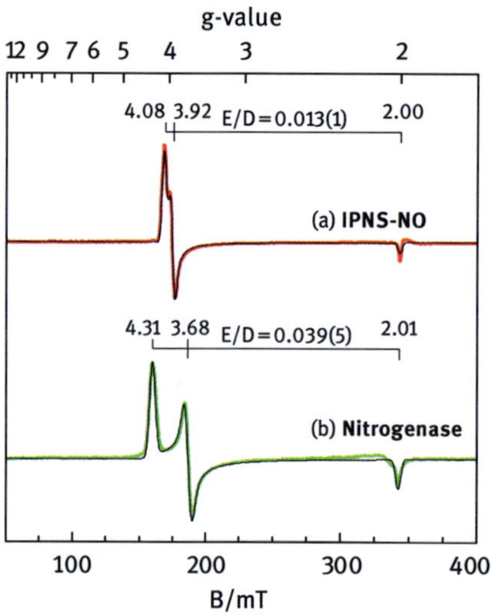


**Fig. 3.15:** EPR spectra (colored lines, 9.62 GHz, 10 K) and simulations (black lines) of $S = 3/2$ centers. (a) Isopenicillin N synthase, (b) the FeMo cluster of nitrogenase. The simulation parameters are as listed. The parenthetical values are the width $\sigma_{E/D}$ of the distribution in $E/D$.

Figure 3.15 shows examples of the $S = 3/2$ spectra and simulations from the Fe center of isopenicillin N synthase bound with NO and the FeMo cluster of nitrogenase [7]. The parameters of the simulation are given in the figure. Both signals originate from transitions within the |±1/2> doublet of the $S = 3/2$ manifold. For $S = 3/2$, the principal *g* values of the |±1/2> doublet are 4, 4, 2 for a metal center in axial symmetry. Diagrams similar to Fig. 3.12 for $S = 3/2$ (and higher) can be produced by SpinCount.

### 3.10.3 Spin systems with *S* = 1, 2, 3, etc.

The focus thus far has been metal centers containing an odd number of unpaired electrons and therefore half-integer spin states. For these odd electron systems, in the absence of a magnetic field (zero-field), the $|\pm m_s>$ spin states will always form degenerate doublets. The applied magnetic field will split a degenerate doublet linearly in energy $\delta E$ with the magnitude of the magnetic field, and the position of EPR signals is given by the resonance condition $\delta E = h\nu = g\beta B$. Metal centers containing an even number of unpaired electrons have integer-spin states. In contrast, all spin levels of an integer-spin system may have different energies in zero-field, with no guarantee of any two spin levels with equal energy. This causes the appearance of EPR spectra to fundamentally change [18]. Figure 3.16 shows the zero-field energies of an $S = 2$ spin system as a function of $E/D$ for $D = 2\ \mathrm{cm}^{-1}$. The states $|\pm m'>$ form doublets that are approximately linear combinations of states $m_s = \pm 1$ or $\pm 2$ [18, 39, 40].

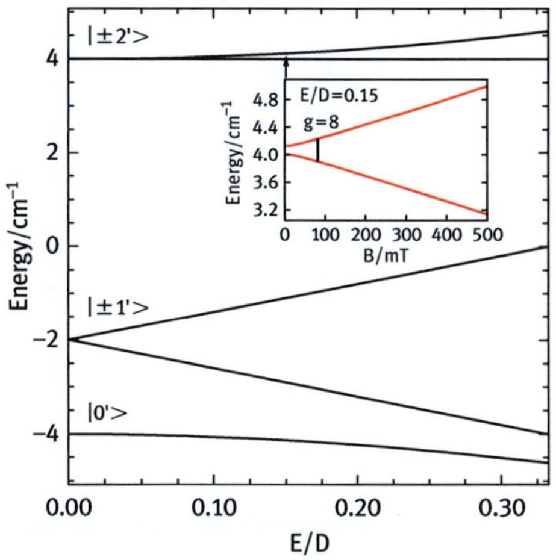


**Fig. 3.16:** Energy as a function of $E/D$ for a $S = 2$ center with $D = 2\ \text{cm}^{-1}$. The inset shows the splitting of the $|\pm 2'\rangle$ doublet for $E/D = 0.15$ as a function of the magnetic field along the $z$-axis of the molecule.

$$|+m'\rangle = (|+m_s\rangle + |-m_s\rangle)/\sqrt{2}$$
$$|-m'\rangle = (|+m_s\rangle - |-m_s\rangle)/\sqrt{2} \qquad (3.11)$$

For $E/D \neq 0$, the $|+m'\rangle$ and $|-m'\rangle$ levels are split in zero-field by an energy $\Delta_1 = 6E$ or $\Delta_2 = 3E^2/D$ for the $|\pm 1'\rangle$ and $|\pm 2'\rangle$ levels, respectively. In the presence of a magnetic field, the resonance condition for integer-spin doublets $|\pm m'\rangle$ is

$$h\nu = [(2\text{m}g_z\beta B \cos\theta)^2 + \Delta_m^2]^{1/2}, \quad m = 1, 2, \qquad (3.12)$$

where $\theta$ is the angle between **B** and the $z$-axis of the molecular frame defined by the **D** tensor. The magnetic field will split the levels of the doublet further apart. An EPR signal may be observed if the energy splitting $\Delta_m$ is less than the microwave quantum $h\nu$. The inset of Fig. 3.16 shows the splitting of the $|\pm 2'\rangle$ doublet for $E/D = 0.15$ as a function of the magnetic field. At a magnetic field of 80 mT, the resonance condition is satisfied.

As is evident from Eq. 3.11, the zero-field states $|+m'\rangle$ and $|-m'\rangle$ do not differ by $m_s = \pm 1$ and cannot obey the standard selection rule of $\Delta m_s = \pm 1$. The appropriate selection rule for the $|\pm m'\rangle$ doublets is $\Delta m_s = \pm 0$. The change in the selection rule affects the polarization direction of the incident microwave magnetic field ($\mathbf{B_1}$), which gives the most intense EPR signals. For half-integer spin centers, the optimal direction has the microwave field oscillating perpendicular to the static magnetic field ($\mathbf{B_1} \perp \mathbf{B}$, perpendicular mode; Figure 3.7), whereas for integer-spin doublets, the optimal orientation has the microwave field oscillating parallel to the static magnetic field ($\mathbf{B_1} \parallel \mathbf{B}$, parallel mode).

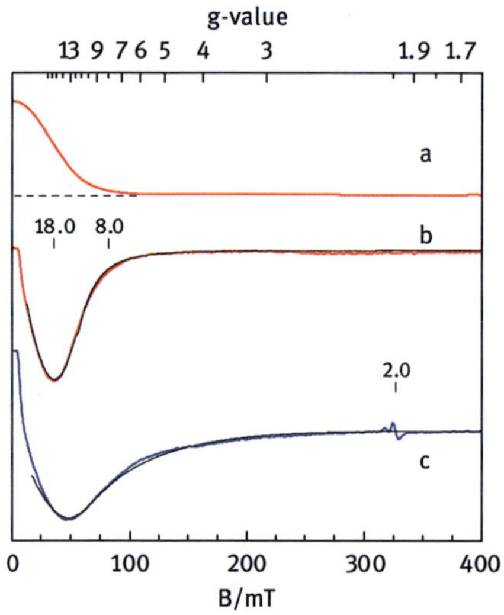


**Fig. 3.17:** EPR spectra and simulations of the reduced ($S = 2$) $Fe_3S_4$ cluster of *D. gigas* ferredoxin II recorded at 4 K with $\mathbf{B_1}$ parallel (b, 9.091 GHz) or perpendicular (c, 9.140 GHz) to **B**. Simulation parameters: $D = -2.5\ cm^{-1}$, $E/D = 0.227$, $\sigma_{E/D} = 0.017$. The absorption spectrum (a) is an integration of (b) demonstrating presence of signal intensity at zero field; a common feature of integer-spin centers but not observed for half-integer spin centers.

Figure 3.17 shows EPR spectra and simulations of the reduced $Fe_3S_4$ cluster of ferredoxin II from *Desulfovibrio gigas* [41]. The energy splitting Δ and change in selection rule result in EPR spectra that are much different in appearance than those of half-integer spin states. The spectra do not fall into one of the four standard types shown in Fig. 3.2. As shown in Fig. 3.17, integer-spin doublets will show signals for both $\mathbf{B_1} \perp \mathbf{B}$ and $\mathbf{B_1} \parallel \mathbf{B}$ orientations, but the $\mathbf{B_1} \parallel \mathbf{B}$ orientation is preferred because (1) the signals are sharper and more intense, (2) overlapping signals from half-integer spin centers with isolated doublets are strictly forbidden, and (3) simulations are less computationally intensive. There are distinguishing features of integer-spin signals evident in the spectra of Fig. 3.17. Integer-spin spectra may be dominated by a downward valley in shape and can have nonzero intensity at very low magnetic fields. This is evident from the integral, Fig. 3.17A, which shows absorption at $B = 0$. As has been discussed above, the zero-field parameters of a molecule are given by a distribution of values having significant spread, and consequently, the parameter Δ has a corresponding distribution in values. The resonance condition (Eq. 3.12) is a function of Δ. For many metal centers, the distribution of Δ straddles the value of the microwave energy $h\nu$, implying that a fraction of molecules with $\Delta = h\nu$ will resonate at $B = 0$. The EPR spectra are often broad owing to the combined broadening effects of the polycrystalline average and the distribution in Δ values. The transition requires a quantum of energy given by the resonance condition (Eq. 3.12), and thus resonances always occur at magnetic fields lower than that expected from $g = 2mg_z$ by an amount related to Δ. For m = 2, $g \approx 8$ ($g_z \approx 2$), signals are to be expected at magnetic fields less than 80 mT for a microwave frequency $\nu \approx 9$ GHz. For ferredoxin, the valley occurs at 36 mT. It is common practice to mark EPR resonances with *g* values in accordance to Eq. 3.5, and for ferredoxin, this position is $g = 18$. However, marking *g* values in this manner for integer-spin spectra is simply a demarcation that does not carry physical significance because the correct resonance condition (Eq. 3.12) is not linear dependent on the magnetic field.

The concentration of species for many types of half-integer spin centers can be obtained from double integration of the EPR signal. This is not true for EPR spectra of integer-spin centers. The intensity of an integer-spin signal is a strong function of Δ, and the common occurrence of resonance into $B = 0$ rules out double integration methods for determination of spin concentrations. A quantitative interpretation of integer-spin signals requires simulation. The simulations of Fig. 3.17 use the parameters given in the caption for the $S = 2$ center. The simulations determine the distribution of the zero-field parameters, and because SpinCount treats the intensity calculation quantitatively, the spin concentration of species can be determined.

Figure 3.18 shows additional examples of integer-spin EPR spectra from iron-sulfur clusters. The Fe-protein of nitrogenase contains a $Fe_4S_4$ cluster. In the fully reduced state of the cluster, all irons are ferrous and form a spin-coupled system with an $S = 4$ state lowest in energy. The EPR signal from the $S = 4$ state (Fig. 3.18a) shows resonances at $g = 16.2$ from the ground doublet and $g = 12.2$ from the first excited doublet [41].

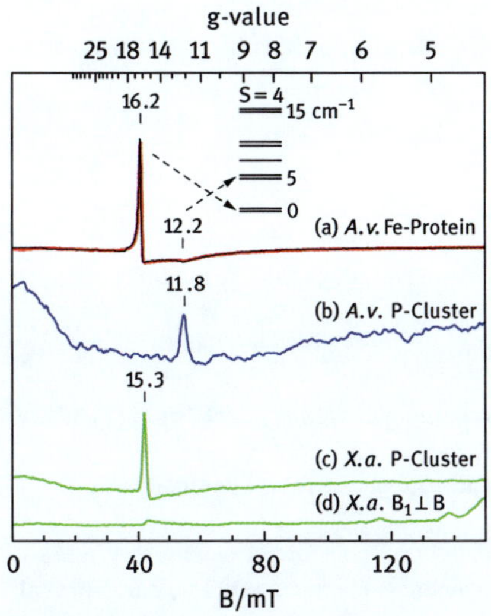


**Fig. 3.18:** (a) Parallel-mode EPR spectra of reduced Fe-protein from *A. vinelandii*. Simulation parameters (black line): $g_y = g_z = 2.05$, $D = -0.8$ cm$^{-1}$, $E/D = 0.32$, $\sigma_{E/D} = 0.08$. Parallel-mode EPR spectra of the P-cluster of nitrogenase from (b) *A. vinelandii* and (c) *X. autotrophicus*. Instrumental parameters: microwave power and frequency, respectively, (a) 13 mW, 9.29 GHz; (b, c) 2 mW, 9.076 GHz; (d) 2 mW, 9.14 GHz; temperature (a, b) 9 K, (c, d) 2 K.

A diagram of the zero-field energies is displayed in the inset. The simulation overlaid on the spectrum quantitatively predicts the intensities of both resonances for the electronic parameters given in the figure. Figure 3.18b and c show EPR spectra recorded in parallel mode from the P-cluster of nitrogenase from *Azotobacter vinelandii* and *Xanthobacter autotrophicus* [42]. In the oxidized state, the P-cluster is essentially composed of two spin $S = 3/2$ $Fe_4S_4$ clusters that are spin coupled to give an $S = 3$ state lowest in energy. Figure 3.18D shows the perpendicular mode spectrum from *X. autotrophicus* on the same relative scale. The signal has significantly less intensity and could easily be overlooked as a baseline impurity.

## 3.11 Conclusion

EPR spectroscopy has been and continues to be of critical importance for the characterization of Fe-S clusters in proteins. This chapter gave a brief introduction to the theory and techniques of EPR as well as introduce the analytical capabilities of SpinCount. The complicated spectroscopy of Fe-S proteins frequently makes it necessary to use more than one spectroscopic technique to fully understand the particular species. Mössbauer spectroscopy is particularly important for Fe complexes and gives information complementary to EPR spectroscopy as will be discussed in the next chapter. This is so important, and we use Mössbauer spectroscopy so frequently, that we have built into SpinCount the same sophisticated ability to interpret Mössbauer spectra. With this, it is now possible to simultaneously calculate and fit both EPR and Mössbauer spectra for the same species and compare with their respective experimental spectra.

## References


[1] Eaton GR, Eaton SS, Salikhov KM. Foundations of modern EPR. Singapore, World Scientific; 1998.

[2] Abragam A, Bleaney B. Electron paramagnetic resonance of transition ions. Oxford: Clarendon Press; 1970.

[3] Carrington A, McLachlan AD. Introduction to magnetic resonance with applications to chemistry and chemical physics. New York: Harper & Row; 1967.

[4] Pake GE, Estle TL. In: Benjamin WA, ed. The physical principles of electron paramagnetic resonance. 2nd ed. Reading (MA); 1973.

[5] Pilbrow JR. Transition ion electromagnetic resonance. New York: Oxford University Press; 1990.

[6] Weil JA, Bolton JR, Wertz JE. Electron paramagnetic resonance: elementary theory and practical applications. New York: Wiley; 1994.

[7] Palmer G. Electron paramagnetic resonance of metalloproteins. In: Que L Jr, ed. Physical methods in bioinorganic chemistry. University Science Books; 2000:121–185.

[8] Gaffney BJ. EPR of Mononuclear non-heme iron proteins. In: Hanson G, Berliner L, eds. Biological magnetic resononance. New York: Springer; 2009:233–268.

[9] Brudvig GW. Electron paramagnetic resonance spectroscopy. In: Sauer K, ed. Methods in enzymology. New York: Academic Press; 1995:536–554.
[10] Hagen WR. Practical approaches to biological inorganic chemistry. In: Crichton RR, Louro RO, eds. EPR spectroscopy. Oxford: Elsevier; 2013:53–75.
[11] Cammack R, MacMillan F. Electron magnetic resonance of iron-sulfur proteins in electron-transfer chains: resolving complexity. In: Hanson G, Berliner L, eds. Metals in biology. New York: Springer; 2010:11–44.
[12] Guigliarelli B, Bertrand P. Application of EPR spectroscopy to the structural and functional study of iron-sulfur proteins. Adv Inorg Chem 1999;47:421–497.
[13] Hagen WR. Probing the iron/sulfur domain with EPR: Pandora's box ajar. ACS Conference Proceedings 1987;459–466.
[14] Cramer WA, Baniulis D, Yamashita E, Zhang H, Zatsman A, Hendrich MP. Cytochrome $b_6f$ complex: structure, spectroscopy, and function of heme $c_n$: n-side electron and proton transfer reactions. In: Fromme P, ed. Photosynthetic protein complexes: a structural approach. Weinheim: Wiley-VCH; 2008:155–179.
[15] Golombek AP, Hendrich MP. Quantitative analysis of dinuclear manganese(II) EPR spectra. J Magn Reson 2003;165:33–48.
[16] Gunderson WA, Zatsman AI, Emerson JP, Farquhar ER, Que L Jr., Lipscomb JD, Hendrich MP. Electron paramagnetic resonance detection of intermediates in the enzymatic cycle of an extradiol dioxygenase. J Am Chem Soc 2008;130:14465–14467.
[17] Gupta R, Fu R, Liu A, Hendrich MP. EPR and Mossbauer spectroscopy show inequivalent hemes in tryptophan dioxygenase. J Am Chem Soc 2010;132:1098–1109.
[18] Hendrich MP, Debrunner PG. Integer-spin electron paramagnetic resonance of iron proteins. Biophys J 1989;56:489–506.
[19] Hendrich MP, Gunderson W, Behan RK, Green MT, Mehn MP, Betley TA, Lu CC, Peters JC. On the feasibility of N2 fixation via a single-site FeI/FeIV cycle: spectroscopic studies of FeI(N2)FeI, FeIV[triple bond]N, and related species. Proc Natl Acad Sci USA 2006;103:17107–17112.
[20] Hendrich MP, Munck E, Fox BG, Lipscomb JD. Integer-spin EPR studies of the fully reduced methane monooxygenase hydroxylase component. J Am Chem Soc 1990;112:5861–5865.
[21] Hendrich MP, Petasis D, Arciero DM, Hooper AB. Correlations of structure and electronic properties from EPR spectroscopy of hydroxylamine oxidoreductase. J Am Chem Soc 2001;123:2997–3005.
[22] Hudder BN, Morales JG, Stubna A, Munck E, Hendrich MP, Lindahl PA. Electron paramagnetic resonance and Mossbauer spectroscopy of intact mitochondria from respiring Saccharomyces cerevisiae. J Biol Inorg Chem 2007;12:1029–1053.
[23] Lacy DC, Gupta R, Stone KL, Greaves J, Ziller JW, Hendrich MP, Borovik AS. Formation, structure, and EPR detection of a high spin Fe(IV)-oxo species derived from either an Fe(III)-oxo or Fe(III)-OH complex. J Am Chem Soc 2010;132:12188–12190.
[24] Lee D, Du Bois J, Petasis D, Hendrich MP, Krebs C, Huynh BH, Lippard SJ. Formation of Fe(III) Fe(IV) species from the reaction between a Diiron(II) complex and dioxygen: relevance to ribonucleotide reductase intermediate X. J Am Chem Soc 1999;121:9893–9894.
[25] Mbughuni MM, Chakrabarti M, Hayden JA, Bominaar EL, Hendrich MP, Münck E, Lipscomb JD. Trapping and spectroscopic characterization of an FeIII-superoxo intermediate from a nonheme mononuclear iron-containing enzyme. Proc Natl Acad Sci USA 2010;107:16788–16793.
[26] Parsell TH, Behan RK, Green MT, Hendrich MP, Borovik AS. Preparation and properties of a monomeric Mn(IV)-oxo complex. J Am Chem Soc 2006;128:8728–8729.
[27] Petasis DT, Hendrich MP. A new Q-band EPR probe for quantitative studies of even electron metalloproteins. J Magn Reson 1999;136:200–206.

[28] Pierce BS, Elgren TE, Hendrich MP. Mechanistic implications for the formation of the Diiron cluster in ribonucleotide reductase provided by quantitative EPR spectroscopy. J Am Chem Soc 2003;125:8748–8759.
[29] Sanakis Y, Petasis D, Petrouleas V, Hendrich M. Simultaneous binding of fluoride and NO to the nonheme iron of photosystem II: quantitative EPR evidence for a weak exchange interaction between the semiquinone QA- and the iron-nitrosyl complex. J Am Chem Soc 1999;121: 9155–9164.
[30] Upadhyay AK, Petasis DT, Arciero DM, Hooper AB, Hendrich MP. Spectroscopic characterization and assignment of reduction potentials in the tetraheme cytochrome c554 from nitrosomonas europaea. J Am Chem Soc 2003;125:1738–1747.
[31] Zatsman AI, Zhang H, Gunderson WA, Cramer WA, Hendrich MP. Heme-heme interactions in the cytochrome b6f complex: EPR spectroscopy and correlation with structure. J Am Chem Soc 2006;128:14246–14247.
[32] Orton JW. Electron paramagnetic resonance: an introduction to transition group ions in crystals. Iliffe; 1968.
[33] Bencini A, Gatteschi D. Electron paramagnetic resonance of exchange coupled systems. Berlin, New York: Springer-Verlag; 1990.
[34] Poole CP. Electron Spin Resonance: A comprehensive treatise on experimental techniques. 2nd ed. New York: Wiley; 1983.
[35] Fox BG, Hendrich MP, Surerus KK, Andersson KK, Froland WA, Lipscomb JD, Munck E. Moessbauer, EPR, and ENDOR studies of the hydroxylase and reductase components of methane monooxygenase from Methylosinus trichosporium OB3b. J Am Chem Soc 1993;115:3688–3701.
[36] Aasa R, Vanngard T. EPR signal intensity and powder shapes. Reexamination. J Magn Reson 1975;19:308–315.
[37] Schulz C, Debrunner PG. Rubredoxin, a simple iron-sulfur protein: its spin Hamiltonian and hyperfine parameters. J Phy Colloq 1976;153–158.
[38] Yoo SJ, Meyer J, Achim C, Peterson J, Hendrich MP, Munck E. Mossbauer, EPR, and MCD studies of the C9S and C42S variants of Clostridium pasteurianum rubredoxin and MCD studies of the wild-type protein. J Biol Inorg Chem 2000;5:475–487.
[39] Hendrich MP, Debrunner PG. EPR of non-Kramers systems in biology. In: Eaton E, Salikhov ed. Foundations of modern EPR. London: World Scientific; 1998:530–547.
[40] Muenck E, Surerus KK, Hendrich MP. Combining Moessbauer spectroscopy with integer spin electron paramagnetic resonance. Methods Enzymol 1993;227:463–479.
[41] Yoo SJ, Angove HC, Burgess BK, Hendrich MP, Muenck E. Moessbauer and integer-spin EPR studies and spin-coupling analysis of the [4Fe-4S]0 cluster of the Fe protein from azotobacter vinelandii nitrogenase. J Am Chem Soc 1999;121:2534–2545.
[42] Surerus KK, Hendrich MP, Christie PD, Rottgardt D, Orme-Johnson WH, Munck E. Moessbauer and integer-spin EPR of the oxidized P-clusters of nitrogenase: POX is a non-Kramers system with a nearly degenerate ground doublet. J Am Chem Soc 1992;114:8579–8590.

# 4 The utility of Mössbauer spectroscopy in eukaryotic cell biology and animal physiology

Mrinmoy Chakrabarti and Paul A. Lindahl

## 4.1 Introduction

In 1957, Rudolf Mössbauer discovered recoilless nuclear fluorescence of gamma rays, an effect that is now given his name. Mössbauer spectroscopy (MBS) is arguably the most powerful method to study iron (Fe); it has been used extensively in fields ranging from geology to biology. For example, Mössbauer (MB) spectrometers were included on Martian exploration rovers to study Fe on that planet [1]. Within the biological realm, MBS has been used extensively to elucidate the magnetic and electronic properties of Fe-containing proteins. Numerous outstanding books and reviews have been written on this subject [2–6]. Thus, it is perhaps surprising that the technique has not been utilized more extensively to address Fe-related problems in cell biology, animal physiology, and biomedicine. There has certainly been some progress in applying MBS to biomedicine [7, 8], but given the importance of this field, we would have expected greater activity. There are undoubtedly many reasons for this, but we suggest that one reason arises from an intellectual barrier of sorts. MBS emerged from physics and has been applied extensively to inorganic chemistry. Accurately and exhaustively interpreting MB spectra requires substantial knowledge of both quantum mechanics and the coordination chemistry of iron, fields in which few biologists and biomedical researchers have been trained. Conversely, few physicists and inorganic chemists have been trained in cell biology, animal physiology, or medicine; thus, they are generally not aware of the critical issues in these fields and how MBS could be applied to them. The aim of this review is to bridge this gap, explaining in plain language the utility of MB spectroscopy for studying Fe metabolism in cells and vertebrate animals. In doing so, we have excluded much of the fundamental and rigorous physics underlying the technique. Instead, we use broad strokes to paint the big picture of this technique and what it can do for these fields.

## 4.2 Transitions associated with MBS

A more descriptive name for MBS is *nuclear gamma-ray resonance*. As the name implies, MBS is somewhat like NMR spectroscopy – both are resonant techniques in which radiation is used to promote nuclear transitions. However, NMR uses radiofrequency radiation to induce transitions whereas MBS uses high-energy gamma-ray radiation. Such radiation is needed because the energies of MB transitions are much greater than those of NMR.

Protons in a nucleus possess a quantum mechanical property called *spin angular momentum*. Nuclei have discrete (not smoothly changing) values of this property. The resulting states are called "nuclear spin states" ($I$). For a single proton, $I$ has the value 1/2. Such states are described by mathematical functions that have solutions only when $I$ and another parameter that reflects magnetic properties, $M_I$, have specific values. There are two "spin functions" associated with $I = 1/2$, namely when $M_I$ equals either +1/2 or −1/2. Spin functions are described using the notation $|I, M_I>$. Thus, for a proton, the two spin functions are called |1/2,+1/2> and |1/2,−1/2>. When a proton is in free space, the two functions have identical energies – they are *degenerate* – and are equally populated (50% will be in one state, 50% in the other). When placed in a magnetic field, the energies associated with the two functions split.

In NMR spectroscopy, the proton is placed in a magnetic field and exposed to radiation of increasing frequency. When the energy of the radiation matches the transition energy, resonance occurs, radiation is absorbed, and an NMR signal is detected. The energy of that resonance (given in parts per million, relative to the resonant energy of a standard compound defined to equal 0 ppm) is sensitive to the chemical environment of the proton. This sensitivity makes NMR spectroscopy useful to chemists even though all the "action" occurs at the nucleus – the portion of the atom generally studied by physicists, not by chemists.

The same basic phenomenon occurs in MBS, but more spin states and transitions are involved and the energies of the transitions are much higher. The $^{57}Fe$ nucleus contains 26 protons and 31 neutrons, each with spin angular momentum that either enhance (→→) or cancel (→←) each other. The result of all 57 nucleons coupling in this way is that the possible values of $I$ include 1/2, 3/2, 5/2, etc. The $I = 1/2$ state is lowest in energy and called the "ground state." The $I = 3/2$ state is the "first excited state." MBS transitions occur between these two states (Fig. 4.1, top panel, center image). These transitions are induced by gamma rays emanating from a $^{57}Co$ source.

As with protons, the two spin functions associated with the $^{57}Fe$ $I = 1/2$ ground state are called |1/2,+1/2> and |1/2,−1/2>. Associated with the $I = 3/2$ excited nuclear spin state are four spin functions designated |3/2,+3/2>, |3/2,+1/2>, |3/2,−1/2>, and |3/2,−3/2>. There are four functions because $M_I$ can have four values when $I = 3/2$. These "rules" arise from the values of $I$ and $M_I$ needed to solve the mathematical function associated with these states.

In $^{57}Fe$ MBS, only six transitions are allowed (indicated by the vertical arrows in Fig. 4.1, top panel, right image). For a free $^{57}Fe$ nucleus in a vacuum, each transition occurs at the same energy (there is a 6-fold degeneracy). However, this degeneracy is generally absent for $^{57}Fe$ nuclei in most biological systems (and in the presence of magnetic fields, see Section 4.7), giving rise to more complicated spectra. The different factors influencing these transition energies will be described in Section 4.7. However, we first need to consider some coordination chemistry of Fe.

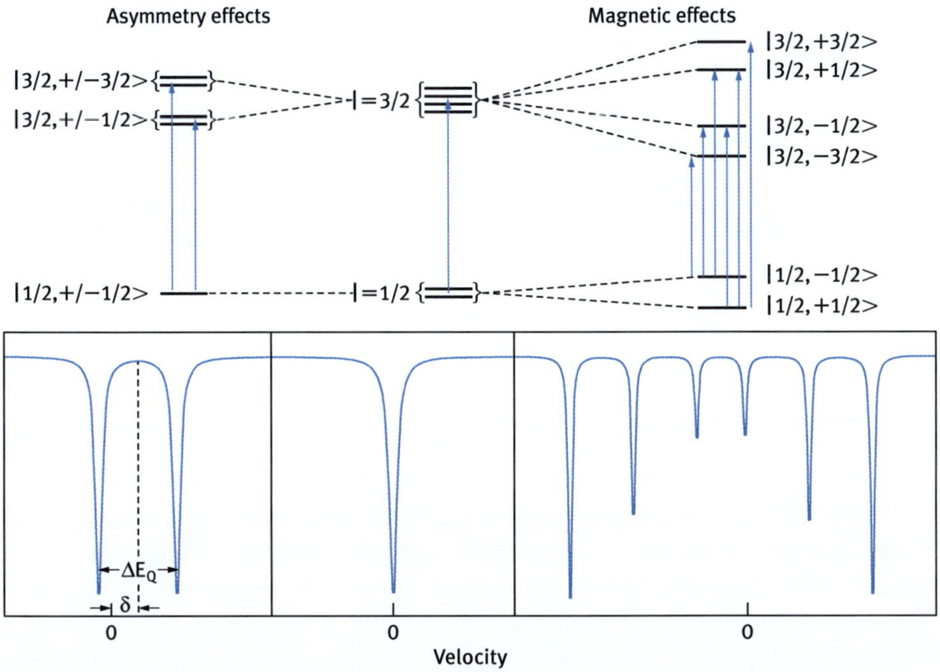


**Fig. 4.1:** Nuclear transitions associated with MB spectroscopy. The basic transitions are shown in the top panel, center image. In the presence of an EFG, the energy levels will shift as shown on the left, giving rise to the quadrupole doublet shown in this figure. In the presence of a magnetic field, levels will shift as shown on the right, often giving rise to the sextet spectral pattern shown in this figure. In many situations, both types of splittings occur.

## 4.3 Coordination chemistry of iron

The elemental state of Fe, as is found in Fe metal and certain organometallic compounds, has no overall charge because Fe atoms contain equal numbers of protons and electrons (26 of each). Up to five electrons can be removed by chemical oxidants, giving rise to the formal oxidation states $Fe^{I}$, $Fe^{II}$, $Fe^{III}$, $Fe^{IV}$, and $Fe^{V}$. Additional electrons are held tightly and cannot be removed under any conditions relevant to biology; sufficiently powerful oxidants are simply not available. Even if they could be generated, higher oxidation states would be so unstable in a water-based environment that the metal ion would immediately grab electrons from other species in the vicinity (e.g. from water), becoming reduced in the process.

In biological systems, Fe in these five oxidation states are not found as autonomous or "free" ions but as complexes coordinated to various ligands (Fig. 4.2). Biologically relevant ligands are molecular species generally containing O, N, or S atoms that "donate" electrons to the metal. The most common ligand is water (or hydroxide

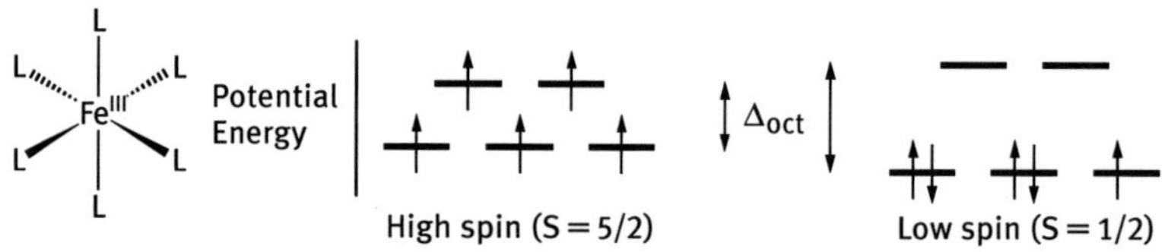


**Fig. 4.2:** Geometry and d-orbital occupation of octahedral $Fe^{III}$ ions. An octahedral complex is shown on the left, with the associated d-orbital splitting pattern shown to the right. The electronic configuration may be HS or LS, depending on the magnitude of the splitting energy, $\Delta_{oct}$.

ion). Within proteins, amino acids such as histidine, cysteine, and aspartic acid can serve as ligands, coordinating to metals. Electron pairs on such atoms are donated to the metal to form a coordinate bond. "Donated" really means "loaned" because ligands retain ownership of the electron pair when they dissociate from the metal.

The $Fe^{II}$ and $Fe^{III}$ states (also called ferrous and ferric, respectively) are most commonly observed in biology. Stabilizing the more extreme oxidation states ($Fe^{I}$, $Fe^{IV}$, and $Fe^{V}$) requires ligands with special properties. Issues of cell biology (Fe trafficking and regulation) generally involve the more stable $Fe^{II}$ and $Fe^{III}$ states, and these will be our focus.

The $Fe^{II}$ and $Fe^{III}$ states have 24 and 23 electrons, respectively, with most of these electrons in inner-shell orbitals. The five to six least tightly held electrons are in the 5 d-type orbitals of the third shell. The 3d-orbitals have unusual shapes, with a node at the nucleus and lobes extending along (or between) particular axes. Orbital geometries give rise to certain geometrical preferences for complexes, including octahedral, tetrahedral, trigonal bipyramidal, etc. For a free $Fe^{II}$ ion in a vacuum, the five 3d-orbitals are degenerate. For an $Fe^{II}$ ion coordinated to four to six ligands, the degeneracy is lifted in ways that depend on the geometry of the complex. In octahedral geometry, two of the 3d-orbitals will have higher energies than the remaining three (Fig. 4.2). The energy separating these two sets of orbitals is called $\Delta_{oct}$. Depending on the geometry of the complex, the orbitals will split into different patterns.

The d-orbitals are responsible for the superior catalytic properties of Fe and other transition metals. Partially filled d-orbitals are involved in bonding, but not strongly so. This allows them to accept and donate electrons without extreme destabilization. Thus, transition metals are relatively stable in multiple oxidation states. Their modest involvement in bonding allows substrate molecules to bind weakly, bringing them close together and thereby increasing the probability of reaction. Such unique catalytic properties make transition metals critical for cellular metabolism. Ironically, these same properties make transition metals like Fe dangerous for cells, as they can also catalyze deleterious reactions that generate reactive oxygen species (ROS) [9]. As a result, Fe trafficking and regulation must be tightly regulated in cells.

## 4.4 Electron spin angular momentum and EPR spectroscopy

Electrons, like protons, possess spin angular momentum. Electronic spin states are similarly designated, except that the letters *S* and $M_S$ replace *I* and $M_I$. A single electron has $S = 1/2$, with two associated spin functions of the form |*S*, $M_S$>, namely ||1/2,+1/2> and |1/2,–1/2>. From an overly simplistic classical perspective, electrons can be viewed as tiny spheres spinning in one direction ($M_S$ = +1/2, spin-up ↑) or the other ($M_S$ = –1/2, spin-down ↓). In the absence of a magnetic field, these two spin functions have the same energy. When a magnetic field is applied to a system with an unpaired electron, the energies of these spin functions split. In EPR spectroscopy, microwave-frequency radiation is used to induce transitions between these states.

## 4.5 High-spin *vs* low-spin $Fe^{II}$ and $Fe^{III}$ complexes

In $Fe^{II}$ and $Fe^{III}$ complexes with more than one unpaired electron, individual spins can couple to give rise to higher overall spin states. Some octahedral $Fe^{III}$ complexes have five unpaired electrons, one in each 3d-orbital. The resulting $S = 5/2$ state has 6 possible $M_S$ values ranging from –5/2 to +5/2, affording six electronic spin functions of the form |*S*, $M_S$>, namely |5/2,+5/2>, |5/2,+3/2> ... |5/2,–5/2>. Other octahedral $Fe^{III}$ complexes have one unpaired electron, and thus, $S = 1/2$. There are two associated spin functions, designated |1/2,+1/2> and |1/2,–1/2>. The former complexes are called "high spin" (HS), whereas the latter are called "low spin" (LS) (Fig. 4.2). Similarly, octahedral $Fe^{II}$ complexes can be HS ($S = 2$) or LS ($S = 0$).

What controls whether a complex is HS or LS? One or two electrons can occupy a given orbital, but if two electrons occupy it, they must be paired, with spins coupled like ↑↓. If $\Delta_{oct}$ for an $Fe^{III}$ complex is less than some critical value (the pairing energy), each of the five electrons will be located in separate d-orbitals, giving rise to the HS configuration. If $\Delta_{oct}$ is greater than the critical pairing energy, four of the electrons will be paired and one will be unpaired, affording the LS configuration. The magnitude of $\Delta_{oct}$ depends on the nature and bond strength of various ligands; thus, the coordinating ligands (and metal oxidation state) together determine whether a given Fe complex will be LS or HS. The geometry of the complex is also critical.

## 4.6 Isomer shift ($\delta$) and quadrupole splitting ($\Delta E_Q$)

One major difference between MBS and NMR spectroscopy is that, in MBS, the four nuclear spin functions associated with the $I = 3/2$ excited state are generally split into two degenerate groups in the absence of a magnetic field. This includes the |3/2,+3/2> and |3/2,–3/2> functions (called the |3/2,±3/2> pair) and the |3/2,+1/2> and |3/2,–1/2> functions (the |3/2,±1/2> pair). In contrast, the ground-state functions do not split in the absence of a magnetic field. Transitions from the ground state to each of these excited state pairs are allowed such that the MB spectrum will show two peaks rather than one (Fig. 4.1, bottom panel, left image). This spectral pattern is called a *quadrupole doublet.*

For reasons that need not concern us, the energy of MB transitions is given in terms of velocity (mm/s) rather than ppm or some other energy unit. The difference velocity of the two transitions of a quadrupole doublet is called the *quadrupole splitting*, $\Delta E_Q$. For biological systems, the magnitude of this parameter ranges from ~0 to ±4 mm/s. The extent of this splitting depends on the asymmetry of the electric field surrounding the $^{57}$Fe nucleus. We need not describe the electric field gradient (EFG) quantitatively, but it should be mentioned that it includes contributions from the asymmetry of the electrons in the iron's d-orbitals as well as from the asymmetry of the ligands coordinating the atom. Compare HS $Fe^{III}$ *vs* $Fe^{II}$ electronic configurations. HS $Fe^{III}$ ions have a symmetrical electronic environment in which each 3d-orbital houses a single electron. For HS $Fe^{II}$ ions, the extra electron breaks the symmetry; thus, one d-orbital will contain two electrons, whereas the others will each contain one electron. As a result, the EFG and thus the $\Delta E_Q$ for HS $Fe^{II}$ complexes are much larger than they are for HS $Fe^{III}$ complexes (*ca* 3 *vs* 0.5 mm/s). Similarly, complexes with less symmetric coordination environments will have larger $\Delta E_Q$ values than those with more symmetric environments.

The average velocity of the two transitions mentioned above is called the *isomer shift*, $\delta$. Such values are reported relative to the centroid of the spectrum of a standard Fe-metal foil at RT, which is defined to be $\delta = 0$ mm/s. $\delta$ basically reflects the oxidation state of an $^{57}$Fe nucleus, but the spin state of the Fe as well as the number and type of ligands (S, N, or O donors) coordinating it also affect this parameter. These two MB parameters, $\Delta E_Q$ and $\delta$, can be used to help identify the oxidation state, electronic configuration, spin state, and coordination environment associated with a mononuclear $Fe^{II}$ or $Fe^{III}$ complex or an Fe/S cluster.

## 4.7 Effects of a magnetic field

When an $^{57}$Fe nucleus is placed in a magnetic field, the degeneracies of the four excited-state functions and two ground-state functions are lifted (Fig. 4.1, right side), such that the energies of each of the six MB transitions are unique. The resulting six-line pattern is called a *sextet*. The splitting of the nuclear states of a diamagnetic (i.e. $S = 0$) $Fe^{II}$ complex is proportional only to the applied magnetic field, whereas for $S \neq 0$ complexes the magnetic splitting is proportional to the sum of external and internal fields. $^{57}$Fe complexes with unpaired electrons possess an internal magnetic field that can affect the nuclear energy levels in the same way as an external field. The presence of an internal magnetic field is due to an interaction between the nuclear spin of the $^{57}$Fe nucleus and the electronic spin of the paramagnetic complex.

At tiny applied magnetic fields (~0.01 T) and higher, HS $Fe^{III}$ $S$ = 5/2 centers and other half-integer spin systems (e.g. $S$ = 1/2 or 3/2) exhibit magnetic splitting due to their internal magnetic field. An $S$ = 0 complex has no internal field, and thus, it exhibits detectable magnetic splitting only when exposed to larger applied fields. Thus, $S$ = 0 complexes exhibit quadrupole doublets when a tiny external field is applied. Curiously, although an integer spin system like an $S$ = 2 $Fe^{II}$ center possesses an internal magnetic field, the magnitude of this field becomes significant only at higher applied magnetic fields. At tiny applied magnetic fields, most integer spin systems will show quadrupole doublets. This is a fundamental (i.e. quantum mechanical) difference between integer and half-integer spin states, and it can be used to experimentally distinguishing half-integer from integer (and diamagnetic) systems. In other words, observing magnetic hyperfine interactions with only a tiny applied field implies a half-integer spin system, whereas the absence of a magnetic spectrum implies an integer or diamagnetic system. Integer and diamagnetic systems can then be distinguished by applying a strong applied magnetic field; in this case, $S$ = 2 $Fe^{II}$ centers would show a significant internal magnetic field, whereas $S$ = 0 $Fe^{II}$ centers would still not possess an internal magnetic field.

The strength of the internal field depends on two factors including (a) the spin state, with higher spin states like $S$ = 5/2 generating larger internal fields and (b) the hyperfine coupling parameter $A$. The entire effect arises because the electronic spin $S$ is coupled to the nuclear spin $I$. The parameter $A$ (which can have orientation dependence, leading to $A_x$, $A_y$, and $A_z$ values) reflects the magnitude of this coupling. Stronger couplings indicate larger $A$ values, which translates into greater internal magnetic fields.

$A$ and $S$ values differ for Fe compounds in different oxidation states and spin states. Thus, the internal magnetic fields generated from these hyperfine interactions can be very helpful in interpreting an MB spectrum. Moreover, these interactions give rise to a synergistic relationship between MB and EPR spectroscopies. MB spectra can be used to quantify internal magnetic fields, whereas EPR spectroscopy can often provide more precise information on $S$ and $A$ than can be unambiguously obtained from MB. With this information on hand, the internal fields that they generate can be more definitively characterized, and in favorable cases, the $A$ values can be interpreted in terms of spin-coupling mechanisms and/or the structure of the Fe complex or cluster giving rise to such interactions.

## 4.8 Slow *vs* fast relaxation limit

The internal magnetic field is different for different spin functions. For $S$ = 1/2, the $|1/2,+1/2\rangle$ function generates an internal field of a particular magnitude and sign, whereas the $|1/2,-1/2\rangle$ function generates an internal field of the same magnitude but of opposite sign. A given $S$ = 1/2 particle fluctuates between these two functions in a temperature-dependent manner. At sufficiently low temperatures (often <10 K), the fluctuation rate (also called the relaxation rate) is slower than the rate of nuclear transitions (technically called the nuclear precession frequency), such that both internal fields are felt by the $^{57}$Fe nucleus. In this so-called slow relaxation limit, the resulting spectrum exhibits magnetic hyperfine interactions arising from both fields (with the influence of each field proportional to its Boltzmann population).

As the temperature is raised, the fluctuation rate increases. Beyond some threshold temperature, fluctuations are faster than the temperature-independent nuclear transitions and the internal fields "average out," such that the $^{57}$Fe nucleus senses a thermally averaged internal field of zero. In this so-called fast relaxation limit, MB spectra exhibit quadrupole doublets rather than magnetic hyperfine interactions; essentially, the magnetic spectrum that was observed at low temperatures "collapses" into a doublet at high temperatures. This phenomenon can be used as evidence for a paramagnetic state, and it can be used to determine $\delta$ and $\Delta E_Q$ for that state. However, be aware that $\delta$ and $\Delta E_Q$ values can change with temperature.

## 4.9 MB properties of individual Fe centers found in biological systems

Virtually every known Fe center in biology has been studied by MBS, and we will not attempt to review such a gigantic effort. Rather, we will summarize the general MB properties of each major group of centers. Such basic information is helpful as we consider the MB properties of more complex biological systems including organelles, cells, and mammalian organs. We will not discuss the more exotic and novel Fe-containing complexes and clusters found in numerous enzyme active sites. The contributions due to such novel (and admittedly very interesting) Fe-containing centers in spectra of complex biological systems such as organelles or whole cells are generally so minor as to be undetectable.

Heme groups are bound in various proteins including hemoglobin and myoglobin. These groups are also found in cytochromes, mitochondrial respiratory complexes, and in various other enzymes. HS and LS $Fe^{II}$ and $Fe^{III}$ states are all common, with general parameters given in Tab. 4.1. Nonheme mononuclear $Fe^{II}/Fe^{III}$ complexes are generally coordinated by 4–6 O, N, and S ligands. Depending on the particular coordinating environment, the center could be HS or LS. As summarized in Tab. 4.1, HS $Fe^{II}$ complexes with N and/or O ligands have $\delta$ = 1.2–1.3 mm/s [2]; loss of a ligand causes $\delta$ to drop by 0.2 mm/s. Replacing N/O with S causes a similar decline of $\delta$.

The MB spectra of spin-coupled systems are complex and beyond the scope of this article. To give readers a basic understanding of these systems, we have listed the MB parameters of some of the commonly observed spin-coupled systems in biology (Tab. 4.1). Fe-O-Fe dimers, $[Fe_2S_2]$, and $[Fe_4S_4]$ clusters contain more than one Fe ion bridged by diamagnetic O or S atoms. In such centers, the local spins of each Fe are

**Tab. 4.1:** Mössbauer and EPR properties of selected iron centers.

| Type of center | Formal oxidation states | System spin state | Isomer shift δ (mm/s) | Quadrupole splitting $\Delta E_Q$ (mm/s) | EPR | Ref. |
|---|---|---|---|---|---|---|
| Heme | $Fe^{II}$ | $S = 2$ | 0.8 (P450)–0.9 | 2.4–2.5 | | [10,11] |
| | | $S = 0$ | 0.49 | 1.1–1.3 | none | [12,13] |
| | $Fe^{III}$ | $S = 5/2$ | 0.42 | 1.24 | $g \sim 6$–2 | [13,14] |
| | $Fe^{III}$ | $S = 1/2$ | 0.2–0.4 | 2.0–2.9 | $g \sim 3$ 2.45, 2.26, 1.91 | [12,13] |
| Nonheme $Fe(O>N)_6$ | $Fe^{II}$ | $S = 2$ | 1.1–1.26 | 2.6–3.5 | | [12] |
| | $Fe^{III}$ | $S = 5/2$ | 0.54 | 0.39 | $g = 4.3$ | [15] |
| | $Fe^{III}$ | $S = 1/2$ | 0.25 | −2.9 | $g = 2.27$, 2.13, 1.97 | [16] |
| $Fe(S)_4$ | $Fe^{II}$ | $S = 2$ | 0.65 | 3.16 | | [17] |
| | $Fe^{III}$ | $S = 5/2$ | 0.25 | 0.74 | | [17] |
| $[Fe\text{-}O\text{-}Fe]_{ox}$ | $[Fe^{III}\ Fe^{III}]$ | $S = 0$ | 0.55 (RNR)<br>0.46 (RNR | −1.62 (RNR)<br>−2.44 (RNR) | None | [18] |
| $[Fe\text{-}O\text{-}Fe]_{semi}$ | $[Fe^{III}\ Fe^{II}]$ | $S = 1/2$ | 0.48, 1.19 | −1.3, 2.4 | $g = 1.95$, 1.86, 1.77 | [19] |
| $[Fe\text{-}O\text{-}Fe]_{red}$ | $[Fe^{II}\ Fe^{II}]$ | $S = 0, 4$ | 1.17, 1.35 (RNR)<br>1.26, 1.27 (RNR) | 3.09–3.11 (RNR)<br>2.93, 3.28 (RNR) | $g = 16$[a] | [18,19] |
| $[Fe_2S_2]^{2+}$ $(SCys)_2$ $(NHis)_2$ | $[Fe^{III}\ Fe^{III}]$ | $S = 0$ | 0.24<br>0.32 | 0.52<br>0.91 | None | [20] |
| $[Fe_2S_2]^{1+}$ $(SCys)_2$ $(NHis)_2$ | $[Fe^{III}\ Fe^{II}]$ | $S = 1/2$ | 0.65–0.74<br>0.22–0.31 | 2.81–3.25<br>0.61–0.63 | $g = 2.02$, 1.90, 1.80 | [20,21] |
| $[Fe_2S_2]^{2+}$ $(SCys)_4$ | $[Fe^{III}\ Fe^{III}]$ | $S = 0$ | 0.28, 0.29<br>0.27, 0.29 | 0.62, 0.44<br>0.50, 0.80 | | [19,22] |
| $[Fe_2S_2]^{1+}$ $(SCys)_4$ | $[Fe^{III}\ Fe^{II}]$ | $S = 1/2$ | 0.32, 0.65 | 0.8, 2.9 | $g = 2.02$, 1.93, 1.93 | [22] |
| $[Fe_3S_4]^{1+}$ $(SCys)_3$ | $[Fe^{III}]$<br>$[Fe^{III}\ Fe^{III}]$ | $S = 1/2$ | 0.27 | 0.71 | $g = 2.01$ | [23–25] |
| $[Fe_3S_4]^{0}$ $(SCys)_3$ | $[Fe^{III}]$<br>$[Fe^{2.5+}\ Fe^{2.5+}]$ | $S = 2$ | 0.32<br>0.46<br>0.46 | 0.52<br>1.47<br>1.47 | | [24] |
| $[Fe_4S_4]^{3+}$ | $[Fe^{III}\ Fe^{III}]$<br>$[Fe^{2.5+}\ Fe^{2.5+}]$ | $S = 1/2$ | 0.29–0.32<br>0.40–0.42 | −0.88 to -0.80<br>−0.93 to 1.03 | $g = 2.12$, 2.04, 2.02 | [26,27] |
| $[Fe_4S_4]^{2+}$ | $[Fe^{2.5+}\ Fe^{2.5+}]$<br>$[Fe^{2.5+}\ Fe^{2.5+}]$ | $S = 0$ | 0.44–0.46 | 0.6–1.5 | None | [28] |
| $[Fe_4S_4]^{1+}$ | $[Fe^{2.5+}\ Fe^{2.5+}]$<br>$[Fe^{II}\ Fe^{II}]$ | $S = 1/2$ or 3/2 | 0.49–0.5<br>0.59–0.62 | 1.23–1.32<br>1.84–2.17 | $g = 2.06$, 1.96, 1.86 | [28] |

Parameters are given relative to α-Fe foil at RT. $\Delta E_Q$ and $\delta$ are given in mm/s, quoted at different temperatures. This is not an exhaustive compilation.

[a] The $S = 4$ state of MMO hydroxylase.

coupled to give a system spin state associated with the entire complex. All the Fe ions in the complex "feel" this system spin. The two irons in Fe-O-Fe dimers can have formal oxidation states of [$Fe^{III}$ $Fe^{III}$], [$Fe^{III}$ $Fe^{II}$], and [$Fe^{II}$ $Fe^{II}$]. For example, each Fe in oxidized [$Fe^{III}$-O-$Fe^{III}$] dimers is HS $Fe^{III}$ with a local spin of $S = 5/2$; these spins often interact to give a system spin of $S_{sys} = 0$. At low applied field and temperature, instead of seeing a magnetically split spectrum as would be observed for mononuclear HS $Fe^{III}$ centers, quadrupole doublets are observed for both $Fe^{III}$ ions. If the coordination environments around the two $Fe^{III}$ ions differ, the respective doublets may have different $\delta$ and $\Delta E_Q$. This type of MB spectrum is observed in oxidized methane monooxygenase reductase, and ribonucleotide reductase (Tab. 4.1). MBS has been used in conjunction with spin labeled EPR and density functional theory to probe the detailed mechanism of these catalytic iron centers [29].

The most commonly observed iron-sulfur clusters (ISCs) in eukaryotic cells have $[Fe_2S_2]^{2+/1+}$, $[Fe_3S_4]^{1+/0}$, and $[Fe_4S_4]^{3+/2+/1+}$ core structures. The formal oxidation states of the iron ions in oxidized $[Fe_4S_4]^{2+}$ clusters are 2$Fe^{II}$ and 2$Fe^{III}$; however, the spins of these irons couple to give $S_{sys} = 0$. The four irons in $[Fe_4S_4]^{2+}$ clusters generally have similar coordination environments. Therefore, at low temperature and low field, all the irons afford the same quadrupole doublet with the same $\delta$ and $\Delta E_Q$. MBS has been used again with other spectroscopic techniques to identify a novel ISC that is critical in cellular iron metabolism [21].

## 4.10 Magnetically interacting Fe aggregates

$Fe^{III}$ in microcrystalline aggregated nanoparticles exhibits internal magnetic fields that may fluctuate or flip directions due to thermal excitation [3]. If the time of measurement (in our case, associated with an MB transition) is longer than the time associated with these fluctuations, the internal magnetic field will fluctuate many times during each measurement. Under these conditions, the internal field gets averaged out and appears to be zero such that a spectrum collected at this condition will show a doublet. The fluctuation rate is temperature-dependent, such that it takes longer to flip as the temperature is lowered. If the temperature is lowered sufficiently, the time required for the flip may be similar to that required for the MB measurement. If a spectrum is collected at temperatures below this critical temperature, the internal magnetic field cannot flip and becomes blocked at one fixed state. In this case, the field does not get averaged and a magnetically split MB spectrum is observed. This is called *superparamagnetic* behavior. The critical temperature is called the blocking temperature $T_B$. The time required to flip also depends on the size of the particle and its magnetic anisotropy (i.e. the directional asymmetry of the internal field).

The Fe in ferritin exhibits superparamagnetic behavior. Ferritin is the major Fe storage protein complex found in the cytosol of higher eukaryotic cells and vertebrate animals. Ferritin is composed of various combinations of 24 heavy and light subunits. Its core is a hollow ~8-nm-diameter sphere that can be filled with $Fe^{III}$ oxyhydroxide nanoparticles in the form of ferrihydrite [30]. A related Fe aggregate is called hemosiderin. This poorly characterized and insoluble form of Fe is found in Fe-overloaded tissues. It is thought to be derived from denatured ferritin [31]. The low-temperature (~5 K) MB spectra of both materials are almost identical, and show magnetically split sextets. At a temperature >200 K, they show quadrupole doublets, again with very similar $\delta$ and $\Delta E_Q$. However, the associated $T_B$ values are very different. Ferritin starts shifting to the superparamagnetic state at a much lower temperature (~50 K) than hemosiderin. At 70 K, hemosiderin shows a magnetically split spectrum, whereas ferritin shows a quadrupole doublet.

$Fe^{III}$ oxyhydroxide (phosphate- or polyphosphate-associated) nanoparticles in yeast [32–35] and human Jurkat cells [36] have blocking temperatures <4.2 K; thus, even at 4.2 K, the majority of the spectra exhibits a quadrupole doublet (in one case, a low-intensity broad magnetic feature was detected near the baseline [32]). The lower $T_B$ associated with these particles indicates that they are smaller than ferritin. This was confirmed for yeast by electron microscopy, which shows clumps of particles with an average particle size of 2–4 nm [34]. X-ray absorption spectroscopy reveals an Fe-P interaction indicating phosphate (or polyphosphate) ions coordinating Fe within these particles. Energy dispersive X-ray analysis imaging confirms the presence of Fe, P, and O; N, S, and C were not detected. The structure and compositions of the nanoparticles in human cells are still being investigated.

Superparamagnetic nanoparticles in yeast give rise to broad, rather isotropic EPR signals centered around $g = 2$. The specific shape of the signal depends on the size and microcrystallinity of the particles. At temperatures lower than $T_B$, such signals exhibit "anti-Curie law" behavior, in that their intensities increase as $T$ increases (normally, EPR signal intensities decline as $T$ increases). At $T > T_B$, the signals follow the Curie law because the magnetic moments of neighboring irons are no longer strongly interacting. Observing this behavior provides evidence for superparamagnetism and associated nanoparticles. However, the value of the $T_B$ associated with this behavior differs according to the spectroscopic method used to measure it.

## 4.11 Insensitivity of MBS and a requirement for $^{57}$Fe enrichment

Only one isotope of iron ($^{57}$Fe) can be used for MB studies, but this isotope is present at only 2% natural abundance. The Fe concentration of normal healthy biological samples is low, ranging from *ca* 0.1 to 1 mM (i.e. 2–20 μM $^{57}$Fe). At these concentrations, data collection times on the order of a month would be required using standard instrumentation. As a result, enriching samples with $^{57}$Fe is essential to obtaining spectra with reasonable signal-to-noise ratios. This requires growing cells in a Fe-deficient medium (to minimize endogenous sources of Fe) while simultaneously supplementing the growth medium with enriched $^{57}Fe$. Even with such enrichment protocols, data collection times as long as 200 h are not uncommon for biological samples whose inherent Fe concentrations are low. Collection times can be minimized using a strong radiation source ($^{57}Co$) along with optimized detector and instrument designs. Another critical procedure is to pack samples (by centrifugation) into MB cups prior to freezing them in liquid $N_2$.

## 4.12 Invariance of spectral intensity among Fe centers

In MBS, each Fe nucleus in a given sample affords about the same relative spectral intensity. This is useful, especially for complex samples that contain multiple species, because it means that the percentage of the total intensity due to a particular spectral feature will be about equal to the percentage of Fe in the sample due to the species giving rise to that feature. This property allows the Fe contents of complex biological samples to be "disentangled" into various groups of Fe in a quantitative manner, and if the total concentration of Fe in the same sample can be determined (see Section 4.12.1–4.12.9), the absolute concentration of each group of Fe within the sample can be determined. We are unaware how such critical information could be obtained by other methods. The following sections highlight some of what has been learned with regard to the cell biology of iron using MBS.

### 4.12.1 Mitochondria

A large proportion of cellular Fe is imported into mitochondria where it is used in the synthesis of ISCs and heme centers; thus, these organelles are a major "hub" for Fe trafficking. Lesuisse *et al.* [32] published the first MB spectra of Fe-overloaded mitochondria. The organelles were isolated from a strain of *Saccharomyces cerevisiae* in which the yeast frataxin homolog 1 protein (Yfh1) was deleted. A deficiency of the mitochondrial protein frataxin in humans causes Friedreich ataxia, a neurodegenerative disease associated with Fe deposits and a deficiency of ISCs and hemes (both of which are synthesized in the mitochondria). Mitochondria isolated from ΔYfh1 cells accumulate large quantities of Fe. MBS of these mitochondria exhibit (at 4.3 K) a single species – a broad quadrupole doublet with $\delta = 0.53$ mm/s and $\Delta E_Q = 0.63$ mm/s. High-field spectra exhibit featureless broadening, which reflects a wide distribution of individual hyperfine fields (i.e. each Fe in a population of these particles feels internal fields of different magnitudes). This indicated a population of inequivalent $Fe^{III}$ ions as is observed for $Fe^{III}$ oxyhydroxide aggregates. The Fe:P molar ratio associated with these particles was ~1:3. These authors concluded that the Fe deposits within Yfh1p-deficient mitochondria consist of $Fe^{III}$ phosphate nanoparticles; no other spectral features were observed.

One of our first objectives upon entering this field was to determine the absolute concentration of Fe (and other metals) in isolated mitochondria. To do this, we pelleted isolated mitochondria into tall, thin glass tubes and marked the heights of the pellets such that the volumes of the packed material could be determined accurately. Samples were transferred to other containers for acid digestion and metal analysis. When multiplied by the dilution factors associated with that transfer, the absolute concentration of Fe in the pelleted sample could be determined. In other experiments, we determined the fraction of the packed volume occupied by the mitochondria themselves (opposed to that due to the buffer). We then divided the measured concentration by this fraction (the so-called packing efficiency) to afford the absolute concentration of Fe in yeast mitochondria [37, 38]. Typical yeast mitochondria contain 700–800 μM Fe, but there can be significant variations (from as low as 140 μM to as high has 20 mM) depending on the concentration of Fe in the growth medium, the type of growth medium, and the time of harvesting.

Despite describing this approach routinely in our publications, we are unaware of another group that measures the absolute Fe concentration of mitochondria using these methods. Virtually all other groups report the "concentration" of Fe in mitochondria in units of "nanomoles of Fe per milligram mitochondrial protein." Such values are actually not concentrations but the *ratio* of two concentrations – "nanomoles of Fe per milliliter" divided by "milligrams of protein per milliliter." Interpreting such ratios as though they were Fe concentrations is fraught with danger, in that an increase could as readily reflect a decline in protein concentration as an increase in Fe concentration. This fundamental problem does not seem to be recognized currently in the field.

Low-temperature, low-field MB spectra of mitochondria isolated from both respiring and fermenting WT *S. cerevisiae* are dominated by a quadrupole doublet located in the central region of the spectrum (Fig. 4.3a and b, respectively) [37, 39]. This so-called central doublet (CD) has $\delta$ = 0.45 mm/s and $\Delta E_Q$ = 1.15 mm/s. These parameters are typical of both $S = 0$ $[Fe_4S_4]^{2+}$ clusters and LS $Fe^{II}$ hemes (Tab. 4.1)] such that the two types of centers cannot be distinguished by MB. However, UV-vis spectroscopy can be used to quantify the heme centers, decomposing them into heme *a*, *b*, and *c* contributions. The MB spectra of mitochondria isolated from human cells are similar (Fig. 4.3d). Also observed by MB were $Fe^{III}$ oxyhydroxide nanoparticles (similar to what was observed by Lesuisse *et al.* [32] but in lesser amounts) and nonheme high-spin (NHHS) $Fe^{II}$ ions. The EPR spectra of such samples affords ~5 signals, including those from the $[Fe_2S_2]^{1+}$ clusters of succinate dehydrogenase and the Rieske Fe/S protein, a $g = 2.0$ radical signal, a signal from the mixed-valence state of the active site of cytochrome c oxidase, and a $g = 4.3$ signal originating from NHHS $Fe^{III}$ ions with rhombic symmetry. Integrating these results, along with Fe concentrations determined by ICP-MS, afforded the first "iron-omic" description of the Fe contents of mitochondria. Admittedly, the achieved level of resolution is far less than most "omic" methods, but it is nonetheless the best obtained currently with regard to Fe.

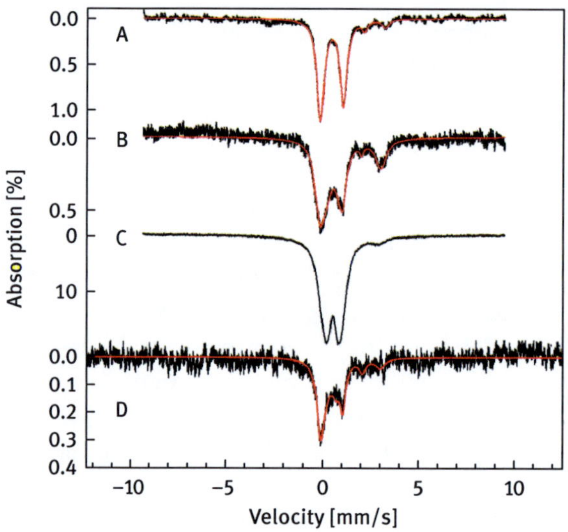


**Fig. 4.3:** (a) MB spectra (5 K, 0.05 T) of respiring yeast mitochondria, (b) fermenting yeast mitochondria, (c) Atm-1-depleted yeast mitochondria, and (d) human Jurkat cell mitochondria. Red lines represent simulations.

The concentration of NHHS $Fe^{II}$ in fermenting mitochondria was unexpectedly large, corresponding to 20%–25% of the MB spectral intensity. For an organelle containing ~700 µM Fe, this translates into 150–175 µM Fe. The concentration of this species was far less in *respiring* mitochondria, which, along with an increased level of ISC Fe, suggested that the detected NHHS $Fe^{II}$ species might represent a “pool” of Fe that is used as feedstock for the synthesis of ISCs [39]. Accordingly, the size of the pool declines during respiration as the rate of ISC synthesis increases.

This ISC-feedstock idea received significant support from an earlier study by Lutz *et al.* [40] in which the rate at which the ISC of a ferredoxin (Yah1) was assembled and installed into the apo-protein was compared before and after adding 1,10-phenanthroline (phen) to a suspension of isolated mitochondria. Phen is a strong $Fe^{II}$ chelator that can penetrate the inner membrane (IM) of mitochondria. Lutz *et al.* reported that ISC assembly was shut down in isolated mitochondria that had been treated with phen. Pandey *et al.* [41] performed similar studies in which the inhibitory effect of phen on ISC assembly was described. These authors concluded that mitochondria must contain an endogenous form of Fe that is used for ISC assembly. Unaware of these studies, Holmes-Hampton *et al.* [42] reported that phen selectively chelates the NHHS $Fe^{II}$ species present in mitochondria. Viewed collectively, these studies provide strong evidence that the NHHS $Fe^{II}$ specie(s) in mitochondria is(are) used as feedstock for ISC assembly. Few techniques besides MBS would allow such conclusions, as NHHS $Fe^{II}$ complexes exhibit neither EPR signals nor intense UV-vis signatures.

We have also obtained MB spectra of various mutant yeast strains, including Yah1p-deficient [33], Atm1p-deficient [34], Aft1-1$^{up}$ [35], and ΔMtm1 [43]. Yah1p is a [$Fe_2S_2$]-containing ferredoxin in the mitochondria that functions as a redox mediator in ISC biogenesis [44]. Atm1p is a transporter on the IM of the mitochondria that may export a sulfur-containing species used by proteins in the cytosol to synthesis ISCs [45]; however, that species has not been identified. The absence of either protein causes a massive increase of mitochondrial Fe along with a deficiency of Fe in the cytosol. MBS was used to investigate the nature of the Fe that accumulated in mitochondria. Aft1p is a transcription factor that, during Fe-deficient conditions, stimulates the so-called iron regulon, a group of ~20 genes that are involved in Fe import and homeostasis [46]. Aft1p in the Aft1-1$^{up}$ strain is mutated such that it constitutively stimulates the iron regulon under all concentrations of Fe in the medium. Like Atm1p, Mtm1p is a transporter on the IM of the mitochondria but the identity of the transported species is unknown. The lack of Mtm1p causes Fe to accumulate in the mitochondria and is associated with a decline of superoxide dismutase activity [47].

MBS was used to characterize the Fe that accumulated in these mitochondria. The low-temperature, low-field spectrum of mitochondria isolated from Yah1p-deficient, Atm1p-deficient, ΔGgc1 [48], Aft1-1$^{up}$, and ΔMtm1 cells are remarkably similar. All are dominated by a single intense species – a broad quadrupole doublet due to superparamagnetic $Fe^{III}$ oxyhydroxide phosphate-associated nanoparticles (Fig. 4.3c). The parameters associated with this doublet and the behavior of the material at high field were indistinguishable from those of ΔYfh1 cells. A broad NHHS $Fe^{II}$ doublet, representing ~2% of the spectrum, was also observed. Also associated with this so-called standard mitochondrial Fe accumulation phenotype is a decline in the level of ISC and heme groups and an $O_2$ dependency (i.e. Fe does not accumulate and ISC/hemes do not decline when cells are grown anaerobically). The similar phenotype of iron nanoparticle accumulation generated by a diverse array of mutations indicates that these phenotypic characteristics are actually secondary effects of these mutations and that the primary function of each protein cannot be directly derived from these properties.

### 4.12.2 Vacuoles

Vacuoles are acidic organelles in yeast that store and sequester Fe from other cellular compartments so as to avoid unwanted reactions [49]. Iron accumulates in the vacuoles of cells grown on medium containing more than *ca* 1 μM Fe. Cockrell *et al.* [15] isolated vacuoles and characterized their Fe content using MBS, EPR, and ICP-MS. Low-field 5-K MB spectra exhibited a sextet characteristic of mononuclear NHHS $Fe^{III}$ complexes (Fig. 4.4a). The hyperfine coupling constant *A* was large, a characteristic of "hard" inorganic O donor ligands (e.g. phosphate or polyphosphate

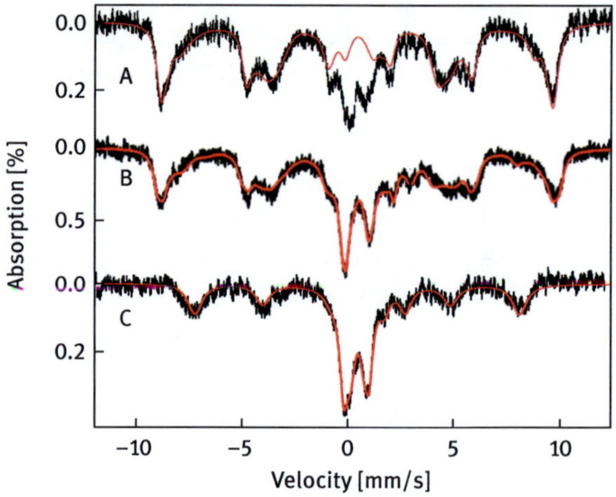


**Fig. 4.4:** (a) MB spectra (5 K, 0.05 T) of fermenting yeast vacuoles, (b) whole fermenting yeast cells, and (c) whole human Jurkat cellsJurkat cells. Red lines represent simulations.

ligands). This species afforded a $g = 4.3$ EPR signal that indicates an $S = 5/2$ state with rhombic (very low) symmetry. Considered collectively, these data indicate that Fe in these organelles is present as nonheme mononuclear HS $Fe^{III}$ ions coordinated to ligands related to polyphosphate. The MBS of many batches also exhibited a broad quadrupole doublet typical of $Fe^{III}$ oxyhydroxide nanoparticles. There appears to be a pH-dependent equilibrium in which nanoparticles form at high pH and mononuclear $Fe^{III}$ species are present at low pH.

### 4.12.3 Whole yeast cells

Fermenting WT yeast cells were grown in minimal medium under Fe-deficient (~1 μM $^{57}Fe$), Fe-replete (10–40 μM), and Fe-overload (100–10,000 μM) conditions and harvested during late exponential phase [50]. The Fe content of Fe-deficient cells was dominated by the CD, HS heme, and nonheme $Fe^{II}$ centers. The CD and heme centers arise primarily from mitochondria, whereas the majority of the NHHS $Fe^{II}$ species appears to be non-mitochondrial (they may be cytosolic). This is the "essential iron-ome" of the cell. The vacuoles accumulate Fe in cells grown with 1→10 μM Fe in the medium and are completely filled in cells grown on 40 μM Fe. Under these conditions, nearly 3/4 of total cellular Fe is found in vacuoles; the remainder is mainly mitochondrial (Fig. 4.4b). Fe-deficient cells lack $Fe^{III}$ oxyhydroxide nanoparticles; these particles accumulate as the Fe concentration in the medium increases. Cells grown in excess medium Fe (and harvested at the end of exponential phase) are remarkably similar to those grown under Fe-replete conditions, indicating tight regulation of Fe import during exponential growth.

Yah1p-deficient [33], Atm1p-deficient [34], Aft1-1$^{up}$ [35], and ΔMtm1 [43] whole cells have also been examined by MBS. As mentioned above, the mitochondria from these cells all exhibit the standard Fe accumulation phenotype, but some differences are evident on the whole-cell level. The MBS of Yah1p- and Atm1p-deficient cells are both dominated by a nanoparticle doublet; whole-cell spectra are essentially indistinguishable from the spectra of the mitochondria isolated from these cells. In contrast, the MBS of Aft1-1$^{up}$ and ΔMtm1 whole cells contain a second major species – namely the nonheme mononuclear HS $Fe^{III}$ species located in vacuoles. Although further studies are required to explain this difference, it appears that the vacuoles in Yah1p- and Atm1p-deficient cells either do not import Fe or that mononuclear vacuolar HS $Fe^{III}$ has been converted into nanoparticles that are not distinguishable from those located in mitochondria. In contrast, the vacuoles in Aft1-1$^{up}$ and ΔMtm1 cells appear to import Fe and maintain it in the standard mononuclear NHHS $Fe^{III}$ state.

### 4.12.4 Human mitochondria and cells

$^{57}$Fe-enriched human Jurkat cells, and mitochondria isolated from such cells have been studied by MBS [36]. Using a packing efficiency of 0.81, the Fe concentration in such cells was determined to be *ca* 400 μM. Approximately half of this is found in the mitochondria (mainly as respiratory complexes, and nanoparticles). Another 160 μM is present as ferritin; the remainder is some combination of nonmitochondrial NHHS $Fe^{II}$, Fe/S clusters, and hemes (Fig. 4.4c). At this level of analysis, the Fe content of these human cells and their mitochondria (Fig. 4.5) are rather similar to that of yeast, except that human cells store Fe in ferritin rather than in vacuoles. The presence of $Fe^{III}$ oxyhydroxide nanoparticles was unexpected, and further studies are underway to determine whether the presence of these particles depends on the type of Fe in the growth medium.

### 4.12.5 Blood

Oshtrakh and Semionkin [51] used MBS to examine erythrocytes from healthy human patients and patients with erythremia [a malignant blood disease characterized by an overproduction of red blood cells (RBCs)]. In both cases, MBS of the blood is dominated by two spectral features including oxyhemoglobin and deoxyhemoglobin. Ortalli *et al.* [52] used MBS to examine hemoglobin from patients with leukemia and Hodgkin disease. The spectra of RBCs exhibited two quadrupole doublets, one from deoxyhemoglobin ($\delta$ = 0.96 mm/s and $\Delta E_Q$ = 2.35 mm/s) and the other from oxyhemoglobin ($\delta$ = 0.26 mm/s and $\Delta E_Q$ = 2.1 mm/s). No significant differences were observed between healthy and diseased states. Oshtrakh [53] also

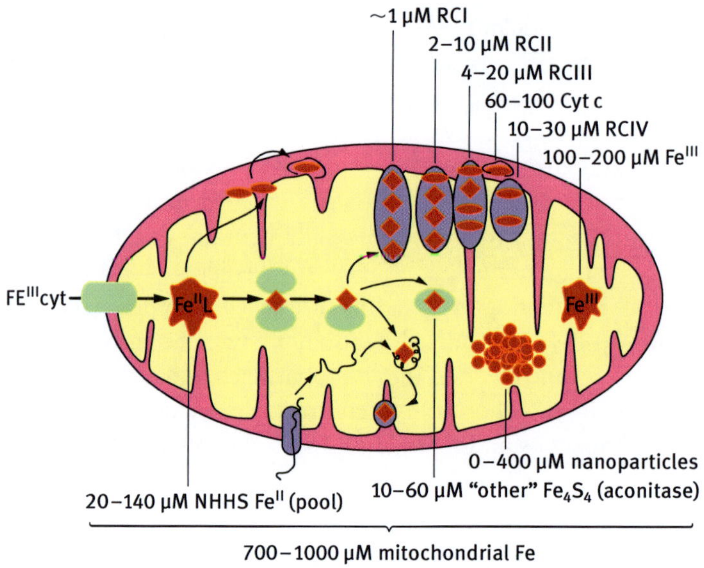


**Fig. 4.5:** Approximate iron distribution and concentrations in human mitochondria based on integrative biophysical methods [36]. A similar distribution is observed for yeast mitochondria [39], but they lack respiratory complex I (RCI). Fe enters the matrix as a pool of $Fe^{II}$ (L in the figure represents an unknown ligand environment of this complex). $Fe^{II}L$ is used as feedstock for ISC assembly (diamonds represent clusters) and perhaps heme biosynthesis (ovals represent hemes). Most proteins enter mitochondria as unfolded polypeptides *via* TOM/TIM complexes (represented by the purple oval at the bottom). Once in the matrix, the signal sequence is clipped and metals are installed as the proteins fold. Clusters are installed most predominantly into the respiratory complexes (RCI-RCIV). Under certain conditions, some $Fe^{II}$ is oxidized to mononuclear $Fe^{III}$. A portion of this can form oxyhydroxide nanoparticles (clump of circles).

obtained MBS of fetal blood. Two doublets were obtained, one with $\delta$ = 0.27 mm/s and $\Delta E_Q$ = 2.1 mm/s, arising from oxyhemoglobin, and the other with $\delta$ = 0.94 mm/s and $\Delta E_Q$ = 2.28 mm/s, arising from deoxyhemoglobin (HS $Fe^{II}$). Shahal *et al.* [54] used MBS to study the oxidative stress response of neonatal *vs* adult RBCs. Neonatal blood was obtained from umbilical veins, and samples were treated with phenylhydrazine, which causes oxidative damage. There was more oxidative damage in treated prenatal RBCs, suggesting an association with the formation of an unidentified Fe degradation product at 20% spectral intensity ($\delta$ = 0.44 and $\Delta E_Q$ = 0.94 mm/s). This product was present exclusively in neonatal RBCs. Ni *et al.* [55] examined the RBC of patients with liver cancer and cirrhosis. Again, the same two doublets were observed, but the deoxyhemoglobin doublet intensity was reduced in the patient's samples. Human serum transferrin has also been examined by MBS [56]. This blood plasma protein binds 1 or 2 $Fe^{III}$ ions and is used to transfer Fe from the blood to organs.

### 4.12.6 Heart

Chua-anusorn *et al.* [57] used MBS to examine autopsied heart tissue from patients with β-thalassemia/hemoglobin E. This disease results from impaired hemoglobin synthesis and causes Fe to accumulate in the heart and other organs. These researchers found that the Fe accumulates as $Fe^{III}$ aggregates, including as ferritin and hemosiderin. More recently, Whitnall *et al.* [58] used MBS to examine the Fe that accumulates in the hearts of mice with a conditional knockdown of frataxin. These mutant mice express frataxin normally except for in heart and skeletal muscle. As mentioned above, a deficiency of frataxin is responsible for the disease Friedreich ataxia. This group found nanoparticles of Fe, P, and S in heart mitochondria (Fig. 4.6a). These aggregates afforded a broad quadrupole doublet with parameters ($\delta$ = 0.48 mm/s and $\Delta E_Q$ = 0.71 mm/s) similar to what we have observed in human Jurkat cells ($\delta$ = 0.48 mm/s and $\Delta E_Q$ = 0.57 mm/s) grown in medium supplemented with 100 μM $^{57}Fe^{III}$ citrate (Fig. 4.6b) [36].

### 4.12.7 Liver

Three major groups of Fe have been identified in normal human and rat livers, including ferritin-like, hemosiderin-like, and mononuclear (heme and nonheme) Fe-containing enzymes. Rimbert *et al.* [59] found that the first two groups constituted 90% of the Fe in fresh and lyophilized iron-overloaded liver samples. One type of overloaded state was caused from excessive intestinal absorption; the other by multiple transfusions given to β-thalassemia patients. Besides the species exhibited by

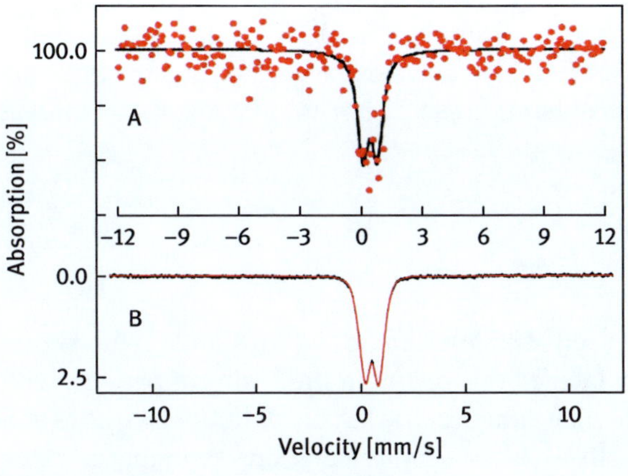


**Fig. 4.6:** (a) MB spectrum (5 K) of the heart of a mouse model of Friedreich ataxia [58]. (b) MB spectrum (5 K, 0.05 T) of whole human Jurkat cells grown in a medium containing 100 μM ferric citrate. Solid lines represent simulations.

normal liver samples, Fe-overloaded tissue from β-thalassemia patients (with ~470 mM Fe) also exhibited features of HS $Fe^{III}$ displaying superparamagnetic behavior above 80 K. The resulting doublet had $\Delta E_Q$ = 0.5 mm/s. Rat Fe-overloaded liver did not show this feature. Whitnall *et al.* [58] examined the livers of mutant mice (where frataxin levels were normal) using MBS. This spectrum was dominated by a sextet due to ferritin. Additional spectral intensity in the center of the spectrum was fitted with a quadrupole doublet, and also assigned to ferritin.

### 4.12.8 Spleen

Human spleen samples were obtained from a patient with primary myelofibrosis and myeloid metaplasia. In this disease, abnormal blood cells and fibers build up inside the bone marrow, which causes the spleen to enlarge and produce blood cells. They also examined the spleen of a patient who had suffered a traumatic spleen lesion (this sample served as control). Samples were rinsed of blood and lyophilized, and MBS were obtained at various temperatures [60]. The diseased spleen contained 10 times more Fe than the normal spleen. However, the Fe in both normal and diseased spleens was present mainly as ferritin. They obtained quadrupole doublets with $\delta$ ranging from 0.14 to 0.46 mm/s (average ~0.36) and $\Delta E_Q$ from 0.5 to 1.9 mm/s (average 0.68).

St Pierre *et al.* [61] used MBS to examine the spleens from Thai and Australian patients with β-thalassemia/hemoglobin E. Only the Australian patients had received blood transfusions and chelation therapy. Tissues were lyophilized and ground to a powder. Fe concentrations of these tissues were substantially higher than normal. Spectra at 78 K were dominated by a doublet with parameters characteristic of either paramagnetic or superparamagnetic HS $Fe^{III}$. Also evident was a sextet, and occasionally a doublet due to heme Fe. Spectra from the Austrialian patients exhibited a stronger sextet at 78 K, which was attributed to polynuclear $Fe^{III}$ oxyhydroxide particles in a goethite-like form (i.e. hemosiderin). At 5 K, a ferritin-like sextet was evident along with the HS $Fe^{III}$ doublet [62].

### 4.12.9 Brain

Galazka-Friedman *et al.* [63] have used MBS to examine the substantia nigra region of the brain. Fe reportedly accumulates in this region in the brains of patients with Parkinson disease. Interestingly, in their samples, the Fe concentration in this region of PD and control brains were both ~1.3 mM (our calculation, assuming a tissue density of 1 gm/mL). The 90-K spectra of quickly prepared fresh-frozen samples were dominated by ferritin-like $Fe^{III}$ in both PD samples and controls. Fe accumulated as ferritin with no other forms of Fe evident. There was no evidence of $Fe^{II}$ in the spectra.

In a more recent study [64], the same group found that the concentration of labile Fe in the brains of PD patients was higher than in controls, but that the overall Fe concentration was not different. Again, MBS detected no $Fe^{II}$ in any samples and only ferritin Fe was observed.

We raised C57BL/6 mice on chow supplemented with 50 mg/kg of $^{57}Fe^{III}$ citrate [65]. MBS was used to examine the Fe content of their brains during development and under Fe-deficient conditions. EPR and UV-visible spectroscopies and ICP-MS were used as auxiliary techniques. Organs were perfused with Ringer's buffer to remove the contribution of blood. Perfused organs were immediately dissected under anaerobic refrigerated conditions, loaded into MB cups, and frozen for later analysis. After subtracting a small contribution of Fe from blood, the brain contained ~180 μM Fe. MB spectra revealed five groups of Fe (Fig. 4.7). In the spectra of brains isolated from 3-week-old animals, a sextet due to ferritin dominated, corresponding to nearly 60% of spectral intensity (*ca* 110 μM). Also observed was the CD (50 μM Fe), HS $Fe^{II}$ hemes (*ca* 10 μM), and NHHS $Fe^{II}$ (*ca* 10 μM). There was no evidence of hemosiderin Fe. The CD and heme contributions primarily arose from mitochondrial respiratory complexes. Viewed simplistically, the 3-week-old mouse brain contains Fe that is stored (ferritin) and Fe that is used (mitochondrial) to generate chemical energy.

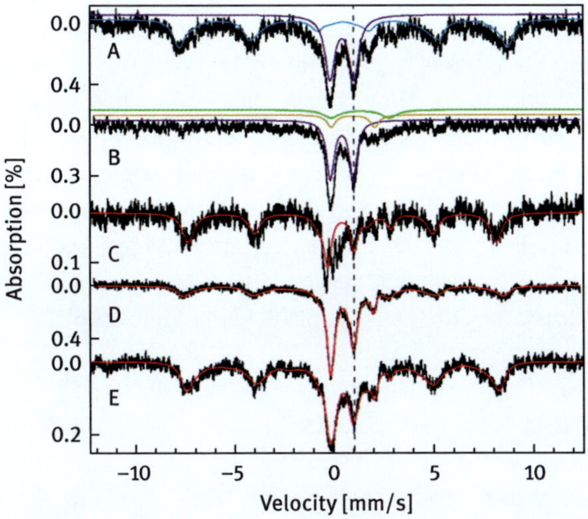


**Fig. 4.7:** MB spectra (5 K, 0.05 T) of mouse brain at different developmental stages: (a) 3 weeks, (b) 3 weeks, iron deficient, (c) 1 week prenatal, (d) 4 weeks, (e) 58 weeks. Solid lines represent simulations. The dashed vertical line shows the high energy line of the CD. Note the ferritin signals surrounding the CD are largely absent in 4-week-old mice, a finding that suggests that storage iron has been consumed to support biogenesis of mitochondria in rapidly growing animals. Modified from a figure in [65]. See the original paper for additional details.

In terms of development, the most intriguing results were that (a) the brains of prenatal mice contained (in terms of percentage and absolute concentrations) mostly ferritin and (b) the mitochondrial Fe components developed dramatically during the first few weeks of life as ferritin levels declined. This burst of mitochondriogenesis might be involved in the increased level of brain activity that occurs during this developmental period. The decline of ferritin was associated with a transient period of Fe deficiency (between 2 and 4 weeks). The rate of Fe import into the brain appears to be too slow during this period to keep up with the rate of mitochondriogenesis. This depletes the stores of ferritin. Within a few weeks, ferritin levels recover. Beyond that age, the Fe content of the brain remains relatively constant throughout the lifetime of the animal.

We have also obtained spectra of other organs (liver, heart, spleen, and kidney), all of which exhibit substantial intensity due to mitochondrial Fe (in addition to ferritin). They also exhibit heme signals due to blood (despite our best efforts to perfuse the organs) and many exhibit low-intensity nonheme HS $Fe^{II}$ doublets. So why have previous investigations not detected such features? Without $^{57}Fe$ enrichment, only tissues with extraordinarily high concentrations of Fe can be examined by MBS, and in these cases, the dominating species tend to be aggregated forms of Fe such as ferritin and hemosiderin. In such spectra, mitochondrial Fe could have easily been obscured by these dominating features. Also puzzling is the absence of blood-related spectral features (oxyhemoglobin and deoxyhemoglobin doublets). Perfusing tissue samples with saline solutions (i.e. removing blood from tissues) is essential for observing other less intense spectral features. Perhaps the human tissues used in other MB studies were perfused before MB sample preparation, but there is typically no mention of this in the experimental sections of the reporting papers. We have not observed spectral evidence of hemosiderin in any of our samples; its prevalence in other studies is troubling because this implies that sample degradation is indeed problematic. Other nondiscussed aspects of sample preparation might also be important. For example, the number of hours between the time of death and freezing MB samples as well as the temperature and the duration of the exposure to air might influence the speciation of Fe in tissues. Finally, variations in age and diet of the donors might influence outcomes. For practical reasons, these factors cannot be controlled in human studies; in contrast, they are easily controlled in rodent studies.

## 4.13 Limitations of MBS and future directions

MBS is admittedly a rather esoteric technique that is only useful, within the context of biology, for the study of Fe. As such, the technique is rarely taught in classes and there are few commercially available manufacturers.[1] Although the initial cost of an instrument is not prohibitive, MB spectrometers are expensive to operate and maintain due to the cost of radioactive $^{57}Co$ sources and liquid helium (LHE). Five-kelvin systems

[1] In the USA, the only manufacturer of MB spectrometers is SEE Co., Edina (MN), www.seeco.us.

that use helium gas refrigerators are available, but these instruments tend to yield slightly broader spectral lines than LHE-based systems and they require more maintenance. The cold-head and compressor associated with these systems are expensive and have limited lifetimes.

Enriching large mammals with $^{57}Fe$ is prohibitively expensive, and unenriched samples of, for example, human tissue inevitably afford noisy spectra from which new insights will be difficult to obtain in the future. The exception is Fe-overloaded human tissues where an adequate signal-to-noise ratio can be achieved; however, in these cases, the spectral features due to the overloaded Fe (ferritin and/or hemosiderin) may not be as interesting as the less intense features that they obscure. Also, studying diseased states inevitably requires studying disease-free controls, which are difficult to obtain in humans. Transgenic mice that model human diseases are increasingly available. All aspects considered, we conclude that MB studies of $^{57}Fe$-enriched rodent tissues hold the most promise to generate major new insights into the Fe content of healthy and diseased states of mammals.

As mentioned in Section 4.1, another limitation of MBS is an intellectual one, in that accurately and exhaustively interpreting MB spectra requires a background in physics and quantum mechanics that few biologists have. Sadly, this requirement discourages biologists and biomedical researchers from using a technique that could potentially be extremely useful for their research. This limitation could be addressed by making instrumentation and analysis software more user-friendly, but it is also important to emphasize that more cursory spectral interpretations can still yield useful insights into cell-biological problems. Indeed, we hope that this review has illustrated this. Interestingly, many biologists routinely use biophysical techniques (e.g. NMR, mass spectra, X-ray diffraction, electron microscopy, UV-vis, fluorescence, IR) without fully understanding the deep physics underlying them; rather, they use these techniques as tools for addressing biological problems. To some extent, we are using MBS in this manner and encourage others, particularly cell biologists and biomedical researchers, to join us. With this shift in attitude along with some user-friendly advances in instrumentation and analysis, MBS might enjoy a renaissance of popularity, facilitating major new insights in the Fe-related metabolism of cells, multicellular organisms, and vertebrate animals. In combination with genetic, biochemical, and molecular-biological experiments, such insights might help generate new treatments and cures for Fe-associated diseases.

## Acknowledgments

We thank current and past members of our research group for their contributions. This project is supported by the National Institutes of Health (GM084266) and the Robert A. Welch Foundation (A1170).

## References

[1] Klingelhofer G, Morris RV, Bernhardt B, et al. Jarosite and hematite at Meridiani Planum from Opportunity's Mössbauer spectrometer. Science 2004;5702:1740–5.
[2] Schunemann V, Winkler H. Structure and dynamics of biomolecules studied by Mössbauer spectroscopy. Rep Prog Phys 2000;63:263–453.
[3] Papaefthymiou GC. The Mössbauer and magnetic properties of ferritin cores. Biochim Biophys Acta 2010;1800:886–97.
[4] Trautwein AX, Bill E, Bominaar EL, Winkler H. Iron-containing proteins and related analogs – complementary Mössbauer, EPR and magnetic susceptibility studies. Structure and Bonding 1991;78:1–95.
[5] Münck E. In: Que L Jr, ed. Physical methods in bioinorganic chemistry: spectroscopy and magnetism. Sausalito (CA): University Science Books; 2000:287–319.
[6] Gutlich P, Link R, Trautwein A. Mössbauer spectroscopy and transition metal chemistry. Springer-Verlag; 1978.
[7] Oshtrakh MI. Mössbauer spectroscopy: application in biomedical research. Hyperfine Interactions 2005;165;313–20.
[8] Croci S, Ortalli I. Mössbauer absorption spectroscopy: bio-medical applications. Riv Nuovo Cimento 2011;34:69–101.
[9] Orino K, Lehman L Tsuji Y, Ayaki H, Torti SV, Torti FM. Ferritin and the response to oxidative stress. Biochem J 2001;357:241–7.
[10] Champion PM, Lipscomb JD, Münck E, Debrunner P, Gunsalus IC. Mössbauer investigations of high-spin ferrous heme proteins. I. Cytochrome P-450. Biochemistry 1975;14:4151–8.
[11] Champion PM, Chiang R, Münck E, Debrunner P, Hager LP. Mössbauer investigations of high-spin ferrous heme proteins. II. Chloroperoxidase, horseradish peroxidase, and hemoglobin. Biochemistry 1975;14:4159–66.
[12] Timoteo CG, Pereira AS, Martins CE, et al. Low-Spin Heme b3 in the catalytic center of nitric oxide reductase from Pseudomonas nautical. Biochemistry 2011;50:4251–62.
[13] Garcia-Serres R, Davydov RM, Matsui T, Ikeda-Saito M, Hoffman BM, Huynh BH. Distinct reaction pathways followed upon reduction of oxy-heme oxygenase and oxy-myoglobin as characterized by Mössbauer spectroscopy. J Am Chem Soc 2007;129:1402–12.
[14] Peisach J Blumberg WE, Ogawa S, Rachmile EA, Oltzik R. Effects of protein conformation on heme symmetry in high-spin ferric heme proteins as studied by electron paramagnetic resonance. J Biol Chem 1971;246:3342–55.
[15] Cockrell AL, Holmes-Hampton GP, McCormick SP, Chakrabarti M, Lindahl PA. Mössbauer and EPR study of iron in vacuoles from fermenting Saccharomyces cerevisiae. Biochemistry 2011;50:10275–83.
[16] Popescu VC, Münck E, Fox BG, et al. Mössbauer and EPR studies of the photoactivation of nitrile hydratase. Biochemistry 2001;40:7984–91.
[17] Lane RW, Ibers JA, Frankel RB, Papaefthymiou GC, Holm RH. Synthetic analogues of the active sites of iron-sulfur proteins. 14. Synthesis, properties, and structures of bis(o-xylyl-α-α'-dithiolato)ferrate(II,III) anions, analogues of oxidized and reduced rubredoxin sites. J Am Chem Soc 1977;99:84–98.
[18] Lynch JB, Juarezgarcia C, Münck E, Que L. Mössbauer and EPR studies of the binuclear iron center in ribonucleotide reductase from E. coli – a new iron-to-protein stoichiometry. J Biol Chem 1989;264:8091–6.
[19] Fox BG, Hendrich MP, Surerus KK, et al. Mössbauer, EPR, and ENDOR studies of the hydroxylase and reductase components of methane monooxygenase from Methylosinus trichosporium OB3b. J Am Chem Soc 1993;115:3688–701.

[20] Fee JA, Findling KL, Yoshida T, et al. Purification and characterization of the Rieske iron-sulfur protein from Thermus thermophilus – evidence for a [2Fe-2S] cluster having non-cysteine ligands. J Biol Chem 1984;259:124–33.
[21] Li H, Mapolelo DT, Dingran NN, et al. The yeast iron regulatory proteins Grx3/4 and Fra2 form heterodimeric complexes containing a [2Fe-2S] cluster with cysteinyl and histidyl ligation. Biochemistry 2009;48:9569–81.
[22] Mitou G, Higgins C, Wittung-Stafshede P, et al. An Isc-type extremely thermostable [2Fe-2S] ferredoxin from Aquifex aeolicus. Biochemical, spectroscopic, and unfolding studies. Biochemistry 2003;42:1354–64.
[23] Kent TA, Dreyer J, Kennedy MC, et al. Mössbauer studies of beef heart aconitase: evidence for facile interconversions of iron-sulfur clusters. Proc Natl Acad Sci USA 1982;79:1096–100.
[24] Papaefthymiou V, Girerd JJ, Moura I, Moura JJG, Münck E. Mössbauer study of D. gigas ferredoxin II and spin-coupling model for $Fe_3S_4$ cluster with valence delocalization. J Am Chem Soc 1987;109:4703–10.
[25] Emptage MH, Kent TA, Kennedy MC, Beinert H, Münck E. Mössbauer and EPR studies of activated aconitase – development of a localized valence state at a subsite of the [4Fe-4S] cluster on the binding of citrate. Proc Natl Acad Sci USA 1983;80:4674–8.
[26] Dilg AW, Grantner K, Iakovleva O, et al. Dynamics of wild-type HiPIPs: a Cys77Ser mutant and a partially unfolded HiPIP. J Biol Inorg Chem 2002;7:691–703.
[27] Dunham WR, Hagen WR, Fee JA, Sands RH, Dunbar JB, Humblet C. An investigation of Chromatium vinosum high-potential iron-sulfur protein by EPR and Mössbauer spectroscopy; evidence for a freezing-induced dimerization in NaCl solutions. Biochim Biophys Acta 1991;1079:253–62.
[28] Middleton P, Dickson DPE, Johnson CE, Rush JD. Interpretation of the Mössbauer spectra of the four-Iron ferredoxin from Bacillus stearothermophilus. Eur J Biochem 1978;88:135–41.
[29] Mbughuni MM, Chakrabarti M, Hayden JA, et al. Trapping and spectroscopic characterization of an $Fe^{III}$-superoxo intermediate from a nonheme mononuclear iron-containing enzyme. Proc Natl Acad Sci USA 2010;107:16788–93.
[30] Theil EC, Behera RK, Tosha T. Ferritins for chemistry and for life. Coord Chem Rev 2013;257:579–86.
[31] Bauminger ER, Cohen SG, Ofer S, Rachmilewitz EA. Quantitative studies of ferritinlike iron in erythrocytes of thalassemia, sickle-cell anemia, and hemoglobin Hammersmith with Mössbauer spectroscopy. Proc Natl Acad Sci USA 1979;76:939–43.
[32] Lesuisse E, Santos R, Matzanke BF, Knight SAB, Camadro JM, Dancis A. Iron use for haeme synthesis is under control of the yeast frataxin homologue (Yfh1). Human Mol Gen 2003;12:879–889.
[33] Miao R, Martinho M, Morales JG, et al. EPR and Mössbauer spectroscopy of intact mitochondria isolated from Yah1p-depleted Saccharomyces cerevisiae. Biochemistry 2008;47:9888–99.
[34] Miao R, Kim H, Koppolu UMK, Ellis EA, Scott RA, Lindahl PA Biophysical characterization of the iron in mitochondria from Atm1p-depleted Saccharomyces cerevisiae. Biochemistry 2009;48:9556–68.
[35] Miao R, Holmes-Hampton GP, Lindahl PA. Biophysical investigation of the iron in Aft1-$1^{up}$ and Gal-YAH1 Saccharomyces cerevisiae. Biochemistry 2011;50:2660–71.
[36] Jhurry ND, Chakrabarti M, McCormick SP, Holmes-Hampton GP, Lindahl PA. Biophysical investigation of the ironome of human Jurkat cells and mitochondria. Biochemistry 2012;51:5276–84.
[37] Hudder BN, Morales JG, Stubna A, Münck E, Hendrich MP, Lindahl PA. Electron paramagnetic resonance and Mössbauer spectroscopy of intact mitochondria from respiring Saccharomyces cerevisiae. J Biol Inorg Chem 2007;12:1029–53.

[38] Lindahl PA, Morales JG, Miao R, Holmes-Hampton GP. Isolation of Saccharomyces cerevisiae mitochondria for Mössbauer, EPR, and electronic absorption spectroscopic analyses. Methods Enzymol 2009;456:267–85.

[39] Morales JG, Holmes-Hampton GP, Miao R, Guo Y, Münck E, Lindahl PA. Biophysical characterization of iron in mitochondria isolated from respiring and fermenting yeast. Biochemistry 2010;49:5436–44.

[40] Lutz T, Westermann B, Neupert W, Herrmann JM. The mitochondria proteins Ssq1 and Jac1 are required for the assembly of iron sulfur clusters in mitochondria. J Mol Biol 2001;307:815–25.

[41] Pandey A, Yoon H, Lyver ER, Dancis A, Pain D. Identification of a Nfs1p-bound persulfide intermediate in Fe-S cluster synthesis by intact mitochondria. Mitochondrion 2012;12:539–49.

[42] Holmes-Hampton GP, Miao R, Morales JG, Guo Y, Münck E, Lindahl PA. A nonheme high-spin ferrous pool in mitochondria isolated from fermenting Saccharomyces cerevisiae. Biochemistry 2010;49:4227–34.

[43] Park J, McCormick SP, Chakrabarti M, Lindahl PA. Insights into the iron-ome and manganese-ome of Δmtm1 Saccharomyces cerevisiae mitochondria. Metallomics 2013;5:656–72.

[44] Lange H, Kaut A, Kispal G, Lill R. A mitochondrial ferredoxin is essential for biogenesis of cellular iron-sulfur proteins. Proc Natl Acad Sci USA 2000;97:1050–5.

[45] Kispal G, Csere P, Prohl C, Lill R. The mitochondrial proteins Atm1p and Nfs1p are essential for biogenesis of cytosolic Fe/S proteins. EMBO J 1999;18:3981–9.

[46] Yamaguchi-Iwai Y, Dancis A, Klausner RD. Aft1 – a mediator of iron regulated transcriptional control in Saccharomyces cerevisiae. EMBO J 1995;14:1231–9.

[47] Yang M, Cobine PA, Molik S, et al. The effects of mitochondrial iron homeostasis on cofactor specificity of superoxide dismutase 2. EMBO J 2006;25:1775–83.

[48] Seguin A, Sutak R, Bulteau AL, et al. Evidence that yeast frataxin is not an iron storage protein in vivo. Biochim Biophys Acta 2010;1802:531–8.

[49] Li L, Chen OS, Ward DM, Kaplan J. CCC1 is a transporter that mediates vacuolar iron storage in yeast. J Biol Chem 2001;276:29515–9.

[50] Holmes-Hampton GP, Jhurry ND, McCormick SP, Lindahl PA. Iron content of Saccharomyces cerevisiae cells grown under iron-deficient and iron-overload conditions. Biochemistry 2013;52:105–14.

[51] Oshtrakh MI, Semionkin VA. Mössbauer study of red blood cells from patients with erythremia. FEBS Letters 1989;257:41–4.

[52] Ortalli I, Pedrazzi G, Varacca V. Mössbauer investigation of hematological diseases. Hyperfine Interactions 1988;42:913–6.

[53] Oshtrakh MI. The features of Mössbauer spectra of hemoglobin in relation to the quadrupole splitting and heme iron stereochemistry. Z Naturforsch A 1998;53:608–14.

[54] Shahal Y, Bauminger ER, Zmora E, et al. Oxidative stress in newborn erythrocytes. Pediatr Res 1991;29:119–22.

[55] Ni X, Hsia Y, Liu R, et al. Mössbauer studies of hemoglobin of the patients with liver cancer and cirrhosis. Hyperfine Interactions 1992;71:1323–6.

[56] Seidel A, Bill E, Häggström L, Nordblad P, Kilár F. Complementary Mössbauer and EPR studies of iron(III) in diferric human serum transferrin with oxalate or bicarbonate as synergistic anions. Arch Biochem Biophys 1994;308:52–63.

[57] Chua-anusorn W, Tran KC, Webb J, Macey DJ, St Pierre TG. Chemical speciation of iron deposits in thalassemic heart tissue. Inorg Chim Acta 2000;300–2:932–6.

[58] Whitnall M, Rahmanto YS, Huang ML-H, et al. Identification of nonferritin mitochondrial iron deposits in a mouse model of Friedreich ataxia. Proc Natl Acad Sci USA 2012;109;20590–5.

[59] Rimbert JN, Dumas F, Kellershohn C, Girot R, Brissot P. Mössbauer spectroscopy study of iron overloaded livers. Biochemie 1985;67:663–8.

[60] Oshtrakh MI, Alenkina IV, Vinogradov A, Konstantinova TS, Kuzmann E, Semionkin VA. Mössbauer spectroscopy of the iron cores in human liver ferritin, ferritin in normal human spleen and ferritin in spleen from patient with primary myelofibrosis: preliminary results of comparative analysis. Biometals 2013;26:229–39.
[61] St Pierre TG, Chua-anusorn W, Webb J, Macey D, Pootrakul P. The form of iron oxide deposits in thalassemic tissues varies between different groups of patients: a comparison between Thai β-thalassemia/hemoglobin E patients and Australian β-thalassemia patients. Biochim Biophys Acta 1998;1407:51–60.
[62] Hackett S, Chua-anusorn W, Pootrakul P, St Pierre TG. The magnetic susceptibilities of iron deposits in thalassaemic spleen tissue. Biochim Biophys Acta 2007;1772:330–7.
[63] Galazka-Friedman J, Bauminger ER, Friedman A, Barcikowska M, Hechel D, Nowik I. Iron in Parkinsonian and control substantia nigra- a Mössbauer spectroscopy study. Mov Disord 1996;11:8–16.
[64] Wypijewska A, Galazka-Friedman J, Bauminger ER, et al. Iron and reactive oxygen species activity in Parkinsonian substantia nigra. Parkinsonism Relat Disord 2010;16:329–33.
[65] Holmes-Hampton GP, Chakrabarti M, Cockrell AL, et al. Changing iron content of the mouse brain during development. Metallomics 2012;4:761–70.

# 5 The interstitial carbide of the nitrogenase M-cluster: insertion pathway and possible function

Yilin Hu and Markus Ribbe

## 5.1 Introduction

Nitrogenase is a complex metalloenzyme of versatile functions. Best known for its role in the biological nitrogen fixation, nitrogenase catalyzes the reduction of nitrogen ($N_2$) to ammonia ($NH_3$), a key step in the global nitrogen cycle [1, 2]. In addition, nitrogenase is capable of reducing a number of "alternative" substrates, such as proton ($H^+$), azide ($N_3^-$), acetylene ($C_2H_2$), cyanide ($CN^-$), and carbon monoxide (CO); most notably, nitrogenase can reduce CO to hydrocarbons of varying lengths, including methane ($CH_4$), ethene ($C_2H_4$), ethane ($C_2H_6$), propene ($C_3H_6$), propane ($C_3H_8$), butene ($C_4H_8$), and butane ($C_4H_{10}$). Interestingly, the reactions of $N_2$- and CO-reduction by nitrogenase parallel two important industrial processes: the Haber-Bosch process, which combines $N_2$ and $H_2$ into ammonia [3], and the Fischer-Tropsch process, which combines CO and $H_2$ into liquid carbon fuels [4]. However, contrary to their industrial counterparts, the nitrogenase-based reactions occur under ambient conditions, making this enzyme an attractive blueprint for future development of cost-efficient approaches for ammonia and carbon fuel production.

Three homologous nitrogenases have been identified so far, which are mainly distinguished by the identity of the heterometal (i.e. Mo, V, or Fe) at the active center [5]. The Mo nitrogenase from *Azotobacter vinelandii* (Fig. 5.1) is the best studied member of this enzyme family. It is a binary system comprising a reductase component and a catalytic component. The reductase component, designated the Fe protein (or NifH), is a $\gamma_2$-homodimer that contains a $[Fe_4S_4]$ cluster at the subunit interface and an ATP binding site within each subunit. The catalytic component, designated the MoFe protein (or NifDK), is an $\alpha_2\beta_2$-heterotetramer that contains a P-cluster ($[Fe_8S_7]$) at each $\alpha/\beta$-subunit interface and an M-cluster ($[MoFe_7S_9C$-homocitrate]) within each $\alpha$-subunit [6–9]. Upon turnover, the two component proteins of nitrogenase form a functional complex, which allows electrons to be transferred from the $[Fe_4S_4]$ cluster of NifH, *via* the P-cluster, to the M-cluster of NifDK, where substrate reduction occurs (Fig. 5.1). The "action central," M-cluster, is arguably one of the most complex metalloclusters identified to date. Ligated to the $\alpha$-subunit of NifDK by only two ligands (i.e. $His^{\alpha442}$ and $Cys^{\alpha275}$), this cluster has a core stoichiometry of 1Mo:7Fe:9S and can be viewed as $[MoFe_3S_3]$ and $[Fe_4S_3]$ subclusters bridged by three $\mu_2$-coordinated sulfide atoms in between (Fig. 5.1b). Additionally, the Mo end of the

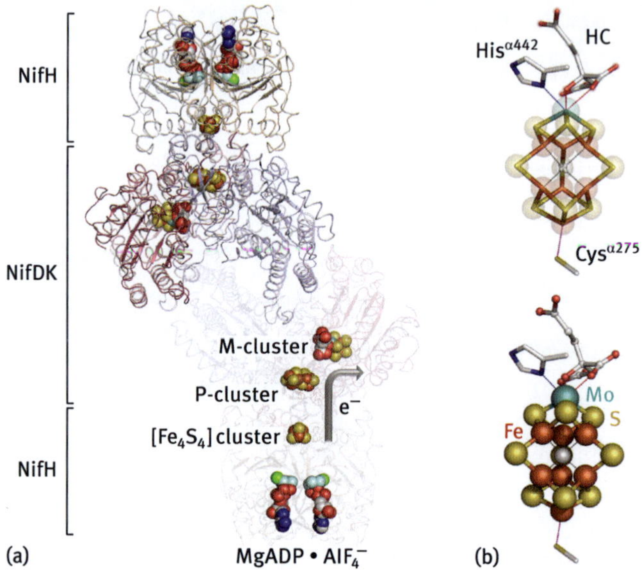


**Fig. 5.1:** Crystal structures of the ADP•$AlF_4^-$-stabilized NifH/NifDK complex (a) and the M-cluster of NifDK (b). (a) One half the NifH/NifDK complex, which consists of the dimeric NifH and one αβ-dimer of the tetrameric NifDK, is shown in the foreground (top), whereas the other equivalent half of the complex is rendered transparent in the background, highlighting the key components that are involved in the electron transfer during substrate turnover (bottom). The two subunits of NifH are colored gray and light brown, and the α- and β-subunits of NifDK are colored red and light blue, respectively. The clusters and ADP•$AlF_4^-$ are shown as space-filling models. (b) The structure of M-cluster is shown in the stick presentation, which highlights its core structure (top), or in the ball-and-stick presentation, which illustrates its elemental composition (bottom). The two ligands ($His^{\alpha442}$ and $Cys^{\alpha275}$) that coordinate the M-cluster in the α-subunit of NifDK are indicated. Atoms are colored as follows: Fe, orange; S, yellow; Mo, cyan; O, red; C, gray; N, dark blue; Mg, green; Al, beige; F, light blue. PYMOL was used to create this figure (PDB IDs: 1N2C, 3U7Q).

cluster is coordinated by a homocitrate (HC) moiety, whereas the central cavity of the cluster is occupied by a $\mu_6$-coordinated light atom (Fig. 5.1b). The discovery of such an interstitial atom in the structure of M-cluster has generated a lot of excitement because of the promising relevance of this atom to the mechanism of nitrogenase [6]. Recently, this atom was identified as a carbide ($C^{4-}$) ion [8, 9], raising questions as to where this carbide originates from and how it is inserted into the M-cluster. Moreover, the newly established function of nitrogenase in reducing CO and coupling CO into hydrocarbon chains prompted additional questions, such as whether the interstitial carbide is involved in C-C coupling and if it plays a role in the catalysis of nitrogenase.

## 5.2 Proposed role of NifB in carbide insertion

A quick examination of the proposed assembly pathway of the M-cluster (Fig. 5.2) provides the initial insights into the carbon insertion process. Assembly of the M-cluster is launched by NifS and NifU, which mobilize Fe and S for the synthesis of $[Fe_4S_4]$ clusters [10]. These small $[Fe_4S_4]$ units are then delivered to NifB and processed into a large FeS cluster (Fig. 5.2). Subsequently, the FeS cluster is transferred

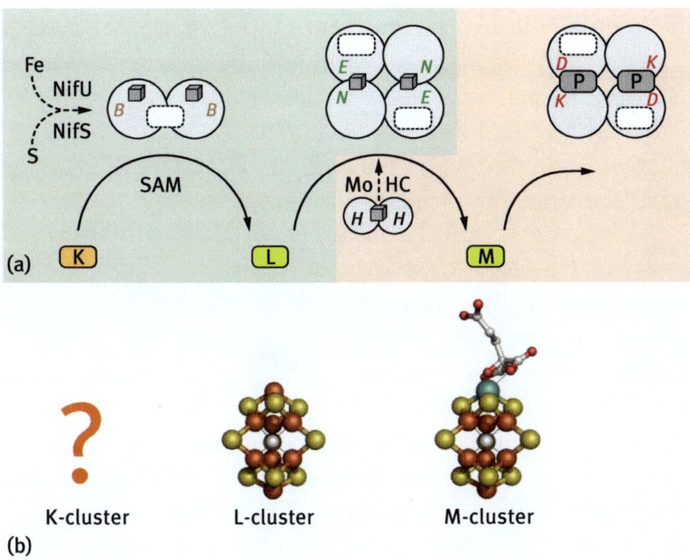


**Fig. 5.2:** The pathway of M-cluster assembly (a) and the cluster intermediates generated along this pathway (b). (a) Assembly of M-cluster is launched by NifS and NifU, which mobilize Fe and S for the synthesis of small FeS clusters. These small units are supplied to NifB and used as a precursor (designated the K-cluster) for the formation of a large FeS core (designated the L-cluster) in a SAM-dependent process. Subsequently, the L-cluster is transferred to NifEN and converted to a mature M-cluster in a NifH-dependent process prior to the transfer of the M-cluster to its target location in NifDK. Deletion of NifDK and NifH eliminates the pathway beyond NifEN (indicated by the light sandy shade), allowing the assembly process to proceed only up to the point of NifEN (indicated by the light blue shade). Consequently, the L-cluster is "backed up" on NifEN and, further upstream, the K-cluster is "backed up" on NifB. The permanent $[Fe_4S_4]$ clusters in NifB and NifEN are shown as gray cubes, and the P-clusters in NifDK are shown as gray rounded rectangles. The cluster binding sites in NifB, NifEN, and NifDK are shown as white rounded rectangles. (b) The L-cluster is an $[Fe_8S_9C]$ cluster that is nearly identical to the M-cluster in structure except for the substitution of Fe for Mo and HC at one end of the cluster. The capture of an all-Fe L-cluster was facilitated by the deletion of NifDK and NifH, which further defined the function of NifH as a Mo/HC insertase, see also (a). The identity of the K-cluster, however, has remained unknown for a long time (indicated by a question mark) due to the instability of NifB in aqueous solutions. Clusters are shown in ball-and-stick presentations, with the atoms colored as those in Fig. 5.1. PYMOL was used to create this figure.

to NifEN, matured to an M-cluster in a NifH-dependent process, and delivered to its final binding site in NifDK (Fig. 5.2) [10]. Of all the biosynthetic components in this proposed pathway, NifB is particularly interesting with regard to the insertion of carbide. NifB plays an essential role in M-cluster assembly, as the NifDK protein generated in a *nifB*-deletion background does not contain an M-cluster at its binding site [11]. Sequence analysis reveals that NifB has a CXXXCXXC signature motif that is characteristic of a family of radical *S*-adenosylmethionine (SAM) enzymes [12, 13]; in addition, it suggests that NifB contains sufficient ligands to accommodate the entire complement of Fe atoms in the M-cluster [14]. These observations suggest that NifB could employ a radical SAM-dependent mechanism to generate a complete FeS core of the M-cluster. Given the well-established role of radical SAM enzymes in mobilizing carbon species [15–18], the interstitial carbide may very well be incorporated upon the formation of the FeS core of M-cluster on NifB. This theory was tested recently by acquiring an intermediate-bound form of NifB (see Section 5.3), which enabled subsequent investigations of the insertion of carbide into the M-cluster (see Section 5.4) and the fate of carbide upon substrate turnover (see Section 5.5).

## 5.3 Accumulation of a cluster intermediate on NifB

To investigate the proposed role of NifB in M-cluster assembly, it is crucial to obtain an intermediate-bound form of NifB. This task can be achieved by deleting NifDK (the downstream receptor for the M-cluster) and NifH (the maturation factor for the M-cluster), which should "back up" the cluster intermediate on NifEN (designated the L-cluster) and further upstream on NifB (designated the K-cluster) (Fig. 5.2). Indeed, the L-cluster was captured on NifEN when this protein was expressed in a *nifHDK*-deletion background [19, 20]. XAS/EXAFS, XES, and crystallographic studies [20–23] identified this cluster as an [$Fe_8S_9C$] cluster that is nearly indistinguishable in structure from the M-cluster except for the replacement of Mo and HC by an Fe atom at one end of the cluster (Fig. 5.2). Biochemical, EPR, and XAS/EXAFS analyses [19, 24–26] further established this cluster as a physiologically relevant precursor to the M-cluster, demonstrating that the L-cluster could be matured into an M-cluster on NifEN upon NifH-mediated insertion of Mo and HC (Fig. 5.2). Together, these studies defined the functions of NifEN and NifH, as well as the sequence of events at/beyond NifEN, in the process of M-cluster assembly (Fig. 5.2). More importantly, they led to the hypothesis that NifB, the protein component that acts prior to NifEN in the assembly process, houses the formation of L-cluster, which represents a complete FeS core of the M-cluster that already has the interstitial carbide in place.

Acquiring proof for this hypothesis, on the other hand, has proven to be challenging. Earlier attempts to characterize NifB from a *nifHDK*-deletion strain of *A. vinelandii* were hampered by the instability of this protein in aqueous solutions. Recently, this problem was circumvented by fusing the 3′ end of *nifB* with the 5′ end of *nifN* and expressing the fused genes in a *nifHDK*-deletion background of *A. vinelandii* [27]. The resultant NifEN-B fusion protein (Fig. 5.3a) consists of both NifEN and NifB and, therefore, should accumulate both the L-cluster and the K-cluster when the biosynthetic flow is interrupted by the deletion of NifH and NifDK (Fig. 5.2). Consistent with this prediction, combined metal, activity, UV-vis, and EPR analyses not only established the presence of the L-cluster on NifEN-B, but also identified the K-cluster as a pair of $[Fe_4S_4]$ clusters (Fig. 5.3b) [27]. Apart from these two transient cluster species, a permanent $[Fe_4S_4]$ cluster (designated the SAM cluster) was identified and assigned to the SAM motif in the NifB entity of this protein [27]. Together with the SAM-cluster,

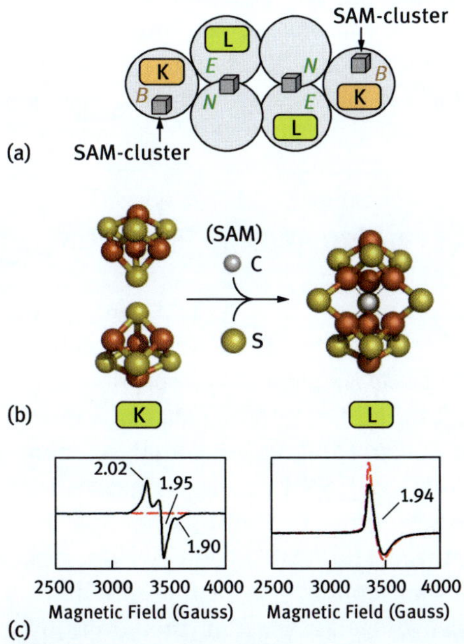


**Fig. 5.3:** Schematic presentation of NifEN-B (a), structural details of K- and L-clusters (b), and spectral changes upon conversion of a K-cluster to an L-cluster (c). (a) NifEN-B contains the L-cluster (lime) in its NifEN entity and the K-cluster (orange) in its NifB entity. In addition, the NifB entity also contains the SAM-cluster (gray), a permanent $[Fe_4S_4]$ cluster that is associated with the SAM domain of NifB. (b) The K-cluster (a $[Fe_4S_4]$ cluster pair) can be converted to an L-cluster (an $[Fe_8S_9]$ cluster) in a SAM-dependent process upon insertion of a carbon atom and a sulfur atom. The clusters are shown in ball-and-stick presentations, with the atoms colored as those in Fig. 5.1. PYMOL was used to create this figure. (c) The K- and SAM-clusters collectively give rise to a SAM-responsive, $S = 1/2$ signal at $g = 2.02$, 1.95 and 1.90 in the dithionite-reduced state (left, black), whereas the L-cluster displays a characteristic $g = 1.94$ signal in the IDS-oxidized state (right, black). In the presence of SAM, the K/SAM-cluster-associated $S = 1/2$ signal disappears (left, red) concomitant with an increase in the magnitude of the L-cluster-specific $g = 1.94$ signal (right, red), which corresponds to the conversion of K-clusters to more L-clusters.

the K-cluster gives rise to an $S = 1/2$ EPR signal at $g = 2.02$, 1.95, and 1.90 (Fig. 5.3c, left, black). This composite signal disappears upon the addition of SAM, implying that (i) the K- and SAM-clusters are located close to each other and (ii) the response of the SAM-cluster to SAM induces the conversion of the nearby K-cluster to an EPR-silent cluster species (Fig. 5.3c, left, red) [27]. Accompanying the disappearance of the K/SAM-cluster-associated $S = 1/2$ signal, there is an increase in the magnitude of the L-cluster-specific $g = 1.94$ signal (Fig. 5.3c, right, red), which is consistent with the formation of "new" L-clusters upon the conversion of K-clusters in the presence of SAM [27]. The combined outcome of these studies suggests that NifB employs a novel synthetic route for the coupling/rearrangement of two $[Fe_4S_4]$ clusters (i.e. the K-cluster) into an $[Fe_8S_9C]$ cluster (i.e. the L-cluster) concomitant with the radical SAM-dependent insertion of carbide, and the addition of "the ninth sulfur" (Fig. 5.3b). Moreover, the NifEN-B fusion protein provided a useful vehicle for the subsequent investigation of carbide insertion into the M-cluster, which led to the identification of the source of carbide and the proposal of a radical SAM-dependent pathway of carbide insertion (see Section 5.4).

## 5.4 Investigation of the insertion of carbide into the M-cluster

The initial insights into the carbide insertion process were gained through the studies of SAM cleavage in the presence of NifEN-B [28]. HPLC and MS analyses showed that *S*-adenosylhomocysteine (SAH) could be generated upon removal of the methyl group of SAM by NifEN-B (Fig. 5.4a, left), suggesting that this methyl group may be mobilized for carbide insertion [28]. In addition, these analyses identified 5′-deoxyadenosine (5′-dAH) as another product of SAM cleavage (Fig. 5.4a, left), which could be enriched by deuterium (i.e. giving rise to 5′-dAD) when the three hydrogen atoms of the methyl group of SAM were substituted by deuterium atoms (Fig. 5.4a, right). This observation implies that mobilization of the SAM-derived methyl group may involve abstraction of a hydrogen atom from this group by a 5′-dA• radical [28]. Interestingly, two radical SAM-dependent RNA methylases, RlmN and Cfr, display the same patterns of SAM cleavage and deuterium substitution [17, 18]. It has been proposed that RlmN and Cfr catalyze the methylation of RNA *via* an initial step of SN2-type methyl transfer from one SAM molecule, followed by the formation of 5′-dA• from a second SAM molecule and the subsequent hydrogen atom abstraction from the methyl group by 5′-dA• [17, 18]. By analogy, NifB could use a similar mechanism to mobilize the methyl group of SAM for carbide insertion during the process of M-cluster assembly. Consistent with this proposal, radiolabeling experiments demonstrated the flow of $^{14}C$ label through the assembly pathway of M-cluster when [$^{14}C$-methyl] SAM was used as the initial carbon source. The $^{14}C$ label first appeared on the L-cluster concomitant with the conversion of K- to L-cluster (Fig. 5.4b, middle) and then on the M-cluster upon the conversion of

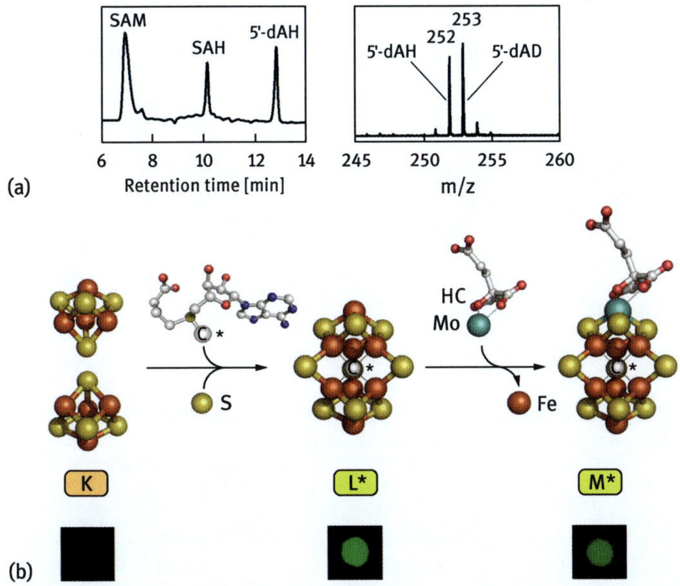


**Fig. 5.4:** Cleavage of SAM by NifEN-B (a) and flow of $^{14}C$ label through the assembly pathway (b). (a) HPLC elution profile showing the cleavage of SAM into SAH and 5′-dAH upon incubation with NifEN-B (left) and LC-MS analysis showing the formation of 5′-dAD, along with 5′-dAH, upon incubation of [methyl-$d_3$] SAM with NifEN-B (right). (b) Schematic presentation (upper) and corresponding autoradiographs (lower) showing the incorporation of $^{14}C$ (derived from [$^{14}C$-methyl] SAM) into the L-cluster and the carry-over of $^{14}C$ into the M-cluster upon incorporation of Mo and HC into the L-cluster. The clusters are shown in ball-and-stick presentations, with the atoms colored as those in Fig. 5.1. PYMOL was used to create this figure.

L- to M-cluster (Fig. 5.4b, right) [28]. The accumulation of $^{14}C$ label on these clusters, along with the absence of $^{14}C$ label from the polypeptides of assembly proteins, suggests a direct transfer of carbon to the cluster intermediates without going through a protein-bound carbon intermediate step [28]. Together, these observations not only established SAM as the source of the interstitial carbide, but also defined the role of NifB as a radical SAM enzyme that catalyzes the insertion of carbon during the K- to L-cluster conversion.

Two reaction pathways have been proposed for the carbide insertion by NifB [28]. One pathway involves the SN2-type methyl transfer from SAM to a sulfide atom of the K-cluster, followed by the generation of a methylene radical *via* hydrogen atom abstraction from the SAM-derived methyl group (Fig. 5.5a), whereas the other pathway involves the formation of a methyl radical *via* the reductive cleavage of SAM, followed by the transfer of methyl radical to an iron atom of the K-cluster and the conversion of this radical to a methylene radical through electron rearrangement and hydrogen atom abstraction (Fig. 5.5b). In both cases, the reaction

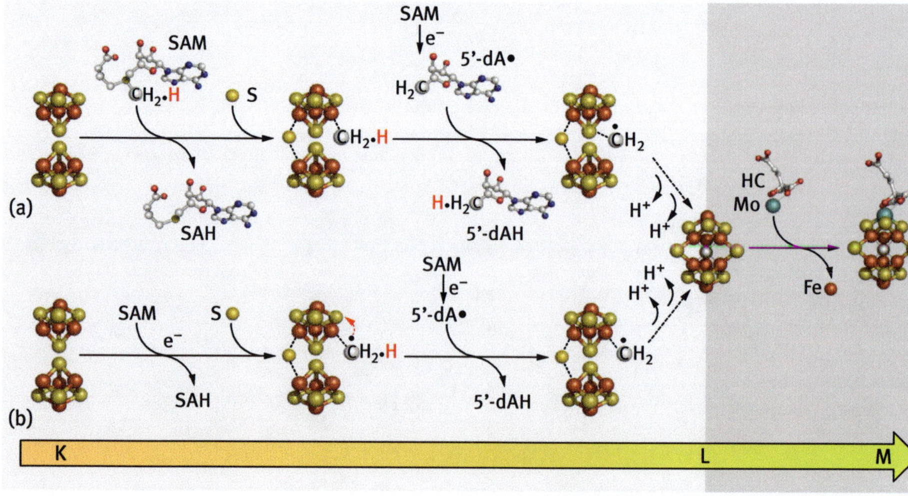


**Fig. 5.5:** Proposed pathways of carbide insertion involving the formation of a methylene (a) or methyl (b) radical. (a) The methyl group of SAM is transferred *via* an SN2 mechanism, followed by the formation of a methylene radical upon hydrogen atom abstraction by 5′-dA• and the transfer of this radical to a sulfide atom of the K-cluster. (b) A methyl radical is formed *via* the reductive cleavage of SAM, followed by the transfer of this radical to an iron atom of the K-cluster and the processing of this intermediate into a methylene radical. Both pathways then continue with multiple deprotonation steps until an interstitial carbide atom is generated, and this process is accompanied by the insertion of a sulfur atom and the restructuring/coupling of the two 4Fe units of the K-cluster into an 8Fe L-cluster. The clusters are shown in ball-and-stick presentations, with the atoms colored as those in Fig. 5.1. PYMOL was used to create this figure.

proceeds with the deprotonation of the carbon intermediate (Fig. 5.5), which could be accomplished either by proton abstraction *via* acid/base chemistry [29–30] or through the transfer of a hydride ($H^-$) to a ferric iron ($Fe^{3+}$) and the subsequent removal of protons [31–32]. Moreover, the processing of the carbon intermediate into a carbide atom is accompanied by the insertion of the ninth sulfur and the restructuring/coupling of the two [$Fe_4S_4$] units of the K-cluster, which eventually lead to the formation of an [$Fe_8S_9C$] L-cluster (Fig. 5.5). The origin of the ninth sulfur is unknown, although SAM has been documented for its ability to serve as a sulfur donor [13–33]. Alternatively, there could be an extra S atom attached to one of the Fe atoms of the K-cluster in a manner analogous to what was observed in the case of (*R*)-2-hydroxyisocaproyl-CoA dehydratase [34]. Although further investigation is required to elucidate the mechanistic details of the carbide insertion process, identification of the source of carbide provided an effective means for the specific labeling of this interstitial atom, which enabled the subsequent tracing of carbide during substrate turnover (see Section 5.5).

## 5.5 Tracing the fate of carbide during substrate turnover

The question of whether the interstitial atom is involved in substrate turnover was tackled even before the unveiling of the identity of this atom. Previous ENDOR/ESEEM analyses suggested that if the interstitial atom was nitrogen, it could not be exchanged upon turnover [35]; however, it did not directly address the question of whether an interstitial carbide atom could be exchanged during catalysis. Facilitated by the recent identification of SAM as the source of carbide, the fate of this atom in substrate turnover could be directly traced by labeling it with either [$^{14}$C-methyl] or [$^{13}$C-methyl] SAM [28]. The $^{14}$C-labeled M-cluster could be subjected to the turnover of "fast" substrates, such as $C_2H_2$ and $N_2$, which would allow a fast exchange of the interstitial carbide with substrates and, consequently, a quick "dilution" of the $^{14}$C label in the M-cluster, whereas the $^{13}$C-labeled M-cluster could be subjected to the turnover of "slow" substrates, such as CO, which would prevent a quick exchange of the interstitial carbide with unlabeled species, thereby enabling a steady enrichment of the $^{13}$C label in the products [36]. As it turned out, the intensity of the $^{14}$C label in M-cluster remained unchanged after extended turnover with $N_2$ or $C_2H_2$ (Fig. 5.6b), even when the substrates were present in large

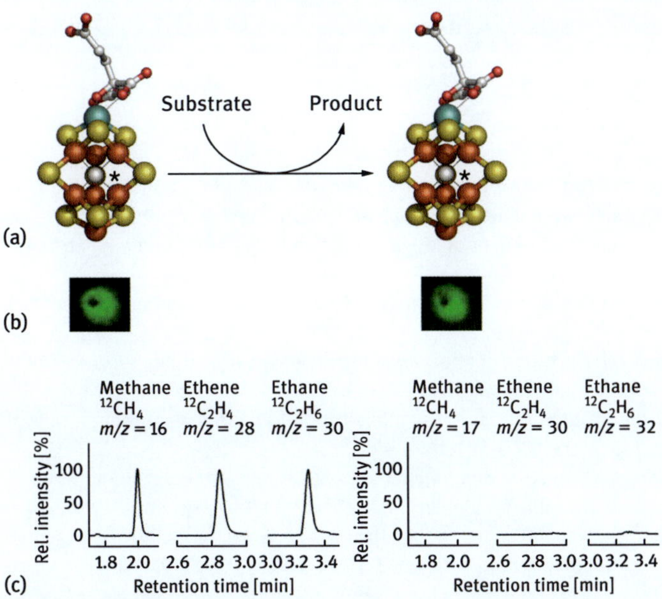


**Fig. 5.6:** Schematic presentation of substrate turnover by labeled M-cluster (a), $^{14}$C experiments with "fast" substrates (b), and $^{13}$C experiments with "slow" substrates (c). (a) The interstitial carbide of the M-cluster was specifically labeled with $^{14}$C or $^{13}$C using [$^{14}$C-methyl] or [$^{13}$C-methyl] SAM. (b) Autoradiograph of $^{14}$C-labeled M-cluster before (left) and after (right) turnover of $C_2H_2$. (c) GC-MS analysis of products generated from the turnover of $^{12}$CO by $^{13}$C-labeled M-cluster.

molar excess to the interstitial carbide [36]. Likewise, no $^{13}C$ label could be detected in the hydrocarbon products when $^{12}CO$ was turned over by the $^{13}C$-labeled M-cluster (Fig. 5.6c), even when the total amount of carbon in these products was kept at a sub-molar ratio to the total amount of the interstitial carbide [36]. Together, these observations provided direct proof that the interstitial carbide can neither be exchanged during turnover nor used as a substrate and incorporated into the products.

A structural role can be proposed for the interstitial carbide in light of these results, one that is required to provide certain "rigidity" to the metal-sulfur core by symmetrically coordinating the six core Fe atoms at the center of the M-cluster (Fig. 5.1b). However, a possible function of this interstitial atom in nitrogenase catalysis cannot be excluded. Previous density functional theory (DFT) calculations suggested that a more stable structure of the M-cluster could be achieved by having a N or O species instead of a C species in the center of the cluster [37, 38]. A recent study of $N_2$ activation on iron metallaboratranes provided further proof for the "flexibility" of the carbide-containing M-cluster, showing that the Fe-C bond distances could be varied during substrate turnover [39]. Thus, the interstitial carbide may participate in catalysis by indirectly tuning the structure and reactivity of the M-cluster. At present, the exact function of this atom in nitrogenase mechanism remains unknown. Further research will hopefully unveil the role of this enigmatic atom and provide new insights into the structure-function relationship of the fascinating nitrogenase system.

## References

[1] Burgess BK. Mechanism of molybdenum nitrogenase. Chem Rev 1996;96:2983–3012.
[2] Howard JB, Rees DC. Structural basis of biological nitrogen fixation. Chem Rev 1996;96:2965–82.
[3] Schlögl R. Angew. Catalytic synthesis of ammonia – a "never-ending story"? Chem Int Ed Engl 2003;42:2004–8.
[4] Rofer-DePoorter CK. A comprehensive mechanism for the Fischer-Tropsch synthesis. Chem Rev 1981;81:447–74.
[5] Eady RR. Structure-function relationships of alternative nitrogenases. Chem Rev 1996; 96:3013–30.
[6] Einsle O, Tezcan FA, Andrade SL, et al. Nitrogenase MoFe-protein at 1.16 Å resolution: a central ligand in the FeMo-cofactor. Science 2002;297:1696–700.
[7] Kim J, Rees DC. Crystallographic structure and functional implications of the nitrogenase molybdenum iron protein from Azotobacter vinelandii. Nature 1992;360:553–60.
[8] Lancaster KM, Roemelt M, Ettenhuber P, et al. X-ray emission spectroscopy evidences a central carbon in the nitrogenase iron-molybdenum cofactor. Science 2011;334:974–7.
[9] Spatzal T, Aksoyoglu M, Zhang L, et al. Evidence for interstitial carbon in nitrogenase FeMo cofactor. Science 2011;334:940.
[10] Hu Y, Ribbe MW. Biosynthesis of the iron-molybdenum cofactor of nitrogenase. J Biol Chem 2013;288:13173–7.
[11] Schmid B, Ribbe MW, Einsle O, et al. Structure of a cofactor-deficient nitrogenase MoFe protein. Science 2002;296:352–6.
[12] Sofia HJ, Chen G, Hetzler BG, Reyes-Spindola JF, Miller NE. Radical SAM, a novel protein superfamily linking unresolved steps in familiar biosynthetic pathways with radical

mechanisms: functional characterization using new analysis and information visualization methods. Nucleic Acids Res 2001;29:1097–106.
[13] Frey PA, Hegeman AD, Ruzicka FJ. The radical SAM superfamily. Crit Rev Biochem Mol Biol 2008;43:63–88.
[14] Schwarz G, Mendel RR, Ribbe MW. Molybdenum cofactors, enzymes and pathways. Nature 2009;460:839–47.
[15] Yan F, LaMarre JM, Röhrich R, et al. RlmN and Cfr are radical SAM enzymes involved in methylation of ribosomal RNA. J Am Chem Soc 2010;132:3953–64.
[16] Yan F, Fujimori DG. RNA methylation by radical SAM enzymes RlmN and Cfr proceeds via methylene transfer and hydride shift. Proc Natl Acad Sci USA 2011;108:3930–4.
[17] Grove TL, Benner JS, Radle MI, et al. A radically different mechanism for S-adenosylmethionine-dependent methyltransferases. Science 2011;332:604–7.
[18] Boal AK, Grove TL, McLaughlin MI, Yennawar NH, Booker SJ, Rosenzweig AC. Structural basis for methyl transfer by a radical SAM enzyme. Science 2011;332:1089–92.
[19] Hu Y, Fay AW, Ribbe MW. Identification of a nitrogenase iron-molybdenum cofactor precursor on NifEN complex. Proc Natl Acad Sci USA 2005;102:3236–41.
[20] Corbett MC, Hu Y, Fay AW, Ribbe MW, Hedman B, Hodgson KO. Structural insights into a protein-bound iron-molybdenum cofactor precursor. Proc Natl Acad Sci USA 2006;103:1238–43.
[21] Kaiser JT, Hu Y, Wiig JA, Rees DC, Ribbe MW. Structure of precursor-bound NifEN: a nitrogenase FeMo cofactor maturase/insertase. Science 2011;331:91–4.
[22] Fay AW, Blank MA, Lee CC, et al. Spectroscopic characterization of the isolated iron-molybdenum cofactor (FeMoco) precursor from the protein NifEN. Angew Chem Int Ed Engl 2011;50:7787–90.
[23] Lancaster KM, Hu Y, Bergmann U, Ribbe MW, Debeer SJ. X-ray spectroscopic observation of an interstitial carbide in NifEN-bound FeMoco precursor. Am Chem Soc 2013;136:610–2.
[24] Hu Y, Corbett MC, Fay AW, et al. FeMo cofactor maturation on NifEN. Proc Natl Acad Sci USA 2006;103:17119–24.
[25] Hu Y, Corbett MC, Fay AW, et al. Nitrogenase Fe protein: a molybdate/homocitrate insertase. Proc Natl Acad Sci USA 2006;103:17125–30.
[26] Yoshizawa JM, Blank MA, Fay AW, et al. Optimization of FeMoco Maturation on NifEN. J Am Chem Soc 2009;131:9321–5.
[27] Wiig JA, Hu Y, Ribbe MW. NifEN-B complex of Azotobacter vinelandii is fully functional in nitrogenase FeMo cofactor assembly. Proc Natl Acad Sci USA 2011;108:8623–7.
[28] Wiig JA, Hu Y, Lee CC, Ribbe MW. Radical SAM-dependent carbon insertion into nitrogenase M-cluster. Science 2012;337:1672–5.
[29] Van der Kamp MW, Żurek J, Manby FR, Harvey JN, Mulholland AJ. Testing high-level QM/MM methods for modeling enzyme reactions: acetyl-CoA deprotonation in citrate synthase. J Phys Chem B 2010;114:11303–14.
[30] Hartman FC, Lee EH. Examination of the function of active site lysine 329 of ribulose-bisphosphate carboxylase/oxygenase as revealed by the proton exchange reaction. J Biol Chem 1989;246:11784–9.
[31] Pavlov M, Siegbahn PEM, Blomberg MRA, Crabtree RH. Mechanism of H-H activation by nickel-iron hydrogenase. J Am Chem Soc 1998;120:548–55.
[32] Yang X, Hall MB. Monoiron hydrogenase catalysis: hydrogen activation with the formation of a dihydrogen, Fe-$H^{\delta-}\cdots H^{\delta+}$-O, bond and methenyl-$H_4MPT^+$ triggered hydride transfer. J Am Chem Soc 2009;131:10901–8.
[33] Hutcheson RU, Broderick JB. Radical SAM enzymes in methylation and methylthiolation. Metallomics 2012;4:1149–54.
[34] Knauer SH, Buckel W, Dobbek HJ. Structural basis for reductive radical formation and electron recycling in (R)-2-hydroxyisocaproyl-CoA dehydratase. Am Chem Soc 2011;133:4342–7.

[35] Lee HI, Benton PM, Laryukhin M, et al. The interstitial atom of the nitrogenase FeMo-cofactor: ENDOR and ESEEM show it is not an exchangeable nitrogen. J Am Chem Soc 2003;125:5604–5.
[36] Wiig JA, Lee CC, Hu Y, Ribbe MW. Tracing the interstitial carbide of the nitrogenase cofactor during substrate turnover. J Am Chem Soc 2013;135:4982–3.
[37] Hinnemann B, Nørskov JK. Modeling a central ligand in the nitrogenase FeMo cofactor. J Am Chem Soc 2003;125:1466–7.
[38] Xie H, Wu R, Zhou Z, Cao Z. Exploring the interstitial atom in the FeMo cofactor of nitrogenase: insights from QM and QM/MM calculations. J Phys Chem B 2008;112:11435–9.
[39] Moret ME, Peters JC. $N_2$ functionalization at iron metallaboratranes. J Am Chem Soc 2011;133:18118–21.

# 6 The iron-molybdenum cofactor of nitrogenase

Thomas Spatzal, Susana L. A. Andrade and Oliver Einsle

## 6.1 Introduction

Biological nitrogen fixation is an essential process for sustaining life on earth. As a building block of all classes of biomolecules, the element nitrogen is a crucial nutrient for all organisms, although its limited availability frequently becomes a growth-limiting factor [1–3]. This is because of all modifications of nitrogen that cycle the biosphere, only a single one – molecular dinitrogen, $N_2^-$ is of such outstanding stability that it constitutes a sink for more than 99% of all of the element at any given point in time. At the same time, atmospheric $N_2$ is a virtually unlimited reservoir, and the ability to access this as a source of nitrogen to sustain organismal growth is a fundamental advantage in the struggle for life. This task, however, is not easy: The triple bond of the $N_2$ molecule has an enthalpy of –942 kJ/mol, making it by far the most stable chemical bond to be cleaved in all of biochemistry. Despite the obvious advantages that come with the ability to break (or "fix") $N_2$, only one single enzymatic system has evolved to carry out this reaction [4]. This by itself highlights the difficulty involved in the process, and indeed, the enzyme system in question, nitrogenase, is a complicated two-component machinery that we still currently struggle to understand. The overall reaction of biological nitrogen fixation,

$$N_2 + 10\,H^+ + 8e^- + 16\,ATP \rightarrow 2\,NH_4^+ + H_2 + 16\,ADP + 16\,P_i$$

occurs under ambient conditions, but at the price of a significant investment of metabolic energy in the form of ATP. Nitrogen fixation is a costly process, and bacteria invest a major part of their energy budget for obtaining the desired nutrient, ammonium. Biological nitrogen fixation is frequently compared with its anthropogenic equivalent, the industrial Haber-Bosch process, designed by chemist Fritz Haber (Nobel Prize 1918) and engineer Carl Bosch (Nobel Prize 1931). Here, a structured surface of metallic iron acts as a catalyst to promote reaction of the gases $N_2$ and $H_2$ under high temperatures and pressures (720 K, 300 bar). The process was developed on an industrial scale at the beginning of the twentieth century, enabling the generation of ammonium not only for the production of explosives for the imminent World War but also as crop fertilizer that has since supported the unprecedented growth of the human population. With a volume of approximately 160 Mt of fixed nitrogen per year, the Haber-Bosch process is one of the dominant industrial processes of our time, having truly changed the face of the world [5]. Although the catalyst is affordable and the process is straightforward to operate, the energetic cost, in particular connected to the production of the synthetic gas $H_2$, is significant [6]. The energy required for hydrogen production is most frequently obtained from the combustion of fossil fuels, leading to significant emissions of $CO_2$. Moreover, the excessive use of fertilizers, mainly in the industrial countries, has led to eutrophication and to what is commonly referred to as "nitrogen pollution." Only seven decades after its invention, Gerhard Ertl was able to explain the mechanism of heterogeneous catalysis in the Haber-Bosch process and was awarded the Nobel Prize in chemistry in 2007 [7]. On the side of homogeneous catalysis, only few complexes were described to bind $N_2$, and even fewer were found to activate the substrate and stoichiometrically cleave the triple bond. In 2003, Yandulov and Schrock [8] presented the first compound able to catalytically cleave $N_2$. Both the Schrock compound and the molecule presented shortly thereafter by Nishibayashi were based on molybdenum [9], lending new substance to the old hypothesis that the Mo ion present in the enzyme nitrogenase might be the site of catalysis. However, a hydride-bridged complex based on titanium by Hou [10] and, most recently, an iron complex with a flexible Fe-B (boron) bond from Peters' laboratory were also found to mediate the reaction [11]. The periodic table thus seems to hold a variety of possibilities for breaking the $N_2$ triple bond, leading us back to the question as to why nature only adopted one such solution.

## 6.2 The metal clusters of nitrogenase

Nitrogenase is an iron-sulfur enzyme. Its two components are the molybdenum-iron protein (MoFe protein) and the iron protein (Fe protein) (Fig. 6.1a). The latter is the

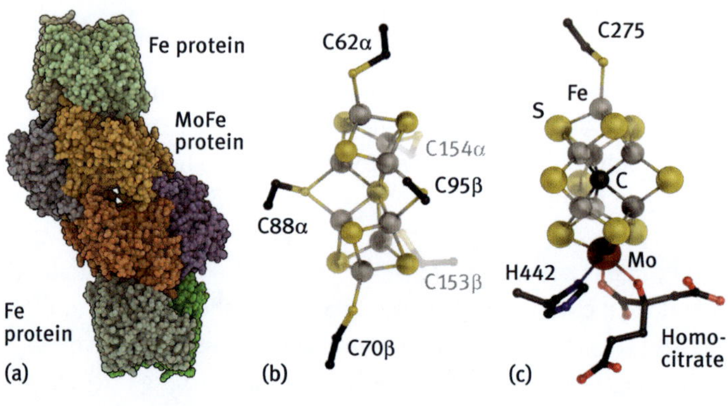


**Fig. 6.1:** The enzyme system nitrogenase consists of two component proteins. (a) During reductive catalysis, the Fe protein, $NifH_2$, transiently forms a complex with the heterotetrameric MoFe protein, $NifD_2K_2$. ATP is hydrolyzed in Fe protein, leading to electron transfer to the active site of MoFe protein and oxidation of the [4Fe:4S] cluster in Fe protein. (b) The [8Fe:7S] P-cluster is an electron transfer site that is reduced by the Fe protein and transfers electrons to the active site of nitrogenase. Upon two-electron oxidation from the all-ferrous $P^N$ state to the $P^{Ox}$ state, two Fe ions of P-cluster undergo a remarkable – and reversible – conformational change. (c) Reduction of substrates by nitrogenase occurs at FeMoco, the unique active site of the enzyme, with a composition of [Mo:7Fe:9S:C]:homocitrate.

site of binding and hydrolysis of ATP, whose free enthalpy is used to transfer an electron from a [4Fe:4S] cluster to MoFe protein. Although it shows the regular cubane structure, the cluster of Fe protein is unusual in at least two aspects. First, it bridges the two subunits of the homodimeric protein, so that each monomer provides two cysteine ligands to the irons of the cluster [12], and second, this cluster is among the few described in literature to undergo two redox transitions within a narrow potential range. This makes three oxidation states, $[4Fe{:}4S]^{+2}$, $[4Fe{:}4S]^{+1}$, and $[4Fe{:}4S]^{+0}$, technically accessible. *In vitro*, the Fe protein was able to mediate both one- and two-electron transfer, and this is of obvious consequence for the overall ATP/$e^-$ stoichiometry of the process, possibly depending on whether the initial electron donor to the enzyme is a ferredoxin or a flavodoxin [13].

MoFe protein contains two further iron-sulfur clusters, both of which are unique to the enzyme. The primary recipient of electrons delivered from ferredoxin is the [8Fe:7S] P-cluster, a moiety generated through a fusion of two cubane-type [4Fe:4S] clusters under abstraction of a single sulfide (Fig. 6.1b) [14]. The P-cluster then passes electrons on to the actual active site of nitrogenase, the iron-molybdenum cofactor (FeMo cofactor, FeMoco). During nitrogenase biosynthesis, FeMoco is synthesized *ex situ* in a complex reaction sequence by multiple maturation factors, and only the finalized moiety is inserted into an apo-version of MoFe protein [15]. FeMoco is the largest and most intricate metal cofactor known to biology (Fig. 6.1c). It consists of an organic molecule, *R*-homocitrate, that is synthesized from acetyl-coenzyme A and 2-oxoglutarate by the maturation factor NifV [16]. Homocitrate is a ligand to the inorganic part of FeMoco, an iron-sulfur cluster containing a molybdenum ion and seven iron ions. Because this cluster is the site of binding and reduction of substrates, it has long been the focus of study for numerous fields of research, including microbiology, inorganic chemistry, theoretical chemistry, spectroscopy, and structural biology. Many proposals for the structure and reactivity of FeMoco were put forward over the years, and yet the first crystal structure, presented by Kim and Rees [17, 18] in 1992, surpassed expectations.

## 6.3 Structure of FeMoco

FeMoco was initially described to consist of two subclusters, [Mo:3Fe:3S] and [4Fe:3S], bridged by three non-protein ligands, in an immediately recognized analogy to P-cluster that appears as two [4Fe:4S] clusters fused *via* one of the sulfides (Fig. 6.1b). In the first analysis, the bridging ligands in FeMoco were not identified but were subsequently assigned to be two sulfides and a third ligand “X” [18]. Most notably, only one of the seven iron atoms of FeMoco, $Fe^{I}$, appeared to show a regular, tetrahedral coordination environment with four sulfido ligands. The other six iron atoms of the cluster were only coordinated to three sulfide ligands each, forming a trigonal prism that surrounded a spacious internal cavity (Fig. 6.2a). Interestingly, the bridging atoms that connected the two triangles of the prism were located on the

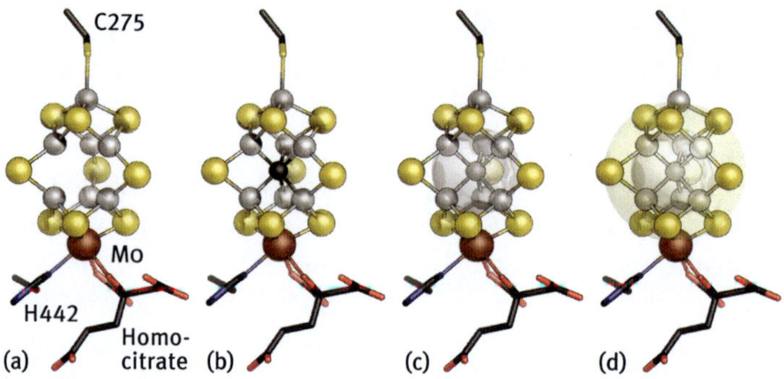


**Fig. 6.2:** FeMoco is the active site of dinitrogen reduction in nitrogenase. (a) In a first structural analysis [18], the central cavity of FeMoco appeared empty, suggesting a substrate binding site surrounded by six coordinatively unsaturated iron ions. (b) Reassessment of the crystal structure at 1.16-Å resolution revealed that a central ligand was previously occluded by Fourier series termination artifacts of the surrounding scatterers. (c) Because of the unique geometry of FeMoco, six iron atoms, $Fe^{II}$–$Fe^{VII}$, are equidistant from the center, placing them on a sphere with a radius $r_{Fe}$ = 2.0 Å. (d) Similarly, all nine sulfides in the cluster structure are equidistant from a center, on a sphere with radius $r_S$ = 3.5 Å.

edges of the prism rather than on its faces, as chemical intuition may (or may not) have suggested. The molybdenum ion formed an integral part of the cofactor, sitting at the apical position of the metal cluster. It attained near-octahedral coordination geometry, with direct interaction to the homocitrate molecule as well as to the imidazole group of a histidine residue, H442 in the *Azotobacter vinelandii* enzyme, in the α-subunit of MoFe protein. Only one further residue, C275, coordinates FeMoco at the distal end from Mo, at $Fe^I$. Arguably, the most outstanding feature of this initial structure of FeMoco was the unprecedented central cavity. On the analytical side, it presented a substantial challenge for theoretical and synthetic chemists who saw themselves forced to assume strong metal-metal bonding forces to explain the structural integrity of the cluster, let alone its extraordinary stability: Although the metal centers of MoFe protein are highly sensitive to $O_2$, FeMoco – but not P-cluster – can be extracted from denatured protein in an intact state, and it can even be reintroduced into a separately generated apo-nitrogenase, where it regains its catalytic capacities [19, 20]. It was difficult to explain these findings in the light of the seemingly fragile structure of the cluster.

Moreover, the six iron atoms, $Fe^{II}$–$Fe^{VII}$, that form the trigonal prism in the center of nitrogenase were coordinatively unsaturated. From the geometry of the cluster, it was obvious that all six irons would have a free coordination site facing the central cavity of the cluster to complete a tetrahedral ligand field (Fig. 6.2a). This was a seemingly unmistakable hint toward a possible substrate binding site, and yet it proved to be impossible to observe any bound molecule. At the time, structural data for

MoFe protein orthologs were available for three organisms, *A. vinelandii, Clostridium pasteurianum* [21], and *Klebsiella pneumoniae* [22], and all showed an identical arrangement for FeMoco. No structural variability and no conformational changes – not even in the form of a partial degradation of the sensitive cluster – were ever observed. For these reasons, the actual binding sites for substrates on FeMoco remained elusive, and the enzymology of nitrogenase reached a hiatus.

## 6.4 Redox properties of FeMoco

The requirement for eight electrons to complete one catalytic turnover necessitates a precise fine-tuning of the electronic states in all metal centers involved, especially at the active site FeMoco. Current mechanistic models assume an accumulation of at least three electrons at the cluster prior to substrate binding [23]. However, the redox chemistry as well as the associated electronic properties of the metal center is scarcely known. Thus far, only three different redox states ($FeMoco^N$, $FeMoco^R$, and $FeMoco^{Ox}$) have been identified spectroscopically, and only two of which ($FeMoco^N$ and $FeMoco^R$) are of undisputed physiological relevance. In MoFe protein, as isolated in the presence of reducing agents, the active site FeMoco is in its paramagnetic $S = 3/2$ resting state ($FeMoco^N$), which is the most studied by far, with a wealth of data that includes vibrational, Mössbauer, electron paramagnetic resonance (EPR)/electron nuclear double resonance (ENDOR), XANES, and EXAFS spectroscopies [4, 24–30]. To date, the integration of these data into a concise functional model is strongly hindered by a lack of detailed understanding of the electronic structure of FeMoco. This is required not only for the interpretation of spectra but also as a solid basis for theoretical approaches that are abundant for the case of this metal center but unfortunately lack the coherence required to provide definitive answers.

From the $FeMoco^N$ state, a one-electron reduction, exclusively mediated by the Fe protein, leads to the $FeMoco^R$ state. This state was only obtained by freeze-trapping the enzyme under turnover conditions [4, 28], and it was thus suggested not to represent a single oxidation state but rather a mixture of different states occurring during catalysis. However, it cannot be excluded that multiple electrons accumulate at the active site in the form of bound reaction intermediates that would not require a change of oxidation state in the metal core [31]. The freeze-trapped $FeMoco^R$ states are EPR silent [32], and the exact redox potential of the $FeMoco^N$/$FeMoco^R$ couple is unknown. In the other direction, a reversible one-electron oxidation of $FeMoco^N$ leads to the $FeMoco^{Ox}$ state [27, 33], a diamagnetic $S = 0$ state [4, 25, 34] with a midpoint redox potential for the $FeMoco^N$/$FeMoco^{Ox}$ couple of −42 mV [35]. Although the $FeMoco^{Ox}$ state is unlikely to play a physiological role, its investigation provided further information about the electronic transitions possible in the cofactor [33].

## 6.5 An overlooked detail: the central light atom

Based on the structural data available at the turn of the millennium, only one little detail could have raised suspicions about the overall picture of FeMoco. The crystallographic structures were resolved to 2.0 Å (*A. vinelandii*) and 1.6 Å (*K. pneumoniae*), and they consistently showed a large and empty cavity in the cluster in the $2F_o$-$F_c$ electron density maps commonly used for model building. While refining atomic models against experimentally observed electron density, the inspection of a further type of map known as $F_o$-$F_c$ difference electron density maps is routinely performed. $F_o$ thereby represents the experimentally observed structure factors, whereas $F_c$ stands for calculated structure factors derived from the molecular model. A positive peak in a difference map would require more experimental information than is actually included in the model. A positive peak is commonly caused by data that still need to be modeled, whereas a negative peak implies that modeling has gone beyond the experimental data. Such peaks provide guidance for modeling, and they usually are overlaid with the $2F_o$-$F_c$ electron density map that represents the actual structure. In case of MoFe protein, the $2F_o$-$F_c$ maps displayed the cavity in FeMoco, but a strong, positive peak appeared in $F_o$-$F_c$ difference electron density maps. The absence of a corresponding peak in the $2F_o$-$F_c$ map constituted evidence that an atom might occupy the cavity, and in retrospect, the overlooked feature was an early indication that something was amiss in the structural model.

Only 10 years after the first structural model for MoFe protein, refined crystallization protocols allowed for much more precise collection of diffraction data, leading to an improved crystal structure at 1.16-Å resolution [36]. This model – at true atomic detail, i.e. with separate electron density maxima for each atom – literally shed new light into the central cavity of FeMoco. It revealed the presence of a light atom in the very center of the cluster, the trigonal prism of iron ions. This atom was lighter than a sulfide, and the data were in best agreement with carbon, nitrogen, or oxygen (Fig. 6.2b). Interestingly, the electron density maximum for the central atom disappeared at resolutions lower than 1.55 Å, leaving behind the empty cavity in the $2F_o$-$F_c$ map and the conspicuous, positive difference electron density peak in the $F_o$-$F_c$ map. This then turned out to be a resolution-dependent artifact induced by the unique geometry of FeMoco itself. In X-ray crystallography, electron density maps are obtained as the Fourier transform of the experimental diffraction data, the structure factors. They represent individual wave functions to be added in a Fourier synthesis, and one of the most amazing properties of this mathematical operation is its robustness: By adding as many wave functions – structure factors – as are available, one obtains an increasingly precise approximation of the actual electron density distribution. The absence of individual terms due to incomplete diffraction data does not lead to missing features in the electron density map, but rather to a gradual deterioration of its overall quality. Limiting the resolution of the diffraction data set is just one such way to reduce completeness, with the result that artifacts appear, betraying the underlying periodic structure of the wave functions that were added. Technically called "Fourier series termination errors," these artifacts are more commonly known as "ripples" for exactly this reason. In protein crystallography, the effect is known to occur when lighter atoms that are coordinated to heavy scatterers such as molybdenum or tungsten, but commonly iron, with only 24 electrons compared with the 38 of $Mo^{+4}$ or the 68 of $W^{+4}$, is already too light to lead to marked distortions in its surroundings.

In FeMoco, however, the central position of the cluster is a very different case. A central light atom is located in the very middle of the trigonal prism formed by $Fe^{II}$–$Fe^{VII}$, with identical distances of $r_{Fe}$ = 2.0 Å to each metal ion (Fig. 6.2c). At the same time, all nine sulfides within FeMoco are equidistant from the center as well, placing them on the surface of a sphere with a radius of $r_S$ = 3.5 Å (Fig. 6.2d). Although the individual ripple effects of a single iron and a single sulfide are small, they are amplified by the number of identical scatterers in the unique geometric arrangement of FeMoco. Consequently, the addition of the individual ripples creates a (fully artificial) resolution-dependent profile of electron density in the cofactor center, and it is exactly in the resolution range of 1.55–2.2 Å (Fig. 6.3a) that the effect adds up to a negative value that is sufficient to conceal the presence of the central atom (Fig. 6.3c).

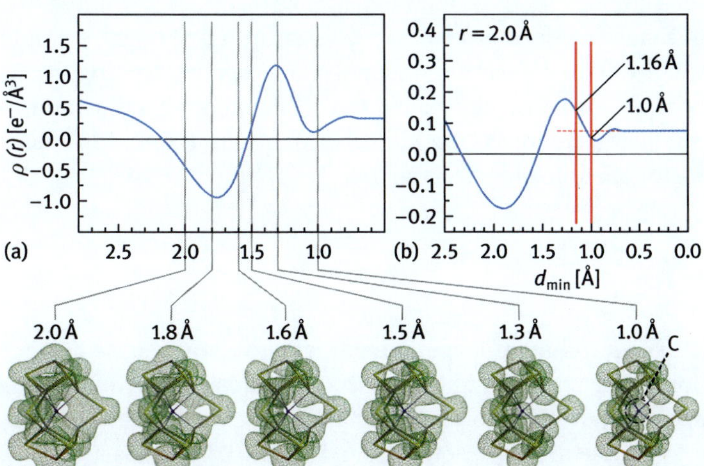


**Fig. 6.3:** Electron density artifacts in the cofactor center. (a) The resolution-dependent electron density profile for the center of FeMoco is the direct result of interfering Fourier series termination artifacts created by the surrounding Fe and S atoms (Fig. 6.2c,d). (b) In the cofactor center, the artifact (blue) leads to an overestimation of the electron density of the central atom at 1.16-Å resolution, whereas the estimate is unbiased at 1.0 Å. (c) Visualization of the electron density artifact at resolutions as indicated in (a). For the Fourier synthesis, $F_{calc}$ from a structural model including a central carbon were used as structure factors at all resolutions. The density maximum for the central atom is well defined at high resolutions but vanishes abruptly at 1.6-Å resolution. Dashed lines indicate the position of the well-defined central atom (c) on the highest-resolution structure in (c).

Once this phenomenon was understood, it was possible to subtract the influence of the density artifact and to quantify the actual electron density present in the cofactor center. According to the observed density peak, the possible candidates for the central atom were C, N, and O, with a most favorable match for nitrogen. Obviously, it then seemed difficult to imagine how a nitrogen atom could be inserted without being in some way connected to the $N_2$-cleaving activity of nitrogenase.

In the following years, based on the discovery of the interstitial ligand and its promising mechanistic implications, a multitude of spectroscopic and theoretical studies were undertaken to clarify the nature of the light atom. On the experimental side, the work was mainly based on resonance techniques such as electron spin echo envelope modulation (ESEEM) and ENDOR spectroscopies that detect characteristic couplings between the electron spin of the metal scaffold and the nuclear spin of the interstitial candidate atom. However, limitations such as the labeling efficiency of the FeMoco with paramagnetic isotopes ($^{13}C$, $^{15}N$, and $^{17}O$), in combination with an ambiguous description of the electronic structure of the metal center prevented a clear assignment [37–39]. Theoretical approaches to elucidate the nature of the light atom largely applied density functional theory (DFT) calculations based on the available structural information [29, 40–50]. Unfortunately, the results of various approaches were not congruent and depended strongly on the input models. Some authors concluded that the central atom would make FeMoco more rigid, whereas others thought it would convey additional flexibility and make conformational rearrangements during catalysis more likely. Thus, despite extensive efforts and the combination of a broad range of analytical and theoretical techniques, the nature of the central atom in FeMoco remained under debate for the following decade.

## 6.6 The nature of X

After the discovery of the central atom, experimental work on nitrogenase FeMoco focused on the question of the chemical nature of the interstitial ligand and on the question of its exchangeability. Hoffman and colleagues [37] showed that enzyme turnover with $^{15}N_2$ did not give rise to new resonances in ENDOR/ESEEM spectroscopy, which would have been expected if a central nitrogen ligand had been exchanged with its paramagnetic counterpart during catalysis. In a follow-up study, they carried out uniform $^{15}N$-labeling of MoFe protein and once more did not see any spectral changes, leading to the conclusion that the central ligand was not a nitrogen species at all [38]. Unfortunately, both results were negative evidence, arguing on the basis of the absence of a signal, so that the authors continued to attempt labeling with $^{13}C$ to probe for carbon. Because of the high cost of $^{13}C$-labeled glucose, however, they limited their study to a labeling efficiency of only 5% and did not see spectral changes under these conditions, so they deemed that carbon was an unlikely candidate for the interstitial atom as well [39].

In 2011, three separate experimental approaches eventually provided direct evidence for the nature of the interstitial atom and settled the debate. Studies based on iron $K_β$ valence-to-core X-ray emission spectroscopy (V2C-XES), on more highly resolved X-ray diffraction data at 1.0-Å resolution and on ESEEM spectroscopy with uniformly $^{15}N$- and $^{13}C$-labeled protein, clarified that the atom in the heart of FeMoco is indeed a carbon [51, 52]. The crystallographic approach mainly relied on a further improvement of data quality, in combination with a novel strategy for data analysis. The increase of resolution from 1.16 Å (PDB 1M1N) to 1.0 Å (PDB 3U7Q) resulted in substantially better electron density maps. This higher precision proved to be a major advance, but two issues still had to be addressed. The first was to incorporate the distorting effects from the surrounding heavy atoms, and the second one was the standing question whether, based on electron density features alone, a reliable discrimination between individual light atom types is at all possible. A simulation of the intrinsic electron density artifact in FeMoco (Fig. 6.3) shows that the ripples generated by the proximity of the surrounding atoms disappear fully only at a resolution of 0.65 Å. At 1.55-Å resolution, the artifact effect turns from negative to positive, with a peak at 1.3 Å, and it is still markedly positive at 1.1 Å, the resolution at which the central atom was initially observed. The electron density in the cofactor center is thus not reduced, but rather enhanced at this resolution. The graph also shows that at 1.0-Å resolution, the influence of the surrounding Fe and S atoms is already negligible (Fig. 6.3b). In consequence, the structural analysis at 1.16 Å is remarkably more biased than at 1.0 Å, where the added ripple effect is insignificant.

The remaining second obstacle then was the unambiguous discrimination among carbon, nitrogen, and oxygen atoms based solely on differences in electron density. These are single-electron differences across the entire electron shell of an atom, making a distinction anything but straightforward. Using atomic resolution data at 1.0 Å, the problem was ultimately overcome by exploring a new strategy of data analysis, focusing on the spatial expansion of shells of electron density. The definition of the extent of the electron shell of an atom is a critical parameter for the evaluation of its total electron density, in particular for a configuration, such as the center of FeMoco, that is surrounded by six heavy atoms. Including electron density from these atoms, even to a minor degree, could significantly distort the results, and we found the analysis of spatial expansion behavior to be a robust way to circumvent these effects. In the analysis, a given atom position was investigated in a series of spheres with radii ranging from 0.2 to 1.4 Å, calculating the average electron density value within each sphere in turn (Fig. 6.4). Note that this is not an analysis of individual shells, but rather of the entire core electron density with an increasing probe radius. A plot of the average density values *vs* the probe sphere radii yielded a characteristic profile that was calculated for each C, N, and O atom within the structure of MoFe protein [52]. With approximately 10,370 carbon, 2,740 nitrogen, and 5,620 oxygen atoms in the structural model of MoFe protein, this then allowed for a solid statistical analysis that proved to be highly robust toward varying environments of an atom,

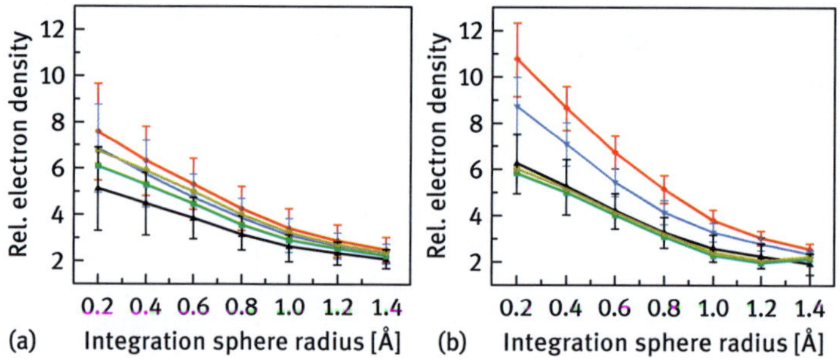


**Fig. 6.4:** Electron density analysis in the center of FeMoco. All instances of carbon (black), nitrogen (blue), and oxygen (red) in the structure of *A. vinelandii* MoFe protein were evaluated in the resolution sphere and the integrated electron density was plotted against the sphere radius. (a) At 1.16-Å resolution (PDB 1M1N), the scattering curves for the individual elements are not ideally separated, and the density profiles for the two copies of FeMoco (green) agreed best with nitrogen. This analysis included 636,000 reflections [36]. (b) In the reanalysis at 1.0 Å (PDB 3U7Q), 960,000 reflections were included, leading to a far better separation of the individual plots for C, N, and O. As the electron density artifact of the cofactor did not have any influence here, the curves for the central atoms (green) clearly showed that the light atom is a carbon species [52].

allowing even for an assessment of the influence of neighboring atoms. The curves obtained for the individual atom types discriminated the different atomic species unequivocally. In the analysis, the averaged curves for all C, N, and O were used as a reference for the electron density profile that was actually observed in the cofactor center. Interestingly, when this analysis was carried out using the data set collected in 2002 at a resolution of 1.16 Å (PDB 1M1N), the profile of the central positions was most similar to nitrogen (Fig. 6.4a). In contrast, the higher resolution of 1.0 Å obtained in 2011 (PDB 3U7Q) yielded profiles that precisely matched carbon [52] (Fig. 6.4b). Two factors account for this unexpected difference: First, the seemingly small increase in resolution of 0.16 Å indeed corresponds to approximately 50% more diffraction maxima in the data set, as the amount of data collected in a diffraction experiment does not scale linearly with resolution. Consequently, the precision of the later analysis was considerably higher than that for the initial data set. Second, the artifactual electron density feature in the cofactor center that is induced by cofactor geometry creates a positive distortion at 1.16 Å that disappears at 1.0 Å (Fig. 6.3b), contributing to an overestimation of the observed electron density in the lower-resolution analysis. With all such effects taken into account, the current crystallographic analysis clearly identifies the central atom as a carbon species.

A second strategy for the identification of the interstitial atom was ESEEM spectroscopy. It detects hyperfine couplings between electron spins and nuclear spins and is therefore a capable tool to identify interactions between the interstitial atom and the metal scaffold. However, ESEEM requires paramagnetic nuclei, and the low natural abundance of the two candidates $^{13}C$ (1.1%) and $^{15}N$ (0.37%) are a significant hindrance for obtaining high data quality. Although enrichment with $^{15}N$ is straightforwardly achieved using the isotope as a sole nitrogen source, either in $(^{15}NH_4)Cl$ or in $^{15}N_2$, the uniform labeling with $^{13}C$ is complicated by the considerable cost of labeled sucrose, the carbon source commonly used for growth of *A. vinelandii*. Earlier studies therefore relied on low labeling fractions of 5%–10%, but in these studies, no unambiguous signals were observed or assigned [39]. This key problem was overcome by a fundamental reorganization of cell growth and protein isolation protocols and a switch to use of glucose as a carbon source, which was far more affordable as a uniformly labeled compound. In the following, this allowed for the quantitative enrichment of MoFe protein with $^{13}C$ and, consequently, a fundamental improvement in signal quality [52]. The ESEEM data brought two major insights. First, the comparison between unlabeled protein and the $^{15}N$-substituted sample confirmed the absence of any additional signals that could be correlated with a $^{15}N$-nucleus strongly coupled to the electron spin of FeMoco. Second, the uniformly $^{13}C$-labeled protein showed an additional spectral feature at the Larmor frequency of $^{13}C$, with a splitting of 2.5 MHz that can only be explained by a $^{13}C$ nuclear spin strongly coupled to a substantial amount of electron spin density, i.e. a carbon in close and direct proximity to the metal scaffold (Fig. 6.5a). Taking into account that no direct metal-carbon interaction is observed in the cluster and the known carbon atoms of H442, C275, and the homocitrate moiety are too far away from the metals to be the origin of the observed strong hyperfine interaction, a central carbon emerged as the only remaining option [52].

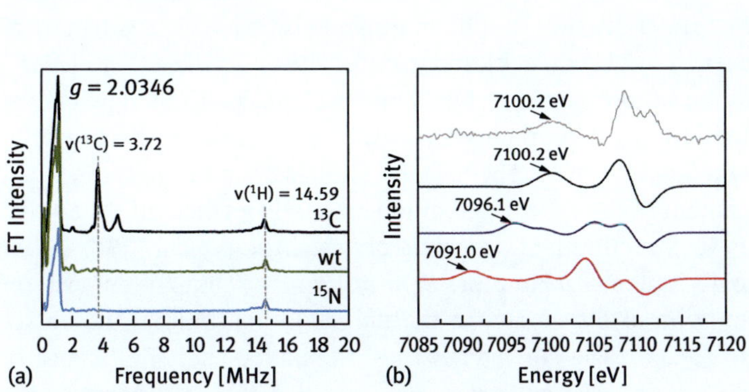


**Fig. 6.5:** Spectroscopic evidence for a central carbon species in FeMoco. (a) Three-pulse ESEEM spectra for MoFe protein [52]. The spectral features for unlabeled ($^{12}C/^{14}N$) protein (green) are identical to those for a uniformly $^{15}N$-labeled sample (blue). In contrast, uniform $^{13}C$ labeling leads to an additional resonance with a Larmor frequency of 3.72 MHz, that is in agreement with a strong Fe-$^{13}C$ coupling. Because there is no other direct iron-carbon bond in FeMoco, this resonance must arise from the central carbon. (b) Experimental V2C XES spectrum of FeMoco after subtraction of the P-cluster contribution (gray). Simulations of the spectrum with an interstitial carbide ($C^{4-}$) are a good match (black), whereas nitride ($N^{3-}$, blue) and oxide ($O^{2-}$, red) produce substantially inferior fits [51].

Finally, a third, complementary approach to identify the central atom was based on iron V2C-XES. The principle of V2C-XES is to monitor the emission of photons during the relaxation of a valence electron into a core hole after an X-ray-induced ionization event. The transition energy is highly element-specific, and it allows a clear assignment of the donating ligand if the transitions can be sufficiently well resolved. In the case of FeMoco, DeBeer and colleagues [51] showed that the 2s and 2p transitions from the interstitial ligand into the Fe 1s-core hole are adequately separated from each other to allow for a distinct assignment. Calculated spectra were in very good agreement with experimental data from model compounds, showing that the transition energies for carbon (7100.2 eV/[2s→1s(Fe)], 7107.9 eV/[2p→1s(Fe)]) are sufficiently separated from nitrogen (7096.1 eV/[2s→1s(Fe)], 7105.1 eV/[2p→1s(Fe)]) and oxygen (7091.0 eV/[2s→1s(Fe)], 7104.0 eV/[2p→1s(Fe)]), respectively (Fig. 6.5b). However, the 2p(ligand) to 1s(Fe) transition strongly interfered with a dominant sulfur 3p orbital transition, which prevented an unambiguous assignment of the former. The 2s→1s transition originating from the light atom, meanwhile, was free from such overlaps and in very good agreement with simulated transition energies for a carbide, $C^{4-}$. Therefore, V2C-XES also confirmed the presence of an interstitial carbon species in the FeMoco.

## 6.7 Insights into the electronic structure of FeMoco

With the atomic structure of FeMoco in place, the next level of detail to be addressed was the electronic structure of the site. Despite multiple efforts, a concise theoretical description of the cluster has not been achieved to date. The crystallizability of MoFe protein was used early on by Hoffmann and colleagues [53] to collect single-crystal EPR data, but unfortunately, no three-dimensional structure of the protein was available at the time to provide a point of reference.These studies were recently repeated, yielding a direct assessment of the orientation of the magnetic **g** tensor of the $S = 3/2$ system with respect to its structure [54]. In single crystals, the typical EPR powder spectra are reduced to a single sharp band per spin system that changes its position with the relative orientation of the cluster in the magnetic field due to the anisotropy of the **g** tensor (Fig. 6.6a). The derived $g^2$ values from a data set of EPR spectra obtained by rotating a crystals in the spectrometer (Fig. 6.6b), the tensor orientation was obtained and matched to the protein orientation derived from diffraction data (Fig. 6.6c). The analysis showed that the longest main axis of the **g** tensor, $g_z = 4.31$, is oriented along the 3-fold symmetry axis of the cluster and that the $g_y = 3.65$ axis attains a distinct position along the cluster edge formed by atoms $Fe^{I}$-$Fe^{III}$-S5-$Fe^{VII}$-Mo, that is stabilized by the surrounding protein (Fig. 6.6c). This underlines a common theme in bioinorganic chemistry, in that the surrounding protein matrix indeed plays a central role in tuning the properties of a metal active site [54].

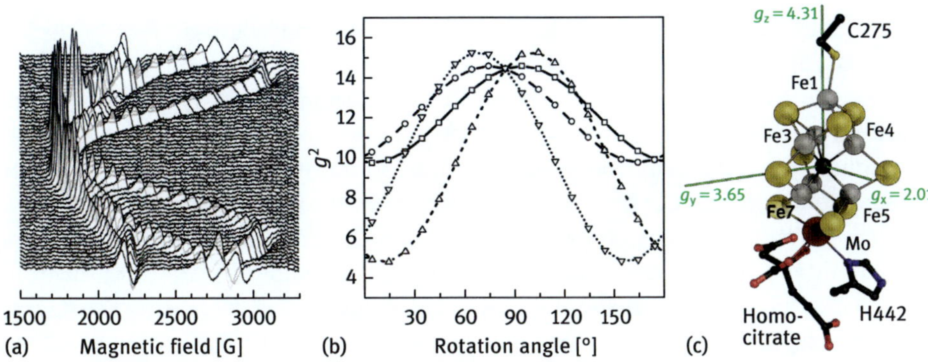


**Fig. 6.6:** Magnetic and electronic properties of FeMoco determined by single-crystal EPR spectroscopy. (a) The characteristic EPR resonance energies of the FeMoco in single crystals of *A. vinelandii* MoFe protein changed with rotation of the crystal in the magnetic field. This was due to the anisotropy of the **g** tensor that reflects the magnetic moment of the $S = 3/2$ spin multiplet of the FeMoco$^N$ ground state of the cluster. (b) A plot of the $g^2$ values for the spectra in (a) revealed four distinct signals rotating in their resonance position that corresponded to the four copies of FeMoco in the unit cell. Together with diffraction data, this information could be used to derive the relative orientation of the **g** tensor and the cluster structure. (c) When superimposed, the $g_z$ axis of the magnetic tensor aligned well with the intrinsic 3-fold axis of FeMoco. The $g_y$ main axis was oriented along the $Fe^{I}$-$Fe^{III}$-$Fe^{VII}$-Mo edge of the cluster. This magnetic anisotropy is most likely induced by the inhomogeneous electrostatic environment of the surrounding protein matrix.

## 6.8 A central carbon – consequences and perspectives

Since the discovery of the interstitial ligand in 2002, a multitude of theoretical and experimental studies were carried out with a primary focus on correctly assigning the chemical nature of the central atom. For nearly a decade, nitrogen was the favored candidate, in line with the available structural data at 1.16 Å, and it was therefore incorporated into most mechanistic considerations for $N_2$ reduction. As a consequence, substantial conformational rearrangements of the FeMoco scaffold would have been required during catalysis. The transient formation of a highly reactive iron surface upon cluster rearrangement was thus proposed as a realistic mechanistic possibility. The identification of the central ligand as a carbon now changes this picture. It necessarily points more toward a structural role for this atom, although even an increase of flexibility due to the interstitial atom has been proposed. Nevertheless, the binding of the substrate $N_2$ or one of its reduction intermediates in the center of FeMoco is now ruled out. Nitrogen binding and reduction may either occur at an iron face or the molybdenum end of the cofactor, and a structural change of the metal scaffold itself seems unlikely. The presence of carbon in the center of the active site conveys stability, not unlike the role that sparse integration of carbon plays in stabilizing the iron scaffold of steel. This is also in agreement with previous observations by

NRVS and EXAFS that revealed only minor changes in bond lengths in FeMoco during catalysis [55].

On a different level, the new focus on iron-carbon chemistry with respect to dinitrogen activation is a promising starting point in the search for new model compounds, in particular due to the fact that, so far, the synthesis of a complete cofactor analogue has not been achieved. In fact, Peters and colleagues [56] showed that the $N_2$ molecule as well as CO can bind to five-coordinated iron complexes that mimic the architecture of FeMoco. The complexes show variations of the iron-carbon distance during substrate binding and reduction by decreasing the degree of covalency of the Fe-C bond, and this may well be in line with FeMoco having to adapt slightly to the different intermediates during turnover. Assuming substrate binding to one or more belt irons, a variation of the Fe-C distances would thus allow for a tuning of the orbital characteristics needed for substrate interaction. A modulation of this kind could therefore be essential for FeMoco to accommodate for substrate binding, activation, and reduction according to current mechanistic considerations involving distal or alternating pathways [56]. Interestingly, the same group showed shortly thereafter that their complex can mediate the catalytic reduction of $N_2$, giving additional significance to this first catalytic iron-based nitrogenase model [11].

With the central ligand identified as carbide, it has become clear that this atom is not introduced as an intermediate of dinitrogen reduction but must be inserted during the biogenesis of the metal center. This realization immediately put the focus on one particular maturation factor – the NifB protein – that assembles a topologically complete, all-iron precursor of FeMoco in an intriguing reaction starting from simple iron-sulfur units synthesized by the general NifS/NifU system. NifB is an enzyme of the radical/SAM family, and shortly after the identification of the interstitial ligand as carbide, Ribbe and colleagues [57] were able to demonstrate that SAM is indeed the source of the interstitial ligand and that it is inserted by NifB. From a mechanistic point of view, it is plausible that the carbon insertion occurs in analogy to the proposed mechanism for RNA methylation. Furthermore, it was shown that the carbon atom can neither be exchanged nor can it directly interact with substrate during turnover [58]. These findings naturally raise questions concerning the electronic state of the central carbon species. SAM donates a methyl group (formal oxidation state – IV), and its insertion into the core of the metal cluster requires multiple deprotonation events but not necessarily a change in redox state. However, in a strongly coupled [Fe:S] system, a purely ionic state can be largely ruled out, in particular, because of the ionic radius of carbide ($r$ = 2.6 Å) that could not be accommodated in the cluster center. To date, an unambiguous identification of the electronic structure of FeMoco – let alone of the central ligand – has not yet been achieved, and further studies will be required to clarify these questions to provide a basis for a sound mechanistic understanding of the nitrogenase system. Undoubtedly, these will require the detailed characterization of the various electronic states of the active site, especially considering the involvement of the central carbon. The highly dynamic process of nitrogenase complex formation significantly restricts the experimental accessibility of intermediates. Theoretical approaches based on precisely refined experimental data will be an essential help in elucidating the multi-electron reduction of dinitrogen, and the field is currently in an unprecedented position to finally tackle the mysteries of biological nitrogen fixation.

## Acknowledgments

The authors thank Doug Rees, Markus Ribbe, Yilin Hu, Frank Neese, and Serena DeBeer for stimulating discussions. This work was supported by Deutsche Forschungsgemeinschaft, Howard Hughes Medical Institute, BIOSS Centre for Biological Signalling Studies, and the European Research Council.

## References

[1] Howard JB, Rees DC. Structural basis of biological nitrogen fixation. Chem Rev 1996;96:2965–82.
[2] Canfield DE, Glazer AN, Falkowski PG. The evolution and future of earth's nitrogen cycle. Science 2010;330:192–6.
[3] Einsle O, Kroneck PMH. Structural basis of denitrification. Biol Chem 2004;385:875–83.
[4] Burgess BK, Lowe DJ. Mechanism of molybdenum nitrogenase. Chem Rev 1996;96:2983–3011.
[5] Smil V. Enriching the earth: Fritz Haber, Carl Bosch, and the transformation of world food production. Boston (MA): MIT Press; 2001.
[6] Schlögl R. Catalytic synthesis of ammonia – a "never-ending story"? Angew Chem Int Edit 2003;42:2004–8.
[7] Ertl G. The arduous way to the Haber-Bosch process. Z Anorg Allg Chem 2012;638:487–9.
[8] Yandulov DV, Schrock RR. Catalytic reduction of dinitrogen to ammonia at a single molybdenum center. Science 2003;301:76–8.
[9] Arashiba K, Miyake Y, Nishibayashi Y. A molybdenum complex bearing PNP-type pincer ligands leads to the catalytic reduction of dinitrogen into ammonia. Nat Chem 2011;3:120–5.
[10] Shima T, Hu SW, Luo G, Kang XH, Luo Y, Hou ZM. Dinitrogen cleavage and hydrogenation by a trinuclear titanium polyhydride complex. Science 2013;340:1549–52.
[11] Anderson JS, Rittle J, Peters JC. Catalytic conversion of nitrogen to ammonia by an iron model complex. Nature 2013;501:84–8.
[12] Georgiadis MM, Komiya H, Chakrabarti P, Woo D, Kornuc JJ, Rees DC. Crystallographic structure of the nitrogenase iron protein from Azotobacter vinelandii. Science 1992;257:1653–9.
[13] Watt GD, Reddy KRN. Formation of an all-ferrous $Fe_4S_4$ cluster in the iron protein component of Azotobacter vinelandii nitrogenase. J Inorg Biochem 1994;53:281–94.
[14] Hu Y, Fay AW, Lee CC, Ribbe MW. P-cluster maturation on nitrogenase MoFe protein. Proc Natl Acad Sci U S A 2007;104:10424–9.
[15] Hu Y, Ribbe MW. Biosynthesis of nitrogenase FeMoco. Coord Chem Rev 2011;255:1218–24.
[16] McLean PA, Dixon RA. Requirement of NifV gene for production of wild-type nitrogenase enzyme in Klebsiella pneumoniae. Nature 1981;292:655–6.
[17] Kim JS, Rees DC. Crystallographic structure and functional implications of the nitrogenase molybdenum iron protein from Azotobacter vinelandii. Nature 1992;360:553–60.

[18] Kim JS, Rees DC. Structural models for the metal centers in the nitrogenase molybdenum-iron protein. Science 1992;257:1677–82.
[19] Robinson AC, Burgess BK, Dean DR. Activity, reconstitution, and accumulation of nitrogenase components in Azotobacter vinelandii mutant strains containing defined deletions within the nitrogenase structural gene-cluster. J Bacteriol 1986;166:180–6.
[20] Curatti L, Ludden PW, Rubio LM. NifB-dependent in vitro synthesis of the iron-molybdenum cofactor of nitrogenase. Proc Natl Acad Sci USA 2006;103:5297–301.
[21] Kim J, Woo D, Rees DC. X-ray crystal-structure of the nitrogenase molybdenum iron protein from Clostridium pasteurianum at 3.0 angstrom resolution. Biochemistry 1993;32:7104–15.
[22] Mayer SM, Lawson DM, Gormal CA, Roe SM, Smith BE. New insights into structure-function relationships in nitrogenase: a 1.6 Å resolution X-ray crystallographic study of Klebsiella pneumoniae MoFe-protein. J Mol Biol 1999;292:871–91.
[23] Hoffman BM, Lukoyanov D, Dean DR, Seefeldt LC. Nitrogenase: a draft mechanism. Acc Chem Res 2013;46:587–95.
[24] Shah VK, Brill WJ. Isolation of a molybdenum-iron cluster from nitrogenase. Proc Natl Acad Sci USA 1981;78:3438–40.
[25] Johnson MK, Thomson AJ, Robinson AE, Smith BE. Characterization of the paramagnetic centers of the molybdenum-iron protein of nitrogenase from Klebsiella pneumoniae using low-temperature magnetic circular-dichroism spectroscopy. Biochim Biophys Acta 1981;671:61–70.
[26] Liu HBI, Filipponi A, Gavini N, et al. EXAFS studies of FeMo-cofactor and MoFe protein – direct evidence for the long-range Mo-Fe-Fe interaction and cyanide binding to the Mo in FeMo-cofactor. J Am Chem Soc 1994;116:2418–23.
[27] Pierik AJ, Wassink H, Haaker H, Hagen WR. Redox properties and EPR spectroscopy of the P-clusters of Azotobacter vinelandii MoFe protein. Eur J Biochem 1993;212:51–61.
[28] Lovell T, Li J, Liu TQ, Case DA, Noodleman L. FeMo cofactor of nitrogenase: a density functional study of states M-N, M-OX, M-R, and M-I. J Am Chem Soc 2001;123:12392–410.
[29] Vrajmasu V, Münck E, Bominaar EL. Density functional study of the electric hyperfine interactions and the redox-structural correlations in the cofactor of nitrogenase. Analysis of general trends in $^{57}$Fe isomer shifts. Inorg Chem 2003;42:5974–88.
[30] Rawlings J, Shah VK, Chisnell JR, et al. Novel metal cluster in iron-molybdenum cofactor of nitrogenase – spectroscopic evidence. J Biol Chem 1978;253:1001–4.
[31] Barney BM, Igarashi RY, Dos Santos PC, Dean DR, Seefeldt LC. Substrate interaction at an iron-sulfur face of the FeMo-cofactor during nitrogenase catalysis. J Biol Chem 2004;279:53621–4.
[32] Huynh BH, Henzl MT, Christner JA, Zimmermann R, Orme-Johnson WH, Münck E. Nitrogenase. 12. Mössbauer studies of the MoFe protein from Clostridum pasteurianum. Biochim Biophys Acta 1980;623:124–38.
[33] Yoo SJ, Angove HC, Papaefthymiou V, Burgess BK, Munck E. Mössbauer study of the MoFe protein of nitrogenase from Azotobacter vinelandii using selective $^{57}$Fe enrichment of the M-centers. J Am Chem Soc 2000;122:4926–36.
[34] Newton WE, Gheller SF, Sands RH, Dunham WR. Mössbauer-spectroscopy applied to the oxidized and semi-reduced states of the iron-molybdenum cofactor of nitrogenase. Biochem Biophys Res Commun 1989;162:882–91.
[35] Angove HC, Yoo SJ, Munck E, Burgess BK. An all-ferrous state of the Fe protein of nitrogenase – interaction with nucleotides and electron transfer to the MoFe protein. J Biol Chem 1998;273:26330–7.
[36] Einsle O, Tezcan FA, Andrade SLA, et al. Nitrogenase MoFe-protein at 1.16 Å resolution: a central ligand in the FeMo-cofactor. Science 2002;297:1696–700.

[37] Lee HI, Benton PMC, Laryukhin M, et al. The interstitial atom of the nitrogenase FeMo-cofactor: ENDOR and ESEEM show it is not an exchangeable nitrogen. J Am Chem Soc 2003;125:5604–5.
[38] Yang TC, Maeser NK, Laryukhin M, et al. The interstitial atom of the nitrogenase FeMo-cofactor: ENDOR and ESEEM evidence that it is not a nitrogen. J Am Chem Soc 2005;127:12804–5.
[39] Lukoyanov D, Pelmenschikov V, Maeser N, et al. Testing if the interstitial atom, X, of the nitrogenase molybdenum-iron cofactor is N or C: ENDOR, ESEEM, and DFT studies of the S=3/2 resting state in multiple environments. Inorg Chem; 46:11437–49.
[40] Dance I. The consequences of an interstitial N atom in the FeMo cofactor of nitrogenase. Chem Commun 2003;324–5.
[41] Hinnemann B, Norskov JK. Modeling a central ligand in the nitrogenase FeMo cofactor. J Am Chem Soc 2003;125:1466–7.
[42] Lovell T, Liu TQ, Case DA, Noodleman L. Structural, spectroscopic, and redox consequences of central ligand in the FeMoco of nitrogenase: a density functional theoretical study. J Am Chem Soc 2003;125:8377–83.
[43] Schimpl J, Petrilli HM, Blochl PE. Nitrogen binding to the FeMo-cofactor of nitrogenase. J Am Chem Soc 2003;125:15772–8.
[44] Hinnemann B, Norskov JK. Chemical activity of the nitrogenase FeMo cofactor with a central nitrogen ligand: density functional study. J Am Chem Soc 2004;126:3920–7.
[45] Huniar U, Ahlrichs R, Coucouvanis D. Density functional theory calculations and exploration of a possible mechanism of $N_2$ reduction by nitrogenase. J Am Chem Soc 2004;126:2588–601.
[46] Noodleman L, Lovell T, Han WG, Li J, Himo F. Quantum chemical studies of intermediates and reaction pathways in selected enzymes and catalytic synthetic systems. Chem Rev 2004;104:459–508.
[47] Cao ZX, Jin X, Zhang QN. Density functional study of the structure of the FeMo cofactor with an interstitial atom and homocitrate ligand ring opening. J Theor Comput Chem 2005;4:593–602.
[48] Dance I. The correlation of redox potential, HOMO energy, and oxidation state in metal sulfide clusters and its application to determine the redox level of the FeMo-co active-site cluster of nitrogenase. Inorg Chem 2006;45:5084–91.
[49] Kästner J, Blöchl PE. Ammonia production at the FeMo cofactor of nitrogenase: results from density functional theory. J Am Chem Soc 2007;129:2998–3006.
[50] Pelmenschikov V, Case DA, Noodleman L. Ligand-bound S = 1/2 FeMo-cofactor of nitrogenase: hyperfine interaction analysis and implication for the central ligand X identity. Inorg Chem 2008;47:6162–72.
[51] Lancaster KM, Roemelt M, Ettenhuber P, et al. X-ray emission spectroscopy evidences a central carbon in the nitrogenase iron-molybdenum cofactor. Science 2011;334:974–7.
[52] Spatzal T, Aksoyoğlu M, Zhang LM, et al. Evidence for interstitial carbon in nitrogenase FeMo cofactor. Science 2011;334:940.
[53] Gurbiel RJ, Bolin JT, Ronco AE, Mortenson L, Hoffman BM. Single-crystal EPR and ENDOR Study of Nitrogenase from Clostridium pasteurianum. J Magn Reson 1991;91:227–40.
[54] Spatzal T, Einsle O, Andrade SL. Analysis of the magnetic properties of nitrogenase FeMo cofactor by single-crystal EPR spectroscopy. Angew Chem 2013;52:10116–9.
[55] George SJ, Barney BM, Mitra D, et al. EXAFS and NRVS reveal a conformational distortion of the FeMo-cofactor in the MoFe nitrogenase propargyl alcohol complex. J Inorg Biochem 2012;112:85–92.
[56] Rittle J, Peters JC. Fe-$N_2$/CO complexes that model a possible role for the interstitial C atom of FeMo-cofactor (FeMoco). Proc Natl Acad Sci USA 2013;110:15898–903.
[57] Wiig JA, Hu YL, Lee CC, Ribbe MW. Radical SAM-dependent carbon insertion into the nitrogenase M-cluster. Science 2012;337:1672–5.
[58] Wiig JA, Lee CC, Hu YL, Ribbe MW. Tracing the interstitial carbide of the nitrogenase cofactor during substrate turnover. J Am Chem Soc 2013;135:4982–3.

# 7 Biotin synthase: a role for iron-sulfur clusters in the radical-mediated generation of carbon-sulfur bonds

Joseph T. Jarrett

## 7.1 Introduction

Biotin is an enzyme cofactor that is involved in carboxylation, transcarboxylation, and decarboxylation reactions [1]. Biotin is required by organisms from all branches of life, although only bacteria, fungi, and some plants can synthesize biotin and animals therefore obtain biotin from dietary sources. In the enzyme acetyl-CoA carboxylase (ACC; ACACA in humans), biotin plays a key role in the capture of carbon dioxide, and the subsequent carboxylation of acetyl-CoA to generate malonyl CoA, a substrate for the fatty acid synthase (FAS) complex [2]. Biotin is also used in gluconeogenesis in the enzyme pyruvate carboxylase, in branched-chain amino acid degradation mediated by the biotin-dependent enzyme propionyl CoA carboxylase (PCCA/B), and in propionic acid metabolism in the enzyme transcarboxylase (also known as methylmalonyl CoA carboxyltransferase from various species of propionibacterium) [2]. Biotin has also been found to play a role in coupling decarboxylation reactions to ion translocation across membranes and the generation of an electrochemical gradient, most likely through protein conformational changes induced by the carboxybiotin intermediate (e.g. glutaconyl CoA decarboxylase in *Acidaminococcus fermentans*) [3].

Biotin is biosynthesized in bacteria, fungi, and plants through a mostly conserved pathway that starts with L-alanine and either pimeloyl CoA or pimeloyl acyl carrier protein (ACP) [4]. Dethiobiotin (DTB), the immediate precursor to biotin, lacks only the sulfur atom that bridges the C6 and C9 carbon atoms to form the five-membered thiophane ring (Fig. 7.1a). The insertion of a sulfur atom is catalyzed by biotin synthase (bacterial, BioB; fungal, Bio2), a member of the radical SAM enzyme superfamily that has recently been extensively reviewed [5, 6]. Radical SAM enzymes contain a $[4Fe\text{-}4S]^{2+}$ cluster that facilitates reductive cleavage of *S*-adenosyl-L-methionine (SAM or AdoMet), generating a highly reactive 5′-deoxyadenosyl radical (5′-dA•) that can abstract a hydrogen atom from unreactive C-H bonds on a substrate or nearby protein residues [7, 8]. In biotin synthase, abstraction of a hydrogen atom from the C9 position of DTB leads to formation of a dethiobiotinyl radical, which reacts with a nearby $[2Fe\text{-}2S]^{2+}$ cluster, generating a new carbon-sulfur bond [5]. A second reaction sequence focused on the C6 position closes the thiophane ring to produce biotin. The chemical details of this enzyme reaction and the properties of biotin synthase essential for catalytic success are further discussed in this chapter.

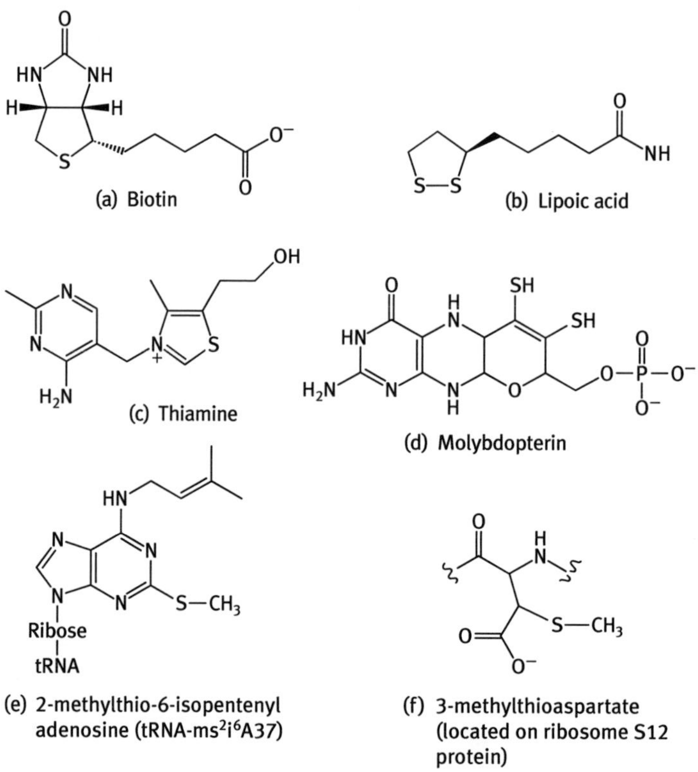


**Fig. 7.1:** A selection of sulfur-containing biomolecules. The cofactors (a) biotin and (b) lipoic acid obtain the sulfur atoms from iron-sulfur clusters in reactions catalyzed by the radical SAM enzymes BioB and LipA. In contrast, (c) thiamine and (d) molybdopterin obtain the sulfur atoms from a peptide thiocarboxylate *via* non-radical polar mechanisms. (e) 2-Methylthio-6-isopentenyladenosine is found as a post-transcriptionally modified base in certain tRNAs; the methylthio group is added by the radical SAM enzyme MiaB. (f) 3-Methylthioaspartate is found as a post-translational modification on the ribosomal S12 protein; the methylthio group is added by the radical SAM enzyme RimO.

## 7.2 Sulfur atoms in biomolecules

A number of cofactors, secondary metabolites, and natural products contain sulfur atoms incorporated at positions that significantly influence their reactivity or structure. Despite sharing a similar valence electronic configuration with oxygen, sulfur is larger, more electron-rich, and can influence chemical reactions through steric effects and through stabilization of electron-deficient enzyme intermediates. Examples of sulfur-containing cofactors include biotin, lipoic acid, thiamine, and molybdopterin (Fig. 7.1). In molybdopterin, the sulfur atoms directly coordinate the molybdenum ion and alter the electronic structure and electrochemical reactivity of this catalytic metal. In thiamine, the sulfur atom sits adjacent to the reactive carbon atom, increasing C-H bond acidity, and in some enzymes, helping to stabilize radical intermediates. In lipoic acid, the sulfur atoms play a direct role in mediating two-electron oxidation-reduction reactions. In biotin, the sulfur atom is distant from the reactive N1′ position, but the steric bulk and hydrophobicity of sulfur play an important steric role in stabilizing enzyme intermediates against attack by bulk water. More recently, sulfur has been found as part of a post-transcriptional modification on adenosine in certain tRNAs [9, 10] (Fig. 7.1e) and a post-translational modification on an aspartate residue in the ribosomal S12 protein [11] (Fig. 7.1f), where the steric bulk of the added methanethiol group is thought to improve the fidelity of the translation process. Various sulfur-containing cofactors and secondary metabolites are found throughout all branches of life and likely offered a significant evolutionary advantage by improving catalytic efficiency, particularly in primary metabolic pathways and protein synthesis.

The incorporation of sulfur into biomolecules generally proceeds through either polar mechanisms, which take advantage of the highly nucleophilic character of sulfur anions, or radical mechanisms, which take advantage of the ability of sulfur to partially stabilize an unpaired valence electron. For example, the sulfur atoms in thiamine and molybdopterin are most likely incorporated through polar mechanisms. During thiamine biosynthesis, a cysteine desulfurase, IscS, catalyzes the transfer of a persulfide sulfur atom from cysteine to the C-terminal carboxylate of the ThiS peptide, generating a thiocarboxylate. The thiocarboxylate is nucleophilic and can attack an electrophilic carbon on an intermediate derived from deoxy-D-xylulose 5-phosphate [12]. Subsequent ring closure through a reaction with dehydroglycine leads to formation of the thiazole ring. In general, polar chemistry would be expected to predominate whenever existing functional groups contain or allow generation of an electrophilic carbon. In contrast, sulfur is occasionally incorporated at positions that are either unreactive aliphatic or electron-rich aromatic carbon positions. In biotin, the biosynthetic precursor DTB has unreactive methyl and methylene groups at the positions where sulfur is be incorporated, precluding use of polar chemistry and requiring that radical chemistry be employed.

## 7.3 Biotin chemistry and biosynthesis

The reactions involving biotin have been most thoroughly studied in the enzyme ACC, in which biotin is attached to a lysine on the biotin carboxyl carrier subunit (AccB) [2]. Electrophilic carbon dioxide is present at very low concentrations (0.1–1% of total dissolved $CO_2$) in slightly alkaline aqueous solutions such as the cytosol of most organisms, whereas the relatively unreactive bicarbonate ion is present at concentrations of 0.1–10 mM, depending on metabolic activity. In the biotin carboxylase subunit of ACC (AccA), bicarbonate is phosphorylated by ATP, and the resulting carboxyphosphate undergoes spontaneous decarboxylation to generate an equivalent of the more reactive carbon dioxide, which is captured by the nucleophilic N1′ position of biotin (Fig. 7.2a). The tethered carboxybiotin intermediate can diffuse to the active site of the

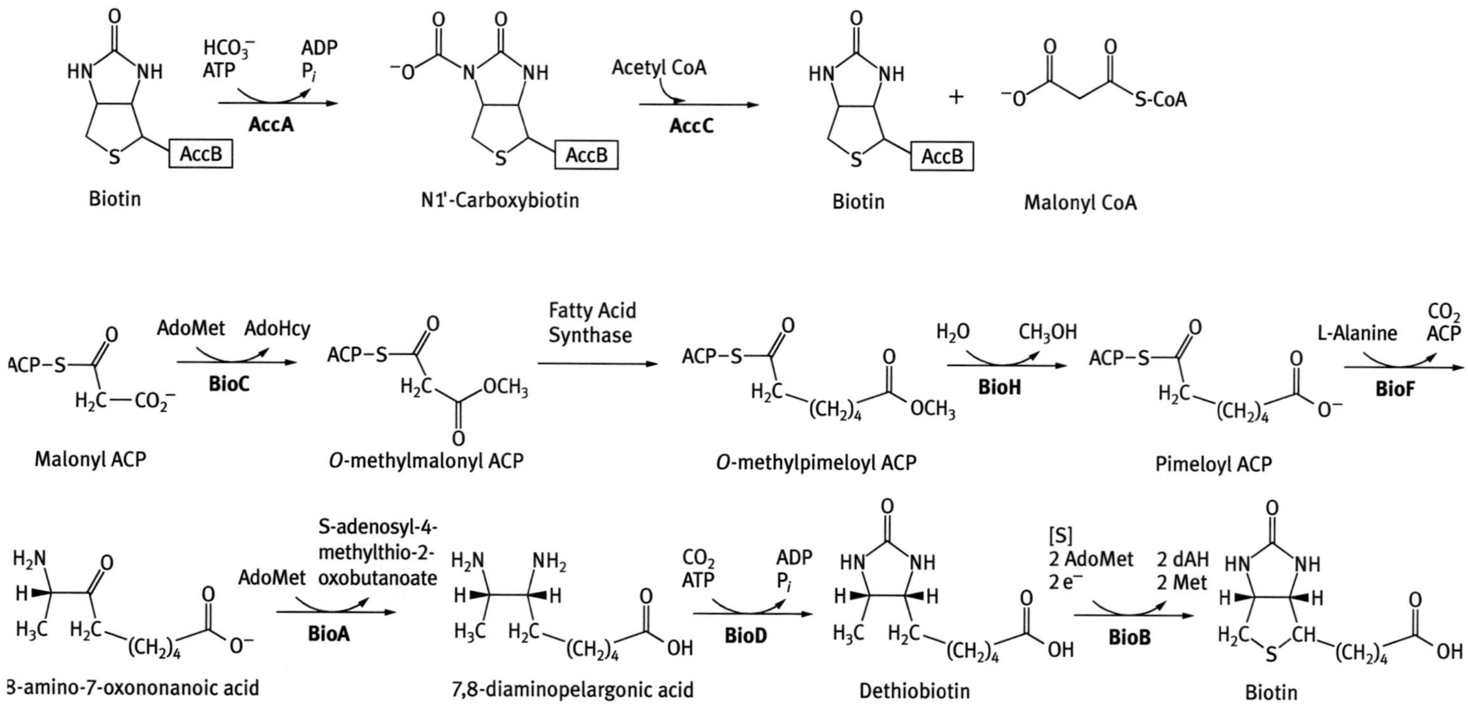


**Fig. 7.2:** (a) Biotin is utilized as a cofactor in enzymes that catalyze carboxylation, decarboxylation, and transcarboxylation reactions; in each of these enzymes, biotin is covalently attached to the protein through a lysine residue. Shown is the reaction from ACC, in which a carboxybiotin intermediate ferries an activated $CO_2$ molecule from the biotin carboxylase subunit (AccA) to the carboxyl transferase subunit (AccC). (b) The biotin biosynthetic pathway in *E. coli*. Pimeloyl ACP is built up by FAS with assistance from BioC and BioH, and the remainder of the cofactor is derived from L-alanine, $CO_2$, and a nitrogen atom from AdoMet in reactions catalyzed by BioF, BioA, and BioD. Incorporation of a sulfur atom requires 2 equiv. AdoMet and two electrons from FldA; the sulfur atom derives from a [2Fe-2S] cluster within biotin synthase (BioB).

carboxylase subunit (AccC), where the enzyme triggers decarboxylation to regenerate carbon dioxide that reacts with an acetyl-CoA enolate to generate the new C-C bond in malonyl CoA. During diffusion between respective subunits, the carboxybiotin intermediate can undergo an energetically wasteful side reaction with bulk water. The presence of a sulfur atom in the thiophane ring blocks the approach of water from one side of the cofactor, resulting in a significant decrease in the spontaneous hydrolysis rate and improved efficiency of biotin-dependent enzymes [13].

The carbon framework of biotin is biosynthesized from L-alanine and either pimeloyl CoA or pimeloyl ACP. Pimelic acid is a seven-carbon dicarboxylic acid that can originate from at least three sources. Many species can scavenge pimelic acid from the environment and couple this to CoA using the ATP-dependent enzyme BioW [14]. In some aerobic species, BioI is a heme monooxygenase that catalyzes multiple hydroxylation reactions on acyl ACPs, resulting in the formation of pimeloyl ACP [15]. Finally, the majority of organisms use BioC and BioH to hijack FAS to produce pimeloyl ACP (Fig. 7.2b) [16]. BioC is an AdoMet-dependent methyltransferase that caps malonyl ACP and makes this a convincing substrate for FAS, as *O*-methylmalonyl ACP is sterically and electrostatically similar to butyryl CoA [17]. Two rounds of FAS catalysis yield *O*-methylpimeloyl ACP, which is then a substrate for the BioH-catalyzed methylesterase hydrolysis reaction [16]. The biotin ureido ring is then begun by coupling L-alanine to pimeloyl CoA (or ACP) in a condensation/decarboxylation reaction, catalyzed by the pyridoxyl phosphate (PLP)-dependent enzyme BioF, generating 8-amino-7-oxononanoic acid (also known as 8-keto-7-aminopelargonic acid) [18]. The remaining ketone undergoes reductive amination through transamination from the methionyl amine of AdoMet, in a reaction catalyzed by the PLP-dependent enzyme BioA, generating 7,8-diaminopelargonic acid [19]. The ureido ring is closed by the action of BioD, an ATP-dependent carboxylase that phosphorylates a spontaneously generated 7-carbamoyl-8-aminopelargonic acid [20], facilitating an intramolecular condensation that generates DTB [21]. Finally, biotin synthase substitutes a sulfur atom in place of hydrogen atoms at the C9 methyl and C6 methylene positions and closes the thiophane ring.

## 7.4 The biotin synthase reaction

Biotin synthase is normally expressed at very low levels in bacteria, and aside from the substrate, product, and the stereochemistry of the sulfur insertion, very little was known about the reaction sequence before studies were performed on recombinant protein. Using recombinant protein in cell-free *Escherichia coli* extracts, Ifuku and colleagues [22, 23] were able to show that AdoMet, $Fe^{2+}$, flavodoxin (FldA), and ferredoxin (FldA):$NADP^+$ oxidoreductase (Fpr) were able to significantly improve the production of biotin. Around this time, the sequence of the *bio* operon by Otsuka *et al.* [24] suggested that BioB was a member of a small but growing family of AdoMet-dependent enzymes that generated 5′-deoxyadenosine (5′-dA) as a product or intermediate, a family that at that time included only lysine-2,3-aminomutase, pyruvate formate-lyase activase, and anaerobic ribonucleotide reductase. Marquet and colleagues went on to confirm that biotin formation is accompanied by the reduction of AdoMet to methionine and 5′-dAH [25] and used isotopic labeling to demonstrate that this product contained two deuterium atoms that originated from both the C6 and C9 positions of deuterated DTB [26]. Finally, a careful analysis of the recombinant *E. coli* BioB protein indicated that it was a dimeric protein that contained approximately two $[2Fe\text{-}2S]^{2+}$ clusters per dimer [27], but under appropriate reducing conditions, the dimeric protein could instead bind two [4Fe-4S] clusters in either the oxidized (+2) or reduced (+1) oxidation states [27–30].

Combining these various observations, we were able to develop mild conditions for generating enzyme with approximately two $[2Fe\text{-}2S]^{2+}$ and two $[4Fe\text{-}4S]^{2+}$ clusters per dimer [29, 31]; this reconstituted protein was able to saturably bind DTB and AdoMet [32], and in the presence of the native reducing system consisting of FldA, Fpr, and NADPH, would produce ~1.8 equivalent (equiv.) of biotin per BioB dimer [31]. Remarkably, biotin production did not require the addition of a sulfur donor [33]. Prior studies had demonstrated that neither AdoMet nor cysteine nor methionine served as a sulfur source, and isotopic labeling of the enzyme suggested that the biotin sulfur atom was derived from an FeS cluster or from a tightly bound sulfide, persulfide, or polysulfide [34, 35]. Consistent with the first possibility, we observed a decrease in absorbance during biotin production, which suggested that reduction or loss of the $[2Fe\text{-}2S]^{2+}$ cluster occurred during catalysis [31]. This observation was later confirmed by both Tse Sum Bui *et al.* [36] and Jameson *et al.* [37], who observed a decrease in the narrow Mössbauer doublet attributable to the $[2Fe\text{-}2S]^{2+}$ cluster and an increase in free $Fe^{2+}$, presumably due to the release of the residual cluster into the solution either immediately prior to or concurrent with biotin formation.

Based on these and other studies of BioB and informed by results emerging from studies of other radical SAM enzymes, we proposed the following reaction sequence (Fig. 7.3) [31]. Biotin synthase initially contains one $[4Fe\text{-}4S]^{2+}$ cluster and one $[2Fe\text{-}2S]^{2+}$ cluster per monomer. AdoMet and DTB bind within the active site, with AdoMet coordinated to the $[4Fe\text{-}4S]^{2+}$ cluster. Flavodoxin transfers an electron into the $[4Fe\text{-}4S]^{2+}$ cluster, which passes this electron into the AdoMet, resulting in spontaneous reductive cleavage to methionine and a 5′-dA• (Fig. 7.4). This high-energy radical abstracts a hydrogen atom from the nearby C9 position of DTB, generating a C9-centered dethiobiotinyl radical, which is then quenched through formation of a bond with the nearby μ-sulfide of the $[2Fe\text{-}2S]^{2+}$ cluster. This quenching reaction requires one-electron oxidation of the sulfide, and inner-sphere electron transfer to the adjacent $Fe^{3+}$ ion generates a reduced $[2Fe\text{-}2S]^{+}$ cluster. The DTB -derived intermediate is monothiolated at the C9 position and remains on the enzyme as either a tightly bound free thiolate or as a thiolate ligand to the $[2Fe\text{-}2S]^{+}$ cluster. 5′-dAH and methionine must dissociate, and a second equivalent of AdoMet must bind, prior to a

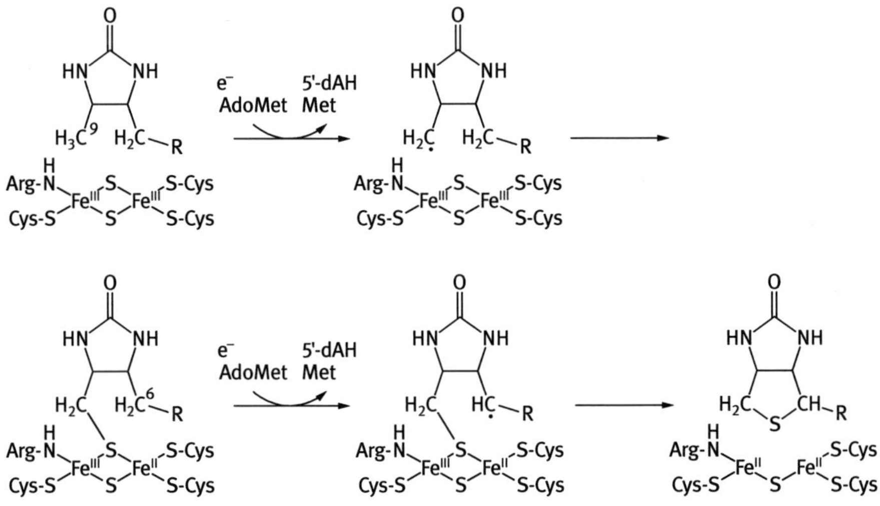


**Fig. 7.3:** The proposed reaction sequence catalyzed by biotin synthase. AdoMet is reductively cleaved by an electron from FldA, generating a 5′-dA• that abstracts a hydrogen atom from the C9 methyl group of DTB. This radical is quenched by the bridging μ-sulfide of the $[2Fe-2S]^{2+}$ cluster, with concomitant transfer of an excess electron from the sulfide into the cluster, generating 9-mercaptodethiobiotin as thiolate ligand to a $[2Fe-2S]^{+}$ cluster. Following exchange of 5′-dAH and methionine for a second equivalent of AdoMet, a similar reaction sequence directed at the C6 methylene group closes the thiophane ring and generates a diferrous cluster that likely dissociates from the enzyme.

second reaction sequence that generates a carbon radical at the C6 position, completing the capture of sulfur (Fig. 7.3). Following the dissociation of biotin, the remnant diferrous cofactor is presumably unstable and dissociates from the enzyme under *in vitro* conditions, rendering the enzyme initially inactive for further turnover. *In vivo*, BioB likely takes advantage of normal FeS cluster repair mechanisms to regenerate the $[2Fe-2S]^{2+}$ cluster of the active enzyme and continue biotin production.

## 7.5 The structure of biotin synthase and the radical SAM superfamily

The structure of *E. coli* biotin synthase containing DTB, AdoMet, and both FeS clusters has been solved to 3.4-Å resolution (Fig. 7.5a) [38]. The protein is a dimer of 38.6-kDa monomers, with the dimer interface consisting of a four-helix bundle that comprises residues from the N-terminus of the primary sequence. For each monomer, the core structure consists of a cylindrical $(\alpha\beta)_8$ barrel (also known as a "TIM" barrel) that encapsulates the active site; this structural topology is now annotated as the BioB/ThiH radical SAM fold (no structure exists of the homologous ThiH enzyme). The $[4Fe-4S]^{2+}$

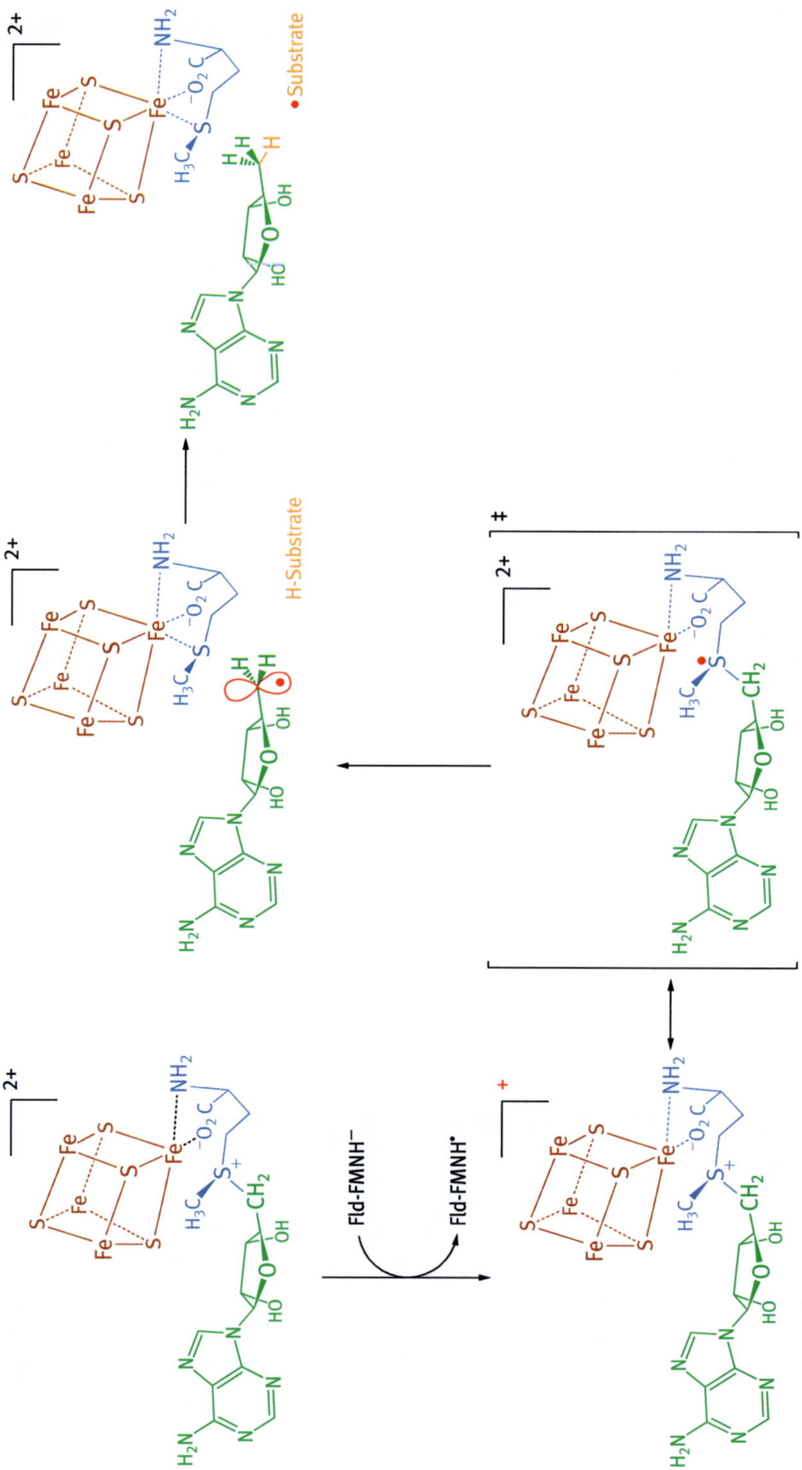


(a)

(Figure Continued)

*(Figure Continued)*

| Gene | Enzyme or pathway | [4Fe-4S] cluster coordniation site |
|---|---|---|
| *Sulfur insertion* | | |
| BioB | Biotin synthase *(E. coli)* | TGACPEDCKYCPQS |
| LipA | Lipoate synthase *(homo sapiens)* | GDTCTRGCRFCSVK |
| MiaB | $ms^2i^6A$-tRNA synthase *(E. coli)* | MEGCNKYCTYCVVP |
| RimO | 3-Methylthioaspartyl synthase *(E. coli)* | SEGCNHRCTFCIIP |
| *Other radical SAM enzymes* | | |
| PflB | Pyruvate formate-lyase activase *(E. coli)* | FQGCLMRCLYCHNR |
| NrdG | Class III ribonucleotide reductase *(E. coli)* | VSGCVHECPGCYNK |
| MoaA | Precursor Z synthase (molybdopterin biosynth.) *(E. coli)* | TDVCNFRCTYCLPD |
| KamA | Lysine-2,3-aminomutase *(C. subterminale)* | TDMCSMYCRHCTRR |
| SplB | Spore photoproduct lyase *(E. subtilis)* | ATGCMGHCHYCYLQ |
| NifB | Nitrogenase FeMo cofactor biosynth. *(A. vinlandii)* | APACNIQCHYCNRK |
| HydG | Hydrogenase Fe-Fe cofactor biosynth. *(C. reinhardtii)* | ANHCMNTCTYCAFR |

(b)

**Fig. 7.4:** (a) The proposed mechanism for the generation of substrate radicals by enzymes in the radical SAM superfamily. AdoMet is coordinated to a $[4Fe\text{-}4S]^{2+}$ cluster through the amine and carboxylate of methionine, resulting in close proximity and possible orbital overlap between the sulfonium and the FeS cluster. Transfer of an electron generates the $[4Fe\text{-}4S]^{+}$ cluster; within this reduced cluster, the lowest-energy excited-state molecular orbital includes the C-S $\sigma^*$ antibonding orbitals. Promotion of an electron to this excited state results in rapid cleavage of the weakened C-S bond and generation of a 5′-dA•. This radical is quenched by abstracting a hydrogen atom from the substrate, generating a substrate radical and 5′-dAH. (b) A selection of enzymes from the radical SAM Superfamily, including all of the known sulfur insertion enzymes. The canonical sequence motif that binds the catalytic [4Fe-4S] cluster is highlighted in green, a semiconserved hydrophobic residue that forms a portion of the adenine binding pocket is in blue.

cluster resides at the C-terminal end of the $\beta_8$ barrel, coordinated to the canonical cysteine motif within a loop between β strand 1 and α helix 1. The $[2Fe\text{-}2S]^{2+}$ cluster is located within the core of the barrel, with metal-ligands consisting of Cys97, Cys128, Cys188, and Arg260, all residues found in disparate β strands within the barrel. The conservation of an arginine (Arg260) as a metal-ligand has been confirmed by HYSCORE (hyperfine sublevel correlation) spectroscopy with $^{15}N$-labeled arginine [41] and is unique among metalloenzymes, although surprisingly, it is not essential for activity [42]. As expected from spectroscopic studies [43], AdoMet is a ligand to the unique Fe position in the $[4Fe\text{-}4S]^{2+}$ cluster and is also held in place by hydrogen bonds between the ribose and Asn153 and Asp155. DTB is positioned between AdoMet and the $[2Fe\text{-}2S]^{2+}$ cluster through hydrogen bonds between the ureido ring and Asn153 and Asn222. The positioning of DTB places the C9 methyl group in an almost direct line among the C5′ position of AdoMet, the site of initial radical generation, and the μ-sulfide of the $[2Fe\text{-}2S]^{2+}$ cluster, the presumed source of the sulfur that is inserted into DTB.

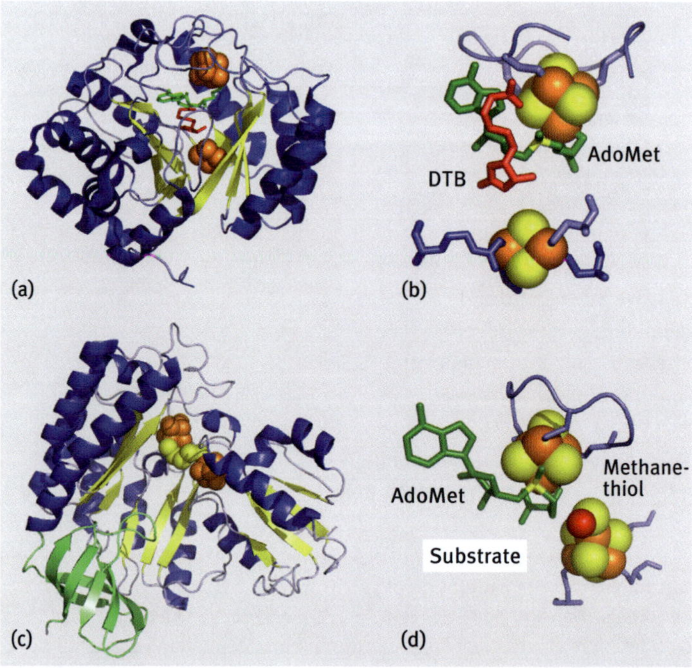


**Fig. 7.5:** (a) Structure of biotin synthase containing $[2Fe\text{-}2S]^{2+}$ and $[4Fe\text{-}4S]^{2+}$ clusters, AdoMet, and DTB [38]. One monomer of the dimeric enzyme is shown, with the core $\beta_8$ barrel in yellow, AdoMet in green, and DTB in red. (b) The active site arrangement shows AdoMet (green) as a ligand to one Fe within the conserved $[4Fe\text{-}4S]^{2+}$ cluster, DTB (red) positioned with the C9 methyl group in van der Waals contact with the AdoMet C5′ methylene, and a $[2Fe\text{-}2S]^{2+}$ cluster nearby that provides the sulfur for sequential formation of 9-MDTB and biotin. Ligands to the $[4Fe\text{-}4S]^{2+}$ cluster are Cys53, Cys57, and Cys60, and ligands to the $[2Fe\text{-}2S]^{2+}$ cluster are Cys97, Cys128, Cys188, and Arg260 (all in dark blue). (c) Structure of the RimO methylthiolation enzyme containing two $[4Fe\text{-}4S]^{2+}$ clusters (orange), which are connected in the published structure by a pentasulfide chain (yellow) [39]. One monomer of the dimeric protein is shown, with the β sheets in yellow, and the RNA-binding TRAM (Trm2 and MiaB) domain in pale green. The published structure does not contain AdoMet or the ribosomal S12 protein substrate and also lacks a small N-terminal putative electron transfer domain (the UPF0004 domain). (d) The active site arrangement of RimO. AdoMet has been modeled into the site using the published structure of MoaA [40] as a guide. Methanethiol has been modeled as a ligand to the second $[4Fe\text{-}4S]^{2+}$ cluster, assuming it binds in the same position as the pentasulfide chain. The presumed substrate binding site can be seen in the protein structure (c) as a solvent-filled pocket surrounded by the FeS clusters, the $\beta_6$ sheet, and the TRAM domain (PDB files: BioB, 1R30; RimO, 4JC0; MoaA, 1TV8).

The topology of biotin synthase differs from that of more typical radical SAM enzymes [44], which usually include only the first six β strands, generating an $(\alpha\beta)_6$ half-barrel that creates a more open active site that allows access to the active site for additional N- or C-terminal domains or for very large protein or RNA substrates. The structure of the methylthiolation enzyme RimO, which contains a core six-stranded β sheet, is shown for comparison (Fig. 7.5c) [39]. The radical SAM superfamily is now estimated to consist of over 48,000 unique enzymes (see the Structure-Function Linkage Database [45] at sfld.rbvi.ucsf.edu/django/superfamily/29/ for an updated tally and classification), of which only 11 enzymes have been structurally characterized [44]. Of these structures, nine contain a concave six-stranded parallel β sheet connected by six α helices – an $(\alpha\beta)_6$ fold – that interacts with additional N- and C-terminal domains to enclose the active site [44]. In those enzymes that have been crystallized in the presence of a substrate, the reactive hydrogen atoms of the substrate are always found within ~3–4 Å of the AdoMet C5′ methylene, in the general vicinity of the location of DTB in the BioB structure, suggesting that these enzymes all catalyze the direct transfer of a hydrogen atom from the substrate to 5′-dAH [46]. In general, the additional domains appear to be largely involved with conferring substrate specificity and steering the substrate to the proper position within the active site. Several of these additional domains also contain FeS clusters, predominantly [4Fe-4S]$^{2+}$ clusters, which are thought to be involved primarily with substrate binding and likely play lesser roles in catalysis [47]. A clear exception is the additional [4Fe-4S]$^{2+}$ clusters found in other radical SAM enzymes that catalyze sulfur insertion, as these clusters are likely to be intimately involved in catalysis.

## 7.6 The [4Fe-4S]$^{2+}$ cluster and the radical SAM superfamily

The diverse enzymes of the radical SAM superfamily catalyze a wide array of transformations, including vitamin $B_{12}$-like rearrangement, dehydration, decarboxylation, and net oxidation and reduction reactions [48, 49]. Each reaction mechanism begins with reductive cleavage of *S*-adenosyl-L-methionine, which in the majority of enzymes, generates an AdoMet-derived radical intermediate, 5′-dA•, and most mechanisms continue with the abstraction of a hydrogen atom from the substrate or protein, generating 5′-dAH and a substrate- or protein-centered radical (Fig. 7.4a). Proteins from the radical SAM superfamily all contain at least one [4Fe-4S]$^{2+}$ cluster that is usually bound within a canonical sequence motif, CxxxCxxC (Fig. 7.4b), found within an extended loop at the C-terminal end of $(\alpha\beta)_8$ barrel or $(\alpha\beta)_6$ three-quarter barrel [46]. The cysteine residues in this motif bind to three of the Fe atoms within the [4Fe-4S]$^{2+}$ cluster, leaving one unique Fe position available for interaction with AdoMet (Fig. 7.5b and d). Detailed ENDOR studies of pyruvate formate lyase (PFL) activase (Fig. 7.4b) with isotopically labeled AdoMet have demonstrated that coordination occurs through the methionyl amine and carboxylate groups, which brings the AdoMet sulfonium to within van der Waals contact (3.3–3.9 Å) of the nearest face of the cuboidal cluster, possibly promoting a sulfide-to-sulfonium nonbonded orbital overlap that may be important for the radical generation reaction [50, 51]. Computational studies indicate that reduction to the [4Fe-4S]$^{+}$ cluster results in the additional electron primarily residing within the sulfide and thiolate sulfur atoms, although proximity of one of the sulfide positions to the sulfonium results in a lowest-energy excited state that includes the sulfonium C-S bond $\sigma^*$ nonbonding orbital [52]. Promotion of an electron into this excited state generates a sulfuranyl radical, the putative transition state, that breaks down through homolytic cleavage of a C-S bond, generating a 5′-dA• radical [52]. This radical is quenched through abstraction of a hydrogen atom from a nearby position on the substrate, generating an oxidized substrate radical. Although the majority of enzymes in the radical SAM superfamily utilize 5′-dA• as a catalytic or stoichiometric oxidant, one radical SAM enzyme, Dph2 from diphthamide biosynthesis, has been proposed to generate a 3-amino-3-carboxypropyl radical that is incorporated into the product [53, 54]. Notably, Dph2 is a noncanonical radical SAM enzyme that shares no sequence or structural homology with the larger radical SAM superfamily [53], and outside of this example, it would appear that the typical radical SAM structure is optimized for generating and utilizing 5′-dA•.

In BioB, the aerobically purified enzyme does not contain a $[4Fe\text{-}4S]^{2+}$ cluster [55]. However, a substoichiometric amount of $[4Fe\text{-}4S]^{2+}$ cluster has been observed *in vivo* in an *E. coli* overexpression system using whole-cell Mössbauer spectroscopy [56]. and anaerobic purification yields protein containing a substoichiometric amount of this FeS cluster. However, the $[4Fe\text{-}4S]^{2+}$ cluster can be fully restored using $Fe^{3+}/Fe^{2+}$, $S^{2-}$, and dithiothreitol (Fig. 7.6a) [29, 30]; alternatively, L-cysteine and cysteine desulfurase can substitute for $S^{2-}$ with similar satisfactory results. The $[4Fe\text{-}4S]^{2+}$ cluster can also be generated in the absence of the $[2Fe\text{-}2S]^{2+}$ cluster by prolonged incubation with sodium dithionite in ≥20% ethylene glycol-containing buffer, with or without added $Fe^{2+}$ or $S^{2-}$, and the reduced $[4Fe\text{-}4S]^{+}$ cluster can be generated under the same conditions, in the absence of ethylene glycol [27–29]. The broad UV-vis absorption maximum at 410 nm, the overlapping Mössbauer doublets ($\delta$ = 0.44, 0.45; $\Delta E_Q$ = 1.03, 1.28), and the single vibrational band observed in resonance Raman ($A_1{}^b$, $\nu$ = 338 $cm^{-1}$) are all fairly typical of a cysteine-ligated cuboidal $[4Fe\text{-}4S]^{2+}$ cluster [28]. The reduced $[4Fe\text{-}4S]^{+}$ cluster has no distinct UV-vis absorption and exhibits an axial EPR spectrum ($g$ = 2.04, 1.94) that is typical of a cuboidal $[4Fe\text{-}4S]^{+}$ cluster [27, 57]. The addition of excess AdoMet results in a change in the $[4Fe\text{-}4S]^{2+}$ cluster vibrational band ($A_1{}^b$, $\nu$ = 342 $cm^{-1}$), splitting of one of the Mössbauer doublets, and a transition to a rhombic EPR spectrum ($g$ = 2, 1.928, 1.854) for the $[4Fe\text{-}4S]^{+}$ cluster [43]. Similar spectroscopic shifts have been observed when AdoMet binds to PFL activase [58].

The $[4Fe\text{-}4S]^{2+}$ cluster in radical SAM enzymes is reduced only at very low electrochemical potentials. In biotin synthase, reduction of the $[4Fe\text{-}4S]^{2+}$ cluster to the $[4Fe\text{-}4S]^{+}$ cluster occurs at $E_h$ = –500 mV, and only ~50% of the cluster is in the reduced state in the presence of dithionite [57]. In other radical SAM enzymes such as PFL activase and lysine-2,3-aminomutase, the substrate-free enzyme is not reduced by dithionite, suggesting a potential much lower than –550 mV. The addition of AdoMet results in a significant shift to higher potentials: in lysine-2,3-aminomutase,

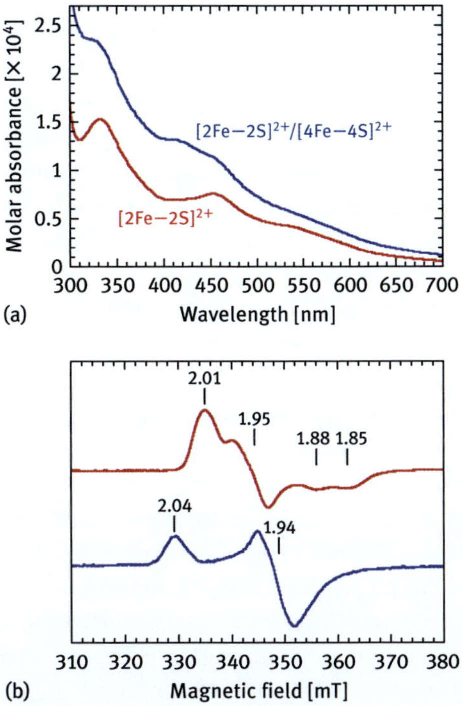


**Fig. 7.6:** Characteristic UV-vis and EPR spectra observed for biotin synthase. (a) The protein is aerobically purified containing one [2Fe-2S]$^{2+}$ cluster per monomer; the brown-colored protein exhibits absorption peaks at 332 and 452 nm (red spectrum, $\varepsilon_{452}$ = 7,500 $M^{-1}$ $cm^{-1}$). Mild reconstitution of the [4Fe-4S]$^{2+}$ cluster leaves the [2Fe-2S]$^{2+}$ cluster intact, providing an active protein with shoulders at 320, 410, and 460 nm (blue spectrum, $\varepsilon_{410}$ = 13,500 $M^{-1}$ $cm^{-1}$). (b) EPR spectrum of BioB frozen during turnover yields a spectrum of the [2Fe-2S]$^{+}$ cluster with 9-MDTB as a bridging thiolate ligand (red spectrum). The spectrum is a composite of two overlapping rhombic signals with an approximate 2:1 ratio. Observed *g* values are indicated, more accurate simulated g values are described in the text (EPR parameters: 9.376 GHz, 40 K, 100 mW). BioB that is reduced with dithionite generates a [4Fe-4S]$^{+}$ cluster (blue spectrum) with observed *g* values typical of a cysteine-coordinated cuboidal cluster (EPR parameters: 9.424 GHz, 20 K, 20 mW).

the potential in the presence of AdoMet but in the absence of lysine is –430 mV, suggesting a 19-kcal/mol stabilization [59, 60]. Due to these very low electrochemical potentials, the resting state for radical SAM enzymes contains the inactive [4Fe-4S]$^{2+}$ cluster, which can be reduced to the active [4Fe-4S]$^{+}$ cluster only after binding of AdoMet and other substrates. In aerobic bacteria such as *E. coli*, reliance on an NADPH-based reducing system, such as the FldA system required by biotin synthase, likely results in only limited generation of active reduced enzyme even in the presence of saturating concentrations of substrate and AdoMet (~1%–30% of total enzyme, depending on the NADPH/NADP$^{+}$ ratio).

## 7.7 The $[2Fe\text{-}2S]^{2+}$ cluster and the sulfur insertion reaction

BioB is aerobically expressed with a $[2Fe\text{-}2S]^{2+}$ cluster that is highly stable to subsequent purification steps [55]. The cluster is bound within the core of $(\alpha\beta)_8$ barrel, with ligands from four separate β strands, and the presence of this cluster within the active site is unique to BioB among all structurally characterized radical SAM enzymes. The $[2Fe\text{-}2S]^{2+}$ cluster has UV-vis absorbance maxima at 332 and 452 nm (Fig. 7.6a), a single narrow Mössbauer doublet ($\delta$ = 0.29; $\Delta E_Q$ = 0.53), and multiple vibrational bands observed in resonance Raman ($\nu$ = 301, 331, 349, 367, 395, and 418 $cm^{-1}$) that are similar to those parameters observed for a ferredoxin-like cysteine-ligated rhomboidal cluster [27, 28]. In WT enzyme, the $[2Fe\text{-}2S]^{2+}$ cluster is not stable to chemical reduction and dissociates from the protein prior to slowly reassembling as a $[4Fe\text{-}4S]^{+}$ cluster in the vacant radical SAM cluster binding site [57]. In N151A, H152A, and N153A mutants reduced with dithionite, Lotierzo *et al.* [61] reported the generation of a short-lived $[2Fe\text{-}2S]^{+}$ cluster, exhibiting an axial EPR spectrum with $g$ = 2.00, 1.91. The instability of the WT enzyme toward reductants has facilitated the quantitative removal of the $[2Fe\text{-}2S]^{2+}$ cluster by reduction with dithionite or deazaflavin in the presence of excess EDTA. Prolonged reconstitution with $Fe^{2+/3+}$, $^{34}S^{2-}$ or $Se^{2-}$, and DTT followed by overnight aerobic dialysis results in a protein that contains ~1 equiv. of a heavy-atom-labeled cluster per monomer. Using BioB labeled in this manner, Marquet and colleagues [34, 62] were able to demonstrate that ~60%–70% of biotin formed in a 2-h assay contained the heavy atom label, suggesting that the $[2Fe\text{-}2S]^{2+}$ cluster is the source of the sulfur in biotin. To avoid the harsh conditions required for stripping and reconstituting the $[2Fe\text{-}2S]^{2+}$ cluster, Farrar *et al.* [63] used natural abundance enzyme containing the as-purified $[2Fe\text{-}2S]^{2+}$ cluster, reconstituted the $[4Fe\text{-}4S]^{2+}$ cluster with $^{34}S^{2-}$, and added 0.5 mM $^{34}S^{2-}$ to the buffer. The biotin formed in the first 10 min contained 81% $^{32}S$; however, after accounting for mixing of the burst phase and steady-state phase, the biotin formed in the burst phase contained only natural abundance sulfur, strongly implicating the $[2Fe\text{-}2S]^{2+}$ cluster as the source for the biotin sulfur atom.

The reaction catalyzed by BioB requires 2 equiv. of AdoMet, despite the enzyme possessing a binding site that can hold only 1 equiv., and thus, the reaction sequence can be divided into two half-turnovers that correspond to the sequence catalyzed by each equivalent of AdoMet. The enzyme most likely proceeds through an intermediate state in which a monothiolated species remains tightly bound during exchange of 5′-dAH and methionine produced in the first half-turnover for AdoMet required for the second half-turnover. Marquet and colleagues [26] used DTB labeled with deuterium at the C9 or C6 positions to demonstrate that hydrogen atoms are transferred from each position to the product 5′-dAH, but they were not able to determine which position reacted first. Because earlier *in vivo* feeding studies had suggested that 9-mercaptodethiobiotin (9-MDTB) could be converted to biotin by *B. sphaericus*, Marquet and colleagues [64] investigated whether 9-MDTB was a substrate for *E. coli* BioB. They found that 9-MDTB is converted to biotin, but the $K_m$ for 9-MDTB is ~200 μM, *ca* 40-fold higher than the $K_m$ for DTB, and the conversion proceeds without loss of the $[2Fe\text{-}2S]^{2+}$ cluster and with the production of a large excess (10–20 equiv.) of 5′-dAH (Farrar and Jarrett, unpublished results). Taylor *et al.* [65] assayed BioB in the presence of a limiting concentration of AdoMet to promote increased formation of the intermediate and were able to detect up to 0.25 equiv. of a monothiolated intermediate whose concentration decayed as biotin was formed. The enzyme intermediate coelutes with a sample of synthetic 9-MDTB. Further evidence for monothiolation at the C9 position was provided in an enzymatic reaction with 9-($^2H_3$-*methyl*)-DTB; formation of 9-($^2H_2$-methylene)-9-MDTB clearly proceeds with transfer of one $^2H$ atom to 5′-dAH. Based on more recent kinetic studies in the presence of stereochemically pure AdoMet, we are able to assign the overall rate of the first turnover as $k_{burst}$ = 0.12 $min^{-1}$, and we observe ~0.05 equiv. 9-MDTB formed in the presence of saturating AdoMet, suggesting that 9-MDTB formation occurs at ~0.13 $min^{-1}$ and is rate-limiting during the first turnover, and that the subsequent reaction at C6 resulting in ring closure proceeds at ~2.2 $min^{-1}$ [63].

## 7.8 Characterization of an intermediate containing 9-MDTB and a $[2Fe\text{-}2S]^+$ cluster

The kinetically competent generation of 9-MDTB and the subsequent conversion to biotin suggests that the enzyme can stabilize a state with tightly bound 9-MDTB. A careful inventory of electrons suggests that the proposed formation of the monothiolated intermediate through a reaction between a dethiobiotinyl carbon radical and the μ-sulfide of a $[2Fe\text{-}2S]^{2+}$ cluster should require one-electron oxidation of the sulfide, which could occur through inner-sphere electron transfer that reduces $Fe^{3+}$ to $Fe^{2+}$ within the $[2Fe\text{-}2S]^{2+}$ cluster (Fig. 7.3). Because the structure of BioB shows the C9 of DTB ~4.6 Å from the μ-sulfide of the $[2Fe\text{-}2S]^{2+}$ cluster [38], one or both of these species must move during formation of a typical 1.9Å C-S bond. At the opposite extremes, the product of this first half-reaction could be a 9-MDTB thiolate ion-paired with a degraded cluster site that contains an $Fe^{2+}$-$Fe^{3+}$ pair or the product could be an intact $[2Fe\text{-}2S]^+$ cluster in which the bridging ligand is now a 9-MDTB thiolate instead of the original sulfide. In either case, a paramagnetic intermediate should form. Huynh and colleagues [37] reported formation of a novel paramagnetic species during catalysis, which, as we later demonstrated, is formed at approximately the same rate as 9-MDTB [66]. The complex EPR spectrum of this species exhibits two overlapping rhombic signals: a major species (~65%–70%) with $g$ = 2.00, 1.94, 1.85, and a minor species (~30%–35%) with $g$ = 2.01, 1.96, 1.88 (there is some uncertainty in these values due to two possible solutions to the data analysis) [66]. Comparison of the electron spin echo envelope modulation spectra of WT enzyme and an Arg260Met mutant as well as the HYSCORE spectra of a $^{15}N$-arginine-labeled WT enzyme indicates that this signal is attributable to a reduced FeS cluster found in the binding site that includes Arg260 [66].

To investigate whether 9-MDTB is coordinated to this paramagnetic species, we selectively labeled DTB with $^{13}C$ in the C9 position, using enzymes from the *E. coli* biotin biosynthetic pathway to synthesize DTB from (3-$^{13}$C)-L-alanine and pimeloyl CoA [41]. EPR spectra were not sufficiently sensitive to detect the hyperfine interaction between the $^{13}C$ and the paramagnetic FeS cluster, but this was likely due to the relatively small expected hyperfine interaction constants (~1–10 MHz) and the broad nature of the $[2Fe\text{-}2S]^+$ cluster spectrum. HYSCORE spectroscopy is a pulsed EPR technique that is able to detect the coupling of nearby nuclear spins with the unpaired electron spin. Using HYSCORE spectroscopy, we were able to detect a strong correlation peak centered at 2.9, 4.7 MHz; this peak corresponds to a correlation of nuclear spin flipping of the $^{13}C$ nucleus in low- and high-energy electron-spin manifolds [41]. The HYSCORE spectra collected at several magnetic field strengths can be simulated using the dominant set of $g$ values from the EPR simulation above, and the axially symmetric hyperfine interaction constants $A$ = 1.2, 1.2, 5.7 MHz, yielding a relatively large $A_{iso}$ = 2.7 MHz, are indicative of close proximity of the C9 position of MDTB with the FeS cluster that could be achieved only through covalent bond formation. Further confirmation that the $[2Fe\text{-}2S]^+$ cluster remains intact comes from the observation of a weaker multinuclear correlation between $^{14}N$ in Arg260 and $^{13}C$ in 9-MDTB at 3.0, 6.3 MHz; this multinuclear coherence arises from correlation between the $^{14}N$ quadrupole spin transition in one electron-spin manifold and the $^{13}C$ spin transition in another electron-spin manifold. Overall, the HYSCORE data support a model in which 9-MDTB remains as a bridging thiolate ligand to an intact $[2Fe\text{-}2S]^+$ cluster, as depicted in Fig. 7.3 [41]. Formation of this intermediate must arise through a significant rearrangement of the substrate binding site, as the C9 position of DTB must move ~2.7 Å from the original position prior to catalysis.

## 7.9 Other important aspects of the biotin synthase reaction

Biotin synthase is a homodimeric enzyme, and a number of studies now suggest that the enzyme exhibits strong anticooperative effects on catalysis. Farrar *et al.* [63] discovered that assays conducted with impure AdoMet exhibited apparent substrate or contaminant inhibition. Further detailed studies demonstrated that *S*-adenosyl-L-homocysteine (AdoHcy), present as a 2%–5% contaminant in most commercial AdoMet samples, was a potent inhibitor of biotin synthase ($K_i$ = 650 nM); however, more surprising was that only 1 equiv. of AdoHcy per dimer was required to completely inhibit both monomers within the dimeric protein [63]. Similarly, sinefungin, a natural product that mimics the structure and charge of AdoMet, was a competitive inhibitor ($K_i$ = 75 μM) that exhibited clear sigmoidal inhibition. Both inhibitors were well-modeled assuming inhibition as a competitive inhibitor in one monomer (increasing the apparent $K_m$) and a noncompetitive inhibitor in the other subunit (decreasing the apparent $k_{cat}$ to zero). We also observed inhibition by the products

5′-dAH and methionine and by the unnatural (*R,S*) diastereomer of AdoMet, although we could not determine whether this inhibition was also cooperative. Cooperative effects had also been observed on the rate of AdoMet and DTB binding, with substrate binding to one monomer ~10-fold faster than binding to the second monomer [32]. Cooperative binding effects, and in particular, the finding that AdoHcy binding in one monomer caused the $k_{cat}$ to decrease to undetectable levels in the other monomer provided the first definitive clues that biotin synthase exhibits half-site activity [63].

In a half-site active enzyme, which must be a multimeric enzyme, only a fraction of the active sites within the multimer undergo turnover at any particular time, due to cooperative effects that significantly increase $K_m$ or decrease $k_{cat}$ in the other active sites. Half-site reactivity can be identified by the minimum stoichiometry of a tight-binding or covalent inhibitor and by the product stoichiometry during the first turnover if the enzyme exhibits burst kinetics. Our prior studies of 9-MDTB formation and decay had suggested that biotin formation in the first turnover (~0.1 $min^{-1}$) was significantly faster than previously reported values for $k_{cat}$ (~0.001–0.005 $min^{-1}$), suggesting that observation of burst kinetics might be possible. After the numerous sources of inhibition had been identified and eliminated from our assay, we were able to demonstrate that BioB exhibits a relatively rapid burst of biotin formation ($k_{burst}$ = 0.12 $min^{-1}$) followed by slower steady-state production of biotin ($k_{SS}$ = 0.009 $min^{-1}$) [63]. Further, the magnitude of the burst corresponds to one equivalent of biotin produced per dimer, suggesting that only one monomer within the dimer is active during the burst phase. More recent studies with differentially $^{32}S$- and $^{34}S$-labeled enzymes provide an even more surprising result. When $^{32}S$ is found only within the $[2Fe\text{-}2S]^{2+}$ cluster, only the first turnover catalyzed by monomer 1 incorporates $^{32}S$ into biotin, and all subsequent turnovers incorporate $^{34}S$ from sulfide in the buffer [63]. However, monomer 2, which did not react during the burst phase, still contains $^{32}S$ in the $[2Fe\text{-}2S]^{2+}$ cluster as demonstrated by extraction and mass spectroscopy, but this monomer remains inactive and the enzyme preferentially uses monomer 1 for all subsequent turnovers (Farrar and Jarrett, unpublished results). Typically, a half-site active enzyme exhibits a stochastic choice of active monomer at the beginning of each turnover, but in biotin synthase, the enzyme appears to remain conformationally locked in a state in which the monomer chosen during the first turnover remains active throughout the assay, whereas the monomer that was inactive during the first turnover remains inactive.

## 7.10 A role for iron-sulfur cluster assembly in the biotin synthase reaction

Several laboratories had observed that catalytic turnover of biotin synthase results in depletion of the $[2Fe\text{-}2S]^{2+}$ cluster and that chemical reconstitution of the $[2Fe\text{-}2S]^{2+}$ cluster was very slow, presumably due to the limited accessibility of the binding site within the $(\alpha\beta)_8$ barrel. When $^{34}S$ sulfide was present in the $[4Fe\text{-}4S]^{2+}$ cluster and the buffer and $^{32}S$ sulfide was present in the $[2Fe\text{-}2S]^{2+}$ cluster, we had observed that after a burst of biotin production using $^{32}S$ from the $[2Fe\text{-}2S]^{2+}$ cluster, subsequent turnover occurred at a much slower rate using $^{34}S$ from the buffer. We suggest that the rate-limiting step during this *in vitro* steady-state turnover was regeneration of the $[2Fe\text{-}2S]^{2+}$ cluster using $Fe^{2+/3+}$ and $S^{2-}$ from the buffer. *In vivo*, one would expect that the iron-sulfur cluster regeneration process would be assisted by the iron-sulfur cluster assembly (ISC) or sulfur utilization factor (SUF) systems, which contain several proteins that coordinate housekeeping and oxidative-damage-repair FeS cluster assembly in *E. coli*.

Lill and colleagues [67] have examined the role of various ISC proteins in mitochondrial FeS enzyme biosynthesis in *S. cerevisiae*, taking advantage of the ability of fermenting yeast to grow under conditions that do not require significant involvement of mitochondria in energy production. The have specifically examined the role of the Isu and Isa proteins (homologues of bacterial IscU and IscA) in assembly of the FeS clusters in biotin synthase. Using Cys to Ala mutants of the respective FeS cluster binding sites, they were able to distinguish effects on $[2Fe\text{-}2S]^{2+}$ cluster and the $[4Fe\text{-}4S]^{2+}$ cluster. They find that Isu1 and Isu2 mutants are unable to assemble the $[2Fe\text{-}2S]^{2+}$ cluster on Bio2, whereas Isa1 and Isa2 mutants are significantly impaired in their ability to assemble the $[4Fe\text{-}4S]^{2+}$ cluster. More recent studies have suggested that Isa1 and Isa2 are specifically involved in assembly of $[4Fe\text{-}4S]^{2+}$ clusters on mitochondrial proteins, including the radical SAM enzyme lipoate synthase (Bio2 was not examined in this study) [68]. The situation in *E. coli* is somewhat murky due to overlapping activities of the ISC and SUF assembly systems. Although an IscU mutant exhibits no detectable phenotype, *in vitro* studies suggest that IscU can bind either $[2Fe\text{-}2S]^{2+}$ or $[4Fe\text{-}4S]^{2+}$ clusters and can deliver $[2Fe\text{-}2S]^{2+}$ clusters to apoferredoxin and apoBioB. Ding and colleagues [69] have demonstrated that although IscA or SufA single mutants show no significant phenotype, an IscA/SufA double mutant exhibits a growth phenotype in minimal media due to a deficiency in branched-chain amino acid biosynthesis, most likely due to a deficiency in dihydroxyacid dehydratase (IlvD), a $[4Fe\text{-}4S]^{2+}$ cluster-containing enzyme. They further showed that IscA containing a reconstituted $[4Fe\text{-}4S]^{2+}$ cluster can transfer this cluster to apoIlvD *in vitro*, generating active enzyme. Ollagnier-de-Choudens *et al.* [70] have demonstrated a similar transfer of FeS clusters from reconstituted IscA into BioB over ~30 min *in vitro*. These studies suggest roles for IscU and IscA in *E. coli* that are similar, although perhaps less specifically targeted, as compared with those proposed in yeast mitochondria.

Since biotin synthase seems likely to depend on the assistance of the ISC system to maintain turnover in *E. coli*, we sought to identify proteins that were most critical in maintaining BioB in the active state containing a $[2Fe\text{-}2S]^{2+}$ cluster. We had previously shown that the apoprotein was rapidly proteolyzed in the *E. coli* cytosol at 37°C, and therefore inefficient FeS cluster assembly would result in significant loss of the enzyme [71]. A screen of ISC and SUF mutants obtained from the KEIO collection (a comprehensive collection of single gene knockouts in *E. coli*) in glucose M9 media containing moderate amounts of iron (0.5–5 μM) demonstrated that only ΔIscU and ΔHscA mutants significantly decreased the stability of overexpressed BioB, as observed in Western blots of whole-cell extracts, with the ΔHscA mutant exhibiting more noticeable effects [72]. We purified BS from a ΔHscA strain and obtained about half as much total protein as compared with WT, and the protein that we did obtain had only one $[2Fe\text{-}2S]^{2+}$ cluster per dimer [72]. *In vitro* studies confirmed a formation of a stoichiometric complex between HscA and apoBioB, with $K_d$ = 1.3 μM. We have proposed that HscA plays an important role in promoting the interaction of IscU with apoBioB. However, addition of HscA, HscB, and reconstituted IscU to biotin synthase assays under a variety of conditions did not enhance steady-state biotin production in a 2-h assay (Reyda and Jarrett, unpublished results), suggesting that other factors must be essential for promoting $[2Fe\text{-}2S]^{2+}$ cluster transfer from IscU to BioB.

## 7.11 Possible mechanistic similarities with other sulfur insertion radical SAM enzymes

As discussed earlier, the insertion of a sulfur atom at a non-electrophilic carbon atom cannot proceed through more common polar mechanisms, but can instead be accomplished through radical chemistry. A subset of radical SAM enzymes appear to catalyze sulfur insertion using the same fundamental mechanism, including BioB, lipoate synthase, and the methylthiolation enzymes RimO [11], MiaB [10], and MtaB [73]. Each of these enzymes contains the consensus radical SAM sequence that binds a $[4Fe\text{-}4S]^{2+}$ cluster and AdoMet, facilitating the generation of 5′-dA• (Fig. 7.4b), and each of these enzymes has also been demonstrated to bind a second $[4Fe\text{-}4S]^{2+}$ cluster to an additional domain located at the N- or C-terminus of the core $(\alpha\beta)_6$ fold. The crystal structure of RimO (Fig. 7.5c, d) shows the proximity of this additional FeS cluster to the radical generation site [77]. Each of these enzymes would be expected to share some mechanistic similarities with BioB; the mechanism of lipoate synthase is discussed in detail elsewhere in this volume and will not be discussed further here.

The methylthiolation enzyme MiaB catalyzes the addition of a methanethiol functional group to C2 of *N*6-isoprenyladenosine-37 of numerous bacterial and eukaryotic tRNAs, generating the hypermodified nucleotide 2-methanethiol-6-isoprenyladenosine [74] MtaB is a related enzyme that acts on N6-threonylcarbamoyladenosine, generating 2-methanethiol-6-threonylcarbamoyladenosine [73] MiaB contains two $[4Fe\text{-}4S]^{2+}$ clusters, including a second cluster bound to Cys10, Cys46, and Cys79 in a domain located at the N-terminus of the core $(\alpha\beta)_6$ radical SAM domain [75]. In contrast with BioB and LipA, MiaB undergoes multiple turnovers in the presence of excess sulfide and AdoMet without destruction of this FeS cluster, and the coproduction of AdoHcy and 5′-dAH suggests that 1 equiv. AdoMet donates a methyl group to sulfide, whereas a second equivalent is involved with abstracting a hydrogen atom, presumably from adenosine

[39, 76]. Consistent with a mechanism that involves methylation of sulfide, methanethiol has been observed as an intermediate during catalysis [76] and will serve as an alternate substrate at sufficiently high concentrations [39, 76]. Using $^{77}$Se-methylselenide (spin = ±1/2) with dithionite-reduced MiaB in which the radical SAM consensus site had been eliminated by site-directed mutagenesis, Forouhar *et al.* [39] used HYSCORE spectroscopy to detect hyperfine coupling between selenium and the second $[4Fe\text{-}4S]^{+}$ cluster, suggesting that methanethiol binds to the unique Fe site on this cluster.

RimO has significant sequence homology with MiaB and shares many mechanistic similarities. The enzyme contains two $[4Fe\text{-}4S]^{2+}$ clusters [11, 73] and catalyzes multiple turnovers of methylthiolation of a peptide substrate corresponding to the Asp88 loop of ribosomal protein S12 [11, 39, 76]. The enzymatic reaction generates both AdoHcy and 5′-dAH in roughly equal amounts, although both products are generated in excess over methylthiolated peptide [39, 76]. Methanethiol is detected by GC-MS as a product formed on the enzyme in the absence of the peptide substrate [76] and is utilized as an alternate substrate that decreases the production of AdoHcy in the presence of the peptide substrate [39, 76]. A structure of RimO with both $[4Fe\text{-}4S]^{2+}$ clusters bound, but in the absence of AdoMet and the substrate, shows a close proximity of the two clusters, separated by ~8 Å edge to edge (Fig. 7.5c and d) [39]. The structure also shows a novel pentasulfide chain that bridges the unique unliganded Fe atoms in the two $[4Fe\text{-}4S]^{2+}$ clusters. Because the radical SAM cluster most likely binds AdoMet in the active state, the catalytic relevance of this full chain is questionable; however, the coordination of this chain to the second $[4Fe\text{-}4S]^{2+}$ cluster may indicate the binding site of the methanethiol intermediate. The structure in Fig. 7.5d shows a methanethiol group modeled into this position on the second $[4Fe\text{-}4S]^{2+}$ cluster as well as AdoMet modeled as bound to the radical SAM $[4Fe\text{-}4S]^{2+}$ cluster, suggesting a possible model for radical activation of the peptide/protein substrate followed by quenching of this radical by the methanethiol ligand. Note, however, that the arrangement of the active site in RimO is quite different than in BioB. Whereas BioB uses the full length of the $(\alpha\beta)_8$ barrel to encompass a linear arrangement of substrates and FeS clusters, which are bound in place by residues donated from the β strands within the barrel [38], RimO appears to use the $(\alpha\beta)_6$ radical SAM domain to hold AdoMet and the radical SAM FeS cluster and an additional αβ domain to hold the second FeS cluster, which approaches the active site from the side rather than from below [39]. This arrangement may be required because of the large dimensions of the natural substrate, which is believed to be the intact fully assembled ribosome.

Perhaps the unifying theme for the radical SAM sulfur insertion enzymes is the employment of an FeS cluster to donate a sulfide or alkylthiol that quenches a carbon radical generated by radical SAM activation chemistry [5]. The unique Fe within the second FeS cluster serves as a Lewis acid that eliminates the proton normally present on sulfide and thiols at physiologic pH. In general, carbon radicals are rapidly quenched by transfer of hydrogen atoms from thiols, so the absence of a hydrogen on the sulfur donor would minimize this nonproductive radical quenching reaction. Further, because the formation of the new C-S bond serves as the radical-chain termination step, an additional electron must be removed from or donated to the substrate or sulfur during C-S bond formation. Coordination of the sulfide or alkylthiol to an FeS cluster facilitates rapid inner-sphere electron transfer into the FeS cluster and production of a stable non-radical-bearing substrate. Following each round of catalysis, one can imagine three scenarios for regenerating the active enzyme. If the redox potentials of the two $[4Fe\text{-}4S]^{2+}$ clusters are approximately matched and the clusters are in close proximity, then electron transfer from the second cluster back into the radical SAM cluster could reactivate the enzyme without need for additional input from an exogenous reductant; this may be the case for MiaB and RimO. If the potentials are not matched, then the second cluster would need to be oxidized and the radical SAM cluster reduced by exogenous proteins to regenerate the active enzyme. Finally, if the sulfur insertion reaction damages the second cluster, as is the case in BioB, then the second FeS cluster would need to be repaired by the ISC or SUF systems prior to further turnover.

## Acknowledgment

This work has been supported by a grant from the National Science Foundation (MCB 09-23829).

## References

[1] Wood HG, Barden RE. Biotin enzymes. Annu Rev Biochem 1977;46:385–413.
[2] Knowles JR. The mechanism of biotin-dependent enzymes. Annu Rev Biochem 1989; 58:195–221.
[3] Kress D, Brugel D, Schall I, Linder D, Buckel W, Essen LO. An asymmetric model for Na+-translocating glutaconyl-CoA decarboxylases. J Biol Chem 2009;284:28401–9.
[4] Marquet A, Tse Sum Bui B, Florentin D. Biosynthesis of biotin and lipoic acid. Vitamins & Hormones 2001;61:51–101.
[5] Fugate CJ, Jarrett JT. Biotin synthase: insights into radical-mediated carbon-sulfur bond formation. Biochim Biophys Acta 2012;1824:1213–22.
[6] Marquet A, Florentin D, Ploux O, Tse Sum Bui B. In vivo formation of C-S bonds in biotin. An example of radical chemistry under reducing conditions. J Phys Org Chem 1998;11:529–35.
[7] Booker SJ. Anaerobic functionalization of unactivated C-H bonds. Curr Opin Chem Biol 2009; 13:58–73.
[8] Shisler KA, Broderick JB. Emerging themes in radical SAM chemistry. Curr Opin Struct Biol 2012;22:701–10.
[9] Atta M, Arragain S, Fontecave M, et al. The methylthiolation reaction mediated by the radical-SAM enzymes. Biochim Biophys Acta 2012;1824:1223–30.
[10] Pierrel F, Douki T, Fontecave M, Atta M. MiaB protein is a bifunctional radical-S-adenosylmethionine enzyme involved in thiolation and methylation of tRNA. J Biol Chem 2004;279:47555–63.
[11] Lee KH, Saleh L, Anton BP, et al. Characterization of RimO, a new member of the methylthiotransferase subclass of the radical SAM superfamily. Biochemistry 2009;48:10162–74.

[12] Hazra A, Chatterjee A, Begley TP. Biosynthesis of the thiamin thiazole in Bacillus subtilis: identification of the product of the thiazole synthase-catalyzed reaction. J Am Chem Soc 2009;131:3225–9.
[13] Tipton PA, Cleland WW. Mechanisms of decarboxylation of carboxybiotin. J Am Chem Soc 1988;110:5866–9.
[14] Ploux O, Soularue P, Marquet A, Gloeckler R, Lemoine Y. Investigation of the first step of biotin biosynthesis in Bacillus sphaericus. Purification and characterization of the pimeloyl-CoA synthase, and uptake of pimelate. Biochem J 1992;287:685–90.
[15] Cryle MJ, De Voss JJ. Carbon-carbon bond cleavage by cytochrome $P450_{Biol}$ (CYP107H1). Chem Comm 2004:86–7.
[16] Lin S, Hanson RE, Cronan JE. Biotin synthesis begins by hijacking the fatty acid synthetic pathway. Nat Chem Biol 2010;6:682–8.
[17] Lin S, Cronan JE. The BioC O-methyltransferase catalyzes methyl esterification of malonyl-acyl carrier protein, an essential step in biotin synthesis. J Biol Chem 2012;287:37010–20.
[18] Webster SP, Alexeev D, Campopiano DJ, et al. Mechanism of 8-amino-7-oxononanoate synthase: spectroscopic, kinetic, and crystallographic studies. Biochemistry 2000;39:516–28.
[19] Sandmark J, Eliot AC, Famm K, Schneider G, Kirsch JF. Conserved and nonconserved residues in the substrate binding site of 7,8-diaminopelargonic acid synthase from Escherichia coli are essential for catalysis. Biochemistry 2004;43:1213–22.
[20] Gibson KJ. Isolation and chemistry of the mixed anhydride intermediate in the reaction catalyzed by dethiobiotin synthetase. Biochemistry 1997;36:8474–8.
[21] Kack H, Sandmark J, Gibson KJ, Schneider G, Lindqvist Y. Crystal structure of two quaternary complexes of dethiobiotin synthetase, enzyme-MgADP-AlF3-diaminopelargonic acid and enzyme-MgADP-dethiobiotin-phosphate; implications for catalysis. Protein Sci 1998;7:2560–6.
[22] Ifuku O, Kishimoto J, Haze S, Yanagi M, Fukushima S. Conversion of dethiobiotin to biotin in cell-free extracts of Escherichia coli. Biosci Biotech Biochem 1992;56:1780–5.
[23] Ifuku O, Koga N, Haze S, Kishimoto J, Wachi Y. Flavodoxin is required for conversion of dethiobiotin to biotin in Escherichia coli. Eur J Biochem 1994;224:173–8.
[24] Otsuka AJ, Buoncristiani MR, Howard PK, et al. The Escherichia coli biotin biosynthetic enzyme sequences predicted from the nucleotide sequence of the bio operon. J Biol Chem 1988;263:19577–85.
[25] Guianvarc'h D, Florentin D, Tse Sum Bui B, Nunzi F, Marquet A. Biotin synthase, a new member of the family of enzymes which uses S-adenosylmethionine as a source of deoxyadenosyl radical. Biochem Biophys Res Commun 1997;236:402–6.
[26] Escalettes F, Florentin D, Tse Sum Bui B, Lesage D, Marquet A. Biotin synthase mechanism: evidence for hydrogen transfer from the substrate into deoxyadenosine. J Am Chem Soc 1999;121:3571–8.
[27] Duin EC, Lafferty ME, Crouse BR, et al. [2Fe-2S] to [4Fe-4S] cluster conversion in Escherichia coli biotin synthase. Biochemistry 1997;36:11811–20.
[28] Cosper M, Jameson G, Hernández H, Krebs C, Huynh B, Johnson M. Characterization of the cofactor composition of Escherichia coli biotin synthase. Biochemistry 2004;43:2007–21.
[29] Ugulava NB, Gibney BR, Jarrett JT. Biotin synthase contains two distinct iron-sulfur cluster binding sites: chemical and spectroelectrochemical analysis of iron-sulfur cluster interconversions. Biochemistry 2001;40:8343–51.
[30] Ugulava NB, Surerus KK, Jarrett JT. Evidence from Mössbauer spectroscopy for distinct $[2Fe\text{-}2S]^{2+}$ and $[4Fe\text{-}4S]^{2+}$ cluster binding sites in biotin synthase from Escherichia coli. J Am Chem Soc 2002;124:9050–1.

[31] Ugulava NB, Sacanell CJ, Jarrett JT. Spectroscopic changes during a single turnover of biotin synthase: destruction of a [2Fe-2S] cluster accompanies sulfur insertion. Biochemistry 2001;40:8352–8.
[32] Ugulava NB, Frederick KK, Jarrett JT. Control of adenosylmethionine-dependent radical generation in biotin synthase: a kinetic and thermodynamic analysis of substrate binding to active and inactive forms of BioB. Biochemistry 2003;42:2708–19.
[33] Sanyal I, Gibson KJ, Flint DH. Escherichia coli biotin synthase: an investigation into the factors required for its activity and its sulfur donor. Arch Biochem Biophys 1996;326:48–56.
[34] Tse Sum Bui B, Florentin D, Fournier F, Ploux O, Mejean A, Marquet A. Biotin synthase mechanism: on the origin of sulphur. FEBS Lett 1998;440:226–30.
[35] Gibson KJ, Pelletier DA, Turner IM, Sr. Transfer of sulfur to biotin from biotin synthase (BioB protein). Biochem Biophys Res Commun 1999;254:632–5.
[36] Tse Sum Bui B, Benda R, Schunemann V, Florentin D, Trautwein AX, Marquet A. Fate of the $(2Fe\text{-}2S)^{2+}$ cluster of Escherichia coli biotin synthase during reaction: a Mössbauer characterization. Biochemistry 2003;42:8791–8.
[37] Jameson GNL, Cosper MM, Hernández HL, Johnson MK, Huynh BH. Role of the [2Fe-2S] cluster in recombinant Escherichia coli biotin synthase. Biochemistry 2004;43:2022–31.
[38] Berkovitch F, Nicolet Y, Wan JT, Jarrett JT, Drennan CL. Crystal structure of biotin synthase, an S-adenosylmethionine-dependent radical enzyme. Science 2004;303:76–9.
[39] Forouhar F, Arragain S, Atta M, et al. Two Fe-S clusters catalyze sulfur insertion by radical-SAM methylthiotransferases. Nat Chem Biol 2013;9:333–8.
[40] Hanzelmann P, Schindelin H. Crystal structure of the S-adenosylmethionine-dependent enzyme MoaA and its implications for molybdenum cofactor deficiency in humans. Proc Natl Acad Sci USA 2004;101:12870–5.
[41] Fugate CJ, Stich TA, Kim EG, Myers WK, Britt RD, Jarrett JT. 9-Mercaptodethiobiotin is generated as a ligand to the $[2Fe\text{-}2S]^{+}$ cluster during the reaction catalyzed by biotin synthase from Escherichia coli. J Am Chem Soc 2012;134:9042–5.
[42] Broach RB, Jarrett JT. Role of the $[2Fe\text{-}2S]^{2+}$ cluster in biotin synthase: mutagenesis of the atypical metal ligand arginine 260. Biochemistry 2006;45:14166–74.
[43] Cosper M, Jameson G, Davydov R, et al. The $[4Fe\text{-}4S]^{2+}$ cluster in reconstituted biotin synthase binds S-adenosyl-L-methionine. J Am Chem Soc 2002;124:14006–7.
[44] Dowling DP, Vey JL, Croft AK, Drennan CL. Structural diversity in the AdoMet radical enzyme superfamily. Biochim Biophys Acta 2012;1824:1178–95.
[45] Pegg SC, Brown SD, Ojha S, et al. Leveraging enzyme structure-function relationships for functional inference and experimental design: the structure-function linkage database. Biochemistry 2006;45:2545–55.
[46] Vey JL, Drennan CL. Structural insights into radical generation by the radical SAM superfamily. Chem Rev 2011;111:2487–506.
[47] Lanz ND, Booker SJ. Identification and function of auxiliary iron-sulfur clusters in radical SAM enzymes. Biochim Biophys Acta 2012;1824:1196–212.
[48] Atta M, Mulliez E, Arragain S, Forouhar F, Hunt JF, Fontecave M. S-Adenosylmethionine-dependent radical-based modification of biological macromolecules. Curr Opin Struct Biol 2010;20:684–92.
[49] Booker SJ, Grove TL. Mechanistic and functional versatility of radical SAM enzymes. F1000 Biol Rep 2010;2:52.
[50] Walsby CJ, Hong W, Broderick WE, et al. Electron-nuclear double resonance spectroscopic evidence that S-adenosylmethionine binds in contact with the catalytically active $[4Fe\text{-}4S]^{+}$ cluster of pyruvate formate-lyase activating enzyme. J Am Chem Soc 2002;124:3143–51.

[51] Walsby CJ, Ortillo D, Broderick WE, Broderick JB, Hoffman BM. An anchoring role for FeS clusters: chelation of the amino acid moiety of S-adenosylmethionine to the unique iron site of the [4Fe-4S] cluster of pyruvate formate-lyase activating enzyme. J Am Chem Soc 2002;124:11270–1.

[52] Kamachi T, Kouno T, Doitomi K, Yoshizawa K. Generation of adenosyl radical from S-adenosylmethionine (SAM) in biotin synthase. J Inorg Biochem 2011;105:850–7.

[53] Zhang Y, Zhu X, Torelli AT, et al. Diphthamide biosynthesis requires an organic radical generated by an iron-sulphur enzyme. Nature 2010;465:891–6.

[54] Zhu X, Dzikovski B, Su X, et al. Mechanistic understanding of Pyrococcus horikoshii Dph2, a [4Fe-4S] enzyme required for diphthamide biosynthesis. Mol Biosyst 2011;7:74–81.

[55] Sanyal I, Cohen G, Flint DH. Biotin synthase: purification, characterization as a [2Fe-2S] cluster protein, and in vitro activity of the Escherichia coli bioB gene product. Biochemistry 1994;33:3625–31.

[56] Benda R, Tse Sum Bui B, Schunemann V, Florentin D, Marquet A, Trautwein AX. Iron-sulfur clusters of biotin synthase in vivo: a Mossbauer study. Biochemistry 2002;41:15000–6.

[57] Ugulava NB, Gibney BR, Jarrett JT. Iron-sulfur cluster interconversions in biotin synthase: dissociation and reassociation of iron is required for conversion of [2Fe-2S] to [4Fe-4S] clusters. Biochemistry 2000;39:5206–14.

[58] Krebs C, Broderick WE, Henshaw TF, Broderick JB, Huynh BH. Coordination of adenosylmethionine to a unique iron site of the [4Fe-4S] of pyruvate formate-lyase activating enzyme: a Mossbauer spectroscopic study. J Am Chem Soc 2002;124:912–3.

[59] Lieder KW, Booker S, Ruzicka FJ, Beinert H, Reed GH, Frey PA. S-Adenosylmethionine-dependent reduction of lysine 2,3-aminomutase and observation of the catalytically functional iron-sulfur centers by electron paramagnetic resonance. Biochemistry 1998;37:2578–85.

[60] Wang SC, Frey PA. Binding energy in the one-electron reductive cleavage of S-adenosylmethionine in lysine 2,3-aminomutase, a radical SAM enzyme. Biochemistry 2007;46:12889–95.

[61] Lotierzo M, Tse Sum Bui B, Leech HK, Warren MJ, Marquet A, Rigby SEJ. Iron-sulfur cluster dynamics in biotin synthase: a new [2Fe-2S]1+ cluster. Biochem Biophys Res Commun 2009;381:487–90.

[62] Tse Sum Bui B, Mattioli TA, Florentin D, Bolbach G, Marquet A. Escherichia coli biotin synthase produces selenobiotin. Further evidence of the involvement of the $[2Fe-2S]^{2+}$ cluster in the sulfur insertion step. Biochemistry 2006;45:3824–34.

[63] Farrar CE, Siu KKW, Howell PL, Jarrett JT. Biotin synthase exhibits burst kinetics and multiple turnovers in the absence of inhibition by products and product-related biomolecules. Biochemistry 2010;49:9985–96.

[64] Tse Sum Bui B, Lotierzo M, Escalettes F, Florentin D, Marquet A. Further investigation on the turnover of Escherichia coli biotin synthase with dethiobiotin and 9-mercaptodethiobiotin as substrates. Biochemistry 2004;43:16432–41.

[65] Taylor AM, Farrar CE, Jarrett JT. 9-Mercaptodethiobiotin Is formed as a competent catalytic intermediate by Escherichia coli biotin synthase. Biochemistry 2008;47:9309–17.

[66] Taylor AM, Stoll S, Britt RD, Jarrett JT. Reduction of the [2Fe-2S] cluster accompanies formation of the intermediate 9-mercaptodethiobiotin in Escherichia coli biotin synthase. Biochemistry 2011;50:7953–63.

[67] Muhlenhoff U, Gerl MJ, Flauger B, et al. The iron-sulfur cluster proteins Isa1 and Isa2 are required for the function but not for the de novo synthesis of the Fe/S clusters of biotin synthase in Saccharomyces cerevisiae. Eukaryotic Cell 2007;6:495–504.

[68] Sheftel AD, Wilbrecht C, Stehling O, et al. The human mitochondrial ISCA1, ISCA2, and IBA57 proteins are required for [4Fe-4S] protein maturation. Mol Biol Cell 2012;23:1157–66.

[69] Lu J, Yang J, Tan G, Ding H. Complementary roles of SufA and IscA in the biogenesis of iron-sulfur clusters in Escherichia coli. Biochem J 2008;409:535–43.

[70] Ollagnier-de-Choudens S, Sanakis Y, Fontecave M. SufA/IscA: reactivity studies of a class of scaffold proteins involved in [Fe-S] cluster assembly. J Biol Inorg Chem 2004;9:828–38.

[71] Reyda MR, Dippold R, Dotson ME, Jarrett JT. Loss of iron-sulfur clusters from biotin synthase as a result of catalysis promotes unfolding and degradation. Arch Biochem Biophys 2008;471:32–41.

[72] Reyda MR, Fugate CJ, Jarrett JT. A complex between biotin synthase and the iron-sulfur cluster assembly chaperone HscA that enhances in vivo cluster assembly. Biochemistry 2009;48:10782–92.

[73] Arragain S, Handelman SK, Forouhar F, et al. Identification of eukaryotic and prokaryotic methylthiotransferase for biosynthesis of 2-methylthio-N6-threonylcarbamoyladenosine in tRNA. J Biol Chem 2010;285:28425–33.

[74] Pierrel F, Hernandez HL, Johnson MK, Fontecave M, Atta M. MiaB protein from Thermotoga maritima. Characterization of an extremely thermophilic tRNA-methylthiotransferase. J Biol Chem 2003;278:29515–24.

[75] Hernandez HL, Pierrel F, Elleingand E, et al. MiaB, a bifunctional radical-S-adenosylmethionine enzyme involved in the thiolation and methylation of tRNA, contains two essential [4Fe-4S] clusters. Biochemistry 2007;46:5140–7.

[76] Landgraf BJ, Arcinas AJ, Lee KH, Booker SJ. Identification of an intermediate methyl carrier in the radical S-adenosylmethionine methylthiotransferases RimO and MiaB. J Am Chem Soc 2013;135:15404–16.

[77] Arragain S1, Garcia-Serres R, Blondin G, et al. Post-translational modification of ribosomal proteins: structural and functional characterization of RimO from Thermotoga maritima, a radical S-adenosylmethionine methylthiotransferase. J Biol Chem 2010;285: 5792–801.

# 8 Molybdenum-containing iron-sulfur enzymes

Russ Hille

## 8.1 Introduction

Molybdenum in biology is frequently associated with iron-sulfur clusters, as perhaps best exemplified by nitrogenase, where molybdenum is part of a $[MoFe_7S_9]$ cluster of the active site. Those molybdenum-containing enzymes other than nitrogenase fall into three groups, as exemplified by the enzymes xanthine oxidase, sulfite oxidase, and DMSO reductase. All the enzymes in xanthine oxidase family possess at least two iron-sulfur clusters in addition to the molybdenum center, and nearly all those in the DMSO reductase family groups possess at least one and usually several iron-sulfur clusters (although no member of the sulfite oxidase family has been identified to date that has an iron-sulfur cluster) [1]. As shown in Fig. 8.1, each molybdenum enzyme family has a distinct metal coordination environment, including at least 1 equivalent (equiv.) of a pyranopterin cofactor (so-called because it consists of a pterin nucleus fused to a pyran ring; Fig. 8.1) that coordinates the metal *via* an enedithiolate side chain. The canonical oxidation state of the cofactor is the tetrahydropterin shown in Fig. 8.1, but as discussed in Section 8.3, there is evidence that the dihydro oxidation state may sometimes be present. Xanthine oxidase and related enzymes have an $LMo^{VI}OS(OH)$ active site with a square-pyramidal coordination geometry. The apical ligand is a Mo=O ligand and the equatorial plane has a Mo=S group, a catalytically labile Mo-OH group and two sulfurs from the enedithiolate side chain of the pyranopterin. Members of the sulfite oxidase family have a related $LMo^{VI}O_2$(S-Cys) active site, again square-pyramidal with an apical Mo=O and a bidentate enedithiolate ligand in the equatorial plane but with a second equatorial Mo=O (rather than Mo-OH) and a cysteine ligand contributed by the protein (rather than a Mo=S) as the final two ligands in the equatorial plane. The final DMSO reductase family possesses 2 equiv. of the pyranopterin cofactor in an $L_2Mo^{VI}Y(X)$ trigonal prismatic coordination geometry. DMSO reductase itself has a catalytically labile Mo=O and a serinate ligand completing the metal coordination sphere of oxidized enzyme. Other family members have S or even Se in place of the Mo=O group, and cysteine, selenocysteine, aspartate, or even hydroxide in place of the serine. Members of the DMSO reductase family, all from bacterial or archaeal sources, are structurally related to the aldehyde:ferredoxin oxidoreductase family of tungsten-containing enzymes [2] (the first protein crystal structure of a pyranopterin-containing enzyme was in fact the tungsten-containing aldehyde:ferredoxin oxidoreductase from *Pyrococcus furiosus* [3]). Although frequently referred to as "molybdopterin" in the literature, in light of the broader distribution of the organic cofactor, "pyranopterin" will be used here.

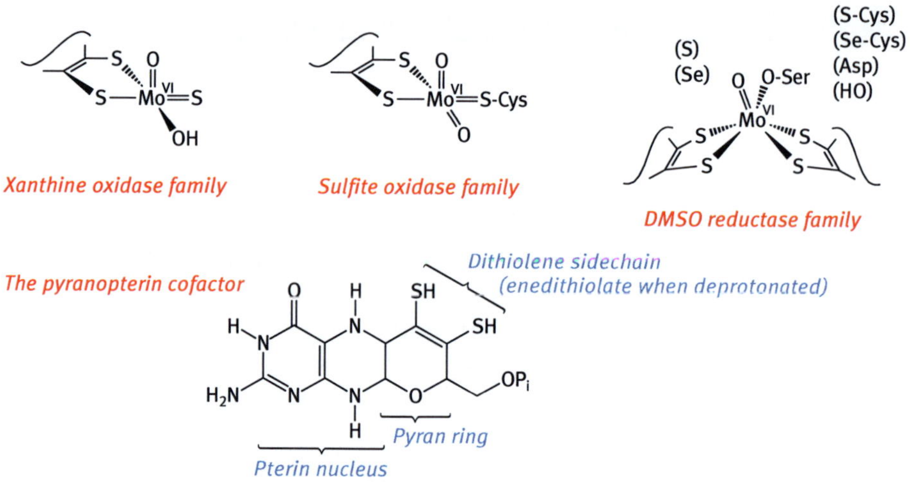


**Fig. 8.1:** The three families of molybdenum-containing enzymes. Structures are for, from left to right, for the xanthine oxidase, sulfite oxidase, and DMSO reductase families. The structure of the pyranopterin cofactor of these enzymes (as well as the tungsten-containing enzymes) is shown at bottom.

All members of the xanthine oxidase family possess multiple iron-sulfur clusters, as do the vast majority of members of the DMSO reductase family – no member of the sulfite oxidase family has yet been identified as having one. What follows is an account of our present understanding of the xanthine oxidase-like and DMSO reductase-like enzymes, with particular attention paid to their iron-sulfur clusters and the relationship of their structures to other iron-sulfur-containing systems.

## 8.2 The xanthine oxidase family

Members of the xanthine oxidase family of molybdenum-containing enzymes usually catalyze the oxidative hydroxylation of a carbon center of their substrates (typically aromatic heterocycles or an aldehydes), with the initial $Mo^{VI}$ state of the active site molybdenum center becoming reduced to $Mo^{IV}$ in the process. Bovine xanthine oxidase, which catalyzes the final two steps of purine metabolism in vertebrates (the oxidation of hypoxanthine to xanthine, and xanthine to uric acid), is the prototypical member and is one of the longest-studied enzymes, having been first purified to homogeneity in 1924 [4]. The enzyme is typically isolated from cow's milk as an oxidase, but the physiologically relevant form expressed in most vertebrate tissues is a dehydrogenase that utilizes $NAD^+$ rather than $O_2$ as oxidizing substrate. Because both forms are products of the same gene, differing only in post-translational modification, the generic term xanthine oxidoreductase is frequently used for the enzyme. These xanthine-oxidizing enzymes are extremely broadly distributed in biology, with only a few organisms oxidizing xanthine to urate by e.g. an $Fe^{2+}$/α-ketoglutarate hydroxylase (in *Aspergillus nidulans* and certain yeasts [5]) or a recently identified third xanthine-oxidizing system (in *Klebsiella* species [6, 7]). In addition to the xanthine-metabolizing molybdenum enzymes, most organisms also encode one or more aldehyde oxidases that are very similar in reaction mechanism and cofactor constitution to their xanthine-oxidizing counterparts.

All members of the xanthine oxidase family have redox-active centers in addition to the molybdenum center: minimally a pair of spinach-ferredoxin-like [2Fe-2S] clusters (i.e. with each iron coordinated by two cysteine residues in addition to the pair of bridging sulfurs), and usually FAD as well. $O_2$ and $NAD^+$ react at the FAD site rather than the molybdenum center, and as a result, the intramolecular electron transfer between molybdenum and FAD *via* the intervening iron-sulfur centers is an obligatory aspect of turnover. As discussed further in Section 8.2.1, the otherwise closely related aldehyde:ferredoxin oxidoreductase from organisms such as *Desulfovibrio gigas* lacks FAD, and the 4-hydroxybenzoyl-CoA reductase from *Thauera aromatica* has an additional redox-active center – a [4Fe-4S] cluster. In all cases, the redox-active centers are found in discretely folded domains or autonomous subunits, with the eukaryotic enzymes being $\alpha_2$-dimers and the bacterial enzymes typically organized as $(\alpha\beta)_2$ or $(\alpha\beta\gamma)_2$. The variation in overall subunit organization notwithstanding, the homologous regions of these enzymes are very similar, as first recognized based on the sequence analysis by Wootton *et al.* [8]. The complex overall architecture of the eukaryotic enzymes appears to have been built up from simpler elements over the course of evolution, from an original $(\alpha\beta\gamma)_2$ form through a $(\alpha\beta)_2$ intermediate in which the iron-sulfur- and FAD-binding domains have been merged into a single subunit, finally to the $(\alpha)_2$ form seen in eukaryotes. The order of the genes in bacterial operons does not always reflect order of the cognate domains in the eukaryotic enzymes, however. The eukaryotic enzymes are invariably organized with the two [2Fe-2S]-containing domains N-terminal to the central FAD domain, followed by the C-terminal molybdenum-binding portion of the protein, whereas the genes encoding the $(\alpha\beta\gamma)_2$ CO dehydrogenase from *Oligotropha carboxidovorans*, for example, are ordered *coxMSL*, encoding FAD-, iron-sulfur-, and molybdenum-binding domains, respectively [9].

### 8.2.1 *D. gigas* aldehyde:ferredoxin oxidoreductase

The aldehyde:ferredoxin oxidoreductase from *D. gigas*, a rod-shaped, anaerobic δ-proteobacterium, was the first member of the xanthine oxidase family to be crystallographically characterized. It was for this enzyme that the overall coordination geometry of the molybdenum coordination sphere was established to be square-pyramidal with the pyranopterin enedithiolate defining the base plane [10] and a highly conserved active site glutamate proposed to act as an active site base (facilitating nucleophilic attack of the equatorial Mo-OH on the substrate carbonyl by abstraction of the Mo-OH proton) [11]. It was subsequently clarified that the catalytically essential

Mo=S group occupied an equatorial rather than apical position [12, 13] and that the equatorial oxygen was a Mo-OH rather than Mo-$OH_2$ as originally proposed. Somewhat surprisingly, it has been suggested based on steady-state assays and the inhibition patterns seen with classic inhibitors of this family of enzymes (e.g. cyanide and ethylene glycol [14]) that the *D. gigas* enzyme is active in the absence of the Mo=S group that in other enzymes of this family is known to be catalytically essential [15]. Aldehydes are in fact generally more susceptible to nucleophilic attack than heterocycles such as xanthine, but the ability of bovine xanthine oxidase to effectively oxidize aldehydes is strictly dependent on the presence of the Mo=S and the conclusion that the *D. gigas* enzyme does not require the Mo=S group may require revision.

The structure of the *D. gigas* enzyme is shown in Fig. 8.2 [10]. The two [2Fe-2S] clusters are found in independently folded N-terminal domains, the first resembling spinach ferredoxin and being comprised principally of β sheet (Fig. 8.2, blue), with the second having a unique four-helix bundle structure (with two short flanking helices) (Fig. 8.2, red). The overall fold of this second iron-sulfur domain is to date unique to the xanthine oxidase family of enzymes but bears some resemblance to zinc finger proteins. There is a 27-residue linker to the molybdenum-binding portion of the protein, which consists of a pair of bilobal two-domain units that lie across one another with the molybdenum center at the interface (Fig. 8.2, light and dark gray). The distal amino group of the pyranopterin cofactor (present as the dinucleotide of cytosine) is within hydrogen-bonding distance (3.1 Å) of the sulfur of Cys139

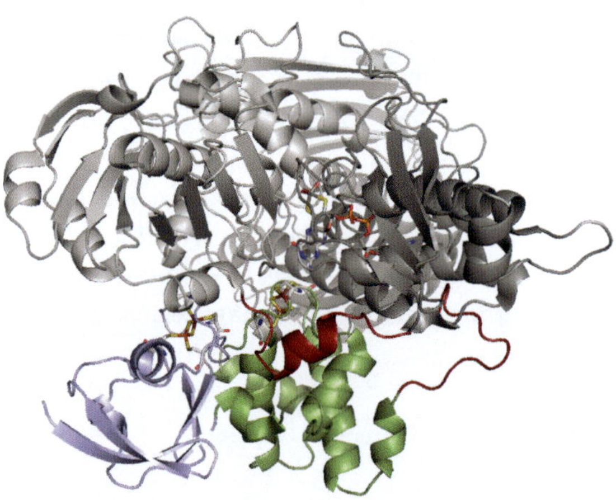
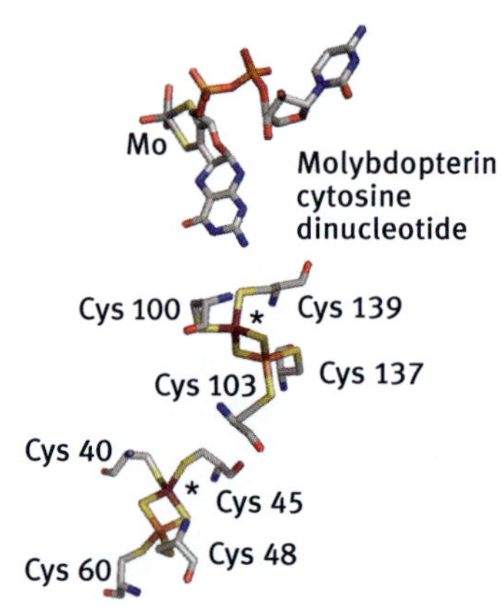


**Fig. 8.2:** The structure of the *D. gigas* aldehyde:ferredoxin oxidoreductase (PDB 1VLB). (left) The overall structure of the protein: the two N-terminal [2Fe-2S]-containing domains are in blue and green, respectively. These are connected *via* a 27-residue long linker (red) to the molybdenum-binding portion of the protein in gray. (right) An enlargement of the redox-active centers, with the molybdenum center (including the pyranoperin cofactor present as the cytosine dinucleotide) at top. The redox-active iron in Fe/S I (which is proximal to the molybdenum center) coordinated by Cys100 and Cys139 and that in Fe/S II coordinated by Cys40 and Cys45 in brown and indicated with asterisks.

that coordinates the nearer (more C-terminal) iron-sulfur cluster. The two iron-sulfur clusters themselves are 9.1 Å apart (edge-to-edge). Unusually for members of the xanthine oxidase family, the *D. gigas* enzyme does not have an FAD-binding domain.

[2Fe-2S] clusters such as are found in members of the xanthine oxidase family are valence-localized, and the redox-active irons have been unambiguously assigned in the crystal structure of the *D. gigas* aldehyde oxidoreductase [16]. As indicated in Fig. 8.2, the redox-active irons in Fe/S I (which is nearer to the molybdenum center) is coordinated by Cys100 and Cys139, whereas that in Fe/S II is coordinated by Cys40 and Cys45. Given the extremely high homology of the iron-sulfur domainsof the *D. gigas* with other members of this family, it is very likely that the corresponding irons are redox-active in all these enzymes.

### 8.2.2 Bovine xanthine oxidoreductase

The subsequent X-ray crystal structure of bovine xanthine oxidoreductase, in both dehydrogenase and oxidase forms (see end of this section), was the first of a family member to contain FAD [17]. The structure of the dimeric bovine xanthine dehydrogenase is shown in Fig. 8.3, with the domains containing each of the redox-active centers color-coded as in Fig. 8.2, with the FAD domain in yellow. The subunit contacts in the dimer are limited and entirely within the molybdenum-binding portion of the protein.

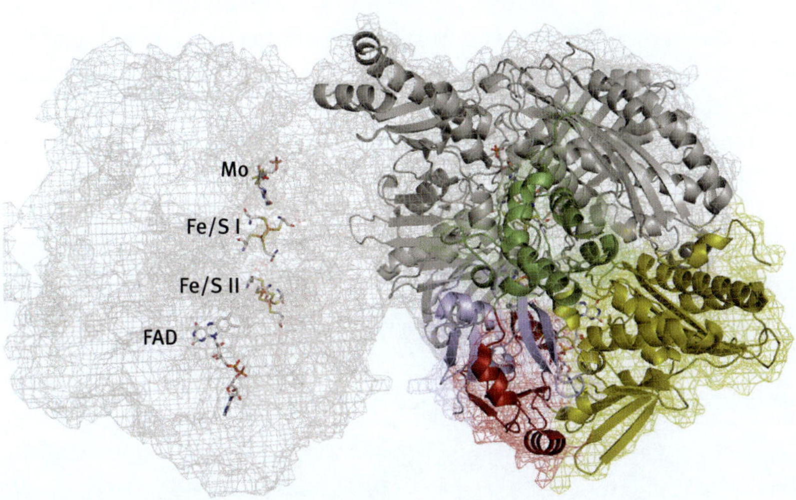


**Fig. 8.3:** The structure of bovine xanthine dehydrogenase (PDB 1FO4). From the N-terminus in the subunit at right, the domains are colored blue and green for the two [2Fe-2S] clusters, yellow for the FAD, and gray for the molybdenum-binding portion of the protein. The linker region between iron-sulfur- and FAD-binding domains is in red at the bottom left. The subunit on the left is rendered in mesh to illustrate the spatial layout of the several redox-active centers within the subunit. The molybdenum centers in the two monomers are 52 Å apart.

The polypeptide strand makes a single pass in going from one domain to the next within each monomer, meaning that not only are the domains themselves autonomous structural elements of the polypeptide but that each domain is encoded by a contiguous stretch of the structural gene for the protein (consistent with the gene duplication/ fusion model by which the protein likely arose). The disposition of the redox-active centers within each protomer defines an approximately linear pathway for electron transfer form the molybdenum center (site of the reductive half-reaction) to the FAD (site of the oxidative half-reaction), with the two iron-sulfur centers intervening. Fe/S II lies approximately 6.4 Å from the C-7 methyl of the FAD, and 7.3 Å from the C-8 methyl. The nearest approach of the electron transfer chains of the two monomers within the dimer is some 52 Å (molybdenum to molybdenum), clearly indicating that inter-subunit electron transfer does not occur.

The two iron-sulfur centers of xanthine oxidase have long been distinguishable on the basis of their electron paramagnetic resonance (EPR) signals [18]. That designated Fe/S I signal has $g_{1,2,3}$ = 2.022, 1.932, and 1.894, with unexceptional linewidths and relaxation properties for a [2Fe-2S] cluster, whereas Fe/S II has $g_{1,2,3}$ = 2.110, 1.991, and 1.902 and unusually broad linewidths and relaxation properties, such that it has been observed only below 25 K [19]. Site-directed mutagenesis studies with rat xanthine oxidoreductase has made possible assignments of the two iron-sulfur clusters [20], with Fe/S I being the cluster in the unusual α-helical domain that is proximal to the molybdenum center, and Fe/S II, that in the more commonly seen ferredoxin-like domain. This assignment is consistent with the known coupling of Fe/S I to the molybdenum center [21, 22]. The pathway for electron transfer within the enzyme is thus Mo → Fe/S I → Fe/S II → FAD. It is noteworthy that the distal amino group of the pyranopterin cofactor is within hydrogen bonding distance to Cys150, coordinating the nearer, presumably redox-active iron of Fe/S I.

Figure 8.4 shows the active site of xanthine oxidoreductase, including several amino acid residues that have been shown to be catalytically important. Phe914 and

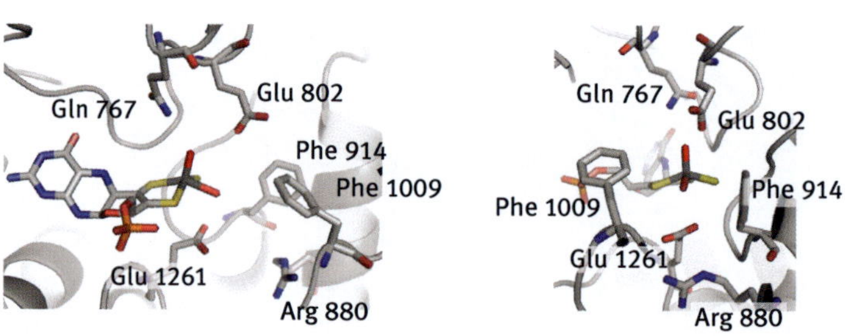


**Fig. 8.4:** The active site of bovine xanthine dehydrogenase (from PDB 1FO4). The several amino acid residues referred to in the text are indicated. The orientation at the right is rotated approximately 90° about the vertical from that at the left and represents the view from the solvent access channel. The PDB file has been modified to indicate that the catalytically essential Mo=S ligand occupies an equatorial position rather than the apical one.

Phe1009 (in the bovine enzyme numbering) are at the end of a 14.5-Å-long substrate access channel and constrain bound substrate to a plane approximately parallel to the apical Mo=O group of the molybdenum center. The equatorial Mo-OH projects directly toward the substrate-binding site. Four other active site residues are universally conserved in xanthine-utilizing enzymes: Glu802, Gln767, Glu1261, and Arg880. These residues, along with the molybdenum center itself, define the structural environment in which catalysis takes place.

Again, xanthine oxidase catalyzes the oxidation of xanthine to uric acid, and is the target of antihyperuricemic drugs such as allopurinol. The overall reaction mechanism of the enzyme is now generally understood to occur as shown in Fig. 8.5 [1, 23] with proton abstraction of the equatorial Mo-OH [24] by the active site glutamate [25], followed by nucleophilic attack on the carbon to be and can be formed very rapidly and noncatalytically simply by mixing substrate with enzyme that has been partially reduced by titration with sodium dithionite [26]. (Although a dead-end intermediate from a kinetic standpoint – the molybdenum center cannot react with substrate until fully reoxidized to the $Mo^{VI}$ state – hydroxylated.) Concomitant hydride transfer to the Mo=S group gives an initial $LMo^{IV}O(SH)(OR)$ intermediate that breaks down by displacement of product from the molybdenum coordination sphere by hydroxide from solvent, with electron transfer from the molybdenum to the other redox-active centers of the enzyme and deprotonation of the Mo-SH to return to the Mo=S of oxidized enzyme. The sequence in which these later events occur depends on the reaction conditions and the substrate utilized – when electron transfer precedes displacement of substrate, an EPR-active $LMo^{V}OS(OR)$ species termed "very rapid" (on the basis of the kinetics of its appearance in the course of the reaction of enzyme with xanthine) is formed. Under most conditions, however, product dissociation precedes electron transfer out of the molybdenum center and the "very rapid" species is bypassed. Not shown in the mechanism are any of the species giving rise to the "rapid" family of EPR signals that appear on approximately the same timescale as the decay of the "very rapid" signal [27–30] and which were long thought to arise from an intermediate lying downstream from the "very rapid" species in the catalytic sequence. The "rapid" signal arises from partially reduced enzyme having bound substrate rather than product, however, the "rapid" species in fact represents a paramagnetic analogue to the Michaelis complex of the enzyme.) A valence bond description of the first step of the reaction has been developed, establishing the interactions of specific atomic and molecular orbitals that lower the barrier to reaction [31]. The upshot is that both Mo=S p → C-H s* and C-H s → Mo=S p* donation contribute to activation of the C-H bond for heterolytic cleavage, along with $O_{eq}$ lp → C-H s* and S lp → C-H s* donation. An $O_{eq}$ lp → Mo + C charge-transfer interaction plays an important role in transition state stabilization, with the electronic delocalization accruing from this interaction reducing electronic repulsion along the reaction coordinate and thus lowering the barrier to reaction.

In the absence of recombinant expression systems that provide substantive amounts of the functional eukaryotic xanthine oxidoreductase, the roles of several highly conserved active site residues have been examined by site-directed

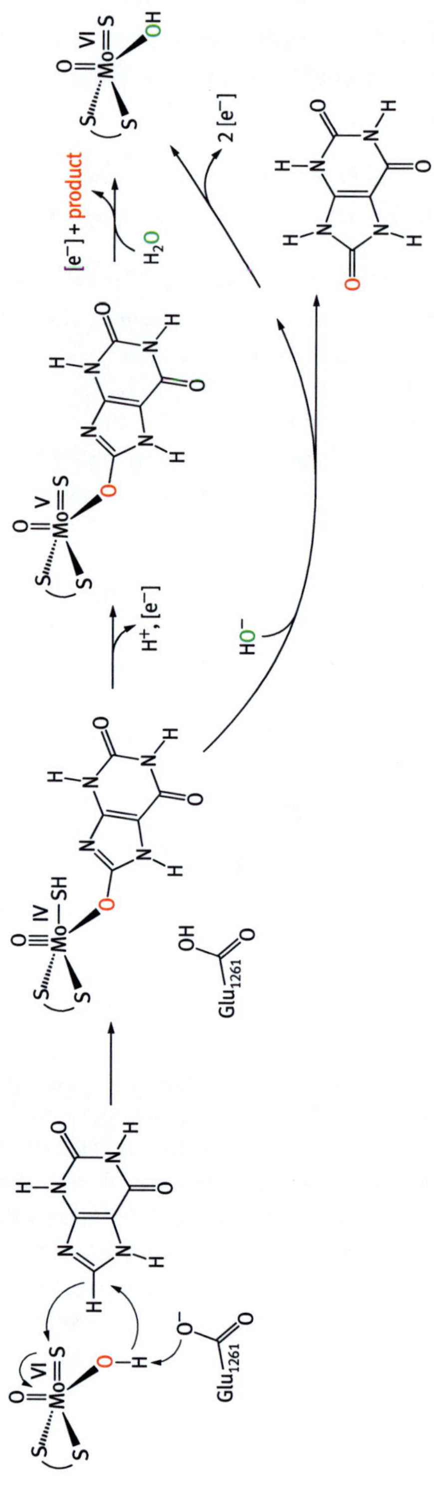


**Fig. 8.5:** The reaction mechanism of xanthine oxidase.

mutagenesis studies of the *Rhodobacter capsulatus* xanthine dehydrogenase, which bears strong structural homology to the bovine enzyme and has a virtually identical active site [32]. Referring to Fig. 8.4, residues important in catalysis include Glu730 (equivalent to Glu1261 in the bovine enzyme), Glu232 (Glu802), Gln197 (Gln767), and Arg310 (Arg880). The substrate-binding site is further comprised of Phe344 and Phe459 (Phe914 and Phe1009 in the bovine enzyme), which again constrain substrate to a plane approximately parallel to the apical Mo=O bond. Mutation of Glu730 in the *R. capsulatus* enzyme to Ala reduces the limiting rate constant for reduction by xanthine, $k_{red}$, by at least seven orders of magnitude [33], corresponding to at least 10 kcal/mol of compromised transition state stabilization with the mutant. Mutation of Glu232 to Ala results in a more modest 12-fold decrease in $k_{red}$ in reductive half-reaction studies as well as a 12-fold increase in $K_d$ [33]. The equivalent E803V mutant of the *Escherichia coli*-expressed human xanthine oxidoreductase exhibits a comparable reduction in the steady-state $k_{cat}$ and increase in $K_m$ [34]. It has been suggested that Glu232/802 accelerates the reaction rate specifically by facilitating a tautomerization of the heterocycle in the course of nucleophilic attack that involves proton transfer from $N_3$ to $N_9$ of the purine ring, thereby compensating for the negative charge accumulating on the imidazole subnucleus of the purine in the course of the reaction [35] (Fig. 8.6). Mutation of Arg310 to the isosteric Met results in a decrease in $k_{red}$ by a factor of approximately $10^4$ [36]; consistent with this, an R881M mutant of the human

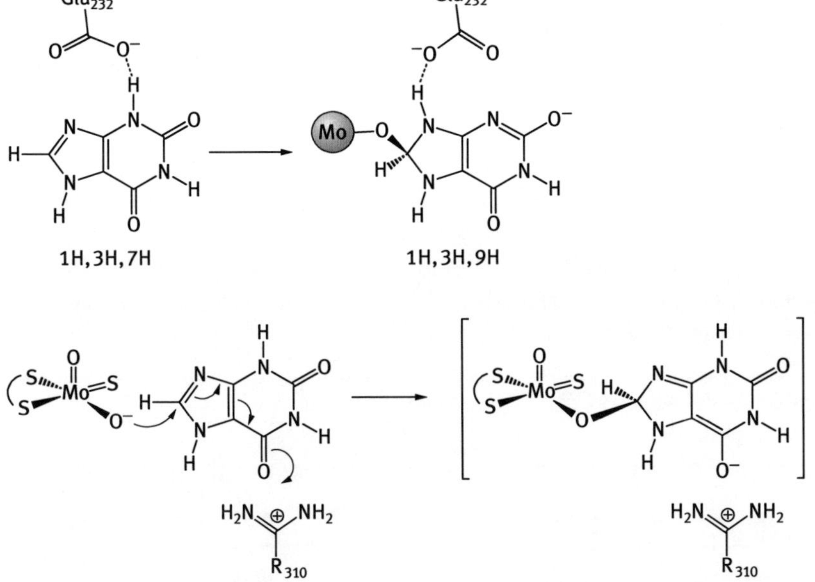


**Fig. 8.6:** The proposed mechanism whereby Glu2332/802 facilitates tautomerization (left) and Arg880 stabilizes charge accumulation on the heterocycle in the course of catalysis.

enzyme has no detectable activity in steady-state assays [34]. This residue is some 8 Å from the Mo-OH oxygen and the site of the hydroxylation chemistry and is proposed to contribute to rate acceleration specifically by stabilizing negative charge accumulation on substrate in the course of nucleophilic attack [36], as shown in Fig. 8.6.

Intramolecular electron transfer is an integral aspect of turnover of all members of the xanthine oxidase family of enzymes, and it has long been recognized that in the course of equilibrium reductive titrations reducing equivalents rapidly distribute themselves among the several redox-active centers according to their reduction potentials and on a timescale that is fast compared with catalysis (with $k_{cat} = 18\ s^{-1}$ at pH 8.5) [37]. The reduction potentials for bovine xanthine oxidase at pH 8.3 are known [38], and these can be used to calculate the distribution of reducing equivalents within enzyme at any given level of reduction (i.e. one-electron reduced, two-electron reduced, and so on) [37], as shown in Tab. 8.1. An examination of Tab. 8.1 indicates that although the FAD/FADH• couple is much lower than the FADH•/$FADH_2$ couple (meaning that the FADH• oxidation state is thermodynamically destabilized and does not accumulate to a significant extent), the midpoint potential for the FAD is approximately equal to the average of the reduction potentials for the two iron-sulfur centers. This leads to an important characteristic of the system, as reflected in the electron distributions between FAD and iron-sulfur clusters for one- and two-electron-reduced enzyme. In one-electron-reduced enzyme, the sole reducing equivalent resides primarily on the highest-potential site, Fe/S II, as expected. In two-electron-reduced enzyme, however, because the midpoint potential of the FAD is equal to the average of the potentials for the two iron-sulfur clusters, the FAD competes very effectively for a pair of electrons – the distribution of reducing equivalents in two-electron-reduced enzyme is approximately evenly split between the FAD and the iron-sulfur clusters. The arrangement is such that, paradoxically, the introduction of a second electron to the one-electron-reduced enzyme results in the net oxidation of the highest-potential site, Fe/S II, whose level of reduction decreases from over 60% to less than 30% in going from one- to two-electron-reduced enzyme (Tab. 8.1). This is because retaining an electron

**Tab. 8.1:** The reduction potentials of xanthine oxidase at pH 8.3 and the distribution of reducing equivalents within one- and two-electron-reduced enzyme.

| Couple | Potential (vs NHE) | Fraction reduced | |
|---|---|---|---|
| | | $XO_{1e}-$ | $XO_{2e}-$ |
| $Mo^{VI/V}$ | −360 | 0.06 | 0.09 |
| $Mo^{V/IV}$ | −321 | 0.00 | 0.02 |
| Fe/S I | −332 | 0.25 | 0.28 |
| Fe/S II | −224 | 0.63 | 0.35 |
| FAD/FADH• | −319 | 0.06 | 0.01 |
| FADH•/$FADH_2$ | −237 | 0.00 | 0.58 |

on Fe/S II in two-electron-reduced enzyme requires that the second electron must go to either Fe/S I or FADH•; both of these have very low reduction potentials, which imparts a thermodynamic "penalty" for maintaining Fe/S II reduced. The upshot is that the FAD does not become reduced to any significant degree until a pair of reducing equivalents are available. Importantly, the arrangement ensures that reducing equivalents are effectively delivered in pairs to the FAD, despite the fact that the iron-sulfur clusters are obligatory one-electron carriers. A great many electron transfer systems, including many containing iron-sulfur clusters such as are considered in Section 8.3, are arranged such that one or another intermediate center has an unusually low (or high) reduction potential, which would ensure the same result: that reducing equivalents are delivered to (or taken from) a terminal redox-active center in pairs.

The explicit rate constants for electron transfer within the bovine xanthine oxidase have been determined using pH-jump [39, 40] and pulse radiolysis [41, 42] methods. In the first type of experiment, advantage is taken of the greater pH dependence of the FAD, and molybdenum reduction potentials relative to those of the iron-sulfur clusters to perturb the distribution of reducing equivalents within partially reduced enzyme. By mixing partially reduced enzyme in dilute buffer at one pH with more concentration buffer at another (under strictly anaerobic conditions), the rate constant for re-equilibration of reducing equivalents between Fe/S I and FAD has been found to range from 155 $s^{-1}$ at pH 6 to 330 $s^{-1}$ at pH 9 [39]. The observed solvent kinetic isotope effect is 6.9, and the linear dependence of the observed rate constant on mole fraction of $D_2O$ indicates that the effect involves a single proton [40]. Knowing the relative reduction potentials of the centers involved, the forward and reverse rate constants for the equilibrium can be calculated explicitly in both $H_2O$ and $D_2O$, with the isotope effect being found to be much larger for the FADH• → Fe/S $I_{ox}$ electron transfer event than the reverse process. It has been concluded that the $N_5$-H proton of the neutral flavin semiquinone is in motion as the system traversed the electron-transfer transition state, i.e. that proton and electron transfer are coupled in xanthine oxidase. This has been attributed to the destabilization of the $FAD^{•-}$ anionic semiquinone by the protein environment, which makes discrete deprotonation of the neutral semiquinone prior to electron transfer thermodynamically unfavorable [40]. This relatively slow electron transfer between flavin and Fe/S I accounts for only about half of the expected spectral change, and in subsequent pulse radiolysis studies, electron transfer between the molybdenum center and Fe/S I (after the former was extremely rapidly reduced with radiolytically generated radical of methylnicotinamide) has been observed with $k_{et}$ = 8,500 $s^{-1}$ [41], with subsequent electron transfer from the iron-sulfur centers on to the flavin occurring at 125 $s^{-1}$ (in good agreement with the pH jump work). It is interesting to note that the distance from the molybdenum atom to the edge of the proximal Fe/S I is 14.5 Å and that the observed rate constant for electron transfer is more or less what one would expect for electron transfer over such a distance with a decay parameter β of 1.6 [43]. The implication is that the intervening pyranopterin does not afford a particularly effective pathway for electron transfer but instead acts simply as a neutral medium. The coupled electron/proton transfer involving the iron-sulfur centers and FAD seen with xanthine oxidase contrasts with that exhibited by other flavin- and iron/sulfur-containing systems such as trimethylamine dehydrogenase (which contains a [4Fe-4S] iron-sulfur cluster and a covalently linked FMN [44]). In this protein, clear evidence is seen for discrete prototropic and redox equilibria with the behavior of the system well accounted for by the following discrete equilibria: $FMNH_2/[4Fe\text{-}4S]_{ox} \rightleftharpoons FMNH^-/[4Fe\text{-}4S]_{ox} \rightleftharpoons FMNH\bullet/[4Fe\text{-}4S]_{red} \rightleftharpoons FMN\bullet^-/[4Fe\text{-}4S]_{red}$ [45]. The different behavior of trimethylamine dehydrogenase and xanthine oxidase has been attributed to the fact that the former enzyme does not thermodynamically destabilize the anionic forms of the flavin, allowing deprotonation to precede electron transfer out of the flavin [42].

Upon reduction of the FAD, reducing equivalents are finally passed on to $O_2$ or $NAD^+$ (depending on the enzyme form) to complete the catalytic sequence. For the dehydrogenase forms, reduction of $NAD^+$ is thought to occur *via* hydride transfer, but the reaction with $O_2$ is more complicated [46, 47]. The reaction of the fully reduced bovine oxidase with $O_2$ occurs in four sequential steps, with the six-electron-reduced enzyme first oxidized to four-electron-reduced enzyme and the four-electron reduced to two-electron-reduced enzyme – both steps involve the quantitative reduction of $O_2$ to $H_2O_2$. Once the two-electron reduced form is generated, however, the remaining two reducing equivalents are lost individually, forming (again quantitatively) 2 equiv. of superoxide, $O_2\bullet^-$. As discussed above, the relative reduction potentials of the redox-active centers are such that in the two-electron-reduced enzyme, the distribution of reducing equivalents at pH 8.5 gives approximately 50% of the enzyme with $FADH_2$ and the remaining 50% with the two iron-sulfur clusters reduced instead (there is little molybdenum reduction in the two-electron-reduced enzyme and little FADH• at the flavin site) [37]. The question is thus why the $FADH_2$ in two-electron-reduced enzyme reacts with $O_2$ to form $O_2\bullet^-$, whereas that in four- and six-electron-reduced enzyme forms $H_2O_2$. It has been suggested that $H_2O_2$ formation involves three discrete steps: (1) an initial one-electron transfer to form a nascent $FADH\bullet...O_2\bullet^-$ complex; (2) regeneration of $FADH_2$ by rapid electron transfer from the iron-sulfur centers to form a $FADH_2...O_2\bullet^-$ complex; and (3) a second (rapid) one-electron transfer to form (with protonation) $FADH\bullet...H_2O_2$. In the absence of a reservoir of reducing equivalents elsewhere in the enzyme upon forming the nascent $FADH\bullet...O_2\bullet^-$ complex in the two-electron-reduced enzyme, the reaction of FADH• with superoxide is sufficiently slow that superoxide has time to dissociate. The one-electron-reduced enzyme thus generated must form $O_2\bullet^-$ [46–48].

Non-mammalian xanthine oxidoreductases exist solely as dehydrogenases. In mammals, however, although the enzyme normally functions as a dehydrogenase, it can be post-translationally converted into an oxidase either by irreversible proteolysis or reversible oxidation of cysteine residues [49–53]. Either modification results in a reorientation of a loop (Gln423-Lys433) in the FAD domain [17], as shown in Fig. 8.7. In the oxidase configuration, this loop is positioned so as to prevent $NAD^+$ from

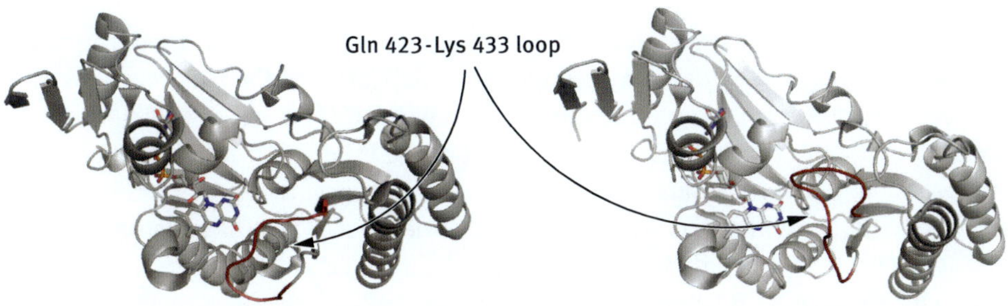


**Fig. 8.7:** Structures of the dehydrogenase (left; PDB 1FO4) and oxidase (right; 1FIQ) forms of bovine xanthine oxidoreductase. The loop that rearranges and prevents $NAD^+$ binding upon proteolysis or cysteine oxidation is indicated in red.

approaching the FAD cofactor, thereby abolishing dehydrogenase activity. The specific sites of proteolysis and formation of disulfide bonds that accompany the irreversible and reversible XDH to XO transition, respectively, have been identified [54]. In the bovine enzyme, proteolysis occurs to the C-terminal side of Lys551 and Lys569 in the linker between the flavin- and the molybdenum-binding domains; the residues involved in disulfide bond formation are Cys535 (also in the linker region) and Cys992 (in the molybdenum-binding domain). Given the copious production of superoxide and hydrogen peroxide (and possibly, indirectly, the hydroxyl radical through Fenton-like chemistry) by the oxidase form of xanthine oxidoreductase, this so-called D-to-O conversion has been implicated in the oxidative stress associated ischemia-reperfusion injury in, for example, heart attack and stroke [55–57].

### 8.2.3 Aldehyde oxidases

Aldehyde oxidases catalyze the oxidation of aliphatic and aromatic aldehydes to the corresponding carboxylic acid. Historically, aldehyde oxidase has been characterized from mammalian liver (most commonly rabbit and rat) and, apart from its substrate specificity, generally closely resembles the better-studied bovine xanthine oxidoreductase in its physicochemical properties [58–68]. It has long been recognized that the molybdenum centers of the two enzymes are fundamentally the same, with aldehyde oxidase exhibiting the same family of "rapid," "slow," and "inhibited" EPR signals seen with xanthine oxidase. Aldehyde oxidase does not, however, exhibit a "very rapid" type of EPR signal [62, 66, 67, 69].

The crystal structure of the mouse AOX3 aldehyde oxidase (mAOX3, encoded by one of four aldehyde oxidase genes in mouse) has recently been reported [70]. As expected, the overall protein architecture generally closely resembles that of previously described members of this family. Surprisingly, given that all the eukaryotic aldehyde-oxidizing enzymes are obligatory oxidases and unable to use $NAD^+$, the configuration of its FAD-binding domain superimposes with the dehydrogenase rather than oxidase configuration of the bovine enzyme (Fig. 8.7). It appears that the lack of reactivity of mAOX3 toward $NAD^+$ is due to the absence of a critical FFP(T)G(S)YR sequence in its residues 396–402, which is known to be important in interacting with the cofactor [71] (unfortunately, these residues are in an unresolved loop of the mAOX3 structure). At the active site, the conserved phenylalanine residues that define the substrate-binding site adjacent to the molybdenum center and the catalytic glutamate occupy essentially identical positions in the mouse AOX3 and the bovine enzyme (Fig. 8.8). Residues that are not conserved include Glu802/Ala807, Arg880/Tyr885 (a methionine in most other aldehyde oxidases), His884/Lys889, and Leu1014/Tyr1019.

In the context of the overall folds of the aldehyde-oxidizing enzymes from *D. gigas* and mouse, it is interesting to note how the FAD-binding domain has been incorporated into the chain trace of the bacterial enzyme in creating the eukaryotic form of the enzyme. Although portions of the two linker regions that connect the C-terminus of the second iron-sulfur domain to the N-terminus of the FAD domain of the mouse aldehyde oxidase and the C-terminus of the FAD domain to the N-terminus of the molybdenum binding portion of the protein are not fully resolved in the structure of the murine (or bovine) enzyme, it is clear that the first of these linkers passes in front of the iron-sulfur domains as shown in Fig. 8.9, and after tracing out the entirety of the FAD domain, the second loops behind the iron-sulfur domains (completing a fifth strand of β sheet in the second of the two molybdenum domains along the way) to connect with the amino terminus of the molybdenum-binding portion of the protein. In the *D. gigas* enzyme, the single 20-amino acid linker that connects the C-terminus of the second iron-sulfur domain with the N-terminus of the first molybdenum domain spans some 25 Å on the surface of the protein, but lies in front of the iron-sulfur domains as shown. From a

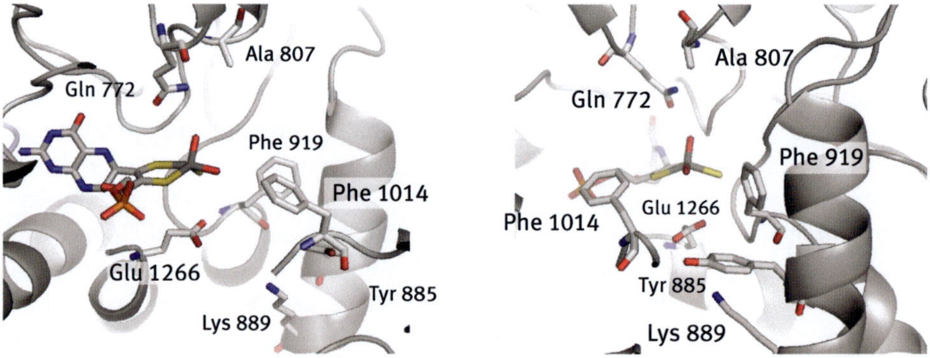


**Fig. 8.8:** The active site of mouse aldehyde oxidase 3 (PDB 3ZYV). Residues in common with the bovine xanthine oxidase include Phe914/919 (bovine/murine numbering), Phe1009/1014, Glu1261/1266, and Gln767/772. Amino acid residues that differ include Glu802/Ala807, Arg880/Tyr885 (a methionine in most other aldehyde oxidases), His884/Lys889, and Leu1014/Tyr1019. Compare with Fig. 8.4.

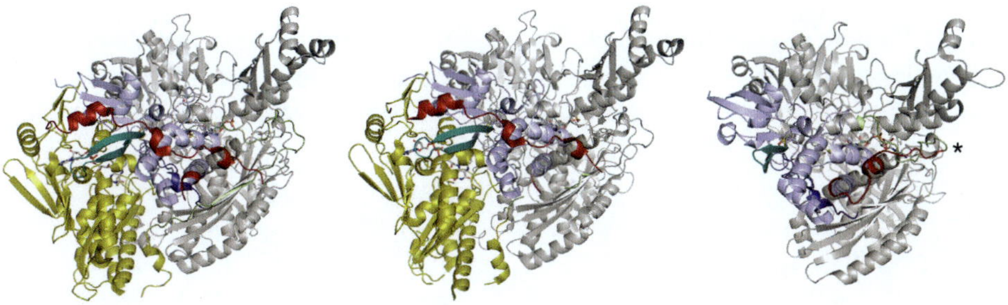

**Fig. 8.9:** A comparison of the polypeptide trace in bovine xanthine dehydrogenase (PDB F1O4), mouse aldehyde oxidase (PDB 3ZYV) and *D. gigas* aldehyde oxidoreductase (PDB 1VLB). The iron-sulfur domains (of one subunit each of the homodimers) are in blue, the FAD domains (when present) in yellow, and the molybdenum domains in gray. The linker between the Fe-S and FAD domains in the first two structures are in red, and the linker between the FAD and Mo domains in green. In the bacterial enzyme at right, the single linker between the Fe-S and the Mo domains is in red and green, with the approximate point of insertion of the FAD domain indicated by an asterisk (far right). The β-turn of the first Fe-S domain that is elongated in the eukaryotic enzymes is shown in teal.

comparison of the two structures, the apparent point at which the FAD domain seen in the eukaryotic enzymes is inserted in the bacterial sequence can be identified as being approximately in the middle of the single prokaryotic linker, as indicated by the asterisk in Fig. 8.9. The point of insertion lies on the opposite side of the two Fe-S domains from the bulk of the FAD domain; thus, although the domains are laid out Fe/S II-Fe/S I-FAD-Mo in the primary sequence, their physical disposition in the protein structure is FAD-Fe/S II-Fe/S I-Mo. The FAD domain of the eukaryotic enzymes has extensive interactions with the Fe-S and Mo domains, including an elongated β turn that protrudes from the first Fe-S domain as compared with the shorter β turn seen in the bacterial enzyme (Fig. 8.9, teal).

Higher plants also encode multiple aldehyde oxidases [72], including enzymes involved in biosynthesis of the crucial plant hormones abscissic acid and indole-3-acetic acid (both reactions involving the oxidative hydroxylation of the respective aldehyde to the carboxylic acid). *Arabidopsis thaliana* has four aldehyde oxidase genes, *AAO1–4* [73, 74], with *AOX1* having a preference for indole-3-acetaldehyde [75] and *AOX3* for abscissic aldehyde [73, 76–78]. The *A. thaliana* AAO1 and AAO3 enzymes have recently been heterologously expressed in *Pichia pastoris* [77], with both isozymes exhibiting the characteristic UV-vis absorption spectra of all members of the xanthine oxidase family, with a broad absorption maximum at ~450 nm and a shoulder at 550 nm and are inhibited by cyanide by removal of the catalytically essential Mo=S ligand. Like the mammalian enzymes [79–82], both *A. thaliana* isozymes generate $O_2{\bullet}^-$ as well $H_2O_2$, which has been implicated in the enzyme's physiological role [77].

The reaction mechanism for aldehyde oxidases, regardless of origin, is believed to involve the same base-assisted proton abstraction from the equatorial Mo-OH group to initiate nucleophilic attack as seen with the xanthine-utilizing enzymes [11]. Concomitant hydride transfer of the aldehyde hydrogen to the Mo=S group occurs through a tetrahedral transition state [83] in which the C-O bond of product is largely formed and the C-H bond of substrate largely broken.

### 8.2.4 CO dehydrogenase

A great many enzymes of the xanthine oxidase family have been biochemically characterized to greater or lesser degrees. Although most of these are likely to be very similar to those already described, several are known to have one or more atypical characteristics that are significant. The first of these to be considered here is the CO dehydrogenase from carboxydotrophic bacteria such as *O. carboxidovorans*. These organisms are aerobes able to grow with CO as sole source of both carbon and energy [84] and are responsible for the annual clearance of ~$2 \times 10^8$ metric tons of CO from the environment [85, 86]. A molybdenum-containing CO dehydrogenase catalyzes the critical first step in this process, the oxidation of CO to $CO_2$ [87], with the reducing equivalents thus generated ultimately being passed on to a CO-insensitive terminal oxidase [88]. A portion of the $CO_2$ thus generated is subsequently fixed non-photosynthetically *via* the reductive pentose phosphate pathway [87]. The Mo-containing CO dehydrogenase from *O. carboxidovorans* and related organisms is distinct from the highly $O_2$-sensitive Ni/Fe-containing CO dehydrogenase from obligate anaerobes such as *Clostridum thermoaceticum* or *Methanosarcina barkerii* [89].

Crystal structures for the CO dehydrogenases from both *O. carboxidovorans* [90] and *Hydrogenophaga pseudoflava* [91] have been reported and are found to be virtually identical. The *O. carboxidovorans* enzyme has a small subunit (CoxS; 18 kDa), with two [2Fe-2S] iron-sulfur clusters, a medium subunit (CoxM; 30 kDa) that possesses FAD, and a large subunit (CoxL; 89 kDa) that has the active site molybdenum center. Each subunit has considerable sequence and structural homology to the corresponding parts of bovine xanthine oxidoreductase (although as noted earlier, the FAD-Fe-S-Mo order of the *coxMSL* genes in the operon differs from the Fe-S-FAD-Mo order seen in the primary sequence of the eukaryotic enzymes). Significantly, the active site of CO dehydrogenase is not a mononuclear molybdenum center but rather a binuclear Mo/Cu center with the structure shown in Fig. 8.10 [92, 93]. The molybdenum has the square-pyramidal coordination geometry seen in other members of this enzyme family, with an apical Mo=O and an equatorial plane consisting of two sulfurs from the pyranopterin cofactor (present as the dinucleotide of cytosine). The remainder of the equatorial plane, however, includes a μ-sulfido bridge to the $Cu^I$ instead of the Mo=S found in other members of this family and a second Mo=O [94] in place of the catalytically labile Mo-OH. The $Cu^I$ ion is also coordinated by Cys388, and a water/hydroxide ligand at a distance of 2.40 Å. The Mo-μS-Cu bond angle is 113°, and the μS-Cu-S(Cys) bond angle is 156°.

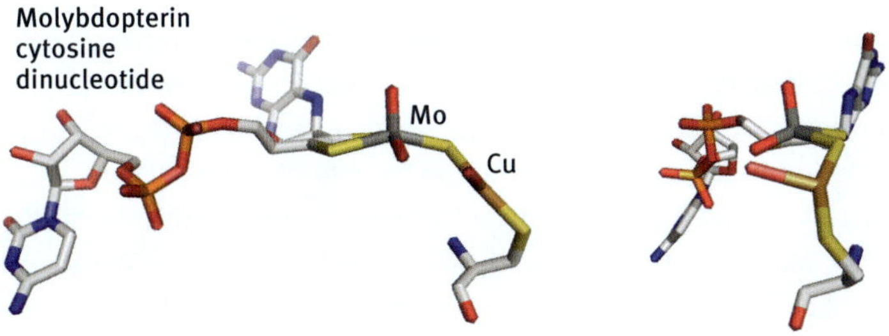


**Fig. 8.10:** The structure of the binuclear $Mo^{VI}/Cu^{I}$ center of CO dehydrogenase. The perspective at the right is rotated approximately 90° about the vertical from that at the left.

CO dehydrogenase is reduced by CO under pseudo first-order conditions with $k_{red}$ = 51 $s^{-1}$ at 25°C [94]. The rate constant is independent of [CO], reflecting a $K_d$ smaller than the ~30-μM lower limit of [CO] that is experimentally accessible and also independent of pH, indicating that there is no acid-base catalysis involved in the rate-limiting step of the reaction. In the course of reaction with CO, an EPR signal clearly attributable to the Mo/Cu binuclear center accumulates (Fig. 8.11, with $g_{1,2,3}$ = 2.0010, 1.9604, and 1.9549 and extremely large hyperfine coupling to the naturally abundant $^{63,65}$Cu nuclei ($I$ = 3/2), with $A_{1,2,3}$ = 117, 164, and 132 MHz [94]. This EPR signal is unchanged on preparation of the sample in $D_2O$, but some line broadening is observed when $^{13}$CO is used as substrate [94].

The $^{13}$C coupling in the EPR-active form of the binuclear cluster of substrate-reduced CO dehydrogenase has been examined by ENDOR spectroscopy [95]. The key observation is that the $^{13}$C hyperfine coupling is essentially entirely isotropic ($a_{iso}$ = 17.3 MHz), which is inconsistent with any structure where there is a direct Mo-C

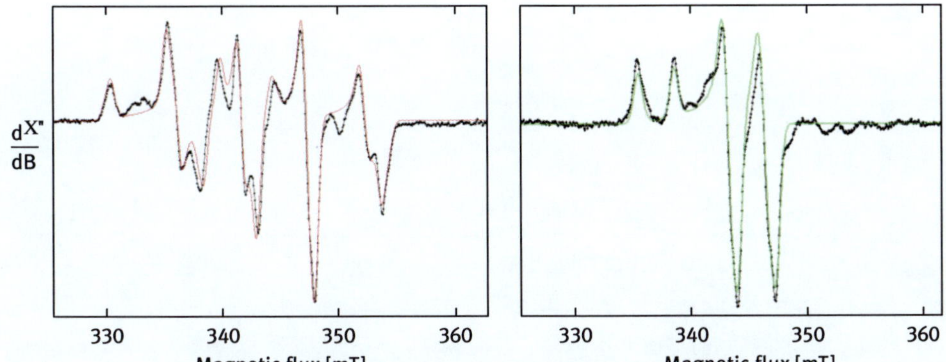


**Fig. 8.11:** EPR of native Mo/Cu CO dehydrogenase (left) and the Ag-substituted enzyme (right). The spectra are in black, and simulations are in color. For the Mo/Cu enzyme, the parameters used were $g_{1,2,3}$ = 2.0010, 1.9604, and 1.9549 and $A_{1,2,3}$ = 117, 164, and 132 MHz; for the Mo/Ag form, $g_{1,2,3}$ = 2.0043, 1.9595, and 1.9540 and $A_{1,2,3}$ = 82.0, 78.9, and 81.9 MHz.

bond or one with only a single atom intervening between the molybdenum and the carbon. The conclusion is that the signal arises from a $Mo^{V}/Cu^{I}$ species having CO bound at the copper of the binuclear center and represents a paramagnetic analogue of the $Mo^{VI}/Cu^{I}$•CO Michaelis complex in the lower mechanism of Fig. 8.12.

A very high-resolution (1.1-Å) structure of CO dehydrogenase in complex with the inhibitor *n*-butylisonitrile has been reported [92], in which the inhibitor is seen to have inserted across the μS-Cu bond (Fig. 8.12, left). A mechanism has been proposed in which CO similarly inserts itself between the bridging sulfur and copper of the binuclear center in the course of the reaction to yield a bridging thiocarbamate and the reduced molybdenum (Fig. 8.12, top right). The thiocarbamate then hydrolyzes with regeneration of the sulfur bridge. In contrast, computational studies have suggested an alternate reaction path, involving nucleophilic attack by the equatorial Mo=O on an initial Cu•CO complex, followed by formation of $CO_2$, and formal reduction of the binuclear cluster (Fig. 8.12, bottom right) [96, 97]. The final step of this alternate mechanism involves reducing equivalents nominally entering the (predominantly Mo-based) redox-active orbital *via* the copper.

A model for the binuclear active site of CO dehydrogenase has recently been synthesized and shown to exhibit very similar EPR characteristics to the enzyme, in particular, the extremely strong Cu hyperfine [98]. Analysis of this model indicates that the redox-active (singly occupied) molecular orbital is extremely delocalized

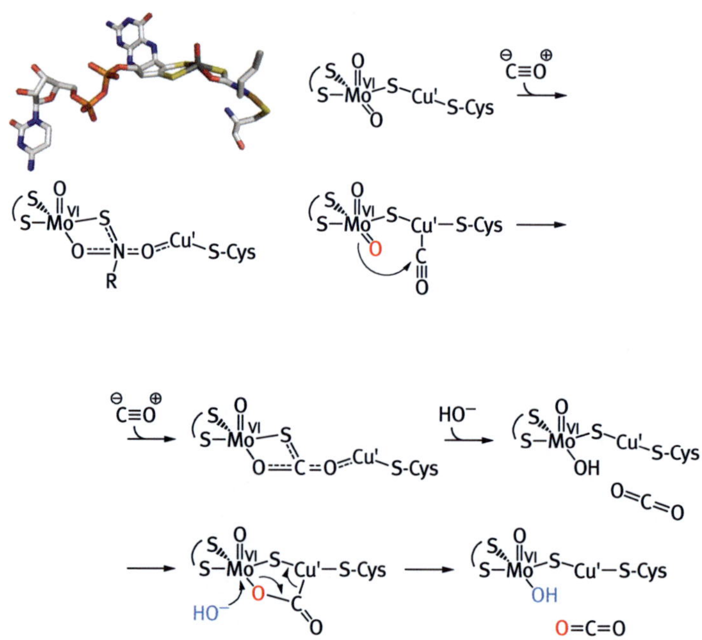

**Fig. 8.12:** Possible reaction mechanisms for CO dehydrogenase.

over the Mo-S-Cu unit and consists of 44% Mo $d_{xy}$ character, with 25% S $p$ character and 21% Cu $d_z2/d_{xy}$ character (along with an undefined amount of Cu $s$ character) [98]. The copper and bridging sulfur thus appear to have the effect of extending the redox-active orbital spatially a considerable distance from the molybdenum, making it possible for the molybdenum to become reduced in the final step depicted in Fig. 8.13, bottom. The binuclear cluster thus appears to be constructed to (1) create a substrate-binding site adjacent to the molybdenum that activates CO for nucleophilic attack and (2) at the same time extend the redox-active Mo $d_{xy}$ orbital such that it can accept an electron pair in the course of the reaction at the more remote site.

The bridging sulfur and copper of CO dehydrogenase can be removed by reaction of the enzyme with cyanide, and a reconstitution protocol has been developed that involves treatment with $Cu^{I}$•thiourea [99]. When the silver salt is used instead, activity is partially recovered [100]. The silver-substituted enzyme thus reactivated is reduced by CO under pseudo first-order conditions with a rate constant of 8.1 $s^{-1}$ (as compared with 51 $s^{-1}$ for the as-isolated enzyme [94]). Significantly, the EPR signal seen upon partial reduction of the enzyme by CO (Fig. 8.11, right) shows the doublets expected for substitution of Ag for Cu ($I$ = 1/2 for the naturally occurring $^{103,105}Ag$), with $g_{1,2,3}$ = 2.043, 1.9595, and 1.9540 (very similar the values seen with the as-isolated enzyme) and $A_{1,2,3}$ = 82.0, 78.9, and 81.9 MHz.

Several quinone species are able to effectively oxidize reduced CO dehydrogenase, and the physiological oxidant for CO dehydrogenase is most likely ubiquinone [101]; the reoxidation reaction occurs at the FAD site, as expected. Quinones are unusual physiological oxidants for this family of enzymes, and an examination of the overall

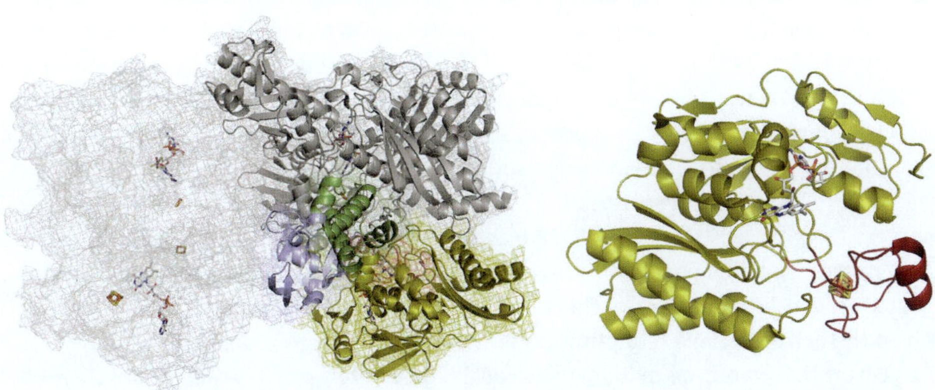

**Fig. 8.13:** The structure of 4-hydroxybenzoyl-CoA reductase from *T. aromatica* (PDB 1RM6). (left) The iron-sulfur- FAD- and molybdenum-containing subunits of the dimer are color-coded as in Fig. 8.3, with the [4Fe-4S]-containing insert in the FAD subunit in red (at the back of the subunit). (right) A closeup of the FAD-containing subunit, more clearly showing the [4Fe-4S]-containing insert.

fold of the FAD-containing domain of CO dehydrogenase indicates that it resembles the dehydrogenase rather than oxidase form of the bovine xanthine oxidoreductase, particularly with regard to the mobile loop referred to above (Fig. 8.7).

### 8.2.5 4-Hydroxybenzoyl-CoA reductase

4-Hydroxybenzoyl-CoA reductase from the archaeon *T. aromatica* [102] catalyzes the reductive dehydroxylation of substrate to benzoyl-CoA, a key metabolic intermediate that is subsequently reductively dearomatized by benzoyl-CoA reductase prior to further degradation. The enzyme plays a critical role in the metabolism of phenolic compounds in this and related obligate anaerobes, which lack the $O_2$-utilizing monooxygenase and dioxygenase generally used by aerobes to cleave the aromatic ring. The reaction is formally the reverse of that catalyzed by xanthine oxidoreductase, with the reducing equivalents required for the reaction provided by a 2*x*[4Fe-4S] bacterial ferredoxin [103]. As shown in Fig. 8.13, the archaeal 4-hydroxybenzoyl-CoA reductase is an $(\alpha\beta\gamma)_2$ hexamer, with separate molybdenum-, FAD-, and 2*x*[2Fe-2S]-containing subunits and an overall protein fold that closely resembles that of other members of this enzyme family [104]. The pyranopterin cofactor of the molybdenum center is present as the dinucleotide of cytosine. 4-Hydroxybenzoyl-CoA reductase is unique, in that it has an additional [4Fe-4S] cluster located in a 41-amino acid insert to the FAD subunit and which is the presumed point of entry of reducing equivalents from ferredoxin [104]. This additional [4Fe-4S] cluster lies unusually far from the isoalloxazine ring of the FAD, 16.5 Å, but this appears to be compensated for by an unusually high peptide packing density and an essentially direct covalent link from Cys122 (coordinating one of the iron atoms of the cluster) through Arg121 to Phe233, which π-stacks onto the *si* face of the isoalloxazine ring [104]. Electron transfer to the FAD is thus likely to be sufficiently fast as to not be rate-limiting to turnover. The reduction potentials of the several redox-active centers of 4-hydroxybenzoyl-CoA reductase have been determined [105], with unusually low potentials seen for the FAD ($\Delta E_{FAD/FADH\bullet} = -250$ mV, $\Delta E_{FADH\bullet/FADH2} = -470$ mV) and molybdenum center ($\Delta E_{Mo^{VI/V}} = -380$ mV, $\Delta E_{Mo^{VI/V}} = -500$ mV) and substantially higher reduction potentials for the two [2Fe-2S] centers (−205 and −255 mV for Fe/S I and Fe/S II, respectively); the [4Fe-4S] cluster also possesses a low potential (−465 mV). Electron flow is in the reverse direction of that seen in other members of this family but is thermodynamically favorable overall given the extremely low reduction potential of the donor ferredoxin.

Given that xanthine oxidase reaction involves obligatory two-electron chemistry [106] and that the enzyme is known catalyze the dehydroxylation of uric acid to xanthine under strongly reducing conditions [107], the reaction mechanism for 4-hydroxybenzoyl-CoA reductase might be thought simply to be the reverse of the hydroxylation pathway for xanthine oxidase, with hydride transfer from an equatorial Mo-SH to C-4 of molybdenum-coordinated substrate, followed directly by dehydroxylation and rearomatization. It has instead been proposed, however, that the enzyme operates *via* a radical-based Birch-like mechanism [105], in which a first reducing equivalent is added to molybdenum-coordinated substrate, followed by protonation at C-4 of substrate and addition of a second reducing equivalent, leading to dehydroxylation and rearomatization. It remains for future work to distinguish between these mechanistic possibilities.

## 8.3 The DMSO reductase family

As mentioned in Section 8.1, the family of molybdenum-containing enzymes epitomized by the DMSO reductase from *Rhodobacter sphaeroides* (the first enzyme of the group to be structurally characterized) is extremely diverse in several regards. In terms of the structure of the molybdenum center, although these enzymes are all thought to have an $L_2MoY(X)$ structure in the oxidized state, there is considerable variability as to the identity of both the amino acid residue X that coordinates the molybdenum and the terminal Mo=Y ligand. The nature of the reaction catalyzed also varies considerably – although most catalyze *bona fide* oxygen atom transfer reactions, others catalyze oxidation/reduction or even hydroxylation/hydration reactions. Finally, the overall architecture, subunit makeup, and constitution of redox-active centers is remarkably diverse. As has been pointed out elsewhere [1], members of this family of molybdenum-containing enzymes are structurally related to the aldehyde:ferredoxin oxidoreductase family of tungsten-containing enzymes, and at least some members are able to accept either metal.

Representative overall architectures of members of the DMSO reductase family of enzymes are illustrated in Fig. 8.14, which illustrates the diversity as regards both subunit constitution and complement of redox-active cofactors seen in these enzymes. The simplest among them are monomeric proteins with an $L_2Mo^{VI}OX$ active site as their sole redox-active center, as exemplified by the periplasmic DMSO reductases from *R. sphaeroides* and *R. capsulatus* (with X = O-Ser and Y = O) [108–111]. Next in complexity are enzymes such as the periplasmic formate dehydrogenase H from *E. coli* [112], which is a monomer with a [4Fe-4S] cluster in addition to the molybdenum center, followed by enzymes such as arsenite oxidase from *Alcaligenes faecalis*, which has a first subunit with the molybdenum center and an iron-sulfur cluster (a [3Fe-4S] cluster in this case), and a second with a Rieske [2Fe-2S] cluster [113]. Similarly, the periplasmic (Nap) and cytoplasmic (Nas) assimilatory nitrate reductases may be either monomers or heterodimers, although they are usually heterodimers with a molybdenum- and [4Fe-4S]-containing subunit (and in the case of the heterodimers a diheme subunit as well) [114]. Greater in complexity are integral membrane enzymes such as the DMSO reductase [115, 116], NarGHI nitrate reductase [117, 118], and formate dehydrogenase N [119] from *E. coli*, which are heterotrimers with multiple redox-active centers in addition to their molybdenum centers. These enzymes may have their catalytic molybdenum-containing subunits localized either in the periplasm (e.g. formate

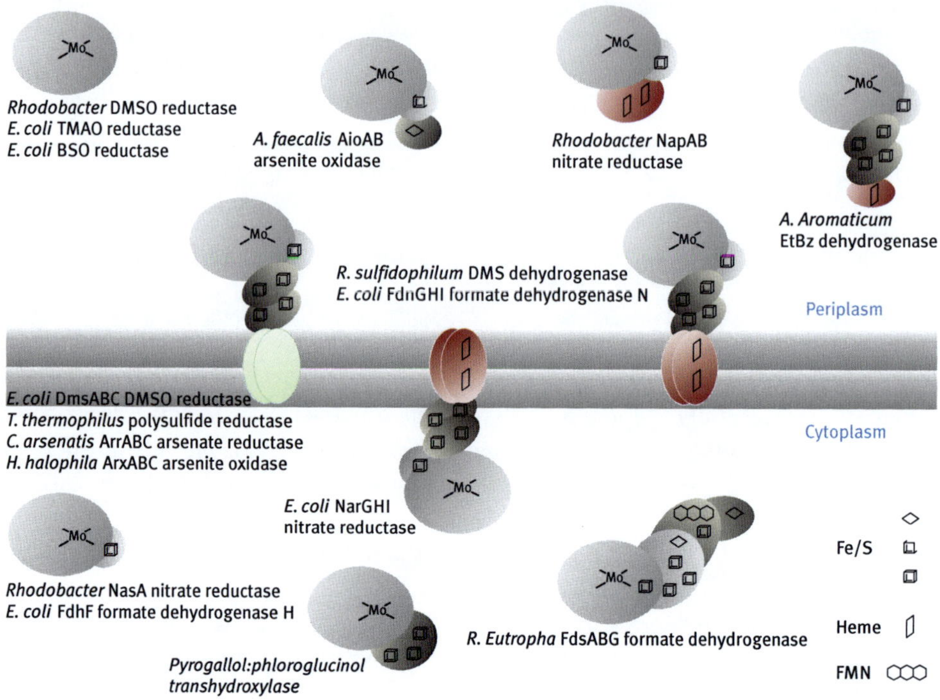


**Fig. 8.14:** Examples of protein architectures seen for members of the DMSO reductase family of enzymes. A key as to the type of redox-active center found in each protein is at bottom right.

dehydrogenase N) or cytosol (the Nar-type nitrate reductases). Finally, most complex are the soluble, cytoplasmic formate dehydrogenases from aerobic bacteria, which are heterotrimers or heterotetramers with a molybdenum center, a minimum of seven iron-sulfur clusters and FMN. As will be discussed in Section 8.3.4.3, these last enzymes are related structurally to NADH dehydrogenases and NiFe hydrogenases.

Several genomic comparisons have been made of DMSO reductase family members, based on the amino acid sequence of either the molybdenum-containing subunit or (in the case of the more complex enzymes) the polyferredoxin subunit, that reveal relationships among its members that might not necessarily have been expected from an examination of Fig. 8.14 [120, 121]. One major clade consists, as expected, of the simple monomeric enzymes having a molybdenum center as their sole redox-active group, but a second includes the Nap and Nas nitrate reductases as well as the FdhF formate dehydrogenase H and arsenite oxidase. A third consists of two subgroups containing enzymes such as polysulfide reductase, and the *E. coli* DmsABC DMSO reductase on the one hand and the NarGHI nitrate reductase, ethylbenzene dehydrogenase, and DMS dehydrogenase on the other hand. The polysulfide reductase clade is interesting in that it includes members from several different phyla of bacteria as well as archaea, a signature of it having evolved prior to the divergence of these groups [122]. The segregation appears to be most closely correlated with the nature of the additional redox-active centers found in these enzymes. Although there is often a correlation with the detailed structure of the molybdenum center, and in particular, the nature of the sixth ligand to the metal provided (usually) by the polypeptide, it is not the case that all members of a given clade possess the same type of ligand. The functionality and nature of the reaction catalyzed are least correlated – the *Rhodobacter* DorA and *E. coli* DmsABC DMSO vreductases are in separate clades, as are the NarGHI and Nap/Nas nitrate reductases; the *E. coli* DmsABC DMSO reductase and *Rhodovulum sulfidophilum* DdhABC DMS dehydrogenase (which catalyzes the opposite reaction) are in the same major clade but in separate subgroups within that clade. The following discussion is organized according to the reaction catalyzed, and the reader is asked to bear in mind that the enzymes under each heading are in fact quite distinct. Given the very large number of enzymes in this family, the discussion here focuses on members that are particularly well characterized from a biochemical standpoint. The interested reader is referred to other reviews dealing with other aspects of enzymes in the DMSO reductase family, particularly more detailed phylogenetic analyses, and discussion of enzymes such as chlorate reductase, selenate reductase, thiosulfate reductase, and tetrathionate reductase that are of considerable microbiological and/or genetic interest [121–123].

### 8.3.1 DMSO reductase and DMS dehydrogenase

#### 8.3.1.1 *Rhodobacter* DMSO reductases

Marine environments generate enormous amounts of DMSO from the degradation of dimethylsulfoniopropionate, the principal osmolyte in seaweeds and phytoplankton. When grown anaerobically on a highly reduced carbon source, and in the presence of DMSO, organisms such as *R. sphaeroides* and *R. capsulatus* express a periplasmically localized, monomeric DMSO reductase that functions as a dissimilatory terminal reductase, reducing DMSO to dimethylsulfide, using reducing equivalents obtaining from the quinone pool (*via* a pentaheme DorC [124]) but without contributing to the transmembrane proton gradient [125–127]. The enzymatically generated DMS is volatile and exchanges readily between ocean and atmosphere, where it plays a central role in cloud nucleation – its concentration has been directly correlated with the number of condensation nuclei in clouds [128]. The *Rhodobacter* DMSO reductases have molybdenum as their sole redox-active centers, and although they lack iron-sulfur clusters, they are mechanistically the best understood members of the family. They thus serve as a paradigm for other enzymes (e.g. the more complex DmsABC DMSO reductase from *E. coli*, considered below) that possess multiple iron-sulfur clusters, which justifies their coverage here.

The crystal structures of the DMSO reductases from both *R. sphaeroides* [110, 111] and *R. capsulatus* [129, 130] have been determined, and that for the *R. sphaeroides* enzyme is shown in Fig. 8.15. The oxidized active site has the $L_2MoO^{VI}$(O-Ser) coordination indicated in Fig. 8.1 [110], with a trigonal prismatic coordination geometry. Reduction results in loss of the Mo=O to give a square-pyramidal coordination geometry with the serine occupying the apical position [110]. The overall protein fold consists of four domains, with the polypeptide trace making multiple passes among the first three of these – only the fourth domain (Fig. 8.15, blue) consists of a contiguous stretch of polypeptide (the C-terminal 150 amino acid residues). As will be discussed further below, in all those proteins possessing an iron-sulfur cluster in their molybdenum-containing subunit, the Fe-S domain is a contiguous insert near the N-terminus, with the cluster in close proximity to the pyranopterin designated Q crystallographically (and hereafter referred to as the Q pterin), and interacting with primarily with the more N-terminal domain (Fig. 8.15, yellow); the other pyranopterin is termed P. (Confusingly, the two pyranopterins have sometimes referred to as the P, for "proximal" and D for "distal" with regard to the iron-sulfur cluster [121]; we will use the original crystallographic designation here, if only because some of these enzymes lack an iron-sulfur and the terms "proximal" and "distal" are not meaningful.) A wide funnel that provides substrate access to the catalytically labile Mo=O of the molybdenum coordination sphere, which has been shown to be transferred from substrate to reduced enzyme in the course of reaction [131]. A cluster of aromatic residues (Tyr165, Trp196, Tyr322, and Tyr360) at the bottom of the funnel constitute the substrate binding pocket. The fully functional enzyme has both equivalents of pyranopterin (present as the dinucleotide

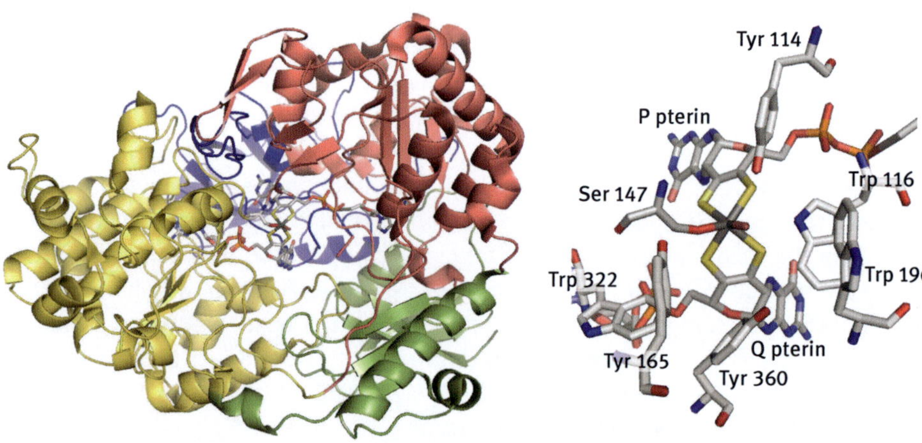


**Fig. 8.15:** The structure of *R. sphaeroides* DMSO reductase (PDB 1EU1). (left) The overall fold of the protein. The four domains of the polypeptide are color-coded, with those in yellow and red related by a pseudo 2-fold axis of symmetry. The Q pyranopterin is associated for the most part with the yellow domain; that designated P is associated with the red domain. (right) A closeup of the active, with active site residues discussed in the text indicated.

of guanine) coordinated to the metal, as shown in Fig. 8.15, but the Q pterin readily dissociates, being replaced by a second Mo=O group to give a catalytically inert from of the enzyme. Accumulation of this Q pterin-dissociated enzyme form caused some confusion in the earlier structural work with the enzyme, but a 1.3-Å-resolution structure of the *R. sphaeroides* enzyme eventually demonstrated the presence of both forms [111]. Conveniently, the Q pterin can be readily re-ligated to the molybdenum by redox cycling the enzyme (reducing with sodium dithionite followed by reoxidation with DMSO) to restore full catalytic activity [132].

A recent survey of crystal structures for enzymes of the DMSO reductase family has noted that the conformations of the P and Q pyranopterins are distinct, with the P pterin always seen in a flatter configuration suggesting that it is a dihydropterin (Fig. 8.16, far left, *top*), whereas the Q pterin always has a conformation closer to that for a tetrahydropterin (Fig. 8.16, far left, bottom) [133]. It has also been noted that it is the tetrahydro Q pterin specifically that is involved in mediating electron transfer into or out of the molybdenum center from/to nearby redox-active centers in members of both the xanthine oxidase and DMSO reductase families of molybdenum enzymes. Indeed, an analysis of the MCD of the paramagnetic "very rapid" species seen with xanthine oxidase has led to the conclusion that electron egress from the molybdenum center to the nearer Fe/S I of the enzyme principally involves $\sigma$, rather than $\pi$, interactions between the molybdenum and the pyranopterin [12], consistent with the pterin being in the tetrahydro oxidation state. Meanwhile, the dihydro-like P pterin in the *R. capsulatus* DMSO reductase has a configuration that more closely resembles that seen in the sulfite oxidase family of molybdenum enzymes. It has been pointed out previously [134] that although the molybdenum centers of the xanthine oxidase and sulfite oxidase families both have a square-pyramidal coordination geometry with an apical Mo=O as well as three sulfurs and an oxygen in the equatorial plane, the orientation of the molybdenum coordination sphere with respect to the pyranopterin cofactor is opposite in the two families: with the pyranopterin group oriented to the left of the metal as shown in Fig. 8.16, the apical Mo=O points *up* for all members of the xanthine oxidase family and *down* for all members of the sulfite oxidase family.

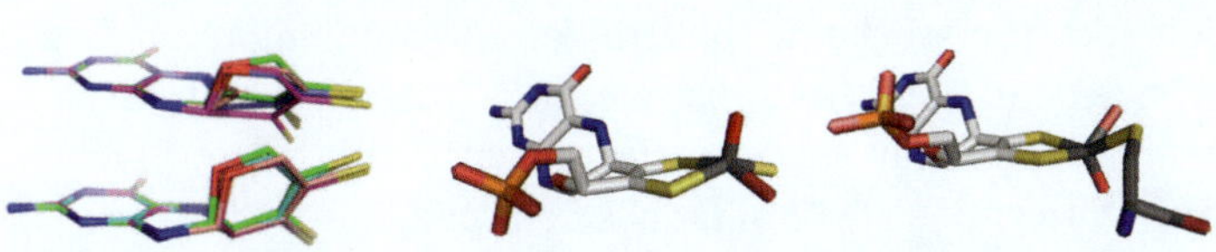

**Fig. 8.16:** A comparison of pyranopterin conformations in molybdenum enzymes. (left) A comparison of the extent of pyranopterin distortion in representative members of the P pterin (top) and Q pterin (bottom) of DMSO reductase family members. (After Rothery RA, Stein B, Solomonson M, Kirk ML, Weiner JH, *Proc Natl Acad Sci USA*, 109, 14773–8, 2012.) Center, the molybdenum center of bovine xanthine oxidase (PDB 1FO4), with the apical oxo group oriented up [135]. (right) The molybdenum center of chicken sulfite oxidase (PDB 1SOX) with the apical oxo group oriented down [136].

The configuration of the pyranopterin in any specific case appears to be dictated by steric and hydrogen-bonding interactions with the polypeptide, which is fixed in the case of each enzyme. It is thus unlikely that the pyranopterin itself is formally redox-active in any given enzyme, a conclusion also drawn from a consideration of the inherent chemistry of metallopterin models [137, 138].

The *Rhodobacter* DMSO reductase has also been examined by both X-ray absorption (XAS) [139] and resonance Raman [132] spectroscopy. The XAS analysis is fully consistent with the crystal structure, indicating a monooxo $Mo^{VI}$ center in oxidized enzyme and a *des*oxo $Mo^{IV}$ center (i.e., lacking a Mo=O group) in reduced. In the XAS data, collected at 10 K, the reduced enzyme shows evidence of a second O/N ligand at 2.16 Å in addition to the oxygen of Ser147 [139], a point considered further later in this section. The Raman work has included an analysis of oxidized and reduced enzyme and an $E_{red}$•DMSO complex formed by treating oxidized enzyme with DMS [132]. The Mo=O stretching mode of oxidized enzyme is seen at 862 $cm^{-1}$, and the oxygen shown to be catalytically labile as oxidation of reduced enzyme with $^{18}O$-labeled DMSO results in a 43-$cm^{-1}$ redshift in the Mo=O mode to 819 $cm^{-1}$. This mode is not seen with the *des*oxo reduced enzyme, as expected, but in the $E_{red}$•DMSO complex a new mode at 862 $cm^{-1}$ is observed that shifts only to 843 $cm^{-1}$ with $^{18}O$ and has been assigned to the S=O stretch of bound DMSO (the latter some 141 $cm^{-1}$ lower in frequency than in free DMSO, reflecting destabilization of the bond). The Mo-O(ser) stretching mode is at 536 $cm^{-1}$ in oxidized enzyme and 513 $cm^{-1}$ in reduced. A careful analysis of the enedithiolate stretching modes, including excitation profiles, clearly indicates that the two pyranopterins are not equivalent, with one being best represented as a discrete enedithiolate and the other as being highly $\pi$-delocalized. The latter more closely resembles that seen previously with sulfite oxidase [140] and has been assigned to the crystallographically identified P pterin, with the discrete enedithiolate being the Q pterin [132]. This assignment is entirely consistent with the structural survey discussed earlier.

Unlike the molybdenum centers of the xanthine oxidase [141] and sulfite oxidase [142, 143] families, that of the DMSO reductase family absorbs extensively throughout the visible and has pronounced spectral changes upon reduction [108]. These absorbance changes have been used to carry out enzyme-monitored turnover experiments in which the absorption spectrum of the *R. sphaeroides* enzyme is monitored as it turns over with either DMSO and TMAO as substrate [144]. The spectra observed in the course of turnover with DMSO can be fit as the weighted sum of four enzyme forms having well-defined absorption spectra: reduced enzyme, the reduced enzyme complexed with DMSO, oxidized enzyme, and the EPR-active "high-*g* split" $Mo^{V}$ form that is an intermediate in the re-reduction of the molybdenum center, as shown in Fig. 8.17. Spectral deconvolution permits the time courses for each of the several catalytically relevant species to be determined "on the fly." Product DMS that accumulates in the course of turnover with DMSO can rebind to the oxidized enzyme and back the reaction up to the $E_{red}$•DMSO species, resulting in a significant decrease in catalytic throughput (Fig. 8.17, right). Such product inhibition does not occur with

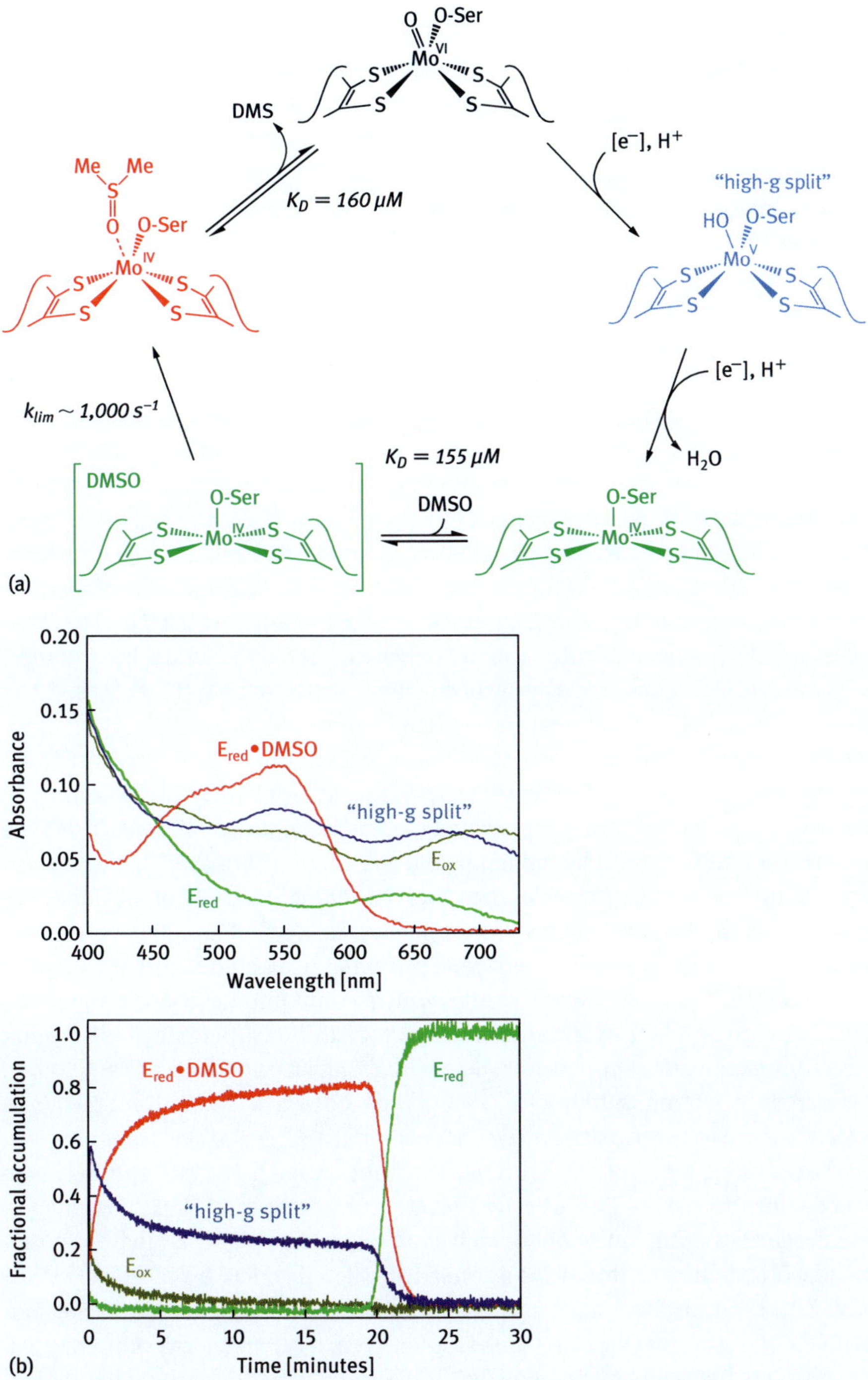


**Fig. 8.17:** The catalytic cycle of DMSO reductase. (left) The overall catalytic cycle. (upper right) The deconvoluted absorption spectra for each of the four spectroscopically distinct species. (lower right) Right, The time course for each species in the course of turnover with DMSO as substrate.

TMAO as substrate, and the predominant species that accumulates in the steady state with this substrate is the $Mo^{V}$ species, whose reduction on to the fully reduced $Mo^{IV}$ species is rate-limiting under the reaction conditions [144].

Mutagenesis studies of Tyr14 [145, 146] and Trp116 [146] of the *R. capsulatus* DMSO reductase have probed the catalytic roles of these residues. Trp116 is within hydrogen-bonding distance of the Mo=O of oxidized enzyme, and Tyr114 (which is a valine in the otherwise closely related TMAO reductase from *Shewanella massilia*) has been proposed to hydrogen bond to the oxygen of DMSO in the course of its reaction with reduced enzyme [145, 147]. The Y114F mutant is in fact somewhat faster in steady-state assays than wild-type enzyme but has a substantially higher $K_m$ for DMSO [145]. The $E_{red}$•DMSO intermediate is considerably less stable in the mutant (as expected given the role of Tyr114 in binding to the substrate S=O group), breaking down faster and being formed to a significantly lesser degree upon rebinding of DMS than seen with wild-type enzyme. As a result, the $E_{red}$•DMSO species does not accumulate as much with the mutant as with wild-type enzyme (the behavior of the mutant with DMSO as substrate in fact resembles that of wild-type enzyme with TMAO, reflecting compromised substrate specificity in the mutant). The Y114F mutant exhibits essentially the same absorption spectrum as wild-type enzyme in the fully oxidized and reduced states and manifests a normal "high-*g* split" EPR signal in the $Mo^{V}$ state, but the spectrum of the $E_{red}$•DMSO complex is perturbed, again consistent with the proposed role of Tyr114 in interacting with substrate. The W116F mutant reacts with DMSO at approximately the same rate as wild-type enzyme but is particularly prone to Q pterin dissociation from the molybdenum. The as-isolated W116F mutant has the Q pterin essentially completely dissociated, but like the wild type, the enzyme can be converted to the functional six-coordinate form by reduction and reoxidation with DMSO; it is this six-coordinate form that is responsible for the observed catalytic activity of the mutant in steady-state assays [146]. Trp116 thus plays a major role in stabilizing the structure of the molybdenum center of wild-type enzyme, presumably by minimizing the binding of water to the reduced molybdenum center, which would initiate dissociation of the Q pterin. The redox-cycled, oxidized W116F mutant exhibits a perturbed absorption spectrum compared with wild-type enzyme, with a long-wavelength band at 680 nm, as compared with 720 nm. Addition of DMS to the oxidized W116F mutant results in conversion to the five-coordinate species rather than the $E_{red}$•DMSO complex.

The essentially quantitative accumulation of the paramagnetic $Mo^{V}$ species in the course of the turnover of the *Rhodobacter* DMSO reductase with TMAO, in conjunction with the absence of other chromophores in the protein, has also enabled an examination the molybdenum center of the enzyme by magnetic circular dichroism (MCD) [148] and XAS [149]. In the MCD work, the two absorption maxima seen at 667 and 540 nm in the $Mo^{V}$ species (Fig. 8.17) are assigned to six discrete but overlapping electronic transitions between specific pairs of frontier molecular orbitals of the molybdenum center. Most of the absorption above 400 nm is due to four specific ligand-to-metal charge-transfer transitions involving one-electron promotion from either of the two highest doubly-occupied orbitals (both of which are principally enedithiolate in character) to the lowest-lying unoccupied orbitals (principally Mo d,p in character). Although the two highest-energy doubly occupied frontier orbitals are approximately equally distributed over the enedithiolates of the P and Q pterins, the same is not true of the two lowest-lying unoccupied orbitals: although these are approximately isoenergetic, the lower of the two is principally P pterin in character, whereas the next lowest is principally Q. Somewhat surprisingly, the XAS results suggest that although the signal-giving species has a *bis*enedithiolate-coordinated molybdenum, Ser147 has dissociated from the metal and been replaced by a Mo-OH ligand. The signal-giving species has long been associated with functional enzyme and is an unquestionable part of the catalytic cycle (in the case of the XAS work, the sample was generated by catalytic turnover), but it is possible that using sodium dithionite as a nonphysiological reductant (in a reaction that is quite sluggish) leads to a different form of the partially reduced enzyme than seen with the physiological DorC cytochrome. It is certainly the case, however, that the fully reduced form of the functional enzyme has Ser147 (re)coordinated to the molybdenum.

Studies of $L_2MoO(OR)$ and $L_2WO(OR)$ compounds have shown that the preferred coordination geometry in the oxidized state is octahedral, rather than the trigonal prismatic geometry seen in the oxidized *R. sphaeroides* DMSO reductase [150–152]. The most important difference is the dihedral angle between the two enedithiolate ligands, being 90° in the model compounds and ~0° in the enzyme. This implies that the polypeptide imposes significant structural constraints on the metal center and imposes a geometry closer to that seen in both the reduced models and enzyme (where the dihedral angle in the square-pyramidal coordination geometry is near 0°). The observed geometry of the oxidized enzyme appears to represent an entatic state that has the effect of labilizing the oxo group for dissociation upon reduction, a conclusion supported by the above MCD study of the partially reduced $Mo^V$ state. A recent XAS/density functional study of a $[Mo^{IV}(OSi)(bdt)_2]/Mo^{VI}O(OSi)(bdt)_2]$ system has shown that the reduced form of the model passes through the same discrete complex with DMSO as seen with the enzyme [153], but this is not inconsistent with the above interpretation because the geometric constraints imposed by the polypeptide in the enzyme-catalyzed reaction are manifested principally in the oxidized rather than reduced state. There are several other important conclusions arising from this study: (1) in agreement with the above MCD study of the enzyme [148], the redox-active orbital is predominantly Mo d in character and the enedithiolate ligands to the metal are "innocent" (i.e. are not directly involved in the oxidation-reduction chemistry of the reaction); (2) oxygen atom transfer is a concerted process wherein lengthening of the S-O bond stabilizes the S-O p* orbital, facilitating concomitant electron transfer from the molybdenum; and (3) the enedithiolate ligands directly stabilize a singlet rather than triple ground state in the reduced complex, which facilitates the overall two-electron reaction. In all, the mechanistic enzymology and model compound work have led to a very clear picture of the chemical course of the reaction catalyzed by DMSO reductase, with chemistry likely to be directly relevant to other oxotransferase members of the DMSO reductase family. Thus, although the *Rhodobacter* DMSO reductase lacks iron-sulfur clusters, it serves as the mechanistic paradigm for the several enzymes considered in the following sections that have an abundance of these clusters.

#### 8.3.1.2 *E. coli* DMSO reductase

As indicated in Fig. 8.14, the DMSO reductase from *E. coli*, encoded by the *dmsABC* gene cluster, is minimally an αβγ heterotrimer (and probably organized into higher order as seen in the polysulfide reductase, NarGHI nitrate reductase, and FdnGHI formate dehydrogenase systems considered in Section 8.3.4) and as such is much more complex than the *Rhodobacter* enzymes [154–156]. The enzyme is a terminal respiratory oxidase, using reducing equivalents from the quinone pool to reduce nitrate to nitrite, contributing to the transmembrane proton gradient in the process through the vectoral release of protons from menaquinol to the periplasm. Even the molybdenum-containing DmsA subunit of the *E. coli* enzyme is more complex in that it has an N-terminal insert that likely possesses a [4Fe-4S] cluster, designated FS0, in addition to its $MGD_2Mo^{VI}(O\text{-}Ser_{176})$ molybdenum center. Although an EPR signal attributable to this cluster has not been observed, several lines of evidence point to its existence: (1) the presence of a cluster of four cysteine residues in the amino terminal domain of DmsA (Cys34, Cys38, Cys42, and Cys75) that is highly homologous to those found in other enzymes (of this family) that been shown crystallographically to coordinate [4Fe-4S] clusters; (2) mutation of Cys38 in this group of cysteines to Ser or Ala results in the appearance of a [3Fe-4S] cluster in DmsA [157]; and (3) the observation of spin-spin interaction between an iron-sulfur cluster in DmsABC with its molybdenum center that is perturbed in the DmsA C38S mutant [158]. DmsA also has an N-terminal *tat* signal sequence and, like the *Rhodobacter* enzymes, is translocated to the periplasm *via* the TAT system [159, 160]. Given the close similarity in amino acid sequence between the molybdenum-binding portion of DmsA and the *Rhodobacter* DMSO reductases, the overall chemistry of oxygen atom transfer is expected to be the same. Indeed, XAS analysis of the *E. coli* enzyme indicates that its the molybdenum center is essentially identical to that seen in the *Rhodobacter* enzymes (with the notable difference that there is no indication that the Q pterin of the *E. coli* enzyme tends to dissociate from the molybdenum) [161]. Like the *Rhodobacter* enzymes, the *E. coli* DMSO reductase is able to reduce a wide range of S- and N-oxides.

DmsB contains four [4Fe-4S] clusters and is co-translated to the periplasm with DmsA after maturation in the cytosol. Based on sequence homology to proteins of known crystal structure, the subunit is believed to consist of a tandem repeat of a 2 × [4Fe-4S] bacterial ferredoxin. As indicated in Fig. 8.18, coordination of the clusters involves pairs of four-cysteine tetrads in which the fourth Cys in each is swapped out

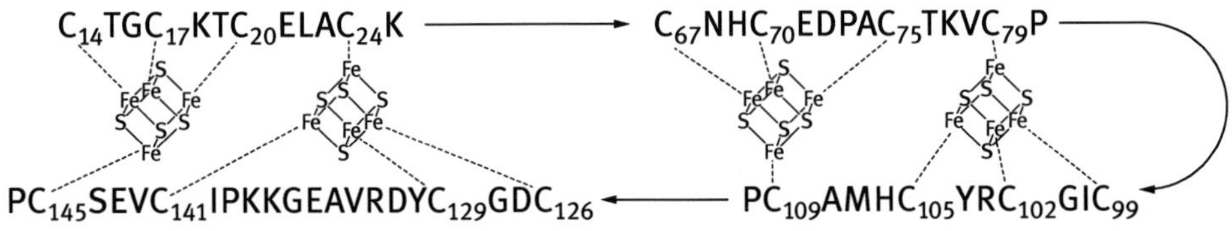


**Fig. 8.18:** Cysteine coordination of the four [4Fe-4S] clusters of DmsB.

and coordinates the cluster associated with the other tetrad in the pair. The arrangement with the first and fourth cysteine tetrads forming one pair of clusters and the second and third tetrads the other is highly conserved in a large group of iron-sulfur containing enzymes, including NADH dehydrogenase and NiFe hydrogenases in addition to the molybdenum-containing enzymes considered here [121, 162]. As will become clear in considering those enzymes that have been crystallographically characterized, the [4Fe-4S] clusters, designated FS1-FS4, are coordinated by tetrads I, IV, II, and III, respectively, and are arranged approximately linearly in the structure of the subunit. FS1 is expected to lie closest to the putative FS0 of the DmsA subunit, and FS4 closest to the DmsC subunit. Like the all remaining DMSO reductase family members considered in subsequent sections, the *E. coli* DMSO reductase is a representative of the "complex iron-sulfur molybdoenzyme" subfamily whose phylogenetic relationships have recently been reviewed [121].

DmsC is an eight-helix membrane-integral anchor subunit that although it has no hemes or other redox-active prosthetic groups, it has a menaquinol-binding site consisting minimally of His65 [116, 163] and Glu87 [164] as defined by mutagenesis studies. Unlike the *Rhodobacter* enzymes, which simply dissipate "extra" reducing equivalents under certain growth conditions, the *E. coli* enzyme contributes to the membrane protonmotive force *via* the vectoral generation of protons in the periplasm in the course of menaquinol oxidation on the periplasmic side of the subunit [154, 156]. The overall structure of DmsC is thought to closely resemble the PsrC subunit of polysulfide reductase, to which it is closely related (see Section 8.3.2).

Although the FS0 cluster of native DmsA has proven refractory by EPR, the FS1-FS4 clusters of DmsB are all EPR detectable, making it possible to determine their reduction potentials [165]. The reduction potentials for the purified protein at pH 6.8 are −240, −330, −120, and −50 mV for FS0-FS4, respectively [116, 121, 157, 158, 165]. The highest-potential FS4 is the only center for which a cleanly resolved EPR signal is observed, with $g_{1,2,3}$ = 2.01, 1.93, and 1.866, but the overall EPR results are consistent with the presence four [4Fe-4S] clusters [165]. As with the bacterial ferredoxins, there is clear evidence of spin-spin interactions among the several centers as they become progressively reduced, complicating the assignment of discrete features in the observed EPR to specific clusters. Although the clusters do not appear to be arranged in uniform thermodynamic order for electron transfer into the molybdenum center in the course of catalysis given the structure for the closely related *T. thermophilus* polysulfide reductase (Section 8.3.2), it is likely that they are sufficiently close together that electron transfer among them is rapid compared with turnover (with electron distributions among the several redox-active centers in partially reduced enzyme best treated using a rapid equilibrium model [37]). The $Mo^{VI/V}$ and $Mo^{V/IV}$ potentials are quite condition-sensitive but are in the range of +0 and –140 mV *vs* SHE, indicating considerable stabilization of the paramagnetic $Mo^{V}$ state; the observed signal [165] is of the type subsequently categorized by Bray and colleagues as “high-*g* unsplit” [166], with $g_{1,2,3}$ = 1.987, 1.976, and 1.960 and no evidence of coupling to protons.

A protein film voltammetric (PFV) study of *E. coli* DMSO reductase has revealed tunnel diode behavior in the enzyme, with the catalytic wave passing through a maximum as the poised potential is increased, and catalytic velocity decreasing beyond this point despite the stronger thermodynamic driving force [167]. The range of poised system potential for optimal catalytic throughput coincides with that for maximum accumulation of the $Mo^{V}$ state. It has been suggested that this behavior is due to protonation events at the molybdenum center (which are know to occur upon reduction [168]), with the kinetics of protonation of the $Mo^{V}$ state being faster than for the $Mo^{IV}$ state [167]. The model not only accounts for the diode-like behavior of the enzyme at high pH but also for the loss of this behavior at lower pH. It should be borne in mind, however, that reduction of immobilized enzyme on the surface of an electrode may well be considerably different than reduction with the enzyme’s physiological reductant, menaquinol, and it is unknown at present whether the enzyme manifests such behavior during turnover with menaquinol.

#### 8.3.1.3 *R. sulfidophilum* DMS dehydrogenase

The purple photosynthetic bacterium *R. sulfidophilum* encodes a heterotrimeric DdhABC dimethylsulfide dehydrogenase that physiologically catalyzes the reverse of the reaction catalyzed by the *E. coli* DMSO reductase, the oxidation of DMS to DMSO, with the reducing equivalents thus obtained going into the quinone pool [169–171]. The *ddhABDC* operon also encodes a DdhD gene product has high sequence homology to TorD, which is known to be a chaperone involved in the maturation of TAT-targeted members of the DMSO reductase family that are localized in the periplasm. Thus, like the *E. coli* enzyme, the DdhAB catalytic core is localized in the periplasm but is membrane-anchored by the DdhC subunit. A detailed analysis of the predicted amino acid sequences of DdhABC indicates that it is more closely related to the *E. coli* NarGHI nitrate reductase (Section 8.3.5.1) and *A. aromaticum* ethylbenzene dehydrogenase (Section 8.3.7) than to the *E. coli* DMSO reductase, however. The differences include an aspartate ligand to the molybdenum a histidine-coordinate [4Fe-4S] cluster in DdhA, a distal [3Fe-4S] (rather than [4Fe-4S]) cluster in the DdhB subunit and the presence of a *b*-type cytochrome in DdhC (coordinated by His81 and Met147) [169]. DdhABC exhibits three distinct $Mo^{V}$ EPR signals at 120 K, attributed to Mo-OH, Mo-$OH_2$, and Mo-X species (with X being an anion such as chloride). The aquo species has $g_{1,2,3}$ = 1.9650, 1.9846, and 2.0006, with coupling to two equivalent protons (4 × $10^{-4}$ $cm^{-1}$), whereas the hydroxy species has $g_{1,2,3}$ = 1.9627, 1.98, and 1.9914 and coupling to three protons, two with stronger coupling comparable to that seen in the aquo species and one with weaker coupling. The anion-complexed species has $g_{1,2,3}$ = 1.9600, 1.9805, and 1.9989 and no proton coupling. These signals are most similar to the $Mo^{V}$ signals seen with the NarGHI nitrate reductase (Section 8.3.5.1), suggesting aspartate as the sixth ligand to the molybdenum coordination sphere [172]. The low-temperature EPR of DdhABC is consistent with the expected constitution of iron-sulfur centers, with a [3Fe-4S] signal in oxidized enzyme with $g_{1,2,3}$ = 1.9650, 1.9870, and 2.0180 and a low-spin *b*-type heme. Upon reduction, signals attributable to the three [4Fe-4S] clusters of DdhB are seen, but (as with the *E. coli* DmsABC), no signal attributable to an iron-sulfur cluster in DdhA is seen [173]. The EPR signal of the [3Fe-4S] FS4 in oxidized enzyme is well defined with $g_{1,2,3}$ = 2.02, 1.99, and 1.97; the signals attributable to FS1-FS4 are in the $g$ = 2.05–1.86 range, but are overlapping and complicated by spin-spin interactions [173]. The reduction potentials for DdhABC have been determined, with the enzyme being found (consistent with the reaction catalyzed) to operate in a quite high oxidation-reduction regime. The reduction potentials for the $Mo^{VI/V}$ and $Mo^{V/IV)}$ couples are +123 and +55 mV *vs* NHE; those for FS1-FS4 (FS4 being the [3Fe-4S] cluster) are +175, −337, +66, and +292 mV; and that for the *b*-type cytochrome in DdhC is +324 mV, reflecting the overall favorability of electron transfer from the molybdenum center to the heme [173]. Interestingly, and as discussed further below, FS2 also has an unusually low reduction potential in both NarGHI and ethylbenzene dehydrogenase.

### 8.3.2 Polysulfide reductase

Polysulfide reductases catalyze the respiratory reduction of inorganic sulfur $(S_n)^{2-}$ to sulfide and $(S_{n-1})^{2-}$, a critical reaction in the biogeochemical cycle of sulfur. The reaction is fundamentally different from the oxygen atom transfer reactions catalyzed by many members of the DMSO reductase family of enzymes, involving simple reduction of substrate to cleave the terminal S-S bond. Its overall subunit organization is very similar to that of the *E. coli* DmsABC DMSO reductase discussed in Section 3.1.2 (Fig. 8.14), including the presence of a cofactorless membrane-integral PsrC subunit [174]. The crystal structure for the enzyme from *Thermus thermophilus* has been reported, and as shown in Fig. 8.19, the enzyme is organized as an $(\alpha\beta\gamma)_2$ heterotrimer with the catalytic PsrA subunit in the periplasm. The 2 equiv. of the pyranopterin cofactor in the molybdenum center of PsrA are present as the guanine dinucleotide. The molybdenum ligand corresponding to Ser176 in the *E. coli* DMSO reductase is Cys173 in the *Thermus* polysulfide reductase, and the additional covalency of the Mo-S bond serves to lengthen the Mo-O distance of the sixth ligand to 2.19 (±0.05) Å, indicating that the oxygen has been protonated at least to a hydroxide (and possibly water, but not a Mo=O as originally

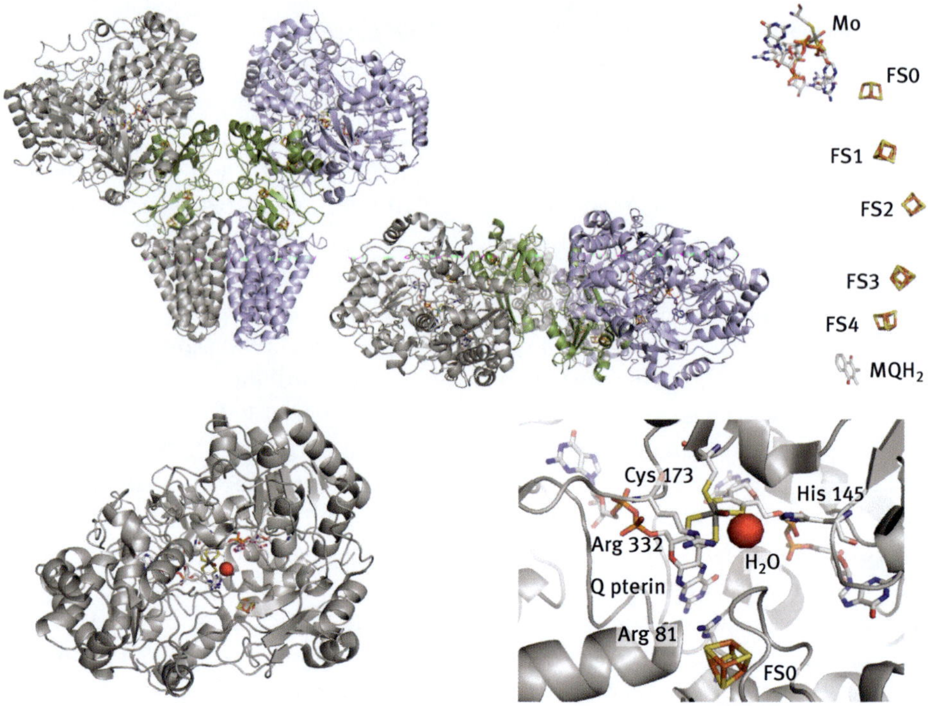


**Fig. 8.19:** The structure of the PsrABC polysulfide reductase from *T. thermophilus*. (top left and center) The overall organization of the subunits in the $(\alpha\beta\gamma)_2$ oligomer. One protomer is in gray and the other blue, with both PsrB subunits in green. (top right) An enlargement of the electron transfer chain, from the molybdenum center (top) to menaquinol (bottom). The right-hand protomer is colored PsrA in blue, PsrB in gray, and PsrC in red. (bottom left) The PsrA subunit, looking down the solvent access channel to the active site. (bottom right) The active site molybdenum center, with Cys173 coordinating the molybdenum and Arg81 intervening between the Q pterin and the FS0. Arg332 and His145 hydrogen bond to a bound water molecule (red sphere).

assigned crystallographically). As is found in all these more complex enzymes, the [4Fe-4S] FS0 cluster is adjacent to the Q pterin of the molybdenum center. Although the overall fold of the molybdenum-binding portion of the PsrA subunit closely resembles that first seen in the *R. sphaeroides* DMSO reductase, the broad funnel providing access to the active site is considerably more constricted in PsrA.

The PsrB subunit contains four [4Fe-4S] clusters as expected on the basis of sequence analysis, and these are organized as two pairs, each of which is similar to the eight-iron bacterial ferredoxins, despite FS1 and FS2 being in discontinuous strands of the polypeptide (Fig. 8.18). The arrangement of the four [4Fe-4S] clusters in PsrB also closely resembles that seen for the iron-sulfur subunits of most of the other molybdenum-containing systems considered in subsequent sections [175, 176] and also the NiFe hydrogenases [177] and NADH dehydrogenase [178–180]. As discussed in Section 8.3.5.3, it is evidently an evolutionarily ancient motif [121]. As discussed in

Section 8.3.1.2 in the case of the *E. coli* DMSO reductase, the cysteine clusters that coordinate these iron-sulfur clusters are organized I-IV from the amino terminus but, as illustrated in Fig. 8.19, are arranged FS1-FS4 progressively away from the FS0 cluster of PsrA with intervening distances are 9–11 Å. This same arrangement is seen in all homologues whose crystal structures have been determined (including the FdnHI formate dehydrogenase and NarGHI nitrate reductase discussed in Sections 8.3.4.2 and 8.3.5.3, respectively). The iron-sulfur cluster FS4 of PsrB lies at the subunit interface immediately adjacent to the membrane-integral PsrC subunit. A comparison of the structure of PsrB with the corresponding subunits of other enzymes considered in subsequent sections is shown in Fig. 8.20, with the conserved polyferredoxin core shown in blue. It can be seen that the layout of the several redox-active centers is highly conserved, but several types of inserts are seen (mostly involving interaction with other subunits of the trimeric protomer).

As shown in Fig. 8.21, the membrane-integral PsrC subunit has a core that consists of eight transmembrane helices arranged as a pair of four-helix bundles; the N-terminal 14 amino acid residues extend into the periplasm and span the length of PsrB. Although there are no redox-active centers in PsrC, menaquinol binds at a site within 9 Å of FS3 in PsrB. The generally hydrophobic binding pocket for menaquinol includes Tyr130 and His21. The role of the second four-helix bundle of PsrC, which is conserved in other cofactorless membrane subunits of similar enzymes, is suggested by a set of hydrophilic amino acid residues (from the cytosolic side of the subunit,

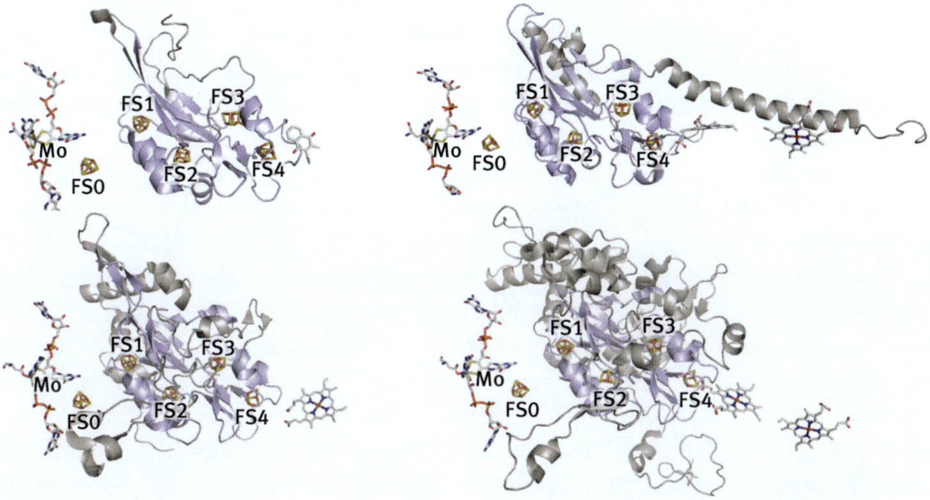


**Fig. 8.20:** A comparison of the iron-sulfur-containing subunits of polysulfide reductase (upper left), ethylbenzene dehydrogenase (lower left), formate dehydrogenase N (upper right), and the Nar nitrate reductase (lower right). The conserved polyferredoxin core of each subunit is indicated in blue. In the second and fourth enzymes, FS4 is a [3Fe-4S] cluster. Each is in a separate clade of these subunits [121]. For all these proteins, the intercluster distances are 9–11 Å edge to edge.

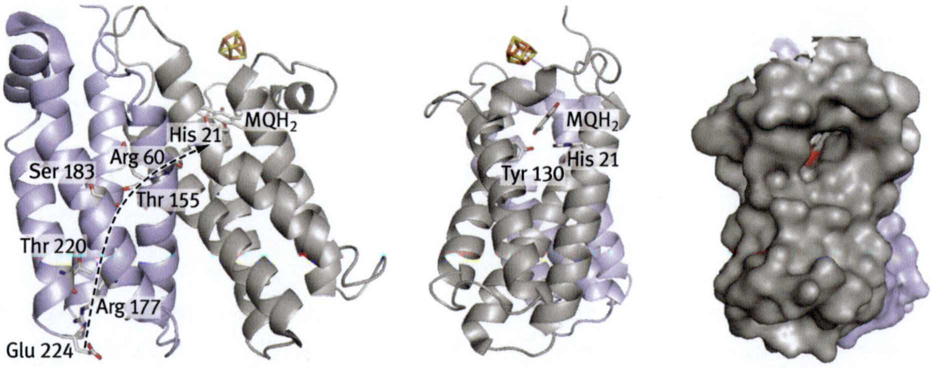


**Fig. 8.21:** The membrane-integral PsrC subunit of *T. thermophilus* polysulfide reductase. The N- and C-terminal four-helix bundles referred to in the text are shown in gray and blue, respectively, from the side (left) and end (center). (right) The channel providing menaquinol access to the binding site. The proximal FS3 of PsrB is also shown, indicating its proximity to the bound menaquinol. In the orientation on the left, a putative proton channel through the second four-helix bundle is indicated, involving Glu224, Arg177, Arg239, Thr220, Ser183, and Thr155, leading to Asp60 and His21 in the first bundle – which may be involved in proton pumping, indicated by the dashed arrow.

Glu224, Arg177, Arg239, Thr220, Ser183, Thr155, Arg60, and His21; Fig. 8.21, dashed arrow) that extends through the core to the menaquinone-binding site in the other bundle that may constitute part of a proton pump contributing to the transmembrane proton gradient [174]. With polysulfide reduction in the periplasm necessarily consuming protons, the loss of protons on oxidation of menaquinol to the periplasm would entail proton-neutral chemistry in the course of turnover, yet a net gain to the proton gradient of 0.5 $H^+/e^-$ during turnover of polysulfide reductase is reported [181, 182]. A proton-pumping role for the second four-helix bundle of PsrC would account for its strict conservation, and mutation of several of the proposed residues in the highly homologous *Wollinella succinogenes* polysulfide reductase results in a loss of activity of the enzyme [182]. Still, definitive evidence for the operation of such a pump is not presently available.

In the active site of the PsrA subunit, a second water molecule is found bound by Arg332 and His145, which constitute a putative substrate-binding site. A reaction mechanism has been proposed based on that of peroxidases (which catalyze similar chemistry in reducing peroxide to water, cleaving an O-O bond), with the terminal sulfur of substrate coordinating the reduced molybdenum center, displacing the bound water/hydroxide and placing the penultimate sulfur in the position of the crystallographically observed water between Arg332 and His145. This putative enzyme•substrate complex has molybdenum coordinated by six sulfur atoms, a structure for which there is chemical precedent [183]. Consistent with this, the *W. succinogenes* polysulfide reductase, which is essentially identical to the *T. thermophilus* enzyme, exhibits several $Mo^V$ signals with very high *g* values depending on conditions (e.g. $g_{1,2,3}$ = 2.0165, 2.0025, and 1.9874 for the "very high-*g* polysulfide" signal) that have been interpreted as reflecting a sulfur-saturated molybdenum coordination sphere [184].

### 8.3.3 Ethylbenzene dehydrogenase

*Aromatoleum aromaticum* (formerly *Azoarcus* strain EbN1) is a β-proteobacterium and an obligate anaerobe that grows on ethylbenzene (a major component of crude oil) as its sole carbon source. Ethylbenzene dehydrogenase, encoded by the *ebdABC* operon, catalyses the first step of this degradative pathway, the hydroxylation of ethylbenzene to (*S*)-1-phenylethanol. The reaction involves the hydroxylation of an unactivated aromatic hydrocarbon without utilizing $O_2$ (i.e. the enzyme is neither a monooxygenase nor a dioxygenase) [185, 186] and contrasts with the reaction catalyzed by enzymes of the xanthine oxidase family that hydroxylate only activated heterocycles or aldehydes. Ethylbenzene dehydrogenase acts on a wide range of alkylaromatics and alkylheterocyclics but requires an ethyl (or substituted ethyl) side chain; toluene and related compounds are inhibitors rather than substrates [187]. The enzyme also catalyzes the dehydrogenation of reduced, bicyclic aromatics such as indane to conjugated products (indene in the case of the reaction with indane), possibly by hydroxylating then dehydrating the substrate. The physiological electron acceptor for ethylbenzene dehydrogenase is unknown.

Ethylbenzene dehydrogenase is a soluble, periplasmically localized αβγ heterotrimer whose crystal structure has been determined. Its overall architecture (Fig. 8.22) generally resembles that of an individual protomer of PsrABC discussed in Section 8.3.2 [188]. The EbdA subunit has a *bis*pyranopterin active site (with the pterin present as the guanine dinucleotide) with Asp223 and an acetate ligand from the crystallization mother liquor coordinated to the (presumably reduced) molybdenum in a distorted trigonal prismatic coordination geometry. Asp223 is hydrogen-bonded to Lys450, an arrangement similar to the Arg-Asp predicted to be conserved in selenate reductase [189] and chlorate reductase [190] on the basis of sequence homology. Unusually, the P pyranopterin has a ring-opened pyran ring (as also seen in the NarGHI nitrate reductase, see Section 8.3.3.1). EbdA also has a [4Fe-4S] cluster proximal to the Q pyranopterin, as seen in the PsrA subunit of polysulfide reductase, but with a

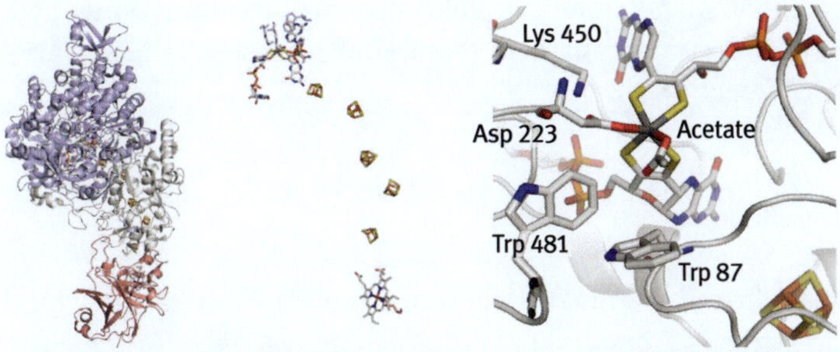


**Fig. 8.22:** The ethylbenzene dehydrogenase from *Aromatoleum aromaticum* (PDB 2IVF). (left) The overall structure of the heterotrimeric enzyme, with the EbdA, EbdB, and EbdC subunits in blue, gray, and red, respectively. (center) An enlargement of the enzyme's electron transfer chain, with the molybdenum center at top and the heme at bottom. (right) The active site of the enzyme, with residues referred to in the text indicated.

histidine ligand replacing one of the cysteine ligands to the cluster (another feature in common with NarGHI). The substrate access tunnel providing access to the active site is principally hydrophobic, as might be expected given the nature of the substrate. The EbdB subunit has three additional [4Fe-4S] clusters and one [3Fe-4S] cluster (with the last being distal to the EbdA subunit, again as seen in NarGHI). As in PsrB, the iron-sulfur clusters of EbdB are organized as two pairs, with a similar cysteine coordination scheme as found in DmsB and PsrB. The EbdC subunit of the soluble ethylbenzene dehydrogenase is distinct from the membrane-integral subunit of PsrABC, with a secondary structure dominated by two sandwiched five-stranded β sheets. It has a single *b*-type cytochrome, coordinated by Met108 and Lys201, in an otherwise very hydrophobic environment. The overall fold is related to the heme-binding domain of the flavocytochrome cellobiose dehydrogenase [191].

The product (*S*)-1-phenylethanol has been modeled into the active site of ethylbenzene dehydrogenase, with the catalytically introduced hydroxyl group coordinated to the molybdenum in place of the acetate ligand [188]. Assuming a Mo=O group in place of the bound acetate in oxidized enzyme, the reaction has been proposed to proceed with proton abstraction from the methyl group of substrate by Asp223 and concomitant hydride transfer to the Mo=O to give a $Mo^{IV}(OH)$ intermediate with a carbocation at C-2 of substrate [187, 188]. This subsequently breaks down by hydroxyl transfer from the molybdenum coordination sphere to give the hydroxylated product, with His192 facilitating the hydroxide transfer. Although an initial quantum mechanical study of the reaction mechanism explicitly comparing carbocation and radical mechanisms favored the carbocation-based mechanism [192], a more detailed density functional study has suggested that a two-step radical-based mechanism is in fact preferred. This would be consistent with the observed pH-dependent kinetic isotope effect of 3–10 (with the larger values observed at higher pH, where the initial hydrogen atom transfer from substrate to the molybdenum center becomes rate-limiting) with 2-$^2$H-ethylbenzene as substrate [193, 194]. The implication is that, in contrast to xanthine oxidase [106], C-H bond cleavage in the ethylbenzene dehydrogenase reaction seems to be homolytic rather than heterolytic. It thus appears that carbon center hydroxylation by members of the xanthine oxidase and DMSO reductase families takes place in fundamentally different ways.

### 8.3.4 Formate dehydrogenases

The bacterial formate dehydrogenases are distinct from the $NAD(P)^+$-dependent (and cofactorless) eukaryotic enzymes from yeasts and plants. *E. coli* encodes three different molybdenum-containing formate dehydrogenases: formate dehydrogenase H (FdhF, the product of the *fdhF* gene, but sometimes referred to in the literature as $FDH_H$ or, confusingly, FdhH), which is a part of the formate hydrogen:lyase complex [195]; formate dehydrogenase N (FdnGHI, or $FDH_N$), a product of the *fdnGHI* operon and expressed along with *narGHI* to form an anaerobic formate:nitrate respiratory chain [196]; and formate dehydrogenase O (FdoGHI, product of the *fdoDEGHI operon*), which is expressed in concert with *narZYX* to form a formate:nitrate respiratory system during the transition from aerobic to anaerobic growth. Each of these systems has broadly distributed cognates in other bacteria but share the common feature of being sensitive to inactivation by $O_2$. In addition, many aerobic bacteria encode a cytoplasmic $NAD^+$-dependent FdsABG(D) formate dehydrogenase, which, in contrast to the eukaryotic enzymes, has multiple redox-active centers, including a molybdenum center. In addition to their intrinsic biological interest, several of these last enzymes readily catalyze the reverse reaction, the reduction of $CO_2$ to formate, a reaction of potential industrial interest as a convenient storage form of $H_2$ [197–199].

#### 8.3.4.1 Formate dehydrogenase H

The first of the formate dehydrogenases to be structurally characterized was the 79-kDa FdhF from *E. coli*, which is localized in the cytoplasm. In the 2.9-Å structure (Fig. 8.23), the monomeric enzyme is seen to have an overall fold similar to the *T. thermophilus* PsrA subunit of polysulfide reductase discussed in Section 8.3.2 [112]. The active site was originally interpreted as having an $L_2Mo(OH)$(Se-Cys) center, with Sec140 in place of the serine seen in the DMSO reductases (the pyranopterin is again present as the guanine dinucleotide). However, a recent examination of the FdhD gene product has demonstrated that the protein is a sulfur transferase that inserts a cyanolyzable sulfur (derived from IscS) into the molybdenum coordination sphere of FdhF [200]. Given that FdhD is required for proper maturation of all formate dehydrogenases (at least in *E. coli*), it seems clear that all have a Mo=S ligand rather than Mo=O or Mo-OH as the sixth ligand to molybdenum in the oxidized enzyme.

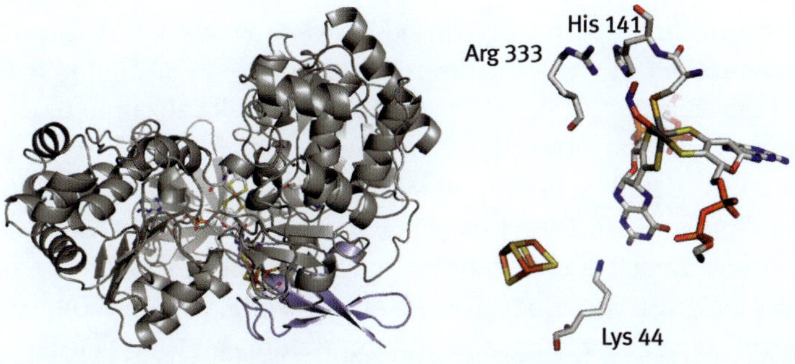


**Fig. 8.23:** The structure of *E. coli* formate dehydrogenase H (FdhF). (left) The overall protein fold, with the N-terminal [4Fe-4S]-containing domain in blue (1FDI). The orientation shown is approximately the same as that in Fig. 8.15 for the *R. sphaeroides* DMSO reductase. (right) A closeup of the active site, showing the inhibitor nitrite bound at the molybdenum, with Arg333 nearby.

As with all other enzyme of this family that contain a [4Fe-4S] cluster, that in FdhF is found adjacent to the Q pterin of the molybdenum center. The distal amino group of the Q pyranopterin of the molybdenum center is 6.4 Å from the nearest sulfur of the [4Fe-4S] cluster, with a highly conserved lysine residue (Lys44 in the *E. coli* FdhF) lying in between the two. Reduction of FdhF with formate to the $Mo^{IV}$ state results in loss of the terminal ligand to give a square-pyramidal $L_2Mo$(Se-Cys) center, analogous to the change in coordination geometry upon reduction first seen with the *R. sphaeroides* DMSO reductase [110]. A structure for the complex of oxidized enzyme with the inhibitor nitrite has also been obtained [112], with Arg333 interacting with the anion as shown in Fig. 8.23 (right). This complex provides a basis for modeling substrate formate into the active site, and doing so places the $C_\alpha$ hydrogen of formate within 1.5 Å of the selenium of Sec140. XAS analysis of FdhF largely was interpreted as largely confirming the molybdenum coordination sphere as defined in the original crystallographic work [201], but with four sulfurs already in the molybdenum coordination sphere, a fifth terminal Mo=S could easily have been missed. There is also evidence for at least a partial Se-S bond between the selenocysteine and a sulfur in the oxidized forms of both the *E. coli* [201] and *D. desulfuricans* enzymes [202], which, with the advantage of hindsight, is likely the terminal Mo=S rather than one of the enedithiolate sulfurs. At least with the *D. desulfuricans* enzyme, the Se-S distance increases from 2.12 in oxidized enzyme to 2.57 in reduced, suggesting some ligand-based oxidation-reduction chemistry in the molybdenum center, at least with dithionite as reductant (see later in this section). Mutation of Sec140 to Cys (U140C) results in a shortening of the Mo-O bond in oxidized enzyme, from 2.1 to 1.7 Å, consistent with deprotonation of a Mo-OH to a Mo=O [201]. This substitution reduced $k_{cat}$ by a factor of 100 in steady-state assays [203].

Reduction of FdhF by formate give rise to a $Mo^V$ EPR signal with $g_{1,2,3}$ = 2.094, 2.001, and 1.989 and a reduced [4Fe-4S] signal with $g_{1,2,3}$ = 2.045, 1.957, and 1.840 [204, 205]. The $Mo^V$ signal is observed at 120 K, whereas the [4Fe-4S] signal is seen only below 50 K; the former signal exhibits coupling to a proton with $A_{1,2,3}$ = 7.5, 19, and 21 MHz and, when generated in $^{77}$Se-labeled enzyme, exhibits very strong and anisotropic coupling to the $^{77}$Se nucleus ($I$ = 1/2) with $A_{1,2,3}$ = 13.2, 75, and 240 MHz. The magnitude of the coupling reflects considerable spin-delocalization of the unpaired electron onto the selenium [204, 206]. When deuterated formate is used to reduce the enzyme, the proton coupling in the $Mo^V$ signal is initially absent but grows in over 30–300 s, depending on the pH (exchange being slower at higher pH) [206]. On the basis of these observations, the signal-giving species has been interpreted as an $L_2Mo^V$(Se-Cys) species lacking a coordinated water or hydroxide, with the coupled (and substrate-derived, but solvent-exchangeable) proton residing on His141 [206].

On the basis of the crystal structure, particularly that of the nitrite complex with oxidized enzyme, a reaction mechanism has been proposed for FdhF that involves initial coordination of formate to the molybdenum, displacing the Mo-OH of oxidized enzyme [112]. Subsequent oxidation of formate to $CO_2$ may occur either by direct electron transfer to the molybdenum with protonation of His141 (consistent with the pH dependence of catalytic activity, reflecting general base catalysis with a p$K_a$ of ~7 [203]) or alternatively by direct hydride transfer involving Sec140. Consistent with this latter proposal, it has been shown that oxidation of formate to $CO_2$ does not entail incorporation of oxygen derived from solvent (as would be expected were the enzyme to oxidatively hydroxylate formate to bicarbonate, followed by dehydration) [206].

The electron density and chain trace of formate-reduced FdhF has recently been reassessed, and an alternate orientation identified for the critical loop containing Sec140. In this alternate configuration, Sec140 (which is unequivocally coordinated to the molybdenum in oxidized enzyme) is oriented away from the molybdenum center; it has also been suggested that the axial ligand to the molybdenum is Mo=S rather than Mo-OH [207]. With regard to the peptide loop in question (residues 138–147), the newly proposed orientation does indeed improve the crystallographic $R/R_{free}$ factor for this region of the polypeptide trace, albeit only modestly. In both the old and new structures, $R/R_{free}$ for the loop in question is substantially higher than for the structure overall, an inevitable consequence of the poorer quality of the electron density map in this region. Concerning the nature of the molybdenum ligand, absent Sec in the metal coordination sphere the electron density better refined using sulfur rather than oxygen, although at the resolution of the electron density (2.3 Å), the improvement in fit was not considered definitive [207]. Again, given the likely sulfurase role of the FdhD gene product, it is very likely that the terminal ligand to the molybdenum is indeed a Mo=S [200]. An alternate mechanism has been proposed based on this structural reinterpretation in which formate binds to the molybdenum and displaces the selenocysteine rather than the putative Mo-OH, with the now-dissociated Sec140 serving as a general base to abstract the $C_\alpha$ proton of formate. Such a mechanism is difficult to reconcile with the structure of the nitrite complex of FdhF [112], however, and is consistent with the available XAS data on FdhF only if it is assumed that dithionite- and formate-reduced FdhF have fundamentally different molybdenum centers. The poor electron density associated with the loop in the formate-reduced enzyme in fact suggests another interpretation: that the electron density is poor because the loop exists in at least two different configurations. The ambiguity is reminiscent of that seen in the crystal structure work with the *R. sphaeroides* DMSO reductase discussed above, where only quite high-resolution data was able to identify the presence of two alternate structures for the molybdenum center. If this is the case with FdhF, then the question is (as in the case with the DMSO reductase) whether there is a catalytically relevant conformational change or whether one or the other of the two configurations is not catalytically relevant. A third mechanism, also based on the alternate crystal structure involving Sec140 dissociation from the molybdenum and Mo=S coordination, has recently been considered in a density function theory-based computational study [208]. This mechanism involves formate coordination to the molybdenum, with insertion of the sulfur into the Se-Mo bond, to yield a Se-S-Mo moiety with the metal formally reduced to $Mo^{IV}$. The Se-S bond is then cleaved, leaving the selenate anion of

Sec140 hydrogen-bonded to His141, which then abstracts the $C_α$ proton from formate. This ultimately yields thioformate coordinated to the (reduced) molybdenum in a bidentate fashion, which then decays to a $Mo^{IV}$=S species and product $CO_2$. With $CO_2$ release, Sec140 then deprotonates and reorients the metal, with a transient Se-S bond formed in the process. It is evident that additional structural, and mechanistic work is needed to resolve the presently ambiguous situation.

Physiologically, as part of the formate:hydrogen lyase complex, FdhF passes the reducing equivalents obtained from oxidation of formate to one of two hydrogenases (Hyd3 or Hyd4, depending on growth conditions) that reduce protons to $H_2$ [195]. Hyd3 is encoded by the complex *hycABCDEFGHI* operon and consists minimally of the HycBCDEFG structural genes. At least, the HycB and HycF subunits possess iron-sulfur clusters that shuttle reducing equivalents to the NiFe active site in the HycE subunit. Hyd3 is localized on the cytosolic side of the cell membrane, and with FdhF in the periplasm, the formate:hydrogen lyase complex contributes to the transmembrane proton gradient by releasing protons to the periplasm and consuming them in the cytosol. Hyd4 is encoded by the similarly complex *hyfABCDEFGHIJR* operon, the enzyme itself being HyfACEFGHI with the active site Ni/Fe center present in the HyfG subunit. Hyd3 and Hyd4, along with the primary Hyd1 and Hyd2 hydrogenases of *E. coli* and other organisms, have been reviewed elsewhere [195].

#### 8.3.4.2 Formate dehydrogenases N and O

The *E. coli* formate dehydrogenase N (FdnGHI) is coexpressed with the NarGHI nitrate reductase (Section 8.3.5) to form a formate:nitrate oxidoreductase system under appropriate growth conditions (specifically, in the absence of $O_2$ and presence of formate and $NO_3^-$). Both enzymes have integral membrane diheme-containing subunits, and the menaquinone pool mediates electron transfer between them. Importantly, as indicated in Fig. 8.14, the catalytic subunits of FdnGHI and NarGHI lie on opposite sides the cell membrane, with NarGH being cytosolic [209] and FdnGH being periplasmic (by virtue of the presence of a *tat* signal sequence on FdnG) [119]. As a result, protons are generated in the periplasm with the oxidation of formate and consumed in the cytosol with the reduction of nitrate so that, like the formate:hydrogen lyase system, the FdnGHI:NarGHI complex contributes to the transmembrane proton gradient [210].

The 1.6-Å crystal structure of FdnGHI has been reported [119] and the enzyme is found to be considerably more complex than the monomeric FdhF. It is organized as an $(αβγ)_3$ trimer of trimers as shown in Fig. 8.24, with the overall organization of the αβγ protomer resembling that seen in PsrABC and EbdABC (despite these being organized as a dimer of trimers or monomer of trimer rather than a trimer of trimers). FdnG is some 200 residues longer than the FdnF formate dehydrogenase H, due in part to a small C-terminal addition that interacts extensively with the FdnH subunit. The substrate access channel is itself is more restricted than seen in the *Rhodobacter* DMSO reductases and is predominantlypositively charged. The molybdenum center of

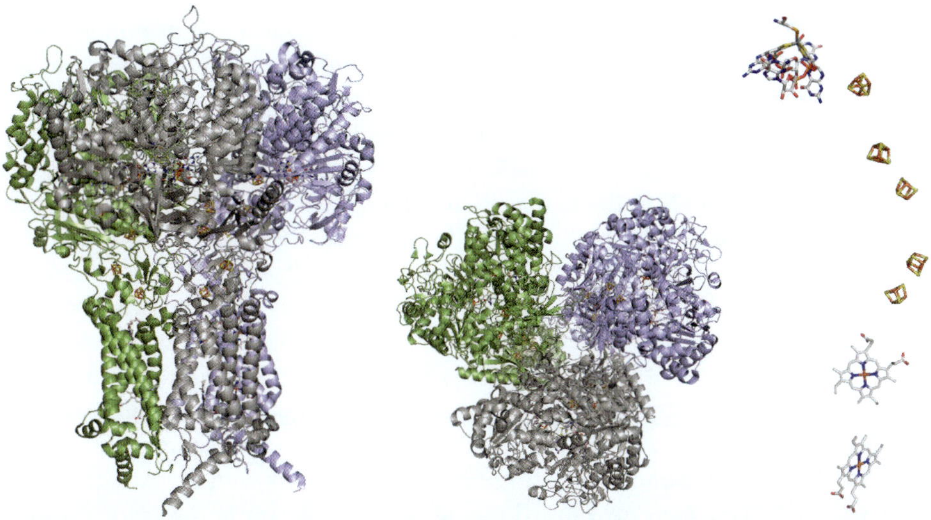

**Fig. 8.24:** The structure of *E. coli* formate dehydrogenase N (FdnGHI). (left) A side view from the (αβγ$_3$ protein illustrating its trimeric nature. A top view of the complex (center) and the arrangement of redox-active centers (right) in one αβγ protomer of the enzyme, illustrating the approximately linear electron transfer chain leading from the membrane-integral hemes at the bottom to the molybdenum center (site of nitrate reduction) at the top.

oxidized FdnG again has the guanine dinucleotide form of the pyranopterin cofactor, with the molybdenum additionally coordinated by Sec196. A sixth ligand in oxidized enzyme was initially modeled as a Mo-OH (at 2.2 Å), but in light of the recent work demonstrating that FdhD inserts a Mo=S group into the molybdenum centers of the formate dehydrogenases [200], the sixth ligand is most likely a terminal sulfido rather than a hydroxy ligand. Overall, the molybdenum coordination sphere closely resembles that seen in oxidized FdnF [112]. Given the overall hydrogen bonding network observed crystallographically in the vicinity of His197 of FdnG, the orientation of its imidazole ring is unambiguous, with its $N_{\delta 1}$ oriented toward the substrate-binding site. The opposite orientation was assigned in the original FdnF structure, but the FdnG structure clearly positions His197 appropriately to abstract the $C_\alpha$ proton of substrate in the course of formate oxidation [119].

As with the PsrB and EbdB subunits of polysulfide reductase and ethylbenzene dehydrogenase discussed above, the four [4Fe-4S] clusters FS1-FS4 of FdnH are arranged in pairs, each of which is similar to the eight-iron bacterial ferredoxins. Again, the arrangement of the four [4Fe-4S] clusters in FdnH also closely resembles that seen in other systems [175–180]. FdnH has a unique C-terminal transmembrane anchoring α helix not seen in otherwise similar systems that helps anchor it to the FdnI subunit and the membrane [119]. The FdnI membrane anchor has two *b*-type cytochromes embedded in a four-helix bundle, with a fifth C-terminal α helix lying approximately parallel to the surface of the membrane on the cytosolic side of the membrane. With the C-terminal membrane-integral helix of FdnH and 1 equiv. of cardiolipin at the subunit interface, the three FdnI subunits form a tightly packed trimer within the membrane that holds the $(\alpha\beta\gamma)_3$ structure together. The two cytochromes are stacked vertically in the center of the four-helix bundle at an angle of approximately 45° to one another, with one closer to the periplasm and FS4 of FdnH (heme $b_P$) and the other closer to the cytoplasm (heme $b_C$); the latter is the site of menaquinone reduction [119]. In both hemes, the planes of the two ligating histidine residues lie at approximately 45° to one another. Again, there is significant homology to the corresponding HyaC subunit of the *E. coli* [NiFe] hydrogenase [176]. The arrangement of the hemes in FdnI also resembles, at least superficially, that seen in the four-helix bundle core of the cytochrome $bc_1$ complex [211], although the ratio of proton translocation to electron transfer in FdnGHI indicates that it does not operate *via* a Q cycle [210, 212].

In addition to the FdnGHI nitrate reductase coexpressed with the NarGHI nitrate reductase, *E. coli* encodes an FdoGHI formate dehydrogenase in the *fdoGHI* operon (also referred to as FDH-Z or FDH-O) that is coexpressed with the NarZYX nitrate reductase in the transition from aerobic to anaerobic growth [213]. FdoGH has ~75% sequence identity to the corresponding FdnGH subunits, and FdoI has 45% identity with FdnI [214]. The key difference between the two systems appears to lie not in their structure or activity, but the conditions under which they are expressed. Specifically, FdnGHI is under the regulatory control of the FNR $O_2$ sensor and is not expressed in the presence of even low concentrations of $O_2$, whereas FdoGHI is not regulated by FNR. This arrangement allows *E. coli* to express a low level of nitrate reductase activity under aerobic conditions (in the presence of nitrate) during the transition from aerobic to anaerobic growth [213].

#### 8.3.4.3 NAD⁺-dependent bacterial formate dehydrogenases

In addition to the (generally) anaerobically expressed and $O_2$-sensitive formate dehydrogenases considered in Sections 8.3.4 through 8.3.4.2, a number of aerobic bacteria express an $O_2$-tolerant and cytoplasmic $NAD^+$-dependent formate dehydrogenase that has a molybdenum center as the active site and is distinct from both these extremely $O_2$-sensitive systems and the cofactor-less eukaryotic $NAD^+$-dependent formate dehydrogenases. In *Ralstonia eutropha*, for example, genetic analysis has predicted that the *fdsGBACD* operon encodes a complex trimeric FdsABG enzyme (sometimes referred to as S-FDH in the literature, for *soluble*) that contains a molybdenum center, multiple iron-sulfur clusters, and FMN, with each subunit bearing strong sequence homology to subunits of NADH dehydrogenase [215–217]. The 105-kDa FdsA subunit also exhibits a 51%–62% sequence similarity to the FdhF and FdnG catalytic subunits of the formate dehydrogenases discussed in the two previous sections, with Cys378 and His379 occupying the positions equivalent to Sec140 and His141 in FdnG. Coordination with Cys378 rather than a selenocysteine residue presumably contributes to the air stability of the molybdenum center. In addition, FdsA also has a 240-amino acid N-terminal extension that has homology to regions of the HoxU and HndD subunits of the *R. eutropha* NAD(P)$^+$-reducing hydrogenases and the C-terminal portion of the NuoG subunit of *E. coli* NADH dehydrogenase (equivalent to Nqo3 in the crystallographically characterized *T. thermophilus* NADH dehydrogenase [178, 180]). This N-terminal region of FdsA is predicted to have four [4Fe-4S] clusters and one [2Fe-2S] cluster; the final (C-terminal) [4Fe-4S] cluster in this N-terminal extension is conserved with NuoG/Nqo3 and occupies a position in the amino acid sequence of FdsA equivalent to FS0 in PsrA, FdnF, and FdnG [217]. The 55-kDa FdsB subunit of has some 45% sequence identity to the FMN- and [4Fe-4S]-containing NuoF subunit of *E. coli* NADH dehydrogenase (Nqo1 of the *Thermus* enzyme) and HoxF of NAD(P)$^+$-dependent hydrogenase, and includes the NAD$^+$-binding site. The 19-kDa FdsG subunit has 34% sequence identity to the NuoE subunit of NADH dehydrogenase (Nqo2 of the *Thermus* enzyme), and is predicted to have a [2Fe-2S] cluster. FdsC and FdsD are not part of the holoenzyme enzyme but appear to be involved in its maturation, possibly being involved in insertion of a Mo=S into the molybdenum coordination sphere as several of the NAD$^+$-dependent formate dehydrogenases have been shown to contain a cyanolyzable sulfur that is presumably a Mo=S group such as found in xanthine oxidase and related enzymes [215, 218, 219].

The NAD$^+$-dependent formate dehydrogenases exhibit the spectroscopic properties expected for the constitution of redox-active centers predicted on the basis of the above sequence analysis. The *R. eutropha* [215] and *M. trichosporium* [218] enzymes absorb throughout the visible, with absorption maxima (or well-resolved shoulders) at 450 nm, which is indicative of the FMN cofactor. The *M. trichosporium* enzyme, which is organized as $(\alpha\beta\delta\gamma)_2$, also exhibits at least five readily resolvable EPR signals attributable to at least one [2Fe-2S] and four [4Fe-4S] clusters, with evidence of magnetic interactions among them such as seen (not surprisingly) in NADH dehydrogenase [218]. A $Mo^V$ signal with $g_{1,2,3}$ = 2.005, 1.091, and 1.984 and evident $^{95,97}Mo$ hyperfine is also observed, but no FMN• signal is seen (simply indicating that the FMN/FMNH• couple has a lower potential than the FMNH•/$FMNH_2$ couple).

Based on the observation that the C-terminal [4Fe-4S] cluster (N7) of subunit NuoG/Nqo3 of the bacterial NADH dehydrogenases is equivalent to FS0 in FdsA [217] and the observation of both sequence [220] and structural [178] similarities of the (cofactorless) C-terminal region of Nqo3 of the *T. thermophilus* NADH dehydrogenase [178, 180] to *bis*MGD molybdenum enzymes, a model of FdsABC can be assembled that is based on the structures of Nqo1-3 of the *Thermus thermophilus* NADH dehydrogenase [178, 180] and FdhF [112], in which the respective N7 cluster and FS0 domains of each are simply overlaid, as shown in Fig. 8.25. Proper orientation is assured by including the entire C-terminal domain of Nqo3, which, as noted previously [121, 178, 180] and illustrated in Fig. 8.25 (lower left), bears a strong structural homology to the molybdenum-binding portion of FdhF. It is noteworthy that cluster N7 of the *T. thermophilus* Nqo3 is some 20 Å removed from the nearest iron-sulfur cluster (N4), but that the alignment with FdsA provides clear evidence that an intervening [4Fe-4S] cluster is present in FdsA; the approximate position of this additional cluster in the structure of Nqo3 is defined by a residual helix-turn motif, which, in Nqo3, has only a single cysteine providing the final ligand to the N4 [4Fe-4S] cluster, but FdsA clearly has a complete cysteine tetrad for coordination of a [4Fe-4S] cluster. The position of this motif, reflecting the position of the additional iron-sulfur cluster that is present in FdsA, is indicated by the brown sphere in Fig. 8.25, lower right. It is thus evident that FdsABG possesses a fully functional electron transfer pathways between the molybdenum and FMN. The homology between FdsA and Nqo3 extends to the presence of a histidine ligand to the N5 [4Fe-4S] cluster (at extreme bottom in Fig. 8.25, lower left). The overall oblong shape of the FdsABG model is consistent with sedimentation studies of the closely related $NAD^+$-dependent formate dehydrogenase from *M. trichosporium*, indicating that the protein is distinctly nonspherical [218]. Although the detailed nature of the structure of FdsABG must await crystallographic work, the model shown in Fig. 8.25 provides important insight into the likely disposition of the several redox-active centers of the enzyme. The structural homologies that make such a model feasible underscore the fact that the $NAD^+$-dependent hydrogenases and NADH dehydrogenase have both evolved from a common ancestor that possessed a molybdenum center as well as a [4Fe-4S] cluster intervening between the crystallographically observed N7 and N4. Both of these sites were subsequently lost in the course of evolution, isolating the N7 cluster and resulting in the diversion of the electron transfer pathway in NADH dehydrogenase (Fig. 8.25, lower left).

Although little mechanistic work has been done with any soluble formate dehydrogenase, formate oxidation at the molybdenum center may proceed analogously to that seen FdhF (with all the issues discussed in Section 8.3.4.1 regarding structural ambiguities, further compounded by the caveat that the less covalent Mo-S-Cys as

**Fig. 8.25:** A model for the structure of the FdsABG formate dehydrogenases. The model was obtained by superimposing the FS0 [4Fe-4S] cluster of FdhF from *E. coli* (PDB 1AA6) with the N7 [4Fe-4S] cluster of the Nqo3 subunit of *T. thermophilus* NADH dehydrogenase (PDB 3IAM), with the Nqo1 and Nqo2 subunits (which have strong homologies to FdsB and G, respectively, included in the model. (upper left) The structures of FdhF (with the molybdenum-binding portion of the protein in gray) and Nqo1-3 (in yellow, gray, and green/red, respectively), with the putative overlap region in red from which the model was constructed. This region contains iron-sulfur cluster FS0 in FdhF, and N7 in the Nqo3 subunit of NADH dehydrogenase. (upper left) The model for FdsABG. (lower left) An overlay of the molybdenum-binding portion of FdhF (gray) with the C-terminal domain of Nqo3 (in blue, upper right), with the domains containing the FS0 and N7 [4Fe-4S] clusters again in red. The rms deviation is 2.7 Å over 428 $C_\alpha$ atoms [178]. (lower right) The disposition of the redox-active centers in the model, with the approximate position of the additional iron-sulfur cluster known to be present in the *R. eutropha* enzyme indicated by the brown sphere. The orientation of the overall complex is the same as in upper right, with the direction of electron transfer in the FdsABG formate dehydrogenase and NADH dehydrogenase indicated.

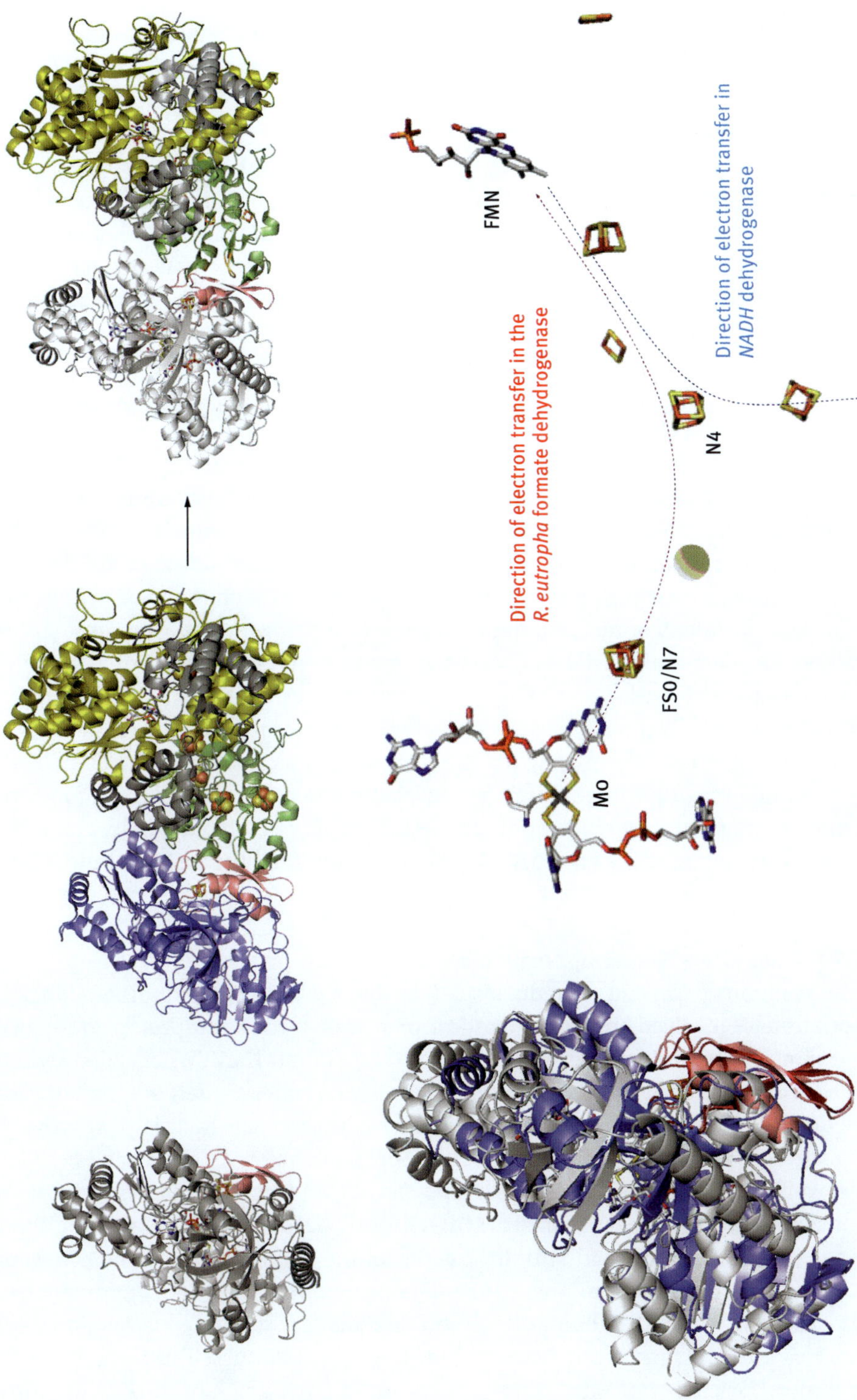
FMN
N4
FS0/N7
Mo
Direction of electron transfer in the
R. eutropha formate dehydrogenase
Direction of electron transfer in
NADH dehydrogenase

compared with the Mo-Se-Cys seen in FdhF may alter the chemical course of the reaction). Although the physiological direction of the reaction catalyzed by FdsABG is formate oxidation to $CO_2$, it is likely that at least some of these air-stable enzymes may also catalyze the reverse reaction, the reduction of $CO_2$ to formate as is seen with the molybdenum- and selenocysteine-containing FdhF from *Clostridium carboxidivorans* [198], and the tungsten- and selenocysteine-containing formate dehydrogenase from *Syntrophobacter fumaroxidans* [199] (both of which are $O_2$-sensitive). As mentioned earlier in this section, such reversibility in an air-stable enzyme would have considerable industrial potential, using formate as a way to store chemical energy (in the form of reducing equivalents).

### 8.3.5 Bacterial nitrate reductases

Three different types of nitrate reductases are found in bacteria, all of which are members of the DMSO reductase family of molybdenum enzymes and catalyze the same reaction, i.e. the reduction of nitrate to nitrite. As a group, they are distinct from the assimilatory nitrate reductases of eukaryotes, which are members of the sulfite oxidase family of molybdenum enzymes (none of which possess iron-sulfur clusters). The bacterial nitrate reductases differ in physiological function (the genetically designated Nar enzymes being respiratory and utilizing menaquinol as reducing substrate, the Nap enzymes dissimilatory but also utilizing menaquinol and the Nas enzymes assimilatory and utilizing ferredoxin, as discussed further below), as well as in their subcellular location (membrane-associated, periplasmic, or cytoplasmic, respectively) and in the detailed structure of the molybdenum center (particularly with regard to the amino acid residue coordinating the molybdenum) [221].

#### 8.3.5.1 Respiratory Nar nitrate reductase

Of the respiratory Nar nitrate reductases (i.e. those using menaquinol as reducing substrate and involved in the generation of a transmembrane proton gradient), the best characterized are those from *E. coli*, NarGHI and NarZYW, encoded by the *narKGHJI* and *narUZYWV* operons, respectively. The enzymes are very similar but, like the FdnGHI and FdoGHI formate dehydrogenases with which each is, respectively, associated (Section 8.3.4.2), have distinct physiological roles. As shown in Fig. 8.26, the crystal structure of the *E. coli* NarGHI enzyme at 1.9-Å resolution shows the protein organized as a $(\alpha\beta\gamma)_2$ dimer of trimers (PDB 1Q16) [117, 222]. The NarG catalytic subunit has the molybdenum center (with the 2 equiv. of the pyranopterin cofactor present as the guanine dinucleotide) and a [4Fe-4S] cluster adjacent to its Q pyranopterin (very similar to the PsrA subunit of polysulfide reductase discussed in Section 8.3.2 as well as the FdhF and FdnG formate dehydrogenases discussed in Section 8.3.4). The iron-sulfur cluster of NarG differs, however, in having one of the coordinating

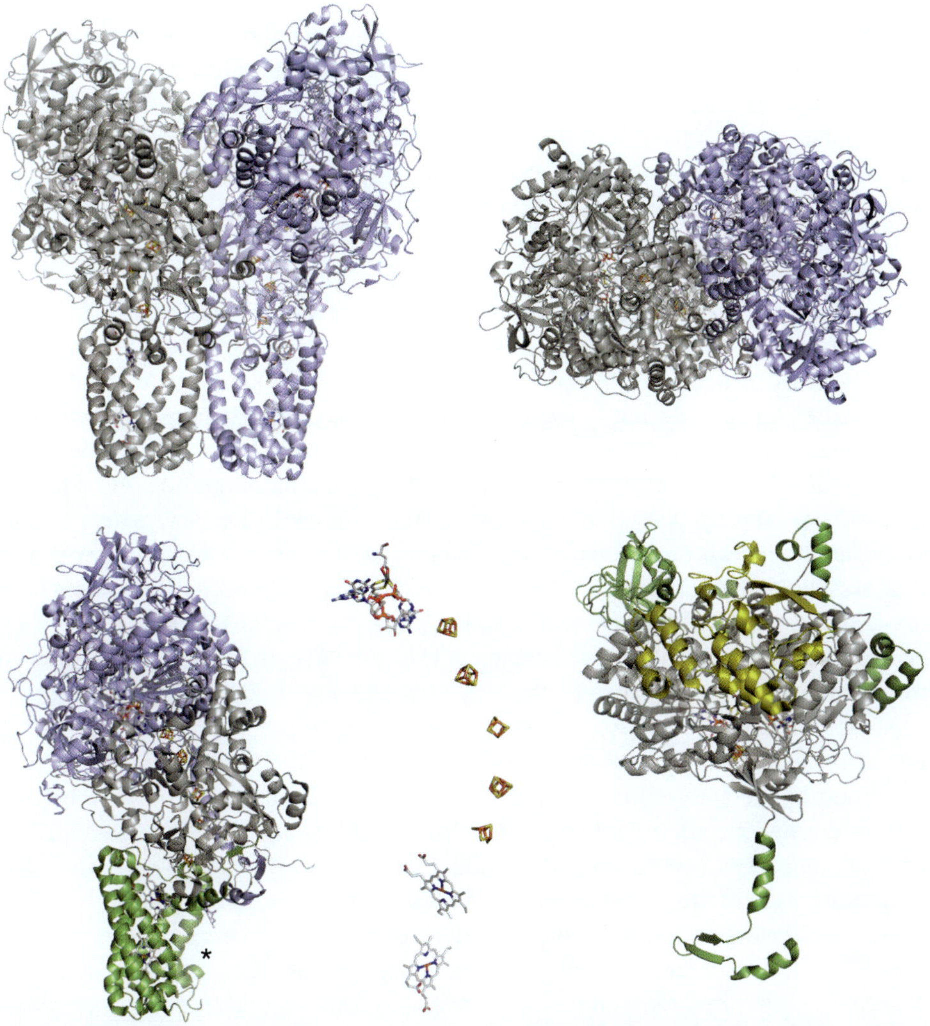


**Fig. 8.26:** NarGHI from *E. coli* (PDB 1Q16). (top) The overall of the organization of the $(\alpha\beta\gamma)_2$ enzyme from the side (with the membrane-integral NarI at bottom) and from the top. (bottom left) The left-hand protomer from above, with the catalytic NarG in blue, the iron-sulfur containing NarH in gray, and the membrane integral NarI in green. The approximate position of the menaquinol-binding site near the distal heme is indicated by the red asterisk [222]. (bottom center) The layout of the eight redox-active centers in the protomer (the perspective is rotated approximately 90° about the vertical relative to that at the left). (bottom right) The NarG subunit, with the inserts discussed in the text rendered in green and yellow.

cysteines replaced with a histidine as seen in the EbdB subunit of ethylbenzene dehydrogenase (Section 8.3.3). Similar His-substituted clusters are observed in the [NiFe] hydrogenase from *D. gigas* [177] and Fe-only hydrogenase from *C. pasteurianum* [223]. NarH harbors four additional iron-sulfur clusters analogous to PsrB and FdnH but with the distal cluster a [3Fe-4S] rather than a [4Fe-4S] cluster (as also seen in EbdB). The interaction between NarG and NarH involves two N-terminal helices in NarG, one of which extends the entire length of NarH. A comparison of NarH with the corresponding subunits of other enzymes considered here is shown in Fig. 8.20, where it can be seen that there are also several inserts to the polyferredoxin core (Fig. 8.20, blue) that make additional contacts with both the NarG and the NarI subunits of the enzyme.

The cytoplasmically exposed NarGH subunits connect to the membrane-integral NarI, which has two *b*-type cytochromes. The hemes are oriented quite differently than seen in FdnI (Section 8.3.4.2), being perpendicular to one another and displaced such that the iron of one heme does not lie in-plane with the other heme. In addition, the histidine ligands of each heme are aligned almost perpendicular to one another rather than at the 45° angle seen in FdnI. A structure of NarGHI in complex with the quinone analogue pentachlorophenol places the menaquinol-binding site adjacent to the distal heme of NarI (indicated by the red asterisk in Fig. 8.22, *center*) [222]. The structure reveals an approximately 75-Å-long electron transfer chain that clearly indicates the path of electron flow (Fig. 8.26, *bottom right*). Thus, electrons flow from the distal heme of NarI (the site of menaquinol reduction) to the proximal heme (5.4 Å apart, edge to edge) to the [3Fe-4S] cluster of NarH (8.9 Å apart, edge to edge), then on through the four [4Fe-4S] clusters (three in NarH and one in NarG, each 9–11 Å apart).

Although the core fold of the NarG subunit closely resembles that seen in the *Rhodobacter* DMSO reductase (Fig. 8.15), it is substantially larger even without consideration of the [4Fe-4S]-containing domain. As originally noted [117], there are several inserts, including residues 1–40, 116–150, 339–472, 616–638, 667–690, and 843–990 (Fig. 8.26, bottom right). The first of these is an N-terminal extension that spans the length of the NarH subunit, and the last (in yellow), a separate subdomain at the interface between the two NarG subunits of the dimer. The molybdenum center of NarG is also distinctive in two other important ways. First, as shown in Fig. 8.27, Asp222 is coordinated to the metal in a bidentate fashion, and there is no terminal

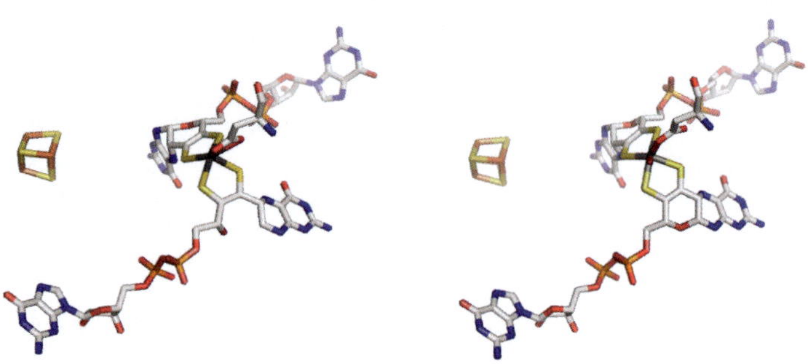

**Fig. 8.27:** Alternate coordination modes for Asp222 in *E. coli* NarG. (left) The bidentate mode with a Mo=O ligand as seen in PDB 1QI6 [117] and the monodentate mode seen in PDB 1R27 [118].

oxo or sulfido ligand. Second, the pyran ring of the P pyranopterin (the one away from the proximal FS0) has opened; possibly allowing the pyran ring to participate in a proton shuttle [224]. A separate structure of just the soluble NarGH subunits (PDB 1R27) has also been reported, showing a very similar overall protein architecture. At the molybdenum center, however, Asp222 (Fig. 8.27) is clearly coordinated in a monodentate fashion as seen in ethylbenzene dehydrogenase [188] (Section 8.3.3), with a terminal oxo group completing the sixth coordination position in a trigonal prismatic geometry similar to that seen in most other members of the DMSO reductase family [118]. In addition, the pyran ring of the P pterin has closed. Rationalizing these structures, it is likely that the NarGHI crystal had become reduced in the synchrotron beam, whereas the NarGH crystal remained oxidized.

Nar nitrate reductases typically exhibit “low-pH” ($g_{1,2,3}$ = 2.001, 1.986, and 1.964) and “high-pH” $Mo^V$ EPR signals ($g_{1,2,3}$ = 1.987, 1.981, and 1.962) [225–230] that differ in the magnitude of the observed hyperfine splitting ($a_{av}$ = 9.6 G for the low-pH form and $a_{av}$ = 3.4 G for the high-pH form); in both signals, the protons are solvent-exchangeable [228]. The different hyperfine coupling may be due either to different orientations of the proton *via* hydrogen bonding with nearby amino acid residue and/or a network created by water molecules (as seen in sulfite oxidase) or, alternatively, from different binding modes for substrate, as has been observed crystallographically [230]. A combined EPR and EXAFS study of the NarGHI from *E. coli* [231] has demonstrated that the high-pH $Mo^V$ species exhibits a small hyperfine coupling to $^{17}O$ ($a_{av}$ ~2.38 G) that has been attributed to a $Mo{=}^{17}O$ unit. For the low-pH species, the presence of a coordinated hydroxyl group is considered responsible for the more strongly coupled proton (although no $^{17}O$ hyperfine has been observed). In the reduced $Mo^{IV}$ state, a substantial amount of a *des*oxo molybdenum species is seen, and in the oxidized $Mo^{VI}$ state, only a single terminal oxo group is observed at a distance of 1.73 Å. Thus, the low-pH form may be a *des*oxo Mo-OH species, whereas the high-pH species has one terminal oxo group.

The His-coordinated FS0 cluster of NarG is unusual in having an $S = 3/2$ ground state in the reduced $[4Fe\text{-}4S]^{1+}$ form, with EPR features at $g$ = 5.023 and 5.556 and a reduction potential of –55 mV *vs* NHE at pH 8.0 [232]. Mutation of the cluster-coordinating His50 to Cys results in a 500-mV decrease in the reduction potential of the cluster with an accompanying loss of activity, and mutation to Ser results in the failure to incorporate either FS0 or the molybdenum center into apo-protein [233]. The four iron-sulfur clusters of NarB have more conventional EPR spectra, with the oxidized [3Fe-4S] FS4 cluster exhibiting an EPR signal with $g_{1,2,3}$ = 2.02, 2.00, and 1.98. The higher-potential [4Fe-4S] FS1 cluster has $g_{1,2,3}$ = 2.05, 1.95, and 1.87 seen on partial reduction of NarGHI; discrete assignments for FS2 and FS3 are complicated by extensive spin-spin interactions among the three paramagnetic centers of fully reduced enzyme [234–237]. The low- and high-potential hemes are of the highly anisotropic, low-spin variety, with $g_1$ values of 3.36 and 3.76, respectively [238].

The NarH FS1 and FS4 have high reduction potentials (+130 and +180 mV, respectively), whereas FS2 and FS3 have much lower potentials (–420 and –55 mV, respectively) [209, 234, 235, 237], and it is thus evident that they are not organized sequentially in order of increasing redution potentials in the direction of physiological electron flow (i.e. from FS4 to FS1). The lower-potential sites are found in more hydrophobic and solvent-shielded environments (FS2 particularly so), whereas the higher-potential clusters are in more polar environments with their coordinating cysteine residues participating in more hydrogen-bonding interactions that would better be able to accommodate the change in charge upon reduction [117]. The reduction potentials of the two *b*-type cytochromes of the NarI subunit have also been determined, being 20 and 120 mV, respectively, for the low- and high-potential hemes (distal and proximal with respect to the NarH subunit) [239]. In the absence of NarH, the reduction potential for the proximal heme drops to –180 mV, suggesting that it is more solvent-exposed in the absence of the partner subunit.

The NarGH fragment from *P. pantotrophus* exhibits $Mo^{VI/V}$ and $Mo^{V/IV}$ reduction potentials of +470 and –50 mV, respectively [240]. The [3Fe-4S] cluster has the highest reduction potential among the iron-sulfur clusters at +24 mV, and the reduction potential for the FS0 cluster ($g \sim 1.833$) is –34 mV. The *E. coli* Nar GHI complex has also been investigated by PFV [241] and is found to behave similarly to the *P. pantotrophus* enzyme over the pH range of 5.0–9.0, with catalytic activity being a function of applied potential (with higher activity at –25 mV and lower activity at -400 mV) and the concentration of nitrate. Two discrete potential regimes are identified, a relatively narrow high-potential one (~–25 mV) at lower (nitrate) and a second at much lower potential (<–300 mV) at millimolar levels of nitrate. Nitrate is proposed to bind to the $Mo^{V}$ state of the active site more effectively than the $Mo^{IV}$ so that low concentrations of substrate are effectively reduced at higher poised potentials. It has been suggested that the redox switch that modulates catalytic activity may involve the participation of the pterin ring *via* a pyran ring opening-closing mechanism. The PFV work has also provided evidence for the involvement of a proton in the catalytic process (associated with the $Mo^{VI}$ state) with a $pK_a$ of 7.8.

#### 8.3.5.2 Nap enzymes

Both the monomeric NapA and heterodimeric NapAB dissimilatory (nonrespiratory) nitrate reductases possess a molybdenum center and a [4Fe-4S] cluster in their NapA subunits, and several crystal structures are now available for both forms [242–244]. In both cases, the reducing equivalents required for turnover come from the quinone pool *via* a membrane-integral NapC *c*-type cytochrome [245, 246]. The first crystal structure of any nitrate reductase was reported for the monomeric NapA from *Desulfovibrio desulfuricans* [114]. As shown in Fig. 8.28, the overall fold of the periplasmically localized protein is very similar to that seen in the *E. coli* FdhF discussed in Section 8.3.3.1 (Fig. 8.19) [112]. The two enzymes have a high degree of sequence similarity, and each can catalyze the reaction of the other enzyme (although less efficiently). In NapA, the distal

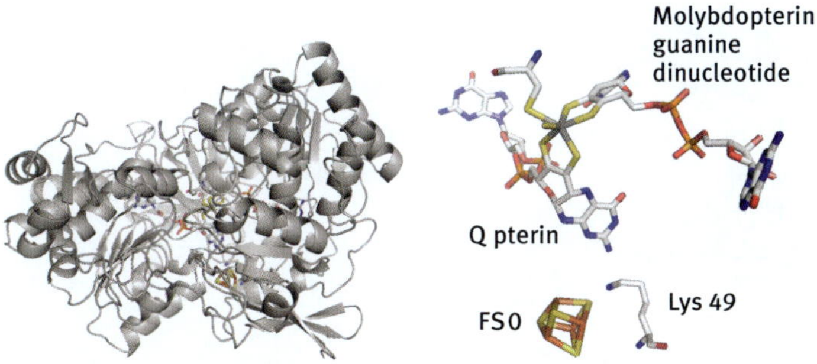


**Fig. 8.28:** The structure of the periplasmic Nap nitrate reductase from *D. desulfuricans*. (left) The overall structure PDB 2NAP), demonstrating the close similarity in overall fold to formate dehydrogenase-H, FdhF (Fig. 8.23). (right) A closeup of the structure of the molybdenum center, showing Lys49 that intervenes between the Q pterin and the FS0.

amino group of the Q pyranopterin of the molybdenum center is 6.4 Å from the nearest sulfur of FS0 (Fig. 8.28) comparable to the distance seen in FdhF, with Lys49 intervening. The molybdenum is coordinated by 2 equiv. of the pyranopterin cofactor (present as the guanine dinucleotide) and Cys140. The sixth coordination position was originally assigned as a terminal oxo group, but the long molybdenum-ligand distance, low thermal (B) parameter for the ligand, and anomalous scattering parameters suggest that the ligand is a terminal sulfur rather than oxygen (as shown in Fig. 8.28) [243]. The terminal sulfur is proposed to form a partial persulfide bond with the cysteine sulfur because the S...S distance of ~2.2 Å is shorter than the sum of the van der Waals radii (Fig. 8.29). Structures with different substrates (e.g. nitrate, perchlorate, cyanide) have also been considered. A more recent crystal structure of the dimeric NapAB from *Cupriavidus necator* (also known as *R. eutropha*) has provided additional support for the overall description of the molybdenum coordination sphere (Fig. 8.30) [247]. The *C. necator* NapAB structure also confirms the disposition of the redox-active centers in a heterodimeric NapAB (as originally seen in the lower-resolution structure of the NapAB from *R. sphaeroides* [242]). The NapB subunit harbors two hemes. As in the monomeric NapA form, the molybdenum center is ~6.5 Å from the closest sulfur of FS0, with the cluster 9.1 Å (edge to edge) from the nearer heme of NapB; a highly conserved

**Fig. 8.29:** Alternate structures proposed for the molybdenum center of Nap enzymes. (left) Originally proposed on the basis of the initial structure of the *D. desulfuricans* NapA [114]. (right) Later suggested [243] after reexamination of anomalous scattering parameters.

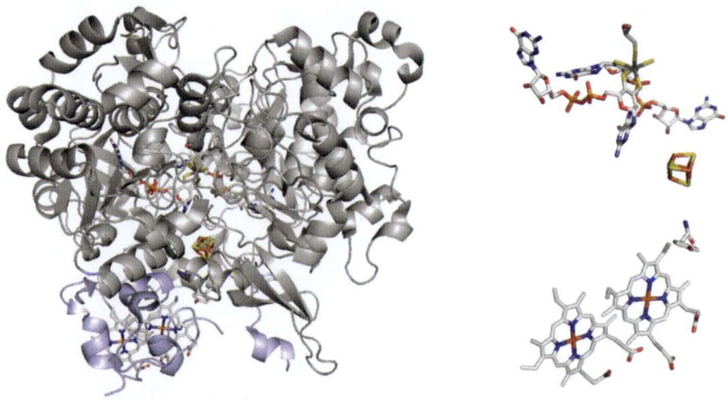

**Fig. 8.30:** The structure of the NapAB from *C. necator* (PDB 3ML1). (left) The overall fold, with the molybdenum-containing NapA in gray and the di-heme NapB in blue. A closeup of the redox-active centers of the enzyme (rotated approximately 45° counterclockwise about the vertical relative to the perspective at left), showing the position of Tyr58 of NapB that intervenes between the proximal heme and the [4Fe-4S] cluster of NapA.

tyrosine (Tyr58 from the *C. necator* NapB) intervenes. The proximal heme is parallel to and in van der Waals contact with the distal heme, although the two are displaced with respect to one another such that the iron-iron distance is 10.0 Å. Both hemes are coordinated by a pair of histidine residues as is typical of *b*-type cytochromes, rather than the histidine-methionine coordination more commonly seen in c-type cytochromes [248]. Also, in both hemes of NapB, the histidines are parallel to one another, which tunes the electronic structure of the heme and hence its reduction potential [249, 250]. The reduction potentials for the proximal and distal hemes of the *C. necator* NapAB, are +50 and +160 mV, respectively.

The Nap nitrate reductases from a number of organisms have been investigated by EPR [242–244, 246, 251–257], with the NapAB from *P. pantothrophus* perhaps being the best understood [246, 251]. Three distinct EPR signals are observed, with the "high-*g*" signal (also called "high-*g* split" and "high-*g* resting") having $g_{1,2,3}$ = 1.999, 1.991, and 1.981 being the most commonly observed signal and considered the most catalytically relevant [246, 258]. The signal exhibits strong coupling to non-exchangeable protons that has been attributed to the β-methylene protons of the cysteine coordinating the molybdenum [259]. XAS analysis indicates the presence of at least one terminal oxo group in the signal-giving species [246]. A QM/MM investigation, meanwhile, has suggested a $MoS_6$ core in the "high-*g*" species, which has been suggested to be a more reduced state where the putative persulfide bond between the sulfido and the S-Cys of the molybdenum is reduced, whereas the "low-*g*" signal (with $g_{1,2,3}$ = 1.996, 1.969, and 1.961) is due to a species coordinated by only 1 equiv. of the pyranopterin cofactor. The reduced [4Fe-4S] cluster has $g_{1,2,3}$ = 2.03, 1.94, and 1.89, and the oxidized low-spin heme $g_{1,2,3}$ = 2.9, 2.2, and 1.5.

Initially, it was thought that the reaction mechanism for the Nap enzymes involved straightforward oxygen atom transfer in which nitrate binds to the reduced (and five-coordinate) molybdenum center *via* one of its oxyanion groups to give a six-coordinate intermediate [114]. Reduction to nitrite results in transfer of the bridging oxygen to the now-oxidized molybdenum as a terminal oxo group, and re-reduction of the molybdenum with reducing equivalents from menaquinol results in protonation of the Mo=O to water, leading to its dissociation to regenerate the pentacoordinate reduced $Mo^{IV}$ center. The recent conclusion that the terminal ligand to molybdenum in oxidized NapA of both *D. desulfuricans* and *C. necator* is a sulfido rather than oxo group, however, requires revision of this mechanism. Specifically, a putative persulfide bond between the terminal sulfide and the cysteine sulfur has been proposed that reacts with substrate to generate an oxo-sulfido intermediate and a sulfide species, with Cys140 (in the *D. Desulfuricans* enzyme) dissociated from the molybdenum center. It remains for future mechanistic work to further clarify the reaction mechanism of the Nap nitrate reductases and permit a comparison with that of the Nar enzymes.

#### 8.3.5.3 Assimilatory Nas nitrate reductase

The assimilatory Nas nitrate reductases are cytosolic and expressed under aerobic conditions; although found in a variety of bacteria [260–266], the majority of work has focused on the genetic aspects of these enzymes. No structure is yet available for any Nas nitrate reductase, although on the basis of sequence homologies, they are expected to be closely related to the periplasmic Nap enzymes and have a cysteine coordinating the molybdenum [121, 267, 268]. The Nas proteins give EPR signals similar to those observed with the Nap proteins, with "very high-g" ($g_{1,2,3}$ = 2.023, 1.998, and 1.993) and "high-*g* split" ($g_{1,2,3}$ = 1.997, 1.990, 1.982, and 1.990) being reported for both the *A. vinelandii* [269] and *Synechococcus* sp. PCC 7942 [267] enzymes. Both signals exhibit splitting due to strongly coupled and non-exchangeable protons that, by analogy to the Nap nitrate reductases, arises from the β-methylene protons of the coordinating cysteine. Here, also the "very high-g" signal is thought to be due to an inactive form of the enzyme, and the "high-g" signal to the functional enzyme. The reduced *Synechococcus* enzyme (and presumably also that from *A. vinelandii*, although it was not reported) exhibits a EPR signal below 20 K attributable to a [4Fe-4S] cluster, with $g_{1,2,3}$ = 2.058, 1.951, and 1.908 [267]. The reduction potential for the cluster is –190 mV *vs* NHE, the lower potential being consistent with sequence analysis indicating that Nas has a [4Fe-4S] cluster coordinated by four cysteines as seen in Nap, rather than a higher-potential cluster (–34 mV) with three cysteines and a histidine as seen in the Nar nitrate reductases [267]. For the molybdenum center, the $Mo^{VI/V}$ and $Mo^{V/IV}$ couples associated with appearance and disappearance of the catalytically relevant "high-g" $Mo^V$ EPR signal are –10 and <–500 mV, with a midpoint potential estimated at ~–350 mV, indicating essentially quantitative accumulation of the paramagnetic $Mo^{V}$ state over the potential range –150 to –300 mV [267]. The enzyme is most active below –200 mV, meaning the substrate most likely interacts with enzyme that has a reduced iron-sulfur cluster and a molybdenum center in the $Mo^{V}$ state. By contrast, the Nar and Nap systems have maximum activities at +20 [240, 241] and –100 mV [270], respectively. Thus, although the EPR signals of the Nas nitrate reductases are very similar to those of the Nap, the cytosolic Nas enzymes operate at much lower reduction potentials than the periplasmically localized Nap enzymes, consistent with their utilizing reducing equivalents provided by low-potential flavodoxins or ferredoxins in turnover [269, 271].

### 8.3.6 Arsenite oxidase and arsenate reductase

It has only recently been recognized that enzymes able to oxidize arsenite and/or reduce arsenate have been recognized as being broadly distributed in biology [122, 272]. These include not only detoxifying enzymes found in all three domains of life but also respiratory enzymes that use the oxidation of arsenite or the reduction of arsenate to generate energy. Both systems are thought to be evolutionarily ancient, possibly present in the last universal common ancestor to all present life forms [122, 273, 274]. These enzymes fall into three distinct groups: the Aio arsenite oxidases (formerly referred to variously as Aox, Aro, or Aso [275]), which provide reducing equivalents for the respiratory chain (and simultaneously transform arsenite to the much less cytotoxic arsenate); the Arr arsenate reductases, which utilize arsenate as a terminal respiratory substrate; and the recently identified Arx arsenite oxidases, which are closely related to the Arr enzymes but function physiologically to oxidize arsenite rather than reduce arsenate.

#### 8.3.6.1 AioAB arsenite oxidase

The AioAB arsenite oxidases are broadly distributed and encoded by a minimal *aioBA* operon that frequently also includes genes encoding a *c*-type cytochrome, a two-component AioS/AioR sensor histidine kinase and regulatory element, a periplasmic AioX arsenite-binding protein, and/or an ArsA ATPase arsenite pump [276]. The enzyme is membrane-associated but on the periplasmic side. Unusually, the twin arginine leader sequence that targets the protein for *tat* export to the periplasm is on the AioB rather than AioA subunit. The crystal structure of the *A. faecalis* AioAB arsenite oxidase [113] is shown in Fig. 8.31 (an unpublished structure of the very similar AioAB from *Rhizobium* strain NT-26 has also been deposited in the Protein Data Bank, PDB 4AAY). The AioA subunit has a [3Fe-4S] cluster in its N-terminal domain and an active site molybdenum center in a four-domain motif resembling that first seen in the *R. sphaeroides* DMSO reductase (Fig. 8.15). The molybdenum center (which has

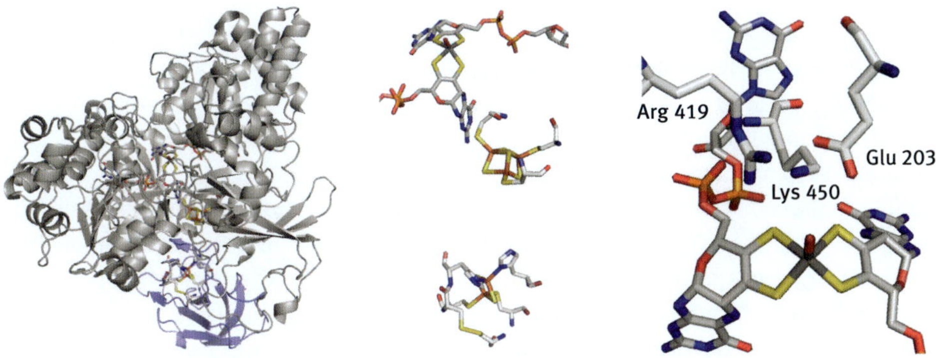


**Fig. 8.31:** The structure of the *A. faecalis* AioAB arsenite oxidase (PDB 1G8J). (left) The overall fold of the dimeric enzyme, with the AioA subunit in gray and the AioB subunit in blue. (center) An enlargement of the structure from the same perspective showing the disposition of the molybdenum center (top), [3Fe-4S] cluster (center), and Rieske [2Fe-2S] cluster (bottom) indicated. Also shown in the disulfide bond between Cys65 and Cys80 in AioB, adjacent to the Rieske cluster. (right) The environment of the molybdenum center, showing Arg419, Lys450, and Glu203, which defines the substrate-binding site.

apparently become reduced in the synchrotron beam) has 2 equiv. of the pyranopterin, present as the guanine dinucleotide, and an apical Mo=O (at a distance of 1.6 Å) in a square-pyramidal coordination geometry. Unique among the molybdenum-containing enzymes, no molybdenum ligand is provided by the polypeptide (the position homologous to the molybdenum-binding amino acid residues in other enzymes of the DMSO reductase family is Ala199, which lies on a polypeptide strand that has looped away from the molybdenum center). Lys450, which lies very close to the apical oxo group, Arg419, and Glu203 define the substrate-binding site (Fig. 8.31, right). Again, the iron-sulfur cluster is adjacent to the Q pyranopterin. X-ray absorption analysis of the oxidized *A. faecalis* arsenite oxidase identifies four sulfurs with Mo-S distances in the 2.37–2.47 range, consistent with the crystallographically observed *bis*enedithiolate coordination of the metal, and two additional oxygen atoms [277]. The data were best fit with a Mo=O at 1.72 Å and a Mo-O(H) at 1.84 Å, but the possibility of two Mo=O at 1.78 Å could not be excluded [277]. The AioB subunit is oblong-shaped, with a six-strand β-barrel at one end and a four-stranded sheet containing a Rieske [2Fe-2S] center at the other. The Rieske-containing subdomain inserts into a shallow depression on the surface of AioA near its [3Fe-4S] cluster, with the histidine-liganded iron of the Rieske cluster nearer to the [3Fe-4S] cluster of AioA; the edge-to-edge distance between the two iron-sulfur clusters is approximately 10 Å. Topologically, the Rieske center lies immediately beneath a disulfide bond between Cys65 and Cys80, both of which are on the protein surface.

The *A. faecalis* arsenite oxidase is able to use either azurin or a *c*-type cytochrome as oxidizing substrate [278], but the *Ralstonia* enzyme appears to be specific for soluble *c*-type cytochromes [279]. The purified *A. faecalis* enzyme absorbs broadly throughout the visible, with pronounced shoulders at 680 and 450 nm that bleach upon reduction. The oxidized enzyme manifests the expected EPR signal of a [3Fe-4S] cluster ($g_{1,2,3}$ = 2.03, 2.01, and 2.00) and the reduced of a Rieske-type center ($g_{1,2,3}$ = 2.03, 1.89, and 1.76) [278]. In the course of reductive titrations, the [3Fe-4S] signal disappears before that for the Rieske center emerges, reflecting the relative reduction potentials of the two centers. To date, no EPR signal attributable to a $Mo^{V}$ state of the molybdenum center has been observed for any arsenite oxidase. The electrochemical behavior of the enzyme has been examined by protein film voltammetry, with reduction potentials for the [3Fe-4S] and Rieske cluster determined to be +260 and +160 mV, respectively [280]. The midpoint potential for the molybdenum center is +292 mV at pH 6.0, and it is seen to behave as an obligate two-electron system, consistent with the failure to observe a $Mo^{V}$ EPR signal. Although these reduction potentials indicate that electron transfer out of the molybdenum center is thermodynamically uphill, they are considerably more positive than that for the arsenite/arsenate couple at +60 mV. The pH dependence of the midpoint potential for the molybdenum center is consistent with the uptake of two protons with the two electrons, consistent with the oxidized enzyme having an $L_2MoO_2$ structure, with one of the oxo groups being protonated and dissociating upon reduction of the molybdenum. This is also consistent with the observation that redox cycling the enzyme by reducing with arsenite followed by reoxidation in $H_2{}^{18}O$ results in incorporation of oxygen into a Mo=O group of the enzyme, as reflected in a 39-$cm^{-1}$ redshift in the 822-$cm^{-1}$ Mo=O stretch in the resonance Raman spectrum of oxidized enzyme to 784 $cm^{-1}$. The implication is that the enzyme operates *via* a simple oxygen atom transfer mechanism as envisaged for DMSO reductase, with a substrate lone pair attacking a catalytically labile Mo=O to initiate catalysis. In PFV experiments performed in the presence of substrate, the catalytic wave in the voltammogram correlates with the $Mo^{VI/IV}$ potential, indicating that electron transfer among the redox-active centers is rapid and that catalytic electron transport is entirely limited by turnover at the molybdenum center [280]. By contrast, the *Rhizobium* arsenite oxidase [281] behaves somewhat differently than the *A. faecalis* enzyme, most significantly in that the pH dependence of the catalytic wave for this enzyme is –30 mV/pH unit, indicating a two-electron/one-proton process. Such an interpretation would be more consistent with an $L_2MoO(OH)$ formulation for the oxidized molybdenum center, the preferred interpretation of the XAS analysis described above. Like the *A. faecalis* enzyme, that from *Rhizobium* does not exhibit a $Mo^{V}$ EPR signal under any conditions thus far examined.

*Ralstonia* sp. S22 possesses an AioAB arsenite oxidase, and an *E. coli* heterologous expression system for recombinant expression system has recently been reported, with the recombinant enzyme exhibiting EPR properties similar to those of the *A. faecalis* enzyme [282]. Interestingly, the reduction potential for the Rieske cluster of the *Ralstonia* enzyme is considerably higher than seen in the *A. faecalis* enzyme at +210 mV [273]. The *Ralstonia* enzyme possesses the disulfide bond adjacent to the Rieske cluster as also seen in the *A. faecalis* enzyme (but not the *Rhizobium* enzyme), and a C106A mutant has been prepared that exhibits similar EPR and reduction potential to the wild-type *Ralstonia* enzyme. This indicates that the presence of the disulfide bond itself does not influence the reduction potential of the cluster, as has long been assumed [282]. Regarding subcellular localization, the overexpressed *Ralstonia* enzyme spontaneously associates with the membrane, indicating that the protein does not need a membrane-anchoring protein to so associate; a Δ*tat* AioB construct has been prepared that, as expected, yields protein that is localized to the cytosol rather than periplasm.

#### 8.3.6.2 Arr arsenate reductase and Arx alternate arsenite oxidase

Although not as well-characterized from a biochemical standpoint as the AioAB arsenite oxidases discussed above, the Arr arsenate reductases have been extensively analyzed genetically and phylogenetically [122]. The operons encoding Arr arsenate reductases are generally organized *arrCABD* (although other arrangements are sometimes observed), with the ArrD gene product predicted to be a Tat-associated chaperone [122, 283]. ArrABC is predicted to have strong sequence homology to the PsrABC polysulfide reductase discussed in Section 8.3.2 and presumably has a similar complement of redox-active centers: a molybdenum- and [4Fe-4S]-containing ArrA (and a conserved Cys homologous to Cys173 of PsrA that coordinates the molybdenum), a 4 × [4Fe-4S]-containing ArrB and a cofactorless membrane-integral ArrC possessing a site for oxidation of menaquinol adjacent to the nearest [4Fe-4S] cluster of ArrB [122]. Given the conserved and distinctive double four-helix bundle structure of PsrC and the role of this subunit in dimerization of polysulfide reductase, it is likely that ArrABC also dimerizes into an $(\alpha\beta\gamma)_2$ oligomer. As with the AioAB arsenite oxidases, the catalytic ArrA subunit is periplasmically localized, although the TAT leader sequence in this case is found in the ArrA subunit rather than ArrB [283]. ArrABC is expressed only in the presence of arsenate and in most organisms is also under the regulatory control of FNR, indicating that it is expressed only under anaerobic conditions [284]. The *arrA* gene has been shown to be a reliable marker for the ability of a microorganism to respire on arsenate and has been used successfully to greatly expand the number of organisms known to be able to do so [285]. As seen with the NarGHI system, a soluble ArrAB unit can be isolated that has electron donor:arsenate oxidoreductase activity; both forms are specific for arsenate as substrate and do not utilize, e.g. nitrate effectively [286, 287]. Unlike the AioAB arsenite oxidases, the ArrABC arsenate reductases appear are to catalyze the reverse of their physiological reaction, the oxidation of arsenite to arsenate, although it is unlikely that they do so in the cell [288].

Quite recently, an alternate ArxABC arsenite oxidase has been identified that more closely resembles the ArrABC arsenate reductases than the AioAB arsenite oxidases described in the previous section [289–291]. Interestingly, the operons from three different organisms, minimally organized as *arxB'ABC*, possess a redundant copy of the *arxB'* gene encoding an alternate 4 × [4Fe-4S]-containing subunit; in two of the three organisms, an additional *arxD* gene encoding a putative TAT-associated chaperone is also present. The ArxABC enzyme has been shown to function physiologically as an arsenite oxidase during anaerobic respiration on nitrate [292] or selenate [293] and under photoautotrophic conditions as well [294]. It has been suggested that, depending on growth conditions, a given organism will express either a respiratory ArrABC enzyme that reduces arsenate utilizing menaquinol as substrate or an ArxABC enzyme that oxidizes arsenite and reduces ubiquinone [122]. The similarity in protein architecture and reversible arsenite/arsenate interconversion notwithstanding, sequence analysis indicates that the ArrA and ArxA catalytic subunits occupy distinct clades within the polysulfide reductase group of molybdenum-containing enzymes (although the two clades are more closely related to one another than to the AioA clade) [291], and as the proteins themselves become better characterized, important differences between them are likely to be established.

### 8.3.7 Pyrogallol:phloroglucinol transhydroxylase

*Pelobacter acidigallici* and related anaerobes are able to grow on polyhydroxylated aromatics such as phloroclucinol (1,3,5-trihydroxybenzene) or pyrogallol (1,2,3-trihydroxybenzene) as sole carbon source, ultimately reductively cleaving the aromatic ring prior to further degradation to acetate [295, 296]. The first step in the degradation of pyrogallol is the transhydroxylation reaction shown in Fig. 8.32, in which the 2-hydroxyl group of the cosubstrate 1,2,3,5-tetrahydroxybenzene is transferred to the 5-position of pyrogallol to generate a new equivalent of the tetrahydroxybenzene and phloroglucinol. The hydroxyl group that is transferred is not derived from solvent, nor does it equilibrate with solvent in the course of turnover [297]. The crystal structure of the transhydroxylase has been determined, and the enzyme found to be an αβ heterodimer, with a molybdenum-containing large subunit and a smaller subunit with three [4Fe-4S] clusters [298], as shown in Fig. 8.33. The large subunit is arranged in four domains as seen in the *Rhodobacter* DMSO reductases (Fig. 8.15), although segments of the chain trace (mostly involving regions involved in substrate binding) differ significantly from that seen in DMSO reductase. The pyranopterin cofactors of the

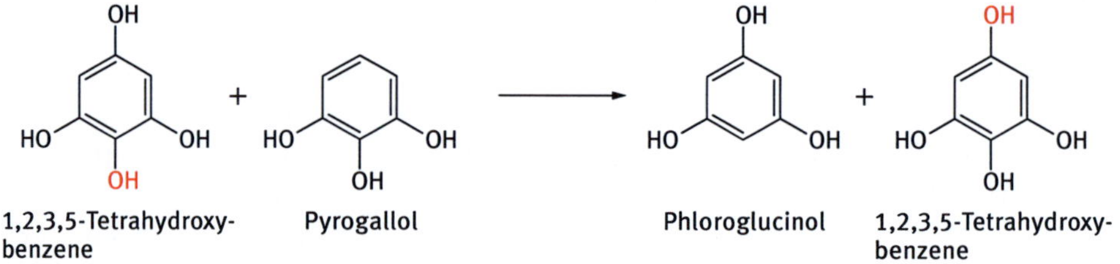


**Fig. 8.32:** The reaction catalyzed by pyrogallol:phloroglucinol transhydroxylase.

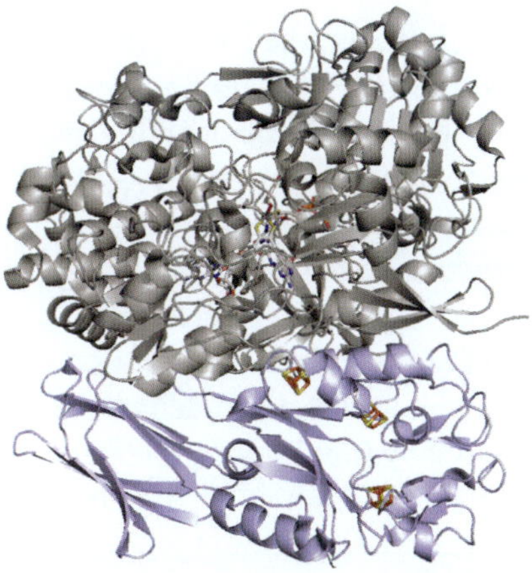

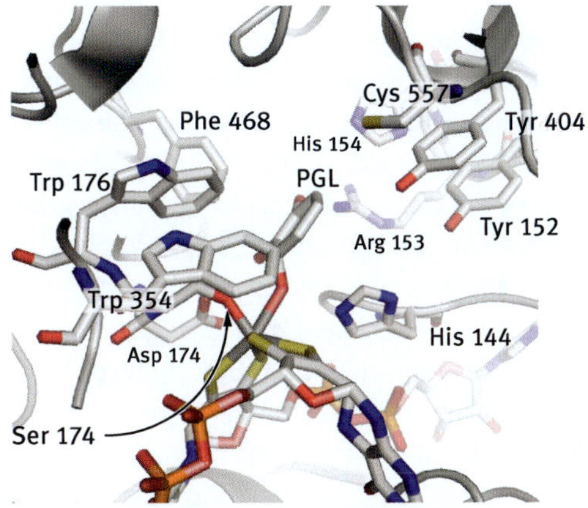


**Fig. 8.33:** The structure of pyrogallol:phloroglucinol transhydroxylase from Pelobacter acidigallici (PDB 1TI4). (left) The large molybdenum-containing subunit is in gray and the smaller iron-sulfur containing subunit in blue. (right) A closeup of the active site of reduced enzyme in complex with pyrogallol.

molybdenum center are present as the guanine dinucleotide, and the molybdenum (which under the strongly reducing crystallization conditions is likely $Mo^{IV}$) is also coordinated by Ser175 and an acetate ion from the crystallization mother liquor; this position is thought normally to be occupied by a water/hydroxide ligand. The small subunit consists of three domains: a normal ferredoxin domain, with two [4Fe-4S] clusters; a second ferredoxin-like domain, with only a single iron-sulfur cluster; and a third consisting of a seven-stranded β-barrel as seen in fibronectin, which is thought to be involved in attaching the enzyme to the inner surface of the cell membrane. The layout of the iron-sulfur clusters in the small subunit resembles that seen in the tungsten-containing formate dehydrogenase from *D. gigas* [299]. The reduced enzyme exhibits two iron-sulfur EPR signals with $g_{1,2,3}$ = 2.08, 1.94, and 1.87 and $g_{1,2,3}$ = 2.06, 1.95, and 1.87 in an approximately 2:1 ratio [300], consistent with the makeup of the enzyme. Without an FS0 cluster in the large subunit, the nearest [4Fe-4S] cluster of the small subunit is 12.4 Å from the distal amino group of the Q pyranopterin, but because the reaction does not involve electron transfer *per se*, very large a distance does not appear likely to compromise turnover.

The crystal structure of the complex of reduced enzyme with the substrate pyrogallol has also been determined [298]. As shown in Fig. 8.33 (right), pyrogallol coordinates to the molybdenum *via* its C1 hydroxyl group, displacing the acetate seen in the uncomplexed structure. Trp176, Trp354, and Phe468 provide a generally hydrophobic environment for substrate; Arg153 hydrogen bonds to the 3-hydroxyl group of substrate and His144 to the coordinating C1 hydroxyl. Interestingly, C1 of the bound pyrogallol is clearly $sp^3$-hybridized and tetrahedral. A model of the ternary complex of enzyme with both substrates places phloroglucinol among His154, Cys557, and

Tyr404 [298] and positioned such that its 2-hydroxyl group is adjacent to C5 of pyrogallol. The proposed reaction mechanism involves transient oxidation of pyrogallol by the molybdenum center to give the orthoquinone, followed by nucleophilic attack of the C5-hydroxyl of phloroglucinol to give a diphenylether intermediate. The cleavage of the C-O bond of phloroglucinol substrate occurs by the formation of an orthoquinone on the pyrogallol moiety, which, upon reduction by the molybdenum center, yields the rearomatized product phloroglucinol (from the aromatic ring of the pyrogallol substrate). Such a mechanism is quite distinct from a previous proposal that involves a simpler hydroxyl group transfer directly involving the molybdenum center, in which the reduced enzyme transiently accepts the C2-hydroxyl from 1,2,3,5-tetrahydroxybenzene to afford phloroglucinol and oxidized enzyme, which then reductively hydroxylates pyrogallol to phloroglucinol [297].

## 8.4 Prospectus

All members of the xanthine oxidase family of molybdenum-containing enzymes possess at least two iron-sulfur clusters, and although a few members of the DMSO reductase family lack iron-sulfur clusters, most have at least one; indeed, most have three, and the soluble formate dehydrogenase has seven (six of which form a wire connecting the enzyme's molybdenum center and FMN, the sites of electron entry and exit in the enzyme during catalysis). In all these enzymes, the iron-sulfur clusters serve to mediate electron transfer into or out of the molybdenum center in the course of catalysis. Our understanding of these enzymes has improved enormously over the past 10 to 15 years. The number of protein crystal structures reported has greatly expanded our understanding of the physical and chemical diversity of these enzymes. Genomics analyses have identified new enzymes, established their widespread presence in all three kingdoms of life, and established deep relationships among them that had not previously been recognized. In a few cases, the genetic context of the structural genes encoding these proteins has provided important clues as to the nature of the enzymes themselves (e.g. with regard to the presence of Mo=S groups in the formate dehydrogenases). At the same time, studies of the reaction mechansms of several of these enzymes has led to a thorough understanding of their reaction mechanisms, to the point that we can speak with confidence of the specific manner in which various active site amino acid residues contribute to transition state stabilization and rate acceleration. For many other of these enzymes, however, even some for which crystal structures exist, many of the most important structural and mechanistic questions remain unanswered or controversial, principally due to the dearth of hard mechanistic information concerning their reactivity. There is much work to do, for example, in clarifying the reaction mechanisms of each of the formate dehydrogenases and nitrate reductases considered here, and given the differences in their active sites, it would be unwise to assume that each group operates by the same mechanism. Also, enzymes such as ethylbenzene dehydrogenase, pyrogallol:phloroglucinol transhydroxylase, and 4-hydroxybenzoylCoA reductase, enzymes that catalyze reactions that are both very interesting and, involved as they are in the metabolism of aromatic hydrocarbons, have considerable industrial potential, our understanding of the manner in which these enzymes act on unreactive carbon centers is still far from complete. Similarly, with cytosolic formate dehydrogenase, the ability to store high-energy reducing equivalents in the form of formate holds considerable industrial promise. For all the progress shown in the past 10 to 15 years, there are still many more enzymes to be discovered, structures to be determined, and reaction mechanisms to be established.

## References

[1] Hille R. The mononuclear molybdenum enzymes. Chem Rev 1996;96:2757–816.
[2] Johnson MK, Rees DC, Adams MWW. Tungstoenzymes. Chem Rev 1996;96:2817–39.
[3] Chan MK, Mukund S, Kletzin A, Adams MWW, Rees DC. Structure of a hyperthermophilic tungstopterin enzyme, aldehyde ferredoxin oxidoreductase. Science 1995;267:1463–9.
[4] Dixon M, Thurlow S. Studies on xanthine oxidase. II. The dynamics of the oxidase system. Biochem J 1924;18:976–88.
[5] Montero-Moran GM, Li M, Rendon-Huerta E, et al. Purification and characterization of the Fe-II- and alpha-ketoglutarate-dependent xanthine hydroxylase from Aspergillus nidulans. Biochemistry 2007;46:5293–304.
[6] Pope SD, Chen L-L, Stewart V. Purine utilization by Klebsiella oxytoca m5al: genes for ring-oxidizing and -opening enzymes. J Bacteriol 2009;191:1006–17.
[7] de la Riva L, Badia J, Aguilar J, Bender RA, Baldoma L. The hpx genetic system for hypoxanthine assimilation as a nitrogen source in Klebsiella pneumoniae: gene organization and transcriptional regulation. J Bacteriol 2008;190:7892–903.
[8] Wootton JC, Nicolson RE, Cock JM, et al. Enzymes depending on the pterin molybdenum cofactor – sequence families, spectroscopic properties of molybdenum and possible cofactor-binding domains. Biochim Biophys Acta 1991;1057:157–85.
[9] Santiago B, Schubel U, Egelseer C, Meyer O. Sequence analysis, characterization and co-specific transcription of the cox gene cluster on the megaplasmid PHCG3 of Oligotropha carboxidovorans. Gene 1999;236:115–24.
[10] Romao MJ, Archer M, Moura I, et al. Crystal-structure of the xanthine oxidase-related aldehyde oxidoreductase from D gigas. Science 1995;270:1170–6.
[11] Huber R, Hof P, Duarte RO, et al. A structure-based catalytic mechanism for the xanthine oxidase family of molybdenum enzymes. Proc Natl Acad Sci USA 1996;93:8846–51.
[12] Jones RM, Inscore FE, Hille R, Kirk ML. Freeze-quench magnetic circular dichroism spectroscopic study of the "very rapid" intermediate in xanthine oxidase. Inorg Chem 1999;38:4963–70.
[13] Bonin I, Martins BM, Purvanov V, Fetzner S, Huber R, Dobbek H. Active site geometry and substrate recognition of the molybdenum hydroxylase quinoline 2-oxidoreductase. Structure 2004;12:1425–35.
[14] Santos-Silva T, Ferroni F, Thapper A, et al. Kinetic, structural, and EPR studies reveal that aldehyde oxidoreductase from Desulfovibrio gigas does not need a sulfido ligand for catalysis and give evidence for a direct Mo-C interaction in a biological system. J Am Chem Soc 2009;131:7990–8.
[15] Doonan CJ, Stockert A, Hille R, George GN. Nature of the catalytically labile oxygen at the active site of xanthine oxidase. J Am Chem Soc 2005;127:4518–22.

[16] More C, Asso M, Roger G, et al. Study of the spin-spin interactions between the metal centers of Desulfovibrio gigas aldehyde oxidoreductase: identification of the reducible sites of the [2Fe-2s](1+2) clusters. Biochemistry 2005;44:11628–35.
[17] Enroth C, Eger BT, Okamoto K, Nishino T, Pai EF. Crystal structures of bovine milk xanthine dehydrogenase and xanthine oxidase: structure-based mechanism of conversion. Proc Natl Acad Sci USA 2000;97:10723–8.
[18] Ormejohn Wh, Beinert H. Heterogeneity of paramagnetic species in 2 iron-sulfur proteins – Clostridium pasteurianum ferredoxin and milk xanthine oxidase. Biochem Biophys Res Commun 1969;36:337.
[19] Hille R, Hagen WR, Dunham WR. Spectroscopic studies on the iron-sulfur centers of milk xanthine-oxidase. J Biol Chem 1985;260:569–75.
[20] Iwasaki T, Okamoto K, Nishino T, Mizushima J, Hori H. Sequence motif-specific assignment of two [2Fe-2S] clusters in rat xanthine oxidoreductase studied by site-directed mutagenesis. J Biochem 2000;127:771–8.
[21] Lowe DJ, Bray RC. Magnetic coupling of molybdenum and iron-sulfur centers in xanthine-oxidase and xanthine dehydrogenases. Biochem J 1978;169:471–9.
[22] Coffman RE, Buettner GR. General magnetic dipolar interaction of spin-spin coupled molecular dimers – application to an EPR-spectrum of xanthine-oxidase. J Phys Chem 1979;83:2392–400.
[23] Xia M, Dempski R, Hille R. The reductive half-reaction of xanthine oxidase – reaction with aldehyde substrates and identification of the catalytically labile oxygen. J Biol Chem 1999;274:3323–30.
[24] Greenwood RJ, Wilson GL, Pilbrow JR, Wedd AG. Molybdenum(v) sites in xanthine-oxidase and relevant analog complexes – comparison of O-17 hyperfine coupling. J Am Chem Soc 1993;115:5385–92.
[25] Albert A, Brown DJ. Purine studies. 1. Stability to acid and alkali – solubility – ionization – comparison with pteridines. J Chem Soc 1954:2060–71.
[26] Hille R, Kim JH, Hemann C. Reductive half-reaction of xanthine-oxidase – mechanistic role of the species giving rise to the rapid type-1 molybdenum(V) electron-paramagnetic resonance signal. Biochemistry 1993;32:3973–80.
[27] Palmer G, Bray RC, Beinert H. Direct studies on electron transfer sequence in xanthine oxidase by electron paramagnetic resonance spectroscopy. I. Techniques + description of spectra. J Biol Chem 1964;239:2657.
[28] Bray RC, Vanngard T. Rapidly appearing molybdenum electron-paramagnetic-resonance signals from reduced xanthine oxidase. Biochem J 1969;114:725.
[29] Bray RC, Palmer G, Beinert H. Direct studies on electron transfer sequence in xanthine oxidase by electron paramagnetic resonance spectroscopy. 2. Kinetic studies employing rapid freezing. J Biol Chem 1964;239:2667.
[30] Edmondso D, Ballou D, Vanheuve A, Palmer G, Massey V. Kinetic studies on substrate reduction of xanthine-oxidase. J Biol Chem 1973;248:6135–44.
[31] Sempombe J, Stein B, Kirk ML. Spectroscopic and electronic structure studies probing covalency contributions to C-H bond activation and transition-state stabilization in xanthine oxidase. Inorg Chem 2011;50:10919–28.
[32] Truglio JJ, Theis K, Leimkuhler S, Rappa R, Rajagopalan KV, Kisker C. Crystal structures of the active and alloxanthine-inhibited forms of xanthine dehydrogenase from Rhodobacter capsulatus. Structure 2002;10:115–25.
[33] Leimkuhler S, Stockert AL, Igarashi K, Nishino T, Hille R. The role of active site glutamate residues in catalysis of Rhodobacter capsulatus xanthine dehydrogenase. J Biol Chem 2004;279:40437–44.

[34] Yamaguchi Y, Matsumura T, Ichida K, Okamoto K, Nishino T. Human xanthine oxidase changes its substrate specificity to aldehyde oxidase type upon mutation of amino acid residues in the active site: roles of active site residues in binding and activation of purine substrate. J Biochem 2007;141:513–24.
[35] Ilich P, Hille R. Tautomerization of the substrate heterocycle in the course of the reaction of xanthine oxidase. Inorg Chim Acta 1997;263:87–93.
[36] Pauff JM, Hemann CF, Juenemann N, Leimkuehler S, Hille R. The role of arginine 310 in catalysis and substrate specificity in xanthine dehydrogenase from Rhodobacter capsulatus. J Biol Chem 2007;282:12785–90.
[37] Olson JS, Ballou DP, Palmer G, Massey V. Mechanism of action of xanthine-oxidase. J Biol Chem 1974;249:4363–82.
[38] Porras AG, Palmer G. The room-temperature potentiometry of xanthine-oxidase – pH-dependent redox behavior of the flavin, molybdenum, and iron-sulfur centers. J Biol Chem 1982;257: 1617–26.
[39] Hille R, Massey V. The equilibration of reducing equivalents within milk xanthine-oxidase. J Biol Chem 1986;261:1241–7.
[40] Hille R. Electron-transfer within xanthine-oxidase – a solvent kinetic isotope effect study. Biochemistry 1991;30:8522–9.
[41] Hille R, Anderson RF. Electron-transfer in milk xanthine-oxidase as studied by pulse-radiolysis. J Biol Chem 1991;266:5608–15.
[42] Hille R, Anderson RF. Coupled electron/proton transfer in complex flavoproteins – solvent kinetic isotope effect studies of electron transfer in xanthine oxidase and trimethylamine dehydrogenase. J Biol Chem 2001;276:31193–201.
[43] Page CC, Moser CC, Chen XX, Dutton PL. Natural engineering principles of electron tunnelling in biological oxidation-reduction. Nature 1999;402:47–52.
[44] Lim LW, Shamala N, Mathews FS, Steenkamp DJ, Hamlin R, Xuong NH. 3-Dimensional structure of the iron-sulfur flavoprotein trimethylamine dehydrogenase at 2.4-Å resolution. J Biol Chem 1986;261:5140–6.
[45] Rohlfs RJ, Huang LX, Hille R. Prototropic control of intramolecular electron-transfer in trimethylamine dehydrogenase. J Biol Chem 1995;270:22196–207.
[46] Hille R, Massey V. Studies on the oxidative half-reaction of xanthine-oxidase. J Biol Chem 1981;256:9090–5.
[47] Porras AG, Olson JS, Palmer G. The reaction of reduced xanthine-oxidase with oxygen – kinetics of peroxide and superoxide formation. J Biol Chem 1981;256:9096–103.
[48] Olson JS, Ballou DP, Palmer G, Massey V. Reaction of xanthine-oxidase with molecular-oxygen. J Biol Chem 1974;249:4350–62.
[49] Dellacor E, Stirpe F. Regulation of rat-liver xanthine-oxidase – involvement of thiol groups in conversion of enzyme-activity from dehydrogenase (type D) into oxidase (type O) and purification of enzyme. Biochem J 1972;126:739.
[50] Stirpe F, Dellacor E. Regulation of rat liver xanthine oxidase – conversion in vitro of enzyme activity from dehydrogenase (type D) to oxidase (type O). J Biol Chem 1969;244:3855.
[51] Battelli MG, Lorenzon E, Stirpe F. Milk xanthine-oxidase type D (dehydrogenase) and type O (oxidase) – purification, interconversion and some properties. Biochem J 1973;131:191–8.
[52] Waud WR, Rajagopalan KV. Mechanism of conversion of rat-liver xanthine dehydrogenase from an NAD+-dependent form (type D) to an O-2-dependent form (type O). Arch Biochem Biophys 1976;172:365–79.
[53] Nakamura M, Yamazaki I. Preparation of bovine-milk xanthine-oxidase as a dehydrogenase form. J Biochem 1982;92:1279–86.

[54] Nishino T, Okamoto K, Kawaguchi Y, et al. Mechanism of the conversion of xanthine dehydrogenase to xanthine oxidase – identification of the two cysteine disulfide bonds and crystal structure of a non-convertible rat liver xanthine dehydrogenase mutant. J Biol Chem 2005;280:24888–94.
[55] Hille R, Nishino T. Flavoprotein structure and mechanism. 4. Xanthine-oxidase and xanthine dehydrogenase. FASEB J 1995;9:995–1003.
[56] McCord JM. Oxygen-derived free-radicals in postischemic tissue-injury. N Engl J Med 1985;312:159–63.
[57] Berry CE, Hare JM. Xanthine oxicloreductase and cardiovascular disease: molecular mechanisms and pathophysiological implications. J Physiol London 2004;555: 589–606.
[58] Rajagopalan KV, Handler P, Fridovich I. Hepatic aldehyde oxidase. 1. Purification and properties. J Biol Chem 1962;237:922.
[59] Rajagopalan KV, Handler P. Hepatic aldehyde oxidase. 3. Substrate-binding site. J Biol Chem 1964;239:2027.
[60] Rajagopalan KV, Handler P. Hepatic aldehyde oxidase. 2. Differential inhibition of electron transfer to various electron acceptors. J Biol Chem 1964;239:2022.
[61] Barber MJ, Coughlan MP, Rajagopalan KV, Siegel LM. Properties of the prosthetic groups of rabbit liver aldehyde oxidase – a comparison of molybdenum hydroxylase enzymes. Biochemistry 1982;21:3561–8.
[62] Turner NA, Doyle WA, Ventom AM, Bray RC. Properties of rabbit liver aldehyde oxidase and the relationship of the enzyme to xanthine-oxidase and dehydrogenase. Eur J Biochem 1995;232:646–57.
[63] Branzoli U, Massey V. Preparation of aldehyde oxidase in its native and deflavo forms – comparison of spectroscopic and catalytic properties. J Biol Chem 1974;249:4339–45.
[64] Hall WW, Krenitsky TA. Aldehyde oxidase from rabbit liver – specificity toward purines and their analogs. Arch Biochem Biophys 1986;251:36–46.
[65] Krenitsky TA. Aldehyde oxidase and xanthine-oxidase – functional and evolutionary relationships. Biochem Pharmacol 1978;27:2763–4.
[66] Rajagopa Kv, Handler P, Palmer G, Beinert H. Studies of aldehyde oxidase by electron paramagnetic resonance spectroscopy. I. Spectra at equilibrium states. J Biol Chem 1968;243:3784.
[67] Rajagopa Kv, Handler P, Palmer G, Beinert H. Studies of aldehyde oxidase by electron paramagnetic resonance spectroscopy. 2. Kinetic studies by rapid freezing. J Biol Chem 1968;243:3797.
[68] Felsted RL, Chu AEY, Chaykin S. Purification and properties of aldehyde oxidases from hog and rabbit livers. J Biol Chem 1973;248:2580–7.
[69] Bray RC, George GN, Gutteridge S, Norlander L, Stell JGP, Stubley C. Studies by electron-paramagnetic-resonance spectroscopy of the molybdenum center of aldehyde oxidase. Biochem J 1982;203:263–7.
[70] Coelho C, Mahro M, Trincao J, et al. The first mammalian aldehyde oxidase crystal structure insights into substrate specificity. J Biol Chem 2012;287:40690–702.
[71] Nishino T. The nicotinamide adenine dinucleotide-binding site of chicken liver xanthine dehydrogenase – evidence for alteration of the redox potential of the flavin by NAD binding or modification of the NAD-binding site and isolation of a modified peptide. J Biol Chem 1989;264:5468–73.
[72] Garattini E, Fratelli M, Terao M. Mammalian aldehyde oxidases: genetics, evolution and biochemistry. Cell Mol Life Sci 2008;65:1019–48.

[73] Seo M, Peeters AJM, Koiwai H, et al. The Arabidopsis aldehyde oxidase 3 (AA03) gene product catalyzes the final step in abscisic acid biosynthesis in leaves. Proc Natl Acad Sci USA 2000;97:12908–13.
[74] Hoff T, Frandsen GI, Rocher A, Mundy J. Biochemical and genetic characterization of three molybdenum cofactor hydroxylases in Arabidopsis thaliana. Biochim Biophys Acta Gene Struct Express 1998;1398:397–402.
[75] Akaba S, Seo M, Dohmae N, et al. Production of homo- and hetero-dimeric isozymes from two aldehyde oxidase genes of Arabidopsis thaliana. J Biochem 1999;126:395–401.
[76] Seo M, Aoki H, Koiwai H, Kamiya Y, Nambara E, Koshiba T. Comparative studies on the Arabidopsis aldehyde oxidase (AAO) gene family revealed a major role of AAO3 in ABA biosynthesis in seeds. Plant Cell Physiol 2004;45:1694–703.
[77] Zarepour M, Simon K, Wilch M, et al. Identification of superoxide production by Arabidopsis thaliana aldehyde oxidases AAO1 and AAO3. Plant Mol Biol 2012;80:659–71.
[78] Koiwai H, Nakaminami K, Seo M, Mitsuhashi W, Toyomasu T, Koshiba T. Tissue-specific localization of an abscisic acid biosynthetic enzyme, AAO3, in Arabidopsis. Plant Physiol 2004;134:1697–707.
[79] Kundu TK, Hille R, Velayutham M, Zweier JL. Characterization of superoxide production from aldehyde oxidase: an important source of oxidants in biological tissues. Arch Biochem Biophys 2007;460:113–21.
[80] Kundu TK, Velayutham M, Zweier JL. Aldehyde oxidase functions as a superoxide generating NADH oxidase: an important redox regulated pathway of cellular oxygen radical formation. Biochemistry 2012;51:2930–9.
[81] Badwey JA, Robinson JM, Karnovsky MJ, Karnovsky ML. Superoxide production by an unusual aldehyde oxidase in guinea-pig granulocytes – characterization and cytochemical-localization. J Biol Chem 1981;256:3479–86.
[82] Wright RM, Clayton DK, Riley MG, McManaman JL, Repine JE. cDNA cloning, sequencing, and characterization of male and female rat liver aldehyde oxidase (rAOX1) – differences in redox status may distinguish male and female forms of hepatic aox. J Biol Chem 1999;274:3878–86.
[83] Alfaro JF, Joswig-Jones CA, Ouyang W, Nichols J, Crouch GJ, Jones JP. Purification and mechanism of human aldehyde oxidase expressed in escherichia coli. Drug Metab Dispos 2009;37:2393–8.
[84] Meyer O, Frunzke K, Mördorf G. Biochemistry of the aerobic utilization of carbon monoxide. Microbial Growth on C1 Compounds 1993:27.
[85] Moxley JM, Smith KA. Factors affecting utilisation of atmospheric co by soils. Soil Biol Biochem 1998;30:65–79.
[86] Moersdorf G, Frunzke K, Gadkari D, Meyer O. Microbial growth on carbon monoxide. Biodegradation 1992;3:61–82.
[87] Jacobitz S, Meyer O. Removal of co dehydrogenase from Pseudomonas carboxydovorans cytoplasmic membranes, rebinding of co dehydrogenase to depleted membranes, and restoration of respiratory activities. J Bacteriol 1989;171:6294–9.
[88] Cypionka H, Meyer O. Carbon-monoxide insensitive respiratory-chain of Pseudomonas carboxydovorans. J Bacteriol 1983;156:1178–87.
[89] Ragsdale SW, Kumar M. Nickel-containing carbon monoxide dehydrogenase/acetyl-CoA synthase. Chem Rev 1996;96:2515–39.
[90] Dobbek H, Gremer L, Meyer O, Huber R. Crystal structure and mechanism of co dehydrogenase, a molybdo iron-sulfur flavoprotein containing S-selanylcysteine. Proc Natl Acad Sci USA 1999;96:8884–9.
[91] Hanzelmann P, Dobbek H, Gremer L, Huber R, Meyer O. The effect of intracellular molybdenum in Hydrogenophaga pseudoflava on the crystallographic structure of the seleno-molybdo-iron-sulfur flavoenzyme carbon monoxide dehydrogenase. J Mol Biol 2000;301:1221–35.

[92] Dobbek H, Gremer L, Kiefersauer R, Huber R, Meyer O. Catalysis at a dinuclear CuSMo(==O) OH cluster in a CO dehydrogenase resolved at 1.1-angstrom resolution. Proc Natl Acad Sci USA 2002;99:15971–6.
[93] Meyer-Klaucke W, Gnida M, Ferner R, Gremer L, Meyer O. Co-dehydrogenase – a new active site motive characterized by XAS. J Inorg Biochem 2001;86:339.
[94] Zhang B, Hemann CF, Hille R. Kinetic and spectroscopic studies of the molybdenum-copper co dehydrogenase from Oligotropha carboxidovorans. J Biol Chem 2010;285:12571–8.
[95] Shanmugam M, Wilcoxen J, Habel-Rodriguez D, Kirk ML, Hoffman BM, Hille, R. 13C and 63,65Cu ENDOR studies of CO dehydrogenase from Oligotropha carboxidovorans. Experimental evidence in support of a copper-carbonyl intermediate. J Am Chem Soc 2013;135:17775–17782.
[96] Siegbahn PEM, Shestakov AF. Quantum chemical modeling of co oxidation by the active site of molybdenum co dehydrogenase. J Comput Chem 2005;26:888–98.
[97] Hofmann M, Kassube JK, Graf T. The mechanism of Mo-/Cu-dependent CO dehydrogenase. J Biol Inorg Chem 2005;10:490–5.
[98] Gourlay C, Nielsen DJ, White JM, Knottenbelt SZ, Kirk ML, Young CG. Paramagnetic active site models for the molybdenum-copper carbon monoxide dehydrogenase. J Am Chem Soc 2006;128:2164–5.
[99] Resch M, Dobbek H, Meyer O. Structural and functional reconstruction in situ of the cusmoo2 active site of carbon monoxide dehydrogenase from the carbon monoxide oxidizing eubacterium Oligotropha carboxidovorans. J Biol Inorg Chem 2005;10:518–28.
[100] Wilcoxen J, Snider S, Hille R. Substitution of silver for copper in the binuclear Mo/cu center of carbon monoxide dehydrogenase from Oligotropha carboxidovorans. J Am Chem Soc 2011;133:12934–6.
[101] Wilcoxen J, Zhang B, Hille R. Reaction of the molybdenum- and copper-containing carbon monoxide dehydrogenase from Oligotropha carboxydovorans with quinones. Biochemistry 2011;50:1910–6.
[102] Gibson J, Dispensa M, Harwood CS. 4-hydroxybenzoyl coenzyme a reductase (dehydroxylating) is required for anaerobic degradation of 4-hydroxybenzoate by rhodoPseudomonas palustris and shares features with molybdenum-containing hydroxylases. J Bacteriol 1997;179:634–42.
[103] Boll M, Fuchs G. Identification and characterization of the natural electron donor ferredoxin and of fad as a possible prosthetic group of benzoyl-CoA reductase (dearomatizing), a key enzyme of anaerobic metabolism. Eur J Biochem 1998;251:946–54.
[104] Unciuleac M, Warkentin E, Page CC, Boll M, Ermler U. Structure of a xanthine oxidase-related 4-hydroxybenzoyl-CoA reductase with an additional 4Fe-4S cluster and an inverted electron flow. Structure 2004;12:2249–56.
[105] Boll M, Fuchs G, Meier C, et al. Redox centers of 4-hydroxybenzoyl-CoA reductase, a member of the xanthine oxidase family of molybdenum-containing enzymes. J Biol Chem 2001;276: 47853–62.
[106] Stockert AL, Shinde SS, Anderson RF, Hille R. The reaction mechanism of xanthine oxidase: evidence for two-electron chemistry rather than sequential one-electron steps. J Am Chem Soc 2002;124:14554–5.
[107] Bergmann F, Dikstein S. Studies on uric acid and related compounds. 3. Observations on the specificity of mammalian xanthine oxidases. J Biol Chem 1956;223:765–80.
[108] Bastian NR, Kay CJ, Barber MJ, Rajagopalan KV. Spectroscopic studies of the molybdenum-containing dimethyl-sulfoxide reductase from Rhodobacter sphaeroides f sp denitrificans. J Biol Chem 1991;266:45–51.
[109] Schneider F, Lowe J, Huber R, Schindelin H, Kisker C, Knablein J. Crystal structure of dimethyl sulfoxide reductase from Rhodobacter capsulatus at 1.88 angstrom resolution. J Mol Biol 1996;263:53–69.

[110] Schindelin H, Kisker C, Hilton J, Rajagopalan KV, Rees DC. Crystal structure of DMSO reductase: redox-linked changes in molybdopterin coordination. Science 1996;272:1615–21.
[111] Li HK, Temple C, Rajagopalan KV, Schindelin H. The 1.3 angstrom crystal structure of Rhodobacter sphaeroides dimethyl sulfoxide reductase reveals two distinct molybdenum coordination environments. J Am Chem Soc 2000;122:7673–80.
[112] Boyington JC, Gladyshev VN, Khangulov SV, Stadtman TC, Sun PD. Crystal structure of formate dehydrogenase h: catalysis involving Mo, molybdopterin, selenocysteine, and an fe4s4 cluster. Science 1997;275:1305–8.
[113] Ellis PJ, Conrads T, Hille R, Kuhn P. Crystal structure of the 100 kDa arsenite oxidase from Alcaligenes faecalis in two crystal forms at 1.64 angstrom and 2.03 angstrom. Structure 2001;9:125–32.
[114] Dias JM, Than ME, Humm A, et al. Crystal structure of the first dissimilatory nitrate reductase at 1.9 angstrom solved by mad methods. Structure 1999;7:65–79.
[115] Rothery RA, Weiner JH. Alteration of the iron-sulfur cluster composition of Escherichia coli dimethyl-sulfoxide reductase by site-directed mutagenesis. Biochemistry 1991;30: 8296–305.
[116] Rothery RA, Weiner JH. Interaction of an engineered 3Fe-4S cluster with a menaquinol binding site of Escherichia coli DMSO reductase. Biochemistry 1996;35:3247–57.
[117] Bertero MG, Rothery RA, Palak M, et al. Insights into the respiratory electron transfer pathway from the structure of nitrate reductase a. Nat Struct Biol 2003;10:681–7.
[118] Jormakka M, Richardson D, Byrne B, Iwata S. Architecture of nargh reveals a structural classification of Mo-bisMGD enzymes. Structure 2004;12:95–104.
[119] Jormakka M, Tornroth S, Byrne B, Iwata S. Molecular basis of proton motive force generation: structure of formate dehydrogenase-N. Science 2002;295:1863–8.
[120] McEwan AG, Ridge JP, McDevitt CA, Hugenholtz P. The DMSO reductase family of microbial molybdenum enzymes; molecular properties and role in the dissimilatory reduction of toxic elements. Geomicrobiol J 2002;19:3–21.
[121] Rothery RA, Workun GJ, Weiner JH. The prokaryotic complex iron-sulfur molybdoenzyme family. Biochim Biophys Acta Biomemb 2008;1778:1897–929.
[122] van Lis R, Nitschke W, Duval S, Schoepp-Cothenet B. Arsenics as bioenergetic substrates. Biochim Biophys Acta Bioenerg 2013;1827:176–88.
[123] Gonzalez PJ, Rivas MG, Mota CS, Brondino CD, Moura I, Moura JJG. Periplasmic nitrate reductases and formate dehydrogenases: biological control of the chemical properties of Mo and w for fine tuning of reactivity, substrate specificity and metabolic role. Coord Chem Rev 2013;257:315–31.
[124] Shaw AL, Hochkoeppler A, Bonora P, Zannoni D, Hanson GR, McEwan AG. Characterization of dorc from Rhodobacter capsulatus, a c-type cytochrome involved in electron transfer to dimethyl sulfoxide reductase. J Biol Chem 1999;274:9911–4.
[125] Yamamoto I, Wada N, Ujiiye T, et al. Cloning and nucleotide-sequence of the gene encoding dimethyl-sulfoxide reductase from Rhodobacter sphaeroides f sp denitrificans. Biosci Biotechnol Biochem 1995;59:1850–5.
[126] Barber MJ, Vanvalkenburgh H, Trimboli AJ, Pollock VV, Neame PJ, Bastian NR. The amino-acid-sequence of Rhodobacter sphaeroides dimethyl-sulfoxide reductase. Arch Biochem Biophys 1995;320:266–75.
[127] Satoh T, Kurihara FN. Purification and properties of dimethylsulfoxide reductase containing a molybdenum cofactor from a photodenitrifier, RhodoPseudomonas sphaeroides fs denitrificans. J Biochem 1987;102:191–7.
[128] Turner SM, Nightingale PD, Spokes LJ, Liddicoat MI, Liss PS. Increased dimethyl sulphide concentrations in sea water from in situ iron enrichment. Nature 1996;383:513–7.

[129] McAlpine AS, McEwan AG, Shaw AL, Bailey S. Molybdenum active centre of DMSO reductase from Rhodobacter capsulatus: crystal structure of the oxidised enzyme at 1.82-angstrom resolution and the dithionite-reduced enzyme at 2.8-angstrom resolution. J Biol Inorg Chem 1997;2:690–701.
[130] McAlpine AS, McEwan AG, Bailey S. The high resolution crystal structure of DMSO reductase in complex with DMSO. J Mol Biol 1998;275:613–23.
[131] Schultz BE, Hille R, Holm RH. Direct oxygen-atom transfer in the mechanism of action of Rhodobacter sphaeroides dimethyl-sulfoxide reductase. J Am Chem Soc 1995;117:827–8.
[132] Garton SD, Hilton J, Oku H, Crouse BR, Rajagopalan KV, Johnson MK. Active site structures and catalytic mechanism of Rhodobacter sphaeroides dimethyl sulfoxide reductase as revealed by resonance raman spectroscopy. J Am Chem Soc 1997;119:12906–16.
[133] Rothery RA, Stein B, Solomonson M, Kirk ML, Weiner JH. Pyranopterin conformation defines the function of molybdenum and tungsten enzymes. Proc Natl Acad Sci USA 2012;109:14773–8.
[134] Schwarz G, Mendel RR. Molybdenum cofactor biosynthesis and molybdenum enzymes. Annu Rev Plant Biol 2006;57:623–47.
[135] Wagner UG, Stupperich E, Kratky C. Structure of the molybdate/tungstate binding protein mop front sporomusa ovata. Structure 2000;8:1127–36.
[136] Solomon EI, Baldwin MJ, Lowery MD. Electronic-structures of active-sites in copper proteins – contributions to reactivity. Chem Rev 1992;92:521–42.
[137] Basu P, Burgmayer SJN. Pterin chemistry and its relationship to the molybdenum cofactor. Coord Chem Rev 2011;255:1016–38.
[138] Williams BR, Fu YC, Yap GPA, Burgmayer SJN. Structure and reversible pyran formation in molybdenum pyranopterin dithiolene models of the molybdenum cofactor. J Am Chem Soc 2012;134:19584–7.
[139] George GN, Hilton J, Rajagopalan KV. X-ray absorption spectroscopy of dimethyl sulfoxide reductase from Rhodobacter sphaeroides. J Am Chem Soc 1996;118:1113–7.
[140] Garton SD, Garrett RM, Rajagopalan KV, Johnson MK. Resonance Raman characterization of the molybdenum center in sulfite oxidase: identification of Mo=o stretching modes. J Am Chem Soc 1997;119:2590–1.
[141] Ryan MG, Ratnam K, Hille R. The molybdenum centers of xanthine-oxidase and xanthine dehydrogenase – determination of the spectral change associated with reduction from the Mo(VI) to the Mo(IV) state. J Biol Chem 1995;270:19209–12.
[142] Johnson JL, Rajagopalan KV. Tryptic cleavage of rat-liver sulfite oxidase – isolation and characterization of molybdenum and heme domains. J Biol Chem 1977;252:2017–25.
[143] Hemann C, Hood BL, Fulton M, et al. Spectroscopic and kinetic studies of Arabidopsis thaliana sulfite oxidase: nature of the redox-active orbital and electronic structure contributions to catalysis. J Am Chem Soc 2005;127:16567–77.
[144] Cobb N, Conrads T, Hille R. Mechanistic studies of Rhodobacter sphaeroides $Me_2SO$ reductase. J Biol Chem 2005;280:11007–17.
[145] Johnson KE, Rajagopalan KV. An active site tyrosine influences the ability of the dimethyl sulfoxide reductase family of molybdopterin enzymes to reduce S-oxides. J Biol Chem 2001;276:13178–85.
[146] Cobb N, Hemann C, Polsinelli GA, Ridge JP, McEwan AG, Hille R. Spectroscopic and kinetic studies of y114f and w116f mutants of $Me_2SO$ reductase from Rhodobacter capsulatus. J Biol Chem 2007;282:35519–29.
[147] Ridge JP, Aguey-Zinsou KF, Bernhardt PV, Brereton IM, Hanson GR, McEwan AG. Site-directed mutagenesis of dimethyl sulfoxide reductase from Rhodobacter capsulatus: characterization of a Y114 -> F mutant. Biochemistry 2002;41:15762–9.

[148] Mtei RP, Lyashenko G, Stein B, Rubie N, Hille R, Kirk ML. Spectroscopic and electronic structure studies of a dimethyl sulfoxide reductase catalytic intermediate: implications for electron- and atom-transfer reactivity. J Am Chem Soc 2011;133:9762–74.

[149] Pushie MJ, Cotelesage JJH, Lyashenko G, Hille R, George GN. X-ray absorption spectroscopy of a quantitatively Mo(v) dimethyl sulfoxide reductase species. Inorg Chem 2013;52:2830–7.

[150] Sung KM, Holm RH. Oxo transfer reactions mediated by bis(dithiolene)tungsten analogues of the active sites of molybdoenzymes in the DMSO reductase family: comparative reactivity of tungsten and molybdenum. J Am Chem Soc 2001;123:1931–43.

[151] Sung KM, Holm RH. Substitution and oxidation reactions of bis(dithiolene)tungsten complexes of potential relevance to enzyme sites. Inorg Chem 2001;40:4518–25.

[152] Wang JJ, Kryatova OP, Rybak-Akimova EV, Holm RH. Comparative kinetics and mechanism of oxygen and sulfur atom transfer reactions mediated by bis(dithiolene) complexes of molybdenum and tungsten. Inorg Chem 2004;43:8092–101.

[153] Tenderholt AL, Wang J-J, Szilagyi RK, et al. Sulfur K-edge X-ray absorption spectroscopy and density functional calculations on Mo(IV) and Mo(VI)=O bis-dithiolenes: insights into the mechanism of oxo transfer in DMSO reductase and related functional analogues. J Am Chem Soc 2010;132:8359–71.

[154] Sambasivarao D, Scraba DG, Trieber C, Weiner JH. Organization of dimethyl-sulfoxide reductase in the plasma-membrane of Escherichia coli. J Bacteriol 1990;172:5938–48.

[155] Weiner JH, Rothery RA, Sambasivarao D, Trieber CA. Molecular analysis of dimethylsulfoxide reductase – a complex iron-sulfur molybdoenzyme of Escherichia coli. Biochim Biophys Acta 1992;1102:1–18.

[156] Weiner JH, Shaw G, Turner RJ, Trieber CA. The topology of the anchor subunit of dimethyl-sulfoxide reductase of Escherichia coli. J Biol Chem 1993;268:3238–44.

[157] Trieber CA, Rothery RA, Weiner JH. Engineering a novel iron-sulfur cluster into the catalytic subunit of Escherichia coli dimethyl-sulfoxide reductase. J Biol Chem 1996;271:4620–6.

[158] Rothery RA, Trieber CA, Weiner JH. Interactions between the molybdenum cofactor and iron-sulfur clusters of Escherichia coli dimethylsulfoxide reductase. J Biol Chem 1999;274:13002–9.

[159] Stanley NR, Sargent F, Buchanan G, et al. Behaviour of topological marker proteins targeted to the tat protein transport pathway. Mol Microbiol 2002;43:1005–21.

[160] Rothery RA, Kalra N, Turner RJ, Weiner JH. Sequence similarity as a predictor of the transmembrane topology of membrane-intrinsic subunits of bacterial respiratory chain enzymes. J Mol Microbiol Biotechnol 2002;4:133–50.

[161] Graham N, Doonan CJ, Rothery RA, Boroumand N, Weiner JH. X-ray absorption spectroscopic characterization of the molybdenum site of Escherichia coli dimethyl sulfoxide reductase. Inorg Chem 2007;46:2–4.

[162] Magalon A, Fedor JG, Walburger A, Weiner JH. Molybdenum enzymes in bacteria and their maturation. Coord Chem Rev 2011;255:1159–78.

[163] Zhao ZW, Weiner JH. Interaction of 2-N-heptyl-4-hydroxyquinoline-N-oxide with dimethyl sulfoxide reductase of Escherichia coli. J Biol Chem 1998;273:20758–63.

[164] Geijer P, Weiner JH. Glutamate 87 is important for menaquinol binding in dmsc of the DMSO reductase (DmsABC) from Escherichia coli. Biochim Biophys Acta Biomemb 2004;1660: 66–74.

[165] Cammack R, Weiner JH. Electron-paramagnetic resonance spectroscopic characterization of dimethyl-sulfoxide reductase of Escherichia coli. Biochemistry 1990;29:8410–6.

[166] Bennett B, Benson N, McEwan AG, Bray RC. Multiple states of the molybdenum center of dimethylsulfoxide reductase from Rhodobacter capsulatus revealed by ER spectroscopy. Eur J Biochem 1994;225:321–31.

[167] Heffron K, Leger C, Rothery RA, Weiner JH, Armstrong FA. Determination of an optimal potential window for catalysis by E coli dimethyl sulfoxide reductase and hypothesis on the role of Mo(V) in the reaction pathway. Biochemistry 2001;40:3117–26.
[168] Stiefel EI. Proposed molecular mechanism for action of molybdenum in enzymes – coupled proton and electron-transfer. Proc Natl Acad Sci USA 1973;70:988–92.
[169] McDevitt CA, Hugenholtz P, Hanson GR, McEwan AG. Molecular analysis of dimethyl sulphide dehydrogenase from Rhodovulum sulfidophilum: its place in the dimethyl sulphoxide reductase family of microbial molybdopterin-containing enzymes. Mol Microbiol 2002;44:1575–87.
[170] Hanlon SP, Holt RA, Moore GR, McEwan AG. Isolation and characterization of a strain of Rhodobacter sulfidophilus – a bacterium which grows autotrophically with dimethylsulfide as electron-donor. Microbiology UK 1994;140:1953–8.
[171] Hanlon SP, Toh TH, Solomon PS, Holt RA, McEwan AG. Dimethylsulfide:acceptor oxidoreductase from Rhodobacter sulfidophilus – the purified enzyme contains b-type haem and a pterin molybdenum cofactor. Eur J Biochem 1996;239:391–6.
[172] McDevitt CA, Hanson GR, Noble CJ, Cheesman MR, McEwan AG. Characterization of the redox centers in dimethyl sulfide dehydrogenase from Rhodovulum sulfidophilum. Biochemistry 2002;41:15234–44.
[173] Creevey NL, McEwan AG, Hanson GR, Bernhardt PV. Thermodynamic characterization of the redox centers within dimethylsulfide dehydrogenase. Biochemistry 2008;47:3770–6.
[174] Jormakka M, Yokoyama K, Yano T, et al. Molecular mechanism of energy conservation in polysulfide respiration. Nat Struct Mol Biol 2008;15:730–7.
[175] Berks BC, Ferguson SJ, Moir JWB, Richardson DJ. Enzymes and associated electron transport systems that catalyse the respiratory reduction of nitrogen oxides and oxyanions. Biochim Biophys Acta Bioenerg 1995;1232:97–173.
[176] Berks BC, Page MD, Richardson DJ, et al. Sequence-analysis of subunits of the membrane-bound nitrate reductase from a denitrifying bacterium – the integral membrane subunit provides a prototype for the diheme electron-carrying arm of a redox loop. Mol Microbiol 1995;15:319–31.
[177] Volbeda A, Charon MH, Piras C, Hatchikian EC, Frey M, Fontecillacamps JC. Crystal-structure of the nickel-iron hydrogenase from Desulfovibrio gigas. Nature 1995;373:580–7.
[178] Sazanov LA, Hinchliffe P. Structure of the hydrophilic domain of respiratory complex i from thermus thermophilus. Science 2006;311:1430–6.
[179] Hunte C, Zickermann V, Brandt U. Functional modules and structural basis of conformational coupling in mitochondrial complex i. Science 2010;329:448–51.
[180] Efremov RG, Baradaran R, Sazanov LA. The architecture of respiratory complex i. Nature 2010;465:441–U61.
[181] Hedderich R, Klimmek O, Kroger A, Dirmeier R, Keller M, Stetter KO. Anaerobic respiration with elemental sulfur and with disulfides. FEMS Microbiol Rev 1998;22:353–81.
[182] Dietrich W, Klimmek O. The function of methyl-menaquinone-6 and polysulfide reductase membrane anchor (psrc) in polysulfide respiration of wolinella succinogenes. Eur J Biochem 2002;269:1086–95.
[183] Nagarajan K, Joshi HK, Chaudhury PK, et al. Structural and functional analogue of the active site of polysulfide reductase from wolinella succinogenes. Inorg Chem 2004;43:4532–3.
[184] Prisner T, Lyubenova S, Atabay Y, MacMillan F, Kroger A, Klimmek O. Multifrequency CW-EPR investigation of the catalytic molybdenum cofactor of polysulfide reductase from wolinella succinogenes. J Biol Inorg Chem 2003;8:419–26.
[185] Kniemeyer O, Heider J. Ethylbenzene dehydrogenase, a novel hydrocarbon-oxidizing molybdenum/iron-sulfur/heme enzyme. J Biol Chem 2001;276:21381–6.
[186] Johnson HA, Pelletier DA, Spormann AM. Isolation and characterization of anaerobic ethylbenzene dehydrogenase, a novel Mo-Fe-S enzyme. J Bacteriol 2001;183:4536–42.

[187] Szaleniec M, Hagel C, Menke M, Nowak P, Witko M, Heider J. Kinetics and mechanism of oxygen-independent hydrocarbon hydroxylation by ethylbenzene dehydrogenase. Biochemistry 2007;46:7637–46.
[188] Kloer DP, Hagel C, Heider J, Schulz GE. Crystal structure of ethylbenzene dehydrogenase from aromatoleum aromaticum. Structure 2006;14:1377–88.
[189] Schroder I, Rech S, Krafft T, Macy JM. Purification and characterization of the selenate reductase from thauera selenatis. J Biol Chem 1997;272:23765–8.
[190] Thorell HD, Stenklo K, Karlsson J, Nilsson T. A gene cluster for chlorate metabolism in ideonella dechloratans. Appl Environ Microbiol 2003;69:5585–92.
[191] Hallberg BM, Bergfors T, Backbro K, Pettersson G, Henriksson G, Divne C. A new scaffold for binding haem in the cytochrome domain of the extracellular flavocytochrome cellobiose dehydrogenase. Structure 2000;8:79–88.
[192] Szaleniec M, Witko M, Heider J. Quantum chemical modelling of the C-H cleavage mechanism in oxidation of ethylbenzene and its derivates by ethylbenzene dehydrogenase. J Mol Catal A Chem 2008;286:128–36.
[193] Szaleniec M, Borowski T, Schuehle K, Witko M, Heider J. Ab initio modeling of ethylbenzene dehydrogenase reaction mechanism. J Am Chem Soc 2010;132:6014–24.
[194] Szaleniec M, Salwinski A, Borowski T, Heider J, Witko M. Quantum chemical modeling studies of ethylbenzene dehydrogenase activity. Int J Quant Chem 2012;112:1990–9.
[195] Bagramyan K, Trchounian A. Structural and functional features of formate hydrogen lyase, an enzyme of mixed-acid fermentation from Escherichia coli. Biochemistry Moscow 2003;68:1159–70.
[196] Jormakka M, Byrne B, Iwata S. Formate dehydrogenase – a versatile enzyme in changing environments. Curr Opin Struct Biol 2003;13:418–23.
[197] Ruschig U, Muller U, Willnow P, Hopner T. $CO_2$ reduction to formate by NADH catalyzed by formate dehydrogenase from Pseudomonas oxalaticus. Eur J Biochem 1976;70:325–30.
[198] Alissandratos A, Kim H-K, Matthews H, Hennessy JE, Philbrook A, Easton CJ. Clostridium carboxidivorans strain p7t recombinant formate dehydrogenase catalyzes reduction of $CO_2$ to formate. Appl Environ Microbiol 2013;79:741–4.
[199] Reda T, Plugge CM, Abram NJ, Hirst J. Reversible interconversion of carbon dioxide and formate by an electroactive enzyme. Proc Natl Acad Sci USA 2008;105:10654–8.
[200] Thome R, Gust A, Toci R, et al. A sulfurtransferase is essential for activity of formate dehydrogenases in Escherichia coli. J Biol Chem 2012;287:4671–8.
[201] George GN, Colangelo CM, Dong J, et al. X-ray absorption spectroscopy of the molybdenum site of Escherichia coli formate dehydrogenase. J Am Chem Soc 1998;120:1267–73.
[202] George GN, Costa C, Moura JJG, Moura I. Observation of ligand-based redox chemistry at the active site of a molybdenum enzyme. J Am Chem Soc 1999;121:2625–6.
[203] Axley MJ, Bock A, Stadtman TC. Catalytic properties of an Escherichia coli formate dehydrogenase mutant in which sulfur replaces selenium. Proc Natl Acad Sci USA 1991;88:8450–4.
[204] Gladyshev VN, Khangulov SV, Axley MJ, Stadtman TC. Coordination of selenium to molybdenum in formate dehydrogenase-H from Escherichia coli. Proc Natl Acad Sci USA 1994;91:7708–11.
[205] Gladyshev VN, Boyington JC, Khangulov SV, Grahame DA, Stadtman TC, Sun PD. Characterization of crystalline formate dehydrogenase h from Escherichia coli – stabilization, EPR spectroscopy, and preliminary crystallographic analysis. J Biol Chem 1996;271:8095–100.
[206] Khangulov SV, Gladyshev VN, Dismukes GC, Stadtman TC. Selenium-containing formate dehydrogenase h from Escherichia coli: a molybdopterin enzyme that catalyzes formate oxidation without oxygen transfer. Biochemistry 1998;37:3518–28.

[207] Raaijmakers HCA, Romao MJ. Formate-reduced E coli formate dehydrogenase H: the reinterpretation of the crystal structure suggests a new reaction mechanism. J Biol Inorg Chem 2006;11:849–54.
[208] Mota CS, Rivas MG, Brondino CD, et al. The mechanism of formate oxidation by metal-dependent formate dehydrogenases. J Biol Inorg Chem 2011;16:1255–68.
[209] Blasco F, Guigliarelli B, Magalon A, Asso M, Giordano G, Rothery RA. The coordination and function of the redox centres of the membrane-bound nitrate reductases. Cell Mol Life Sci 2001;58:179–93.
[210] Jones RW, Lamont A, Garland PB. The mechanism of proton translocation driven by the respiratory nitrate reductase complex of Escherichia coli. Biochem J 1980;190:79–94.
[211] Iwata S, Lee JW, Okada K, et al. Complete structure of the 11-subunit bovine mitochondrial cytochrome bc(1) complex. Science 1998;281:64–71.
[212] Jones RW. Proton translocation by the membrane-bound formate dehydrogenase of Escherichia coli. FEMS Microbiol Lett 1980;8:167–71.
[213] Abaibou H, Pommier J, Benoit S, Giordano G, Mandrandberthelot MA. Expression and characterization of the Escherichia coli fdo locus and a possible physiological-role for aerobic formate dehydrogenase. J Bacteriol 1995;177:7141–9.
[214] Plunkett G, Burland V, Daniels DL, Blattner FR. Analysis of the Escherichia coli genome. 3. DNA-sequence of the region from 87.2 to 89.2 minutes. Nucl Acids Res 1993;21:3391–8.
[215] Friedebold J, Bowien B. Physiological and biochemical-characterization of the soluble formate dehydrogenase, a molybdoenzyme from Alcaligenes eutrophus. J Bacteriol 1993;175:4719–28.
[216] Friedebold J, Mayer F, Bill E, Trautwein AX, Bowien B. Structural and immunological studies on the soluble formate dehydrogenase from Alcaligenes eutrophus. Biol Chem Hoppe-Seyler 1995;376:561–8.
[217] Oh JI, Bowien B. Structural analysis of the fds operon encoding the NAD(+)-linked formate dehydrogenase of Ralstonia eutropha. J Biol Chem 1998;273:26349–60.
[218] Jollie DR, Lipscomb JD. Formate dehydrogenase from Methylosinus trichosporium OB3B – purification and spectroscopic characterization of the cofactors. J Biol Chem 1991;266:21853–63.
[219] Muller U, Willnow P, Ruschig U, Hopner T. Formate dehydrogenase from Pseudomonas oxalaticus. Eur J Biochem 1978;83:485–98.
[220] Finel M. Organization and evolution of structural elements within complex I. Biochim Biophys Acta Bioenerg 1998;1364:112–21.
[221] Stolz JF, Basu P. Evolution of nitrate reductase: molecular and structural variations on a common function. Chembiochem 2002;3:198–206.
[222] Bertero MG, Rothery RA, Boroumand N, et al. Structural and biochemical characterization of a quinol binding site of Escherichia coli nitrate reductase a. J Biol Chem 2005;280:14836–43.
[223] Peters JW, Lanzilotta WN, Lemon BJ, Seefeldt LC. X-ray crystal structure of the Fe-only hydrogenase (CPL) from Clostridium pasteurianum to 1.8 angstrom resolution. Science 1998;282:1853–8.
[224] Enemark JH, Garner CD. The coordination chemistry and function of the molybdenum centres of the oxomolybdoenzymes. J Biol Inorg Chem 1997;2:817–22.
[225] Bray RC, Vincent SP, Lowe DJ, Clegg RA, Garland PB. Electron-paramagnetic-resonance studies on molybdenum of nitrate reductase from Escherichia coli K12. Biochem J 1976;155:201–3.
[226] Clegg RA. Purification and some properties of nitrate reductase (EC 1.7.99.4) from Escherichia coli K12. Biochem J 1976;153:533–41.
[227] Hettmann T, Anemuller S, Borcherding H, Mathe L, Steinrucke P, Diekmann S. Pseudomonas stutzeri soluble nitrate reductase alpha beta-subunit is a soluble enzyme with a similar

electronic structure at the active site as the inner membrane-bound 47 holoenzyme. Febs Letters 2003;534:143–50.

[228] George GN, Bray RC, Morpeth FF, Boxer DH. Complexes with halide and other anions of the molybdenum center of nitrate reductase from Escherichia coli. Biochem J 1985;227:925–31.

[229] Vincent SP, Bray RC. Electron-paramagnetic-resonance studies on nitrate reductase from Escherichia coli K12. Biochem J 1978;171:639–47.

[230] Correia C, Besson S, Brondino CD, et al. Biochemical and spectroscopic characterization of the membrane-bound nitrate reductase from marinobacter hydrocarbonoclasticus 617. J Biol Inorg Chem 2008;13:1321–33.

[231] George GN, Turner NA, Bray RC, Morpeth FF, Boxer DH, Cramer SP. X-ray-absorption and electron-paramagnetic-resonance spectroscopic studies of the environment of molybdenum in high-pH and low-pH forms of Escherichia coli nitrate reductase. Biochem J 1989;259:693–700.

[232] Rothery RA, Bertero MG, Cammack R, et al. The catalytic subunit of Escherichia coli nitrate reductase a contains a novel 4Fe-4S cluster with a high-spin ground state. Biochemistry 2004;43:5324–33.

[233] Rothery RA, Bertero MG, Spreter T, Bouromand N, Strynadka NCJ, Weiner JH. Protein crystallography reveals a role for the fs0 cluster of Escherichia coli nitrate reductase a (narghi) in enzyme maturation. J Biol Chem 2010;285:8801–7.

[234] Guigliarelli B, Asso M, More C, et al. Epr and redox characterization of iron-sulfur centers in nitrate reductases-A and reductases-Z from Escherichia coli – evidence for a high-potential and a low-potential class and their relevance in the electron-transfer mechanism. Eur J Biochem 1992;207:61–8.

[235] Guigliarelli B, Magalon A, Asso M, et al. Complete coordination of the four Fe-S centers of the beta subunit from Escherichia coli nitrate reductase. Physiological, biochemical, and EPR characterization of site-directed mutants lacking the highest or lowest potential 4Fe-4S clusters. Biochemistry 1996;35:4828–36.

[236] Magalon A, Rothery RA, Giordano G, Blasco F, Weiner JH. Characterization by electron paramagnetic resonance of the role of the Escherichia coli nitrate reductase (narghi) iron-sulfur clusters in electron transfer to nitrate and identification of a semiquinone radical intermediate. J Bacteriol 1997;179:5037–45.

[237] Rothery RA, Magalon A, Giordano G, Guigliarelli B, Blasco F, Weiner JH. The molybdenum cofactor of Escherichia coli nitrate reductase a (narghi) – effect of a mobab mutation and interactions with Fe-S clusters. J Biol Chem 1998;273:7462–9.

[238] Magalon A, LemesleMeunier D, Rothery RA, Frixon C, Weiner JH, Blasco F. Heme axial ligation by the highly conserved his residues in helix II of cytochrome b (nari) of Escherichia coli nitrate reductase a (narghi). J Biol Chem 1997;272:25652–8.

[239] Rothery RA, Blasco F, Magalon A, Asso M, Weiner JH. The hemes of Escherichia coli nitrate reductase a (narghi): potentiometric effects of inhibitor binding to nari. Biochemistry 1999;38:12747–57.

[240] Anderson LJ, Richardson DJ, Butt JN. Catalytic protein film voltammetry from a respiratory nitrate reductase provides evidence for complex electrochemical modulation of enzyme activity. Biochemistry 2001;40:11294–307.

[241] Elliott SJ, Hoke KR, Heffron K, et al. Voltammetric studies of the catalytic mechanism of the respiratory nitrate reductase from Escherichia coli: how nitrate reduction and inhibition depend on the oxidation state of the active site. Biochemistry 2004;43:799–807.

[242] Arnoux P, Sabaty M, Alric J, et al. Structural and redox plasticity in the heterodimeric periplasmic nitrate reductase. Nat Struct Biol 2003;10:928–34.

[243] Najmudin S, Gonzalez PJ, Trincao J, et al. Periplasmic nitrate reductase revisited: a sulfur atom completes the sixth coordination of the catalytic molybdenum. J Biol Inorg Chem 2008;13: 737–53.

[244] Jepson BJN, Mohan S, Clarke TA, et al. Spectropotentiometric and structural analysis of the periplasmic nitrate reductase from Escherichia coli. J Biol Chem 2007;282:6425–37.
[245] Breton J, Berks BC, Reilly A, Thomson AJ, Ferguson SJ, Richardson DJ. Characterization of the paramagnetic iron-containing redox centers of Thiosphaera pantotropha periplasmic nitrate reductase. Febs Letters 1994;345:76–80.
[246] Butler CS, Charnock JM, Bennett B, et al. Models for molybdenum coordination during the catalytic cycle of periplasmic nitrate reductase from Paracoccus denitrificans derived from EPR and exafs spectroscopy. Biochemistry 1999;38:9000–12.
[247] Coelho C, Gonzalez PJ, Moura JJG, Moura I, Trincao J, Romao MJ. The crystal structure of Cupriavidus necator nitrate reductase in oxidized and partially reduced states. J Mol Biol 2011;408:932–48.
[248] Zoppellaro G, Bren KL, Ensign AA, et al. Studies of ferric heme proteins with highly anisotropic/highly axial low spin (s=½) electron paramagnetic resonance signals with bis-histidine and histidine-methionine axial iron coordination. Biopolymers 2009;91:1064–82.
[249] Basu P, Shokhirev NV, Enemark JH, Walker FA. NMR-studies of hindered ligand rotation, magnetic-anisotropy, curie behavior, proton spin relaxation, and ligand-exchange in some novel oxomolybdenum(V) iron(III) porphyrinate complexes. J Am Chem Soc 1995;117:9042–55.
[250] Walker FA. Models of the bis-histidine-ligated electron-transferring cytochromes. Comparative geometric and electronic structure of low-spin ferro- and ferrihemes. Chem Rev 2004;104: 589–615.
[251] Bennett B, Berks BC, Ferguson SJ, Thomson AJ, Richardson DJ. Mo(v) electron-paramagnetic-resonance signals from the periplasmic nitrate reductase of Thiosphaera pantotropha. Eur J Biochem 1994;226:789–98.
[252] Gonzalez PJ, Rivas MG, Brondino CD, Bursakov SA, Moura I, Moura JJG. Epr and redox properties of periplasmic nitrate reductase from Desulfovibrio desulfuricans atcc 27774. J Biol Inorg Chem 2006;11:609–16.
[253] Gonzalez PJ, Correia C, Moura I, Brondino CD, Moura JJG. Bacterial nitrate reductases: molecular and biological aspects of nitrate reduction. J Inorg Biochem 2006;100:1015–23.
[254] Simpson PJL, Richardson DJ, Codd R. The periplasmic nitrate reductase in shewanella: the resolution, distribution and functional implications of two nap isoforms, NapEDABC and NapDAGHB. Microbiology 2010;156:302–12.
[255] Simpson PJL, McKinzie AA, Codd R. Resolution of two native monomeric 90 kda nitrate reductase active proteins from shewanella gelidimarina and the sequence of two NapA genes. Biochem Biophys Res Commun 2010;398:13–8.
[256] Fourmond V, Burlat B, Dementin S, et al. Major Mo(v) EPR signature of Rhodobacter sphaeroides periplasmic nitrate reductase arising from a dead-end species that activates upon reduction. Relation to other molybdoenzymes from the DMSO reductase family. J Phys Chem B 2008;112:15478–86.
[257] Bertrand P, Frangioni B, Dementin S, et al. Effects of slow substrate binding and release in redox enzymes: theory and application to periplasmic nitrate reductase. J Phys Chem B 2007;111:10300–11.
[258] Butler CS, Charnock JM, Garner CD, et al. Thiocyanate binding to the molybdenum centre of the periplasmic nitrate reductase from Paracoccus pantotrophus. Biochem J 2000;352:859–64.
[259] Butler CS, Fairhurst SA, Ferguson SJ, et al. Mo(v) co-ordination in the periplasmic nitrate reductase from Paracoccus pantotrophus probed by electron nuclear double resonance (endor) spectroscopy. Biochem J 2002;363:817–23.
[260] Olmo-Mira MF, Cabello P, Pino C, et al. Expression and characterization of the assimilatory NADH-nitrite reductase from the phototrophic bacterium Rhodobacter capsulatus e1f1. Arch Microbiol 2006;186:339–44.
[261] Lin JT, Stewart V. Nitrate assimilation by bacteria. In: Poole RK, ed. Adv Microb Physiol 1998; 39:1.

[262] Wang BM, Pierson LS, Rensing C, Gunatilaka MK, Kennedy C. Nast-mediated antitermination plays an essential role in the regulation of the assimilatory nitrate reductase operon in Azotobacter vinelandii. Appl Environ Microbiol 2012;78:6558–67.
[263] Ramos F, Blanco G, Gutierrez JC, Luque F, Tortolero M. Identification of an operon involved in the assimilatory nitrate-reducing system of Azotobacter vinelandii. Mol Microbiol 1993;8:1145–53.
[264] Gates AJ, Luque-Almagro VM, Goddard AD, Ferguson SJ, Dolores Roldan M, Richardson DJ. A composite biochemical system for bacterial nitrate and nitrite assimilation as exemplified by Paracoccus denitrificans. Biochem J 2011;435:743–53.
[265] Luque-Almagro VM, Gates AJ, Moreno-Vivian C, Ferguson SJ, Richardson DJ, Roldan MD. Bacterial nitrate assimilation: gene distribution and regulation. Biochem Soc Trans 2011;39:1838–43.
[266] Ogawa KI, Akagawa E, Yamane K, et al. The nasB operon and nasA gene are required for nitrate/nitrite assimilation in Bacillus subtilis. J Bacteriol 1995;177:1409–13.
[267] Jepson BJN, Anderson LJ, Rubio LM, et al. Tuning a nitrate reductase for function – the first spectropotentiometric characterization of a bacterial assimilatory nitrate reductase reveals novel redox properties. J Biol Chem 2004;279:32212–8.
[268] Richardson DJ, Berks BC, Russell DA, Spiro S, Taylor CJ. Functional, biochemical and genetic diversity of prokaryotic nitrate reductases. Cell Mol Life Sci 2001;58:165–78.
[269] Gangeswaran R, Lowe DJ, Eady RR. Purification and characterization of the assimilatory nitrate reductase of Azotobacter vinelandii. Biochem J 1993;289:335–42.
[270] Frangioni B, Arnoux P, Sabaty M, et al. In Rhodobacter sphaeroides respiratory nitrate reductase, the kinetics of substrate binding favors intramolecular electron transfer. J Am Chem Soc 2004;126:1328–9.
[271] Hirasawa M, Rubio LM, Griffin JL, et al. Complex formation between ferredoxin and Synechococcus ferredoxin : nitrate oxidoreductase. Biochim Biophys Acta Bioenerg 2004;1608:155–62.
[272] Heinrich-Salmeron A, Cordi A, Brochier-Armanet C, et al. Unsuspected diversity of arsenite-oxidizing bacteria as revealed by widespread distribution of the aoxB gene in prokaryotes. Appl Environ Microbiol 2011;77:4685–92.
[273] Duval S, Santini JM, Nitschke W, Hille R, Schoepp-Cothenet B. The small subunit aroB of arsenite oxidase lessons on the 2Fe-2S Rieske protein superfamily. J Biol Chem 2010;285:20442–51.
[274] Lebrun E, Brugna M, Baymann F, et al. Arsenite oxidase, an ancient bioenergetic enzyme. Mol Biol Evol 2003;20:686–93.
[275] Lett MC, Muller D, Lievremont D, Silver S, Santini J. Unified nomenclature for genes involved in prokaryotic aerobic arsenite oxidation. J Bacteriol 2012;194:207–8.
[276] Osborne TH, Jamieson HE, Hudson-Edwards KA, et al. Microbial oxidation of arsenite in a subarctic environment: diversity of arsenite oxidase genes and identification of a psychrotolerant arsenite oxidiser. Bmc Microbiology 2010;10.
[277] Conrads T, Hemann C, George GN, Pickering IJ, Prince RC, Hille R. The active site of arsenite oxidase from Alcaligenes faecalis. J Am Chem Soc 2002;124:11276–7.
[278] Anderson GL, Williams J, Hille R. The purification and characterization of arsenite oxidase from Alcaligenes faecalis, a molybdenum-containing hydroxylase. J Biol Chem 1992;267:23674–82.
[279] Lieutaud A, van Lis R, Duval S, et al. Arsenite oxidase from Ralstonia sp 22 characterization of the enzyme and its interaction with soluble cytochromes. J Biol Chem 2010;285:20433–41.
[280] Hoke KR, Cobb N, Armstrong FA, Hille R. Electrochemical studies of arsenite oxidase: an unusual example of a highly cooperative two-electron molybdenum center. Biochemistry 2004;43:1667–74.

[281] Bernhardt PV, Santini JM. Protein film voltammetry of arsenite oxidase from the chemolithoautotrophic arsenite-oxidizing bacterium nt-26. Biochemistry 2006;45:2804–9.
[282] van Lis R, Nitschke W, Warelow TP, Capowiez L, Santini JM, Schoepp-Cothenet B. Heterologously expressed arsenite oxidase: a system to study biogenesis and structure/function relationships of the enzyme family. Biochim Biophys Acta Bioenerg 2012;1817:1701–8.
[283] Saltikov CW, Newman DK. Genetic identification of a respiratory arsenate reductase. Proc Natl Acad Sci USA 2003;100:10983–8.
[284] Saltikov CW, Cifuentes A, Venkateswaran K, Newman DK. The ars detoxification system is advantageous but not required for As(V) respiration by the genetically tractable shewanella species strain Ana-3. Appl Environ Microbiol 2003;69:2800–9.
[285] Malasarn D, Saltikov W, Campbell KM, Santini JM, Hering JG, Newman DK. Arra is a reliable marker for As(V) respiration. Science 2004;306:455.
[286] Krafft T, Macy JM. Purification and characterization of the respiratory arsenate reductase of Chrysiogenes arsenatis. Eur J Biochem 1998;255:647–53.
[287] Malasarn D, Keeffe JR, Newman DK. Characterization of the arsenate respiratory reductase from Shewanella sp strain ana-3. J Bacteriol 2008;190:135–42.
[288] Richey C, Chovanec P, Hoeft SE, Oremland RS, Basu P, Stolz JF. Respiratory arsenate reductase as a bidirectional enzyme. Biochem Biophys Res Commun 2009;382:298–302.
[289] Zargar K, Hoeft S, Oremland R, Saltikov CW. Identification of a novel arsenite oxidase gene, arxA, in the haloalkaliphilic, arsenite-oxidizing bacterium Alkalilimnicola ehrlichii strain MLHE-1. J Bacteriol 2010;192:3755–62.
[290] Sultana M, Vogler S, Zargar K, et al. New clusters of arsenite oxidase and unusual bacterial groups in enrichments from arsenic-contaminated soil. Arch Microbiol 2012;194:623–35.
[291] Zargar K, Conrad A, Bernick DL, et al. Arxa, a new clade of arsenite oxidase within the DMSO reductase family of molybdenum oxidoreductases. Environ Microbiol 2012;14:1635–45.
[292] Hoeft SE, Blum JS, Stolz JF, et al. Alkalilimnicola ehrlichii sp nov., a novel, arsenite-oxidizing haloalkaliphilic gammaproteobacterium capable of chemoautotrophic or heterotrophic growth with nitrate or oxygen as the electron acceptor. Int J Syst Evol Microbiol 2007;57:504–12.
[293] Fisher JC, Hollibaugh JT. Selenate-dependent anaerobic arsenite oxidation by a bacterium from Mono Lake, California. Appl Environ Microbiol 2008;74:2588–94.
[294] Kulp TR, Hoeft SE, Asao M, et al. Arsenic(III) fuels anoxygenic photosynthesis in hot spring biofilms from Mono Lake, California. Science 2008;321:967–70.
[295] Schnell S, Brune A, Schink B. Degradation of hydroxyhydroquinone by the strictly anaerobic fermenting bacterium Pelobacter massiliensis sp nov. Arch Microbiol 1991;155:511–6.
[296] Brune A, Schnell S, Schink B. Sequential transhydroxylations converting hydroxyhydroquinone to phloroglucinol in the strictly anaerobic, fermentative bacterium Pelobacter massiliensis. Appl Environ Microbiol 1992;58:1861–8.
[297] Reichenbecher W, Schink B. Towards the reaction mechanism of pyrogallol-phloroglucinol transhydroxylase of Pelobacter acidigallici. Biochim Biophys Acta Protein Struct Mol Enzymol 1999;1430:245–53.
[298] Messerschmidt A, Niessen H, Abt D, Einsle O, Schink B, Kroneck PMH. Crystal structure of pyrogallol-phloroglucinol transhydroxylase, an Mo enzyme capable of intermolecular hydroxyl transfer between phenols. Proc Natl Acad Sci USA 2004;101:11571–6.
[299] Raaijmakers H, Macieira S, Dias JM, et al. Gene sequence and the 1.8 angstrom crystal structure of the tungsten-containing formate dehydrogenase from Desulfolvibrio gigas. Structure 2002;10:1261–72.
[300] Kisker C, Schindelin H, Baas D, Retey J, Meckenstock RU, Kroneck PMH. A structural comparison of molybdenum cofactor-containing enzymes. FEMS Microbiol Rev 1998;22:503–21.

# 9 The role of iron-sulfur clusters in the biosynthesis of the lipoyl cofactor

Nicholas D. Lanz and Squire J. Booker

## 9.1 Introduction

Lipoic acid (LA) is a simple, yet essential cofactor found in all domains of life. It consists of an eight-carbon fatty acid chain containing, in its reduced form, sulfhydryl groups at C6 and C8. It is used in nature both as a redox cofactor and a prosthetic arm that can shuttle intermediates between subunits of large multiprotein complexes. LA is synthesized in *Escherichia coli* and many other organisms in a two-step process that bifurcates from the type II fatty acid biosynthetic pathway. In the first step, catalyzed by the octanoyltransferase, LipB, an octanoyl chain from octanoyl-acyl carrier protein (ACP) is appended onto a specific lysyl residue of a lipoyl carrier protein (LCP). In the second step, catalyzed by lipoyl synthase (LipA in *E. coli*), sulfur atoms are inserted at C6 and C8 of the appended octanoyl chain *via* radical-dependent processes. LipA is a member of the radical *S*-adenosylmethionine (SAM) superfamily of enzymes, which use a [4Fe-4S] cluster to cleave SAM reductively to methionine and a 5′-deoxyadenosyl 5′-radical (5′-dA•). The 5′-dA•, a potent oxidant, abstracts hydrogen atoms from C6 and C8 of the octanoyl substrate to initiate sulfur insertion. Intriguingly, lipoyl synthases contain one additional [4Fe-4S] cluster ligated by cysteines residing in a $Cx_4Cx_5C$ motif located near the N-termini of their primary structures. This cluster has been proposed to be the source of the inserted sulfur atom and therefore sacrificed during turnover. Thus, within a single polypeptide, lipoyl synthases demonstrate two novel and distinctly different roles for iron-sulfur clusters.

## 9.2 Discovery of LA

In the late 1940s, O'Kane and Gunsalus [1] reported that an unknown factor found in yeast extract was required to oxidize pyruvate in *Streptococcus (Enterococcus) faecalis*. The factor was named "pyruvate oxidation factor" due to the inability of this organism to oxidize pyruvate to acetate and carbon dioxide in its absence [2]. Similarly, an unknown factor required for lactic acid bacteria to grow in the absence of acetate, designated acetate-replacing factor, was first reported by Guirard *et al.* [3] and further characterized by Skeggs *et al.* [4]. Reed *et al.* [5] finally isolated and characterized this factor in 1951. Based upon the solubility of the factor in hydrophobic solvents, its acidity and its involvement in the formation of acetate through the oxidative decarboxylation of pyruvate, the compound was renamed lipoic acid.

Initially, acid hydrolysis of 10 tons of bovine liver was required to isolate just 30 mg of LA [6]. Once isolated, further study led to the elucidation of the structure and properties of the molecule. Elemental analysis revealed that the cofactor contains eight carbons and two sulfurs [6, 7]. When reduced with potassium cyanide, a nitroprusside assay indicated the presence of free thiols [6–8]. The thiols could also be reversibly oxidized to a disulfide [7]. A single titratable acid group with a p$K_a$ of 4.76 was found, suggesting that LA contains an aliphatic carboxyl group separated from any unsaturated or polar groups by at least two methylene carbons [6, 8]. Desulfurization by Raney nickel resulted in *n*-octanoic acid (OA), and the lack of a methyl group in the infrared spectrum of LA indicated that one of the sulfur atoms resides on the terminal carbon. These studies lead to the conclusion that LA is 4,8-, 5,8-, or 6,8-dithio-*n*-OA [6]. Brockman and colleagues [9] determined 6,8-dithio-*n*-OA to be the biologically active form. Finally, the dextrorotatory (+) stereoisomer was established as the active form by Walton *et al.* [10], allowing Mislow and Meluch [11] to assign (+)-α-LA as 6*R*-6,8-dithio-*n*-OA (Fig. 9.1a) [10, 11].

A second form, β-LA, also exists in nature, which is the thiosulfinate of LA (Fig. 9.1b) [8]. This form is less prevalent and less biologically active than α-LA but can rescue *E. coli* LA auxotrophs [12, 13].

## 9.3 Functions of the lipoyl cofactor

### 9.3.1 Primary metabolism

LA is best known for its role as a cofactor in primary metabolism, in which, in its functional form, LA is covalently attached to a specific lysyl residue of a subunit of one

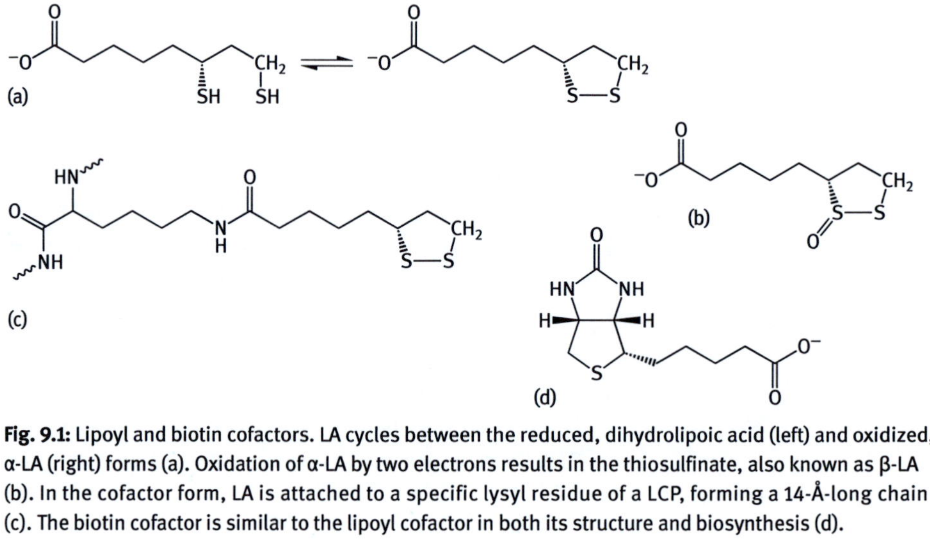


**Fig. 9.1:** Lipoyl and biotin cofactors. LA cycles between the reduced, dihydrolipoic acid (left) and oxidized, α-LA (right) forms (a). Oxidation of α-LA by two electrons results in the thiosulfinate, also known as β-LA (b). In the cofactor form, LA is attached to a specific lysyl residue of a LCP, forming a 14-Å-long chain (c). The biotin cofactor is similar to the lipoyl cofactor in both its structure and biosynthesis (d).

of several large protein complexes. Linked *via* an amide bond to the ε-amino group of the lysyl residue, the cofactor has a reach of approximately 14 Å (Fig. 9.1c) [14]. LA serves both as a redox cofactor, cycling between the five-membered dithiolane ring (α-LA) and the two-electron reduced dihydrolipoamide forms (DHLA) and as a "swinging arm" by which reaction intermediates are shuttled between subunits. It is an essential cofactor in the pyruvate dehydrogenase (PDC), α-ketoglutarate dehydrogenase (KDC), branched-chain α-keto acid dehydrogenase (BCKDC), and acetoin dehydrogenase complexes (ADCs) as well as the glycine cleavage system (GCS) [15–22].

The PDC catalyzes the conversion of pyruvate to acetyl-CoA. The complex consists of a thiamine pyrophosphate (TPP)-dependent dehydrogenase ($E_1$), a lipoyl-dependent dihydrolipoyl transacetylase ($E_2$), and a flavin adenine dinucleotide (FAD)-dependent dihydrolipoyl dehydrogenase ($E_3$) [23]. In the first step of the reaction, the $E_1$ subunit catalyzes the oxidative decarboxylation of pyruvate [14]. The resulting TPP-bound hydroxyethyl group is then transferred to the $E_2$-bound lipoyl cofactor with concomitant reduction of the cofactor's intramolecular disulfide bond. Coenzyme A (CoA) is then used to cleave the resulting thioester bond, forming acetyl-CoA and the dihydrolipoyl cofactor [24]. The dihydrolipoyl cofactor must then be reoxidized by the $E_3$ subunit before the next round of catalysis can take place. In this instance, electrons are transferred to the FAD cofactor to yield $FADH_2$, which is subsequently used to reduce $NAD^+$ to NADH. The KDC and BCKDC catalyze very similar reactions and have the same general subunit composition [23, 25]. In the case of the KDC, α-ketoglutarate is converted to succinyl-CoA and $CO_2$ as part of the TCA cycle, whereas the BCKDC catalyzes the decarboxylation of α-keto acids derived from branched-chain amino acids, forming acyl-CoA derivatives.

In the GCS, four proteins catalyze the reversible decarboxylation of glycine to form $N^5$,$N^{10}$-methylene-tetrahydrofolate (THF), ammonia, and $CO_2$ [26]. The four proteins, P, H, T, and L (GcvP, GcvH, GcvT, and GcvL in *E. coli*), consist of a PLP-dependent decarboxylase, a LCP, a THF-dependent aminomethyl transferase, and a dihydrolipoyl dehydrogenase, respectively. Here, GcvP catalyzes the oxidative decarboxylation of glycine, releasing $CO_2$ and reductively transferring the methylamine moiety to the lipoyl cofactor located on GcvH [15, 27–33]. GcvT then catalyzes the transfer of the methylene group to THF, forming $N^5$,$N^{10}$-methylene-THF and releasing ammonia [31, 34]. The dihydrolipoyl group must then be reoxidized by GcvL in a reaction identical to that catalyzed by the $E_3$ subunit of the PDH [35, 36].

Certain Gram-negative bacteria that respire anaerobically (e.g. *Alcaligenes eutrophus, Clostridium magnum, Klebsiella pneumonia*, and *Pelobacter carbinolicus*) possess an ADC that is used to convert acetoin (3-hydroxy-2-butanone) to acetaldehyde and acetyl-CoA [21, 22, 37, 38]. This complex permits these bacteria to use acetoin as their sole source of carbon. The ADC is very similar to the PDC, KDC, and BCKDC complexes in its subunit composition. Unlike the aforementioned complexes, the ADC does not catalyze a decarboxylation reaction. Instead, it catalyzes the cleavage of acetoin, a four-carbon compound, into acetyl-CoA and acetaldehyde. By analogy to the PDC, the hydroxyethyl product generated on $E_1$ in the TPP-dependent step is successively transferred to the lipoyl cofactor of the $E_2$ subunit and then to CoA. Oxidation of the resulting DHLA by the $E_3$ subunit proceeds as described for the α-keto acid dehydrogenases as well.

### 9.3.2 Antioxidant

More recently, LA has been noted for its role as an intracellular antioxidant. LA has a lower redox potential than many common cellular reducing agents. Two groups have measured the reversible midpoint potential under slightly different conditions and reported it to be −0.325 V at pH 7 and 25°C and −0.294 V at pH 6.8 and 22°C [39, 40]. Comparatively, glutathione has a potential of −0.25 V at pH 6.8 and 30°C, and thioredoxin has a potential of −0.27 V at pH 7 and 25°C [41–43]. The midpoint potential of LA is also quite similar to that of the $NAD^+$/NADH (−0.318 V) and $NADP^+$/NADPH (−0.317 V) redox couples [44]. In particular, LA has been implicated in the recycling of vitamins C and E as well as glutathione [45–47]. DHLA may directly recycle vitamin E (tocopherol), but, more likely, it reduces vitamin E *via* vitamin C (ascorbic acid), glutathione, NADH, NADPH, or ubiquinol [45, 46, 48–52]. Additionally, DHLA and α-LA are scavengers of reactive oxygen species (ROS), including singlet oxygen, hypochlorous acid, and superoxide, hydroxide, and peroxyl radicals [46, 53–60]. DHLA is a stronger reducing agent; however, α-LA scavenges singlet oxygen, whereas DHLA does not [50, 55]. Furthermore, α-LA acid is also an excellent scavenger of hypochlorous acid [55]. The α-LA is oxidized by an additional two electrons by hypochlorous acid to a β-LA (Fig. 9.1b) [12]. Not only can LA react with a broad range of ROS and cellular reducing agents, its water and lipid solubility, along with its ability to cross the blood-brain barrier, allow LA to access a variety of cellular targets [61].

In addition to its role as an antioxidant, LA is also involved in the regulation of several cellular signaling pathways [62, 63]. In particular, LA can modulate the activity of NF-κβ, a transcription factor that regulates the expression of genes involved in inflammation and defense against pathogens [64, 65]. NF-κβ itself is regulated by the redox state of the cell, most likely *via* a disulfide bond in the protein [66, 67]. LA is thought to reduce the disulfide in a manner similar to that of thioredoxin [67]. This regulation is particularly important in the pathology of human immunodeficiency virus (HIV). NF-κβ binds HIV proviral DNA and promotes transcription of the viral genome. LA can inhibit NF-κβ, and consequently, HIV replication [68, 69]. LA has also been used to treat several other conditions associated with oxidative stress or ROS [70, 71]. These include diabetic neuropathy, Alzheimer disease, heavy metal poisoning, mushroom poisoning, and liver cirrhosis [72–74].

## 9.4 Pathways for lipoyl cofactor biosynthesis

### 9.4.1 Exogenous pathway

*E. coli* and many other organisms use two separate pathways to generate the lipoyl cofactor (Fig. 9.2)] [75, 76]. In the exogenous or salvage pathway, LA is scavenged from the environment or dietary sources and attached to LCPs in a process that takes place in two steps. In the first step, free LA is activated in an ATP-dependent reaction to generate lipoyl-AMP and pyrophosphate. In the second step, the lipoyl group is transferred to the LCP with concomitant release of AMP [17, 77]. In *E. coli*, both reactions are catalyzed by one enzyme, lipoate-protein ligase A (LplA) [78, 79]. In mammals and some other organisms, two separate enzymes are required to attach exogenous LA to LCPs [80, 81]. The adenylation of the lipoyl group is catalyzed by lipoate-activating enzyme (LAE), whereas lipoyl transferase (LT) catalyzes the subsequent transfer of the lipoyl group to LCPs. Notably, a number of human pathogens lack this pathway for

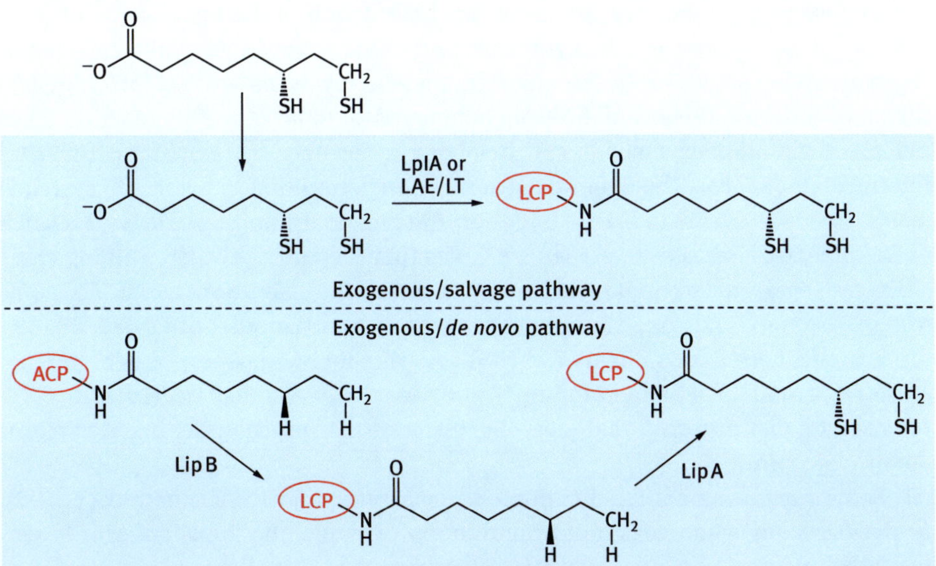


**Fig. 9.2:** Pathways for LA biosynthesis. Nutritionally or environmentally available LA can be incorporated into LCPs *via* an exogenous pathway (top). LA is taken up and attached to an LCP in a two-step process catalyzed by LplA in some prokaryotes or by the combined efforts of LAE and LT in eukaryotes and a few prokaryotes. Alternatively, LA can be synthesized *de novo via* the endogenous pathway (bottom). LipB transfers an octanoyl chain to an LCP, which is followed by insertion of the two sulfur atoms by LipA. The indicated *pro*-R hydrogen at C6 is abstracted during the LipA reaction, and sulfur is inserted with inversion of stereochemistry. The *de novo* biosynthetic pathway is the only mechanism for lipoyl cofactor biosynthesis in some organisms, including a number of human pathogens.

LA salvage. *Mycobacterium tuberculosis*, *Mycobacterium leprae*, and *Corynebacterium diptheriae*, the causative agents of tuberculosis, leprosy, and diphtheria, respectively, do not contain an *lplA* gene [82]. Among the Gram-negative pathogens, strains of *Rickettsia prowazekii* (epidemic typhus), *Rickettsia rickettsii* (Rocky Mountain spotted fever), *Burkholderia pseudomallei* (melioidosis), *Neisseria gonorrhoeae* (gonorrhea), *Neisseria meningitidis* (bacterial meningitis), *Legionella pneumophila* (Legionnaires disease), and *Helicobacter pylori* (linked to gastric ulcers and stomach cancer) also do not encode for a lipoate protein ligase [82]. Each of these organisms appears to contain an endogenous pathway for lipoyl cofactor biosynthesis and one or more lipoyl-dependent complexes, suggesting that LA may still be essential for survival [82]. Unfortunately, for most of the organisms listed above, very little is known about the lipoyl cofactor biosynthetic pathways outside of the reported genomic analyses.

### 9.4.2 Endogenous pathway

The endogenous, or *de novo*, pathway for construction of the lipoyl cofactor is an offshoot of type II fatty acid biosynthesis, and like the exogenous pathway, it requires two enzyme activities. In the first step, an octanoyl transferase (LipB) transfers the eight-carbon fatty acyl chain from octanoyl-ACP to an LCP. This reaction takes place *via* a ping-pong mechanism, involving a covalent octanoyl-LipB thioester intermediate [83–85]. The octanoyl group is transferred first to a specific cysteinyl residue of LipB (Cys169 in *E. coli*) and then attached to the appropriate lysyl residue of the target LCP in a second step [86, 87]. The final step in the two-step endogenous pathway is catalyzed by lipoyl synthase (LipA) and is the insertion of two sulfur atoms into the octanoyl-LCP, one stereoselectively at C6 and the other at C8 [85, 88, 89]. LipA, a member of the radical SAM (RS) superfamily of enzymes, uses a [4Fe-4S] cluster and *S*-adenosyl-L-methionine to generate the free radical necessary to drive the reaction and an additional [4Fe-4S] cluster as the source of the inserted sulfur atoms (*vide infra*).

Some organisms exhibit modified pathways for lipoyl cofactor biosynthesis. In *Bacillus subtilis* and yeast mitochondria, for example, the lipoyl cofactor is synthesized only on Gcv3 (analogous to GcvH from *E. coli*), and an amidotransferase is required to transfer the lipoyl group to other LCPs [90–92]. Schonauer *et al.* [90] demonstrated that, in yeast mitochondria, an LplA homologue (Lip3) and Gcv3 are absolutely required for lipoylation of the $E_2$ subunits of PDC (Lat1) and KDC (Kgd2). More recently, work in the Cronan laboratory has established that the *B. subtilis* endogenous pathway is similar to that in yeast mitochondria [91–93]. A *B. subtilis* gene formerly annotated as encoding a lipoate protein ligase (*lipM*) was actually found to encode an octanoyl transferase [93]. However, the specificity of LipM is limited to GcvH [91]. *B. subtilis* also encodes an amidotransferase (LipL) that is required to transfer either the octanoyl or lipoyl moiety from GcvH to the $E_2$ subunits of the PDC and

KDC [91, 92]. It is yet unknown whether LipL acts before or after insertion of the two sulfur atoms by LipA.

## 9.5 Characterization of LipA

### 9.5.1 Discovery of LipA

Twenty years after the discovery of LA, the *lipA* gene was mapped in *E. coli* by Herbert and Guest *via* chemical mutagenesis and complementation [94]. The authors identified strains that could grow anaerobically in media containing glucose but which failed to grow aerobically in the same media. The strains could be rescued to grow aerobically on glucose if supplemented with LA or two products of the PDC and KDC, acetate and succinate, respectively. These mutants allowed the authors to map the *lip* locus of the *E. coli* genome. The Ashley and Cronan laboratories finally cloned and sequenced the *lipA* gene in the early 1990s [95, 96].

### 9.5.2 *In vivo* characterization of LipA

Before the prevalence of genome sequencing and recombinant gene expression and protein overproduction, results from *in vivo* metabolic feeding studies told much of the story of lipoyl cofactor biosynthesis. Studies by Reed *et al.* [97] determined that OA could act as a precursor to LA. In these studies, greater than 98% of LA extracted from *E. coli* cells cultured in media supplemented with [1-$^{14}$C]-OA was found to contain $^{14}$C at C1. Furthermore, radiolabel from [1-$^{14}$C]-hydroxyoctanoic acid, [1-$^{14}$C]-hexanoic acid, and [1,6-$^{14}$C]-adipate was not incorporated into LA, indicating that OA is converted into LA as an intact unit rather than from smaller fragments.

To determine whether biosynthesis of LA from OA proceeds *via* desaturated intermediates, OA specifically tritiated at C5, C6, C7, or C8 was synthesized and administered to growing *E. coli* along with [1-$^{14}$C]-OA to assess the changes in the ratios of the two radioisotopes in isolated LA [98]. Little to no tritium was lost from the OA labeled at C5, C7, and C8, whereas approximately half of the tritium was lost from the OA labeled at C6. Given that the labeled OAs were a racemic mixture, loss of 50% of the label is consistent with stereoselective removal of hydrogen at C6. The insignificant loss of tritium from C5 and C7 suggests that OA is not desaturated during sulfur insertion. Finally, the retention of tritium at C8 was suggested to be due to significant isotopic discrimination against removing a tritium.

Both [(6*S*)-6-$^{3}$H]-OA and [(6*R*)-6-$^{3}$H]-OA were also synthesized to determine whether the pro-*R* or pro-*S* hydrogen is removed from OA during LA formation [99]. Radiolabel was retained in isolated LA when [(6*S*)-6-$^{3}$H]-OA was fed to *E. coli*, whereas radiolabel was not found in LA when [(6*R*)-6-$^{3}$H]-OA was used, suggesting that the pro-*R* hydrogen is removed from the octanoyl substrate. In addition, given that the natural stereoisomer of LA is the *R* configuration at C6, insertion of the sulfur atom must occur with inversion of configuration when taking into account the change in stereochemical priority rules upon sulfur insertion. This observed inversion of configuration raised the possibility that sulfur insertion at C6 takes place in two steps, possibly *via* the intermediacy of a hydroxylated species that is subsequently activated and displaced by a sulfur nucleophile in an $S_N2$ fashion. If hydroxylation were to occur with retention of configuration, subsequent $S_N2$ displacement would give rise to the observed inversion.

The deuterium tracer studies of Robert White yielded similar conclusions to those discussed above [100, 101]. In these studies, *E. coli* was supplemented with either [2-$^2H_3$]-acetate or $^2H_{15}$-OA, and the deuterium content of isolated LA was measured by GC-MS. A mass shift of +13 for LA generated by cells supplemented with $^2H_{15}$-OA was consistent with OA being a precursor to LA. The result also demonstrated that only two hydrogens are lost from the octanoyl substrate and that insertion of the sulfur atoms does not require intermediate desaturation. Supplementation with [2-$^2H_3$]-acetate yielded LA with a distribution of deuterium atoms that would be expected for LA synthesized from saturated fatty acids. By comparison to the biosynthesis of palmitic acid, deuterium incorporated at C6 would be in the *S* configuration. The distribution of deuterium atoms incorporated into LA was consistent with retention of this deuterium, confirming that LA is synthesized with inversion of configuration.

A second study by White [102] assessed whether hydroxyoctanoic acid is an intermediate in LA biosynthesis. In this study, *E. coli* was supplemented with [8,8-$^2H_2$]-6,8-dihydroxy-OA, [8,8-$^2H_2$]-8-hydroxy-OA, or [6-$^2H_1$]-6-hydroxy-OA. LA extracted from these cultures did not contain the deuterium label, indicating that *E. coli* is unable to utilize the hydroxylated forms of OA. The deuterium label was detected in cells supplemented with [8,8-$^2H_2$]-8-mercaptooctanoic acid (MOA) and, to a lesser extent, [6-$^2H_1$]-6-MOA. Although these results suggest that LA is synthesized *via* a thiolated intermediate rather than a hydroxylated intermediate, 8-MOA could not be extracted from *E. coli* [103].

Studies of two strains of *E. coli* that are auxotrophic for LA yielded similar results [104]. The two strains, W1485-*lip2* and JRG33-*lip9*, have mutations that result, respectively, into S308F and E196K substitutions in LipA. Although these strains are LA auxotrophs, they will grow when supplemented with high concentrations of 8-MOA. The growth response to 8-MOA was approximately 20-fold higher for W1485-*lip2* than JRG33-*lip9*, suggesting that W1485-*lip2* is defective for insertion of sulfur at C8 but not C6. Neither strain could be rescued with 6-MOA supplementation. One hypothesis that was advanced to explain these observations was that 8-MOA is an intermediate in LA biosynthesis, implying that sulfur insertion at C8 precedes insertion at C6. It should be noted that a strain containing a transposon insertion in the *lipA* gene was also rescued with 8-MOA or 6-MOA, suggesting that a redundant pathway for insertion of the second sulfur exists in *E. coli* [96].

### 9.5.3 LipA is an iron-sulfur enzyme

Upon sequencing of the *lipA* gene and determination of the LipA primary structure, the presence of a motif ($Cx_3Cx_2C$) that had the potential to bind metals was noted [95]. This same motif was also present in biotin synthase (BioB), an enzyme that catalyzes sulfur insertion into dethiobiotin, the final step in biotin biosynthesis [96]. Biotin is similar in structure to LA (Fig. 9.1d), and as will be described later, the final step in its biosynthesis is quite similar to the reaction catalyzed by LipA. Busby *et al.* [105] first reported the purification of LipA and noted that the enzyme contained iron and acid-labile sulfide (Fe/S). Although the total iron and sulfide content of the protein was low, 1.8 iron ions and 2.2 sulfide ions per monomer, resonance Raman and UV-vis spectroscopic studies revealed that the iron was in the form of a [4Fe-4S] cluster. The authors determined *via* molecular sieve chromatography that LipA is primarily a dimer and thus inferred that the [4Fe-4S] cluster bridges the dimer interface. Soon after, Ollagnier-de Choudens and Fontecave also reported that LipA could bind an Fe/S cluster [106]. In this instance, the protein was purified and refolded from insoluble aggregates. It could then be reconstituted with up to 1.8 to 2.3 iron and sulfide ions per polypeptide. Although the UV-vis spectrum was similar to spectra of [2Fe-2S] cluster-containing enzymes, the electron paramagnetic resonance (EPR) spectrum of the reduced protein suggested the presence of a $[4Fe\text{-}4S]^{+}$ cluster. The axial EPR signal, exhibiting *g* values of 2.03 and 1.98, displayed temperature dependence and power saturation properties that were more consistent with a [4Fe-4S] cluster than a [2Fe-2S] cluster, but its spin concentration only accounted for 30% of the iron in the sample.

A subsequent study by the Fontecave group and its collaborators [107] reported higher iron (3.7–4.2) and sulfide content. Again, an EPR signal was observed in the presence of a low-potential reductant, and the results of temperature dependence and power saturation studies were consistent with the presence of a $[4Fe\text{-}4S]^{+}$ cluster. Mössbauer spectroscopic analysis of the reconstituted enzyme, however, indicated the presence of several species. The first species, observed as a quadrupole doublet with $\Delta E_q = 1.20$ mm/s and $\delta = 0.44$ mm/s and accounting for 50% of the total iron, was assigned as a $[4Fe\text{-}4S]^{2+}$ cluster. The other quadrupole doublet ($\Delta E_q = 0.6$ mm/s and $\delta = 0.3$ mm/s) accounted for 10% to 15% of the iron and was assigned as a $[2Fe\text{-}2S]^{2+}$ cluster. The remainder of the iron was present as paramagnetic species that were poorly defined. Exposure of the protein to oxygen followed by gel filtration resulted in degradation of the [4Fe-4S] cluster to a [2Fe-2S] cluster. In addition to containing only 1.9 to 2.1 iron and sulfide ions per monomer, UV-vis, EPR, and Mössbauer spectroscopic analyses all indicated that the iron was in the form of a [2Fe-2S] cluster. The authors concluded that LipA contains one [4Fe-4S] cluster per monomer, but that interconversion between cluster types may occur depending upon the redox state of the cell. The true cluster content of LipA was not determined until new methods for overproduction, purification, and activity determination were developed (*vide infra*).

### 9.5.4 LipA is an RS enzyme

It was first noted that two of the products of the activating enzyme of pyruvate formate-lyase are 5′-dA and methionine as opposed to *S*-adenosylhomocysteine (SAH), the product of SAM when used in its traditional role as methyl donor [108]. Further experiments demonstrated the transfer of tritium between the C5′ hydrogens of SAM and lysine in the reaction of lysine 2,3-aminomutase, suggesting direct hydrogen abstraction by a 5′-dA• [109, 110]. These enzymes, and others discovered later, use an electron donated from a [4Fe-4S]$^{1+}$ cluster to cleave SAM reductively, generating methionine and a 5′-dA• (Fig. 9.3). In a groundbreaking bioinformatics study, Sofia *et al.* [111] uncovered a superfamily of over 600 related enzymes that were proposed to use SAM to carry out radical-mediated catalysis, and gave this superfamily the name "radical SAM." In the last 10–15 years since that study, the number of RS proteins has grown to at least 44,680 individual sequences representing 3,142 genomes and at least 40 different types of reactions [112]. The superfamily is identified by an almost universally conserved $Cx_3Cx_2C$ motif containing cysteines that ligate three iron ions of a [4Fe-4S] cluster. Additionally, RS enzymes often have an aromatic residue adjacent to the third cysteine that may lower the redox potential of the cluster as well as a second glycine- and proline-rich motif that is similar to SAM-binding sites in methyltransferases [111]. These features allow SAM to bind to the unique iron site of the [4Fe-4S] cluster of RS enzymes *via* its α-amino and α-carboxylate groups positioning its

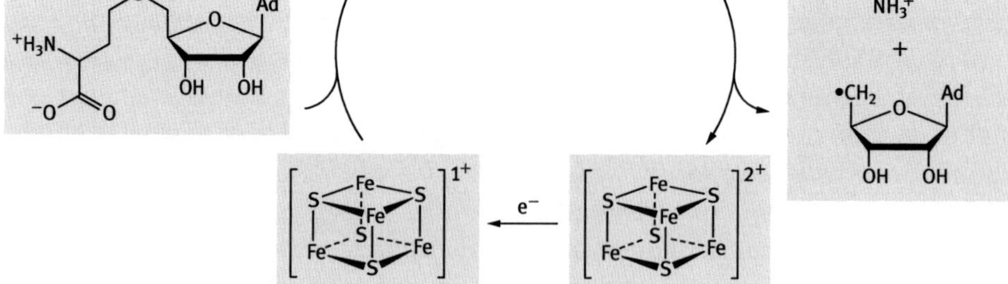


**Fig. 9.3:** Mechanism of SAM cleavage by RS enzymes. Binding of SAM to a reduced [4Fe-4S]$^{1+}$ cluster facilitates transfer of an electron into the sulfonium ion of SAM, resulting in reductive cleavage of SAM to methionine and a 5′-deoxyadenosyl 5′-radical. The [4Fe-4S]$^{2+}$ cluster must be reduced by one electron for subsequent rounds of turnover.

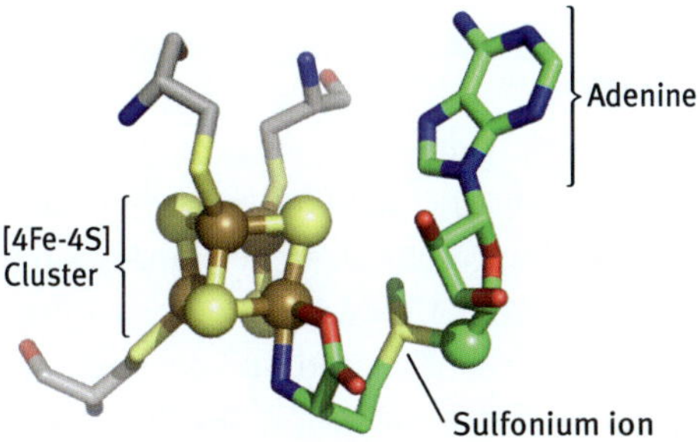


**Fig. 9.4:** SAM bound to the [4Fe-4S] cluster of PFL activase (3CB8). SAM binds the unique iron ion of the RS cluster *via* its α-carboxy- and α-amino groups. The carbon backbones of SAM and the cysteine ligands to the cluster are shown in pale green and gray, respectively. Sulfur, iron, oxygen, and nitrogen are depicted in pale yellow, tan, red, and blue, respectively. The 5′-carbon that bears the radical upon homolytic cleavage, immediately adjacent to the sulfonium ion, is depicted as a pale green sphere. The structures were prepared using PyMOL Molecular Graphics software (www.pymol.org).

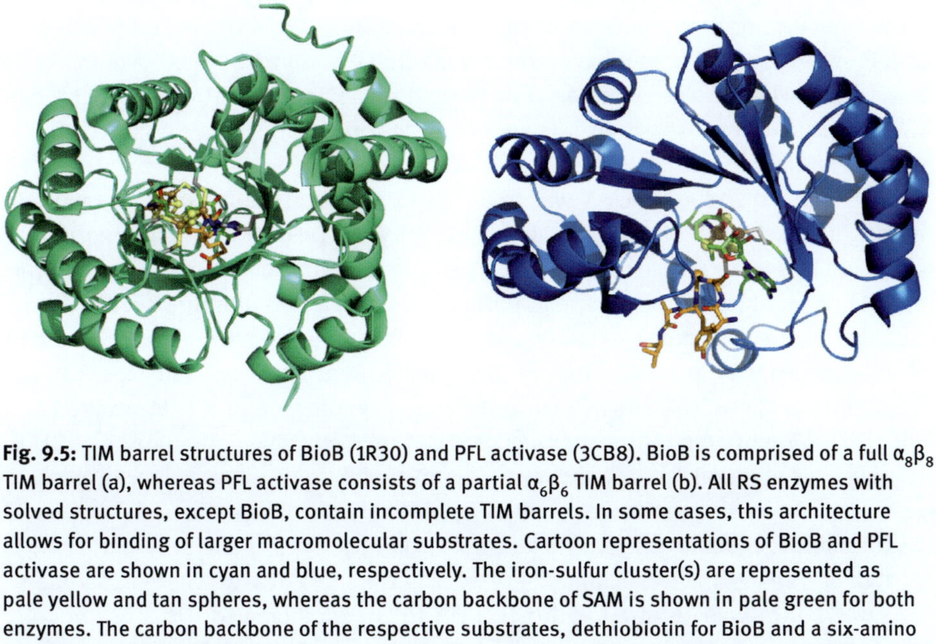

**Fig. 9.5:** TIM barrel structures of BioB (1R30) and PFL activase (3CB8). BioB is comprised of a full $\alpha_8\beta_8$ TIM barrel (a), whereas PFL activase consists of a partial $\alpha_6\beta_6$ TIM barrel (b). All RS enzymes with solved structures, except BioB, contain incomplete TIM barrels. In some cases, this architecture allows for binding of larger macromolecular substrates. Cartoon representations of BioB and PFL activase are shown in cyan and blue, respectively. The iron-sulfur cluster(s) are represented as pale yellow and tan spheres, whereas the carbon backbone of SAM is shown in pale green for both enzymes. The carbon backbone of the respective substrates, dethiobiotin for BioB and a six-amino acid peptide for PFL activase, are shown in orange. The structures were prepared using PyMOL Molecular Graphics software (www.pymol.org).

sulfonium group approximately 4 Å from the iron-sulfur cluster (Fig. 9.4) [113–117]. This binding configuration presumably allows for facile transfer of an electron from the cluster to cleave SAM. X-ray crystallographic structures of a number of RS enzymes have shown that the core RS fold is comprised of either a full ($\alpha_8\beta_8$) or partial ($\alpha_n\beta_n$) triose phosphate isomerase (TIM) barrel (Fig. 9.5) [113–120].

The first demonstration of LipA turnover was published by Miller *et al.* [121], who developed a two-step coupled assay to measure the enzyme's activity. In the first step, LipA, octanoyl-ACP (then believed to be the physiological substrate for LipA), SAM, and dithionite (a requisite low-potential reductant) were allowed to react, and then LipB and apo-PDC were added to allow transfer of the synthesized lipoyl cofactor to the PDC. The activity of the PDC was then determined in a spectrophotometric assay in which the lipoyl-PDC was used to reduce an analogue of nicotinamide adenine dinucleotide, 3-acetylpyridine adenine dinucleotide, in a thiamine-, pyruvate-, CoA-, and cysteine-dependent reaction. A maximum of $0.032 \pm 0.010$ equivalent (equiv.) of product per equivalent of LipA were produced in a 1-h reaction, which was absolutely dependent on the presence of SAM, reductant, octanoyl-ACP, and LipA. The presence of the lipoyl group attached to the $E_2$ subunit of the PDC was confirmed by matrix-assisted laser desorption/ionization mass spectrometry. The authors could not readily explain the miniscule amount of product produced but did note that only a small fraction of the [4Fe-4S] cluster was reduced by sodium dithionite and that the amount of lipoyl cofactor produced correlated with the amount of $[4Fe\text{-}4S]^+$ cluster detected by EPR. Furthermore, the presumed product of the first reaction, lipoyl-ACP, could not be detected. However, it has since been determined that octanoyl-LCP, and not octanoyl-ACP, is the substrate for LipA [85]. Therefore, in the enzymatic assay described above, the LipA reaction did not begin until the addition of LipB.

The true substrate for LipA was revealed in a second study by the same laboratories using both *in vivo* and *in vitro* methods [85]. Despite multiple reports that octanoyl-LCPs were not substrates for LipA, the inability to detect lipoyl-ACP warranted further study [96, 121]. The authors found that a *lipB* mutant strain of *E. coli* could be rescued by supplementation with OA, but only if intact copies of the *lplA* and *lipA* genes were present. This result suggested that LplA can transfer OA to the $E_2$ subunits of the PDC and KDC. To confirm this hypothesis, $[1\text{-}^{14}C]$-OA was fed to a *lipA lipB fadE* mutant strain that expressed the lipoyl-carrying domain of the PDH $E_2$ subunit. This strain lacks the endogenous pathway for lipoyl cofactor biosynthesis as well as the ability to degrade fatty acids. The octanoylated $E_2$ domain was detected by electrophoretic mobility shift and mass spectrometry, again indicating that LplA is capable of transferring the octanoyl moiety of OA to LCPs. In a separate experiment, the same strain was cultured with $[^2H_{15}]$-OA to produce labeled $[^2H_{15}]$-octanoyl-$E_2$. After washing the cells to remove free $^2H_{15}$-OA and resuspending them in fresh media, a cosmid encoding *lipA* was transduced into the cells using λ phage. The authors then detected $^2H_{13}$-LA, demonstrating that LipA can utilize octanoyl-LCPs as a substrate.

*In vitro* assays confirmed that LipA catalyzes cleavage of SAM to generate 5′-deoxyadenosine (5′-dA) and uses an octanoyl-LCP as a substrate (85). In a reaction containing reduced LipA, LipB, octanoyl-ACP, apo-PDC $E_2$ domain, and *S*-$[2,8\text{-}^3H]$-adenosylmethionine, formation of 5′-$[2,8\text{-}^3H]$-dA was detected. The authors estimated that 10 to 40 molecules of SAM were cleaved for each molecule of lipoyl product generated. SAM was consumed, and 5′-dA was produced only when reduced LipA, octanoyl-PDH $E_2$ domain, and labeled SAM were present. Alternatively, when octanoyl-PDC $E_2$ domain was replaced with octanoyl-ACP, 50-fold less 5′-dA was produced, even after a considerably longer incubation time. To confirm this result with a physiological substrate, octanoyl-PDH was purified and incubated with reduced LipA and SAM. Formation of lipoyl-PDH required both the presence of SAM and LipA, and its extent of formation depended on the LipA concentration. Together, these results indicated that octanoyl-PDH, and presumably other octanoyl-LCPs, are the preferred substrates for LipA.

Shortly after, a second report detailed the utilization of a second LCP, the H protein of the GCS (GcvH), as a substrate for LipA [88]. An *in vitro* coupled enzyme assay, in which lipoyl-GcvH reduces $NAD^+$ *via* lipoamide dehydrogenase, was used to measure the amount of lipoyl-GcvH produced by LipA. In this assay, the rate of reduction of $NAD^+$ by lipoamide dehydrogenase is a function of the concentration of lipoyl-GcvH synthesized by LipA, which could be assessed from a standard curve of reduction rates in assays containing graded concentrations of authentic lipoyl-GcvH. Using this method of quantification of lipoyl product, along with quantification of SAM and 5′-dA by HPLC, the authors found that approximately 2.4 to 2.7 equiv. of 5′-dA were produced for each equivalent of lipoyl product. The amount of 5′-dA produced above 2 equiv. was attributed to abortive cleavage of SAM.

In a second experiment, in which octanoyl-$^2H_{15}$-GcvH was used as the substrate, GC-MS analysis revealed that mono-deuterated 5′-dA was the primary product, with a small amount of unlabeled 5′-dA generated as well. Di-deuterated 5′-dA was not detected. The generation of labeled 5′-dA is evidence for direct hydrogen removal from the substrate by the 5′-dA•, and the lack of multiply deuterated 5′-dA or deuterated SAM is further evidence that SAM is not regenerated during the reaction. Curiously, when the octanoyl-$^2H_{15}$-GcvH substrate was used, production of a monothiolated species was observed but not the lipoyl product, and the amount of 5-dA generated decreased approximately 2-fold. This result suggests that there is a significant isotope effect associated with abstraction of one of the two deuterium atoms of the substrate.

Bryant *et al.* [122] simplified the assay for LipA using a synthetic tetrapeptide substrate containing an octanoyllysine moiety, which allowed direct detection by HPLC. The tetrapeptide sequence was derived from the sequence of the $E_2$ subunit of the *Sulfolobus solfataricus* PDC flanking the conserved lysine that bears the lipoyl group. The viability of this substrate was assessed using LipA from *S. solfataricus*, and conversion of ~55% of the octanoyl peptide to the lipoyl peptide was observed by HPLC during a 30-min incubation. Similarly, an octanoyl tripeptide derived from the sequence of the $E_2$ subunit of the *E. coli* PDC was synthesized, and the ability of *E. coli* LipA to convert this peptide to the lipoyl form was measured. Approximately 31% of the peptide was converted to the lipoyl-peptide in 30 min. This simplified assay provided a means to probe the mechanistic details of the LipA reaction more easily (*vide infra*).

### 9.5.5 Product inhibition of LipA

*S*-Adenosylhomocysteine/methylthioadenosine nucleosidase (SAHN) is an important enzyme for degrading adenosine-containing products of SAM-dependent reactions [123]. These products, 5′-dA, SAH, and 5′-methylthioadenosine (MTA), frequently inhibit SAM-dependent enzymes; however, SAHN can relieve this inhibition (Fig. 9.6).

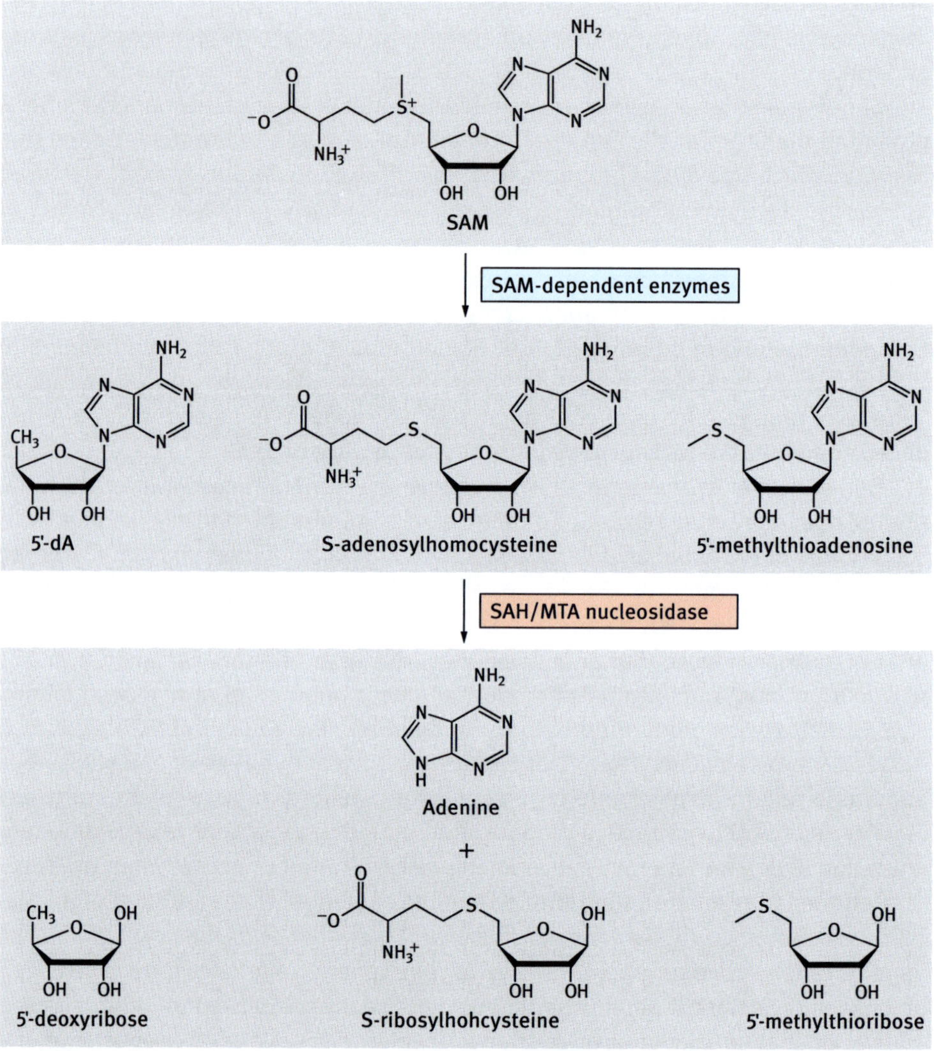


**Fig. 9.6:** Typical adenine-containing products of SAM-dependent enzymes and their further metabolism by SAH/MTA nucleosidase (SAHN). Catalysis by SAM-dependent enzymes most often affords 5′-dA, SAG, or 5′-MTA as by-products. SAHN metabolizes these compounds into adenine plus 5′-deoxyribose, *S*-ribosylhomocysteine, or 5′-methylthioribose, respectively. Further downstream processing converts these compounds into usable cellular metabolites.

Biotin synthase was the first RS enzyme shown to be inhibited by 5′-dA [124]. A strain of *E. coli* containing a null mutation in the gene that encodes for SAHN (*pfs*) was unable to produce biotin but was able to make the immediate precursor to biotin, dethiobiotin, suggesting that the activity of BioB was inhibited in this strain by the accumulation of the cleavage products of SAM. Moreover, growth of this strain was impaired in the presence of exogenously added 5′-dA but could be rescued by cosupplementation with biotin and LA, two metabolites known to be required for aerobic growth of *E. coli*, which are also dependent on RS enzymes for their biosynthesis.

Challand *et al.* [125] further studied the *in vitro* inhibition of RS enzymes by 5′-dA and methionine. Addition of methionine to LipA, BioB, or tyrosine lyase assays had little to no effect on the activity of the enzymes, but significant inhibition was observed for all three enzymes in the presence of 5′-dA. The addition of both 5′-dA and methionine had the most profound effect on the activity of each enzyme, with $IC_{50}$ values in the range of 100 and 500 μM for each product. Inclusion of SAHN in the assay relieved the inhibition and, to varying degrees, improved the activity of each enzyme compared with the uninhibited control reaction.

### 9.5.6 LipA contains two [4Fe-4S] clusters

LipA contains two tri-cysteine motifs. The canonical $Cx_3Cx_2C$ RS motif is comprised of cysteines 94, 98, and 101 (*E. coli* numbering). The second cysteine motif, in a $Cx_4Cx_5C$ configuration, is strictly conserved only among LipA proteins. The presence of these two motifs is consistent with LipA containing two sites at which it can bind metal cofactors; however, early evidence indicated the presence of only one cluster. It was hypothesized that the endogenous iron-sulfur cluster assembly machinery in *E. coli* was not capable of manufacturing clusters at a rate sufficient for incorporation into heterologously produced LipA. Kriek *et al.* [126] attempted to remedy this situation by overexpressing the *E. coli isc* operon (*iscSUA-hscBA-fdx*) – which encodes iron-sulfur cluster assembly proteins – simultaneously with *lipA*. Different expression vectors were created containing either the full *isc* operon or various truncated segments of it engineered downstream of the *lipA* gene such that the operon is controlled by the same promoter that controls *lipA* expression. Iron content increased significantly from 2.7 to approximately 3.5 iron ions per monomer, and sulfide content increased by a similar proportion in the as-isolated enzyme. Curiously, it was necessary for the entire operon to be present to improve the cluster content, and the presence of smaller segments of the operon had a deleterious effect. Although the improvement in cluster content was significant, the iron and sulfide content could still account only for one cluster per monomer.

A similar approach to determining the true cluster content of *E. coli* LipA was taken by Cicchillo and colleagues [127–129] using a plasmid containing the *isc* operon from *Azotobacter vinelandii*. In this study, the *A. vinelandii isc* operon, under the control of an arabinose-inducible promoter, was co-transformed into *E. coli* along with the expression vector for *E. coli lipA*. Gene expression from the *isc*-containing plasmid was induced first and was allowed to proceed for one doubling time before expression of the *lipA* gene located on the second plasmid was initiated. This strategy allowed for production of iron-sulfur cluster assembly proteins before expression of *lipA*. The as-isolated LipA contained 6.9 iron and 6.4 sulfide ions per polypeptide and could be chemically reconstituted to contain 13.8 iron and 13.1 sulfide ions per polypeptide. Analysis both by Mössbauer and EPR spectroscopies indicated that approximately 97% of the iron in the as-isolated enzyme and 67% of the iron in the reconstituted enzyme was in the form of $[4Fe\text{-}4S]^{2+}$ clusters. In conjunction with the stoichiometry of iron and sulfide ions per polypeptide, this analysis indicated that ~1.7 [4Fe-4S] clusters per polypeptide were present in the as-isolated enzyme and that ~2.3 [4Fe-4S] clusters were present in the reconstituted enzyme.

The observed stoichiometry of two [4Fe-4S] clusters per polypeptide correlated with the presence of the two predicted cluster-binding motifs, and therefore, variants were constructed in which the three cysteines of either motif were substituted with alanine. The variant containing C→A substitutions at the cysteines of the RS motif, C94A-C98A-C101A, contained 4.2 iron and 4.7 sulfide ions per polypeptide, whereas the variant containing C→A substitutions at the cysteines of the second motif, C68A-C73A-C79A, contained 3.0 iron and 3.6 sulfide ions per polypeptide. In each case, Mössbauer spectroscopy showed that greater than 90% of the iron was in the form of $[4Fe\text{-}4S]^{2+}$ clusters. Therefore, the C94A-C98A-C101A variant contained ~1.0 [4Fe-4S] cluster per protein, whereas the C68A-C73A-C79A variant contained ~0.7 clusters per protein. Both sets of substitutions resulted in loss of one [4Fe-4S] cluster, supporting the hypothesis that the two cysteine-containing motifs each bind an Fe/S cluster. Furthermore, both lipoyl-GcvH and 5′-dA were produced by the wild-type enzyme, whereas neither was detected in activity determinations of the two variants. Thus, both clusters are essential for activity of the enzyme.

### 9.5.7 Two distinct roles for the iron-sulfur clusters

The role of the RS cluster in facilitating SAM cleavage has been well documented, and the loss of activity of the variant in which this cluster is missing was expected. SAM binds to the unique iron site of the cluster *via* its α-amino and α-carboxylate groups, positioning the sulfonium ion in close proximity to the Fe/S cluster for facile electron transfer [130]. The coordination of SAM to the cluster was originally determined by $^{15}N$- and $^{17}O$-electron nuclear double resonance and Mössbauer spectroscopies [131–133]. This coordination was confirmed by X-ray crystallographic studies of several RS enzymes (Fig. 9.4) [113–116, 134]. The energetics of SAM cleavage has been characterized for the RS enzyme lysine 2,3-aminomutase [135]. The midpoint potential of the RS cluster can vary, but with SAM bound, it was determined to be –430 mV [136]. In contrast, the potential required to reduce the sulfonium group of SAM in solution is –1.8 V [137, 138]. The difference of ~1.4 V means that the equilibrium constant would be too small for the reaction to occur at a reasonable rate. However, binding of lysine lowers the midpoint potential of the cluster by ~170 mV, whereas the binding of SAM to the cluster raises the potential required to cleave SAM by approximately 810 mV, dropping the difference to ~400 mV [135]. As of yet, the 5′-dA• has not been observed spectroscopically due to its high energy. An analogue of SAM, *S*-3′,4′-anhydroadenosyl-L-methionine was used to stabilize the product of its cleavage by lysine 2,3-aminomutase [139, 140]. Characterization of the resulting 5′-deoxy-3′,4′-anhydroadenosine 5′-radical by EPR spectroscopy confirmed that a radical intermediate is formed by RS enzymes.

The role of the auxiliary cluster in catalysis was unknown; however, it was speculated that the second cluster may act as a sacrificial donor of sulfur [127]. An accumulation of evidence now suggests that the auxiliary cluster of LipA provides the sulfur required for the reaction. Multiple studies have provided evidence that formation of the lipoyl cofactor requires only the presence of LipA, octanoyl-LCP, SAM, and reductant [88, 121, 122]. In one labeling study in particular, it was determined that both inserted sulfur atoms are derived from the same polypeptide of LipA itself [141]. In this study, LipA was overproduced in *E. coli* supplemented with either natural abundance $Na_2S$ or $Na_2{}^{34}S$, and the resulting proteins were mixed in a 1:1 ratio for further studies. If each LipA polypeptide donates only one sulfur atom, then a 1:2:1 ratio of $^{32}S$-$^{32}S$, $^{32}S$-$^{34}S$, and $^{34}S$-$^{34}S$ product would be expected. However, the lipoyl product observed was almost exclusively $^{32}S$-$^{32}S$ or $^{34}S$-$^{34}S$, indicating that both sulfur atoms are supplied by the same polypeptide. Given that LipA contains only six conserved cysteines and that each is used as ligands to two [4Fe-4S] clusters, the most likely source of the inserted sulfur atoms would be one of the [4Fe-4S] clusters. Results of a recent study using Mössbauer spectroscopy to follow cluster configuration during turnover is consistent with the degradation of one of the [4Fe-4S] clusters during turnover [N.D. Lanz et al., unpublished]. The starting protein used in this study contained 1.8 $[4Fe\text{-}4S]^{2+}$ clusters, whereas the protein obtained after reaction with SAM, dithionite, and a synthetic octanoyl-peptide substrate contained 0.7 $[4Fe\text{-}4S]^{2+}$ clusters per polypeptide as well as a substantial amount of mononuclear $Fe^{2+}$. Although the spectroscopic analysis cannot distinguish which cluster is degraded, the stoichiometry suggests that one of the clusters is left intact, whereae the other is destroyed. This observation argues that both sulfur atoms are derived not only from the same polypeptide but also from the same [4Fe-4S] cluster.

### 9.5.8 A unique intermediate

Douglas *et al.* [89] determined that the sulfur atoms of the lipoyl cofactor are inserted in a stepwise manner. *In vitro* LipA reactions in the presence of an octanoyllysine-containing tripeptide substrate were quenched prior to completion, and analysis by

LC-MS revealed the presence of four tripeptide species: unreacted octanoyllysine substrate, lipoyllysine product, reduced dihydrolipoyllysine product, and a monothiolated octanoyllysine species, proposed to be an intermediate in the reaction. To determine at which carbon atom sulfur is inserted first, a tripeptide substrate containing an [8,8,8-$^2H_3$]-octanoyllysine amino acid was synthesized for use in deuterium tracer experiments. Insertion of a sulfur atom at C8 would result in a product having two deuterium atoms, whereas insertion of a sulfur atom at C6 would result in a product having one deuterium atom, thereby distinguishing between the two possibilities. The results showed that the sulfur atom was inserted exclusively at C6 and also that insertion at C8 took place with a substantial deuterium kinetic isotope effect, which could bias the sequence of sulfur insertions. Therefore, a reaction was conducted with unlabeled substrate, and the monothiolated species was purified, alkylated with iodoacetamide, and subjected to two-dimensional correlation nuclear magnetic resonance spectroscopic analysis to assess whether the species contained sulfur at C6 or C8. The spectrum confirmed that the 6-mercaptooctanoyl peptide is the exclusive species. The generation of the monothiolated intermediate in the context of the auxiliary cluster serving as the sulfur donor suggests that the intermediate could be covalently cross-linked to LipA *via* the sacrificial [4Fe-4S] cluster.

A kinetic and spectroscopic analysis of the formation and decay of the 6-mercaptooctanoyl intermediate was recently conducted [N.D. Lanz et al., unpublished]. Because incorporation of three deuteriums at C8 essentially arrests the reaction after insertion of one sulfur atom [88, 89], the possible presence of a LipA/H protein cross-linked species upon initial sulfur insertion could be investigated by conducting the reaction with [8,8,8-$^2H_3$]octanoyl-H protein [N.D. Lanz et al., unpublished]. Upon incubation of LipA and [8,8,8-$^2H_3$]octanoyl-H protein under turnover conditions and purification of the protein-containing reaction components by anion-exchange chromatography with detection by polyacrylamide gel electrophoresis, three species were observed: unreacted LipA, which eluted essentially in the void volume of the column; unreacted H protein, which eluted near the end of the gradient; and a species containing both LipA and H protein, which eluted near the middle of the gradient. The purified cross-linked species was analyzed by Mössbauer spectroscopy and was found to contain 0.4 $[4Fe\text{-}4S]^{2+}$ clusters, 1.3 $[3Fe\text{-}4S]^{0}$ clusters, and mononuclear $Fe^{2+}$. Because the ratio of the $[4Fe\text{-}4S]^{2+}$ clusters to the $[3Fe\text{-}4S]^{0}$ cluster was much lower than the 1:1 predicted ratio, it was hypothesized that some of the $[4Fe\text{-}4S]^{2+}$ clusters were adventitiously oxidized to $[3Fe\text{-}4S]^{0}$ clusters during purification by anion-exchange chromatography. Therefore, gentler methods for isolating the cross-linked species were employed. In these experiments, LipA from *Thermus thermophilus* (*Tt*LipA) or *E. coli* (*Ec*LipA) was reacted with a biotin-tagged [8,8,8-$^2H_3$] octanoyl peptide, and the cross-linked species was analyzed directly without purification (*Ec*LipA) or after purification on a streptavidin mutein column. In both samples, the total iron in clusters was split among $[3Fe\text{-}4S]^{0}$ clusters (*Tt*LipA 36%, *Ec*LipA 36%), $[4Fe\text{-}4S]^{2+}$ clusters (*Tt*LipA 40%, *Ec*LipA 32%), and $[4Fe\text{-}4S]^{2+}$ clusters with SAM- or methionine-bound (*Tt*LipA 24%,

*Ec*LipA 32%). In these experiments, the amount of observed $[3Fe\text{-}4S]^0$ cluster correlates with the yield of monothiolated peptide quantified by LC-MS.

In an experiment designed to test the intermediacy of the cross-linked species, the cross-linked species between *Ec*LipA and an eight-amino acid peptide containing an octanoyllysine residue was generated by inclusion of only one equivalent of SAM in the reaction. This method exploits the stepwise manner of sulfur insertion by LipA and the requirement of two molecules of SAM to generate one molecule of lipoyl product. Because only one equivalent of SAM is present, the hydrogen at C6 can be removed and the first carbon-sulfur bond formed; however, the reaction is then arrested until additional SAM is supplied. Approximately 0.9 equiv. of the monothiolated species was produced in this manner. Mössbauer spectroscopic analysis revealed the same three species described above in similar amounts as observed for the previous experiments (27% $[3Fe\text{-}4S]^0$, 24% SAM- or methionine-bound $[4Fe\text{-}4S]^{2+}$, 28% $[4Fe\text{-}4S]^{2+}$). Excess peptide substrate was removed by size-exclusion chromatography, and SAM and reductant were added back to allow the reaction to proceed to completion. LC-MS analysis indicated a substantial loss of the monothiolated species and the formation of the lipoyl peptide. Furthermore, the rate of lipoyl product formation from the cross-linked species was somewhat faster than that of the overall reaction, demonstrating the kinetic competence of the intermediate. The Mössbauer spectrum of the fully reacted sample lacked the signal corresponding to the $[3Fe\text{-}4S]^0$ cluster and showed the presence of only $[4Fe\text{-}4S]^{2+}$ clusters and $Fe^{2+}$, consistent with degradation of the auxiliary cluster during turnover.

### 9.5.9 A proposed mechanism for the biosynthesis of the lipoyl cofactor

The summation of the decades of work described here has resulted in a proposal for a mechanism for sulfur insertion to generate the lipoyl cofactor (Fig. 9.7) [88, 89, 142]. First, an electron from the reduced $[4Fe\text{-}4S]^+$ RS cluster is used to cleave SAM reductively to generate a 5′-dA•. The 5′-dA• then abstracts a hydrogen atom from C6 of the octanoyl group of an LCP, creating a carbon-centered radical that attacks a bridging μ-sulfido ion of the auxiliary cluster of LipA. The product of the first half-reaction is a cross-linked species in which the auxiliary [4Fe-4S] cluster of LipA is covalently attached to the octanoyl moiety of the LCP. Methionine and 5′-dA then vacate the active site to allow a second molecule of SAM to bind and be reductively cleaved. The second 5′-dA• then abstracts a hydrogen atom from C8 of the octanoyl group, and the resulting carbon-centered radical attacks a second bridging μ-sulfido ion of the auxiliary [4Fe-4S] cluster. Protonation of this species results in the release of the reduced dihydrolipoyl-containing LCP. Presumably, a mechanism exists *in vivo* to reassemble the auxiliary cluster for further rounds of turnover, although the nature of that process is yet unknown.

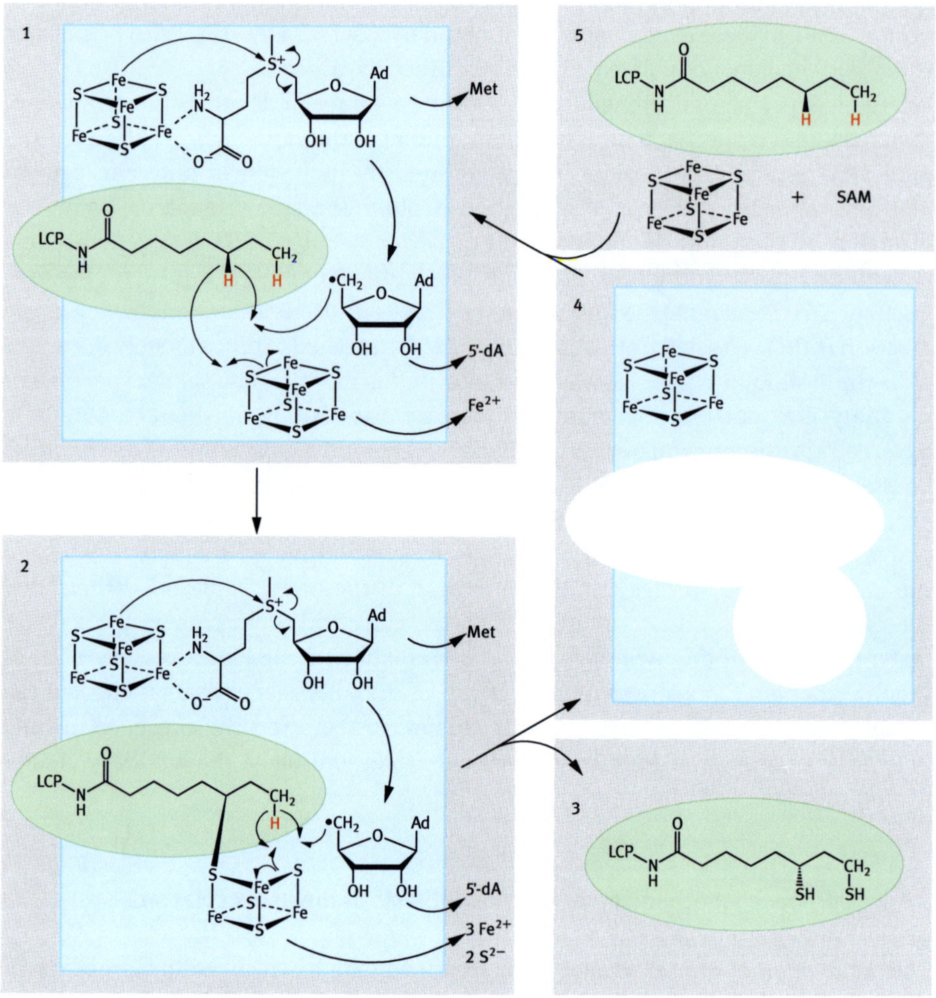


**Fig. 9.7:** Mechanism of lipoyl cofactor biosynthesis by LipA. The stepwise insertion of two sulfur atoms into the octanoyl chain by LipA requires two equivalents of the 5′-dA• to abstract hydrogen atoms from C6 and C8. The hydrogen atom from C6 is abstracted in the first step (panel 1), followed by abstraction of the hydrogen atom from C8 in the second step (panel 2). The carbon-centered radical generated in each instance recombines with a separate bridging μ-sulfido ion of the auxiliary [4Fe-4S] cluster of LipA to generate the lipoyl cofactor. Two equivalents of 5′-dA, methionine, and $S^{2-}$ as well as 4 equiv. of $Fe^{2+}$ are generated for each equivalent of lipoyl product. Release of the lipoyl product (panel 3), generates an inactive form of LipA lacking one of its [4Fe-4S] clusters (panel 4). Reassembly of the auxiliary [4Fe-4S] cluster and binding of the SAM and octanoyl-LCP substrates are required for the enzyme to catalyze multiple rounds of turnover (panel 5).

## 9.6 Conclusions

The reaction catalyzed by LipA is remarkable in that it requires the removal of hydrogen atoms from two completely unactivated carbon centers and the subsequent insertion of sulfur atoms. Indeed, the carbon-hydrogen bonds at C8 and C6 that must be cleaved by LipA have homolytic bond dissociation energies of 98 and 95 kcal/mol, respectively [143]. To accomplish this feat, LipA employs two [4Fe-4S] clusters for two distinct purposes. One cluster interacts with SAM to facilitate its reductive cleavage to a 5′-dA•, whereas the second cluster is believed to serve as the sulfur donor. The requirement of radical chemistry to remove the C6 and C8 hydrogens in the reaction necessitates that the addition of sulfur also involve a homolytic mechanism. A homolytic mechanism would require that the inserted sulfur atom be bonded to a species that could stabilize an unpaired electron. The Fe-S bonds in an Fe/S cluster would seem to be the most appropriate choice, given that homolytic Fe-S bond cleavage can take place concomitant with reduction of the iron ion. The sacrifice of an Fe/S cluster to incorporate sulfur atoms into organic molecules necessitates a mechanism to rebuild them after each turnover. This aspect of sulfur insertion by RS enzymes is relatively poorly understood but is currently moving to the forefront of studies.

## Acknowledgment

This work was supported by NIH Grant GM-63847 (S.J.B.)

## References


[1] O'Kane DJ, Gunsalus IC. Pyruvic acid metabolism. A factor required for oxidation by Streptococcus faecalis. J Bacteriol 1948;56:499–506.

[2] Gunsalus IC, Dolin MI, Struglia L. Pyruvic acid metabolism. III. A manometric assay for pyruvate oxidation factor. J Biol Chem 1952;194:849–57.

[3] Guirard BM, Snell EE, Williams RJ. The nutritional role of acetate for lactic acid bacteria. I. The response to substances related to acetate. Arch Biochem 1946;9:361–79.

[4] Skeggs HR, Wright LD, Cresson EL, et al. Discovery of a new acetate-replacing factor. J Bacteriol 1956;72:519–24.

[5] Reed LJ, Busk BGD, Gunsalus IC, Schnakenberg GHF. Crystalline a-lipoic acid: a catalytic agent associated with pyruvate dehydrogenase. Science 1951;114:93.

[6] Reed LJ, Gunsalus IC, Schnakenberg GHF, et al. Isolation, characterization and structure of α-lipoic acid. J Am Chem Soc 1953;75:1267–70.

[7] Brockman JA, Stokstad ELR, Patterson EL, Pierce JV, Macchi M, Day FP. Structure of protogen-A. J Am Chem Soc 1952;74:1868.

[8] Reed LJ, DeBusk BG, Gunsalus IC, Schnakenberg GHF. Chemical nature of α-lipoic acid. J Am Chem Soc 1951;73:5920.

[9] Bullock MW, Brockman JA, Patterson EL, Pierce JV, Stokstad ELR. Synthesis of compounds in the thioctic acid series. J Am Chem Soc 1952;74:3455.
[10] Walton E, Wagner AF, Peterson LH, Holly FW, Folkers K. The synthesis of (+)-α-lipoic acid and its optical antipode. J Am Chem Soc 1954;76:4748.
[11] Mislow K, Meluch WC. The stereochemistry of a-lipoic acid. J Am Chem Soc 1956;78:5920.
[12] Biewenga GP, Dejong J, Bast A. Lipoic acid favors thiolsulfinate formation after hypochlorous acid scavenging: a study with lipoic acid derivatives. Arch Biochem Biophys 1994;312:114–20.
[13] Herbert AA, Guest JR. Lipoic acid content of Escherichia coli and other microorganisms. Arch Biochem Biophys 1975;106:259–66.
[14] Reed LJ. Multienzyme complexes. Acc Chem Res 1974;7:40–6.
[15] Robinson JR, Klein SM, Sagers RD. Glycine metabolism. Lipoic acid as the prosthetic group in the electron transfer protein P2 from Peptococcus glycinophilus. J Biol Chem 1973;248:5319–23.
[16] Koike M, Reed LJ. α-Keto acid dehydrogenation complexes. II. The role of protein-bound lipoic acid and flavin adenine dinucleotide. J Biol Chem 1960;235:1931–8.
[17] Reed LJ, Koike M, Levitch ME, Leach FR. Studies on the nature and reactions of protein-bound lipoic acid. J Biol Chem 1958;232:143–58.
[18] Ambrose-Griffen MC, Danson MJ, Griffen WG, Hale G, Perham RN. Kinetic analysis of the role of lipoic acid residues in the pyruvate dehydrogenase multienzyme complex of Escherichia coli. Biochem J 1980;187:393–401.
[19] Angelides KJ, Hammes GG. Mechanism of action of the pyruvate dehydrogenase multienzyme complex from Escherichia coli. Proc Natl Acad Sci USA 1978;75:4877–80.
[20] Pettit FH, Yeaman SJ, Reed LJ. Purification and characterization of branched chain α-keto acid dehydrogenase complex of bovine kidney. Proc Natl Acad Sci USA 1978;75:4881–5.
[21] Krüger N, Oppermann FB, Lorenzl H, Steinbüchel A. Biochemical and molecular characterization of the Clostridium magnum acetoin dehydrogenase enzyme system. J Bacteriol 1994;176:3614–30.
[22] Priefert H, Hein S, Krüger N, Zeh K, Schmidt B, Steinbüchel A. Identification and molecular characterization of the Alcaligenes eutrophus H16 aco operon genes involved in acetoin catabolism. J Bacteriol 1991;173:4056–71.
[23] Perham RN. Domains, motifs, and linkers in 2-oxo acid dehydrogenase multienzyme complexes: a paradigm in the design of a multifunctional protein. Biochemistry 1991;30:8501–12.
[24] Voet DJ, Voet JG, Pratt CW. Principles of biochemistry. 2nd ed. Hoboken, NJ: John Wiley & Sons; 2006.
[25] Hackert ML, Xu WX, Oliver RM, et al. Branched-chain alpha-keto acid dehydrogenase complex from bovine kidney: radial distribution of mass determined from dark-field electron micrographs. Biochemistry 1989;28:6816–21.
[26] Sagers RD, Gunsalus IC. Intermediary metabolism of Diplococcus glycinophilus. I. Glycine cleavage and one-carbon interconversions. J Bacteriol 1961;81:541–9.
[27] Fujiwara K, Okamura-Ikeda K, Motokawa Y. Mechanism of the glycine cleavage reaction. Further characterization of the intermediate attached to H-protein and of the reaction catalyzed by T-protein. J Biol Chem 1984;259:10664–8.
[28] Hiraga K, Kikuchi G. The mitochondrial glycine cleavage system. Purification and properties of glycine decarboxylase from chicken liver mitochondria. J Biol Chem 1980;255:11664–70.
[29] Klein SM, Sagers RD. Glycine metabolism. IV. Effect of borohydrine reduction on the pyridoxal phosphate-containing glycine decarboxylase from Peptococcus glycinophilus. J Biol Chem 1967;242:301–5.
[30] Fujiwara K, Okamura K, Motokawa Y. Hydrogen carrier protein from chicken liver: purification, characterization, and role of its prosthetic group, lipoic acid, in the glycine cleavage reaction. Arch Biochem Biophys 1979;197:454–62.
[31] Kochi H, Kikuchi G. Mechanism of the reversible glycine cleavage reaction in Arthrobacter globiformis. I. Purification and function of protein components required for the reaction. J Biochem 1974;75:1113–27.

[32] Kochi H, Kikuchi G. Mechanism of reversible glycine cleavage reaction in Arthrobacter globiformis. Function of lipoic acid in the cleavage and synthesis of glycine. Arch Biochem Biophys 1976;173:71–81.
[33] Motokawa Y, Kikuchi G. Glycine metabolism by rat liver mitochondria. IV. Isolation and characterization of hydrogen carrier protein, and essential factor for glycine metabolism. Arch Biochem Biophys 1969;135:402–9.
[34] Motokawa Y, Kikuchi G. Glycine metabolism by rat liver mitochondria. II. Methylene tetrahydrofolate as the direct one carbon donor in the reaction of glycine synthesis. J Biochem 1969;65:71–5.
[35] Baginsky ML, Huennekens FM. Further studies on the electron transport proteins involved in the oxidative decarboxylation of glycine. Arch Biochem Biophys 1967;120:703–11.
[36] Klein SM, Sagers RD. Glycine metabolism. III. A flavin-linked dehydrogenase associated with the glycine cleavage system in Peptococcus glycinophilus. J Biol Chem 1967;242:297–300.
[37] Deng WL, Chang HY, Peng HL. Acetoin catabolic system of Klebsiella pneumoniae CG43: sequence, expression, and organization of the aco operon. J Bacteriol 1994;176:3527–35.
[38] Oppermann FB, Steinbuchel A. Identification and molecular characterization of the aco genes encoding the Pelobacter carbinolicus acetoin dehydrogenase enzyme system. J Bacteriol 1994;176:469–85.
[39] Sanadi DR, Langley M, Searls RL. alpha-Ketoglutaric dehydrogenase. VI. Reversible oxidation of dihydrothioctamide by diphosphopyridine nucleotide. J Biol Chem 1959;234:178–82.
[40] Ke B. The polarographic behavior of alpha-lipoic acid. Biochim Biophys Acta 1957;25:650–1.
[41] Lin TY, Kim PS. Urea dependence of thiol-disulfide equilibria in thioredoxin: confirmation of the linkage relationship and a sensitive assay for structure. Biochemistry 1989;28:5282–7.
[42] Krause G, Lundstrom J, Barea JL, Pueyo de la Cuesta C, Holmgren A. Mimicking the active site of protein disulfide-isomerase by substitution of proline 34 in Escherichia coli thioredoxin. J Biol Chem 1991;266:9494–500.
[43] Scott EM, Duncan IW, Ekstrand V. Purification and properties of glutathione reductase of human erythrocytes. J Biol Chem 1963;238:3928–33.
[44] Dawson RMC, Elliott DC, Elliott WH, Jones KM. Data for biochemical research. New York: Oxford University Press; 1989.
[45] Scholich H, Murphy ME, Sies H. Antioxidant activity of dihydrolipoate against microsomal lipid peroxidation and its dependence on alpha-tocopherol. Biochim Biophys Acta 1989;1001:256–61.
[46] Kagan VE, Shvedova A, Serbinova E, et al. Dihydrolipoic acid – a universal antioxidant both in the membrane and in the aqueous phase. Reduction of peroxyl, ascorbyl and chromanoxyl radicals. Biochem Pharmacol 1992;44:1637–49.
[47] Rosenberg HR, Culik R. Effect of α-lipoic acid on vitamin C and vitamin E deficiencies. Arch Biochem Biophys 1959;80:86–93.
[48] Kagan V, Serbinova E, Packer L. Antioxidant effects of ubiquinones in microsomes and mitochondria are mediated by tocopherol recycling. Biochem Biophys Res Commun 1990;169:851–7.
[49] Kagan VE, Serbinova EA, Forte T, Scita G, Packer L. Recycling of vitamin E in human low density lipoproteins. J Lipid Res 1992;33:385–97.
[50] Packer L, Witt EH, Tritschler HJ. α-Lipoic acid as a biological antioxidant. Free Radic Biol Med 1995;19:227–50.
[51] Gotz ME, Dirr A, Burger R, et al. Effect of lipoic acid on redox state of coenzyme Q in mice treated with 1-methyl-4-phenyl-1,2,3,6-tetrahydropyridine and diethyldithiocarbamate. Eur J Pharmacol 1994;266:291–300.
[52] Bast A, Haenen GR. Regulation of lipid peroxidation by glutathione and lipoic acid: involvement of liver microsomal vitamin E free radical reductase. Adv Exp Med Biol 1990;264:111–6.

[53] Suzuki YJ, Tsuchiya M, Packer L. Thioctic acid and dihydrolipoic acid are novel antioxidants which interact with reactive oxygen species. Free Radic Res Commun 1991;15:255–63.
[54] Scott BC, Aruoma OI, Evans PJ, et al. Lipoic and dihydrolipoic acids as antioxidants. A critical evaluation. Free Radic Res 1994;20:119–33.
[55] Haenen GR, Bast A. Scavenging of hypochlorous acid by lipoic acid. Biochem Pharmacol 1991;42:2244–6.
[56] Stevens B, Perez SR, Small RD. The photoperoxidation of unsaturated organic molecules – IX. Lipoic acid inhibition of rubrene autoperoxidation. Photochem Photobiol 1974;19:315–16.
[57] Suzuki YJ, Tsuchiya M, Packer L. Antioxidant activities of dihydrolipoic acid and its structural homologues. Free Radic Res Commun 1993;18:115–22.
[58] Devasagayam TP, Di Mascio P, Kaiser S, Sies H. Singlet oxygen induced single-strand breaks in plasmid pBR322 DNA: the enhancing effect of thiols. Biochim Biophys Acta 1991;1088:409–12.
[59] Devasagayam TP, Subramanian M, Pradhan DS, Sies H. Prevention of singlet oxygen-induced DNA damage by lipoate. Chem Biol Interact 1993;86:79–92.
[60] Stary FE, Jindal SJ, Murray RW. Oxidation of alpha-lipoic acid. J Org Chem 1975;40:58–62.
[61] Packer L, Tritschler HJ, Wessel K. Neuroprotection by the metabolic antioxidant alpha-lipoic acid. Free Radic Biol Med 1997;22:359–78.
[62] Kiemer AK, Diesel B. Activation of cytoprotective signaling pathways by alpha-lipoic acid. In: Patel MS, Packer L, eds. Lipoic acid: energy production, antioxidant activity and health effects. New York: CRC Press; 2008:439–60.
[63] Han D, Hamilton RT, Lam PY, Packer L. Modulation of cellular redox and metabolic status by lipoic acid. In: Patel MS, Packer L, eds. Lipoic acid: energy production, antioxidant activity and health effects. New York: CRC Press; 2008:293–314.
[64] Ghosh S, May MJ, Kopp EB. NF-kappa B and Rel proteins: evolutionarily conserved mediators of immune responses. Annu Rev Immunol 1998;16:225–60.
[65] Hofmann MAB, Ziegler A, Wahl R, Nawroth P, Tritschler HJ. Lipoate Effects on Atherogenesis. In: Fuchs J, Packer L, Zimmer G, eds. Lipoic acid in health and disease. New York: Marcel Dekker; 1997:321–36.
[66] Schreck R, Albermann K, Baeuerle PA. Nuclear factor kappa B: an oxidative stress-responsive transcription factor of eukaryotic cells (a review). Free Radic Res Commun 1992;17:221–37.
[67] Matthews JR, Wakasugi N, Virelizier JL, Yodoi J, Hay RT. Thioredoxin regulates the DNA binding activity of NF-kappa B by reduction of a disulphide bond involving cysteine 62. Nucleic Acids Res 1992;20:3821–30.
[68] Suzuki YJ, Aggarwal BB, Packer L. Alpha-lipoic acid is a potent inhibitor of NF-kappa B activation in human T cells. Biochem Biophys Res Commun 1992;189:1709–15.
[69] Baur A, Harrer T, Peukert M, Jahn G, Kalden JR, Fleckenstein B. Alpha-lipoic acid is an effective inhibitor of human immuno-deficiency virus (HIV-1) replication. Klin Wochenschr 1991;69:722–4.
[70] Halliwell B, Gutteridge JMC. Free radicals in biology and medicine. Oxford: Oxford University Press; 1999.
[71] Biewenga GP, Haenen GR, Bast A. The pharmacology of the antioxidant lipoic acid. Gen Pharmacol 1997;29:315–31.
[72] Biewenga GP, Haenen GRMM, Bast A. An overview of lipoate chemistry. In: Fuchs J, Packer L, Zimmer G, eds. Lipoic acid in health and disease. New York: Marcel Dekker; 1997:1–32.
[73] Borcea V, Nourooz-Zadeh J, Wolff SP, et al. alpha-Lipoic acid decreases oxidative stress even in diabetic patients with poor glycemic control and albuminuria. Free Radic Biol Med 1999;26:1495–500.
[74] Tirosh O, Sen CK, Roy S, Kobayashi MS, Packer L. Neuroprotective effects of alpha-lipoic acid and its positively charged amide analogue. Free Radic Biol Med 1999;26:1418–26.

[75] Morris TW, Reed KE, Cronan JE Jr. Lipoic acid metabolism in Escherichia coli: the lplA and lipB genes define redundant pathways for ligation of lipoyl groups to apoprotein. J Bacteriol 1995;177:1–10.
[76] Cronan JE Jr, Zhao X, Jiang Y. Function, attachment and synthesis of lipoic acid in Escherichia coli. Adv Microb Physiol 2005;50:103–46.
[77] Reed LJ, Leach FR, Koike M. Studies on a lipoic acid-activating system. J Biol Chem 1958;232:123–42.
[78] Morris TW, Reed KE, Cronan JE Jr. Identification of the gene encoding lipoate-protein ligase A of Escherichia coli. Molecular cloning and characterization of the lplA gene and gene product. J Biol Chem 1994;269:16091–100.
[79] Green DE, Morris TW, Green J, Cronan JE Jr, Guest JR. Purification and properties of the lipoate protein ligase of Escherichia coli. Biochem J 1995;309:853–62.
[80] Fujiwara K, Okamura-Ikeda K, Motokawa Y. Purification and characterization of lipoyl-AMP:$N^{\varepsilon}$-lysine lipoyltransferase from bovine liver mitochondria. J Biol Chem 1994;269:16605–9.
[81] Fujiwara K, Takeuchi S, Okamura-Ikeda K, Motokawa Y. Purification, characterization, and cDNA cloning of lipoate-activating enzyme from bovine liver. J Biol Chem 2001;276:28819–23.
[82] Spalding MD, Prigge ST. Lipoic acid metabolism in microbial pathogens. Microbiol Mol Biol Rev 2010;74:200–28.
[83] Jordan SW, Cronan JE Jr. A new metabolic link. The acyl carrier protein of lipid synthesis donates lipoic acid to the pyruvate dehydrogenase complex in Escherichia coli and mitochondria. J Biol Chem 1997;272:17903–6.
[84] Brody S, Oh C, Schweizer UHE. Mitochondrial acyl carrier protein is involved in lipoic acid synthesis in Saccharomyces cerevisiae. FEBS Lett 1997;408:217–20.
[85] Zhao X, Miller JR, Jiang Y, Marletta MA, Cronan JE Jr. Assembly of the covalent linkage between lipoic acid and its cognate enzymes. Chem Biol 2003;10:1293–302.
[86] Zhao X, Miller JR, Cronan JE Jr. The reaction of LipB, the octanoyl-[acyl carrier protein]:protein N-octanoyltransferase of lipoic acid synthesis, proceeds through an acyl-enzyme intermediate. Biochemistry 2005;44:16737–46.
[87] Nesbitt NM, Baleanu-Gogonea C, Cicchillo RM, et al. Expression, purification, and physical characterization of Escherichia coli lipoyl(octanoyl)transferase. Protein Expr Purif 2005;39:269–82.
[88] Cicchillo RM, Iwig DF, Jones AD, et al. Lipoyl synthase requires two equivalents of S-adenosyl-L-methionine to synthesize one equivalent of lipoic acid. Biochemistry 2004;43:6378–86.
[89] Douglas P, Kriek M, Bryant P, Roach PL. Lipoyl synthase inserts sulfur atoms into an octanoyl substrate in a stepwise manner. Angew Chem Int Ed Engl 2006;45:5197–9.
[90] Schonauer MS, Kastaniotis AJ, Kursu VA, Hiltunen JK, Dieckmann CL. Lipoic acid synthesis and attachment in yeast mitochondria. J Biol Chem 2009;284:23234–42.
[91] Christensen QH, Martin N, Mansilla MC, de Mendoza D, Cronan JE Jr. A novel amidotransferase required for lipoic acid cofactor assembly in Bacillus subtilis. Mol Microbiol 2011;80:350–63.
[92] Martin N, Christensen QH, Mansilla MC, Cronan JE Jr, de Mendoza D. A novel two-gene requirement for the octanoyltransfer reaction of Bacillus subtilis lipoic acid biosynthesis. Mol Microbiol 2011;80:335–49.
[93] Christensen QH, Cronan JE Jr. Lipoic acid synthesis: a new family of octanoyltransferases generally annotated as lipoate protein ligases. Biochemistry 2010;49:10024–36.
[94] Herbert AA, Guest JR. Biochemical and genetic studies with lysine + methionine mutants of Escherichia coli: lipoic acid and α-ketoglutarate dehydrogenase-less mutants. J Gen Microbiol 1968;53:363–81.
[95] Hayden MA, Huang I, Bussiere DE, Ashley GW. The biosynthesis of lipoic acid: cloning of lip, a lipoate biosynthetic locus of Escherichia coli. J Biol Chem 1992;267:9512–5.

[96] Reed KE, Cronan JE Jr. Lipoic acid metabolism in Escherichia coli: sequencing and functional characterization of the lipA and lipB genes. J Bacteriol 1993;175:1325–36.
[97] Reed LJ, Okaichi T, Nakanishi I. Studies on the biosynthesis of lipoic acid. Abstr Int Symp Chem Nat Prod (Kyoto) 1964:218–20.
[98] Parry RJ. Biosynthesis of lipoic acid. 1. Incorporation of specifically tritiated octanoic acid into lipoic acid. J Am Chem Soc 1977;99:6464–6.
[99] Parry RJ, Trainor DA. Biosynthesis of lipoic acid. 2. Stereochemistry of sulfur introduction at C-6 of octanoic acid. J Am Chem Soc 1978;100:5243–4.
[100] White RH. Stable isotope studies on the biosynthesis of lipoic acid in Escherichia coli. Biochemistry 1980;19:15–9.
[101] White RH. Stoichiometry and stereochemistry of deuterium incorporated into fatty acids by cells of Escherichia coli grown on [methyl-$^{2}H_{3}$]acetate. Biochemistry 1980;19:9–15.
[102] White RH. Biosynthesis of lipoic acid: extent of incorporation of deuterated hydroxy- and thiooctanoic acids into lipoic acid. J Am Chem Soc 1980;102:6605–7.
[103] Parry RJ. Biosynthesis of some sulfur-containing natural products investigations of the mechanism of carbon-sulfur bond formation. Tetrahedron 1983;39:1215–38.
[104] Hayden MA, Huang IY, Iliopoulos G, Orozco M, Ashley GW. Biosynthesis of lipoic acid: characterization of the lipoic acid auxotrophs Escherichia coli W1485-lip2 and JRG33-lip9. Biochemistry 1993;32:3778–82.
[105] Busby RW, Schelvis JPM, Yu DS, Babcock GT, Marletta MA. Lipoic acid biosynthesis: lipA is an iron-sulfur protein. J Am Chem Soc 1999;121:4706–7.
[106] Ollagnier-de Choudens S, Fontecave M. The lipoate synthase from Escherichia coli is an iron-sulfur protein. FEBS Lett 1999;453:25–8.
[107] Ollagnier-de Choudens S, Sanakis Y, Hewitson KS, et al. Iron-sulfur center of biotin synthase and lipoate synthase. Biochemistry 2000;39:4165–73.
[108] Knappe J, Neugebauer FA, Blaschkowski HP, Gänzler M. Post-translational activation introduces a free radical into pyruvate formate-lyase. Proc Natl Acad Sci USA 1984;81:1332–5.
[109] Moss M, Frey PA. The role of S-adenosylmethionine in the lysine 2,3-aminomutase reaction. J Biol Chem 1987;262:14859–62.
[110] Baraniak J, Moss ML, Frey PA. Lysine 2,3-aminomutase. Support for a mechanism of hydrogen transfer involving S-adenosylmethionine. J Biol Chem 1989;264:1357–60.
[111] Sofia HJ, Chen G, Hetzler BG, Reyes-Spindola JF, Miller NE. Radical SAM. a novel protein superfamily linking unresolved steps in familiar biosynthetic pathways with radical mechanisms: functional characterization using new analysis and information visualization methods 2001;29:1097–106.
[112] Booker SJ. Radical SAM enzymes and radical enzymology. Biochim Biophys Acta 2012;1824: 1151–3.
[113] Layer G, Moser J, Heinz DW, Jahn D, Schubert WD. Crystal structure of coproporphyrinogen III oxidase reveals cofactor geometry of radical SAM enzymes. EMBO J 2003;22:6214–24.
[114] Berkovitch F, Nicolet Y, Wan JT, Jarrett JT, Drennan CL. Crystal structure of biotin synthase, an S-adenosylmethionine-dependent radical enzyme. Science 2004;303:76–9.
[115] Hänzelmann P, Schindelin H. Crystal structure of the S-adenosylmethionine-dependent enzyme MoaA and its implications for molybdenum cofactor deficiency in humans. Proc Natl Acad Sci USA 2004;101:12870–5.
[116] Lepore BW, Ruzicka FJ, Frey PA, Ringe D. The x-ray crystal structure of lysine-2,3-aminomutase from Clostridium subterminale. Proc Natl Acad Sci USA 2005;102:13819–24.
[117] Jarrett JT. The novel structure and chemistry of iron-sulfur clusters in the adenosylmethionine-dependent radical enzyme biotin synthase. Arch Biochem Biophys 2005;433:312–21.
[118] Boal AK, Grove TL, McLaughlin MI, Yennawar NH, Booker SJ, Rosenzweig AC. Structural basis for methyl transfer by a radical SAM enzyme. Science 2011;332:1089–92.

[119] Suzuki Y, Noma A, Suzuki T, et al. Crystal structure of the radical SAM enzyme catalyzing tricyclic modified base formation in tRNA. J Mol Biol 2007;372:1204–14.
[120] Nicolet Y, Rubach JK, Posewitz MC, et al. X-ray structure of the [FeFe]-hydrogenase maturase HydE from Thermotoga maritima. J Biol Chem 2008;283:18861–72.
[121] Miller JR, Busby RW, Jordan SW, et al. Escherichia coli LipA is a lipoyl synthase: in vitro biosynthesis of lipoylated pyruvate dehydrogenase complex from octanoyl-acyl carrier protein. Biochemistry 2000;39:15166–78.
[122] Bryant P, Kriek M, Wood RJ, Roach PL. The activity of a thermostable lipoyl synthase from Sulfolobus solfataricus with a synthetic octanoyl substrate. Anal Biochem 2006;351:44–9.
[123] Parveen N, Cornell KA. Methylthioadenosine/S-adenosylhomocysteine nucleosidase, a critical enzyme for bacterial metabolism. Mol Microbiol 2011;79:7–20.
[124] Choi-Rhee E, Cronan JE Jr. A nucleosidase required for in vivo function of the S-adenosyl-L-methionine radical enzyme, biotin synthase. Chem Biol 2005;12:589–93.
[125] Challand MR, Ziegert T, Douglas P, et al. Product inhibition in the radical S-adenosylmethionine family. FEBS Lett 2009;583:1358–62.
[126] Kriek M, Peters L, Takahashi Y, Roach PL. Effect of iron-sulfur cluster assembly proteins on the expression of Escherichia coli lipoic acid synthase. Protein Expr Purif 2003;28:241–5.
[127] Cicchillo RM, Lee KH, Baleanu-Gogonea C, Nesbitt NM, Krebs C, Booker SJ. Escherichia coli lipoyl synthase binds two distinct [4Fe-4S] clusters per polypeptide. Biochemistry 2004;43:11770–81.
[128] Dos Santos P, Johnson D, Ragle B, Unciuleac M, Dean D. Controlled expression of nif and isc iron-sulfur protein maturation components reveals target specificity and limited functional replacement between the two systems. J Bacteriol 2007;189:2854–62.
[129] Johnson DC, Unciuleac M-C, Dean DR. Controlled expression and functional analysis of iron-sulfur cluster biosynthetic components within Azotobacter vinelandii. J Bacteriol 2006;188:7551–61.
[130] Frey PA, Hegeman AD, Ruzicka FJ. The radical SAM superfamily. Crit Rev Biochem Mol Biol 2008;43:63–88.
[131] Chen D, Walsby C, Hoffman BM, Frey PA. Coordination and mechanism of reversible cleavage of S-adenosylmethionine by the (4Fe-4S) center in lysine 2,3-aminomutase. J Am Chem Soc 2003;125:11788–9.
[132] Walsby CJ, Hong W, Broderick WE, et al. Electron-nuclear double resonance spectroscopic evidence that S-adenosylmethionine binds in contact with the catalytically active $(4Fe\text{-}4S)^{+}$ cluster of pyruvate formate-lyase activating enzyme. J Am Chem Soc 2002;124:3143–51.
[133] Krebs C, Broderick WE, Henshaw TF, Broderick JB, Huynh BH. Coordination of adenosylmethionine to a unique iron site of the (4Fe-4S) of pyruvate formate-lyase activating enzyme: a Mössbauer spectroscopic study. J Am Chem Soc 2002;124:912–3.
[134] Cosper MM, Cosper NJ, Hong W, et al. Structural studies of the interaction of S-adenosylmethionine with the [4Fe-4S] clusters in biotin synthase and pyruvate formate-lyase activating enzyme. Protein Sci 2003;12:1573–7.
[135] Wang SC, Frey PA. Binding energy in the one-electron reductive cleavage of S-adenosylmethionine in lysine 2,3-aminomutase, a radical SAM enzyme. Biochemistry 2007;46:12889–95.
[136] Hinckley GT, Frey PA. Cofactor dependence of reduction potentials for $[4Fe\text{-}4S]^{2+/1+}$ in lysine 2,3-aminomutase. Biochemistry 2006;45:3219–25.
[137] Colichman EL, Love DL. Polarography of sulfonium salts. J Org Chem 1952;18:40–6.
[138] Saeva FD, Morgan BP. Mechanism of one-electron electrochemical reductive cleavage reactions of sulfonium salts. J Am Chem Soc 1984;106:4121–5.
[139] Magnusson OT, Reed GH, Frey PA. Spectroscopic evidence for the participation of an allylic analogue of the 5′-deoxyadenosyl radical in the reaction of lysine 2,3-aminomutase. J Am Chem Soc 1999;121:9764–5.

[140] Magnusson OT, Reed GH, Frey PA. Characterization of an allylic analogue of the 5′-deoxyadenosyl radical: an intermediate in the reaction of lysine 2,3-aminomutase. Biochemistry 2001;40:7773–82.
[141] Cicchillo RM, Booker SJ. Mechanistic investigations of lipoic acid biosynthesis in Escherichia coli: both sulfur atoms in lipoic acid are contributed by the same lipoyl synthase polypeptide. J Am Chem Soc 2005;127:2860–1.
[142] Booker SJ, Cicchillo RM, Grove TL. Self-sacrifice in radical S-adenosylmethionine proteins. Curr Opin Chem Biol 2007;11:543–52.
[143] McMillen DF, Golden DM. Hydrocarbon bond dissociation energies. Annu Rev Phys Chem 1982;33:493–532.

# 10 Iron-sulfur clusters and molecular oxygen: function, adaptation, degradation, and repair

Yvain Nicolet and Juan C. Fontecilla-Camps

## 10.1 Introduction

Iron-sulfur clusters (Fe-S) are ubiquitous and the fundamental protein cofactors involved in processes such as electron transfer, redox reactions, and Lewis acid-base chemistry. Remarkably, a $Fe_4S_4$ cluster can be used to generate a radical species that initiates numerous difficult and central reactions of the cell. Having evolved before the oxygenation of earth's atmosphere by photosynthetic organisms, Fe-S clusters are generally oxygen sensitive. Consequently, microbes have adopted several strategies to protect themselves from oxygen toxicity. The simplest strategy is restriction to colonization of the remaining anoxic environments such as the intestinal tracts of animals and the bottom of lakes. Other coping mechanisms have involved (i) rapid reduction of oxygen and its derived species, (ii) Fe-S cluster repair mechanisms, and (iii) their replacement by oxygen-resistant cofactors. The unusual $Fe_3S_3$ cluster of $O_2$-tolerant hydrogenases, which is capable of performing two successive one-electron oxidations, represents a remarkable example of the former strategy. Because oxygen is a very convenient final electron acceptor, some originally anaerobic organisms developed the possibility of switching between anaerobic and aerobic metabolisms. In this case, nature has used the intrinsic oxygen lability of Fe-S clusters to regulate that transition. In this chapter, we discuss the effects of oxygen on Fe-S clusters and adaptations that have developed during evolution, using the available structural data whenever possible.

Iron-sulfur (Fe-S) clusters are among the most abundant prosthetic groups found in proteins, and they are, by far, the most common metal-containing clusters. The most common types are the $Fe_2S_2$ and the $Fe_4S_4$ centers, but they can also be more complex, as exemplified by the FeMo cofactor of nitrogenase, or the presence of other transition metals such as nickel in the active site of anaerobic carbon monoxide dehydrogenase. Although it is often possible to insert $Fe_2S_2$ or $Fe_4S_4$ clusters into a suitable "apo"-protein by simply mixing iron and sulfide ions and a reducing agent, it is known that in living organisms, the assembly of Fe-S clusters requires the mobilization of complex machineries involving numerous proteins (as discussed in Chapters 16 and 22 by Wayne Outten and Hong Ye, respectively). These machineries not only provide correctly assembled clusters but also tightly control their trafficking and delivery to target proteins. Such complexity represents a non-negligible metabolic cost to the cell and underscores the importance of Fe-S clusters for life (indeed, any defect in assembly may result in severe disease if is not directly lethal). Fe-S clusters are also among the main targets of reactive oxygen (ROS) and nitrogen species, which are, for instance, produced by macrophages to fight pathogenic microbial infections. Pathogenic prokaryotes, whose survival depends on the activity of several key Fe-S cluster-containing proteins, have developed a series of protective mechanisms in response to host defenses. In this chapter, we will succinctly discuss the biological roles of Fe-S clusters and will elaborate on the molecular mechanisms involved in their protection and their degradation by ROS.

## 10.2 Fe-S clusters – reasons for their abundance

### 10.2.1 Origin of Fe-S clusters

The autotrophic theory of the origin of life posits that primordial metabolic processes, such as the Krebs cycle and the pentose pathway, along with their amino acid-generating side reactions [1], evolved in connection with minerals such as the clay montmorillonite or, more relevant to our discussion, transition metal sulfides such as FeS and NiS [2]. The central feature of the latter proposition is the "pyrite-pulled" metabolism proposed by G. Wächterhäuser:

$$FeS + H_2S \rightarrow FeS_2\,(\text{pyrite}) + H_2 \tag{1}$$

Reaction (1) generates the energy-rich dihydrogen molecule that is supposed to have been subsequently used in prebiotic reactions such as carbon fixation. Mineral surfaces can readily adsorb many metabolites and, under the appropriate conditions, facilitate the polymerization of both amino acids and nucleotides and catalyze many other reactions that may be relevant for the origin of life. Using a related approach, Russell and Hall [3] have proposed that primordial pre-cellular reactions took place within FeS-coated bubbles in hydrothermal settings. According to this view, the last universal common ancestor (LUCA) would have had mineral boundaries, later evolving into protoarchaeal and protobacterial cells. Thus, LUCA would not have been a cell but rather a set of interconnected, self-regenerating mineral-based interface reactions. Both the mineral surface and FeS-bubble metabolic theories try to solve serious problems related to other origin-of-life theories, which include the problem of large-scale diffusion in the supposedly vast water masses and unfavorable peptide synthesis in water. However, the surface metabolism approach has been criticized because of the difficulty in de-adsorbing polymers such as polypeptides and nucleic acids from mineral surfaces [4]. Structural work by us and others has shown that many fundamental, and probably very ancient, reactions involve metalloenzymes that contain active sites reminiscent of transition metal sulfides, including hydrogen metabolism by both NiFe- and FeFe-hydrogenases [5], carbon fixation by CO dehydrogenase and acetyl-CoA synthase [6, 7] and nitrogen fixation by nitrogenase [8]. It is not clear, however, whether these similarities result from the capture of Fe- and S-containing active sites from the inorganic world [9] or from biological evolution to metal-based gas metabolisms.

### 10.2.2 Functions of Fe-S clusters

Fe-S clusters in proteins are very often involved in electron transfer. These metal centers, of which the most common representatives are the $Fe_4S_4$ clusters and $Fe_2S_2$ clusters, usually switch between +2 and +1 redox states, as in hydrogenases and ferredoxins, respectively. The former clusters can also be found at more positive potentials such as in the high potential iron-sulfur proteins (HiPiPs), which operate between +2 and +3 redox states [10]. The clusters' Fe ions can be formally defined as being either ferrous (+2) or ferric (+3) depending on their oxidation state: in $Fe_4S_4]^{1+}$ clusters there are 1 $Fe^{3+}$ and 3 $Fe^{2+}$ ions, whereas $Fe_4S_4]^{2+}$ centers contain 2 $Fe^{3+}$ and 2 $Fe^{2+}$ ions and the superoxidized $Fe_4S_4]^{3+}$ state has 3 $Fe^{3+}$ and 1 $Fe^{2+}$. This formal assignment does not represent the actual electronic structure of the clusters because the Fe ions can be magnetically coupled; in their +2 redox state, the metal ions form two $Fe^{2.5+}$ $Fe^{2.5+}$ mixed-valence pairs rather than isolated 2 $Fe^{3+}$ and 2 $Fe^{2+}$ ions [11].

Less frequently found are $Fe_3S_4$ clusters, which function between 0 and +1 redox states and have, like in NiFe-hydrogenases, generally more positive redox potentials than their $Fe_4S_4$ counterparts. Protein structures often show arrangements of $Fe_4S_4$ clusters, which are separated by 10 to 14 Å and form "wires" connecting the active site to the molecular surface. An example of such a wire is the $Fe_4S_4$-$Fe_3S_4$-$Fe_4S_4$ motif of NiFe-hydrogenases [12]. The position of the central $Fe_3S_4$ cluster is not easy to rationalize; when this cluster was transformed into a $Fe_4S_4$ center, which introduced a potential change of greater than 200 mV, the observed catalytic rate was comparable to the one obtained for the wild-type enzyme. Nevertheless, the mutated hydrogenase displayed an elevated sensitivity to damage caused by oxygen exposure [13].

Fe-S clusters can also participate in protein-based redox catalysis. Because standard Fe-S clusters, such as $Fe_2S_2$ and $Fe_4S_4$ centers, are normally involved in one-electron processes (but see Section 10.4.2) they can be central to reactions involving radical chemistry such as those carried out by radical *S*-adenosyl-L-methionine (SAM)-dependent enzymes (SDEs). Initially, SDEs were found to constitute a superfamily of proteins sharing a TIM barrel-like three-dimensional fold [14–16]. They perform the reductive cleavage of SAM using an electron coming from a conserved $Fe_4S_4]^{1+}$ cluster, which results in the production of methionine and the 5′-deoxyadenosyl radical species (5′-dA$^{\bullet}$) (Fig. 10.1).

5′-dA$^{\bullet}$ is used by SDEs to catalyze difficult reactions such as methylations, cyclizations, sulfur atom insertions, oxidations, and transfer of radicals to other proteins [17, 18]. Their substrates vary from small organic molecules to nucleic acids and proteins. Lately, there has been a growing interest in characterizing these proteins, especially because they are ubiquitous and usually involved in key processes such as the biosynthesis of essential cofactors and antibiotics [18]. SDEs are also involved in RNA (thio)-methylation, which can increase the specificity of translation [19, 20] or cause antibiotic resistance [21–24]. The role played by the conserved $Fe_4S_4$ cluster in generating the 5′-dA$^{\bullet}$ species of SDEs has been extensively studied [17, 18]. These

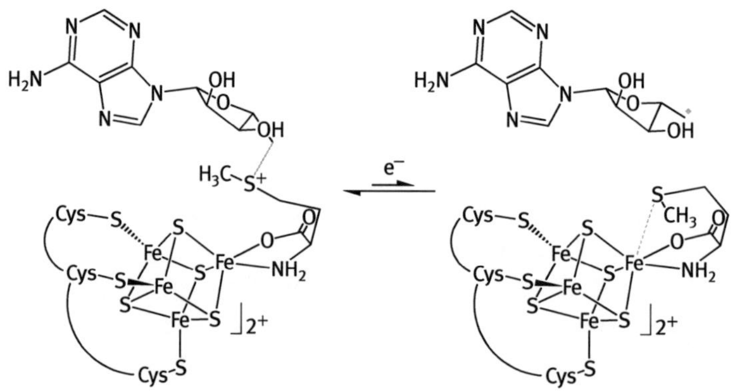


**Fig. 10.1:** SAM cleavage by radical SAM enzymes.

enzymes usually coordinate three irons of the cluster with a three-cysteine sequence consensus motif [14]. Conversely, the methionine fragment of SAM ligates the remaining cluster iron ion through a bidentate amino carboxylate linkage [25–27]. The nature of the radical reaction and the affinity of the protein matrix for the resulting SAM cleavage products determines whether SAM will be used as substrate, and cleavage products will be released, or they will remain bound to the enzyme [28]. In the first case, at the end of the radical-based reaction, a fresh SAM molecule will bind to the enzyme to restore its activity, whereas in the second case SAM will be regenerated. These different outcomes led to the proposition of two alternative mechanisms involving (i) sulfur-based electron transfer between the Fe-S cluster and SAM when the latter is substrate and (ii) iron-based electron transfer when it is a cofactor [17]. However, as indicated by the available SDE X-ray structures and ENDOR data [25, 27], SAM always binds the $Fe_4S_4$ cluster in the same way [29]. Furthermore, our own theoretical calculations have shown that electron transfer between the $Fe_4S_4$ cluster and the SAM sulfonium ion depends only on the energy levels of their respective molecular orbitals, and consequently, both outcomes should use the same mechanism, which we have concluded is iron based [29].

Recently, other members with novel protein folds have been added to the SDE superfamily [30]. One example is ThiC, an enzyme involved in thiamine biosynthesis. Although it has neither the standard SDE fold nor the canonical $Fe_4S_4$ cluster cysteine-binding motif, ThiC still generates the 5′-dA$^•$ radical [31]. A more extreme case is Dph2, which uses both unique $Fe_4S_4$ cluster and SAM binding sites to generate a 3-amino-3-carboxypropyl radical rather than 5′-dA$^•$ [32]. Recent theoretical calculations indicate that as a function of the orientation of SAM relative to the unique iron atom of the $Fe_4S_4$ cluster, a different $S^+$-C bond can be cleaved [33]. Thus, different relative bindings of the $Fe_4S_4$ cluster and SAM result in an ample variety of radical-based reactions.

In (*E*)-4-hydroxy-3-methyl-but-2-enyl diphosphate reductase, an enzyme involved in the biosynthesis of isoprenoid, a $Fe_4S_4]^+$ cluster carries out a two-electron reduction, generating a $[Fe_4S_4]^{3+}$ HiPiP-like intermediate [34]. Along with $O_2$-tolerant NiFe-hydrogenases (see Section 10.4.2), this is one of few examples of two-electron redox chemistry catalyzed by $Fe_4S_4$ clusters. Fe-S clusters modified with additional organic ligands or other metal ions are present in FeFe-hydrogenase [5], nitrogenase [8], and CODH-acetyl-CoA synthase [6, 7].

Non-redox chemistry is also common among Fe-S cluster-containing proteins with one of their Fe ions acting as a Lewis acid. One of the best-studied examples is aconitase, which converts citrate into isocitrate [35, 36]. Similar activities are displayed by other dehydratases such as IspG [37], fumarase A [38, 39], and quinolinate synthase [40, 41]. Despite their different folds, these enzymes share an Fe-S cluster-based Lewis chemistry, a commonality that most probably emerged through convergent evolution.

## 10.3 Oxygen and Fe-S clusters

It is now widely accepted that a major evolutionary breakthrough took place on earth with the appearance of oxygenic photosynthesis in cyanobacteria over 2.2 billion years ago. This major evolutionary development enabled living organisms to utilize an abundant source of electrons from water and reducing power from the sun. However, the disadvantage for living organisms present at that time was the concomitant production of molecular oxygen, which progressively accumulated in the atmosphere. Molecular oxygen has been certainly beneficial in allowing the evolution of the eukaryotic cell and multicellular organisms. Indeed, respiration using molecular oxygen as the final electron acceptor increases the ATP production efficiency up to 15-fold when compared with fermentative processes. Furthermore, the presence of $O_2$ and $O_3$ in the atmosphere ensures a relative protection against solar UV radiation and may have allowed organisms to leave the sea and colonize the land. However, molecular oxygen is both toxic and mutagenic for cells, and its increased abundance forced living organisms to adapt for survival before being able to take advantage of it.

### 10.3.1 Properties of molecular oxygen and its partially reduced species

Most of the properties of molecular oxygen and of its ROS result from their reduction potentials and their electronic structures, which dictate their respective reactivity. Ground-state oxygen (also called $O_2$ triplet) is a stable biradical (Fig. 10.2). Each of the two oxygen atoms has a spin state of 1/2 for a total spin $S = 1$, making ground-state oxygen ($O_2$) a triplet ($2S + 1 = 3$). Organic molecules usually undergo two-electron redox chemistry, transferring paired electrons with opposing spins as required by the Pauli exclusion principle. However, molecular oxygen, which has two electrons with parallel spins and consequently occupying different orbitals, cannot accept these electrons

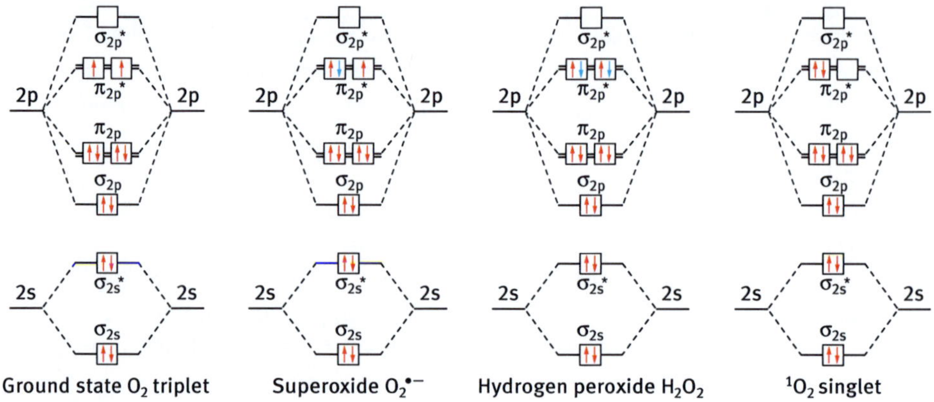


**Fig. 10.2:** Molecular orbital diagrams of dioxygen and its reduction products.

without a spin flip of one electron, which requires about 23 kcal/mol. This energy prerequisite makes the reaction unlikely because of this high activation energy barrier. However, under certain circumstances, for example, when photo-excited, molecular oxygen converts to the singlet state, which can react with many double bond-containing organic molecules. Conversely, although triplet molecular oxygen is not a strong univalent oxidant, it readily reacts with unpaired electron-containing organic radicals and transition metals according to reactions (2) and (3).

$$Fe^{2+} + O_2 \rightarrow (Fe^{2+}\text{-}O_2 \leftrightarrow Fe^{3+}\text{-}O_2^{\bullet-}) \rightarrow Fe^{3+} + O_2^{\bullet-} \quad (2)$$

$$R^{\bullet} + O_2 \rightarrow R\text{-}O\text{-}O^{\bullet} \quad (3)$$

The superoxide anion ($O_2^{\bullet-}$) is a much stronger univalent oxidant than molecular oxygen, but its anionic nature reduces its propensity to react with electron-rich molecules. This feature favors inner-sphere electron transfer (whereas molecular oxygen can perform outer-sphere electron transfer). Conversely, the protonated superoxide radical species ($HO_2^{\bullet}$) ($pK_a = 4.8$) is much more reactive. One of the best-characterized reactions of superoxide is dismutation, which produces molecular oxygen and hydrogen peroxide according to reactions (4) or (4′). Conversely, superoxide anions can reduce transition metals as in reaction (2) when going from right to left.

$$O_2^{\bullet-} + O_2^{\bullet-} + 2\,H^+ \rightarrow H_2O_2 + O_2 \text{ (slow)} \quad (4)$$

$$HO_2^{\bullet} + O_2^{\bullet-} + H^+ \rightarrow H_2O_2 + O_2 \text{ (fast)} \quad (4')$$

Hydrogen peroxide is also a strong oxidant, but the electronic structure of the molecule makes it moderately stable because about 38 kcal/mol is required to disrupt it.

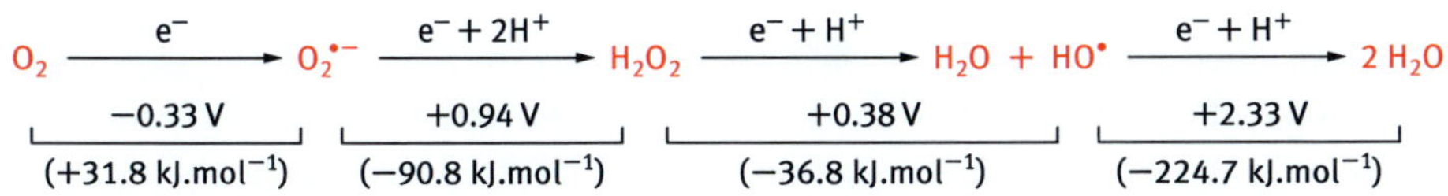


**Fig. 10.3:** Potential diagram for oxygen at pH 7 in the standard state [42].

Perhaps the best-characterized reaction of hydrogen peroxide is its reduction by ferrous ions (or other reduced transition metal ions), known as the Fenton's reaction (5), which generates the very reactive hydroxyl radical species. The exact nature of the reaction is still under debate but it appears that it proceeds through an inner-sphere mechanism [43].

$$Fe^{2+} + H_2O_2 \rightarrow Fe^{3+} + HO^- + HO^\bullet \tag{5}$$

The hydroxyl radical is much more reactive than the other ROS described above. In fact, this species is the second most reactive oxidant known after fluorine. It displays indiscriminate diffusion-limited reactivity with fatty acids, proteins, or DNA, making it a major threat for living organisms. Figure. 10.3 depicts the reduction potentials of the different oxygen species.

### 10.3.2 Oxidative damage to Fe-S clusters

It has been known for a long time that oxygen and ROS have an impact on the activity of Fe-S cluster-containing proteins. Indeed, early reports on dihydroxyacid dehydratase demonstrated that the superoxide ion impairs its enzymatic activity [44, 45]. Subsequently, Flint and colleagues [39, 46] demonstrated that upon oxidation by either superoxide ion or hydrogen peroxide, the $Fe_4S_4]^{2+}$ cluster of dihydroxyacid dehydratase was transiently converted into a metastable $Fe_4S_4]^{3+}$ cluster, which rapidly released one ferrous ion according to reactions (6) [or (6′)] and (7). Because the lost iron is essential for catalysis, the oxidized enzyme is inactive. Oxygen itself can directly oxidize a $Fe_4S_4$ cluster according to reaction (8), as has been observed in the fumarate nitrate reduction regulator (FNR) (see Section 10.3.4) [47–49]. This reaction produces a superoxide radical species, which in turn causes additional oxidative damage.

$$Fe_4S_4]^{2+} + O_2^{\bullet -} + 2\,H^+ \rightarrow Fe_4S_4]^{3+} + H_2O_2 \tag{6}$$

$$Fe_4S_4]^{2+} + H_2O_2 + H^+ \rightarrow Fe_4S_4]^{3+} + H_2O + HO^\bullet \tag{6′}$$

$$Fe_4S_4]^{3+} \rightarrow Fe_3S_4]^{+} + Fe^{2+} \tag{7}$$

$$Fe_4S_4]^{2+} + O_2 \rightarrow Fe_4S_4]^{3+} + O_2^{\bullet -} \tag{8}$$

In most cases, the result of ROS-induced damage to a $Fe_4S_4$ cluster is a stable $Fe_3S_4]^+$ center. Consequently, it is not uncommon to get mainly a $Fe_3S_4]^+$ species when purifying a protein with a solvent-exposed $Fe_4S_4$ cluster [27]. When this happens, *in vitro* Fe-S cluster reconstitution is required. Cluster degradation can naturally proceed further as in the case of the well-studied FNR, the protein switch between anaerobic and aerobic metabolisms in facultative anaerobic microorganisms [47–55]. Under anaerobic conditions, FNR is a homodimer in which each monomer contains a $Fe_4S_4$ cluster. In this oligomeric state, FNR is capable of binding specific palindromic DNA sequences to either activate or repress a vast number of genes. However, when exposed to dioxygen, i.e. when the organism reaches an aerobic environment, the $Fe_4S_4$ cluster is rapidly converted into a $Fe_2S_2$ cluster *via* a transient unstable $Fe_3S_4$ cluster [56–58]. The $Fe_2S_2$ cluster-containing form of FNR is monomeric and cannot bind DNA, resulting in extensive metabolic remodeling [53, 59]. FNR dimer dissociation is most likely provoked by the structural rearrangements required to change the Fe-S cluster binding site geometry. This is so because $Fe_4S_4$ clusters are cubes, whereas $Fe_2S_2$ clusters are rhombs (see Fig. 10.6). A more detailed description of FNR function and mechanism is given in Chapter 13 (Patricia Kiley). In conclusion, FNR uses the reactivity of a $Fe_4S_4$ cluster toward ROS to sense environmental $O_2$ levels and regulate gene expression accordingly.

### 10.3.3 Molecular mechanisms of oxidative damage to $Fe_4S_4$ clusters

As mentioned earlier, the first step in $Fe_4S_4]^{2+}$ cluster degradation upon ROS exposure is the oxidation of the cluster to the $Fe_4S_4]^{3+}$ state. However, there are different kinds of $Fe_4S_4$ clusters depending on the number and nature of their ligands. Dehydratases such as aconitase [35, 36] and, as mentioned above, radical SAM enzymes such as lysine 2,3-aminomutase [17] use a solvent exposed $Fe_4S_4$ cluster either as a Lewis acid to bind and activate a substrate or as a redox center to reductively produce a 5′-dA• radical species from SAM (Fig. 10.1). Both proteins coordinate their $Fe_4S_4$ cluster with only three protein ligands and thus, respectively, keep the fourth iron atom available for either substrate or cofactor binding. Conversely, redox-active Fe-S clusters, like in ferredoxin, do not need to keep the fourth iron open for ligation with substrates or cofactors. Ferredoxins may have four buried protein ligands and be protected by the protein matrix. Consequently, ROS will not react the same way, or at all, with each type of $Fe_4S_4$ cluster. It has been proposed that buried $Fe_4S_4$ clusters with four protein ligands are less sensitive to oxidation by superoxide and hydrogen peroxide. Nevertheless, they can still be oxidized by molecular oxygen, most probably *via* outer-sphere electron transfer, according to reaction (8), as reported for FNR [47–49]. Conversely, in solvent-exposed $Fe_4S_4$ clusters with three protein ligands, the fourth iron atom is potentially accessible to superoxide and hydrogen peroxide. In this case, there is most probably inner-sphere electron transfer, involving an iron-centered oxidation mechanism. Subsequently, the resulting $Fe_4S_4]^{3+}$ cluster spontaneously decays, as described in reaction (7).

### 10.3.4 $Fe_3S_4$ to $Fe_2S_2$ cluster conversion in FNR

Two sets of experiments conducted on FNR gave apparently contradictory results about the mechanism of cluster conversion. Kinetic analyses concluded that the $Fe_3S_4$ to $Fe_2S_2$ cluster conversion was first order with concomitant release of one ferric and two sulfide ions [47–49] as described in reaction (9). Crack and colleagues also proposed that, *in vitro*, depending on the buffer used [for instance, phosphate or 2-amino-2-hydroxymethyl-1,3-propanediol (Tris)], ferric ion can be rapidly reduced, most probably by the concomitantly released sulfide ions, explaining apparent discrepancies between different studies [48, 55]. They thus put forward that an amplification system operates *in vivo* because one oxygen molecule can oxidize four $Fe_4S_4$-FNR complexes according to reaction (10), which is the combination of reactions (4) (catalyzed by endogenous superoxide dismutase) and (8) [49].

$$Fe_3S_4]^{+} \rightarrow Fe_2S_2]^{2+} + Fe^{3+} + 2\,S^{2-} \tag{9}$$

$$4\,Fe_4S_4]^{2+} + O_2 + 4\,H^{+} \rightarrow 4\,Fe_2S_2]^{2+} + 4\,Fe^{2+} + 4\,Fe^{3+} + 8\,S^{2-} + 2\,H_2O \tag{10}$$

Conversely, Raman resonance spectroscopy data showed that $Fe_2S_2$ cluster formation in FNR involves production of cysteine persulfide ligands, thereby supporting a sulfur-centered oxidation mechanism [reaction (11)] [52]. Also, upon addition of ferrous ions and DTT, it was possible to convert the cysteine persulfide-bound $Fe_2S_2$ cluster back to a $Fe_4S_4$ center. This result supports the existence of a hitherto unidentified cellular Fe-S cluster repair mechanism that does not require additional sulfide ions but instead uses those liberated upon persulfide reduction [52]. Reaction (11) contradicts the previous kinetic analysis, which supported a non-oxidative first order reaction [reaction (9)]. However, reaction (11) agrees with previous reports, which suggested that cysteine persulfides are formed during oxidative Fe-S cluster degradation in other proteins, such as aconitase and biotin synthase.

$$2\,Cys\text{-}S^{-} + Fe_3S_4]^{+} + O_2 + 4\,H^{+} \rightarrow 2\,Cys\text{-}S\text{-}S^{-} + Fe_2S_2]^{2+} + Fe^{3+} + 2\,H_2O \tag{11}$$

### 10.3.5 X-ray crystallographic studies

HydE is a radical SAM enzyme involved in the assembly of the active site of FeFe-hydrogenase [5, 60], that, in addition to its canonical $Fe_4S_4$ cluster, has three cysteine residues capable of binding a solvent-exposed second Fe-S cluster of unknown function [61]. In an independent study using HydE from *Thermotoga maritima*, we were able to trap several X-ray structures of intermediates in the time course of $Fe_4S_4$ cluster degradation upon oxygen exposure [62]. Some of these intermediate species turned out to be very valuable to further our understanding of the changes that occur during the oxidative damage of a $Fe_4S_4$ cluster. Unexpectedly, a regular $Fe_2S_2$ cluster bound to one cysteine thiolate and two cysteine persulfide ligands (species **3** in Fig. 10.4) [62] was found in a structure that was solved from crystals grown in the glove box. The fourth ligand is arginine, a residue that has already been observed to coordinate a $Fe_2S_2$ cluster in biotin synthase, another radical SAM enzyme [63]. The starting protein material, which contained two $Fe_4S_4$ clusters per protein molecule, may have been exposed to a very limited amount of molecular oxygen during sample transfer between different glove boxes before setting up the crystallization experiments. This would explain the presence of cysteine persulfide ligands of a $Fe_2S_2$ cluster in HydE. The direct observation of these ligands hints at their possible biological relevance and strengthens the proposal that they come from $Fe_4S_4$ cluster oxidative damage.

In full agreement with what was observed in FNR [52], addition of a reductant such as DTT to our HydE crystals released sulfide ions from the cysteine persulfide ligands, leading to a regular $Fe_2S_2$ cluster bound to three cysteine residues and an

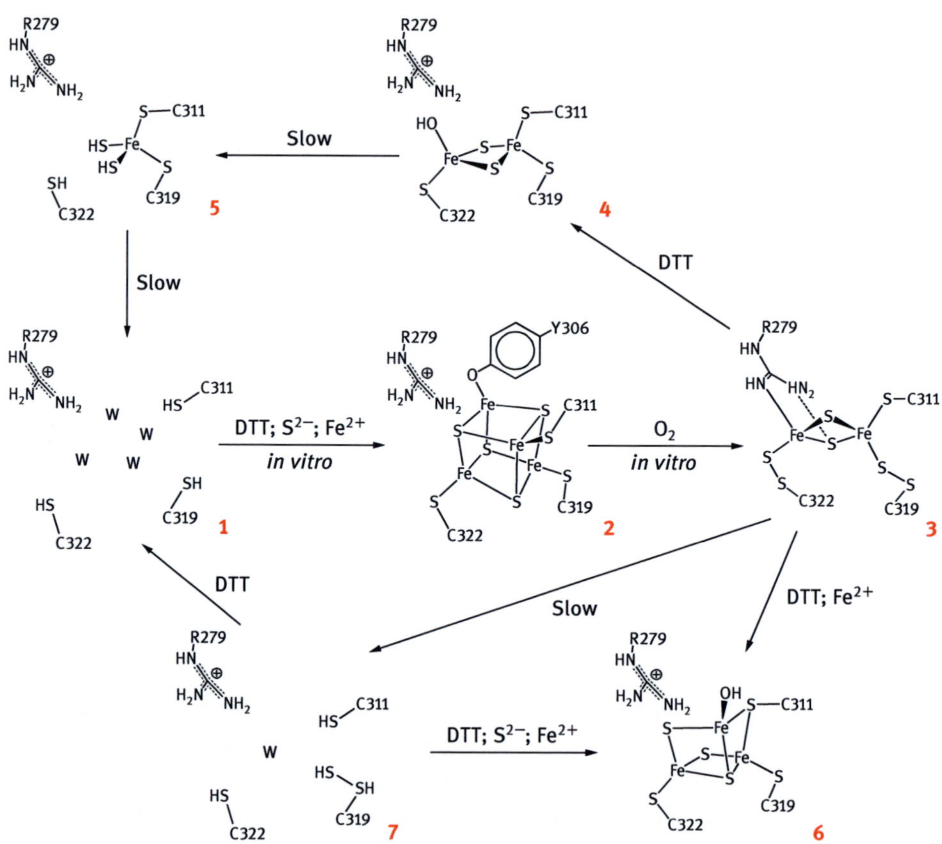


**Fig. 10.4:** Postulated degradation of a FeS cluster based on the TmHydE crystal structures. Unless specified otherwise, the reactions were performed in protein crystals [62].

exchangeable fourth ligand. The latter was modeled as a water molecule but may correspond to a hydrogen sulfide ion released from the cluster after reduction (species **4** in Fig. 10.4) [62]. In the crystals, these sulfide ions were readily mobilized to constitute an unusual $Fe_3S_3$ cluster (conversion from **3** to **6** in Fig. 10.4) because packing constraints prevented the formation of a regular $Fe_4S_4$ cluster [62].

Thus, the first step of $O_2$-induced $Fe_4S_4$ cluster degradation corresponds to an outer-sphere iron oxidation [reaction (8)] followed by reaction (7) [49]. Conversely, the presence of cysteine persulfide ligands in species **3** (Fig. 10.4), also detected by Raman resonance spectroscopy in FNR, demonstrates that subsequent oxidative reactions are sulfur centered. Furthermore, the observation that some FNR molecules exhibit only one cysteine persulfide supports that oxidation steps occur successively [52]. However, molecular oxygen only marginally reacts with sulfur atoms when compared with superoxide, and hydrogen peroxide. Because $Fe^{3+}$-S-R is isoelectronic with $Fe^{2+}$ •S-R, instead of adding a new oxygen molecule to the $Fe_3S_4]^+$ as shown in reaction (11), an alternative mechanistic proposition would favor the reaction of the superoxide anion released during reaction (8) with one of the neighboring cysteine ligands of the cluster. The resulting cysteine persulfenate, instead of getting converted into a cysteine sulfinate as expected, reacts with one of the sulfide ligands from the sulfur-rich environment of the adjacent iron to generate one cysteine persulfide with concomitant release of hydrogen peroxide and one ferrous ion. These observations are in agreement with previous reports on FNR [47, 55]. Reaction of hydrogen peroxide with a second cysteine residue will produce cysteine sulfenate according to reaction (12). Furthermore, it is well known that cysteine sulfenate can react with different thiols, such as glutathione [reaction (13)]. In a sulfide-rich environment, it probably converts into a second cysteine persulfide ligand. If superoxide or hydrogen peroxide were the primary oxidants [reactions (6) and (6′)] one would expect the production of only one cysteine persulfide per damaged Fe-S cluster.

$$\text{R-SH} + H_2O_2 \rightarrow \text{R-S-OH} + H_2O \tag{12}$$

$$\text{R-S-OH} + {}^{-}\text{S-R}' \rightarrow \text{R-S-S-R}' + OH^{-} \tag{13}$$

### 10.3.6 Alternative reactions can occur and compete

Aging a HydE-bound $Fe_2S_2$ cluster (with or without cysteine persulfide ligands) under anaerobic conditions led to its slow decay, probably *via* a non-redox process, with slow release of Fe and S ions. This process started with the iron atom that was most exposed to solvent (species **5** in Fig. 10.4) and ended with the complete disappearance of the cluster. This observation fully supports an Fe-S cluster degradation mechanism involving iron depletion, as described for bacterial aconitase [64]. When we crystallized HydE without the SAM cofactor a slow transformation of the conserved $Fe_4S_4$ cluster to a stable $Fe_3S_4$ form was observed. Thus, species **6** in Fig. 10.4 [62] probably represents an intermediate in the inter-conversion of $Fe_3S_3$ and $Fe_2S_2$ clusters (Fig. 10.5). These non-redox Fe-S cluster degradation processes agree well with experiments showing that in FNR, a first-order kinetic reaction [reaction (9)] occurs when converting the $Fe_3S_4$ cluster into a $Fe_2S_2$ center in the presence of ferrous or ferric ion chelators [47]. In fact, this "non-redox" or "iron-depletion" process should compete with the sulfur-centered oxidation reaction described earlier (Fig. 10.5). The *in vivo* balance between the two pathways will certainly depend on the efficiency of the rescue systems. Indeed, efficient "free" iron trapping as well as good "ROS-neutralizing" enzymes such as efficient catalases and superoxide dismutases will certainly favor the non-redox path. Conversely, by storing the sulfur ions, the sulfur-centered oxidation path has the great advantage of allowing *in situ* repair without requiring a complete and energetically expensive *de novo* Fe-S cluster synthesis. Such a repair system has not yet been characterized.

### 10.3.7 Structural changes

As was mentioned earlier, FNR uses the oxygen sensitivity of a $Fe_4S_4$ cluster to sense the $O_2$ concentration in the cell and thereby acts as a molecular switch between anaerobic and aerobic metabolisms. We have already discussed the chemical mechanisms involved in this process, but the question of how $Fe_4S_4$ cluster degradation induces FNR dimer dissociation remains. Obviously, as $Fe_4S_4$ clusters are cubane, whereas $Fe_2S_2$ are rhombic, their coordination is different (tetrahedral *vs* planar), and consequently, the dissimilar cluster stereochemistry should dictate the conformational changes responsible for dimer dissociation. Although no three-dimensional structural information on FNR is available, it is possible to assess the amplitude of such conformational changes by comparing HydE structures with either $Fe_4S_4$ or $Fe_2S_2$ second clusters. Indeed, a new crystal form of HydE containing a second $Fe_4S_4$ cluster, suggested that converting it to a $Fe_2S_2$ center would require one cysteine ligand to move about 6 Å (Cα-to-Cα distance) (Fig. 10.6). This, in turn, should induce a drastic rearrangement of one of the HydE Cys-containing loops [62]. An equivalent major conformational change should promote dimer dissociation in FNR and thus trigger the switch between aerobic, and anaerobic metabolisms. Thus, FNR is a remarkable product of evolution that takes advantage of the intrinsic lability of an Fe-S cluster, converting a potential evolutionary vulnerability into a regulatory asset. As Fe-S clusters are major targets of oxidative damage, the protein uses one $Fe_4S_4$ cluster in a structural environment that makes it highly sensitive to oxidation, enabling FNR to induce a cellular response to the oxidative stress.

## 10.4 Adaptation to oxygen

As mentioned above, Fe-S clusters present today in living organisms most probably appeared before the emergence of oxygenic photosynthesis. Thereafter, these

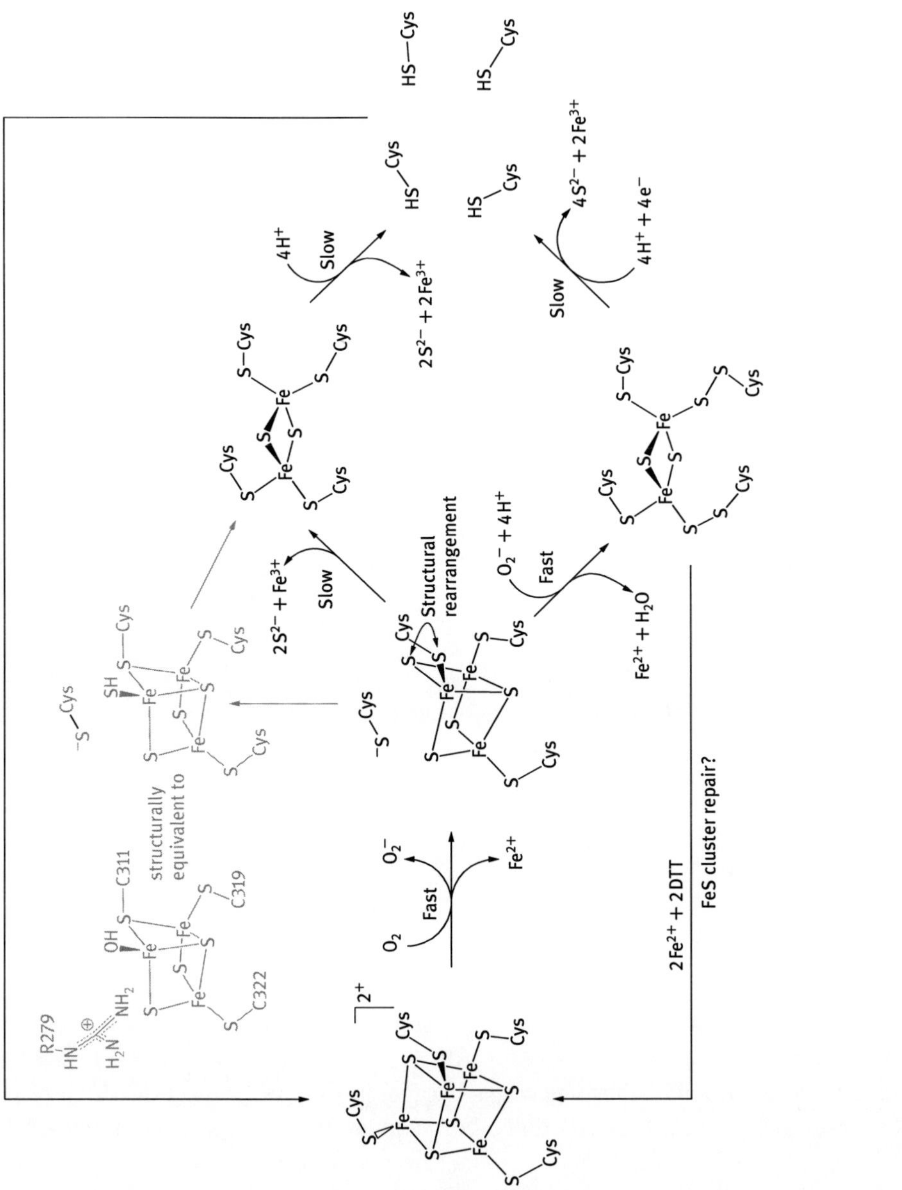


**Fig. 10.5:** Alternative mechanisms for FeS cluster degradation. The $Fe_3S_4$ cluster can either slowly decompose to a $Fe_2S_2$ cluster *via* a ligand exchange process (top) or be further damaged by ROS *via* sulfur-centered oxidative reactions leading to the formation of cysteine persulfide ligands, as observed in FNR and in TmHydE (bottom) [52, 62]. A postulated intermediate during the required structural rearrangements that is supported by a species observed in the TmHydE structures is depicted in gray (Fig. 10.4 and [62]).

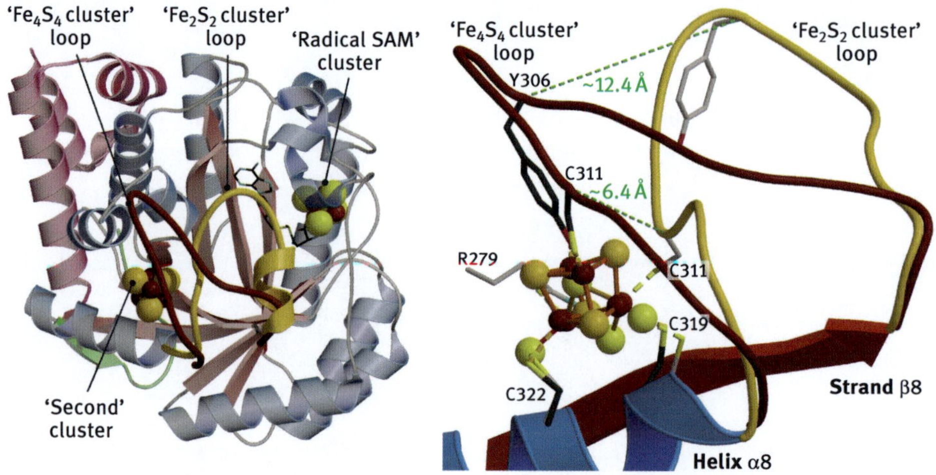


**Fig. 10.6:** Structural differences between SAM-independent $Fe_4S_4$ and $Fe_2S_2$ cluster-containing TmHydE. (left) View of the whole structure. The loop containing Cys311 covers the barrel cavity when the protein coordinates a $Fe_2S_2$ cluster (gold) and moves away from it when the protein binds a $Fe_4S_4$ cluster (brown). (right) Superposition between species 2 and 3 (Fig. 10.4) [62].

organisms had to deal with the major change induced by the presence of atmospheric oxygen that affected Fe-S clusters as well as many other metal-containing active centers. Different adaptations were developed and some of them are briefly discussed here.

### 10.4.1 Switch between metabolisms or restriction to niches

The easiest solution to the oxygen problem was to avoid being exposed to it so that no metabolic changes were required. This choice implied that, to survive, organisms could colonize a few specific niches in which they could conserve a strict anaerobic way of life. Conversely, other organisms became facultative, conserving the ability to live under strict anaerobic conditions, but also developing alternative pathways suitable for aerobic life; this is, for example, the case of the gut bacterium *Escherichia coli*, which lives in the human digestive tract where oxygen concentrations are highly variable. During the anaerobic-to-aerobic transition, which is controlled by FNR and other regulators and takes place at the distal end of the mammalian rectum, *E. coli* uses several enzymes such as the $O_2$-tolerant NiFe-hydrogenase Hyd-1 [65] and cytochrome *bd* oxidases [66] to consume $O_2$ and thereby allow anaerobic metabolic pathways to continue to function. As described in detail in the next section, Hyd-1 has a peculiar proximal (relative to its active site [12]) Fe-S cluster, which protects the NiFe active site from oxygen-induced damage.

### 10.4.2 $O_2$-tolerant NiFe hydrogenases

During the long evolutionary history of [NiFe]-hydrogenases, at least three different approaches have been developed to cope with the damaging effects of $O_2$: (i) a hydrophobic tunnel leading to the active site that constitutes a filter in the regulatory NiFe-hydrogenases, which allows the passage of $H_2$ while blocking passage of the bulkier $O_2$ [67–69]; (ii) a SeCys, which can reduce protons at low $O_2$ levels [70] replaces Cys as one of the terminal Ni ligands in $O_2$-resistant [NiFeSe]-hydrogenases [71], which can reduce protons at low $O_2$ levels [70]; and (iii) the presence of two supernumerary Cys residues found near the proximal Fe-S cluster of some $O_2$-tolerant [NiFe]-hydrogenases, which oxidize dihydrogen at 21% $O_2$ [72–74]. The latter are membrane-bound periplasmic enzymes typified by *Ralstonia eutropha* (*Re*MBH) [75], *Escherichia coli* hydrogenase 1 (*Ec*Hyd-1) [76], and *Aquifex aeolicus* hydrogenase 1 (*Aa*Hase-1) [74]. This NiFe-hydrogenase subclass has already been used in proof-of-concept technological applications such as a biofuel cell [77]. Upon oxidation, standard (i.e. $O_2$-sensitive) NiFe-hydrogenases display two EPR-active redox states called Ni-A and Ni-B. The latter, described as $Ni^{III}$-$Fe^{II}$ with a bridging $OH^-$ ion, undergoes rapid reductive reactivation [75]. Conversely, the Ni-A form, a $Ni^{III}$-$Fe^{II}$ center with either a putative bridging (hydro)peroxo [78] or, equivalently, a cysteine S=O and a $OH^-$ Ni-Fe bridging ligand, needs a prolonged reactivation process, which may not normally take place *in vivo*. Except for the Ni-A species, the redox states described in standard NiFe-hydrogenases are also present in $O_2$-tolerant NiFe-hydrogenases [74, 79]. As mentioned earlier, in standard NiFe-hydrogenases, electron transfer between the active site and the molecular surface takes place through a "wire" made of the proximal and distal (relative to the active site) $Fe_4S_4$ clusters and the medial $Fe_3S_4$ center [12]. A recent EPR study using the $O_2$-tolerant *Re*MBH, which detected coupling between an additional paramagnetic species and both the active site and the medial cluster, suggested the presence of an atypical modification at or near the proximal cluster [79]. In addition, +2/+1 and +3/+2 redox transitions, which were assigned to this cluster, have been reported for three of these $O_2$-tolerant enzymes (Fig. 10.7) [80–82]. The +2/+3 superoxidation, which can furnish an additional electron to the active site, was proposed to be a redox switch against oxidative stress [82]. Site-directed mutagenesis studies found Cys19 and Cys120, the two supernumerary conserved cysteine residues (*Ec*Hyd-1 numbering herein), to be required for the unusual two-electron redox behavior of the proximal cluster and the associated $O_2$ tolerance of the enzyme [83, 84]. Remarkably, the nonphysiological super-reduction of HiPIP $[Fe_4S_4]^{3+}$ cluster to the +1 state occurs at a potential about 1,000 mV more negative than that of the +3/+2 redox couples recorded for the $O_2$-tolerant NiFe-hydrogenases (Fig. 10.7) [85, 86].

The crystal structures of the oxidized and $H_2$-reduced forms of *Hydrogenovibrio marinus* (*Hm*) MBH [87] and *Ec*Hyd-1 [65] and of the reduced form of *Re*MBH [88] revealed that the proximal cluster, which is a symmetrically arranged standard cuboidal $Fe_4S_4(Cys)_4$ cluster in standard hydrogenases, is substituted by a $Fe_4S_3(Cys)_6$ cluster. Remarkably, the absence of one sulfide ion is compensated by both the

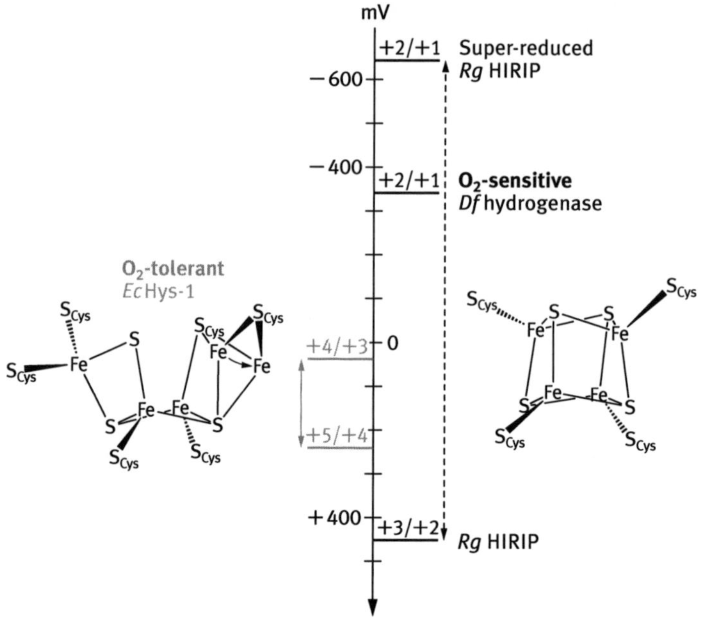


**Fig. 10.7:** Redox potential range for (right) a symmetrical $[Fe_4S_4](Cys)_4$ cluster using the example of the $O_2$-sensitive hydrogenase from ***Desulfovibrio fructosovorans*** [13] for the +2/+1 redox couple and HiPIPs for the +3/+2 redox couple [85] (the difference between these two redox pairs is greater than 700 mV) and (left) the asymmetrical $Fe_4S_3(Cys)_6$ cluster with the +4/+3 and the +5/+4 redox couples of EcHyd-1 [80]. Rg = *Rhodopila globiformis*.

bridging Cys19 thiolate and the terminally bound Cys120, which binds a Fe atom that ligates to another cysteine thiolate. The absence of one of the sulfide ligands also implies that the standard $Fe_4S_4(Cys)_4$ cluster redox states +1, +2, and +3 must now be, respectively, corrected to +3, +4, and +5 to account for the loss of two negative charges in the cluster. The $Fe_4S_3(Cys)_6$ arrangement results in an unprecedented asymmetric Fe-S cluster (Fig. 10.7), which becomes further distorted upon superoxidation when the Fe coordinated by Cys19 and Cys20 shifts away from the remaining $Fe_3S_3$ fragment to bind the deprotonated amide N of Cys20 [65, 87]. The crystal structures of the as-isolated and $H_2$-reduced *Ec*Hyd-1 led some of us to propose that the conserved Glu76 deprotonates the Cys20 amide N and allows formation of the Fe-N bond [65]. In that work, quantum calculations defining electronic/spin states that fit available structural and spectroscopic data were also reported. It was concluded that the interplay between the strong asymmetry of the $Fe_4S_3$ cluster and the resulting Fe valence/spin localization in the +3, +4, and +5 redox states is the key feature that explains its facile superoxidation [89].

Indeed, one very important point, frequently ignored when analyzing the significant potential differences between the normal $Fe_4S_4$ centers, found in $O_2$-sensitive hydrogenases (Fig. 10.7, right), and the $O_2$-tolerant hydrogenase proximal cluster (Fig. 10.7, left) has to do with the extent of electronic/spin-delocalization at mixed-valence iron pairs [90, 91]. Whenever full electronic delocalization takes place within a single mixed-valence pair of a $Fe_4S_4$ cluster, which changes the pair from localized $Fe^{3+}$-$Fe^{2+}$ to delocalized $Fe^{2.5+}$-$Fe^{2.5+}$ valences, the bonding energy E is stabilized by approximately 350 mV ($E^{deloc} = E^{loc}-350$ mV) [11]. In standard $Fe_4S_4$ clusters, the 1+ and 3+ redox states contain one $Fe^{2.5+}$-$Fe^{2.5+}$ delocalized pair, and the 2+ state two such pairs. Unexpectedly, in the $O_2$-tolerant NiFe-hydrogenases $Fe_4S_3$ clusters, the corresponding +3, +4, and +5 redox states were found to be fully localized [89]. The potential $E^0$ of a given redox pair corresponds to the difference between the two redox state energies: $E^0 = E(\text{ox})-E(\text{red})$. Therefore, in the case of standard $Fe_4S_4$ clusters, the electronic delocalization contribution to the (2+/1+) redox pair will be ~−350 mV and the equivalent contribution to the (3+/2+) redox pair will be positive at ~ +350 mV. Consequently, without taking into account other less remarkable effects on redox potentials [92, 93], the electronic delocalization within mixed-valence pair(s) generates the observed ~700-mV difference between the two redox pairs of HiPiP (Fig. 10.7, right). This implies that in $O_2$-sensitive NiFe-hydrogenases, once the first electron leaves the symmetrical reduced $Fe_4S_4(Cys)_4$ proximal cluster oxidizing it from 1+ to 2+, the potential gap for the second (2+ → 3+) oxidation is not energetically favorable. Strikingly, computational studies have shown that in the asymmetrical $Fe_4S_3(Cys)_6$ proximal cluster, the +3, +4, and +5 redox states consist only of valence-localized $Fe^{3+}$ and $Fe^{2+}$ ions [89]. This means there will be no significant delocalization contribution to $E^0$(4+/3+) and $E^0$(5+/4+). Thus, it is the absence of a potential gap between the (4+/3+) and (5+/4+) redox pairs that allows for the second oxidation of the $Fe_4S_3(Cys)_6$, asymmetrical proximal cluster to occur.

The ability of the proximal cluster of $O_2$-tolerant NiFe-hydrogenases to provide two electrons in rapid succession to the active site is most likely responsible for the resistance of these enzymes to $O_2$-induced oxidative stress [82, 83]. Under physiological conditions, the $O_2$-tolerant NiFe-hydrogenases are most probably in a reduced state, such as the hydride-bound Ni-C or Ni-R forms [74, 79, 94]. The ready availability of four electrons, two from the active site and two from the proximal cluster prevents the accumulation of the Ni-A state [78]. Figure 10.8 shows a tentative mechanism for protection against oxidative stress that takes into account the following observations: (i) $O_2$ binds the $Ni^{3+}$-containing Ni-C-$H^-$ (or the isoelectronic $Ni^I$ + $H^+$ (described as the photolyzed Ni-L* state in $O_2$-sensitive hydrogenases) (state **1**) and forms either a (hydro)peroxo species or an electronically equivalent Cys S=O and bridging $OH^-$; **2**; in the former case, molecular oxygen could insert into the nickel-hydride bond [95]; ii) The electron resulting from the +3 to +4 oxidation as well as a proton from Glu76 leads to species **3**. In $O_2$-sensitive hydrogenases, **3** rearranges to Ni-SU **4′**, which is an inactive oxidized state, one electron more reduced than Ni-A and subsequently oxidizes to Ni-A **5′**. In $O_2$-tolerant hydrogenases, upon the deprotonation of Glu76, the unique iron moves toward the amide of Cys20 resulting in N pyramidalization [96]

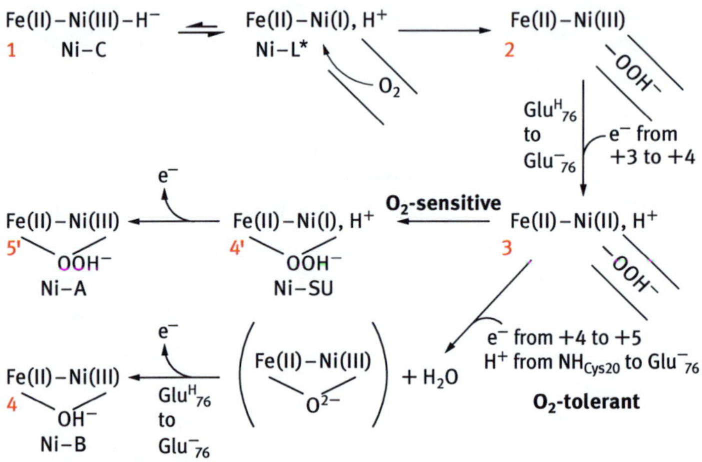


**Fig. 10.8:** Proposed reaction mechanism against $O_2$-induced oxidative stress at the active site of NiFe-hydrogenases. The active site is represented by the Ni-Fe center and events occurring at the proximal cluster are depicted in gray. The partially reduced $O_2$ species in 4′ and 5′ is shown as hydroperoxide but the isoelectronic Cys sulfenate (S=O) and bridging $OH^-$ are a possible alternative.

and amide proton acidification [97]. This in turn allows the deprotonation of amide of Cys20 and the subsequent superoxidation of the cluster from +4 to +5. The extra proton and electron permit the rapid reduction of the oxo species at the active site to $H_2O$, leading to the easily activated Ni-B state **4**.

In conclusion, the crucial feature that distinguishes $O_2$-sensitive from $O_2$-tolerant hydrogenases is not located at the active site nor is it found in the hydrophobic channel network that allows for $H_2/O_2$ diffusion but is rather attributable to the properties of the asymmetric proximal $Fe_4S_3$ cluster [65, 82, 84, 87, 88]. The cluster asymmetry allows for (1) the movement of one unique ferrous ion, which results in amide deprotonation, and (2) the full iron valence localization in the three redox states that significantly facilitates the +4 to +5 superoxidation. By comparison, the $Fe_4S_4$ proximal clusters of $O_2$-sensitive NiFe-hydrogenases are dense and rigid, and the complete electronic delocalization of the +2 state rules out two successive electron transfers.

### 10.4.3 Protective systems against ROS

To survive in aerobic environments, cells had to prevent the deleterious effects of ROS and consequently they developed protective mechanisms. The best-known protective enzymes are superoxide dismutase and catalase. The former catalyzes the dismutation of superoxide ions into oxygen and hydrogen peroxide according to reaction (4) and plays an important role in the cell's defense against free radicals [98]. Catalase converts hydrogen peroxide into water, and oxygen and consequently prevents oxidative damage [99, 100]. Another adaptation to oxygen is the SUF (*sulfur* mobilization) Fe-S cluster assembly machinery that functions mainly when bacteria are exposed to stress [101]. The SUF machinery is found in plastids and bacteria [102, 103]. More details on this machinery may be found in Chapters 16 and 22 (by Wayne Outten and Hong Ye, respectively).

Recently, it has been observed that superoxide-exposed cells recover their Fe-S cluster-dependent activities without requiring *de novo* protein synthesis, suggesting that a dedicated repair system for damaged Fe-S clusters that is neither ISC nor SUF exists [104]. A putative repair system would theoretically represent a much better solution than complete *de novo* Fe-S cluster reassembly after cells are exposed to a major oxidative stress. Based on a transcriptomic analysis of *E. coli* exposed to a nitric oxide stress, one of the highly induced genes was found to be *ytfE* [105–107]. Deletion of *ytfE* resulted in a decrease in the activity of the major Fe-S cluster-containing enzymes [108]. In addition, cells displayed greater sensitivity to oxidative damage and an inability to recover when the source of oxidative stress was removed [108, 109]. Conversely, the addition of purified YtfE protein to a crude extract of an *E. coli*$^{\Delta yftE}$ strain previously exposed to oxidative stress complemented the gene deletion and led to the recovery of major Fe-S cluster-containing proteins, such as aconitase B and fumarase A, within minutes [107]. Therefore, YtfE seems to be a major component of a new machinery dedicated to the repair of damaged Fe-S clusters, which has been named repair of iron centers (RIC) [109, 110]. The *ytfE* gene is found in many pathogenic organisms and appears to be an important virulence factor because it confers resistance to the bactericidal oxidative bursts produced by macrophages. Although paralogues of *ytfE* have been found in eukaryotes, the function of their gene products remains uncharacterized [109, 111]. The 25-kDa RIC proteins form homodimers through intermolecular disulfide bridges. However, dimerization does not seem to be required for RIC activity [108, 109]. Amino acid sequence analysis suggests that the protein is homologous to the oxygen transporter hemerythrin, which consists of a four-helical bundle with an additional specific domain at its N-terminus [109, 111]. Biochemical and spectroscopic characterization indicated that RIC, like hemerythrin, contains two iron atoms per monomer, arranged in a nonheme binuclear center with histidine and carboxylate ligands [108, 112]. It is not known how RIC mobilizes iron for Fe-S cluster repair. Clearly, further characterization will be required to determine the molecular bases of its activity.

### 10.4.4 Evolutionary replacement of Fe-S clusters to keep essential functions in aerobic organisms

Many Fe-S cluster-containing proteins, which evolved when the atmosphere on earth was anoxic, play essential roles in cells. Replacing them with counterparts that do not contain Fe-S clusters may be considered as an ultimate adaptation against oxygen toxicity. One interesting example is found in the enzymes responsible for the *de novo* production of the four deoxyribonucleotide triphosphates, which constitute the building blocks of DNA. The conversion of ribonucleotides into deoxyribonucleotides is catalyzed by three evolutionarily unrelated ribonucleotide reductases (RNR) [113, 114]. Each of the three enzymes requires a system to generate a free-radical species at its active site. Class III RNRs, which are considered the most ancient RNRs, use a radical SAM enzyme, and are therefore Fe-S dependent (Fig. 10.1) [114, 115]. Class II RNRs use a different cofactor, 5′-deoxyadenosylcobalamin, which is much less oxygen sensitive than $Fe_4S_4$ clusters [116]. Class I RNRs are found in eukaryotes and have further evolved because they use a binuclear iron center and molecular oxygen (!) to generate the necessary radical species [114].

Another example corresponds to ThiH and ThiO, two enzymes that produce the dehydroglycine precursor involved in the biosynthesis of the thiazole ring of thiamine. Under anoxic conditions, ThiH, which also belongs to the radical SAM protein superfamily, uses its oxygen-sensitive $Fe_4S_4$ cluster to produce dehydroglycine from tyrosine [117, 118]. Under aerobic conditions, dehydroglycine is produced by ThiO, which is a FAD-dependent glycine oxidase [119, 120]. A third interesting example is the replacement of the anaerobic sulfatase-maturating enzyme (anSME), yet another SAM radical protein [121, 122], by an oxygen-dependent formylglycine-generating enzyme [123]. The former oxidizes either a serine or a cysteine residue, whereas the formylglycine-generating enzyme oxidizes only a cysteine residue. anSME uses three $Fe_4S_4$ clusters, whereas the formylglycine-generating enzyme resorts to generating cystine by the oxidation of two cysteine residues, with the absolute requirement of molecular oxygen to oxidize cysteine from the sulfatase to formylglycine [123–125]. Here, we have discussed only three examples that illustrate how dependence on an Fe-S cluster can be bypassed, but many others might be described.

## 10.5 Conclusions

The increasing concentrations of atmospheric molecular oxygen that commenced about 2.2 billion years ago led to several adaptive changes in extant microorganisms. Adaptations consisted basically in preserving the existing biochemical processes either by protecting and repairing the crucial Fe-S clusters of many of the involved proteins or by substituting Fe-S clusters with aerobically stable, functionally equivalent, counterparts. Only in a subsequent step did organisms evolve to directly benefit from $O_2$, using it as oxidant in monooxygenase or dioxygenases or, much more importantly, as the final electron acceptor in respiration. In addition to their involvement in respiration, Fe-S clusters participate in many other key metabolic pathways, such as the Krebs cycle and photosynthesis. Because of the energetic advantages of aerobic respiration, some microorganisms have developed the capacity of living in both anaerobic and aerobic settings, and they are able to use the Fe-S cluster's sensitivity to molecular oxygen to directly sense its levels in their environment. This is best illustrated by

FNR. Furthermore, in some NiFe-hydrogenases, the modification of a regular cuboid Fe-S cluster has rendered it capable of providing two electrons to successively reduce molecular oxygen at the enzyme's active site. We thus conclude that although Fe-S clusters are fossils from a time when molecular oxygen was not present on earth, they are still among the most widespread and fundamental prosthetic groups found in living organisms. Finally, the study of Fe-S cluster-containing proteins is also a very rich source of information about the processes of evolution in general.

## References


[1] Davis BK. Evolution of the genetic code. Prog Biophys Mol Biol 1999;72:157–243.

[2] Wachtershauser G. Before enzymes and templates – theory of surface metabolism. Microbiol Rev 1988;52:452–84.

[3] Russell MJ, Hall AJ. The emergence of life from iron monosulphide bubbles at a submarine hydrothermal redox and pH front. J Geol Soc 1997;154:377–402.

[4] Lambert JF. Adsorption and polymerization of amino acids on mineral surfaces: a review. Orig Life Evol Biosph 2008;38:211–42.

[5] Fontecilla-Camps JC, Volbeda A, Cavazza C, Nicolet Y. Structure/function relationships of [NiFe]- and [FeFe]-hydrogenases. Chem Rev 2007;107:4273–303.

[6] Darnault C, Volbeda A, Kim EJ, et al. Ni-Zn- Fe-4-S-4 and Ni-Ni- Fe-4-S-4 clusters in closed and open subunits of acetyl-CoA synthase/carbon monoxide dehydrogenase. Nat Struct Biol 2003;10:271–9.

[7] Kung Y, Drennan CL. A role for nickel-iron cofactors in biological carbon monoxide and carbon dioxide utilization. Curr Opin Chem Biol 2011;15:276–83.

[8] Hoffman BM, Lukoyanov D, Dean DR, Seefeldt LC. Nitrogenase: a draft mechanism. Acc Chem Res 2013.

[9] Nicolet Y, Piras C, Legrand P, Hatchikian CE, Fontecilla-Camps JC. Desulfovibrio desulfuricans iron hydrogenase: the structure shows unusual coordination to an active site Fe binuclear center. Structure 1999;7:13–23.

[10] Carter CW Jr, Kraut J, Freer ST, et al. A comparison of Fe4S4 clusters in high-potential iron protein and in ferredoxin. Proc Natl Acad Sci USA 1972;69:3526–9.

[11] Mouesca JM, Chen JL, Noodleman L, Bashford D, Case DA. Density functional Poisson-Boltzmann calculations of redox potentials for iron sulfur clusters. J Am Chem Soc 1994;116:11898–914.

[12] Volbeda A, Charon MH, Piras C, Hatchikian EC, Frey M, Fontecilla-Camps J. Crystal structure of the nickel-iron hydrogenase from Desulfovibrio gigas. Nature 1995;373:580–7.

[13] Rousset M, Montet Y, Guigliarelli B, et al. [3Fe-4S] to [4Fe-4S] cluster conversion in Desulfovibrio fructosovorans [NiFe] hydrogenase by site-directed mutagenesis. Proc Natl Acad Sci USA 1998;95:11625–30.

[14] Sofia HJ, Chen G, Hetzler BG, Reyes-Spindola JF, Miller NE. Radical SAM, a novel protein superfamily linking unresolved steps in familiar biosynthetic pathways with radical mechanisms: functional characterization using new analysis and information visualization methods. Nucleic Acids Res 2001;29:1097–106.

[15] Nicolet Y, Drennan CL. AdoMet radical proteins – from structure to evolution – alignment of divergent protein sequences reveals strong secondary structure element conservation. Nucleic Acids Res 2004;32:4015–25.

[16] Vey JL, Drennan CL. Structural insights into radical generation by the radical SAM superfamily. Chem Rev 2011;111:2487–506.
[17] Frey PA, Hegeman AD, Ruzicka FJ. The radical SAM superfamily. Crit Rev Biochem Mol Biol 2008;43:63–88.
[18] Challand MR, Driesener RC, Roach PL. Radical S-adenosylmethionine enzymes: Mechanism, control and function. Nat Prod Rep 2011;28:1696–721.
[19] Hernandez HL, Pierrel F, Elleingand E, et al. MiaB, a bifunctional radical-S-adenosylmethionine enzyme involved in the thiolation and methylation of tRNA, contains two essential [4Fe-4S] clusters. Biochemistry 2007;46:5140–7.
[20] Perche-Letuvee P, Kathirvelu V, Berggren G, et al. 4-Demethylwyosine synthase from Pyrococcus abyssi is a radical-S-adenosyl-L-methionine enzyme with an additional [4Fe-4S] +2 cluster that interacts with the pyruvate co-substrate. J Biol Chem 2012;287:41174–85.
[21] Boal AK, Grove TL, McLaughlin MI, Yennawar NH, Booker SJ, Rosenzweig AC. Structural basis for methyl transfer by a radical SAM enzyme. Science 2011;332:1089–92.
[22] Grove TL, Benner JS, Radle MI, et al. A radically different mechanism for S-adenosylmethionine-dependent methyltransferases. Science 2011;332:604–7.
[23] Booth MP, Challand MR, Emery DC, Roach PL, Spencer J. High-level expression and reconstitution of active Cfr, a radical-SAM rRNA methyltransferase that confers resistance to ribosome-acting antibiotics. Protein Expr Purif 2010;74:204–10.
[24] Yan F, LaMarre JM, Rohrich R, et al. RlmN and Cfr are radical SAM enzymes involved in methylation of ribosomal RNA. J Am Chem Soc 2010;132:3953–64.
[25] Walsby CJ, Hong W, Broderick WE, et al. Electron-nuclear double resonance spectroscopic evidence that S-adenosylmethionine binds in contact with the catalytically active [4Fe-4S] + cluster of pyruvate formate-lyase activating enzyme. J Am Chem Soc 2002;124:3143–51.
[26] Walsby CJ, Ortillo D, Broderick WE, Broderick JB, Hoffman BM. An anchoring role for FeS clusters: chelation of the amino acid moiety of S-adenosylmethionine to the unique iron site of the [4Fe-4S] cluster of pyruvate formate-lyase activating enzyme. J Am Chem Soc 2002;124:11270–1.
[27] Walsby CJ, Ortillo D, Yang J, et al. Spectroscopic approaches to elucidating novel iron-sulfur chemistry in the “radical-SAM” protein superfamily. Inorg Chem 2005;44:727–41.
[28] Challand MR, Ziegert T, Douglas P, et al. Product inhibition in the radical S-adenosylmethionine family. FEBS Lett 2009;583:1358–62.
[29] Nicolet Y, Amara P, Mouesca JM, Fontecilla-Camps JC. Unexpected electron transfer mechanism upon AdoMet cleavage in radical SAM proteins. Proc Natl Acad Sci USA 2009;106:14867–71.
[30] Dowling DP, Vey JL, Croft AK, Drennan CL. Structural diversity in the AdoMet radical enzyme superfamily. Biochim Biophys Acta 2012;1824:1178–95.
[31] Chatterjee A, Li Y, Zhang Y, et al. Reconstitution of ThiC in thiamine pyrimidine biosynthesis expands the radical SAM superfamily. Nat Chem Biol 2008;4:758–65.
[32] Zhang Y, Zhu X, Torelli AT, et al. Diphthamide biosynthesis requires an organic radical generated by an iron-sulphur enzyme. Nature 2010;465:891–6.
[33] Kampmeier JA. Regioselectivity in the homolytic cleavage of S-adenosylmethionine. Biochemistry 2010;49:10770–2.
[34] Wang W, Wang K, Span I, et al. Are free radicals involved in IspH catalysis? An EPR and crystallographic investigation. J Am Chem Soc 2012;134:11225–34.
[35] Flint DH, Allen RM. Iron-sulfur proteins with nonredox functions. Chem Rev 1996;96: 2315–34.
[36] Beinert H, Kennedy MC, Stout CD. Aconitase as iron-sulfur protein, enzyme, and iron-regulatory protein. Chem Rev 1996;96:2335–74.
[37] Lee M, Grawert T, Quitterer F, et al. Biosynthesis of isoprenoids: crystal structure of the [4Fe-4S] cluster protein IspG. J Mol Biol 2010;404:600–10.

[38] Flint DH, Emptage MH, Guest JR. Fumarase a from Escherichia coli: purification and characterization as an iron-sulfur cluster containing enzyme. Biochemistry 1992;31:10331–7.
[39] Flint DH, Tuminello JF, Emptage MH. The inactivation of Fe-S cluster containing hydrolyases by superoxide. J Biol Chem 1993;268:22369–76.
[40] Ollagnier-de Choudens S, Loiseau L, Sanakis Y, Barras F, Fontecave M. Quinolinate synthetase, an iron-sulfur enzyme in NAD biosynthesis. FEBS Lett 2005;579:3737–43.
[41] Rousset C, Fontecave M, Ollagnier de Choudens S. The [4Fe-4S] cluster of quinolinate synthase from Escherichia coli: investigation of cluster ligands. FEBS Lett 2008;582:2937–44.
[42] Wood PM. The potential diagram for oxygen at pH 7. Biochem J 1988;253:287–9.
[43] Pryor WA, Houk KN, Foote CS, et al. Free radical biology and medicine: it's a gas, man! Am J Physiol Regul Integr Comp Physiol 2006;291:R491–511.
[44] Boehme DE, Vincent K, Brown OR. Oxygen and toxicity: inhibition of amino acid biosynthesis. Nature 1976;262:418–20.
[45] Kuo CF, Mashino T, Fridovich I. α,β-Dihydroxyisovalerate dehydratase: a superoxide-sensitive enzyme. J Biol Chem 1987;262:4724–7.
[46] Imlay JA. Pathways of oxidative damage. Annu Rev Microbiol 2003;57:395–418.
[47] Crack J, Green J, Thomson AJ. Mechanism of oxygen sensing by the bacterial transcription factor fumarate-nitrate reduction (FNR). J Biol Chem 2004;279:9278–86.
[48] Crack JC, Gaskell AA, Green J, Cheesmant MR, Le Brun NE, Thomson AJ. Influence of the environment on the 4Fe-4S2+ to 2Fe-2S2+ cluster switch in the transcriptional regulator FNR. J Am Chem Soc 2008;130:1749–58.
[49] Crack JC, Green J, Cheesman MR, Le Brun NE, Thomson AJ. Superoxide-mediated amplification of the oxygen-induced switch from 4Fe-4S to 2Fe-2S clusters in the transcriptional regulator FNR. Proc Natl Acad Sci USA 2007;104:2092–7.
[50] Crack JC, Green J, Le Brun NE, Thomson AJ. Detection of sulfide release from the oxygen-sensing 4Fe-4S cluster of FNR. J Biol Chem 2006;281:18909–13.
[51] Jervis AJ, Crack JC, White G, et al. The O2 sensitivity of the transcription factor FNR is controlled by Ser24 modulating the kinetics of 4Fe-4S to 2Fe-2S conversion. Proc Natl Acad Sci USA 2009;106:4659–64.
[52] Zhang B, Crack JC, Subramanian S, et al. Reversible cycling between cysteine persulfide-ligated [2Fe-2S] and cysteine-ligated [4Fe-4S] clusters in the FNR regulatory protein. Proc Natl Acad Sci USA 2012;109:15734–9.
[53] Khoroshilova N, Popescu C, Munck E, Beinert H, Kiley PJ. Iron-sulfur cluster disassembly in the FNR protein of Escherichia coli by O2: 4Fe-4S to 2Fe-2S conversion with loss of biological activity. Proc Natl Acad Sci USA 1997;94:6087–92.
[54] Lazazzera BA, Beinert H, Khoroshilova N, Kennedy MC, Kiley PJ. DNA binding and dimerization of the Fe-S-containing FNR protein from Escherichia coli are regulated by oxygen. J Biol Chem 1996;271:2762–8.
[55] Sutton VR, Mettert EL, Beinert H, Kiley PJ. Kinetic analysis of the oxidative conversion of the 4Fe-4S2+ cluster of FNR to a 2Fe-2S2+ cluster. J Bacteriol 2004;186:8018–25.
[56] Fleischhacker AS, Kiley PJ. Iron-containing transcription factors and their roles as sensors. Curr Opin Chem Biol 2011;15:335–41.
[57] Crack JC, Green J, Thomson AJ, Le Brun NE. Iron-sulfur cluster sensor-regulators. Curr Opin Chem Biol 2012;16:35–44.
[58] Green J, Crack JC, Thomson AJ, LeBrun NE. Bacterial sensors of oxygen. Curr Opin Microbiol 2009;12:145–51.
[59] Kiley PJ, Beinert H. Oxygen sensing by the global regulator, FNR: the role of the iron-sulfur cluster. FEMS Microbiol Rev 1998;22:341–52.
[60] Mulder DW, Shepard EM, Meuser JE, et al. Insights into [FeFe]-hydrogenase structure, mechanism, and maturation. Structure 2011;19:1038–52.

[61] Nicolet Y, Rubach JK, Posewitz MC, et al. X-ray structure of the [FeFe]-hydrogenase maturase HydE from Thermotoga maritima. J Biol Chem 2008;283:18861–72.
[62] Nicolet Y, Rohac R, Martin L, Fontecilla-Camps JC. X-ray snapshots of possible intermediates in the time course of synthesis and degradation of protein-bound Fe4S4 clusters. Proc Natl AcadSci USA 2013;110:7188–92.
[63] Berkovitch F, Nicolet Y, Wan JT, Jarrett JT, Drennan CL. Crystal structure of biotin synthase, an S-adenosylmethionine-dependent radical enzyme. Science 2004;303:76–9.
[64] Varghese S, Tang Y, Imlay JA. Contrasting sensitivities of Escherichia coli aconitases A and B to oxidation and iron depletion. J Bacteriol 2003;185:221–30.
[65] Volbeda A, Amara P, Darnault C, et al. X-ray crystallographic and computational studies of the O2-tolerant [NiFe]-hydrogenase 1 from Escherichia coli. Proc Natl Acad Sci USA 2012;109: 5305–10.
[66] Alvarez AF, Malpica R, Contreras M, Escamilla E, Georgellis D. Cytochrome d but not cytochrome o rescues the toluidine blue growth sensitivity of arc mutants of Escherichia coli. J Bacteriol 2010;192:391–9.
[67] Volbeda A, Montet Y, Vernede X, Hatchikian EC, Fontecilla-Camps JC. High-resolution crystallographic analysis of Desulfovibrio fructosovorans [NiFe] hydrogenase. Int J Hydrogen Energy 2002;27:1449–61.
[68] Buhrke T, Lenz O, Krauss N, Friedrich B. Oxygen tolerance of the H2-sensing [NiFe] hydrogenase from Ralstonia eutropha H16 is based on limited access of oxygen to the active site. J Biol Chem 2005;280:23791–6.
[69] Duche O, Elsen S, Cournac L, Colbeau A. Enlarging the gas access channel to the active site renders the regulatory hydrogenase HupUV of Rhodobacter capsulatus O2 sensitive without affecting its transductory activity. FEBS J 2005;272:3899–908.
[70] Parkin A, Goldet G, Cavazza C, Fontecilla-Camps JC, Armstrong FA. The difference a Se makes? Oxygen-tolerant hydrogen production by the [NiFeSe]-hydrogenase from Desulfomicrobium baculatum. J Am Chem Soc 2008;130:13410–6.
[71] Garcin E, Vernede X, Hatchikian EC, Volbeda A, Frey M, Fontecilla-Camps JC. The crystal structure of a reduced [NiFeSe] hydrogenase provides an image of the activated catalytic center. Structure 1999;7:557–66.
[72] Cracknell JA, Vincent KA, Ludwig M, Lenz O, Friedrich B, Armstrong FA. Enzymatic oxidation of H2 in atmospheric O2: the electrochemistry of energy generation from trace H2 by aerobic microorganisms. J Am Chem Soc 2008;130:424–5.
[73] Lenz O, Ludwig M, Schubert T, et al. H2 conversion in the presence of O2 as performed by the membrane-bound [NiFe]-hydrogenase of Ralstonia eutropha. ChemPhysChem 2010;11:1107–19.
[74] Pandelia ME, Fourmond V, Tron-Infossi P, et al. Membrane-bound hydrogenase I from the hyperthermophilic Bacterium Aquifex aeolicus: enzyme activation, redox intermediates and oxygen tolerance. J Am Chem Soc 2010;132:6991–7004.
[75] Ludwig M, Cracknell JA, Vincent KA, Armstrong FA, Lenz O. Oxygen-tolerant H2 oxidation by membrane-bound [NiFe] hydrogenases of Ralstonia Species coping with low level H2 in air. J Biol Chem 2009;284:465–77.
[76] Lukey MJ, Parkin A, Roessler MM, et al. How Escherichia coli is equipped to oxidize hydrogen under different redox conditions. J Biol Chem 2010;285:3928–38.
[77] Vincent KA, Cracknell JA, Clark JR, et al. Electricity from low-level H2 in still air – an ultimate test for an oxygen tolerant hydrogenase. Chem Commun 2006:5033–5.
[78] Volbeda A, Martin L, Cavazza C, et al. Structural differences between the ready and unready oxidized states of [NiFe] hydrogenases. J Biol Inorg Chem 2005;10:239–49.
[79] Saggu M, Zebger I, Ludwig M, et al. Spectroscopic insights into the oxygen-tolerant membrane-associated [NiFe] hydrogenase of Ralstonia eutropha H16. J Biol Chem 2009; 284:16264–76.

[80] Roessler MM, Evans RM, Davies RA, Harmer J, Armstrong FA. EPR spectroscopic studies of the FeS clusters in the O0032-tolerant [NiFe]-hydrogenase Hyd-1 from Escherichia coli and characterization of the unique [4Fe3S] cluster by HYSCORE. J Am Chem Soc 2012;134:15581–94.
[81] Knüttel K, Schneider K, Erkens A, et al. Redox properties of the metal centers in the membrane-bound hydrogenase from Alcaligenes eutrophus CH34. Bull Pol Acad Sci Chem 1994;42:495–511.
[82] Pandelia ME, Nitschke W, Infossi P, Giudici-Orticoni MT, Bill E, Lubitz W. Characterization of a unique [FeS] cluster in the electron transfer chain of the oxygen tolerant [NiFe] hydrogenase from Aquifex aeolicus. Proc Natl Acad Sci USA 2011;108:6097–102.
[83] Goris T, Wait AF, Saggu M, et al. A unique iron-sulfur cluster is crucial for oxygen tolerance of a [NiFe]-hydrogenase. Nature Chem Biol 2011;7:310–8.
[84] Lukey MJ, Roessler MM, Parkin A, et al. Oxygen-tolerant [NiFe]-hydrogenases: the individual and collective importance of supernumerary cysteines at the proximal Fe-S cluster. J Am Chem Soc 2011;133:16881–92.
[85] Heering HA, Bulsink YBM, Hagen WR, Meyer TE. Reversible super reduction of the cubane [4Fe-4S] (3+,2+,1+) in the high-potential iron-sulfur protein under non-denaturating conditions. Eur J Biochem 1995;232:811–7.
[86] Cammack R. “Super reduction” of chromatium high-potential iron-sulphur protein in the presence of dimethyl sulphoxide. Biochem Biophys Res Commun 1973;54:548–54.
[87] Shomura Y, Yoon KS, Nishihara H, Higuchi Y. Structural basis for a [4Fe-3S] cluster in the oxygen-tolerant membrane-bound [NiFe]-hydrogenase. Nature 2011;479:253–6.
[88] Fritsch J, Scheerer P, Frielingsdorf S, et al. The crystal structure of an oxygen-tolerant hydrogenase uncovers a novel iron-sulphur centre. Nature 2011;479:249–52.
[89] Mouesca JM, Fontecilla-Camps JC, Amara P. The Structural plasticity of the proximal 4Fe3S cluster is responsible for the O2 tolerance of membrane-bound NiFe hydrogenases. Angew Chem Int Ed 2013;52:2002–6.
[90] Walters EM, Garcia-Serres R, Jameson GNL, et al. Spectroscopic characterization of site-specific Fe4S4 cluster chemistry in ferredoxin: thioredoxin reductase: implications for the catalytic mechanism. J Am Chem Soc 2005;127:9612–24.
[91] Orio M, Mouesca JM. Variation of average g values and effective exchange coupling constants among [2Fe-2S] clusters: a density functional theory study of the impact of localization (trapping forces) versus delocalization (double-exchange) as competing factors. Inorg Chem 2008;47:5394–416.
[92] Dey A, Francis EJ, Adams MWW, et al. Solvent tuning of electrochemical potentials in the active sites of HiPIP versus ferredoxin. Science 2007;318:1464–8.
[93] Capozzi F, Ciurli S, Luchinat C. Coordination sphere versus protein environment as determinants of electronic and functional properties of iron-sulfur proteins. Struct Bonding 1998;90:127–60.
[94] Pandelia ME, Infossi P, Stein M, Giudici-Orticoni MT, Lubitz W. Spectroscopic characterization of the key catalytic intermediate Ni-C in the O2-tolerant NiFe hydrogenase I from Aquifex aeolicus: evidence of a weakly bound hydride. Chem Commun 2012;48:823–5.
[95] Schlangen M, Schwarz H. Insertion of molecular oxygen in transition-metal hydride bonds, oxygen-bond activation, and unimolecular dissociation of metal hydroperoxide intermediates. Helv Chim Acta 2008;91:379–86.
[96] Nemirovskiy OV, Gross ML. Complexes of iron(II) with cysteine-containing peptides in the gas phase. J Am Soc Mass Spectrom 1996;7:977–80.
[97] Sigel H, Martin RB. Coordinating properties of the amide bond – stability and structure of metal-ion complexes of peptides and related ligands. Chem Rev 1982;82:385–426.
[98] Miller AF. Superoxide dismutases: ancient enzymes and new insights. FEBS Lett 2012;586:585–95.

[99] Nicholls P. Classical catalase: ancient and modern. Arch Biochem Biophys 2012;525:95–101.
[100] Mishra S, Imlay J. Why do bacteria use so many enzymes to scavenge hydrogen peroxide? Arch Biochem Biophys 2012;525:145–60.
[101] Johnson DC, Dean DR, Smith AD, Johnson MK. Structure, function, and formation of biological iron-sulfur clusters. Annu Rev Biochem 2005:247–81.
[102] Kessler D, Papenbrock J. Iron-sulfur cluster biosynthesis in photosynthetic organisms. Photosynth Res 2005;86:391–407.
[103] Balk J, Pilon M. Ancient and essential: the assembly of iron-sulfur clusters in plants. Trends Plant Sci 2011;16:218–26.
[104] Djaman O, Outten FW, Imlay JA. Repair of oxidized iron-sulfur clusters in Escherichia coli. J Biol Chem 2004;279:44590–9.
[105] Mukhopadhyay P, Zheng M, Bedzyk LA, LaRossa RA, Storz G. Prominent roles of the NorR and Fur regulators in the Escherichia coli transcriptional response to reactive nitrogen species. Proc Natl Acad Sci USA 2004;101:745–50.
[106] Justino MC, Vicente JB, Teixeira M, Saraiva LM. New genes implicated in the protection of anaerobically grown Escherichia coli against nitric oxide. J Biol Chem 2005;280:2636–43.
[107] Justino MC, Almeida CC, Teixeira M, Saraiva LM. Escherichia coli di-iron YtfE protein is necessary for the repair of stress-damaged iron-sulfur clusters. J Biol Chem 2007;282:10352–9.
[108] Justino MC, Almeida CC, Goncalves VL, Teixeira M, Saraiva LM. Escherichia coli YtfE is a di-iron protein with an important function in assembly of iron-sulphur clusters. FEMS Microbiol Lett 2006;257:278–84.
[109] Justino MC, Baptista JM, Saraiva LM. Di-iron proteins of the RIC family are involved in iron-sulfur cluster repair. Biometals 2009;22:99–108.
[110] Vine CE, Justino MC, Saraiva LM, Cole J. Detection by whole genome microarrays of a spontaneous 126-gene deletion during construction of a ytfE mutant: confirmation that a ytfE mutation results in loss of repair of iron-sulfur centres in proteins damaged by oxidative or nitrosative stress. J Microbiol Meth 2010;81:77–9.
[111] Overton TW, Justino MC, Li Y, et al. Widespread distribution in pathogenic bacteria of di-iron proteins that repair oxidative and nitrosative damage to iron-sulfur centers. J Bacteriol 2008;190:2004–13.
[112] Todorovic S, Justino MC, Wellenreuther G, et al. Iron-sulfur repair YtfE protein from Escherichia coli: structural characterization of the di-iron center. J Biol Inorg Chem 2008;13:765–70.
[113] Torrents E, Aloy P, Gibert I, Rodriguez-Trelles F. Ribonucleotide reductases: divergent evolution of an ancient enzyme. J Mol Evol 2002;55:138–52.
[114] Nordlund N, Reichard P. Ribonucleotide reductases. Annu Rev Biochem 2006:681–706.
[115] Logan DT, Andersson J, Sjoberg BM, Nordlund P. A glycyl radical site in the crystal structure of a class III ribonucleotide reductase. Science 1999;283:1499–504.
[116] Sintchak MD, Arjara G, Kellogg BA, Stubbe J, Drennan CL. The crystal structure of class II ribonucleotide reductase reveals how an allosterically regulated monomer mimics a dimer. Nat Struct Biol 2002;9:293–300.
[117] Challand MR, Martins FT, Roach PL. Catalytic activity of the anaerobic tyrosine lyase required for thiamine biosynthesis in Escherichia coli. J Biol Chem 2010;285:5240–8.
[118] Kriek M, Martins F, Challand MR, Croft A, Roach PL. Thiamine biosynthesis in Escherichia coli: identification of the intermediate and by-product derived from tyrosine. Angew Chem Int Ed 2007;46:9223–6.
[119] Park JH, Dorrestein PC, Zhai H, Kinsland C, McLafferty FW, Begley TP. Biosynthesis of the thiazole moiety of thiamin pyrophosphate (Vitamin B1). Biochemistry 2003;42:12430–8.

[120] Settembre EC, Dorrestein PC, Park JH, Augustine AM, Begley TP, Ealick SE. Structural and mechanistic studies on ThiO, a glycine oxidase essential for thiamin biosynthesis in Bacillus subtilis. Biochemistry 2003;42:2971–81.
[121] Benjdia A, Leprince J, Sandstrom C, Vaudry H, Berteau O. Mechanistic investigations of anaerobic sulfatase-maturating enzyme: direct C-beta H-atom abstraction catalyzed by a radical AdoMet enzyme. J Am Chem Soc 2009;131:8348–+.
[122] Goldman PJ, Grove TL, Sites LA, McLaughlin MI, Booker SJ, Drennan CL. X-ray structure of an AdoMet radical activase reveals an anaerobic solution for formylglycine posttranslational modification. Proc Natl Acad Sci USA 2013;110:8519–24.
[123] Dierks T, Dickmanns A, Preusser-Kunze A, et al. Molecular basis for multiple sulfatase deficiency and mechanism for formylglycine generation of the human formylglycine-generating enzyme. Cell 2005;121:541–52.
[124] Roeser D, Preusser-Kunze A, Schmidt B, et al. A general binding mechanism for all human sulfatases by the formylglycine-generating enzyme. Proc Natl Acad Sci USA 2006;103:81–6.
[125] Bojarova P, Williams SJ. Sulfotransferases, sulfatases and formylglycine-generating enzymes: a sulfation fascination. Curr Opin Chem Biol 2008;12:573–81.

# 11 A retrospective on the discovery of [Fe-S] cluster biosynthetic machineries in *Azotobacter vinelandii*

Patricia C. Dos Santos and Dennis R. Dean

## 11.1 Introduction

Iron-sulfur clusters ([Fe-S]) clusters are inorganic prosthetic groups that participate in many essential life processes. Despite their relatively simple structures, the biological assembly of [Fe-S] clusters is complicated and the defining features of the process have only emerged over the past two decades. The biosynthetic process involves two principle interacting proteins, an L-cysteine desulfurase and a molecular assembly scaffold. In this chapter, a retrospective is provided that describes the initial discovery of the mechanism for the assembly of [Fe-S] clusters in the nitrogen-fixing organism *Azotobacter vinelandii*. Why and how the study of biological nitrogen fixation ultimately contributed the key insights into the general process of [Fe-S] cluster assembly is described.

Beinert, Holm, and Munck, pioneers in the field of bioinorganic chemistry, have aptly described biological iron-sulfur clusters ([Fe-S] clusters) as nature's modular, multipurpose structures [1]. What are [Fe-S] clusters, why do they exist in nature, and why are they so important to so many biological processes? [Fe-S] clusters were originally discovered as inorganic prosthetic groups composed exclusively of iron and sulfide contained in a class of redox-active proteins denominated as ferredoxins [2, 3]. Ferredoxins are electron/proton carrier proteins that contain [Fe-S] clusters, usually in the form of rhombic [2Fe-2S] or cubane [4Fe-4S] clusters (Fig. 11.1). [Fe-S] clusters are most often covalently attached to their cognate protein partners through cysteinyl thiolate ligation to their metal sites, although other ligation modes are also known to exist and not all of the metal sites are necessarily coordinated by a protein-donated ligand [4]. Ferredoxins represent a specialized class of a wide variety of proteins, now generically designated as [Fe-S] proteins, that contain one or more [Fe-S] clusters. It is the capacity of [Fe-S] clusters to exist in multiple oxidation states that endows [Fe-S] proteins with their ability to serve as electron/proton carriers. Indeed, the reversibility of redox properties of [Fe-S] proteins is an integral aspect of essential energy transducing processes in nitrogen fixation, photosynthesis, and respiration. The key involvement of [Fe-S] clusters in such life-sustaining processes is intimately linked to the wide range of redox potentials they can attain as a consequence of their respective polypeptide environments [5]. However, the ability of proteins to tune the redox potentials of their cognate [Fe-S] clusters is not the only feature of [Fe-S] clusters exploited by nature. The chemical and structural versatility of [Fe-S]

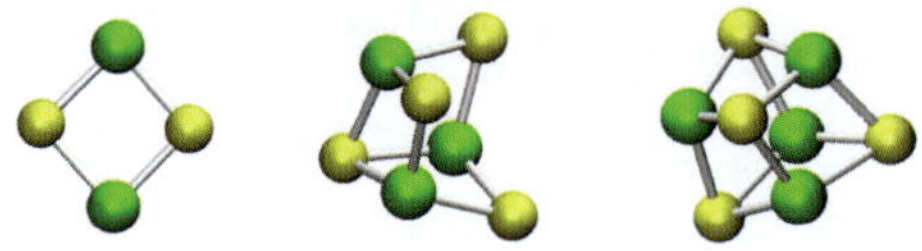

[2Fe–2S] [3Fe–4S] [4Fe–4S]

**Fig. 11.1:** Simple [Fe-S] cluster types found in nature.

clusters, uniquely achieved by combining the individual chemical properties of Fe and S, has enabled [Fe-S] proteins to fulfill other important biological roles including activation of substrates for chemical transformations, serving as environmental sensors, providing structural determinants within proteins, and functioning as agents of gene regulation [4, 6].

Although [Fe-S] clusters play a central role in many biological processes, there is also a significant penalty associated with their use. Indeed, $Fe^{2+}$ and $S^{2-}$, necessary for both the assembly and disassembly of [Fe-S] clusters, are toxic to a wide variety of cellular processes in aerobic organisms. How then did [Fe-S] clusters become so pervasive in biology? The answer to this question is not known for certain but could be linked to the "iron-sulfur world" theory, which proposes that chemical reactions associated with formation of the organic building blocks necessary for the emergence of life on earth occurred on metal-sulfur surfaces in the highly reducing environment of prebiotic earth [7, 8]. Certain aspects of such chemistries might have been captured in primordial organisms in the form of [Fe-S] clusters, particularly when considering that facile "spontaneous" assembly of [Fe-S] clusters occurs under reducing conditions. This possibility finds some credence when it is considered that life likely originated in an anoxic environment in which free $Fe^{2+}$ and free $S^{2-}$ did not pose the same risk as is associated with the oxygen-saturated environment that dominates life on earth today. For example, the potential for lethal formation of reactive oxygen species (ROS) through Fenton chemistry did not exist for emergent life forms. This luxury was eliminated, however, by the advent of oxygenic photosynthesis, which gradually converted the biosphere from a reducing environment to a primarily oxidizing environment. Given that fundamental life sustaining processes requiring [Fe-S] clusters were almost certainly well developed by the time photosynthesis emerged, the spontaneous assembly of [Fe-S] clusters from free $Fe^{2+}$ and $S^{2-}$ could not possibly continue, even if that was the case in primordial life. What this means is that at some point during evolution and certainly upon transition of earth to an oxidizing atmosphere, living organisms needed to develop a way to construct [Fe-S] clusters such that free $Fe^{2+}$ and free $S^{2-}$ were not required [9, 10]. In other words, these elements needed to be trafficked and combined in nontoxic forms or living organisms needed to evolve alternative strategies to replace the many functions supplied by [Fe-S] proteins.

Considering the importance of [Fe-S] clusters in sustaining essential life processes [11] as well as the striking structural simplicity of most [Fe-S] clusters, it might seem curious that so little was known about their biological assembly until relatively recently. However, as is discussed in this narrative, there are very good reasons why a fundamental understanding of the assembly of [Fe-S] clusters took so long to develop when compared with our understanding of the biosynthetic pathways for formation of many other organic cofactors. One reason is that the critical importance of [Fe-S] clusters to so many biological processes prevented fortuitous discovery of assembly factors in genetic screens because assembly factors were necessary for survival. Another reason is that the structural simplicity of [Fe-S] clusters, and the ability to form them spontaneously *in situ* from free $Fe^{2+}$ and $S^{2-}$, failed to inspire serious inquiry into potential mechanisms for their biological formation. In this chapter, we provide a retrospective about the work using the bacterium *Azotobacter vinelandii*, and studies on the specialized process of biological $N_2$ fixation, that ultimately led to some of the key insights about the [Fe-S] cluster assembly process.

## 11.2 An introduction to nitrogenase

Serendipity is defined as the accident of finding something good or useful while not specifically searching for it. Initial work on nitrogenase was not aimed at understanding how simple [Fe-S] clusters are formed in biological systems. Instead, the focus was directed at understanding the genetic determinants and chemical mechanism of biological $N_2$ fixation, an essential contributor to the biogeochemical nitrogen cycle. Here we discuss some of the general features of the enzyme nitrogenase, the catalytic component of $N_2$ fixation, and the initial approaches that were used to explore its function [12, 13]. It will be seen that, by both chance and design, these approaches inexorably led to the discovery of unifying features involved in the assembly of simple [Fe-S] clusters necessary to sustain cellular metabolism.

Simple rhombic [2Fe-2S] and cubane [4Fe-4S] clusters (Fig. 11.1), composed only of Fe and S, represent the dominant forms of [Fe-S] clusters in biological systems [4]. However, there are also other types of [Fe-S] clusters that have higher nuclearity, those that contain another metal in addition to Fe, and those that also contain organic constituents [13, 14]. A broad spectrum of [Fe-S] cluster types can be found in nitrogenase, the enzyme that catalyzes the nucleotide-dependent reduction of $N_2$ to yield two molecules of ammonia ($NH_3$) (Fig. 11.2). Nitrogenase is a two component enzyme that contains a canonical [4Fe-4S] cluster, involved in electron transfer, a novel [8Fe-7S] cluster (P-cluster) that also serves as an agent of electron transfer, and a [7Fe-9S-Mo-C-homocitrate] cluster (FeMo cofactor, or FeMoco), which contains molybdenum (Mo) and provides the site for $N_2$ activation and reduction [15]. One component of nitrogenase is called the Fe protein, a name derived from the observation that it contains a single [4Fe-4S] cluster bridged between two identical subunits. Each Fe protein subunit also contains a nucleotide-binding site. The other component of nitrogenase is called the MoFe protein, a name derived from the fact it contains FeMoco. The MoFe

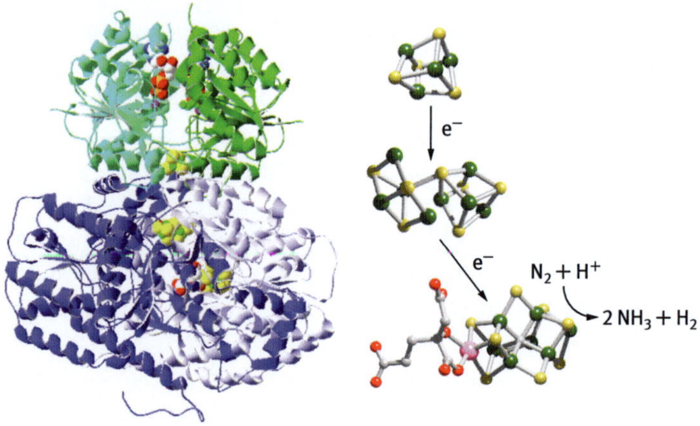


**Fig. 11.2:** Mo-dependent nitrogenase and its associated [Fe-S] clusters. (left) Ribbon representation of a Mo-dependent nitrogenase catalytic unit. The MoFe protein α-subunit is shown in light blue, and the β-subunit is shown in dark blue. The Fe protein is shown in a complex with the MoFe protein, and the identical Fe protein subunits are indicated in light and dark green. Nucleotides bound to the Fe protein and the associated [Fe-S] clusters are shown as space-filling models. (right) Ball-and-stick representations of the nitrogenase-associated [Fe-S] clusters: [4Fe-4S] cluster (top), [8Fe-7S] P-cluster (center), [7Fe-9S-C-Mo-homocitrate] FeMoco structure from PDB 3U7Q (bottom). The electron flow direction and site of nitrogen reduction is indicated.

protein is an $\alpha_2\beta_2$ heterotetramer that, in combination with two Fe proteins, appears to contain two independent catalytic units. One P-cluster is located at the interface of each MoFe protein αβ interface and one FeMoco is buried entirely within each MoFe protein α subunit. During catalysis, the Fe Protein and MoFe protein cycle through a series of sequential single electron transfer events and each cycle is coupled to the binding and hydrolysis of two ATPs. FeMoco becomes activated for $N_2$ binding and reduction once the MoFe protein has accumulated a sufficient number of electrons that have been delivered by the Fe protein [16].

Models for how intermolecular electron transfer is coupled to nucleotide hydrolysis, where and how nitrogenase substrates interact with FeMoco, and how FeMoco is activated for substrate binding have been proposed [17–19]. These models were developed through application of the biochemical-genetic approach described in a following section, advanced spectroscopic analyses, and consideration of high-resolution crystallographic structures of the Fe protein and MoFe protein. However, when the biochemical-genetic approach for analysis of nitrogenase structure and function was initiated, the structures of nitrogenase proteins and their associated metalloclusters were not known, nor was the identity, organization, and function of genes associated with formation of an active nitrogenase understood. During the late 1970s, a variety of approaches were undertaken with the goal of understanding how nitrogenase functions at the mechanistic level and how the complex [Fe-S] clusters associated with its activity are formed [20, 21]. There were two practical reasons for wanting to understand the mechanism of nitrogenase at the chemical level as well as for understanding the pathways involved in assembly of its associated [Fe-S] clusters. First, a mechanistic understanding of nitrogenase might be useful for the development of synthetic biomimetic catalysts for $N_2$ reduction. Second, attempts to endow higher plants with the capacity to produce their own nitrogenous fertilizer necessarily depend on a fundamental understanding of the genetic and biochemical determinants required for formation of an active nitrogenase.

Attempts to elucidate the organization and function of genes associated with nitrogenase predated the availability of genomic sequences. The pioneering work in this area of research was performed in the late 1970s and early 1980s and used the $N_2$-fixing facultative anaerobe *Klebsiella pneumoniae* as an experimental organism. As a starting point, this work involved the isolation of mutant strains having point mutations or insertion mutations that resulted in a defective capacity for $N_2$ fixation [22, 23]. The chromosomal position of each mutation was subsequently determined by classical genetic mapping technologies using phage transduction and three factor crosses, and the individual proteins affected were identified by protein mobility shifts that could be recognized by analysis of mutant crude extracts using two-dimensional gel electrophoresis [24–28]. In retrospect, it is amazing that, without the benefit of DNA sequencing technology, nearly every gene associated with $N_2$ fixation in *K. pneumonia*, collectively designated as *nif* genes (for nitrogen fixation), was correctly identified in this way, as were their relative chromosomal locations, and their organization into transcriptional units (Fig. 11.3) [29, 30].

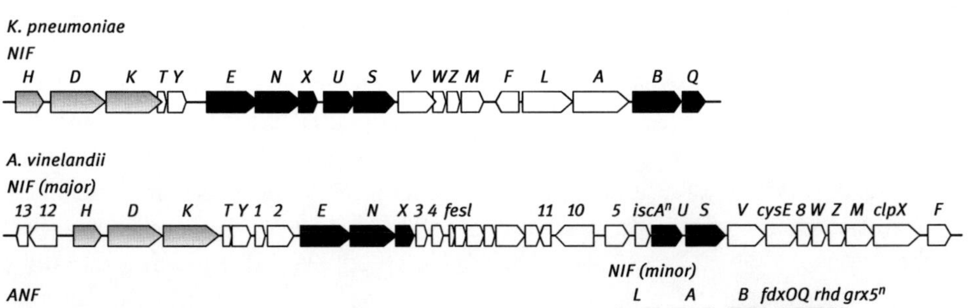


**Fig. 11.3:** Nitrogen fixation-associated gene regions in *K. pneumoniae* and *A. vinelandii*. The organization of gene regions for the three nitrogen fixation systems, Nif, Anf, and Vnf, is indicated. Genes encoding nitrogenase catalytic components are shown in light gray; genes known or suspected to be involved in [Fe-S] cluster assembly or trafficking are shown in black.

Once the fundamental genetic determinants necessary for $N_2$ fixation by *K. pneumoniae* were established, a variety of investigators turned their attention to *A. vinelandii* as the experimental organism because (i) *A. vinelandii* fixes $N_2$ under aerobic growth conditions, making large-scale growth possible without requiring specialized fermentation facilities, (ii) this bacterium abundantly produces nitrogenase when grown under conditions that impose a need for $N_2$ fixation, (iii) it is tractable to facile genetic manipulation, and (iv) it produces multiple, genetically distinct, forms of nitrogenase (discussed later). *A. vinelandii* turned out to be an excellent choice as an experimental organism and proved to be instrumental in advancing the current mechanistic understanding of $N_2$ fixation. For example, the highest resolution structures currently available were produced using nitrogenase isolated from *A. vinelandii* [31–33]. Also, current mechanistic models for nitrogenase are based on interpretation of biochemical and biophysical analyses of altered nitrogenases produced by mutant strains derived from *A. vinelandii* [15]. Finally, a comprehensive understanding of the mechanism for assembly of FeMoco, which provides the $N_2$ activation and reduction site, has emerged from studies on components isolated from *A. vinelandii* [34, 35].

Foundational work using *K. pneumoniae* proved to be valuable when investigators initiated serious studies on the biochemical-genetic analysis of $N_2$ fixation in *A. vinelandii* because it provided a genetic roadmap and a baseline analysis for establishing gene-product relationships. Early efforts to construct a physical map of genes involved in $N_2$ fixation in *A. vinelandii* revealed a much more complex picture than that established for *K. pneumoniae* (Fig. 11.3) [36, 37]. For example, not all *A. vinelandii* genes associated with $N_2$ fixation are physically linked on the genome, as was true for *K. pneumoniae*. Also, there are many more $N_2$ fixation-associated genes in *A. vinelandii* when compared with *K. pneumoniae*. Finally, it was discovered that *A. vinelandii* encodes three genetically distinct $N_2$-fixing systems, two of which do not require Mo for their respective activities [38]. One of these is designated as the "V-dependent" nitrogenase [genetically indicated as *vnf* because its counterpart to the FeMoco contains the transition metal, vanadium (V) in place of Mo]. The other is designated as "Fe-only" nitrogenase (genetically indicated as *anf*) because it represents a counterpart to the FeMoco in which Fe substitutes for the Mo atom [39–41]. There is a hierarchy of physiological expression of the different nitrogenases. Notably, the V-dependent enzyme is expressed only when Mo is not available in the growth medium but V is available. The Fe-only nitrogenase is expressed only when neither Mo nor V is available in the growth medium. Such hierarchical expression of genes that encode the different nitrogenases might reflect the catalytic efficiencies of the corresponding enzyme systems, with the Mo-dependent enzyme being the most efficient and the Fe-only enzyme being the least efficient [42]. Although the catalytic components of the different nitrogenases are genetically distinct, they all appear to share common mechanistic features. The entire suite of genes associated with all three systems was ultimately established by determination of the genome sequence of *A. vinelandii* and transcriptome analyses of gene expression under conditions favoring each of the respective systems (Fig. 11.3) [43–46].

## 11.3 Approaches to identify gene-product and product-function relationships

Two related biochemical-genetic approaches were initially used to establish gene-product and product-function relationships in $N_2$ fixation [12] prior to the advent of the field now known as genomics. The first of these involved searching for mutant strains defective in $N_2$ fixation followed by identification of the affected gene, and subsequent characterization of the specific biochemical phenotype associated with loss of that function. The second approach used *in vitro* biochemical complementation experiments, involving the mixture of extracts, or purified proteins, prepared from separately isolated mutant strains that were each defective in a particular function, in an attempt to rescue a biochemical function. The power of such classical approaches, first developed in the 1940s, has been tremendously augmented over the past several decades through application of recombinant DNA technologies and the capacity for rapid and accurate DNA sequence analyses. Among the many tools now available to researchers that have been applied to the analysis of $N_2$ fixation are included: (1) prediction of possible functional features based on deduction of primary structures from DNA sequences; (2) heterologous production of proteins at prodigious levels using recombinant expression plasmids; (3) directed *in vitro* mutagenesis and gene replacement; and (4) placement of affinity tags within a gene that encodes a protein of particular interest to facilitate its purification for subsequent biochemical analysis. We now describe how these technologies, in combination with the classical biochemical-genetic approach, were applied to provide an understanding of some of the basic features associated with $N_2$ fixation-specific [Fe-S] cluster formation.

## 11.4 FeMoco and development of the scaffold hypothesis for complex [Fe-S] cluster formation

Two aspects of $N_2$ fixation make it particularly amenable to application of the biochemical-genetic approach for the study of [Fe-S] cluster biogenesis. First, the ability to fix $N_2$ is dispensable for the growth of $N_2$-fixing organisms when an alternative nitrogen source, for example, ammonia, is supplemented in the growth medium. In other words, mutant strains defective in $N_2$ fixation grow normally and are indistinguishable from wild-type strains, unless they are cultured under conditions that demand $N_2$ fixation [47]. The other reason $N_2$ fixation is amenable to biochemical-genetic analysis is that nitrogenase is a very poor catalyst [48], and therefore, it must be produced at very high levels to sustain growth under conditions that demand $N_2$

fixation. Consequently, $N_2$-fixing cells also have a high demand for the acquisition of the Fe and S needed for assembly of nitrogenase-associated [Fe-S] clusters. It is the conditional dispensability of the capacity for physiological $N_2$ reduction as well as the high levels of nitrogenase production in $N_2$ fixing cells that proved to be important aspects in the discovery of how Fe and S are combined for formation of both complex and simple [Fe-S] clusters.

It has already been noted that the FeMoco provides the site for $N_2$ binding and reduction. This feature was firmly established using a variety of experimental methods, but there were two, simple, but elegant, approaches that first led to discovery of FeMoco and subsequently to the concept of a molecular scaffold involvement in complex [Fe-S] cluster assembly. In the early analysis of mutant strains defective in $N_2$ fixation, several mutants that were specifically defective in MoFe protein activity but lacked mutations in the structural genes that encode the MoFe protein, *nifD* or *nifK*, were identified (Tab. 11.1). Rather, these mutant strains had mutations in genes designated *nifE*, *nifN*, or *nifB* [12, 30]. It was then discovered that extraction of purified MoFe protein using a chaotropic organic solvent, *N*-methyl formamide, released a factor that contained Fe and Mo, which could be added to extracts of *nifB*, *nifE*, or *nifN* mutants to reconstitute MoFe protein activity [49]. Because the released factor contained both Fe and Mo, it was designated FeMoco, and *nifB*, *nifE*, and *nifN* were recognized to be necessary for FeMoco formation.

**Tab. 11.1:** Gene-product relationships and phenotypes associated with selected *A. vinelandii nif* genes.

| Gene | Product and function | Fe protein[a] | MoFe protein[b] | Required for V- and Fe-only nitrogenases |
|---|---|---|---|---|
| *nifH* | Fe protein subunit | NA | NA | No |
| *nifD* | MoFe protein α-subunit | A | NA | No |
| *nifK* | MoFe protein β-subunit | A | NA | No |
| *nifE* | FeMo-cofactor assembly scaffold α-subunit | A | NA | No |
| *nifN* | FeMo-cofactor assembly scaffold β-subunit | PA | NA | No |
| *nifB* | SAM-dependent C-insertion into FeMoco | A | NA | Yes |
| *nifU* | Simple [Fe-S] cluster assembly scaffold | PA | PA | Yes |
| *nifS* | Cysteine desulfurase | PA | PA | Yes |
| *nifV* | Homocitrate synthase | A | PA | Yes |
| *nifM* | Fe protein maturase *cis-trans* Prolyl-isomerase | NA | PA | Yes |

[a] Fe protein activity (nmol of ethylene/min/mg of crude extracts) after addition of saturating levels of purified MoFe protein.
[b] MoFe protein activity (nmol of ethylene/min/mg of crude extracts) after addition of saturating levels of purified Fe protein.
A, activity detected (>50% of wild-type levels); NA, no activity (<1%); PA, partial activity detected (5%–40%).

It is now well understood that the biosynthesis of FeMoco involves the products of *nifE*, *nifN*, and *nifB* as well as the participation of other *nif* gene products, for example, *nifV*, which encodes a homocitrate synthase (Tab. 11.1) [50, 51]. However, none of the specific functions of these proteins was known during the early development of the scaffold hypothesis for complex [Fe-S] cluster assembly. With respect to assembly of FeMoco, one possibility was that Fe, S, Mo, and organic constituents could be separately inserted into the MoFe protein. Such an assembly path is analogous to on-site construction of a house within a neighborhood using brick, wood, and other construction materials separately delivered to the construction site. The other possibility was that FeMoco could be preassembled and then inserted into MoFe protein as an intact entity. This latter type of construction is analogous to the prefabrication of a house, followed by subsequent delivery of the house to the neighborhood. To differentiate between these possibilities, extracts of mutant strains that do not produce MoFe protein but express functional products of all of the other *nif* genes were mixed with extracts of strains that are defective in *nifE*, *nifN*, or *nifB*, which all produce an inactive FeMoco-less MoFe protein [52, 53]. In this experiment, MoFe protein activity was reconstituted, demonstrating that FeMoco can be assembled in the absence of MoFe protein (produced in extracts that do not contain MoFe protein) and therefore must be pre-constructed and subsequently inserted intact into a cofactorless form of MoFe protein (produced in extracts of strains lacking NifE, NifN, or NifB function).

The above experiments were executed in the absence of any knowledge about the primary structure of NifE, NifN, or NifB and without direct knowledge about the distribution of the [Fe-S] clusters contained within the MoFe protein. However, there were already some clues available that made it possible to advance the “scaffold hypothesis” for formation of FeMoco. The main premise of the scaffold hypothesis was that because FeMoco is synthesized separately from the MoFe protein, it could be anticipated that some aspect of assembly might occur on a molecular “scaffold” that duplicates certain structural features of the MoFe protein, particularly within FeMoco-binding regions [54]. This hypothesis was initially developed on the foundation of an apparent correlation between NifD (MoFe protein α-subunit) and NifE and between NifK (MoFe protein β-subunit) and NifN, respectively. Namely, it was already known from genetic mapping and two-dimensional gel electrophoresis analyses that *nifD* and *nifK* were co-transcribed and expressed at approximately equal levels to each other and that *nifE* and *nifN* were separately co-transcribed and expressed at approximately equal levels to each other, suggesting that the structural genes (NifD and NifK) were in a different operon from the cofactor assembly genes (NifE and NifN). Furthermore, NifD was similar in size and isoelectric point to NifE, and NifK was similar to NifN [28, 55]. Elucidation of primary structures, deduced from DNA sequence analysis, did, in fact, reveal significant conservation in primary structure between NifD and NifE and between NifK and NifN, respectively [54, 56]. Particularly rich sequence conservation between a segment of NifD and NifE, was observed in the region predicted to provide at least a portion of the FeMoco-binding pocket, on the basis of the NifD primary structure [57]. These observations provided powerful support for the scaffold hypothesis and indicated that NifE and NifN might form a heterotetrameric complex that provides a transient site for an intermediate stage of FeMoco formation. The primary structural comparisons between NifD and NifE also suggested that FeMoco might be located entirely within the MoFe protein α-subunit and that the NifEN complex might contain another cluster, analogous to the MoFe protein P-cluster, bridged between the NifE and NifN subunits. All of these predictions turned out to be correct [58–60].

## 11.5 An approach for the analysis of *nif* gene product function

Details concerning other specialized aspects of FeMoco assembly, for example, the formation and attachment of the organic constituent, homocitrate, and the insertion of the central carbide atom [35], will not be discussed in detail here, although some of these are summarized in Tab. 11.1. The topic of FeMoco assembly has been exhaustively reviewed [34, 35] and is discussed elsewhere in this volume. Nevertheless, as will be discussed below, efforts to understand the biochemical and regulatory functions of all the *nif*-associated gene products ultimately led to the discovery of general principles concerning how simple [Fe-S] clusters can be assembled on molecular scaffolds. In this regard, it is interesting and instructive that analysis of the assembly of nature's most complex [Fe-S] cluster known to date, FeMoco, provided the conceptual starting point for understanding how simple [Fe-S] clusters are assembled on molecular scaffolds.

The size and complexity of the *nif* regulon, together with the fact that nitrogenase, and some of its assembly factors, are oxygen sensitive, prohibited transferring the entire system to a tractable host such as *Escherichia coli* for heterologous expression and subsequent analysis. Instead, the ability to efficiently transform *A. vinelandii* using isolated DNA, and the inherent capacity of this organism for frequent homologous recombination, was exploited to place in-frame deletions within the *A. vinelandii* genomic region for each of the *nif*-associated genes, separately, and in various combinations [61]. This simple strategy (Fig. 11.4) involved the use of restriction enzymes to delete known

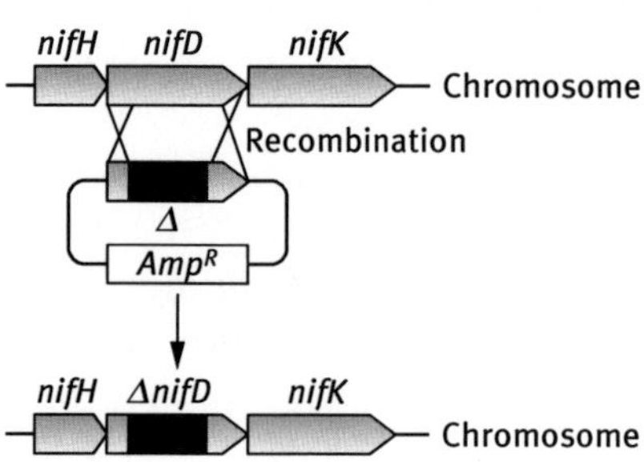


**Fig. 11.4:** Diagrammatic representation of *A. vinelandii* mutant strain construction. A specific deletion carried within a plasmid can be transferred to the chromosome by reciprocal recombination.

segments of *nif* genes contained in recombinant plasmids and subsequent transfer of such deletions to the corresponding region of the chromosome through transformation and reciprocal recombination. The values of this approach were that the native organism could be analyzed without disrupting the coordination and balance of *nif* gene expression and the precise nature of every genetic lesion was defined.

Because nitrogenase activity requires both the Fe protein and the MoFe protein and because neither protein exhibits any catalytic activity in the absence of the other, mutants that are specifically defective in one component or the other can be readily identified by adding either purified Fe protein or MoFe protein to mutant strain extracts in an attempt to reconstitute activity. One example of this type of experiment is that extracts of a mutant strain deleted for *nifB* exhibit no nitrogenase activity, but activity can be detected when isolated MoFe protein, or FeMoco extracted from MoFe protein, is added to the crude extract [12]. Hence, NifB is required for formation of an active MoFe protein and, more specifically, for formation of FeMoco. It was this type of analysis that launched the discovery of the roles of NifS and NifU as representing nature's minimum tool kit for formation of simple [Fe-S] clusters.

### 11.5.1 Phenotypes associated with loss of *NifS* or *NifU* function indicate their involvement in nitrogenase-associated [Fe-S] cluster formation

Strains in which either *nifS* or *nifU* were deleted revealed phenotypes that, in aggregate, distinguished them from all other $N_2$ fixation deficient mutant strains (Tab. 11.1). These phenotypes included the following: (1) negligible growth when cultured in the absence of a fixed nitrogen source under conditions that require expression of the Mo-dependent nitrogenase, the V-dependent nitrogenase, or the Fe-only nitrogenase; (2) significant descrease in the activities of both the Fe protein and the MoFe protein in crude extracts; (3) loss of the dark brown color characteristic of extracts prepared from $N_2$-fixing wild-type cells in extracts prepared from *nifU*- and *nifS*-deletion strains [62–65].

A common feature of all three nitrogenase systems is that they require [Fe-S] clusters for their respective activities. Similarly, the common feature of the Fe protein and MoFe protein is that they also both require [Fe-S] clusters to sustain a functional nitrogenase. These considerations indicated that NifS and NifU might have complementary functions related to the mobilization of the S and Fe necessary to assemble both the simple and the complex [Fe-S] clusters associated with nitrogenase. Because the characteristic dark brown color of extracts prepared from $N_2$-fixing cells is derived primarily from nitrogenase-associated [Fe-S] clusters, the relatively light color of NifS- and NifU-deficient extracts also implied that [Fe-S] cluster assembly was defective in *nifS* and *nifU* mutant strains. However, loss of NifS or NifU function does not completely eliminate nitrogenase activity [65] and therefore does not eliminate the assembly of nitrogenase-associated [Fe-S] clusters. This observation was interpreted to indicate that separate cellular functions, independent of $N_2$ fixation, might replace the functions of NifS and NifU but very inefficiently. In aggregate, these possibilities resulted in the design of experiments that led to the discovery that NifS encodes a cysteine desulfurase involved in mobilization of S for [Fe-S] cluster assembly [66] and that NifU provides a molecular scaffold for the assembly of simple [Fe-S] clusters [67, 68]. They also led to an appreciation that the functions of NifS and NifU represent general features of [Fe-S] cluster formation throughout nature [6].

### 11.5.2 *NifS* is a cysteine desulfurase

One technical problem associated with functional analysis of proteins involved in maturation of nitrogenase components is that such proteins are only produced at very low levels because they engage in multiple turnovers. Therefore, as a way to gain insight into the possible function of its product, the *nifS* gene was heterologously expressed at very high levels in *E. coli* using a recombinant plasmid construct. Purification and initial analysis of NifS produced in this way revealed that it was a homodimer that exhibited a visible spectrum characteristic of pyridoxal-phosphate-dependent (PLP) enzymes [66]. Subsequent chemical analysis indeed confirmed that NifS contained PLP and, therefore, the possibility that L-cysteine is a substrate for NifS was investigated for the following reasons: (i) PLP enzymes are frequently involved in the elimination or rearrangement reactions involving amino acid substrates, (ii) a gene encoding a CysE homolog, which catalyzes the rate-limiting step in cysteine biosynthesis, is co-transcribed with *nifS*, and (iii) involvement for NifS in the mobilization of S was predicted based on the various phenotypes of a NifS-deficient strain. This possibility was confirmed by showing that, of all the amino acids, only the addition of L-cysteine to NifS resulted in a transient shift in its visible spectrum. Furthermore, prolonged incubation of L-cysteine with NifS resulted in the precipitation of colloidal sulfur and accumulation of L-alanine, demonstrating a role for NifS in S mobilization [69].

Basic features of the NifS L-cysteine desulfurization mechanism were elucidated using a combination of approaches [66, 69]. First, the production of colloidal S from L-cysteine indicated the likely participation of a protein thiol group in the S mobilization reaction, consistent with the observation that NifS catalytic activity is extremely sensitive to thiol alkylating reagents, such as *N*-ethylmaleimide. Second, phylogenetic comparisons of NifS primary structures from a variety of $N_2$-fixing microorganisms revealed that residue Cys325 is strictly conserved. Third, substitution of Cys325 by Ala325 resulted in complete loss of NifS activity and an *A. vinelandii* mutant strain that produces the Ala325-substituted protein exhibits the same phenotype as a strain deleted for *nifS*. Fourth, treatment of NifS with the L-cysteine analogues, L-allylglycine or vinylglycine, were shown to irreversibly inactivate NifS by respective formation of a γ-methylcystathionyl or cystathionyl residue derived from Cys325. Fifth, the formation of an enzyme-bound cysteinyl persulfide, located at the Cys325 position, was identified as an intermediate in the NifS-catalyzed L-cysteine desulfurization reaction.

Finally, evidence was obtained for the occurrence of an enamine intermediate in the NifS-catalyzed formation of L-alanine. These observations indicated that a cysteinyl persulfide could be formed on NifS through nucleophilic attack by the active site Cys325 thiol on a cysteine-PLP ketamine adduct and that formation of a NifS-bound cysteinyl persulfide could represent a benign way to activate and traffic S for nitrogenase [Fe-S] cluster formation. Subsequent to the initial characterization of the NifS cysteine desulfurase reaction, other studies established that NifS represents a broad class of enzymes that use L-cysteine and PLP chemistry to accomplish the activation and mobilization of S for biosynthesis of many S-containing biomolecules (Fig. 11.5) [70, 71]. Analysis of other cysteine desulfurases, including crystallographically determined high-resolution structures [72–76], has provided detailed insight into the probable NifS mechanism (Fig. 11.6).

The overall function of cysteine desulfurases can be considered to occur in two stages, formation of the enzyme-bound persulfide and subsequent delivery of S to an acceptor molecule. All cysteine desulfurases are thought to share similar mechanistic features in the first stage, whereas the second stage appears to vary with overall functionality of the cysteine desulfurase type and its corresponding delivery target (Fig. 11.5). The proposed intermediates formed during the first stage are shown in Fig. 11.6 [69, 77]. In the NifS resting state, PLP is covalently bound to the conserved Lys202 residue through a Schiff base (intermediate 1). The binding of cysteine to PLP is proposed to occur through formation of a tetrahedral C4′ intermediate, geminal diamine (intermediates 2–3). This results in the displacement of Lys202 from PLP and formation of an external Cys-aldimine Schiff base (intermediate 4). A key step involves proton abstraction by a protein residue serving as a general base (intermediate 5). Because a monoprotic residue facilitates intermediate 5 formation, this function is unlikely to involve the active site Lys202 residue [77]. Also, formation of the proposed quinonoid (intermediate 6) is independent of the active site cysteine residue because the NifS Ala325-substituted variant is still able to bind cysteine and to form intermediates 2–5. It appears likely that a histidine residue conducts the nucleophilic attack on the α-proton and either His100 or His201 are candidates to perform this role.

NifS inactivation studies using the suicide inhibitors, allylglycine and vinylglycine, have eliminated the possible involvement of Cys325 during conversion to the Cys-quinonoid form (intermediate 6) [69]. Instead, inhibition profiles favor the involvement of the active site Cys325 residue serving as a general acid during the protonation of C4′ of the Cys-PLP quinonoid adduct. Hydrogen bonding between the imine hydrogen and the phenolate oxygen allows expansion of the π-orbital system leading to a highly conjugated structure of PLP, therefore permitting electron delocalization. The committed step of the NifS reaction is characterized by nucleophilic attack of the deprotonated active site cysteine thiol on the thiol group of the substrate (intermediate 7). This event leads to formation of the persulfide bond, and conversion of cysteine to an alanine-enamine intermediate (intermediate 8). *In vitro* reactions performed in $^2H_2O$ confirmed that α and β hydrogens are exchanged during the reaction, supporting

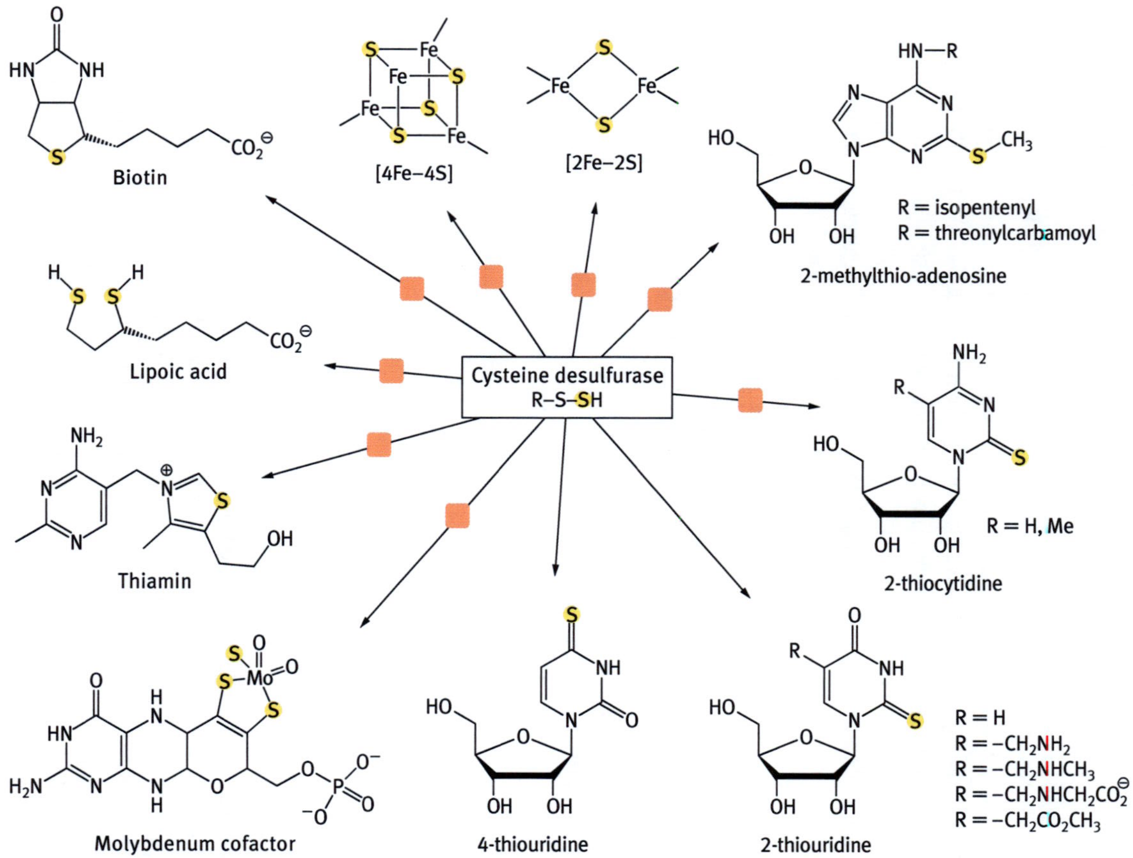


**Fig. 11.5:** Cysteine desulfurases are involved in the first step of sulfur mobilization in the biosynthesis of many thio-containing cofators. Biosynthetic pathways requiring participation of at least one [Fe-S] protein are indicated by a red box.

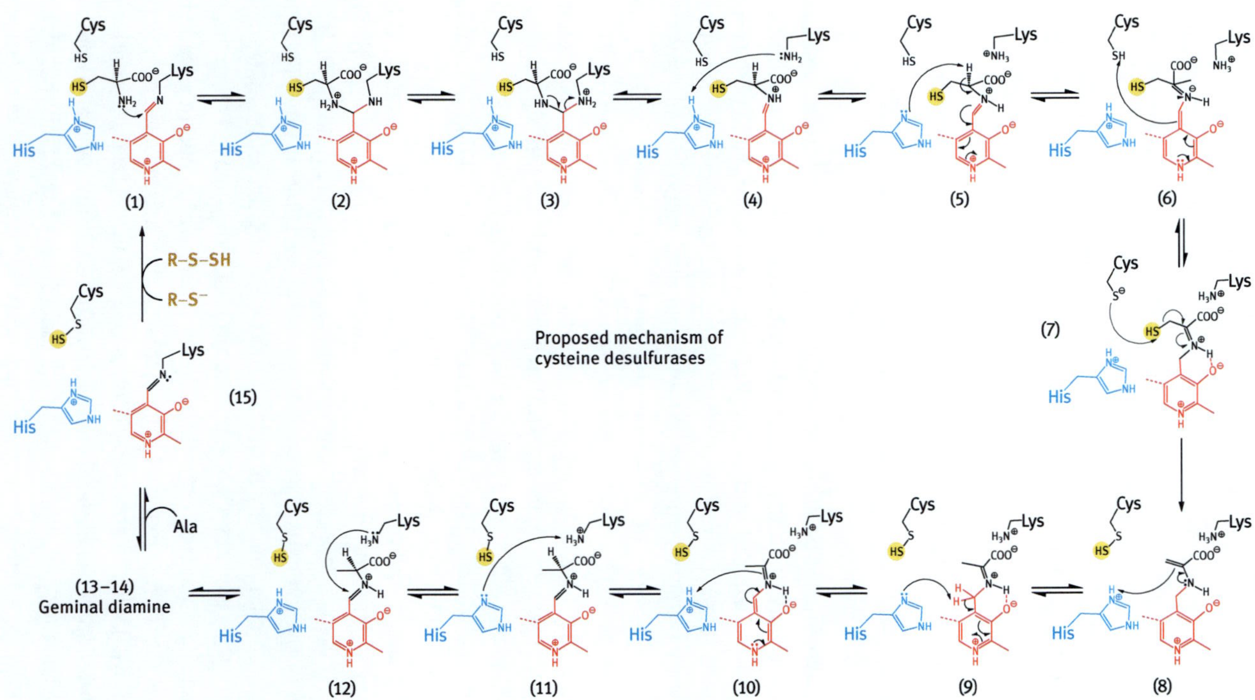


**Fig. 11.6:** Proposed L-cysteine desulfurase mechanism. The reaction involves two committed steps: the formation of the enzyme-bound persulfide bond and the subsequent transfer of the terminal persulfide sulfur to an acceptor molecule (R-S–). The proposed mechanism includes the PLP-coordinating lysine PLP (Lys202 of *A. vinelandii* NifS) and at least one histidine (shown in blue) serving as a general acid general base; His100 and His201 of *A. vinelandii* NifS are candidates for this role.

the occurrence of intermediate 8 [69]. Furthermore, the subsequent release of alanine is proposed to follow the same steps but in reverse order: ketamine (intermediate 9), quinonoid (intermediate 10), aldimine (intermediates 11–12), and gem-diamine (intermediates 13–14) to restore the internal Schiff base with Lys202 and form the persulfurated Cys325 residue. Formation of persulfurated NifS form (intermediate 15) marks the end of the first half of the cysteine desulfurase reaction, apparently common to all cysteine desulfurases. The second step, involving delivery of S for [Fe-S] cluster formation, is not yet clearly understood, but the mechanism of S delivery appears to differ among the different cysteine desulfurase types and their various targets (Fig. 11.5). The involvement of NifS in providing S for formation of [Fe-S] clusters on NifU is discussed in Section 11.5.3.

### 11.5.3 Extension of the scaffold hypothesis to NifU function

The apparent involvement of NifS in the mobilization of S for nitrogenase-associated [Fe-S] cluster assembly naturally led to an investigation of a possible complementary role for NifU in mobilizing Fe and/or serving as a scaffold for [Fe-S] cluster formation [78, 79]. A capacity for NifU to bind [Fe-S] clusters had in fact already been suspected on the basis of its primary structure [80] because phylogenetic comparisons indicated that NifU is comprised of three modules, and each module contains conserved cysteine residues (Fig. 11.7). The prediction that NifU might contain an [Fe-S] cluster was confirmed by analysis of the isolated protein heterologously produced in *E. coli*. This analysis, together with analysis of different recombinant NifU forms, each having one of the nine-conserved cysteine residues substituted by alanine, showed that NifU contains a redox-active [2Fe-2S] cluster, attached to the central domain through thiolate ligands provided by NifU residues Cys137, Cys139, Cys172, and Cys175.

Once it was discovered that overexpressed and purified NifU contained two [2Fe-2S] clusters, one cluster within each central domain of a NifU dimer [78], the possibility that such clusters represent precursors for nitrogenase-associated [Fe-S] clusters became an obvious consideration. It was expected that if the [2Fe-2S] clusters contained within recombinantly produced NifU were capable of being transferred to another protein, they would be labile and could easily be removed by treatment with metal chelating reagents. However, the [2Fe-2S] clusters attached to as-isolated NifU are quite stable, even when exposed to oxygen, and they are released only very slowly by treatment of NifU with Fe chelators, such as α-α′ dipyridyl [78]. Alternative possibilities for the function of the [2Fe-2S] clusters contained in the central domain of NifU, now designated as the “permanent clusters,” include a redox role in the release of the persulfide from NifS, a redox role in the assembly of “transient” clusters contained in other NifU domains, or as a structural element. The function of the NifU permanent [2Fe-2S] clusters remains elusive. However, given the conservation of cysteines in the N- and C-terminal domains of NifU and the precedence of the role of NifEN as

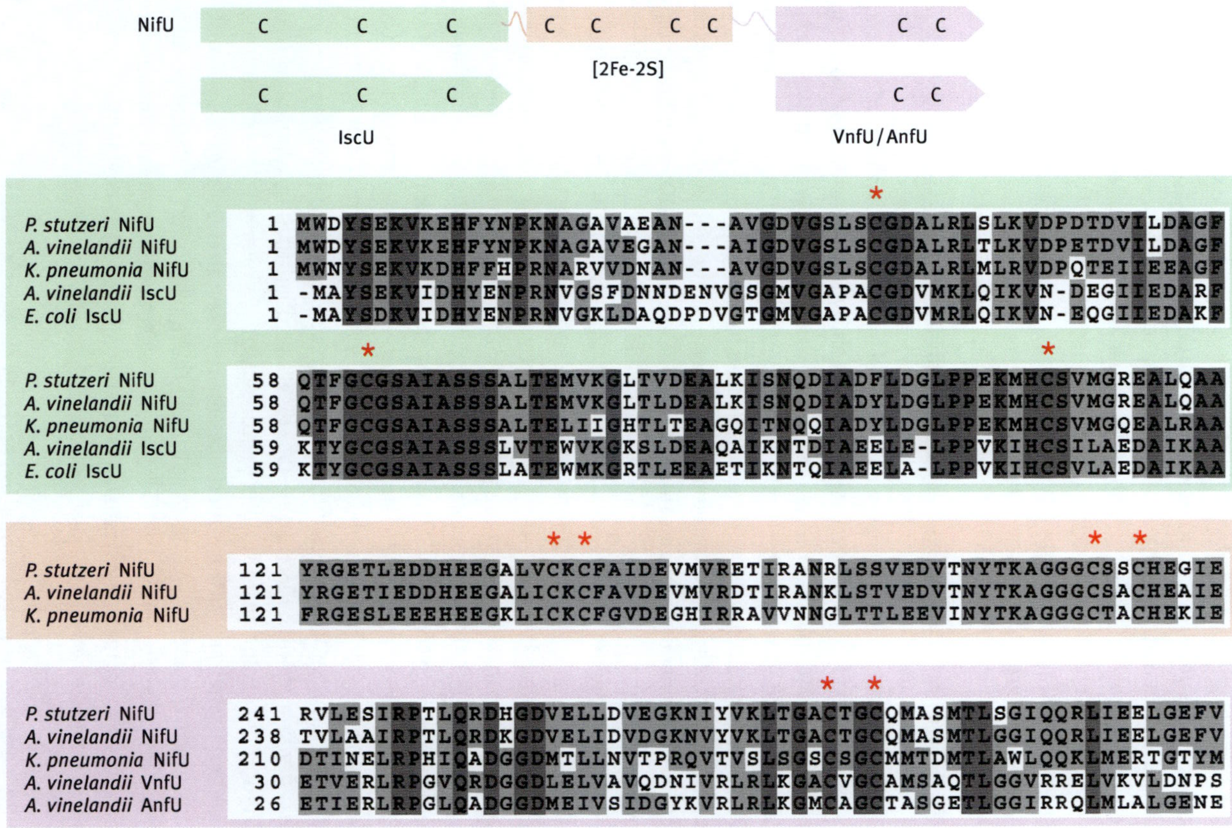


**Fig. 11.7:** NifU is a modular protein that contains conserved cysteine residues in each module. (top) Diagrammatic representation of NifU domains including the relative position of conserved cysteine residues and location of the permanent [2Fe-2S] cluster. (bottom) Sequence alignments comparing IscU and the N-terminal domain of NifU sequences (green box), central domain of NifU sequences (pink box), and VnfU, AnfU, and C-terminal domain of NifU sequences (violet box). Conserved cysteine residues are indicated with a star.

an assembly scaffold for FeMoco, the possibility that NifU serves as a scaffold for the formation of transient [Fe-S] clusters was an attractive one. Evidence supporting this possibility was obtained using a combination of approaches: (i) physiological analysis of defects associated with substitution of each of the conserved cysteine residues contained in the N- and C-terminal domains of NifU [68, 79], (ii) NifS-assisted assembly of [Fe-S] clusters within isolated NifU and NifU fragments [67], (iii) activation of a cluster-less nitrogenase Fe protein using either isolated NifU or NifU fragments that contain [Fe-S] clusters formed *in vitro* [68, 81], (v) isolation of NifU or NifU fragments produced *in vivo* that contain labile [Fe-S] clusters having the same biophysical features of [Fe-S] clusters formed on NifU *in vitro*.

#### 11.5.3.1 Physiological characterization of *nifU* mutants

An important catalytic role for Cys325 was indicated for NifS by showing that the Ala325 substitution resulted in the same physiological phenotype as a strain deleted for *nifS*. However, a similar approach in the assessment of NifU function was much more complicated because nine conserved cysteines were identified, rather than the single conserved cysteine contained in NifS. Functional roles for conserved NifU cysteines were evaluated by substituting each of them, separately, and, in combination, by alanine. It was found that individual substitution of Cys35, Cys62, or Cys106 by alanine or the combined substitution of all three residues, resulted in the same moderate impairment in $N_2$-fixing growth when compared with the wild-type strain [68]. In other words, no cumulative effect was manifested upon combined substitution of conserved cysteines in the N-terminal domain, indicating these residues might all participate in the same quasi-dispensable function. Given the potential for cysteinyl residues to provide thiolate ligands for [Fe-S] clusters, the results suggested the possibility that, in aggregate, NifU residues Cys35, Cys62, and Cys106, could provide a nucleation site for the formation of "transient" [Fe-S] clusters destined for incorporation into nitrogenase. Similar to the situation with the N-terminal domain, there was only a very minor effect on the capacity for $N_2$-fixing growth when either of the two conserved cysteine residues in the C-terminal domain of NifU, Cys272 and Cys275, were substituted by alanine. However, a combination of alanine substitutions for conserved cysteines located in both the N- and C-terminal domain of NifU resulted in a severe defect in the ability for the respective strains to grow under $N_2$-fixing conditions. These results led to the proposal that the N- and C-terminal domains of NifU could have either redundant or complementary functions related to formation of transient [Fe-S] clusters at both sites [68].

#### 11.5.3.2 *In vitro NifS*-assisted formation of [Fe-S] clusters on NifU

As a way to test if NifU can serve as an assembly scaffold for [Fe-S] cluster formation, isolated recombinant NifU, containing only the intact permanent [2Fe-2S] clusters, was incubated under anoxic conditions in the presence of $Fe^{2+}$, L-cysteine, and NifS. The results of these experiments demonstrated that NifS facilitated the

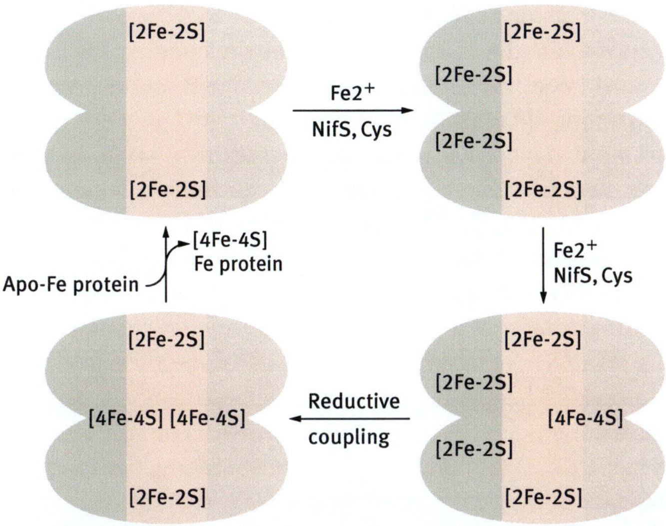


**Fig. 11.8:** NifU is an [Fe-S] cluster assembly scaffold that can be used to activate the nitrogenase Fe protein. As-isolated NifU dimer contains a redox-active [2Fe-2S] cluster located within the central domain of each NifU monomer (shown in pink). Upon incubation with NifS, cysteine, and $Fe^{2+}$, NifU undergoes sequential assembly of [2Fe-2S] clusters at the N-terminal IscU domain (shown in green). Prolonged incubation results in the formation of [4Fe-4S] at the C-terminal NfU domain (shown in violet) and subsequent appearance of a [4Fe-4S at the N-terminal domain.

formation of [Fe-S] clusters within NifU in a $Fe^{2+}$- and L-cysteine-dependent manner (Fig. 11.8). An analysis of the time-dependent [Fe-S] cluster formation process using spectroscopic and analytical methods indicated that, ultimately [4Fe-4S] clusters were assembled in both the N- and C-terminal NifU domains and that [2Fe-2S] clusters were initial intermediates in that process. Details of the process were further explored by testing if isolated NifU fragments, containing either the N- or C-terminal domain, could serve as *in vitro* assembly scaffolds [68, 81]. The results of these experiments revealed that NifS-dependent formation of the [4Fe-4S] cluster within the N-terminal NifU domain followed formation of [2Fe-2S] cluster intermediates, although no intermediates could be detected in formation of the [4Fe-4S] cluster within the C-terminal domain. It is not clear why there are two, apparently separate, [Fe-S] cluster-assembly sites located within NifU. However, it is noteworthy that there are genes associated with the *vnf* and *anf* systems that encode paralogues to the NifU C-terminal domain (Fig. 11.7) [44]. Given that NifU is required for all three nitrogenase systems and that the N-terminal domain of NifU apparently plays the dominant functional role, it could be that the NifU C-terminal domain and its corresponding VnfU and AnfU paralogues could function primarily as agents of delivery rather than assembly. In other words, [Fe-S] clusters formed on the N-terminal domain could be transferred to the C-terminal NifU domain types to provide target specificity.

#### 11.5.3.3 Activation of Fe protein using cluster-loaded *NifU*

Assembly of [Fe-S] clusters within the N- and C-terminal domains of NifU, and demonstration that conserved cysteine residues within these domains are required for such assembly provided strong circumstantial evidence that NifU provides a scaffold for [Fe-S] cluster assembly. Nevertheless, two other experimental approaches were necessary to establish the role of NifU in simple [Fe-S] cluster formation. First, the capacity for [Fe-S] cluster-loaded NifU to activate a nitrogenase target protein was needed. Second, it was necessary to establish that [Fe-S] clusters assembled on NifU *in vitro* faithfully duplicated an *in vivo* process.

It was known that binding of MgATP produced a conformational change in the Fe protein that exposed its [4Fe-4S] cluster to solvent, and this property was exploited to produce apoprotein needed for [Fe-S] cluster transfer experiments. Solvent exposure of the cluster makes it susceptible to removal by treatment of metal chelators such as α-α′ -dipyridyl [82, 83]. Chelation of the [4Fe-4S] cluster, followed by a desalting step, performed in the presence of a reductant, such as dithionite, removed the released $Fe^{2+}$ and $S^{2-}$. The resulting apo-Fe protein was then used as a target for cluster transfer experiments by mixing with cluster-loaded NifU prepared *in vitro*.

The initial [Fe-S] cluster-transfer experiments involved mixing NifU, NifS, $Fe^{2+}$, and L-cysteine to accomplish cluster assembly on the NifU scaffold. After removal of the excess $Fe^{2+}$ and L-cysteine, the sample was mixed with apo-Fe protein, and activation was evaluated by subsequent measurement of recovered activity. The results showed that mixing with the reconstituted holo-NifU resulted in rapid and nearly complete restoration of Fe protein activity, whereas omission of any of the assembly ingredients resulted in no recovery of Fe protein activity in control experiments. In addition, removal of NifS from the reaction mixture, post-assembly, had only a very modest effect on the capacity for apo-Fe protein activation, indicating that, although NifS is required for *in vitro* assembly of [Fe-S] clusters on the NifU scaffold, transfer of the [Fe-S] cluster does not require the participation of NifS [68].

In light of the results from genetic and *in vitro* assembly experiments, indicating the capacity for [Fe-S] cluster formation on both the N- and C-terminal domains, the ability to transfer [Fe-S] clusters assembled on either NifU domain was evaluated. One approach involved using a full length NifU, in which either the N-terminal domain or the C-terminal domain was inactivated. In these experiments, activation of the Fe protein could be accomplished, whereas no activation occurred when a full length NifU inactivated in both the N-terminal and the C-terminal domains was used. A second series of experiments involved successful apo-Fe protein activation using preparations of [Fe-S] cluster-loaded N- or C-terminal fragments. In these experiments, the various NifU samples used for cluster transfer contained spectroscopically defined [Fe-S] cluster species shown primarily to be in the form of [4Fe-4S] clusters, suggesting that [4Fe-4S] clusters are transferred intact from NifU to the apo-Fe protein [81].

#### 11.5.3.4 [Fe-S] clusters are assembled on NifU *in vivo*

A final approach used to establish that NifU functions as an [Fe-S] cluster assembly scaffold involved testing whether or not a cluster-loaded form of NifU could be produced *in vivo*. This effort faced a number of technical challenges. For example, [Fe-S] clusters assembled on NifU are labile because they must have a capacity for facile release to accomplish their delivery to target proteins. Such lability is manifested by rapid degradation when transient [Fe-S] clusters assembled on NifU *in vitro* are exposed to oxygen, reducing reagents, such as dithionite, or treated with chelating reagents. Consequently, a procedure for gentle and rapid purification of NifU was needed. Another challenge was related to the dynamic nature of *in vivo* cluster assembly and delivery. What this means is that [Fe-S] clusters assembled on NifU must be transiently located on the scaffold as they must be constantly delivered to other target proteins. Consequently, the occupancy level of the NifU cluster assembly sites could not be anticipated. Finally, because NifU is involved in the catalytic activation of other proteins, the physiological requirement for NifU is relatively low and it is accordingly produced in low quantities.

The technical aspects were overcome by decoupling the expression of *nifU* and *nifS* in *A. vinelandii* from nitrogen fixation by placing their expression under control of a strong, inducible promoter [84]. The rationale of the approach was to produce NifU and NifS at high levels in the absence of nitrogenase target proteins, reasoning that the absence of target recipients for [Fe-S] clusters might cause [Fe-S] clusters to accumulate on the NifU scaffold (Fig. 11.9). In addition, a benign affinity tag was placed within NifU such that its

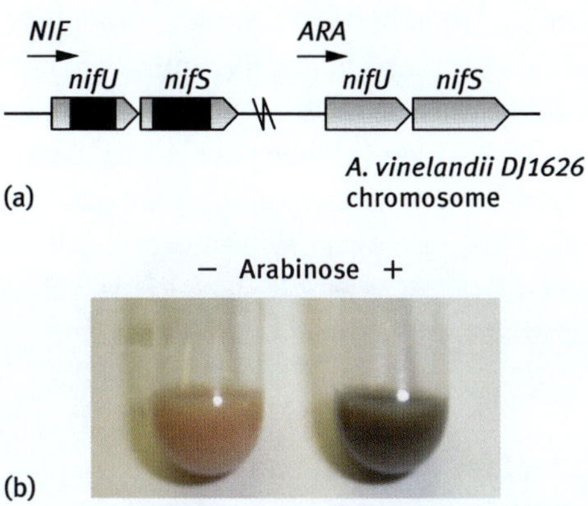


**Fig. 11.9:** *A. vinelandii* arabinose-controlled expression of NifU and NifS. (a) Schematic representation of gene regions containing *nifU* and *nifS* of *A. vinelandii* strain DJ1626. It contains two copies of the *nifU-nifS* genes. The *nif*-regulated copy contains an in frame deletion (black-colored region), whereas the second copy is located at a different chromosomal position. Expression of the second *nifUS* copy is under control of the arabinose regulatory system. (b) Resuspended cell pellet of *A. vinelandii* cells cultured in media supplemented with ammonia in the absence (–) or presence of arabinose (+). Note that the dark color of cells cultured in the presence of arabinose indicated a high level of [Fe-S] cluster accumulation.

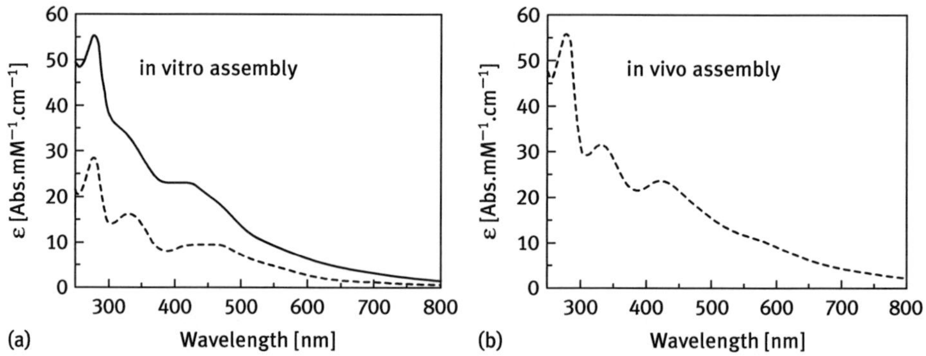


**Fig. 11.10:** UV-vis absorption spectra of NifU. (a) Spectra of NifU before (dashed line) and after (solid line) *in vitro* Fe-S cluster assembly. The dashed-line spectrum represents the as-isolated form of NifU that contains only the permanent [2Fe-2S] clusters. The solid line spectrum represents *in vitro* reconstituted NifU that contains both the permanent [2Fe-2S] clusters and transient [4Fe-4S] clusters. (b) Spectrum of [Fe-S] clusterloaded NifU produced by *A. vinelandii* strain DJ1626 (Fig. 11.8).

rapid and gentle purification could be accomplished. This strategy was successful, as shown in Figs. 11.9 and 11.10. Namely, NifU produced in *A. vinelandii*, under conditions where no other *nif* genes were expressed, and therefore, no acceptor proteins were produced, contained transient [Fe-S] clusters. The nearly identical spectra of cluster-loaded forms of NifU prepared *in vitro* or *in vivo* (Fig. 11.10) indicated that they were likely to be identical. These spectroscopic results, and the fact that cluster-loaded NifU produced *in vivo* can also be used for the effective *in vitro* activation of apo-Fe protein, provided a final compelling argument to support the scaffold hypothesis for the assembly of simple [Fe-S] clusters. It should be noted that the function of NifU is not limited to the activation of Fe protein. Indeed, there is now also *in vitro* evidence that simple [Fe-S] units assembled on NifU provide the building blocks necessary for the assembly of the more complex FeMoco [64]. Also, the scaffold hypothesis, originally developed for FeMoco, has been further developed to include the participation of a series of assembly scaffolds as well as carrier proteins that traffic [Fe-S] cluster intermediates.

### 11.5.4 Discovery of *isc* system for [Fe-S] cluster formation and functional cross-talk among [Fe-S] cluster biosynthetic systems

A variety of considerations indicated that the functions of NifU, and NifS might represent a duplication of "housekeeping" functions involved in the assembly of general [Fe-S] clusters, involved in other aspects of cellular metabolism unrelated to nitrogen fixation. First, there are many cellular processes that depend on [Fe-S] proteins, so there must be a mechanism for the assembly of their associated [Fe-S] clusters [11]. Second, nitrogenase is produced at very high levels in $N_2$ fixing cells, so it seemed reasonable to expect that NifU and NifS could represent a mechanism to specifically satisfy an increased demand for [Fe-S] cluster formation. Third, there was already evidence in the literature that L-cysteine can serve as the ultimate S source for formation of [Fe-S] clusters in *E. coli*, a non-$N_2$-fixing organism [85]. Fourth, *E. coli* extracts exhibited L-cysteine desulfurase activity and, based on this feature, a protein having PLP-dependent L-cysteine desulfurase activity and primary structure similarity to NifS was isolated [86]. Finally, elimination of NifU or NifS function did not fully eliminate the capacity for producing active nitrogenase components, indicating that limited functional replacement of NifS and NifU activities were mediated by some other cellular components [65].

A search for a "housekeeping" [Fe-S] biosynthetic system in *A. vinelandii* was initiated by purification of a protein having L-cysteine desulfurase activity from extracts of a mutant strain in which *nifS* was deleted [87]. Because this effort predated the genomics era, the gene region that encoded the protein with cysteine desulfurase activity was isolated using a reverse genetics approach that relied on using the sequence of peptides generated by trypsin digestion to predict the DNA sequence. Examination of this region revealed that the gene encoding a NifS-like protein was contained in an apparent transcriptional unit that contained a total of seven genes. These genes were, respectively, designated *iscR, iscS, iscU, iscA, hscB, hscA, fdx*, and *iscX* (Fig. 11.11) and their proposed functions are shown in Tab. 11.2. Primary structural comparisons revealed strong similarity between components of the Nif and Isc systems: NifS was homologous to IscS, the N-terminal domain of NifU was homologous to IscU, and a *nif* gene product, now designated $IscA^{Nif}$, was homologous to IscA. Fdx does not bear primary structure similarity to any of the *nif* gene products, but it does contain a [2Fe-2S] cluster that might serve a function similar to the permanent [2Fe-2S] cluster located in the NifU central domain. There is no counterpart to the NifUC-terminal domain encoded within the *isc* gene cluster, but there is a gene, designated NfuA [88], located elsewhere on the genome that shares primary structure similarity when compared with the NifU-terminal domain, including the two conserved cysteinyl residues. There are no *nif* counterparts to *iscR, hscB, hscA*, or *iscX*. A description of the details of the Isc system is beyond the scope of the present discussion, and this topic

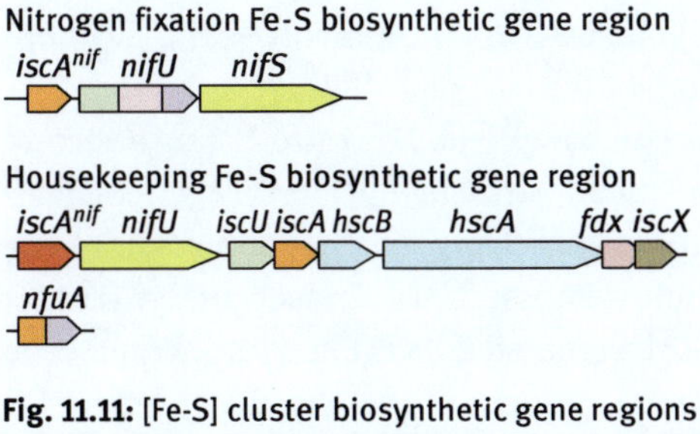


**Fig. 11.11:** [Fe-S] cluster biosynthetic gene regions in *A. vinelandii*. Genes whose products display primary structure similarities are color-coded accordingly.

**Tab. 11.2:** Selected *A. vinelandii* proteins known or suspected to be involved in simple [Fe-S] cluster assembly and their proposed functions.

| Gene product | Proposed function |
|---|---|
| NifU | [Fe-S] cluster assembly scaffold/ [Fe-S] cluster carrier |
| NifS | Cysteine desulfurase |
| $IscA^{nif}$ | [Fe-S] cluster carrier |
| IscR | [Fe-S] cluster transcription regulator |
| IscS | Cysteine desulfurase |
| IscU | [Fe-S] cluster assembly scaffold |
| IscA | [Fe-S] cluster carrier |
| HscB | [Fe-S] cluster delivery chaperone |
| HscA | [Fe-S] cluster delivery chaperone |
| Fdx | Ferredoxin, specific function unknown |
| IscX | Unknown |
| NfuA | [Fe-S] cluster carrier |

is reviewed elsewhere [4, 6]. However, it is noteworthy that, although the Isc system for the assembly and delivery of [Fe-S] clusters appears to be more complicated than the Nif system, both are unified by the involvement of an L-cysteine desulfurase and an assembly scaffold. All of the essential mechanistic elements discovered for the Nif system have also been independently established for the Isc system.

## 11.6 The Isc system is essential in *A. vinelandii*

Identification of the Isc system provided insight into why the mechanism for assembly of such an important and simple prosthetic group took so long to discover. The answer to this question was expected to be associated with the participation of [Fe-S] proteins in so many essential biological processes. In other words, impairment in the ability to assemble [Fe-S] clusters would be manifested in the loss of many, apparently unrelated, essential metabolic processes. Circumstantial evidence for this possibility was found in the instability of mutant strains when attempts were made to incorporate targeted genetic lesions into the *isc* gene region [87]. A more direct and definitive approach to assess the importance of [Fe-S] cluster formation involved the development of a system for the controlled expression of the *isc* genes [89].

The controlled expression of *isc* genes was accomplished by introducing a second copy of the *isc* gene region within the *A. vinelandii* genome [89]. In this genetic construction, the expression of the duplicated *isc* gene copy was decoupled from the normal *isc* regulatory elements. Instead, their expression was placed under control of the sucrose catabolic regulatory elements, designated as *scr* (Fig. 11.12). When cells containing the duplicated genes are grown in the presence of sucrose, they produce Isc components from both the *isc*-regulated genes and the *scr*-regulated genes. However,

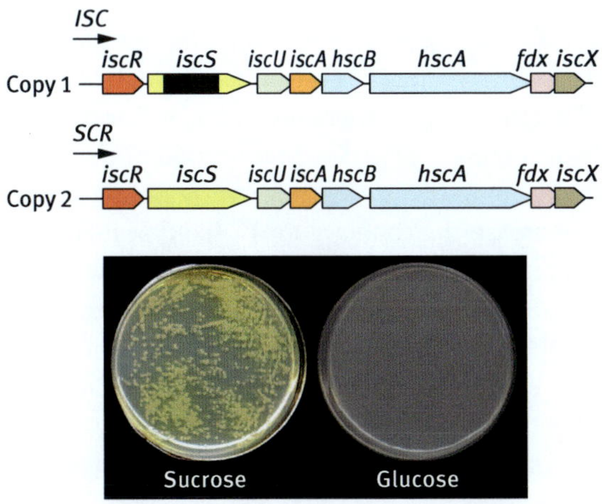


**Fig. 11.12:** Sucrose-dependent expression of *iscS*. (top) Schematic representation of duplicate *isc* gene regions contained in *A. vinelandii* strain DJ1422. The *isc*-regulated copy carries an in-frame deletion within the *iscS* coding sequence (black-colored region within *iscS*), whereas the second copy, whose expression is under control of the scr regulatory elements, is intact. (bottom) *A. vinelandii* strain DJ1422 cultured on Petri plates using sucrose or glucose as the only carbon source.

when such cells are grown in the absence of sucrose, for example, when glucose is used as the sole carbon source, only the *isc*-regulated copy is expressed. The approach of using an inducible copy of each Isc gene enabled researchers to assess whether or not individual Isc components were essential. An example of this approach is illustrated in Fig. 11.12, which shows how growth of a strain deleted for the *isc*-regulated *iscS* gene can be supported by activation of the *scr*-regulated copy of *iscS*. This strain cannot grow when cultured in the absence of sucrose because the capacity to produce IscS is eliminated. In contrast, when sucrose is added to the growth medium robust growth is restored. Using this controlled gene expression approach, IscS, IscU, and Fdx were also demonstrated to be essential in *A. vinelandii* [89].

## 11.7 There is limited functional cross-talk between the Nif and Isc systems

The fact that neither a *nifU* nor a *nifS* deletion eliminates the capacity for assembly of an active nitrogenase represented a major clue that ultimately led to the discovery of the Isc system. On the basis of minimal retention of activity, it was suspected that, under $N_2$-fixing conditions, NifU and NifS might represent a specialized way to meet the increased demand for [Fe-S] clusters in the abundantly produced [Fe-S] cluster-containing nitrogenase components. It was not possible to directly test this possibility by asking if the combined elimination of NifS and IscS function or combined elimination of NifU and IscU function completely eliminated the capacity for producing an active nitrogenase because IscS and IscU are both essential. Instead, an orthogonal approach, which exploited the negative regulation of *isc* genes by IscR, was used. Inactivation of IscR results in highly elevated expression of all the other housekeeping [Fe-S] cluster biosynthetic components. Therefore, the ability of elevated expression of the Isc system to increase the capacity for $N_2$ fixation in strains deleted for *nifU* or *nifS* was tested, and rescue of the very slow $N_2$-fixing growth phenotypes associated with deletion of *nifU* or *nifS* was achieved through high Isc expression [84]. Although this result does not eliminate the possibility that other factors can also replace NifU and NifS function in *A. vinelandii*, it does establish that the capacity for limited functional cross-talk exists between the Nif and the Isc systems.

## 11.8 Closing remarks

In this chapter, we have not attempted to provide a comprehensive summary of the overall process of the assembly of [Fe-S] clusters. This topic represents a very active and rapidly advancing area of research. Among the many important and fascinating features not addressed in this narrative are included the regulation of [Fe-S] cluster formation, trafficking of [Fe-S] clusters, and the repair of damaged [Fe-S] proteins. Here we have only provided a personalized perspective on the conceptual and experimental basis from which the field emerged. Emphasized here is the role of nitrogen fixation research as providing some of the foundational discoveries that underpin the importance of S trafficking and participation of molecular assembly scaffolds for the assembly of [Fe-S] clusters, in most, and perhaps all, biological systems. It is of particular interest that the peculiarities of the $N_2$-fixing system for specialized [Fe-S] cluster formation contributed so significantly to these discoveries.

## Acknowledgments

Work in the laboratory of Dennis R. Dean is supported by NSF MCB-071770, and work in the laboratory of Patricia C. Dos Santos is supported by NSF MCB-1054623.

## References

[1] Beinert H, Holm RH, Munck E. Iron-sulfur clusters: nature's modular, multipurpose structures. Science 1997;277:653–9.
[2] Mortenson LE, Valentine RC, Carnahan JE. An electron transport factor from Clostridium pasteurianum. Biochem Biophys Res Commun 1962;7:448–52.
[3] Tagawa K, Arnon DI. Ferredoxins as electron carriers in photosynthesis and in the biological production and consumption of hydrogen gas. Nature 1962;195:537–43.
[4] Johnson DC, Dean DR, Smith AD, Johnson MK. Structure, function, and formation of biological iron-sulfur clusters. Annu Rev Biochem 2005;74:247–81.

[5] Ewen KM, Kleser M, Bernhardt R. Adrenodoxin: the archetype of vertebrate-type [2Fe-2S] cluster ferredoxins. Biochim Biophys Acta 2011;1814:111–25.
[6] Roche B, Aussel L, Ezraty B, Mandin P, Py B, Barras F. Iron/sulfur proteins biogenesis in prokaryotes: formation, regulation and diversity. Biochim Biophys Acta 2013;1827:455–69.
[7] Wachtershauser G. Groundworks for an evolutionary biochemistry: the iron-sulphur world. Prog Biophys Mol Biol 1992;58:85–201.
[8] Wachtershauser G. Origin of life. Life as we don't know it. Science 2000;289:1307–8.
[9] Peters JW, Williams LD. The origin of life: look up and look down. Astrobiology 2012;12:1087–92.
[10] Imlay JA. Iron-sulphur clusters and the problem with oxygen. Mol Microbiol 2006;59:1073–82.
[11] Py B, Barras F. Building Fe-S proteins: bacterial strategies. Nat Rev Microbiol 2010;8:436–46.
[12] Brill WJ. Biochemical genetics of nitrogen fixation. Microbiol Rev 1980;44:449–67.
[13] Rees DC, Howard JB. The interface between the biological and inorganic worlds: iron-sulfur metalloclusters. Science 2003;300:929–31.
[14] Peters JW, Broderick JB. Emerging paradigms for complex iron-sulfur cofactor assembly and insertion. Annu Rev Biochem 2012;81:429–50.
[15] Seefeldt LC, Hoffman BM, Dean DR. Mechanism of Mo-dependent nitrogenase. Annu Rev Biochem 2009;78:701–22.
[16] Hoffman BM, Lukoyanov D, Dean DR, Seefeldt LC. Nitrogenase: a draft mechanism. Acc Chem Res 2013;46:587–95.
[17] Seefeldt LC, Hoffman BM, Dean DR. Electron transfer in nitrogenase catalysis. Curr Opin Chem Biol 2012;16:19–25.
[18] Duval S, Danyal K, Shaw S, et al. Electron transfer precedes ATP hydrolysis during nitrogenase catalysis. Proc Natl Acad Sci USA 2013;110:16414–9.
[19] Yang ZY, Khadka N, Lukoyanov D, Hoffman BM, Dean DR, Seefeldt LC. On reversible H2 loss upon N2 binding to FeMo-cofactor of nitrogenase. Proc Natl Acad Sci USA 2013;110:16327–32.
[20] Mortenson LE, Thorneley RN. Structure and function of nitrogenase. Annu Rev Biochem 1979;48:387–418.
[21] Dixon RAY, Cannon F, Kondorosi A. Construction of a P plasmid carrying nitrogen fixation genes from Klebsiella pneumoniae. Nature 1976;260:268–71.
[22] Streicher S, Gurney E, Valentine RC. Transduction of the nitrogen-fixation genes in Klebsiella pneumoniae. Proc Natl Acad Sci USA 1971;68:1174–7.
[23] Cannon FC, Dixon RA, Postgate JR. Derivation and properties of F-prime factors in Escherichia coli carrying nitrogen fixation genes from Klebsiella pneumoniae. J Gen Microbiol 1976;93:111–25.
[24] Kennedy C. Linkage map of the nitrogen fixation (nif) genes in Klebsiella pneumoniae. Mol Gen Genet 1977;157:199–204.
[25] Dixon R, Kennedy C, Kondorosi A, Krishnapillai V, Merrick M. Complementation analysis of Klebsiella pneumoniae mutants defective in nitrogen fixation. Mol Gen Genet 1977;157:189–98.
[26] St John RT, Johnston HM, Seidman C, et al. Biochemistry and genetics of Klebsiella pneumoniae mutant strains unable to fix N2. J Bacteriol 1975;121:759–65.
[27] MacNeil T, MacNeil D, Roberts GP, Supiano MA, Brill WJ. Fine-structure mapping and complementation analysis of nif (nitrogen fixation) genes in Klebsiella pneumoniae. J Bacteriol 1978;136:253–66.
[28] Roberts GP, Brill WJ. Gene-product relationships of the nif regulon of Klebsiella pneumoniae. J Bacteriol 1980;144:210–6.
[29] Dixon R, Eady RR, Espin G, et al. Analysis of regulation of Klebsiella pneumoniae nitrogen fixation (nif) gene cluster with gene fusions. Nature 1980;286:128–32.
[30] Roberts GP, Brill WJ. Genetics and regulation of nitrogen fixation. Annu Rev Microbiol 1981;35:207–35.
[31] Spatzal T, Aksoyoglu M, Zhang L, et al. Evidence for interstitial carbon in nitrogenase FeMo cofactor. Science 2011;334:940.

[32] Schindelin H, Kisker C, Schlessman JL, Howard JB, Rees DC. Structure of ADP × AlF4(-)-stabilized nitrogenase complex and its implications for signal transduction. Nature 1997;387:370–6.

[33] Strop P, Takahara PM, Chiu H, Angove HC, Burgess BK, Rees DC. Crystal structure of the all-ferrous $[4Fe\text{-}4S]^0$ form of the nitrogenase iron protein from Azotobacter vinelandii. Biochemistry 2001;40:651–6.

[34] Rubio LM, Ludden PW. Biosynthesis of the iron-molybdenum cofactor of nitrogenase. Annu Rev Microbiol 2008;62:93–111.

[35] Hu Y, Ribbe MW. Biosynthesis of the iron-molybdenum cofactor of nitrogenase. J Biol Chem 2013;288:13173–7.

[36] Jacobson MR, Brigle KE, Bennett LT, et al. Physical and genetic map of the major nif gene cluster from Azotobacter vinelandii. J Bacteriol 1989;171:1017–27.

[37] Joerger RD, Bishop PE. Nucleotide sequence and genetic analysis of the nifB-nifQ region from Azotobacter vinelandii. J Bacteriol 1988;170:1475–87.

[38] Bishop PE, Joerger RD. Genetics and molecular biology of alternative nitrogen fixation systems. Ann Rev Plant Physiol Plant Mol Biol 1990;41:109–25.

[39] Pau RN, Mitchenall LA, Robson RL. Genetic-evidence for an Azotobacter vinelandii nitrogenase lacking molybdenum and vanadium. J Bacteriol 1989;171:124–9.

[40] Joerger RD, Jacobson MR, Premakumar R, Wolfinger ED, Bishop PE. Nucleotide-sequence and mutational analysis of the structural genes (anfhdgk) for the $2^{nd}$ alternative nitrogenase from Azotobacter vinelandii. J Bacteriol 1989;171:1075–86.

[41] Luque F, Pau RN. Transcriptional regulation by metals of structural genes for Azotobacter vinelandii nitrogenases. Mol Gen Genet 1991;227:481–7.

[42] Eady RR. Structure-function-relationships of alternative nitrogenases. Chem Rev 1996;96:3013–30.

[43] Hamilton TL, Ludwig M, Dixon R, et al. Transcriptional profiling of nitrogen fixation in Azotobacter vinelandii. J Bacteriol 2011;193:4477–86.

[44] Setubal JC, dos Santos P, Goldman BS, et al. Genome sequence of Azotobacter vinelandii, an obligate aerobe specialized to support diverse anaerobic metabolic processes. J Bacteriol 2009;191:4534–45.

[45] Dos Santos PC, Fang Z, Mason SW, Setubal JC, Dixon R. Distribution of nitrogen fixation and nitrogenase-like sequences amongst microbial genomes. BMC Genomics 2013;13:162.

[46] O'Carroll IP, Dos Santos PC. Genomic analysis of nitrogen fixation. Methods Mol Biol 2011;766:49–65.

[47] Bishop PE, Brill WJ. Genetic analysis of Azotobacter vinelandii mutant strains unable to fix nitrogen. J Bacteriol 1977;130:954–6.

[48] Thorneley RN, Lowe DJ. Nitrogenase of Klebsiella pneumoniae. Kinetics of the dissociation of oxidized iron protein from molybdenum-iron protein: identification of the rate-limiting step for substrate reduction. Biochem J 1983;215:393–403.

[49] Shah VK, Brill WJ. Isolation of a molybdenum-iron cluster from nitrogenase. Proc Natl Acad Sci USA 1981;78:3438–40.

[50] Hoover TR, Robertson AD, Cerny RL, et al. Identification of the V factor needed for synthesis of the iron-molybdenum cofactor of nitrogenase as homocitrate. Nature 1987;329:855–7.

[51] Zheng LM, White RH, Dean DR. Purification of the Azotobacter vinelandii NifV-encoded homocitrate synthase. J Bacteriol 1997;179:5963–6.

[52] Shah VK, Imperial J, Ugalde RA, Ludden PW, Brill WJ. Invitro synthesis of the iron-molybdenum cofactor of nitrogenase. Proc Natl Acad Sci USA 1986;83:1636–40.

[53] Ugalde RA, Imperial J, Shah VK, Brill WJ. Biosynthesis of iron-molybdenum cofactor in the absence of nitrogenase. J Bacteriol 1984;159:888–93.

[54] Dean DR, Brigle KE. Azotobacter vinelandii nifD and nifE-encoded polypeptides share structural homology. Proc Natl Acad Sci USA 1985;82:5720–3.

[55] Roberts GP, MacNeil T, MacNeil D, Brill WJ. Regulation and characterization of protein products coded by the nif (nitrogen fixation) genes of Klebsiella pneumoniae. J Bacteriol 1978;136:267–79.
[56] Brigle KE, Weiss CM, Newton WE, Dean DR. Products of the iron-molybdenum cofactor-specific biosynthetic genes, nifE and nifN, are structurally homologous to the products of the nitrogenase molybdenum-iron protein genes, nifD and nifK. J Bacteriol 1987;169:1547–53.
[57] Brigle KE, Newton WE, Dean DR. Complete nucleotide-sequence of the Azotobacter vinelandii nitrogenase structural gene-cluster. Gene 1985;37:37–44.
[58] Goodwin PJ, Agar JN, Roll JT, Roberts GP, Johnson MK, Dean DR. The Azotobacter vinelandii NifEN complex contains two identical [4Fe-4S] clusters. Biochemistry 1998;37:10420–8.
[59] Hu Y, Ribbe MW. Biosynthesis of Nitrogenase FeMoco. Coord Chem Rev 2011;255:1218–24.
[60] Kaiser JT, Hu Y, Wiig JA, Rees DC, Ribbe MW. Structure of precursor-bound NifEN: a nitrogenase FeMo cofactor maturase/insertase. Science 2011;331:91–4.
[61] Dos Santos PC. Molecular biology and genetic engineering in nitrogen fixation. Methods Mol Biol 2012;766:81–92.
[62] Kennedy C, Dean D. The nifU, nifS and nifV gene products are required for activity of all three nitrogenases of Azotobacter vinelandii. Mol Gen Genet 1992;231:494–8.
[63] Dos Santos PC, Dean DR, Hu Y, Ribbe MW. Formation and insertion of the nitrogenase iron-molybdenum cofactor. Chem Rev 2004;104:1159–73.
[64] Zhao D, Curatti L, Rubio LM. Evidence for nifU and nifS participation in the biosynthesis of the iron-molybdenum cofactor of nitrogenase. J Biol Chem 2007;282:37016–25.
[65] Jacobson MR, Cash VL, Weiss MC, Laird NF, Newton WE, Dean DR. Biochemical and genetic analysis of the nifUSVWZM cluster from Azotobacter vinelandii. Mol Gen Genet 1989;219:49–57.
[66] Zheng L, White RH, Cash VL, Jack RF, Dean DR. Cysteine desulfurase activity indicates a role for NIFS in metallocluster biosynthesis. Proc Natl Acad Sci USA 1993;90:2754–8.
[67] Yuvaniyama P, Agar JN, Cash VL, Johnson MK, Dean DR. NifS-directed assembly of a transient [2Fe-2S] cluster within the NifU protein. Proc Natl Acad Sci USA 2000;97:599–604.
[68] Dos Santos PC, Smith AD, Frazzon J, Cash VL, Johnson MK, Dean DR. Iron-sulfur cluster assembly: NifU-directed activation of the nitrogenase Fe protein. J Biol Chem 2004;279:19705–11.
[69] Zheng L, White RH, Cash VL, Dean DR. Mechanism for the desulfurization of L-cysteine catalyzed by the nifS gene product. Biochemistry 1994;33:4714–20.
[70] Mueller EG. Trafficking in persulfides: delivering sulfur in biosynthetic pathways. Nat Chem Biol 2006;2:185–94.
[71] Kessler D. Enzymatic activation of sulfur for incorporation into biomolecules in prokaryotes. FEMS microbiology reviews 2006;30:825–40.
[72] Kaiser JT, Clausen T, Bourenkow GP, Bartunik HD, Steinbacher S, Huber R. Crystal structure of a NifS-like protein from Thermotoga maritima: implications for iron sulphur cluster assembly. J Mol Biol 2000;297:451–64.
[73] Tirupati B, Vey JL, Drennan CL, Bollinger JM Jr. Kinetic and structural characterization of Slr0077/SufS, the essential cysteine desulfurase from Synechocystis sp. PCC 6803. Biochemistry 2004;43:12210–9.
[74] Cupp-Vickery JR, Urbina H, Vickery LE. Crystal structure of IscS, a cysteine desulfurase from Escherichia coli. J Mol Biol 2003;330:1049–59.
[75] Shi R, Proteau A, Villarroya M, et al. Structural basis for Fe-S cluster assembly and tRNA thiolation mediated by IscS protein-protein interactions. PLoS Biol 2010;8:e1000354.
[76] Marinoni EN, de Oliveira JS, Nicolet Y, et al. $(IscS-IscU)_2$ complex structures provide insights into $Fe_2S_2$ biogenesis and transfer. Angew Chem Int Ed Engl 2012;51:5439–42.
[77] Behshad E, Bollinger JM Jr. Kinetic analysis of cysteine desulfurase CD0387 from Synechocystis sp. PCC 6803: formation of the persulfide intermediate. Biochemistry 2009;48:12014–23.

[78] Fu W, Jack RF, Morgan TV, Dean DR, Johnson MK. The nifU gene product from Azotobacter vinelandii is a homodimer that contains two identical [2Fe-2S] clusters. Biochemistry 1994;33:13455–63.
[79] Agar JN, Yuvaniyama P, Jack RF, et al. Modular organization and identification of a mononuclear iron-binding site within the NifU protein. J Biol Inorg Chem 2000;5:167–77.
[80] Beynon J, Ally A, Cannon M, et al. Comparative organization of nitrogen fixation-specific genes from Azotobacter vinelandii and Klebsiella pneumoniae: DNA sequence of the nifUSV genes. J Bacteriol 1987;169:4024–9.
[81] Smith AD, Jameson GN, Dos Santos PC, et al. NifS-mediated assembly of [4Fe-4S] clusters in the N- and C-terminal domains of the NifU scaffold protein. Biochemistry 2005;44:12955–69.
[82] Zheng L, Dean DR. Catalytic formation of a nitrogenase iron-sulfur cluster. J Biol Chem 1994;269:18723–6.
[83] Walker GA, Mortenson LE. Effect of magnesium adenosine 5'-triphosphate on the accessibility of the iron of clostridial azoferredoxin, a component of nitrogenase. Biochemistry 1974;13:2382–8.
[84] Dos Santos PC, Johnson DC, Ragle BE, Unciuleac MC, Dean DR. Controlled expression of nif and isc iron-sulfur protein maturation components reveals target specificity and limited functional replacement between the two systems. J Bacteriol 2007;189:2854–62.
[85] White RH. Origin of the labile sulfide in the iron-sulfur proteins of Escherichia coli. Biochem Biophys Res Commun 1983;112:66–72.
[86] Flint DH. Escherichia coli contains a protein that is homologous in function and N-terminal sequence to the protein encoded by the nifS gene of Azotobacter vinelandii and that can participate in the synthesis of the Fe-S cluster of dihydroxy-acid dehydratase. J Biol Chem 1996;271:16068–74.
[87] Zheng L, Cash VL, Flint DH, Dean DR. Assembly of iron-sulfur clusters. Identification of an iscSUA-hscBA-fdx gene cluster from Azotobacter vinelandii. J Biol Chem 1998;273:13264–72.
[88] Bandyopadhyay S, Naik SG, O'Carroll IP, et al. A proposed role for the Azotobacter vinelandii NfuA protein as an intermediate iron-sulfur cluster carrier. J Biol Chem 2008;283:14092–9.
[89] Johnson DC, Unciuleac MC, Dean DR. Controlled expression and functional analysis of iron-sulfur cluster biosynthetic components within Azotobacter vinelandii. J Bacteriol 2006;188:7551–61.

# 12 A stress-responsive Fe-S cluster biogenesis system in bacteria – the *suf* operon of Gammaproteobacteria

F. Wayne Outten

## 12.1 Introduction to Fe-S cluster biogenesis

Iron-sulfur (Fe-S) clusters are minerals captured by the biosphere and integrated into cellular metabolism at the earliest stages of life [1, 2]. The elegant work of Holm and others suggests that early Fe-S cluster incorporation into metabolism may not have required any biogenesis machinery [3–6]. Fe-S clusters are thermodynamically stable, especially in anaerobic environments, and can undergo facile ligand exchange with small molecule (i.e. nonproteinaceous) ligands *in vitro*. Iron and sulfide were both abundant in certain environmental niches of the early anaerobic earth, for example, in hydrothermal vents where some of the most phylogenetically ancient organisms still thrive [7]. The earliest precursors of modern cells may have simply encapsulated preformed Fe-S clusters, concentrating them within membranes to be bound by metalloproteins or other metabolites with suitable ligands and/or binding site geometry (see discussion in Chapter 23 by Boyd *et al.*). It is likely that the first piece of the biogenesis machinery to have evolved is the Fe-S scaffold protein, which allows cells to physically sequester Fe-S clusters and direct them into specific target metalloproteins. A scaffold protein helps to limit nonspecific incorporation of Fe-S clusters into proteins with potential metal-binding amino acid residues. Such cluster misincorporation is still a technical problem that must be overcome by experimentalists characterizing new candidate Fe-S proteins [8].

What makes a good Fe-S cluster scaffold protein? The scaffold structure must promote formation of a stable cluster during assembly but be flexible enough to efficiently release the Fe-S cluster for insertion into a target metalloprotein. These requirements are further complicated by the potential need for the scaffold to accommodate multiple cluster types with very different geometries (planar *vs* cuboidal, for example). Elegant biophysical characterization of the IscU Fe-S cluster scaffold protein from *E. coli* has clearly shown that the scaffold is metamorphic, that is, it can exist in two interconvertible conformations that are similar in their free energies [9–11]. The partially disordered (D) state of IscU promotes sulfur acquisition during cluster assembly as well as transfer of the fully formed cluster. In contrast, the structured (S) conformation of IscU facilitates iron binding and iron-sulfide interactions to initially assemble the *de novo* cluster. Interconversion between these different conformations is partially controlled by *cis-trans* isomerization of two peptidyl-prolyl peptide bonds in IscU as well as by interaction with molecular chaperones that preferentially bind a specific conformation [9].

As Fe-S cluster metabolism became more predominant among early lineages of life, there was likely selection for specific iron and sulfide donation pathways. Organisms that filled some environmental niches may have struggled with limited access to bioavailable iron and sulfur sources, a problem that was certainly exacerbated by the accumulation of oxygen after the development of oxygenic photosynthesis [12–14]. This scarcity selected for the evolution of high affinity uptake transporters and dedicated intracellular homeostasis systems for iron and sulfur, at least partially to ensure continued biogenesis of Fe-S clusters required for core metabolism. We will discuss each in turn, beginning with sulfur trafficking, which is better characterized than *in vivo* iron donation.

## 12.2 Sulfur trafficking for Fe-S cluster biogenesis

Among the Bacteria and Eukarya, the cysteine desulfurase enzyme family controls sulfide donation for Fe-S cluster assembly (Fig. 12.1). Cysteine desulfurases are homodimeric enzymes that utilize pyridoxal-5′-phosphate (PLP) as a cofactor to catalyze the removal of sulfur from L-cysteine substrate [15, 16]. The sulfur is abstracted from L-cysteine by nucleophilic attack from the thiolate anion of a Cys residue in the desulfurase active site on the PLP-L-cysteine protonated quinonoid adduct. This results in the sulfur atom forming a persulfide bond (R-S-SH) at the active site Cys residue of the cysteine desulfurase enzyme. The remaining backbone of L-cysteine is released from PLP as L-alanine. The persulfide sulfur is then transferred from the cysteine desulfurase to a partner protein (usually the scaffold) *via* a thiol exchange reaction in which a Cys residue from the partner protein carries out a nucleophilic attack on the persulfide of the donor (Fig. 12.1). This sulfur trafficking pathway likely limits the release of sulfide *in vivo* during cluster biogenesis. Because sulfide is toxic to most aerobic organisms, the controlled trafficking of a persulfide species avoids deleterious side reactions during cluster assembly. The persulfide is $S^0$ and must be reduced to $S^{2-}$ during the subsequent steps of cluster assembly. Recent evidence suggests that the source of reducing equivalents for this may be provided by partner redox proteins, such as [2Fe-2S] Fdx in the Isc pathway, that directly donate electrons for sulfur reduction [17, 18]. It also has been proposed that oxidation of $Fe^{2+}$ to $Fe^{3+}$ during formation of the first $[2Fe\text{-}2S]^{2+}$ cluster intermediate may provide some of the reducing power for sulfide formation. Despite numerous published results, it is still difficult to establish if persulfide and iron are donated sequentially (and if so, which comes first) or if they must be provided simultaneously for proper cluster assembly. Results obtained *in vitro* are often complicated by the lack of one or more potential partner proteins for the cysteine desulfurase and/or scaffold protein, which may significantly alter the biochemical behavior of the system. Context is everything in Fe-S cluster biogenesis.

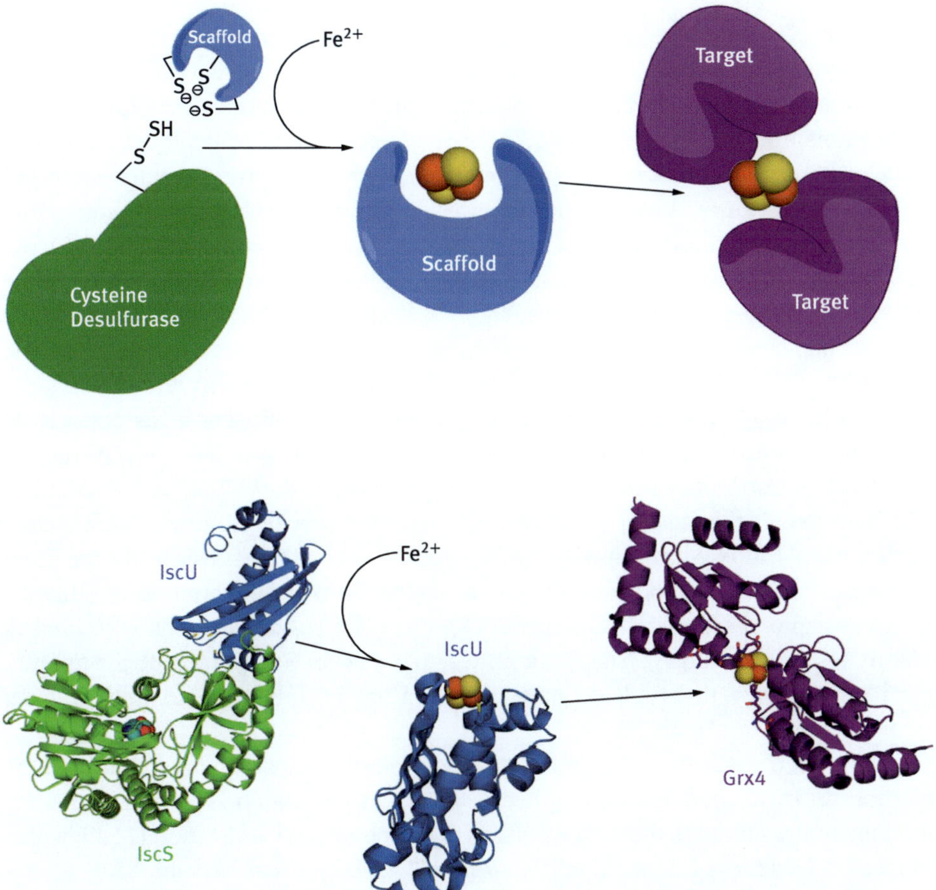


**Fig. 12.1:** Core components of Fe-S cluster assembly. (top) Diagram of persulfide transfer from cysteine desulfurase to scaffold protein. Iron donation and cluster assembly proceed and then cluster transfer occurs to a target Fe-S metalloprotein. (bottom) Structures of IscS cysteine desulfurase (green) and IscU scaffold (blue) are shown (from co-structure of $IscS_2$-$IscU_2$ in PDB 4EB5) [19]. [2Fe-2S] monothiol glutaredoxin 4 (Grx4) is shown in purple as a target protein (from PDB 2WCI) [68].

## 12.3 Iron donation for Fe-S cluster biogenesis

The donation of iron for Fe-S cluster biogenesis is still largely uncharacterized. It is generally accepted (although not proven experimentally) that iron used for Fe-S cluster biogenesis comes from a ferrous pool of "labile" iron. The nature of the labile iron pool is still under investigation, but it is thought to consist of weakly chelated $Fe^{2+}$ that is routed into iron metallocofactor biogenesis. This pool also is sensed by iron metalloregulatory proteins that control cellular iron homeostasis. To direct labile iron into Fe-S cluster biogenesis, the scaffold protein could directly bind iron from this pool. Multiple studies to address this question have not resulted in a clear answer. Although *Azotobacter vinelandii* and *Escherichia coli* IscU scaffolds do not appear to have appreciable iron binding capability, *Thermotoga maritima* SufU can bind either $Fe^{2+}$ or $Fe^{3+}$ with affinities that range from the high nanomolar to low micromolar [20–22].

Alternatively, iron donation may require a separate iron "chaperone" protein that specifically interacts with the scaffold protein. Metallochaperones have been identified for other essential metals, such as copper and cobalt. Examples of well-studied copper chaperones include Atx1 in eukaryotes and CopZ in bacteria. Atx1 efficiently traffics copper and releases it upon direct protein-protein contact with its cognate acceptor Ccc2 [23–25]. An analogous iron chaperone would ensure that a subpopulation of the labile iron pool is dedicated for Fe-S cluster biogenesis as opposed to heme biosynthesis or maturation of mono- and di-nuclear iron enzymes. A number of candidate iron donor proteins have been put forth over the last two decades, of which frataxin is the most thoroughly studied (see Chapter 18 by Tong). There is clear evidence that frataxin is intimately involved in Fe-S cluster biogenesis by the Isc (iron-sulfur cluster) pathway, both in Bacteria as well as in the mitochondria of Eukarya. *E. coli* contains a microbial frataxin (CyaY). CyaY binds two $Fe^{2+}$ ions with a weak affinity ($K_d$ = 3.8 μM) [26]. The weak affinity of CyaY for $Fe^{2+}$ is in contrast to other metallochaperones that typically have metal-binding affinities of picomolar or lower for their target metals [27, 28].

CyaY can also mediate the controlled oxidation of ferrous iron to ferric iron in the presence of $H_2O_2$, thereby avoiding hydroxyl radical generation [26]. This reaction is fundamentally similar to that of the mini-ferritin protein Dps (see Section 12.8.2), which sequesters ferric iron under oxidative stress in *E. coli*. CyaY can accommodate approximately 26 $Fe^{3+}$, likely in the form of polynuclear iron hydroxo(oxo) species that are only partially associated with the protein. Iron associates with CyaY at the carboxyl side chain of conserved aspartate and glutamate residues [29]. Since carboxyl groups are less susceptible to oxidation than the thiol side chain of cysteine ligands that are prevalent in ferrous iron-binding proteins, it is proposed that CyaY sequesters ferric iron in the presence of hydrogen peroxide to help minimize the production of hydroxyl radicals [27]. Based on these studies and analogous work on eukaryotic frataxins, it was originally proposed that CyaY may be an iron donor for the Isc pathway [30].

However, a role for CyaY in iron donation has recently been called into question [31–34]. In *E. coli*, CyaY has been shown to inhibit *in vitro* [2Fe-2S] cluster formation on IscU when using IscS as the sulfur donor. The presence of CyaY does not facilitate iron delivery for cluster reconstitution on IscU, regardless of whether iron is added in solution or bound directly to CyaY. CyaY inhibition likely occurs through direct binding of CyaY to the IscS-IscU co-complex such that CyaY inhibits a step in cluster assembly, such as persulfide transfer from IscS to IscU. It was shown that CyaY residues that bind iron are also involved in binding to IscS because mutation of these residues diminished the ability of CyaY to inhibit IscS. The presence of iron slightly enhanced the inhibitory effect of CyaY on IscS-IscU cluster formation, leading to the proposal that CyaY may act as a gatekeeper to control Fe-S cluster assembly based on *in vivo* iron levels [31]. One could imagine a model where *in vivo* oxidation of $Fe^{2+}$ to $Fe^{3+}$ on CyaY during oxidative stress might enhance CyaY repression of IscS-IscU to limit Isc-mediated cluster assembly. Such regulation would be consistent with other post-transcriptional repression of Isc under certain stress conditions (see below in Section 12.5). Further studies are necessary to confirm the exact function of frataxin.

In addition to frataxin, in eukaryotes the poly(rC)-binding proteins (PCBPs) were shown to be required for efficient *in vivo* iron loading into the ferritin iron storage protein as well as iron- and 2-oxoglutarate-dependent dioxygenases [35–37]. PCBP1 can bind up to 3 iron ions with high nanomolar to low micromolar affinity and was shown to directly interact with ferritin in a $Fe^{2+}$-dependent manner. However, close PCBP homologues are largely absent in currently sequenced microbial genomes; thus, it is not clear if they play any role in iron trafficking in non-eukaryotic organisms.

Unlike copper, specific high-affinity metallochaperones for individual iron-containing enzymes have not been identified in bacteria. This difference may be due to the abundance of iron containing enzymes in the cell, which requires a large labile iron pool and makes a "one chaperone per metalloenzyme" system less tenable. The labile copper pool is quite small (less than femtomolar) due to tight buffering by metallochaperones and glutathione (GSH) [29, 38, 39]. In contrast, the labile iron pool is thought be considerable (low micromolar) in most cell types. It is possible that the use of metallochaperones for other metals besides iron may have arisen to protect weakly chelated iron pools from direct competition by metals like copper and zinc. $Cu^{II}$ and $Zn^{II}$ should bind more tightly to cellular ligands than $Fe^{II}$ based on their relative positions in the Irving-Williams series. This explanation certainly seems plausible in the case of copper, which has been shown to disrupt iron homeostasis when present in excess of the cell's copper buffering capacity [40, 41]. In that particular case, $Cu^{I}$ may be the culprit due to its thiophilic nature, which allows it to compete strongly for binding to Cys thiolates in cluster-binding sites. At present iron donation for Fe-S cluster biogenesis is perhaps the least well-defined step of the entire pathway. We will come back to this topic in more detail later.

## 12.4 Fe-S cluster assembly and trafficking

As iron and persulfide are mobilized to the scaffold protein, the cluster assembly proceeds. In the well characterized U-type scaffolds (such as IscU and the N-terminus of NifU), the first observable intermediate in cluster assembly is a $[2Fe\text{-}2S]^{2+}$ cluster [42–47]. This cluster type can then be transferred from IscU to target apo-proteins with the assistance of heat shock cognate molecular chaperones (HscA and HscB in *E. coli*) [48–51]. Depending on the target apo-protein, the chaperones can enhance cluster transfer rates from $[2Fe\text{-}2S]^{2+}$ IscU up to 700-fold [50]. Structural analysis and biochemical experiments suggest that the molecular chaperones are required to initiate ligand exchange during cluster transfer by promoting dissociation of one of the IscU Cys ligands from the Fe-S cluster [48, 50–53]. Opening this coordinate site presumably allows the target apo-protein to access the cluster and initiate the transfer process. Molecular chaperone proteins also help control the equilibrium between structured and partially disordered conformations of IscU *via* an ATP-dependent binding cycle [49, 54, 55].

With additional iron and sulfide donation, the $[2Fe\text{-}2S]^{2+}$ form of the IscU dimer can further assemble a second $[2Fe\text{-}2S]^{2+}$ cluster. At this stage, the addition of one electron to each cluster will catalyze reductive coupling of the $2 \times [2Fe\text{-}2S]^{+}$ clusters into a single $[4Fe\text{-}4S]^{2+}$ cluster on IscU [44, 48]. The $[4Fe\text{-}4S]^{2+}$ cluster can be transferred to a target apo-protein but does not require molecular chaperones for the transfer [56]. The $[4Fe\text{-}4S]^{2+}$ cluster on IscU is sensitive to oxygen, and upon oxidation, it may "disassemble" back to the $[2Fe\text{-}2S]^{2+}$ form *in vitro*. The oxygen-sensitive cluster conversion process may create an *in vivo* pool of IscU that contains both $[4Fe\text{-}4S]^{2+}$ and $[2Fe\text{-}2S]^{2+}$ clusters, which are, by far, the most common cluster types required by Fe-S metalloproteins. The equilibrium between cluster forms of IscU would be controlled by oxygen and by the specific target apo-proteins that are drawing Fe-S clusters from the biogenesis pool. The number and types of various target Fe-S apo-proteins would in turn be controlled by cellular metabolic requirements.

Recently, it has also become clear that *in vivo* Fe-S cluster metabolism requires an array of Fe-S cluster trafficking proteins, including the A-type carrier (ATC) family [54–63] and the monothiol glutaredoxins [16, 64–68]. ATC proteins directly interact with the scaffold proteins and are highly efficient at accepting the clusters and passing them to target apo-proteins. Meanwhile, retrograde cluster transfer from the carrier back to the scaffold is typically unfavorable. The Fe-S cluster trafficking carrier proteins may provide an additional layer of *in vivo* specificity that would be difficult to encode directly in a single scaffold. For example, both ErpA, and IscA are required for full maturation of the Fe-S subunits of formate-hydrogen lyase complex in *E. coli* [69]. Similarly, Fe-S cluster acquisition by the Fe-S metalloregulatory proteins IscR and NsrR is controlled by different ATC proteins *in vivo* [70, 71]. Another possible use of carrier proteins *in vivo* is to act as a way station where clusters can accumulate while they await cellular demand. If sulfur trafficking, iron donation, and *de novo* cluster assembly are the slowest and/or most sensitive steps of Fe-S cluster biogenesis, there may be a physiological advantage to loading a pool of Fe-S carrier proteins to minimize reliance on the early steps in biogenesis. There is also strong evidence that the ATC proteins can reversibly interconvert $[4Fe\text{-}4S]^{2+}$ and $[2Fe\text{-}2S]^{2+}$ clusters in response to oxygen [72, 73]. Monothiol glutaredoxins are also known to accommodate multiple cluster types [74]. Cluster interconversion could allow them to load Fe-S clusters into multiple types of enzymes. Cluster interconversion may also provide the various trafficking proteins with a mechanism to "protect" sensitive $[4Fe\text{-}4S]^{2+}$ clusters by carrying out a controlled oxidation and disassembly to the $[2Fe\text{-}2S]^{2+}$ form in response to oxygen toxicity. In some cases all-ferric $[2Fe\text{-}2S]^{2+}$ clusters are more resistant to oxidation than $[4Fe\text{-}4S]^{2+}$ clusters that contain a mixture of $Fe^{2+}/Fe^{3+}$.

## 12.5 Iron and oxidative stress are intimately intertwined

From the observations described in Section 12.4, it is clear that oxygen is a major antagonist of Fe-S cluster metabolism, likely through the production of ROS such as $H_2O_2$ and superoxide ($O_2$•). Superoxide is electrostatically attracted to the iron cations of the Fe-S cluster and can directly oxidize $Fe^{2+}$ to $Fe^{3+}$ [75]. Cuboidal [4Fe-4S] clusters in most proteins are not stable in the $[4Fe\text{-}4S]^{3+}$ state. Cluster oxidation leads to ejection of iron and further oxidation of the remaining iron or sulfide ($S^{2-}$) causes cluster disassembly (described in detail in Chapter 10 by Nicolet and Fontecilla-Camps) (Fig. 12.2). Some Fe-S clusters are protected from direct disruption by being buried in a solvent-protected environment within the metalloprotein. However, many Fe-S cluster-dependent dehydratase enzymes require that an open coordination site on one Fe atom be exposed to solvent to bind and activate the enzyme substrate. Fe-S clusters in this class of dehydratase are some of the most sensitive targets of oxidative stress [75–83]. ROS such as $H_2O_2$ also react readily with the thiolate anion form of cysteine, resulting in oxidation of the organosulfur to sulfenic, sulfinic, and sulfonic acids (Fig. 12.2) [84]. Because Cys residues are often used to coordinate Fe-S clusters, oxidative modifications of the active sites of Fe-S metalloproteins (including Fe-S cluster scaffolds) would block binding of Fe-S clusters. The protein-bound persulfide species carried on the cysteine desulfurases and trafficked by Fe-S cluster biogenesis proteins is also sensitive to oxidation providing a further mechanism by which Fe-S cluster biogenesis can be disrupted by oxygen toxicity.

Oxygen also has an indirect effect on the ability of organisms to acquire iron. In the presence of oxygen, $Fe^{2+}$ is readily oxidized to the ferric ion $Fe^{3+}$. In aqueous solutions near neutral pH, $Fe^{3+}$ will hydrolyze $H_2O$ to generate iron$^{III}$ oxide-hydroxide compounds. Iron hydroxides can then polymerize *via* a pH-dependent process known as olation, thereby forming insoluble iron hydroxide precipitates. The aqueous chemistry of ferric iron results in vary low amounts of bioavailable iron in many environments that are otherwise favorable for life and often makes iron a limiting nutrient for growth. Although multiple strategies have evolved to circumvent iron limitation (including use of non-iron alternatives in many metabolic pathways), there appear to be a subset of absolutely essential iron metalloproteins in many organisms [60, 85].

Clearly, iron metalloproteins are a target of oxidative stress; however, iron also directly contributes to oxidative stress. This is because weakly chelated $Fe^{2+}$ participates in the production of hydroxyl radicals *via* Fenton chemistry ($Fe^{2+} + H_2O_2 \rightarrow Fe^{3+} + {-OH} + {\cdot OH}$). To avoid toxicity, bacterial cells dynamically regulate iron distribution to maintain iron homeostasis [86, 87]. Multiple strategies are used to maintain necessary intracellular iron levels: secretion of iron-scavenging siderophores, regulation of iron import systems, utilization of iron storage proteins, and controlled distribution of iron to various pathways [88]. Intracellular iron availability is sensed by several global transcriptional regulators, including the ferric uptake regulator (Fur) and the iron-sulfur cluster regulator IscR (see Chapter 13 by Mettert *et al.*). Fur directly binds

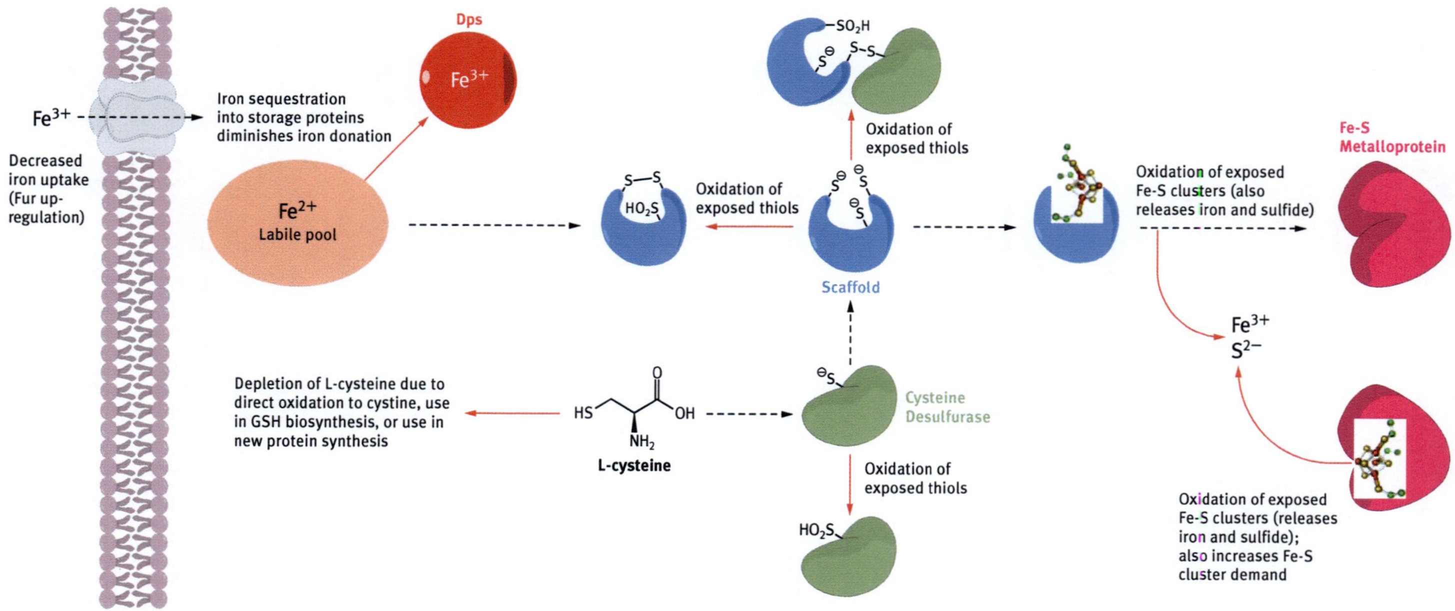


**Fig. 12.2:** Effects of oxidative stress on Fe-S cluster metabolism in *E. coli*, steps of iron or Fe-S homeostasis that may be perturbed by oxidative stress are indicated by dashed arrows. Reactions/processes that occur in response to stress and disrupt iron or Fe-S homeostasis are indicated by red arrows.

iron, and under iron-replete conditions $Fe^{2+}$-Fur represses the transcription of genes that encode iron starvation stress response proteins (such as siderophore biosynthesis systems, iron uptake transporters, and non-iron alternative enzymes) [87, 89]. As cellular $Fe^{2+}$ levels drop, apo-Fur dissociates from its binding site at target promoters and thereby allows RNA polymerase access to initiate transcription. For example, $Fe^{2+}$-Fur represses *suf* transcription when cellular iron levels are adequate, but this repression is abolished under iron starvation conditions [90, 91]. Fur also indirectly regulates many mRNA transcripts in *E. coli* by repressing the transcription of the RyhB small RNA under iron-replete conditions [92–94]. Under low iron conditions, the RyhB small RNA is expressed and binds to target mRNAs to regulate their stability or translation. For example, when RyhB is expressed, the small RNA binds to the *isc* mRNA and promotes RnaseE-mediated degradation of *isc* transcript under iron starvation conditions (Fig. 12.3).

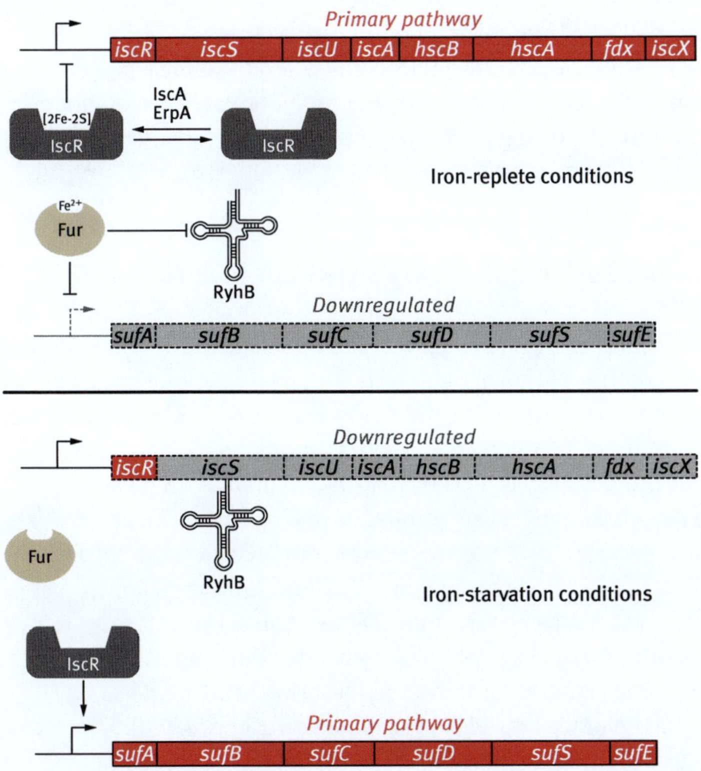


**Fig. 12.3:** Iron-dependent regulation of *isc* and *suf* operons in *E. coli*. Gray dashed boxes indicate transcriptional or post-transcriptional downregulation of mRNA. Red boxes indicate increased expression of mRNA. RyhB interacts with the *iscS* region of the polycistronic *isc* mRNA and does not appear to decrease *iscR* expression.

Deletion of Fur leads to a decrease of total cellular iron (due to the reduced expression of some iron storage proteins) but causes an increase in the labile pool of iron due to increased iron uptake and decreased iron utilization by iron metalloproteins. The increase in labile iron causes increased sensitivity to hydrogen peroxide ($H_2O_2$) and other ROS that react with $Fe^{2+}$ *via* the Fenton reaction [78, 95, 96]. It has also been shown that *fur* expression is upregulated in response to $H_2O_2$ induction *via* the OxyR transcriptional activator that senses peroxide stress. Increased Fur expression helps to replace Fur protein damaged by ROS [97] and allows Fur to both sequester labile iron released from damaged iron metalloproteins and to tightly repress any further iron uptake during the oxidative stress.

These complex regulatory loops show that iron starvation and oxidative stress responses are intimately intertwined in *E. coli*. Both conditions result in disruption of iron metabolism and *de facto* iron limitation for iron cofactor biosynthesis *in vivo*. In the case of oxidative stress, iron limitation occurs due to increased iron demand to replace damaged iron enzymes, decreased iron uptake due to Fur upregulation, and increased iron sequestration in iron storage proteins (such as Dps and Bfr, see Section 12.8.2) (Fig. 12.2). *E. coli* attempts to circumvent iron limitation by expressing non-iron alternative enzymes and by decreasing production of non-essential iron metalloproteins. However, as noted above, there are at least a handful of iron-dependent enzymes that cannot be replaced and their maturation must be maintained for cell survival.

## 12.6 Stress-response Fe-S cluster biogenesis in *E. coli*

Oxygen-dependent disruption of Fe-S cluster metabolism has selected for the evolution of Fe-S cluster biogenesis pathways that are able to maintain cluster biogenesis under stress. The most well-studied example of a stress-response Fe-S biogenesis pathway is the Suf system found in many facultative anaerobes among the Gammaproteobacteria, such as *E. coli*. The Suf pathway is broadly distributed among Archaea and Bacteria, where it is often the primary housekeeping pathway for cluster biogenesis (see Chapter 11 by Dos Santos and Dean for more discussion of housekeeping Suf systems in Gram-positive Bacteria) [98]. However, in some Bacteria, and in a limited set of Eukaryotes, Suf appears to have taken on the role of either a stress-response system or an organelle-specific system for Fe-S cluster biogenesis (for Eukaryotic Suf systems, note the role of Suf in some plant chloroplasts as described in Chapter 22 by Ye). In those cases, Suf augments other housekeeping cluster biogenesis pathways (typically the Isc pathway, see Chapter 13) and is a *bona fide* stress-response system for maintaining Fe-S cluster biogenesis under a variety of stress conditions.

*E. coli* switches between the Isc and the Suf systems based on environmental conditions using multiple transcriptional and post-transcriptional regulatory mechanisms (Fig. 12.3). Transcription of the *iscRSUA-hscBA-fdx-iscX* operon is controlled by the IscR metalloregulatory protein (for a comprehensive description of IscR, see Chapter 13 by

Mettert *et al.*) [99]. Briefly, IscR regulates global Fe-S metabolism in response to Fe-S cluster demand [71, 100–102]. IscR itself binds a [2Fe-2S] cluster and holo-IscR represses *isc* transcription. As Fe-S cluster demand exceeds the ability of the biogenesis machinery to load the [2Fe-2S] cluster into IscR, the apo-IscR predominates, leading to increased *isc* transcription. Interestingly, apo-IscR is not inactive. Instead, loss of the Fe-S cluster allows apo-IscR to acquire a new set of target promoters for which it activates transcription [101, 103]. Although *isc* transcription is generally responsive to a certain window of Fe-S cluster demand, under severe iron limitation, the *isc* mRNA is selectively degraded by RNaseE due to the action of the RyhB small RNA regulator (Fig. 12.3) [104]. Under such conditions, the cell begins to utilize the Suf pathway for Fe-S cluster biogenesis while minimizing its reliance on Isc.

Transcription of the *sufABCDSE* operon is increased under iron starvation stress *via* two mechanisms. First, the *suf* promoter is activated by apo-IscR, which will predominate when iron is limiting for Fe-S cluster biogenesis [101, 103]. Second, under iron-replete conditions, the transcription of *suf* is repressed by $Fe^{2+}$-Fur, an iron metalloregulatory protein (Fig. 12.3) [90, 91]. During iron starvation, $Fe^{2+}$-Fur repression is abolished, allowing *suf* transcription to increase. Transcription of *suf* also increases under oxidative stress by the concerted efforts of the OxyR peroxide stress response regulator and integration host factor (IHF) as well as through separate activation by apo-IscR [90, 105]. Due to this complex regulation, *suf* transcription should generally increase under any stress condition in which iron or Fe-S metabolism is disrupted, including iron limitation and oxidative stress.

The reciprocal regulation of Isc and Suf is also reflected in their respective mutant phenotypes [41, 62, 63, 90, 98, 106–110]. Although deletion of the *isc* locus causes a general defect in Fe-S cluster biogenesis under normal conditions, increased expression of *suf* is able to compensate for the lack of Isc function. Similarly, deletion of the *suf* operon has little effect on Fe-S cluster biogenesis under normal growth conditions, but the *suf* mutants are sensitive to stresses that disrupt Fe-S metabolism. The alternate regulation and phenotypes associated with Isc and Suf raises the question of what are the specific adaptations of the Suf pathway in organisms such as *E. coli* that allow it to maintain Fe-S cluster biogenesis when the housekeeping Isc pathway is compromised?

## 12.7 Sulfur trafficking in the stress-response Suf pathway

SufS is the cysteine desulfurase that mobilizes sulfur from L-cysteine during Fe-S cluster biogenesis. This pyridoxal-5′-phospate (PLP) dependent reaction results in an enzyme bound persulfide (R-S-SH) species on Cys364 of SufS. The steady-state activity of SufS is many-fold lower than that of the IscS cysteine desulfurase used by the Isc pathway. Comparison of the structures of the two provides a partial explanation for this difference (Fig. 12.4). Both cysteine desulfurases are homodimers with PLP binding sites located in the vicinity of the active site Cys residues used for sulfur abstraction from L-cysteine substrate. The active site Cys328 residue of IscS is on a flexible, solvent exposed loop. In most *E. coli* IscS structures, the active site loop is disordered and cannot be resolved. Figure 12.4 shows the *A. fulgidus* IscS co-crystallized with IscU, which has a well-defined active site loop containing Cys321 (analogous to Cys328). In contrast, the SufS active site loop is largely covered by a "lid" structure provided by the neighboring SufS monomer (Fig. 12.4). The active site lid restricts solvent access to SufS Cys364, and this important structural element is not present in IscS (Fig. 12.4). Cysteine desulfurase enzyme turnover requires release of the persulfide from the active site Cys residue, which depends on the ability of a reductant (usually DTT or TCEP *in vitro* or a partner protein thiol *in vivo*) to gain access to the persulfide species in the active site.

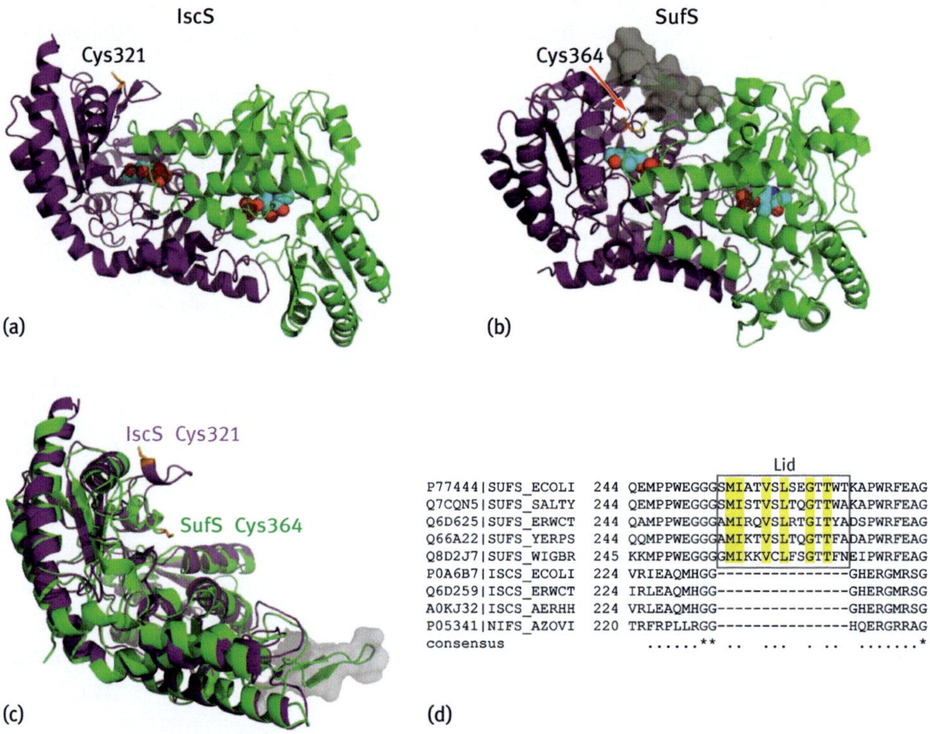


**Fig. 12.4:** Comparison of IscS and SufS cysteine desulfurases. (a) *A. fulgidus* IscS homodimer with one monomer in green and one in violet (taken from IscU-IscS co-structure in PDB 4EB5). Active Cys321 is shown in orange with PLP cofactor in cyan. (b) *E. coli* SufS homodimer with one monomer in green and one in violet (take from PDB 1KMJ) [111]. Active Cys364 is shown in orange with PLP cofactor in cyan. Active site "lid" structure highlighted in surface representation (transparent gray). (c) FATCAT structural alignment of IscS (violet) and SufS (green) monomers from A and B [112]. Active site Cys residues are in orange. PLP is not shown and IscU is omitted for comparison of IscS and SufS structures. (d) Partial amino acid alignment of select cysteine desulfurases generated by FASTA. The lid region conserved in SufS family members is boxed with highly conserved residues highlighted in yellow.

The structural lid over the SufS active site limits access of reductants to the Cys364 persulfide and results in low steady-state activity of SufS compared with IscS [113–119]. At present, it is not clear if the rate at which SufS forms the enzyme persulfide is actually slower than IscS as direct quantitative comparisons have not been done between the two enzymes under pre-steady-state conditions. Possibly SufS reacts quickly to form the persulfide but then turns over at a much slower rate than IscS in the *in vitro* assays due to the failure to release persulfide [113].

Is there an advantage to this structural limitation of SufS? Recently, we determined that SufS Cys364 is more resistant to direct oxidation by $H_2O_2$ compared with IscS Cys328, indicating that the active site lid may prevent disruption of sulfur mobilization by SufS in the presence of hydrogen peroxide [120]. These experiments had to be conducted in the absence of strong reductants such as DTT to prevent "repair" of partially oxidized Cys residues. As one might imagine, DTT and TCEP are not the physiological reductants for the persulfides on IscS or SufS *in vivo*. In both cases, the persulfide species is released from the cysteine desulfurase enzyme by the nucleophilic attack of a thiol residue from a specific partner protein. For IscS, the Fe-S scaffold protein IscU is the partner protein that accepts the persulfide. The *E. coli* Suf system lacks a true IscU homologue (in contrast to Gram-positive Bacteria, as discussed in Chapter 11 by Dos Santos and Dean) but SufS does interact with the SufE protein.

Although there is poor sequence conservation between the two proteins, SufE is structurally similar to IscU [121]. In contrast to IscU, there is no evidence that SufE can act as an Fe-S cluster scaffold because it lacks two of the three conserved Cys residues used by IscU to coordinate its cluster. Instead, SufE has a single conserved Cys residue at position 51 (analogous to IscU Cys63) that is used to accept sulfur from SufS. As a result, addition of SufE to the SufS activity assay results in a dramatic enhancement of basal SufS activity such that it reaches the level of IscS [116, 117, 122]. Addition of IscU can also mildly enhance IscS activity *in vitro* but only if the strong chemical reductants (DTT, TCEP) are omitted from the reaction until it is quenched. If the reductants are present during the reaction, they apparently release IscS persulfide much more efficiently than IscU such that activity enhancement by IscU is not observed [120].

Although the SufE Cys51 residue is on a loop structure near the surface of SufE, the Cys51 side chain is actually oriented into a protected groove in the SufE structure (Fig. 12.5) [121]. The groove is lined with hydrophobic residues and has limited solvent exposure. SufE Cys51 can be oxidized by $H_2O_2$ but is much less sensitive to oxidation than IscU Cys63, which is part of a more extended and exposed Fe-S cluster-binding site in IscU [120]. Recently, a co-structure was solved containing two close homologues of SufS and SufE, named CsdA and CsdE, respectively, which were identified in a third bacterial cysteine desulfurase system known as the Csd (cysteine sulfinate desulfinase) system (Fig. 12.5) [123]. These proteins catalyze a cysteine desulfurase reaction, but the resulting persulfide appears to be donated to a partner protein known as CsdL for the synthesis of an unknown sulfur-containing metabolite or protein [124]. However, the co-structure of CsdA-CsdE is illuminating. In the structure, the SufE homologue CsdE has adopted a confirmation in which the active site Cys residue that accepts persulfide sulfur is flipped out of its groove on the CsdE surface and is extended into the active site of the cysteine desulfurase CsdA [123]. Apparently protein-protein interactions between CsdA and CsdE trigger this conformational change to facilitate sulfur transfer from the desulfurase to the acceptor proteins (Fig. 12.5). One would assume that SufE undergoes a similar conformational change upon interacting with SufS. In fact, using hydrogen deuterium exchange mass spectrometry, we have observed changes in solvent accessibility around SufE Cys51 when it interacts with SufS [125].

Does the presence of the partner proteins alter the sensitivity of SufS or IscS to oxidative stress? We determined that addition of IscU to IscS actually made IscS activity more sensitive to disruption by oxidative stress [120]. In contrast, SufS activity was still enhanced in the presence of SufE and was not sensitive to $H_2O_2$ exposure. SufS-SufE maintained a higher level of absolute activity than IscS or IscS-IscU over the entire range of $H_2O_2$ concentrations tested (ranging from low to mid micromolar concentrations). When we measured the level of Cys oxidation at the active site Cys residues of each protein after $H_2O_2$ exposure, we observed that IscU was readily oxidized at its active site Cys63 and Cys106 residues and formed mixed disulfide species with Cys328 of IscS. In contrast, we observed little oxidation of SufS Cys364 and only mild oxidation of SufE Cys51 in response to $H_2O_2$ exposure. Thus, it appears that the Isc sulfur transfer pathway is inherently more sensitive to oxidative stress than the Suf pathway, providing a partial rationale for the cell to utilize Suf under oxidative stress conditions.

In addition to their differential sensitivity to oxidation, SufS-SufE and IscS-IscU also show differences in their relative activity levels under low L-cysteine concentrations [120]. SufS-SufE are up to 6-fold more active at low L-cysteine concentrations

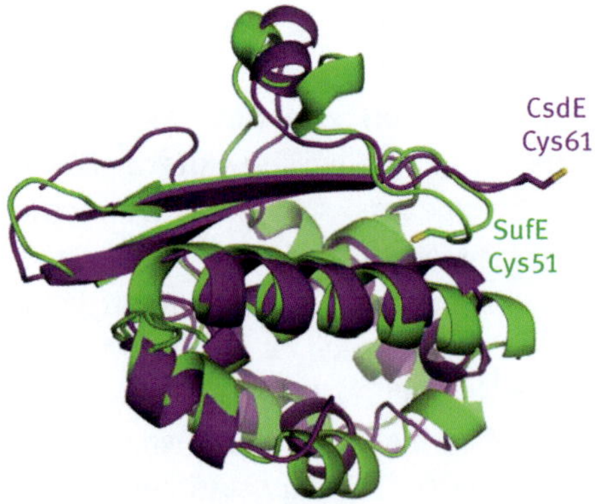


**Fig. 12.5:** Comparison of resting and active SufE/CsdE sulfur transfer shuttles. CsdE from co-structure with CsdA (PDB 4LW4) is shown in purple with active site Cys61 in stick representation [123]. SufE (PDB 1MZG) is shown in green with active site Cys51 in stick representation [121]. Structures were aligned using FATCAT [112].

than IscS-IscU. The greater ability of Suf to maintain sulfide mobilization under low L-cysteine conditions may have *in vivo* relevance. Under oxidative stress, cellular L-cysteine pools may become transiently depleted as L-cysteine is used for GSH synthesis or to replace damaged thiols in oxidized proteins. Recently, it was also shown that L-cysteine biosynthesis is downregulated under iron starvation stress by RyhB to divert L-serine for siderophore production [126]. L-Serine is a common intermediate in both L-cysteine and enterobactin biosynthesis pathways. Thus, Suf may be better adapted than Isc to acquire sulfide under stress conditions in which the L-cysteine substrate is limiting (Fig. 12.2).

## 12.8 Stress-responsive iron donation for the Suf pathway

The complexity of *in vivo* iron homeostasis poses a significant challenge for understanding iron trafficking to Fe-S cluster biogenesis pathways. Iron is a required nutrient in *E. coli* (with total cellular iron content typically in the hundreds of micromolar range depending on environmental iron levels) [127]. Identification of an *in vivo* iron donor for Suf is complicated by a number of obstacles. Confirming a role for an iron donor *in vitro* is difficult because scaffold proteins can easily incorporate weakly chelated iron into nascent clusters. Any passive source of iron ions will "donate" iron to cluster assembly *in vitro*, making it difficult to evaluate the physiological significance of previous studies [27, 31, 128, 129]. Based on the role of Suf during iron starvation and oxidative stress conditions that disrupt iron homeostasis, it is reasonable to posit that *E. coli* Suf must acquire iron from a protected source that is resistant to these stresses. At present, the *in vivo* iron donation process is still under study. Here we will summarize the likely suspects for Suf iron donation and discuss the evidence for and against their role in iron trafficking to the Suf pathway.

### 12.8.1 SufD

The SufB, SufC, and SufD proteins are localized in the cytoplasm and interact as a stable $SufBC_2D$ complex (Fig. 12.6) [58, 116, 130–134]. The $SufBC_2D$ complex forms a novel Fe-S scaffold upon which iron and sulfide assemble into a stable cluster prior to transfer to target Fe-S proteins [58, 134, 135]. SufB is the final destination of iron and the site of cluster assembly, SufC has intrinsic ATPase activity, and the role of SufD in cluster assembly is not fully characterized. SufD is homologous to SufB (26% identical, 45% similar over the C-terminal ~150 residues) and likely resulted from a gene duplication of SufB. However, SufD does not appear to coordinate an Fe-S cluster on its own. $SufB_2C_2$ and $SufC_2D_2$ complexes have also been isolated, but the role of these sub-complexes has not been fully studied [59, 133, 133, 136]. We found that the absence of SufD diminishes *in vivo* iron incorporation into SufB with only a modest effect on

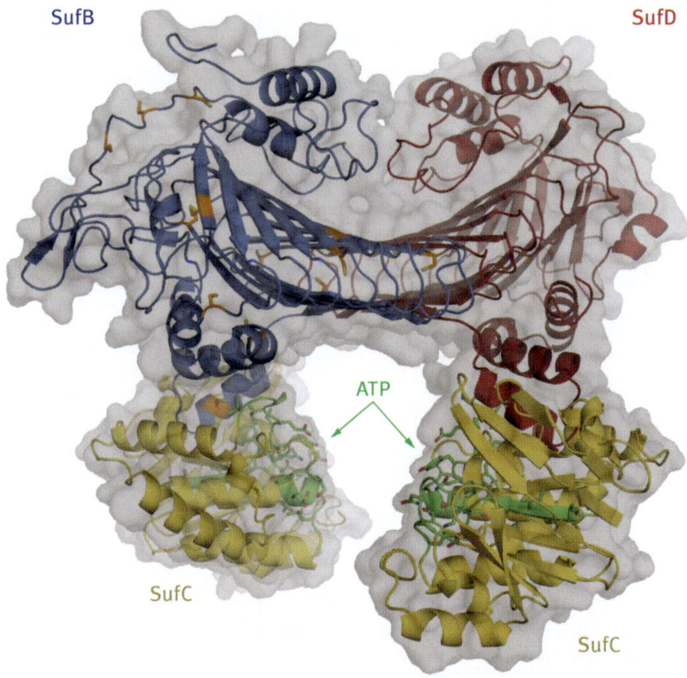


**Fig. 12.6:** Model structure of complete $SufBC_2D$ structure. SufB is in blue, SufD is in red, and SufC monomers are in yellow. ATP binding sites on SufC are highlighted in green in stick representations. All Cys residues in SufB are highlighted in orange in stick representations. Surface representation is in transparent gray. SufB was modeled onto one SufD subunit from the $SufC_2D_2$ complex (PDB 2ZU0) [130].

sulfide, suggesting that SufD may be required for iron acquisition [136]. A selective depletion of iron on SufB was also observed when SufC ATPase activity was altered by a point mutation that did not otherwise alter $SufBC_2D$ interactions. Based on these results, we proposed that SufC ATPase activity works in concert with SufD to facilitate iron delivery into the Suf pathway [136]. The iron donation step may require energy in the form of ATP hydrolysis to obtain iron and concentrate it for cluster assembly against an unfavorable concentration gradient.

The $SufBC_2D$ complex is isolated associated with 1 equiv. of $FADH_2$ per complex [134, 136]. All three proteins are required for the full binding of $FADH_2$ with a $K_d$ for $FADH_2$ of 12 μM. Reduced flavins are very efficient ferric iron-reducing agents and $SufBC_2D$-$FADH_2$ was shown to mobilize iron from ferric citrate and the iron-bound bacterial frataxin CyaY, presumably *via* a reductive mechanism [134]. It was proposed that $SufBC_2D$ uses $FADH_2$ as a redox cofactor to mobilize iron for the Suf pathway although the *in vivo* iron donor remains unclear. CyaY does not appear to play a direct role in Suf mediated Fe-S cluster biogenesis and is instead closely linked to the Isc system (W. Outten, unpublished results) [137–139].

### 12.8.2 Iron storage proteins

*E. coli* contains four potential iron storage proteins that could release iron for Suf cluster assembly: ferritin A (FtnA), ferritin B (FtnB), heme-binding bacterioferritin (Bfr), and "DNA-binding protein from starved cells" (Dps) [140]. FtnA, Bfr, and Dps form multisubunit protein shells that oxidize $Fe^{2+}$ to $Fe^{3+}$ to store iron as ferric hydroxide cores. Stored iron must be reduced to $Fe^{2+}$ for iron to be mobilized for cell use. Bfr and FtnA are similar, and each forms a 12-faced polyhedron structure consisting of 24 subunits. Both proteins also have a similar ferroxidase center that utilizes $O_2$ to catalyze ferrous iron oxidation during storage. One key difference between FtnA and Bfr is that Bfr incorporates twelve inter-subunit heme groups that are not present in FtnA [141–143]. Kinetic profiles of iron release from *E. coli* and *A. vinelandii* Bfr indicate that heme increases iron release by 2- to 4-fold using $FMNH_2$ as the reductant [144]. It is not clear if FtnB is functionally similar to the other ferritins because it lacks the residues critical for ferroxidase activity and those residues important for multimeric protein assembly [145]. However, a role for FtnB in some step of Fe-S cluster repair was observed in *Salmonella* [146].

In contrast to FtnA, and Bfr, Dps quaternary structure only consists of 12 subunits, leading to a smaller overall capacity for iron storage (approximately 500 iron atoms compared with 4,500 iron atoms). Dps also contains a novel ferroxidase center that utilizes $H_2O_2$ to oxidize $Fe^{2+}$ in a controlled reaction that produces $H_2O$ rather than hydroxyl radicals (in contrast to the Fenton reaction) [147–149]. Dps also has the additional ability to bind to DNA, possibly to protect DNA from Fenton chemistry through its iron sequestration and $H_2O_2$ reduction activities [150]. Dps is highly regulated in response to oxidative stress as well as during stationary phase in *E. coli* [151, 152]. At present, it is not clear if iron stored in Dps can be directly used as an iron source for metallocofactor assembly or if Dps is exclusively an iron detoxification system. However, its co-regulation with the Suf pathway under oxidative stress suggests it could serve as an iron source for that pathway.

In *E. coli*, transcription of both *bfr* and *ftnA* increases under iron-replete conditions [153]. For *bfr*, it is not entirely clear how this regulation occurs, although it may be a target of the small RNA RyhB [94]. In the case of *ftnA*, this regulation is mediated through a complex regulatory mechanism whereby $Fe^{2+}$-Fur binds upstream of the *ftnA* promoter and blocks H-NS repression of *ftnA* [154]. By relieving H-NS repression of *ftnA*, $Fe^{2+}$-Fur indirectly activates *ftnA* transcription under high iron conditions. Consequently, FtnA stores up to 50% of total cellular iron under iron-sufficient growth conditions, with the majority of *ftnA* expression occurring during the post-exponential phase of growth [155]. In contrast, Bfr accounts for only 1% of total cellular iron under iron-replete conditions in *E. coli*. However, when cells are grown under more moderate iron levels (6 μM in minimal media), the relationship between Bfr and FtnA changes. Under low to moderate iron conditions, Bfr accounts for approximately 14% of protein-bound iron, whereas FtnA accounts for only 5% [156]. Thus, it appears that FtnA is primarily a storage protein for conditions of iron excess, whereas Bfr may be preferentially used for iron storage in growth conditions with low to moderate iron availability. Bfr expression is also controlled by the nutrient starvation regulator RpoS, such that it is likely to be more highly expressed in stationary or lag phase growth conditions.

The complex regulation of FtnA and Bfr complicates the interpretation of their physiological roles in iron trafficking. However, genetic evidence indirectly links the Suf pathway to Bfr in the Gram-negative plant pathogen *Erwinia chrysanthemi* [157]. In *E. chrysanthemi* strains lacking the *sufC* gene, iron release from Bfr is somehow impaired leading to iron accumulation in Bfr. In contrast, deletions of *sufA*, *sufB*, *sufD*, *sufS*, or *sufE* reduced the amount of iron loaded into Bfr *in vivo*. The results suggest that the Suf proteins directly or indirectly influence iron cycling through Bfr. Because of these phenotypes, it is tempting to speculate that Suf directly acquires iron from Bfr or another iron storage protein to bypass the labile iron pool. A dedicated iron trafficking pathway from storage proteins to Suf could be advantageous under iron starvation and oxidative stress conditions in which the labile pool of iron will be diminished.

### 12.8.3 Other candidates

YtfE contains a di-iron center and plays a role in *in vivo* Fe-S cluster repair after nitrosative and oxidative stresses. Deletion of *ytfE* caused defects in Fe-S enzyme activity in *E. coli*. Addition of holo-YtfE to cell lysates promoted the repair of the [4Fe-4S] clusters of fumarase and aconitase. The results suggest YtfE donates iron for the repair of damaged [4Fe-4S] clusters *in vivo* [158–162]. To date, no results have been published indicating a direct relationship between Suf and YtfE.

YggX (*yggX*) and GSH contribute in an overlapping manner to the maintenance of labile iron homeostasis in *Salmonella enterica* [139, 163–165]. YggX is a small soluble protein that is regulated in response to oxidative stress [166]. Deletion of *yggX* in *E. coli* or *S. enterica* leads to specific defects in iron and Fe-S cluster metabolism [163, 164, 166]. It was hypothesized that YggX may be an intracellular iron chaperone; however, subsequent studies suggest YggX cannot bind iron *in vitro* [163, 167]. It is clear that YggX plays some role in Fe-S metabolism, but the exact function of YggX in this process is not known. GSH is present at millimolar levels in the cell and is considered the main cellular reductant. GSH can chelate metals as well as reduce ferric iron and has been proposed as an iron chelator. It was also proposed that GSH facilitates binding of iron by Fur [165]. A *gshA* deletion mutant that is unable to synthesize GSH showed sensitivity to $H_2O_2$ and impairment of Fe-S cluster repair [165]. The phenotype of a *gshA* deletion is aggravated when combined with the *yggX* gene deletion. An *yggX gshA* double-deletion mutant strain showed enhanced sensitivity to streptonigrin and further increased sensitivity to $H_2O_2$ [165]. These results suggest that the pool of iron available for participation in Fenton chemistry increased. Interestingly, the *yggX gshA cyaY* triple mutant strain has a strong phenotype, with increased nutritional auxotrophies and decreased activity of iron-containing and Fe-S enzymes [139]. At present, it is not clear if YggX (and GSH) are directly important for iron trafficking or if their effects on Fe-S cluster metabolism are indirectly caused by increased oxidative stress.

## 12.9 Unanswered questions about Suf and Isc roles in *E. coli*

There is growing evidence that Suf mobilization of iron and sulfide is more robust than Isc under stress conditions. However, there are other steps of cluster biogenesis where Suf and Isc may diverge in function. The inherent flexibility of the IscU scaffold protein likely renders it sensitive to disruption, especially by oxidation of exposed thiol ligands. It is possible that SufB and IscU are not equally sensitive to stress. SufB does not appear to catalyze cluster assembly *via* a stable $[2Fe\text{-}2S]^{2+}$ intermediate as observed for IscU, although $[4Fe\text{-}4S]^{2+}$ SufB can by converted into the $[2Fe\text{-}2S]^{2+}$ form by air exposure [135, 136]. Interestingly, SufB contains some linear [3Fe-4S] cluster (in addition to $[4Fe\text{-}4S]^{2+}$) after *in vivo* expression with SufCDSE [136]. The linear [3Fe-4S] cluster may represent a cluster intermediate or a cluster degradation product. Regardless, it is easily convertible back to the $[4Fe\text{-}4S]^{2+}$ form by the addition of iron and one electron [168, 169]. Possibly, the ability to stabilize this particular cluster type provides some advantage under oxidative stress conditions.

The Fe-S cluster trafficking steps may also be sensitive to disruption, but here, there is less divergence between Suf and Isc. Both pathways encode an ATC protein (SufA and IscA) that appear to overlap considerably in their *in vivo* function and *in vitro* biochemical characteristics [61, 62, 109, 170–172]. Although the downstream targets of the ATC proteins are likely to be specific for each ATC, the ATCs seem to share some ability to interact with both SufB and IscU scaffolds. Monothiol glutaredoxins may be used for Suf stress-responsive cluster trafficking, but the ability of the Grx4 to interface with Suf and Isc has not been carefully compared [173, 174]. It is likely that Suf and Isc differentially interact with currently uncharacterized pieces of the Fe-S cluster metabolism machinery *in vivo*. Future work in this field should help delineate the complex relationship between housekeeping and stress-responsive Fe-S cluster biogenesis.

## Acknowledgment

Work in the laboratory of F. Wayne Outten was supported by NIH GM 81706.

## References


[1] Dupont CL, Butcher A, Valas RE, Bourne PE, Caetano-Anolles G. History of biological metal utilization inferred through phylogenomic analysis of protein structures. Proc Natl Acad Sci USA 2010;107:10567–72.

[2] Meyer J. Iron-sulfur protein folds, iron-sulfur chemistry, and evolution. J Biol Inorg Chem 2008;13:157–70.

[3] Hagen KS, Reynolds JG, Holm RH. Definition of reaction sequences resulting in self-assembly of $[Fe_4S_4(Sr)4]^{2-}$ clusters from simple reactants. J Am Chem Soc 1981;103:4054–63.

[4] Hagen KS, Watson AD, Holm RH. Synthetic routes to $Fe_2S_2$, $Fe_3S_4$, $Fe_4S_4$, and $Fe_6S_9$ clusters from the common precursor $[Fe(Sc_2H_5)4]^{2-}$ – structures and properties of $[Fe_3S_4(Sr)4]^{3-}$ and $[Fe_6S_9(Sc_2H_5)2]^{4-}$, examples of the newest types of Fe-S-Sr Clusters. J Am Chem Soc 1983;105:3905–13.

[5] Beinert H, Holm RH, Munck E. Iron-sulfur clusters: nature's modular, multipurpose structures. Science 1997;277:653–9.

[6] Venkateswara Rao P, Holm RH. Synthetic analogues of the active sites of iron-sulfur proteins. Chem Rev 2004;104:527–59.

[7] Liu Y, Beer LL, Whitman WB. Methanogens: a window into ancient sulfur metabolism. Trends Microbiol 2012.

[8] Yang J, Bitoun JP, Ding H. Interplay of IscA and IscU in biogenesis of iron-sulfur clusters. J Biol Chem 2006;281:27956–63.

[9] Dai Z, Tonelli M, Markley JL. Metamorphic protein IscU changes conformation by cis-trans isomerizations of two peptidyl-prolyl peptide bonds. Biochemistry 2012;51:9595–602.

[10] Kim JH, Tonelli M, Kim T, Markley JL. Three-dimensional structure and determinants of stability of the iron-sulfur cluster scaffold protein IscU from Escherichia coli. Biochemistry 2012;51:5557–63.

[11] Kim JH, Tonelli M, Markley JL. Disordered form of the scaffold protein IscU is the substrate for iron-sulfur cluster assembly on cysteine desulfurase. Proc Natl Acad Sci USA 2012;109:454–9.

[12] Anbar AD. Oceans elements and evolution. Science 2008;322:1481–3.

[13] Anbar AD, Knoll AH. Proterozoic ocean chemistry and evolution: a bioinorganic bridge? Science 2002;297:1137–42.

[14] Kim KM, Qin T, Jiang YY, et al. Protein domain structure uncovers the origin of aerobic metabolism and the rise of planetary oxygen. Structure 2012;20:67–76.

[15] Zheng L, White RH, Cash VL, Dean DR. Mechanism for the desulfurization of L-cysteine catalyzed by the nifS gene product. Biochemistry 1994;33:4714–20.

[16] Zheng L, White RH, Cash VL, Jack RF, Dean DR. Cysteine desulfurase activity indicates a role for NIFS in metallocluster biosynthesis. Proc Natl Acad Sci USA 1993;90:2754–8.

[17] Kim JH, Frederick RO, Reinen NM, Troupis AT, Markley JL. [2Fe-2S]-ferredoxin binds directly to cysteine desulfurase and supplies an electron for iron-sulfur cluster assembly but is displaced by the scaffold protein or bacterial frataxin. J Am Chem Soc 2013;135:8117–20.

[18] Yan R, Konarev PV, Iannuzzi C, et al. Ferredoxin competes with bacterial frataxin in binding to the desulfurase IscS. J Biol Chem 2013;288:24777–87.

[19] Marinoni EN, de Oliveira JS, Nicolet Y, et al. (IscS-IscU)2 complex structures provide insights into $Fe_2S_2$ biogenesis and transfer. Angew Chem Int Ed Engl 2012;51:5439–42.

[20] Adinolfi S, Rizzo F, Masino L, et al. Bacterial IscU is a well folded and functional single domain protein. Eur J Biochem 2004;271:2093–100.

[21] Nuth M, Yoon T, Cowan JA. Iron-sulfur cluster biosynthesis: characterization of iron nucleation sites for assembly of the $[2Fe\text{-}2S]^{2+}$ cluster core in IscU proteins. J Am Chem Soc 2002;124:8774–5.

[22] Smith AD, Agar JN, Johnson KA, et al. Sulfur transfer from IscS to IscU: the first step in iron-sulfur cluster biosynthesis. J Am Chem Soc 2001;123:11103–4.

[23] Lin SJ, Pufahl RA, Dancis A, O'Halloran TV, Culotta VC. A role for the Saccharomyces cerevisiae ATX1 gene in copper trafficking and iron transport. J Biol Chem 1997;272:9215–20.
[24] Pufahl RA, Singer CP, Peariso KL, et al. Metal ion chaperone function of the soluble Cu(I) receptor Atx1. Science 1997;278:853–6.
[25] Robinson NJ, Winge DR. Copper metallochaperones. Annu Rev Biochem 2010;79:537–62.
[26] Bou-Abdallah F, Adinolfi S, Pastore A, Laue TM, Dennis Chasteen N. Iron binding and oxidation kinetics in frataxin CyaY of Escherichia coli. J Mol Biol 2004;341:605–15.
[27] Banci L, Bertini I, Ciofi-Baffoni S, et al. Human Sco1 functional studies and pathological implications of the P174L mutant. Proc Natl Acad Sci USA 2007;104:15–20.
[28] Changela A, Chen K, Xue Y, et al. Molecular basis of metal-ion selectivity and zeptomolar sensitivity by CueR. Science 2003;301:1383–7.
[29] Nair M, Adinolfi S, Pastore C, Kelly G, Temussi P, Pastore A. Solution structure of the bacterial frataxin ortholog, CyaY: mapping the iron binding sites. Structure 2004;12:2037–48.
[30] Layer G, Ollagnier de Choudens S, Sanakis Y, Fontecave M. Iron-sulfur cluster biosynthesis: characterization of Escherichia coli CyaY as an iron donor for the assembly of [2Fe-2S] clusters in the scaffold IscU. J Biol Chem 2006;281:16256–63.
[31] Adinolfi S, Iannuzzi C, Prischi F, et al. Bacterial frataxin CyaY is the gatekeeper of iron-sulfur cluster formation catalyzed by IscS. Nat Struct Mol Biol 2009;16:390–6.
[32] Prischi F, Konarev PV, Iannuzzi C, et al. Structural bases for the interaction of frataxin with the central components of iron-sulphur cluster assembly. Nat Commun 2010;1:95.
[33] Iannuzzi C, Adinolfi S, Howes BD, et al. The role of CyaY in iron sulfur cluster assembly on the E. coli IscU scaffold protein. PLoS One 2011;6:e21992.
[34] Bridwell-Rabb J, Iannuzzi C, Pastore A, Barondeau DP. Effector role reversal during evolution: the case of frataxin in Fe-S cluster biosynthesis. Biochemistry 2012;51:2506–14.
[35] Leidgens S, Bullough KZ, Shi H, et al. Each member of the poly-r(C)-binding protein 1 (PCBP) family exhibits iron chaperone activity toward ferritin. J Biol Chem 2013;288:17791–802.
[36] Nandal A, Ruiz JC, Subramanian P, et al. Activation of the HIF prolyl hydroxylase by the iron chaperones PCBP1 and PCBP2. Cell Metab 2011;14:647–57.
[37] Shi H, Bencze KZ, Stemmler TL, Philpott CC. A cytosolic iron chaperone that delivers iron to ferritin. Science 2008;320:1207–10.
[38] Rae TD, Schmidt PJ, Pufahl RA, Culotta VC, O'Halloran TV. Undetectable intracellular free copper: the requirement of a copper chaperone for superoxide dismutase. Science 1999;284:805–8.
[39] Rae TD, Schmidt PJ, Pufahl RA, Culotta VC, O'Halloran TV. The function of the copper chaperone for Cu,Zn-superoxide dismutase implies a negligible concentration of intracellular free copper. J Inorg Biochem 1999;74:271.
[40] Macomber L, Imlay JA. The iron-sulfur clusters of dehydratases are primary intracellular targets of copper toxicity. Proc Natl Acad Sci USA 2009;106:8344–9.
[41] Chillappagari S, Seubert A, Trip H, Kuipers OP, Marahiel MA, Miethke M. Copper stress affects iron homeostasis by destabilizing iron-sulfur cluster formation in Bacillus subtilis. J Bacteriol 2010;192:2512–24.
[42] Fu W, Jack RF, Morgan TV, Dean DR, Johnson MK. nifU gene product from Azotobacter vinelandii is a homodimer that contains two identical [2Fe-2S] clusters. Biochemistry 1994;33:13455–63.
[43] Agar JN, Krebs C, Frazzon J, Huynh BH, Dean DR, Johnson MK. IscU as a scaffold for iron-sulfur cluster biosynthesis: sequential assembly of [2Fe-2S] and [4Fe-4S] clusters in IscU. Biochemistry 2000;39:7856–62.
[44] Agar JN, Zheng L, Cash VL, Dean DR, Johnson MK. Role of the IscU Protein in Iron-Sulfur Cluster Biosynthesis: IscS-mediated Assembly of a 2Fe-2S Cluster in IscU. J Am Chem Soc 2000;122:2136–7.
[45] Yuvaniyama P, Agar JN, Cash VL, Johnson MK, Dean DR. NifS-directed assembly of a transient [2Fe-2S] cluster within the NifU protein. Proc Natl Acad Sci USA 2000;97:599–604.

[46] Smith AD, Jameson GN, Dos Santos PC, et al. NifS-mediated assembly of [4Fe-4S] clusters in the N- and C-terminal domains of the NifU scaffold protein. Biochemistry 2005;44:12955–69.
[47] Chandramouli K, Unciuleac MC, Naik S, Dean DR, Huynh BH, Johnson MK. Formation and properties of [4Fe-4S] clusters on the IscU scaffold protein. Biochemistry 2007;46:6804–11.
[48] Cupp-Vickery JR, Peterson JC, Ta DT, Vickery LE. Crystal structure of the molecular chaperone HscA substrate binding domain complexed with the IscU recognition peptide ELPPVKIHC. J Mol Biol 2004;342:1265–78.
[49] Silberg JJ, Tapley TL, Hoff KG, Vickery LE. Regulation of the HscA ATPase reaction cycle by the co-chaperone HscB and the iron-sulfur cluster assembly protein IscU. J Biol Chem 2004;279:53924–31.
[50] Chandramouli K, Johnson MK. HscA and HscB stimulate [2Fe-2S] cluster transfer from IscU to apoferredoxin in an ATP-dependent reaction. Biochemistry 2006;45:11087–95.
[51] Vickery LE, Cupp-Vickery JR. Molecular chaperones HscA/Ssq1 and HscB/Jac1 and their roles in iron-sulfur protein maturation. Crit Rev Biochem Mol Biol 2007;42:95–111.
[52] Bonomi F, Iametti S, Morleo A, Ta D, Vickery LE. Studies on the mechanism of catalysis of iron-sulfur cluster transfer from IscU[2Fe2S] by HscA/HscB chaperones. Biochemistry 2008;47:12795–801.
[53] Bonomi F, Iametti S, Morleo A, Ta D, Vickery LE. Facilitated transfer of IscU-[2Fe2S] clusters by chaperone-mediated ligand exchange. Biochemistry 2011;50:9641–50.
[54] Silberg JJ, Hoff KG, Tapley TL, Vickery LE. The Fe/S assembly protein IscU behaves as a substrate for the molecular chaperone Hsc66 from Escherichia coli. J Biol Chem 2001;276:1696–700.
[55] Silberg JJ, Vickery LE. Kinetic characterization of the ATPase cycle of the molecular chaperone Hsc66 from Escherichia coli. J Biol Chem 2000;275:7779–86.
[56] Unciuleac MC, Chandramouli K, Naik S, et al. In vitro activation of apo-aconitase using a [4Fe-4S] cluster-loaded form of the IscU [Fe-S] cluster scaffolding protein. Biochemistry 2007;46:6812–21.
[57] Chahal HK, Dai Y, Saini A, Ayala-Castro C, Outten FW. The SufBCD Fe-S scaffold complex interacts with SufA for Fe-S cluster transfer. Biochemistry 2009;48:10644–53.
[58] Chahal HK, Outten FW. J Inorg Biochem 2012;116:126–34.
[59] Gupta V, Sendra M, Naik SG, et al. Native Escherichia coli SufA, coexpressed with SufBCDSE, purifies as a [2Fe-2S] protein and acts as an Fe-S transporter to Fe-S target enzymes. J Am Chem Soc 2009;131:6149–53.
[60] Loiseau L, Gerez C, Bekker M, et al. ErpA, an iron sulfur (Fe-S) protein of the A-type essential for respiratory metabolism in Escherichia coli. Proc Natl Acad Sci USA 2007;104:13626–31.
[61] Lu J, Yang J, Tan G, Ding H. Complementary roles of SufA and IscA in the biogenesis of iron-sulfur clusters in Escherichia coli. Biochem J 2008;409:535–43.
[62] Vinella D, Brochier-Armanet C, Loiseau L, Talla E, Barras F. Iron-sulfur (Fe/S) protein biogenesis: phylogenomic and genetic studies of A-type carriers. PLoS Genet 2009;5:e1000497.
[63] Wollenberg M, Berndt C, Bill E, Schwenn JD, Seidler A. A dimer of the FeS cluster biosynthesis protein IscA from cyanobacteria binds a [2Fe-2S] cluster between two protomers and transfers it to [2Fe-2S] and [4Fe-4S] apo proteins. Eur J Biochem 2003;270:1662–71.
[64] Rodriguez-Manzaneque MT, Tamarit J, Belli G, Ros J, Herrero E. Grx5 is a mitochondrial glutaredoxin required for the activity of iron/sulfur enzymes. Mol Biol Cell 2002;13:1109–21.
[65] Achebach S, Tran QH, Vlamis-Gardikas A, Mullner M, Holmgren A, Unden G. Stimulation of Fe-S cluster insertion into apoFNR by Escherichia coli glutaredoxins 1:2 and 3 in vitro. FEBS Lett 2004;565:203–6.
[66] Rouhier N, Unno H, Bandyopadhyay S, et al. Functional, structural, and spectroscopic characterization of a glutathione-ligated [2Fe-2S] cluster in poplar glutaredoxin C1. Proc Natl Acad Sci USA 2007;104:7379–84.

[67] Bandyopadhyay S, Gama F, Molina-Navarro MM, et al. Chloroplast monothiol glutaredoxins as scaffold proteins for the assembly and delivery of [2Fe-2S] clusters. EMBO J 2008;27:1122–33.
[68] Iwema T, Picciocchi A, Traore DA, Ferrer JL, Chauvat F, Jacquamet L. Structural basis for delivery of the intact [$Fe_2S_2$] cluster by monothiol glutaredoxin. Biochemistry 2009;48:6041–3.
[69] Pinske C, Sawers RG. A-type carrier protein ErpA is essential for formation of an active formate-nitrate respiratory pathway in Escherichia coli K-12. J Bacteriol 2012;194:346–53.
[70] Vinella D, Loiseau L, Ollagnier de Choudens S, Fontecave M, Barras F. In vivo [Fe-S] cluster acquisition by IscR and NsrR, two stress regulators in Escherichia coli. Mol Microbiol 2013;87:493–508.
[71] Giel JL, Nesbit AD, Mettert EL, Fleischhacker AS, Wanta BT, Kiley PJ. Regulation of iron-sulphur cluster homeostasis through transcriptional control of the Isc pathway by [2Fe-2S]-IscR in Escherichia coli. Mol Microbiol 2013;87:478–92.
[72] Ollagnier-de-Choudens S, Sanakis Y, Fontecave M. SufA/IscA: reactivity studies of a class of scaffold proteins involved in [Fe-S] cluster assembly. J Biol Inorg Chem 2004;9:828–38.
[73] Mapolelo DT, Zhang B, Naik SG, Huynh BH, Johnson MK. Spectroscopic and functional characterization of iron-sulfur cluster-bound forms of Azotobacter vinelandii (Nif)IscA. Biochemistry 2012;51:8071–84.
[74] Zhang B, Bandyopadhyay S, Shakamuri P, et al. Monothiol glutaredoxins can bind linear [$Fe_3S_4$]$^+$ and [$Fe_4S_4$]$^{2+}$ clusters in addition to [$Fe_2S_2$]$^{2+}$ clusters: spectroscopic characterization and functional implications. J Am Chem Soc 2013;135:15153–64.
[75] Flint DH, Tuminello JF, Emptage MH. The inactivation of Fe-S cluster containing hydro-lyases by superoxide. J Biol Chem 1993;268:22369–76.
[76] Djaman O, Outten FW, Imlay JA. Repair of oxidized iron-sulfur clusters in Escherichia coli. J Biol Chem 2004;279:44590–9.
[77] Imlay JA. Iron-sulphur clusters and the problem with oxygen. Mol Microbiol 2006;59:1073–82.
[78] Keyer K, Imlay JA. Superoxide accelerates DNA damage by elevating free-iron levels. Proc Natl Acad Sci USA 1996;93:13635–40.
[79] Keyer K, Imlay JA. Inactivation of dehydratase [4Fe-4S] clusters and disruption of iron homeostasis upon cell exposure to peroxynitrite. J Biol Chem 1997;272:27652–9.
[80] Varghese S, Tang Y, Imlay JA. Contrasting sensitivities of Escherichia coli aconitases A and B to oxidation and iron depletion. J Bacteriol 2003;185:221–30.
[81] Gardner PR, Fridovich I. Superoxide sensitivity of the Escherichia coli aconitase. J Biol Chem 1991;266:19328–33.
[82] Gardner PR, Fridovich I. Superoxide sensitivity of the Escherichia coli 6-phosphogluconate dehydratase. J Biol Chem 1991;266:1478–83.
[83] Gardner PR, Fridovich I. Inactivation-reactivation of aconitase in Escherichia coli. A sensitive measure of superoxide radical. J Biol Chem 1992;267:8757–63.
[84] Leonard SE, Carroll KS. Chemical 'omics' approaches for understanding protein cysteine oxidation in biology. Curr Opin Chem Biol 2011;15:88–102.
[85] Gari K, Ortiz AML, Borel V, Flynn H, Skehel JM, Boulton SJ. MMS19 links cytoplasmic iron-sulfur cluster assembly to DNA metabolism. Science 2012;337:243–45.
[86] Amir A, Meshner S, Beatus T, Stavans J. Damped oscillations in the adaptive response of the iron homeostasis network of E. coli. Mol Microbiol 2010;76:428–36.
[87] Andrews SC, Robinson AK, Rodriguez-Quinones F. Bacterial iron homeostasis. FEMS Microbiol Rev 2003;27:215–37.
[88] Semsey S, Andersson AM, Krishna S, Jensen MH, Masse E, Sneppen K. Genetic regulation of fluxes: iron homeostasis of Escherichia coli. Nucleic Acids Res 2006;34:4960–7.
[89] Hantke K. Regulation of ferric iron transport in Escherichia coli K12: isolation of a constitutive mutant. Mol Gen Genet 1981;182:288–92.
[90] Outten FW, Djaman O, Storz G. A suf operon requirement for Fe-S cluster assembly during iron starvation in Escherichia coli. Mol Microbiol 2004;52:861–72.

[91] Patzer SI, Hantke K. SufS is a NifS-like protein, and SufD is necessary for stability of the [2Fe-2S] FhuF protein in Escherichia coli. J Bacteriol 1999;181:3307–9.
[92] Masse E, Escorcia FE, Gottesman S. Coupled degradation of a small regulatory RNA and its mRNA targets in Escherichia coli. Genes Dev 2003;17:2374–83.
[93] Masse E, Gottesman S. A small RNA regulates the expression of genes involved in iron metabolism in Escherichia coli. Proc Natl Acad Sci USA 2002;99:4620–5.
[94] Masse E, Vanderpool CK, Gottesman S. Effect of RyhB small RNA on global iron use in Escherichia coli. J Bacteriol 2005;187:6962–71.
[95] Maringanti S, Imlay JA. An intracellular iron chelator pleiotropically suppresses enzymatic and growth defects of superoxide dismutase-deficient Escherichia coli. J Bacteriol 1999;181:3792–802.
[96] Woodmansee AN, Imlay JA. Reduced flavins promote oxidative DNA damage in non-respiring Escherichia coli by delivering electrons to intracellular free iron. J Biol Chem 2002;277:34055–66.
[97] Varghese S, Wu A, Park S, Imlay KR, Imlay JA. Submicromolar hydrogen peroxide disrupts the ability of Fur protein to control free-iron levels in Escherichia coli. Mol Microbiol 2007;64:822–30.
[98] Takahashi Y, Tokumoto U. A third bacterial system for the assembly of iron-sulfur clusters with homologs in archaea and plastids. J Biol Chem 2002;277:28380–3.
[99] Schwartz CJ, Giel JL, Patschkowski T, et al. IscR, an Fe-S cluster-containing transcription factor, represses expression of Escherichia coli genes encoding Fe-S cluster assembly proteins. Proc Natl Acad Sci USA 2001;98:14895–900.
[100] Giel JL, Rodionov D, Liu M, et al. IscR-dependent gene expression links iron-sulphur cluster assembly to the control of $O_2$-regulated genes in Escherichia coli. Mol Microbiol 2006;60:1058–75.
[101] Yeo WS, Lee JH, Lee KC, Roe JH. IscR acts as an activator in response to oxidative stress for the suf operon encoding Fe-S assembly proteins. Mol Microbiol 2006;61:206–18.
[102] Fleischhacker AS, Stubna A, Hsueh KL, et al. Characterization of the [2Fe-2S] cluster of Escherichia coli transcription factor IscR. Biochemistry 2012;51:4453–62.
[103] Nesbit AD, Giel JL, Rose JC, Kiley PJ. Sequence-specific binding to a subset of IscR-regulated promoters does not require IscR Fe-S cluster ligation. J Mol Biol 2009;387:28–41.
[104] Desnoyers G, Morissette A, Prevost K, Masse E. Small RNA-induced differential degradation of the polycistronic mRNA iscRSUA. EMBO J 2009;28:1551–61.
[105] Lee JH, Yeo WS, Roe JH. Induction of the sufA operon encoding Fe-S assembly proteins by superoxide generators and hydrogen peroxide: involvement of OxyR, IHF and an unidentified oxidant-responsive factor. Mol Microbiol 2004;51:1745–55.
[106] Nachin L, Loiseau L, Expert D, Barras F. SufC: an unorthodox cytoplasmic ABC/ATPase required for [Fe-S] biogenesis under oxidative stress. EMBO J 2003;22:427–37.
[107] Tokumoto U, Kitamura S, Fukuyama K, Takahashi Y. Interchangeability and distinct properties of bacterial Fe-S cluster assembly systems: functional replacement of the isc and suf operons in Escherichia coli with the nifSU-like operon from Helicobacter pylori. J Biochem (Tokyo) 2004;136:199–209.
[108] Ranquet C, Ollagnier de Choudens S, Loiseau L, Barras F, Fontecave M. Cobalt stress in Escherichia coli: the effect on the iron-sulfur proteins. J Biol Chem 2007;282:30442–51.
[109] Tan G, Lu J, Bitoun JP, Huang H, Ding H. IscA/SufA paralogues are required for the [4Fe-4S] cluster assembly in enzymes of multiple physiological pathways in Escherichia coli under aerobic growth conditions. Biochem J 2009;420:463–72.
[110] Jang S, Imlay JA. Hydrogen peroxide inactivates the Escherichia coli Isc iron-sulphur assembly system, and OxyR induces the Suf system to compensate. Mol Microbiol 2010;78:1448–67.
[111] Lima CD. Analysis of the E. coli NifS CsdB protein at 2.0 Å reveals the structural basis for perselenide and persulfide intermediate formation. J Mol Biol 2002;315:1199–208.
[112] Ye Y, Godzik A. FATCAT: a web server for flexible structure comparison and structure similarity searching. Nucleic Acids Res 2004;32:W582–5.

[113] Selbach BP, Pradhan PK, Dos Santos PC. Protected sulfur transfer reactions by the Escherichia coli Suf system. Biochemistry 2013;52:4089–96.
[114] Turowski VR, Busi MV, Gomez-Casati DF. Structural and functional studies of the mitochondrial cysteine desulfurase from Arabidopsis thaliana. Mol Plant 2012;5:1001–10.
[115] Tirupati B, Vey JL, Drennan CL, Bollinger JM Jr. Kinetic and structural characterization of Slr0077/SufS, the essential cysteine desulfurase from Synechocystis sp. PCC 6803. Biochemistry 2004;43:12210–19.
[116] Outten FW, Wood MJ, Munoz FM, Storz G. The SufE protein and the SufBCD complex enhance SufS cysteine desulfurase activity as part of a sulfur transfer pathway for Fe-S cluster assembly in Escherichia coli. J Biol Chem 2003;278:45713–9.
[117] Ollagnier-de-Choudens S, Lascoux D, Loiseau L, Barras F, Forest E, Fontecave M. Mechanistic studies of the SufS-SufE cysteine desulfurase: evidence for sulfur transfer from SufS to SufE. FEBS Lett 2003;555:263–7.
[118] Mihara H, Kurihara T, Yoshimura T, Esaki N. Kinetic and mutational studies of three NifS homologs from Escherichia coli: mechanistic difference between L-cysteine desulfurase and L-selenocysteine lyase reactions. J Biochem (Tokyo) 2000;127:559–67.
[119] Mihara H, Maeda M, Fujii T, Kurihara T, Hata Y, Esaki N. A nifS-like gene, csdB, encodes an Escherichia coli counterpart of mammalian selenocysteine lyase. Gene cloning, purification, characterization and preliminary x-ray crystallographic studies. J Biol Chem 1999;274:14768–72.
[120] Dai Y, Outten FW. The E. coli SufS-SufE sulfur transfer system is more resistant to oxidative stress than IscS-IscU. FEBS Lett 2012;586:4016–22.
[121] Goldsmith-Fischman S, Kuzin A, Edstrom WC, et al. The SufE sulfur-acceptor protein contains a conserved core structure that mediates interdomain interactions in a variety of redox protein complexes. J Mol Biol 2004;344:549–65.
[122] Loiseau L, Ollagnier-de-Choudens S, Nachin L, Fontecave M, Barras F. Biogenesis of Fe-S cluster by the bacterial Suf system: SufS and SufE form a new type of cysteine desulfurase. J Biol Chem 2003;278:38352–9.
[123] Kim S, Park S. Structural changes during cysteine desulfurase CsdA and sulfur acceptor CsdE interactions provide insight into the trans-persulfuration. J Biol Chem 2013;288:27172–80.
[124] Trotter V, Vinella D, Loiseau L, Ollagnier de Choudens S, Fontecave M, Barras F. The CsdA cysteine desulphurase promotes Fe/S biogenesis by recruiting Suf components and participates to a new sulphur transfer pathway by recruiting CsdL (ex-YgdL), a ubiquitin-modifying-like protein. Mol Microbiol 2009;74:1527–42.
[125] Singh H, Dai Y, Outten FW, Busenlehner LS. Escherichia coli SufE sulfur transfer protein modulates the SufS cysteine desulfurase through allosteric conformational dynamics. J Biol Chem 2013;288: 36189–200.
[126] Salvail H, Lanthier-Bourbonnais P, Sobota JM, et al. A small RNA promotes siderophore production through transcriptional and metabolic remodeling. Proc Natl Acad Sci USA 2010;107:15223–8.
[127] Outten CE, O'Halloran TV. Femtomolar sensitivity of metalloregulatory proteins controlling zinc homeostasis. Science 2001;292:2488–92.
[128] Ding H, Yang J, Coleman LC, Yeung S. Distinct iron binding property of two putative iron donors for the iron-sulfur cluster assembly: IscA and the bacterial frataxin ortholog CyaY under physiological and oxidative stress conditions. J Biol Chem 2007;282:7997–8004.
[129] Wang W, Huang H, Tan G, et al. In vivo evidence for the iron-binding activity of an iron-sulfur cluster assembly protein IscA in Escherichia coli. Biochemical J 2010;432:429–36.
[130] Nachin L, El Hassouni M, Loiseau L, Expert D, Barras F. SoxR-dependent response to oxidative stress and virulence of Erwinia chrysanthemi: the key role of SufC, an orphan ABC ATPase. Mol Microbiol 2001;39:960–72.
[131] Rangachari K, Davis CT, Eccleston JF, et al. SufC hydrolyzes ATP and interacts with SufB from Thermotoga maritima. FEBS Lett 2002;514:225–8.

[132] Eccleston JF, Petrovic A, Davis CT, Rangachari K, Wilson RJ. The kinetic mechanism of the SufC ATPase: the cleavage step is accelerated by SufB. J Biol Chem 2006;281:8371–8.
[133] Wada K, Sumi N, Nagai R, et al. Molecular dynamism of Fe-S cluster biosynthesis implicated by the structure of SufC2-SufD2 complex. J Mol Biol 2009;387:245–58.
[134] Wollers S, Layer G, Garcia-Serres R, et al. Iron-sulfur (Fe-S) cluster assembly: the SufBCD complex is a new type of Fe-S scaffold with a flavin redox cofactor. J Biol Chem 2010;285:23331–41.
[135] Layer G, Gaddam SA, Ayala-Castro CN, et al. SufE transfers sulfur from SufS to SufB for iron-sulfur cluster assembly. J Biol Chem 2007;282:13342–50.
[136] Saini A, Mapolelo DT, Chahal HK, Johnson MK, Outten FW. SufD and SufC ATPase activity are required for iron acquisition during in vivo Fe-S cluster formation on SufB. Biochemistry 2010;49:9402–12.
[137] Li DS, Ohshima K, Jiralerspong S, Bojanowski MW, Pandolfo M. Knock-out of the cyaY gene in Escherichia coli does not affect cellular iron content and sensitivity to oxidants. FEBS Lett 1999;456:13–6.
[138] Vivas E, Skovran E, Downs DM. Salmonella enterica strains lacking the frataxin homolog CyaY show defects in Fe-S cluster metabolism in vivo. J Bacteriol 2006;188:1175–9.
[139] Thorgersen MP, Downs DM. Oxidative stress and disruption of labile iron generate specific auxotrophic requirements in Salmonella enterica. Microbiology 2009;155:295–304.
[140] Andrews SC. The Ferritin-like superfamily: evolution of the biological iron storeman from a rubrerythrin-like ancestor. Biochim Biophys Acta 2010;1800:691–705.
[141] Cheesman MR, le Brun NE, Kadir FH, et al. Haem and non-haem iron sites in Escherichia coli bacterioferritin: spectroscopic and model building studies. Biochem J 1993;292:47–56.
[142] Le Brun NE, Cheesman MR, Thomson AJ, et al. An EPR investigation of non-haem iron sites in Escherichia coli bacterioferritin and their interaction with phosphate. A study using nitric oxide as a spin probe. FEBS Lett 1993;323:261–6.
[143] Andrews SC, Le Brun NE, Barynin V, et al. Site-directed replacement of the coaxial heme ligands of bacterioferritin generates heme-free variants. J Biol Chem 1995;270:23268–74.
[144] Yasmin S, Andrews SC, Moore GR, Le Brun NE. A new role for heme, facilitating release of iron from the bacterioferritin iron biomineral. J Biol Chem 2011;286:3473–83.
[145] Le Brun NE, Crow A, Murphy ME, Mauk AG, Moore GR. Iron core mineralisation in prokaryotic ferritins. Biochim Biophys Acta 2010;1800:732–44.
[146] Velayudhan J, Castor M, Richardson A, Main-Hester KL, Fang FC. The role of ferritins in the physiology of Salmonella enterica sv. typhimurium: a unique role for ferritin B in iron-sulphur cluster repair and virulence. Mol Microbiol 2007;63:1495–507.
[147] Ilari A, Ceci P, Ferrari D, Rossi GL, Chiancone E. Iron incorporation into Escherichia coli Dps gives rise to a ferritin-like microcrystalline core. J Biol Chem 2002;277:37619–23.
[148] Zhao G, Ceci P, Ilari A, et al. Iron and hydrogen peroxide detoxification properties of DNA-binding protein from starved cells. A ferritin-like DNA-binding protein of Escherichia coli. J Biol Chem 2002;277:27689–96.
[149] Chiancone E, Ceci P. The multifaceted capacity of Dps proteins to combat bacterial stress conditions: detoxification of iron and hydrogen peroxide and DNA binding. Biochim Biophys Acta 2010;1800:798–805.
[150] Martinez A, Kolter R. Protection of DNA during oxidative stress by the nonspecific DNA-binding protein Dps. J Bacteriol 1997;179:5188–94.
[151] Almiron M, Link AJ, Furlong D, Kolter R. A novel DNA-binding protein with regulatory and protective roles in starved Escherichia coli. Genes Dev 1992;6:2646–54.
[152] Altuvia S, Almiron M, Huisman G, Kolter R, Storz G. The dps promoter is activated by OxyR during growth and by IHF and sigma S in stationary phase. Mol Microbiol 1994;13:265–72.
[153] McHugh JP, Rodriguez-Quinones F, Abdul-Tehrani H, et al. Global iron-dependent gene regulation in Escherichia coli. A new mechanism for iron homeostasis. J Biol Chem 2003;278:29478–86.

[154] Nandal A, Huggins CC, Woodhall MR, et al. Induction of the ferritin gene (ftnA) of Escherichia coli by $Fe^{2+}$-Fur is mediated by reversal of H-NS silencing and is RyhB independent. Mol Microbiol 2010;75:637–57.
[155] Abdul-Tehrani H, Hudson AJ, Chang YS, et al. Ferritin mutants of Escherichia coli are iron deficient and growth impaired, and fur mutants are iron deficient. J Bacteriol 1999;181:1415–28.
[156] Sevcenco AM, Pinkse MW, Wolterbeek HT, Verhaert PD, Hagen WR, Hagedoorn PL. Exploring the microbial metalloproteome using MIRAGE. Metallomics 2011;3:1324–30.
[157] Expert D, Boughammoura A, Franza T. Siderophore-controlled iron assimilation in the enterobacterium Erwinia chrysanthemi: evidence for the involvement of bacterioferritin and the Suf iron-sulfur cluster assembly machinery. J Biol Chem 2008;283:36564–72.
[158] Justino MC, Vicente JB, Teixeira M, Saraiva LM. New genes implicated in the protection of anaerobically grown Escherichia coli against nitric oxide. J Biol Chem 2005;280:2636–43.
[159] Justino MC, Almeida CC, Goncalves VL, Teixeira M, Saraiva LM. Escherichia coli YtfE is a di-iron protein with an important function in assembly of iron-sulphur clusters FEMS Microbiol Lett 2006;257:278–84.
[160] Justino MC, Almeida CC, Teixeira M, Saraiva LM. Escherichia coli di-iron YtfE protein is necessary for the repair of stress-damaged iron-sulfur clusters. J Biol Chem 2007;282:10352–9.
[161] Overton TW, Justino MC, Li Y, et al. Widespread distribution in pathogenic bacteria of di-iron proteins that repair oxidative and nitrosative damage to iron-sulfur centers. J Bacteriol 2008;190:2004–13.
[162] Todorovic S, Justino MC, Wellenreuther G, et al. Iron-sulfur repair YtfE protein from Escherichia coli: structural characterization of the di-iron center. J Biol Inorg Chem 2008;13:765–70.
[163] Gralnick JA, Downs DM. The YggX protein of Salmonella enterica is involved in Fe(II) trafficking and minimizes the DNA damage caused by hydroxyl radicals: residue CYS-7 is essential for YggX function. J Biol Chem 2003;278:20708–15.
[164] Skovran E, Lauhon CT, Downs DM. Lack of YggX results in chronic oxidative stress and uncovers subtle defects in Fe-S cluster metabolism in Salmonella enterica. J Bacteriol 2004;186: 7626–34.
[165] Thorgersen MP, Downs DM. Analysis of yggX and gshA mutants provides insights into the labile iron pool in Salmonella enterica. J Bacteriol 2008;190:7608–13.
[166] Pomposiello PJ, Koutsolioutsou A, Carrasco D, Demple B. SoxRS-regulated expression and genetic analysis of the yggX gene of Escherichia coli. J Bacteriol 2003;185:6624–32.
[167] Osborne MJ, Siddiqui N, Landgraf D, Pomposiello PJ, Gehring K. The solution structure of the oxidative stress-related protein YggX from Escherichia coli. Protein Sci 2005;14:1673–8.
[168] Kennedy MC, Emptage MH, Dreyer JL, Beinert H. The role of iron in the activation-inactivation of aconitase. J Biol Chem 1983;258:11098–105.
[169] Kennedy MC, Kent TA, Emptage M, Merkle H, Beinert H, Munck E. Evidence for the formation of a linear [3Fe-4S] cluster in partially unfolded aconitase. J Biol Chem 1984;259:14463–71.
[170] Balasubramanian R, Shen G, Bryant DA, Golbeck JH. Regulatory roles for IscA and SufA in iron homeostasis and redox stress responses in the cyanobacterium Synechococcus sp. strain PCC 7002. J Bacteriol 2006;188:3182–91.
[171] Pinske C, Sawers RG. Delivery of iron-sulfur clusters to the hydrogen-oxidizing [NiFe]-hydrogenases in Escherichia coli requires the A-type carrier proteins ErpA and IscA. PLoS One 2012;7:e31755.
[172] Mapolelo DT, Zhang B, Randeniya S, et al. Monothiol glutaredoxins and A-type proteins: partners in Fe-S cluster trafficking. Dalton Trans 2013;42:3107–15.
[173] Butland G, Babu M, Diaz-Mejia JJ, et al. eSGA: E. coli synthetic genetic array analysis. Nat Methods 2008;5:789–95.
[174] Yeung N, Gold B, Liu NL, et al. The E. coli monothiol glutaredoxin GrxD forms homodimeric and heterodimeric Fe-S cluster containing complexes. Biochemistry 2011;50:8957–69.

# 13 Sensing the cellular Fe-S cluster demand: a structural, functional, and phylogenetic overview of *Escherichia coli* IscR

Erin L. Mettert, Nicole T. Pernaand and Patricia J. Kiley

## 13.1 Introduction

Given the versatile chemical reactivity of iron-sulfur (Fe-S) cluster cofactors, it is perhaps not surprising that in addition to serving structural, catalytic, or electron transfer roles for a variety of enzymes, Fe-S clusters enable many transcriptional regulatory proteins to sense specific environmental signals and respond appropriately by altering gene expression [1–4]. For the bacterium *Escherichia coli*, Fe-S cluster transcription factors are known to differentially regulate gene expression in response to several redox active compounds or gases, such as $O_2$, ROS, NO, and redox-cycling drugs [3, 4]. As such, these transcription factors – by altering mRNA levels through transcriptional regulation – play an important role in managing redox stress in cells.

Perhaps the most well-known example, and the subject of many reviews, is the global regulator FNR, which directly senses $O_2$ through the lability of its $[4Fe\text{-}4S]^{2+}$ cluster [3–8]. During anaerobic growth of *E. coli*, acquisition of the $[4Fe\text{-}4S]^{2+}$ cluster by FNR promotes protein dimerization, enabling FNR to bind site-specifically to DNA sites within promoter regions to control the transcription of genes encoding for proteins involved in metabolism, among numerous other functions. This transcriptional response enables *E. coli* to adapt its metabolism to anaerobic environments. In the presence of $O_2$, the $[4Fe\text{-}4S]^{2+}$ cluster is converted to a $[2Fe\text{-}2S]^{2+}$ cluster, thereby inactivating FNR through loss of dimerization. This mechanism of cluster conversion serves as a paradigm for the $O_2$ sensitivity of Fe-S clusters in general.

Although the sensitivity of the $[4Fe\text{-}4S]^{2+}$ cluster to $O_2$ is optimized for the role of FNR in regulating relevant pathways under anaerobic conditions, the activities of many Fe-S enzymes are essential under both aerobic and anaerobic growth conditions. Thus, the innate instability of some Fe-S clusters to $O_2$ and ROS [9–11] suggests that *in vivo*, the cellular requirements for Fe-S cluster biogenesis should vary with environmental changes in $O_2$ tension. An initial clue that *E. coli* responds to changes in Fe-S cluster demand was based on the finding that expression of the housekeeping Fe-S cluster biogenesis system (the Isc pathway) [12] is upregulated under aerobic conditions compared with anaerobic conditions. Furthermore, we discovered that the transcription factor IscR, the first gene in the operon encoding the Isc proteins (*iscRSUAhscBAfdx*), is responsible for this $O_2$-mediated differential expression by negatively regulating expression of the *isc* operon. Within the last 10 years, extensive studies have focused on how IscR makes use of the ligation state of its own Fe-S cluster to assess the cellular Fe-S demand and, accordingly, globally alter gene transcription to maintain Fe-S cluster homeostasis. These studies have also revealed that, unlike FNR, IscR is a novel type of regulator in that both the Fe-S cluster-containing form and the clusterless, apo-protein form are active to modulate gene expression. Furthermore, these two forms of IscR differentially regulate gene expression by site-specifically binding to two distinct DNA motifs in response to $O_2$ availability. In this chapter, we highlight key structural features of IscR and relate these to the major findings that have established its critical physiological role in Fe-S homeostasis. In addition, we discuss the results of phylogenetic analyses, which have revealed insights into evolutionary aspects of IscR function.

## 13.2 General properties of IscR

IscR is a member of the Rrf2 family of transcription factors that are widely distributed in bacteria and are responsible for sensing a variety of signals [13]. Other members include the nitric oxide sensor NsrR [14], the *Rhizobiaceae* iron-responsive regulator RirA [15], and the global cysteine regulator CymR of *Bacillus subtilis* [13]. Although this protein family has not been extensively studied, the crystal structures of several Rrf2 family members, including IscR, have been solved in recent years (Fig. 13.1) [13, 16, 17]. These structures have revealed some key features common to members of the Rrf2 family, such as the presence of an N-terminal winged helix-turn-helix (wHTH) DNA-binding motif and a long dimerization helix. This long dimerization helix positions the wHTH at the extreme ends of the protein dimer, enabling these relatively small proteins to span 23–27 bp of DNA [13]. In addition, sequence comparisons among Rrf2 members have revealed that a subset contains three conserved cysteine residues in their C-terminal region, suggesting that some family members may ligate Fe-S clusters [13, 18].

Although the apo-protein form of IscR was used to solve its crystal structure [16], several lines of evidence have demonstrated that IscR can indeed ligate an Fe-S cluster. Upon overexpression and isolation under anaerobic conditions, IscR is red in color and exhibits an absorption spectrum typical of a protein containing a [2Fe-2S] cluster [12]. The presence of a mostly reduced $[2Fe\text{-}2S]^{1+}$ cluster in isolated IscR was supported by EPR and Mössbauer spectroscopies, along with sulfide and iron analysis. Furthermore, Mössbauer studies using whole cells indicate that *in vivo*, the IscR [2Fe-2S] cluster is predominantly in the reduced (1+) state [12, 19]. Substitution of all three C-terminal cysteine residues (Cys92, Cys98, and Cys104) or just Cys92 with alanine, followed by purification yields IscR protein that does not contain any Fe-S cluster [18]. Thus, these results indicate that the conserved cysteine residues, in fact, serve as ligands for the IscR [2Fe-2S] cluster.

Until recently, it was unclear what residue provides the fourth ligand to the [2Fe-2S] cluster because IscR only contains the three aforementioned cysteine residues. Analysis of purified [2Fe-2S]-IscR by resonance Raman and NMR spectroscopies revealed that the [2Fe-2S] cluster has partial non-cysteinyl coordination. Indeed, subsequent mutational analysis identified His107 as the fourth ligand of the [2Fe-2S] cluster [19]. Although His107 is highly conserved among IscR orthologues, it is not conserved across all Rrf2 family members [13]. Furthermore, the $(Cys)_3(His)_1$ ligation is atypical because most [2Fe-2S] clusters have all cysteinyl $(Cys)_4$ ligation, or in the case of Rieske proteins, a $(Cys)_2(His)_2$ ligation. Interestingly, the few proteins known to bind a [2Fe-2S] cluster with $(Cys)_3(His)_1$ ligation, such as the yeast Fra2-Grx3/4 protein complexes [20] and the mitochondrial mitoNEET protein [21], appear to have possible roles as sensors. Further work is needed to determine whether the unique cluster ligation scheme allows IscR to more efficiently carry out its function in sensing the cellular Fe-S demand.

The apo-IscR crystal structure revealed that the electron density for residues encompassing the cluster-binding region were either weak or missing (Fig. 13.1) [16], suggesting conformational flexibility of this region in the absence of the [2Fe-2S] cluster. Nevertheless, the location of the weak electron density indicated that the cluster-binding site is exposed to solvent, perhaps allowing for efficient incorporation and/or loss of the [2Fe-2S] cluster. Additionally, the cluster-binding region from one monomer was proximal to the wHTH DNA binding domain of the other monomer, thus raising the possibility that upon [2Fe-2S] cluster ligation, a conformational

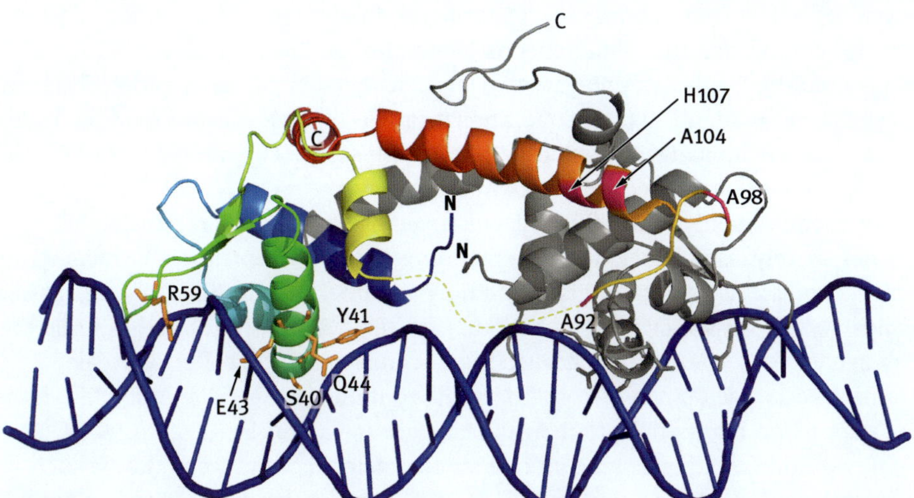


**Fig. 13.1:** Overall structure of an apo-IscR dimer bound to a 29-bp type 2 site derived from the *hyaA* promoter. In this crystal form, an IscR variant with three (Cys92, Cys98, and Cys104) of the four cluster ligands substituted with alanine (shown in magenta) was used. The IscR dimer is shown as a ribbon representation, with one subunit in gray and the other subunit in rainbow colors. Side chains making specific contacts to the *hyaA* DNA are shown as sticks, and the dashed line indicates residues for which electron density was missing.

change within IscR could influence DNA binding. Recent studies to address the mechanism of DNA binding by apo- and holo-IscR are described in further detail later in this chapter.

## 13.3 [2Fe-2S]-IscR represses Isc expression *via* a negative feedback loop

The significance of the [2Fe-2S] cluster for IscR transcriptional activity was recognized by the initial observation that a functional Isc pathway is required for IscR to repress transcription of the *iscRSUAhscBAfdx* operon *in vivo* [12]. Presumably, the Isc pathway provides IscR its [2Fe-2S] cluster and provides a link between IscR [2Fe-2S] cluster occupancy and the level of synthesis of the Isc pathway. Since this discovery, the IscR-dependent autoregulatory mechanism of Isc expression has been investigated in detail to better understand how Isc Fe-S cluster biogenesis machinery is regulated and, accordingly, how Fe-S cluster homeostasis is maintained.

DNase I footprinting experiments revealed that [2Fe-2S]-IscR binds to three individual sites within the *iscR* promoter region ($P_{iscR}$; Fig. 13.2a) [22]; however, only two of these sites, designated site A and site B, are necessary for IscR-mediated repression of $P_{iscR}$ *in vivo* [23]. Because these sites partially overlap promoter elements recognized by RNA polymerase (RNAP), we predicted that [2Fe-2S]-IscR represses $P_{iscR}$ by preventing RNAP from binding to the promoter. In support of this notion, DNase I footprinting revealed that simultaneous incubation of [2Fe-2S]-IscR and RNAP with DNA containing the *iscR* promoter region ($P_{iscR}$) resulted in pattern of protection from DNase I cleavage identical to the one when just [2Fe-2S]-IscR was present (Fig. 13.2b) [24]. This result indicates that binding of [2Fe-2S]-IscR to $P_{iscR}$ prevents RNAP from binding to the promoter and activating transcription.

The requirement of the [2Fe-2S] cluster in IscR-mediated repression of $P_{iscR}$ was definitively established by demonstrating that $P_{iscR}$ repression is relieved in strains expressing IscR mutants containing alanine substitutions of the cluster ligands (Cys92, Cys98, Cys104, and His107) [19, 23]. [2Fe-2S]-IscR appears to be the form needed for DNA binding because fluorescence anisotropy assays revealed that *in vitro*, [2Fe-2S]-IscR binds to DNA fragments containing either site A or site B from $P_{iscR}$ with much higher affinity than the clusterless variant IscR-C92A/C98A/C104A [23]. In addition, strains lacking individual components of the Isc pathway, such as IscS, IscU, or Fdx, yielded similar defects in $P_{iscR}$ repression as a strain lacking the entire Isc pathway (Δ*iscSUAhscBAfdx*) [12, 23, 25]. Although this defect was less severe in a strain lacking the A-type carrier (ATC) protein IscA [23, 25], recent studies demonstrated that another ATC, ErpA, also participates in IscR [2Fe-2S] cluster delivery [25]. Finally, in contrast to the Isc pathway, deletion of the *sufABCDSE* operon, encoding the alternate, stress-induced Suf Fe-S cluster biogenesis pathway, had no effect on the ability of [2Fe-2S]-IscR to repress $P_{iscR}$ *in vivo* [23, 25]. Taken together, these findings

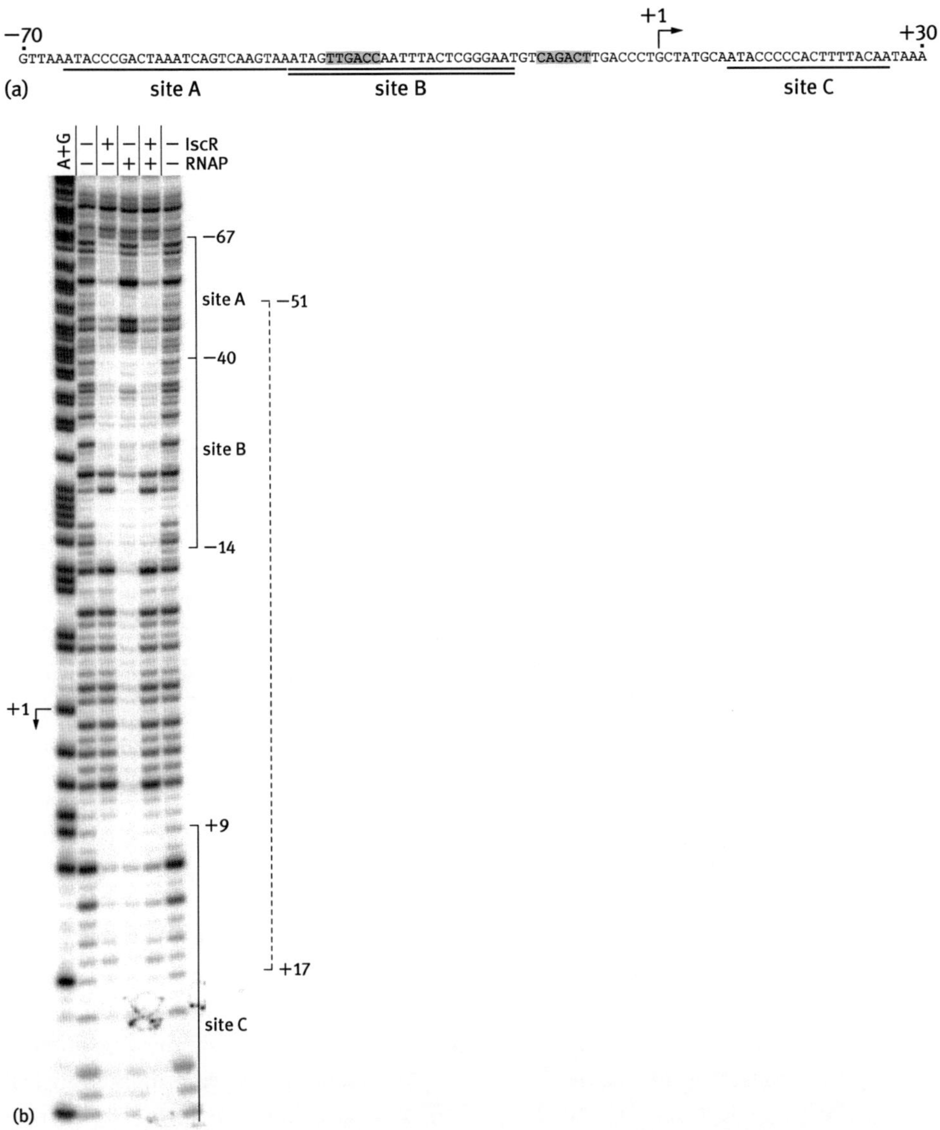


**Fig. 13.2:** IscR prevents binding of RNA polymerase to the *iscR* promoter *in vitro*. (a) Diagram of the *iscR* promoter region with the IscR binding sites A (underlined), B (double underlined), C (underlined), the transcriptional start site (arrow), and the –35 and –10 promoter elements (shaded) indicated. (b) Results of a DNase I footprinting assay, carried out as previously described [22], to evaluate binding of IscR (500-nM dimers) or $\sigma^{70}$-RNAP (100-nM dimers) to the top strand of the *iscR* promoter. IscR, RNAP, or both were incubated with DNA radiolabeled at one end for 30 min at 37°C before the addition of DNase I for 30 s. Electrophoresis of the samples revealed the IscR- or RNAP-specific patterns of protection of the DNA from DNase I cleavage. The presence (+) or absence (–) of the proteins in the reaction and the extent of the IscR (solid line) and $\sigma^{70}$-RNAP (dashed line) footprints relative to the +1 transcription start site are noted. Samples were electrophoresed with a Maxam-Gilbert (A + G) ladder.

support the conclusion that the Isc pathway provides the [2Fe-2S] cluster required for IscR to repress $P_{iscR}$ expression.

Despite the *in vivo* indicators of pathway specificity, IscR was capable of receiving clusters *in vitro* from either IscU or SufB scaffold proteins without the assistance of any ATC. This result was in contrast to those obtained for the Rrf2 paralogue NsrR, which required the ATCs ErpA or SufA for acquisition of its Fe-S cluster *in vitro* and the Isc pathway and ErpA for Fe-S acquisition *in vivo* [25]. Although we know the levels of the Isc components and that they are quite abundant in cells [24, 26, 27], we do not know yet the levels of other Fe-S cluster assembly proteins, which will be critical in understanding the differences between *in vivo* and *in vitro* results.

## 13.4 IscR adjusts synthesis of the Isc pathway based on the cellular Fe-S demand

*E. coli* is predicted to contain greater than 150 proteins that contain Fe-S clusters, some of which are known to be sensitive to destruction or damage by $O_2$ and ROS. Although it is not known in general how levels of these proteins compare between aerobic and anaerobic growth conditions, gene expression studies suggest that Fe-S proteins are highly expressed under both conditions [28]. Thus, to mitigate the damage to Fe-S clusters by $O_2$ or ROS, the need for Fe-S cluster biogenesis or repair is likely increased in the presence of $O_2$. Consistent with this notion, there is 2- to 5-fold less $P_{iscR}$ repression under aerobic conditions than anaerobic conditions. Furthermore, this aerobic derepression of $P_{iscR}$ occurs despite IscR protein levels being ~10-fold higher under aerobic than anaerobic conditions [23]. Hence, we interpret this as indicating that the cellular Fe-S demand appears to be higher when $O_2$ is present.

Recent research has focused on how IscR may sense these changes in cellular Fe-S demand. Based on previous studies with FNR [29–31], it was reasonable to hypothesize that $O_2$ may destabilize the IscR $[2Fe\text{-}2S]^{1+}$ cluster, or as in the case for SoxR in response to redox cycling drugs [32], the oxidation state of the cluster may be altered, thereby affecting holo-IscR activity. However, results from *in vitro* experiments suggest that $O_2$-mediated destruction of the IscR $[2Fe\text{-}2S]^{1+}$ cluster does not occur at a biologically relevant rate [12] (Kiley and Fleischhacker, unpublished results). Furthermore, although Mössbauer spectroscopy revealed that the $[2Fe\text{-}2S]^{+1}$ cluster is indeed subject to immediate oxidation to the (2+) state upon exposure to $O_2$, [2Fe-2S]-IscR was shown to have similar affinity for DNA containing the $P_{iscR}$ site B *in vitro* regardless of the cluster oxidation state [19].

Rather, the current hypothesis is that IscR cluster occupancy differs between aerobic and anaerobic conditions due to competition between IscR and other Fe-S substrates for the Isc machinery (Fig. 13.3). This competition would be elevated under aerobic conditions due to increased rates of $O_2$- or ROS-mediated Fe-S cluster

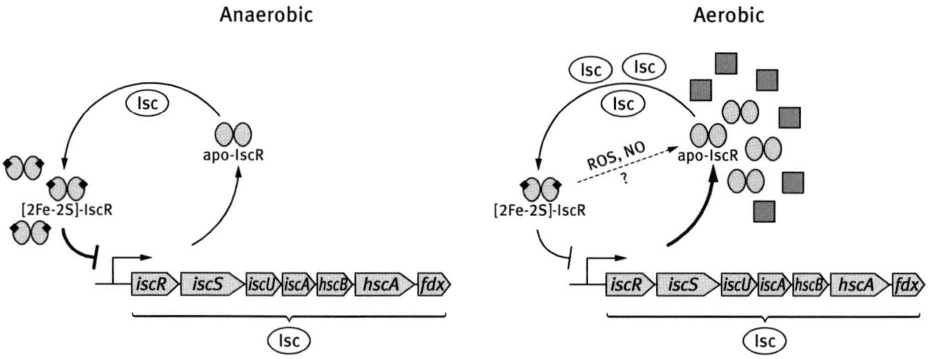


**Fig. 13.3:** Model describing how IscR responds to the differential demand in Fe-S cluster biogenesis between aerobic and anaerobic growth conditions. (Adapted from Giel JL, Nesbit AD, Mettert EL, Fleischhacker AS, Wanta BT, Kiley PJ, *Mol Microbiol*, 87, 478–92, 2013.) When $O_2$ is present, the rate of general Fe-S cluster turnover is predicted to be high, resulting in an increased demand for Fe-S cluster assembly. This leads to competition between IscR and apo-protein substrates (gray squares) for the Isc machinery, resulting in low IscR [2Fe-2S] cluster occupancy and thus derepression of the *iscRSUAhscBAfdx* operon. ROS and NO may also directly destabilize the IscR [2Fe-2S] cluster to generate apo-IscR, although this has yet to be confirmed. In contrast, when $O_2$ is absent, the demand for Fe-S cluster biogenesis is low due to heightened stability of Fe-S clusters. As a result, IscR [2Fe-2S] cluster occupancy is increased, culminating in repression of the Isc pathway. Although not depicted, under iron-limiting conditions, synthesis of [2Fe-2S]-IscR may be decreased, leading to elevated levels of apo-IscR.

turnover for proteins containing labile clusters. In turn, this would lead to low IscR [2Fe-2S] cluster occupancy and thus less repression of the *iscRSUAhscBAfdx* operon. Under anaerobic conditions, Fe-S clusters are predicted to be more stable and thus less competition would exist between IscR and apo-protein substrates. As a result, IscR would be predominantly in the holo-protein form to repress expression of the Isc pathway. Consistent with this model, IscR titration experiments revealed that half-maximal repression of $P_{iscR}$ *in vivo* required ~9-fold more IscR protein under aerobic conditions relative to anaerobic conditions [23]. If the concentration of holo-IscR is presumed to be equivalent to the IscR protein concentration required for 50% $P_{iscR}$ repression, these results indicate that IscR [2Fe-2S] cluster occupancy is lower under aerobic conditions. Although direct measurement of IscR cluster occupancy would further support this hypothesis, we currently lack robust methods to assess cluster occupancy at physiological levels of Fe-S proteins. Nevertheless, overexpression of FNR, or FNR-L28H, which contains a leucine-to-histidine substitution that renders the [4Fe-4S] cluster resistant to oxygen, diminished aerobic $P_{iscR}$ repression, whereas anaerobic $P_{iscR}$ expression remained unaffected [23]. This finding argues that under anaerobic conditions, there is sufficient Isc machinery present to efficiently respond to changes in Fe-S demand, thereby decreasing the competition between IscR and substrate proteins for Fe-S clusters.

Although this model provides an explanation for how *E. coli* exploits changes in IscR [2Fe-2S] cluster occupancy to sense and respond to differential requirements for Fe-S biogenesis, many questions regarding this proposed mechanism still remain. For example, as mentioned earlier in this chapter, it is not known whether the atypical $(Cys)_3(His)_1$ cluster ligation scheme or some other feature makes IscR a less efficient competitor for the Isc machinery compared with other Fe-S substrate proteins. If IscR is less able to acquire its Fe-S cluster, then other important cellular Fe-S cluster requirements would be satisfied before IscR acquires its [2Fe-2S] cluster. Furthermore, whether $(Cys)_3(His)_1$ ligation makes IscR more prone to cluster loss upon exposure to ROS remains to be established. Although expression of the *iscRSUAhscBAfdx* operon was induced upon exposure of cultures to $H_2O_2$- and the $O_2^-$-generating drug paraquat [33–36], further work is needed to establish whether ROS can directly target the stability of the IscR [2Fe-2S] cluster, in addition to increasing the general cellular need for Fe-S synthesis.

It is also possible that [2Fe-2S]-IscR responds to other environmental signals. For example, in response to excess amounts of cobalt, cellular ROS levels and expression of *iscRSUAhscBAfdx* were both increased [37], suggesting that [2Fe-2S]-IscR may indirectly sense cobalt stress. [2Fe-2S]-IscR may also sense NO, known to damage some Fe-S clusters [38, 39], and/or levels of Fe. Indeed, in cells treated with the $Fe^{2+}$-specific chelator 2,2′-dipyridyl, transcription of the *iscRSUAhscBAfdx* operon was upregulated [25, 33, 40]. Under these same growth conditions, however, the small regulatory RNA RyhB was found to target the *iscSUAhscBAfdx* transcript for degradation. In contrast, the *iscR* transcript remained relatively stable, presumably due to secondary structure within the transcript that protected the transcript from degradation by ribonucleases [41]. Thus, this differential transcript degradation could allow for IscR-mediated gene regulation while simultaneously decreasing Isc-mediated Fe-S biogenesis under conditions when Fe is limiting. However, this mechanism could also simply increase expression of the regulator without coupling it to changes in levels of the biogenesis machinery, which are already fairly abundant [24, 26, 27]. To resolve the role of Fe in the regulation of IscR- and Isc-mediated cluster assembly will require determining whether Fe limitation significantly alters IscR protein levels. Ultimately, these and future studies to decipher how IscR is regulated will provide key insights into understanding how Fe-S homeostasis is maintained throughout *E. coli*'s ever-changing environment.

## 13.5 IscR has a global role in maintaining Fe-S homeostasis

Subsequent to discovering that IscR functions as a repressor of the *iscRSUAhscBAfdx* operon [12], genome-wide transcription profiling data, in addition to findings from biochemical and genetic studies, have revealed that IscR either directly or indirectly controls the expression of more than 40 genes in *E. coli* [22, 34, 42–45]. Similar to the *isc* operon, IscR also represses the expression of ErpA and NfuA, proteins that have additional roles in Fe-S cluster assembly [22]. In addition, IscR is predicted to bind the promoter region of *cysE* [22]. CysE is a serine acetyltransferase that catalyzes a rate-limiting step in biosynthesis of cysteine [46], which is needed for Fe-S cluster assembly. Surprisingly, IscR directly activates transcription of the *sufABCDSE* operon that encodes the alternate Suf Fe-S biogenesis pathway [22, 34]. Although Isc is considered to be the housekeeping Fe-S biogenesis pathway, the Suf system is proposed to function primarily under oxidative and nitrosative stress and Fe-limiting conditions [33, 35, 47, 48]. Thus, the role of IscR in maintaining Fe-S cluster homeostasis goes beyond that of simply regulating the Isc pathway.

## 13.6 Fe-S cluster ligation broadens DNA site specificity for IscR

To establish the consensus IscR DNA binding site, sequences of the IscR protected regions in DNase I footprinting experiments were analyzed. This analysis, in addition to further investigation by phylogenetic footprinting, revealed that IscR specifically binds to two classes of DNA target sites, referred to as type 1 and type 2 sites (Fig. 13.4) [22]. However, as later studies revealed, the promoters of genes controlled by type 1 and type 2 sites show differences in their regulation by IscR according to the availability of $O_2$. The different roles of these two sites in allowing IscR to maintain Fe-S homeostasis will be discussed later in the chapter. Because the nucleotide sequences of these sites differ, it was not immediately apparent how an IscR dimer, predicted to have a single wHTH DNA binding domain in each subunit, could distinguish between the two sites. Even more intriguing was the finding that type 1 and type 2 sites are differentially regulated by IscR depending on whether the IscR [2Fe-2S] cluster is present. *In vitro* studies revealed that only [2Fe-2S]-IscR binds with high affinity to a type 1 site (found in the *iscR* promoter), whereas both [2Fe-2S]-IscR and apo-IscR bind with similar high affinity to a type 2 (found in the *hya* and *suf* promoters) site [18, 23]. Because other known Fe-S cluster-containing transcription factors (e.g. FNR, SoxR)

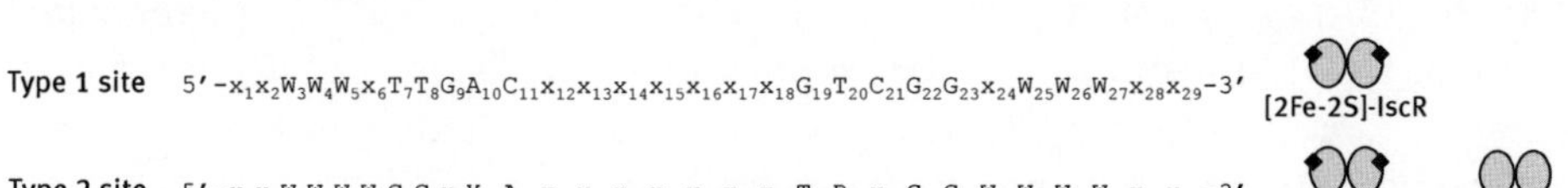


**Fig. 13.4:** The proposed consensus DNA binding sites for IscR are shown. [2Fe-2S]-IscR binds with high affinity to type 1 sites, whereas both [2Fe-2S]-IscR and apo-IscR bind with equally high affinity to type 2 sites. For the type 2 site, underlined is the symmetrical sequence present in each half site shown to be important for base-specific recognition by IscR. A, T, C, and G represent adenine, thymine, cytosine, and guanine, respectively; R represents A or G, Y represents C or T, W represents A or T, and x is undefined. Prototypical promoters that contain type 1 sites are *iscR, erpA*, and *nfuA*, whereas those that contain type 2 sites are *hyaA, sufA*, and *hybO*.

do not regulate transcription when they lack a cluster [6, 49], the finding that both holo- and apo-IscR are both active forms of the protein was truly unexpected. Furthermore, because Fe-S cluster occupancy of IscR is regulated by $O_2$ availability, broadening of IscR target site recognition upon [2Fe-2S] ligation revealed a very unique mechanism for differential gene regulation.

The crystal structure of apo-IscR bound to the type 2 site of the *hyaA* promoter has provided the major framework to understand how IscR distinguishes between two different DNA sites (Fig. 13.1). Although nonspecific interactions to the minor groove and DNA backbone provide much of the binding energy required to form the IscR-*hyaA* complex, the structure also revealed nucleotide-specific contacts between the IscR recognition helix and major groove of the DNA [16]. These findings corroborated earlier mutational analysis of the *hyaA* type 2 site that identified the same nucleotides important for IscR site-specific binding [18]. In addition, alanine substitution of the residues participating in these specific contacts (Ser40, Tyr41, Glu43, Gln44, and Arg59) established their relative importance in IscR recognition of the symmetrical $^{7}$CCxYA$^{11}$ sequence (where x is undefined and Y is a cytosine or thymine) present in each half site of the type 2 motif (Fig. 13.4) [16].

The presence of Glu43 within the IscR-*hyaA* interface was particularly compelling because acidic residues are not typically involved in nucleotide recognition due to potential electrostatic repulsion with the DNA backbone. Nevertheless, Glu43 was determined to be an important specificity-determining contact with the $^{7}$CC$^{8}$ dinucleotides in each type 2 half site. Although mutation of $^{7}$CC$^{8}$ in one half site disrupted *in vitro* binding affinity for wild-type IscR, the binding affinity of the mutant protein IscR-E43A was unaffected. Furthermore, unlike wild-type IscR, IscR-E43A did not discriminate between nucleotide bases at positions 7 or 8 [16]. These results suggest that the Glu43 contacts with the symmetrical $^{7}$CC$^{8}$ dinucleotides are critical for specific recognition of type 2 sites.

In contrast to the symmetrical type 2 site, the type 1 site is asymmetrical, containing $^{7}$CC$^{8}$ in one half site and a $^{7'}$TT$^{8'}$ in the other half site (Fig. 13.4). Because apo-IscR binds only type 2 sites with high affinity, it was proposed that an unfavorable interaction between Glu43 and $^{7'}$TT$^{8'}$ may inhibit apo-IscR from binding type 1 sites. In support of this hypothesis, both apo- and holo-forms of the IscR-E43A mutant bound the *iscRB* type 1 site with similar high affinity *in vitro* [16]. Thus, Glu43 of the wild-type protein is vital for discrimination against apo-IscR binding to the type 1 site. Furthermore, this finding implies that for [2Fe-2S]-IscR to bind to the type 1 site, the electrostatic repulsion between Glu43 and $^{7'}$TT$^{8'}$ must be removed. The current model proposes that ligation of the [2Fe-2S] cluster may induce a conformational change within IscR that repositions Glu43 away from $^{7'}$TT$^{8'}$, allowing for favorable contacts with the type 1 site.

Additional sequence differences in the two IscR binding motifs may also account for the differential regulation of type 1 and type 2 sites. Indeed, simply replacing the $^{7'}$TT$^{8'}$ dinucleotide of the *iscRB* type 1 site with cytosines was not sufficient to allow high-affinity binding by apo-IscR [16]. This suggests that the potentially favorable interaction between Glu43 and the CC dinucleotide may lock the IscR-DNA complex in such a conformation that essential contacts to other nucleotides in the type 1 site are not made. Although residues Ser40, Tyr41, Gln44, and Arg59 were found to be important for [2Fe-2S]-IscR binding to both type 1 and type 2 sites, the relative influence of contacts made by Ser40 and Gln44 varied between the two sites [16]. Thus, it appears that upon [2Fe-2S] ligation, repositioning of Glu43 allows for a more flexible protein conformation such that IscR can make different nucleotide contacts to accommodate binding to type 1 sites. Studies to establish a structural basis for this model and to further characterize important nucleotides within the type 1 site are ongoing.

Although IscR from *E. coli* represents the paradigm for understanding mechanistically how a transcription factor is capable of binding site-specifically to two different sites, this may be an emerging property of other transcription factors. For example, the Fur proteins from *Campylobacter jejuni* and *Bradyrhizobium japonicum* and the Fur homologue BosR from *Borrelia burgdorferi* apparently recognize multiple DNA sequences, although the exact relationship between sequence specificity and affinity has not been established [50–52]. In addition, the functionality of both apo- and holo-forms may not be unique to IscR; whereas Fur requires its $Fe^{2+}$ cofactor to repress transcription in most bacteria, apo-Fur appears to regulate expression of a few genes in *C. jejuni* and *Helicobacter pylori* [50, 53].

## 13.7 Phylogenetic analysis of IscR

To establish the extent to which IscR is conserved among bacteria and, more specifically, how well the key IscR residues that allow differential DNA site regulation are conserved, we conducted evolutionary analyses. The primary IscR amino acid sequence from *E. coli* strain MG1655 was used to query the NCBI reference protein database. This blastp search, plus further verification by sequence alignment, produced 4,527 proteins similar to MG1655 IscR. Among these proteins, we identified four major clades (Fig. 13.5a): the first clade (1,209 proteins) includes *E. coli* IscR; the second clade (1,310 proteins) includes *E. coli* NsrR; and the third clade (1,292 proteins) includes *B. subtilis* CymR. This observation is consistent with the fact that all three of these proteins belong to the Rrf2 protein family. The fourth clade (715 proteins, designated here as the Unknown clade) does not appear to contain any well-characterized transcription factors, and literature in the PubMed database linked to these proteins almost exclusively consisted of genome sequence reports. Only one taxon, *Gemmatimonas aurantiaca* T-27 gi|226227389, could not be definitively assigned to one of these clades. Although taxonomic distribution does not reveal any obviously strict patterns, some interesting trends emerge among the four clades, which encompass a wide range of bacterial diversity (Tab. 13.1). For instance, the Unknown clade is dominated by Alphaproteobacteria and Actinobacteria. Gammaproteobacteria and Betaproteobacteria

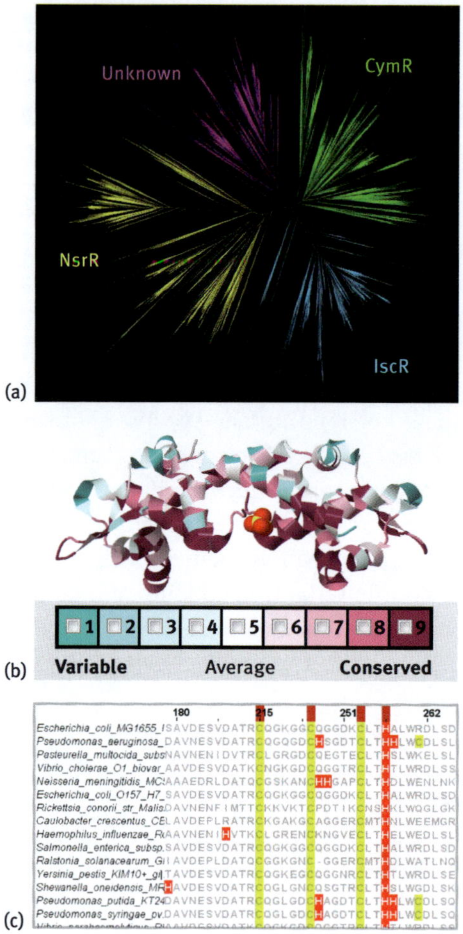


**Fig. 13.5:** (a) Shown is an unrooted tree for 4,527 proteins similar to *E. coli* MG1655 IscR. The MG1655 IscR protein sequence was used to query the reference protein (refseq protein) database using blastp with an *E* value cutoff of 0.0001 and a maximum number of hits (20,000) that exceeded the number of matches retrieved. The search returned 4,529 hits. These, plus MG1655 IscR, were aligned with MAFFT (version 7) using the FFT-NS-1 option (PMID: 23329690). Three sequences from one organism were very poorly aligned (*Saccharopolyspora spinosa*) and were thus removed, and the remaining sequences were realigned using the same options. Phylogenetic analysis was conducted using FastTree2 with default parameters (PMID: 19377059). Trees were visualized using Archeopteryx (PMID: 19860910). (b) Shown is the structure of apo-IscR colorized according to amino acid conservation. The conservation scores and mapping onto the structure (PDB ID: 4hf0) were performed using the ConSurf server (PMID: 20478830) with an MAFFT alignment of the 1,209 IscR clade proteins. (c) Shown is an alignment of a subset of IscR proteins. This is a portion of the 1,209 protein IscR clade MAFFT alignment visualized using Jalview2.8 (PMID: 19151095) with gaps relative to the *E. coli* MG1655 IscR protein hidden. Red blocks mark columns of the alignment corresponding to Cys92, Cys98, Cys104, and His107. All cysteines (yellow) and histidines (orange) in this region are highlighted.

**Tab. 13.1:** Taxonomic distribution of four clades of Rrf2 family proteins.

| Taxonomy[a] | Clade[b] | | | | Total |
|---|---|---|---|---|---|
| | IscR | CymR | NsrR | Unknown | |
| Proteobacteria; Gammaproteobacteria | 733 | 4 | 387 | 17 | 1,140 |
| Proteobacteria; Alphaproteobacteria | 162 | 56 | 276 | 268 | 762 |
| Firmicutes; Clostridia | 5 | 437 | 86 | 5 | 533 |
| Firmicutes; Bacilli | 0 | 224 | 191 | 9 | 424 |
| Proteobacteria; Betaproteobacteria | 259 | 10 | 104 | 0 | 373 |
| Actinobacteria; Actinobacteridae | 0 | 0 | 47 | 218 | 265 |
| Proteobacteria; Deltaproteobacteria | 8 | 91 | 44 | 20 | 163 |
| Cyanobacteria; Oscillatoriophycideae | 0 | 108 | 0 | 0 | 108 |
| Firmicutes; Negativicutes | 0 | 63 | 7 | 0 | 70 |
| Cyanobacteria; Nostocales | 0 | 57 | 0 | 0 | 57 |
| Bacteroidetes; Flavobacteriia | 1 | 1 | 21 | 32 | 55 |
| Proteobacteria; Epsilonproteobacteria | 0 | 22 | 5 | 27 | 54 |
| Actinobacteria; Coriobacteridae | 0 | 34 | 13 | 0 | 47 |
| Bacteroidetes; Cytophagia | 0 | 0 | 12 | 31 | 43 |
| Spirochaetes; Spirochaetales | 9 | 25 | 3 | 0 | 37 |
| Deinococcus-Thermus; Deinococci | 0 | 25 | 0 | 10 | 35 |
| Bacteroidetes; Sphingobacteriia | 0 | 6 | 9 | 17 | 32 |
| Planctomycetes; Planctomycetia | 0 | 0 | 8 | 19 | 27 |
| Thermotogae; Thermotogales | 16 | 1 | 5 | 0 | 22 |
| Fusobacteria; Fusobacteriales | 0 | 16 | 4 | 0 | 20 |
| Cyanobacteria; Stigonematales | 0 | 18 | 0 | 0 | 18 |
| Synergistetes; Synergistia | 0 | 6 | 6 | 4 | 16 |
| Firmicutes; Erysipelotrichia | 1 | 13 | 1 | 0 | 15 |
| Bacteroidetes; Bacteroidia | 0 | 5 | 6 | 3 | 14 |
| Chlorobi; Chlorobia | 0 | 1 | 12 | 0 | 13 |
| Acidobacteria; Acidobacteriales | 0 | 2 | 8 | 1 | 11 |
| Deferribacteres; Deferribacterales | 4 | 0 | 2 | 4 | 10 |
| Nitrospirae; Nitrospirales | 0 | 1 | 8 | 1 | 10 |
| Actinobacteria; Rubrobacteridae | 0 | 2 | 1 | 6 | 9 |
| Cyanobacteria; Pleurocapsales | 0 | 9 | 0 | 0 | 9 |
| Proteobacteria; Zetaproteobacteria | 6 | 0 | 3 | 0 | 9 |
| Verrucomicrobia; Verrucomicrobiae | 0 | 1 | 2 | 6 | 9 |
| Candidate division OP9 | 0 | 5 | 2 | 0 | 7 |
| Aquificae; Aquificales | 0 | 0 | 6 | 0 | 6 |
| Bacteroidetes; Bacteroidetes Order II | 0 | 0 | 6 | 0 | 6 |
| Chloroflexi; Chloroflexales | 0 | 6 | 0 | 0 | 6 |
| Chloroflexi; Ktedonobacteria | 0 | 2 | 3 | 0 | 5 |
| Ignavibacteriae; Ignavibacteria | 0 | 2 | 3 | 0 | 5 |
| Actinobacteria; Acidimicrobidae | 0 | 0 | 2 | 2 | 4 |
| Cyanobacteria | 0 | 4 | 0 | 0 | 4 |
| Dictyoglomi; Dictyoglomales | 0 | 4 | 0 | 0 | 4 |

*(Continued)*

Tab. 13.1: Taxonomic distribution of four clades of Rrf2 family proteins. (*Continued*)

| Taxonomy[a] | Clade[b] | | | | Total |
|---|---|---|---|---|---|
| | IscR | CymR | NsrR | Unknown | |
| Verrucomicrobia; Opitutae | 0 | 0 | 0 | 4 | 4 |
| Acidobacteria; Holophagae | 0 | 3 | 0 | 0 | 3 |
| Acidobacteria; Solibacteres | 0 | 1 | 1 | 1 | 3 |
| Candidate division OD1 | 0 | 2 | 1 | 0 | 3 |
| Candidate division OP8 | 0 | 1 | 1 | 1 | 3 |
| Chloroflexi; Dehalococcoidia | 0 | 3 | 0 | 0 | 3 |
| Cyanobacteria; Gloeobacteria | 0 | 2 | 1 | 0 | 3 |
| Firmicutes; environmental samples | 0 | 3 | 0 | 0 | 3 |
| Thermodesulfobacteria | 1 | 1 | 1 | 0 | 3 |
| Verrucomicrobia | 0 | 0 | 0 | 3 | 3 |
| Armatimonadetes; Chthonomonadetes | 0 | 1 | 1 | 0 | 2 |
| Chloroflexi; Anaerolineae | 0 | 0 | 2 | 0 | 2 |
| Chloroflexi; Sphaerobacteridae | 0 | 2 | 0 | 0 | 2 |
| Firmicutes | 0 | 2 | 0 | 0 | 2 |
| Thermobaculum | 0 | 1 | 0 | 1 | 2 |
| Acidobacteria; Chloracidobacterium | 0 | 1 | 0 | 0 | 1 |
| Acidobacteria; Koribacter | 0 | 0 | 1 | 0 | 1 |
| Actinobacteria; Microthrix | 0 | 0 | 0 | 1 | 1 |
| Caldiserica; Caldisericia | 0 | 0 | 0 | 1 | 1 |
| Caldithrix; | 0 | 0 | 1 | 0 | 1 |
| Candidate division KSB1 | 1 | 0 | 0 | 0 | 1 |
| Candidate division NC10 | 0 | 0 | 0 | 1 | 1 |
| Candidate division NKB19 | 0 | 0 | 1 | 0 | 1 |
| Candidate division OP3 | 0 | 0 | 0 | 1 | 1 |
| Candidate division WS1 | 0 | 0 | 1 | 0 | 1 |
| Chlamydiae; Chlamydiales | 0 | 0 | 1 | 0 | 1 |
| Chloroflexi; Herpetosiphonales | 0 | 1 | 0 | 0 | 1 |
| Chloroflexi; Thermomicrobiales | 0 | 1 | 0 | 0 | 1 |
| Chrysiogenetes; Chrysiogenales | 0 | 1 | 0 | 0 | 1 |
| Cyanobacteria; Prochlorales | 0 | 1 | 0 | 0 | 1 |
| dsDNA viruses; unclassified dsDNA phages | 0 | 1 | 0 | 0 | 1 |
| Euryarchaeota; Thermoplasmata | 0 | 1 | 0 | 0 | 1 |
| Fibrobacteres; Fibrobacterales | 0 | 1 | 0 | 0 | 1 |
| Lentisphaerae; Lentisphaeria | 0 | 0 | 1 | 0 | 1 |
| Metazoa; Ecdysozoa | 1 | 0 | 0 | 0 | 1 |
| Planctomycetes; Phycisphaerae | 0 | 0 | 1 | 0 | 1 |
| Synergistetes | 0 | 1 | 0 | 0 | 1 |
| Verrucomicrobia; Spartobacteria | 0 | 0 | 0 | 1 | 1 |
| Total | 1,207 | 1,292 | 1,308 | 715 | 4,521 |

[a] Taxonomic information was extracted from Reference Protein XML records.

[b] Clades were assigned according to phylogenetic analysis (Fig. 13.5).

tend to have IscR and NsrR, but very few CymR and Unknown clade members. In addition, Bacilli and other Firmicutes tend to have NsrR and CymR, although there is some evidence for a few IscR and/or Unknown clade members in these taxa.

We found that IscR is most prevalent in the Proteobacteria (primarily gamma, then beta and alpha), similar to what was observed previously [54]. However, it is also present in distantly related organisms, such as the Spirochaetes and Thermotogae. Notably, the presence of IscR appears to be widespread among pathogenic bacteria and, accordingly, was demonstrated to be important for virulence by the plant pathogen *Dickeya dadantii* (formerly known as *Erwinia chyrsanthemi*) [55] and the opportunistic human pathogen *Pseudomonas aeruginosa* [56]. Remarkably, the key residues involved in IscR [2Fe-2S] cluster ligation and DNA binding are all highly conserved among members of the IscR clade (Fig. 13.5b and Fig. 13.5c). This finding implies that the mechanism by which IscR regulates two different DNA target sites may also be highly conserved among numerous bacteria. Despite this high degree of conservation, we and Rodionov *et al.* [54] observed an intriguing variation among the Alphaproteobacteria; whereas most IscR proteins possess the [2Fe-2S] cluster ligands, these residues are not conserved in the Rhodobacterales. Not only does the Rhodobacterales IscR lack the cluster ligands, its genomic context is a combination of *isc* (*iscRS*) and *suf* (*sufBCDS*) genes, suggesting that apo-IscR might regulate a hybrid pathway. However, the genomic organization of Fe-S cluster assembly genes of another Alphaproteobacteria, *Caulobacter crescentus*, also has this hybrid operon, but the IscR protein has all of the cluster ligands [57]. Thus, understanding how IscR regulates this hybrid operon among the Alphaproteobacteria will be important in revealing differences in regulation of Fe-S homeostasis within this diverse class of bacteria.

## 13.8 Binding to two classes of DNA sites allows IscR to differentially regulate transcription in response to $O_2$

By identifying which IscR-regulated promoters contain a type 1 site *vs* a type 2 site, it is apparent that the ability of IscR to discriminate between these two sites poses a major advantage for maintaining Fe-S cluster homeostasis. For example, type 1 site-containing promoters include those of the *iscRSUAhscBAfdx* operon, *erpA*, *nfuA*, and *cysE*, all of which are known or are predicted to be repressed by [2Fe-2S]-IscR and encode proteins involved in Fe-S cluster assembly [22]. Because the Fe-S demand under anaerobic conditions is low, it makes physiological sense to have these genes repressed by [2Fe-2S]-IscR (Fig. 13.3). In contrast, under aerobic conditions when the Fe-S demand is high and levels of apo-IscR increase due to competition with other substrate proteins for the Isc machinery, de-repression of *iscRSUAhscBAfdx*, *erpA*, *nfuA*, and *cysE* occurs so that the cellular Fe-S requirements can be met. Interestingly, although all IscR-dependent promoters identified thus far contain a single target site, the one exception is the *iscR* promoter, which contains two type 1 sites [22], perhaps emphasizing the significance of regulating Isc Fe-S biogenesis *in vivo*.

Although promoters containing a type 2 site can be bound by both apo- and holo-IscR *in vitro* [18], transcription profiling results indicate that most of these promoters are limited to regulation by IscR under aerobic growth conditions when apo-IscR is the predominant form of the protein [22]. As previously mentioned, promoters with a type 2 site include that of the *hyaABCDEF* operon, encoding the anaerobic Fe-S enzyme hydrogenase 1. Two other anaerobic Fe-S enzymes, hydrogenase 2 and the periplasmic nitrate reductase, are also regulated by type 2 site-containing promoters, *hybO* and *napF*, respectively. Under aerobic conditions, IscR represses transcription of these enzymes, perhaps directing Fe-S cluster biogenesis toward Fe-S proteins that exhibit essential cellular functions in the presence of $O_2$ [22]. IscR also activates the expression of some promoters that have a type 2 site. These include those of the *sufABCDSE* operon and *sodA*, encoding the $Mn^{2+}$-containing superoxide dismutase ($Mn^{2+}$-SOD) [22, 34]. Thus, IscR-dependent regulation of type 2 sites enables *E. coli* to maintain Fe-S cluster homeostasis not only by directly controlling Fe-S biogenesis, but also by decreasing ROS and possibly protecting Fe-S clusters from destruction under conditions of oxidative stress.

Further investigation of promoters having a type 2 site revealed that several are regulated by additional transcription factors. For instance, *sufA* is regulated by Fur, OxyR, IHF, and NsrR [33, 35, 58], and *sodA* is regulated by Fur, ArcA, IHF, and SoxS [59–61]. Therefore, it was hypothesized that coordinate regulation of type 2 site-containing promoters by multiple transcription factors may at least partially account for the decreased regulation by [2Fe-2S]-IscR under anaerobic conditions of this class of promoters. Indeed, this mechanism of $O_2$-dependent regulation was recently demonstrated for the *hyaA* promoter. Apo-IscR represses $P_{hyaA}$ under aerobic conditions, whereas the anaerobic transcription factors ArcA and AppY promote $P_{hyaA}$ transcription under anaerobic conditions by antagonizing IscR-mediated repression. Although the mechanism by which AppY prevents IscR repression remains to be elucidated, ArcA was shown to compete with IscR in binding $P_{hyaA}$ *in vitro*. Furthermore, it was demonstrated that, *in vivo*, [2Fe-2S]-IscR exhibits weaker repressor activity at *hyaA* compared with apo-IscR [62]. This weaker repressor activity, in combination with antirepression by ArcA and AppY and the fact that anaerobic levels of [2Fe-2S]-IscR are low due to negative autoregulation, would ultimately lead to increased $P_{hyaA}$ expression under anaerobic conditions. Thus, in contrast to type 1 sites where $O_2$-dependent regulation appears to primarily depend on changes in IscR [2Fe-2S] cluster occupancy, $O_2$-dependent regulation of type 2 sites may involve the combinatorial action of multiple transcription factors. Because the identity and number of transcription factors that regulate each type 2 site-containing promoter varies, further investigation is required to dissect the mechanism by which each promoter is coordinately regulated. In summary, the ability of IscR to recognize two classes of

DNA target sites provides a novel and efficient mechanism for sculpting the *E. coli* transcriptome in response to $O_2$.

## 13.9 Roles of IscR beyond Fe-S homeostasis

IscR also regulates type 2 sites in promoters of genes whose functional ties to Fe-S homeostasis are not currently obvious. For example, under iron-limiting conditions, IscR directly activates transcription of *fimE*, encoding the FimE recombinase that decreases expression of type I fimbriae, thus reducing biofilm formation [42]. Also, under iron-limiting conditions, IscR promotes expression of NrdEF, the $Mn^{2+}$-containing ribonucleotide reductase [45]. Because IscR also upregulates $Mn^{2+}$-SOD [22], it is intriguing that IscR upregulates several isoenzymes that contain $Mn^{2+}$, which, unlike their counterparts that contain $Fe^{2+}$, would not be sensitive to oxidation. In conjunction with LexA, IscR represses *cka*, encoding the pore-forming toxin colicin K [44]; this coordinate gene regulation has shown to important for delayed production of colicin K following induction of the SOS response, a global response to DNA damage that prompts DNA repair. In addition, IscR directly represses *rnlA*, an essential gene for the activity of RNase LS, an endoribonuclease important for RNA metabolism and antagonizing bacteriophage T4 [43]. Overall, these findings suggest that IscR may respond to stresses in addition to changes in the cellular Fe-S demand. Further investigation of the direct IscR regulon is needed to gain insight on ancillary regulatory roles of IscR.

## 13.10 Additional aspects of IscR regulation

Although the critical findings establishing IscR's role in Fe-S homeostasis have been reviewed here, further research is needed to shed light on additional factors that enable IscR to carry out this vital function. For example, although IscR is a target of the ClpXP protease [63], it is not known how regulating IscR protein levels may influence expression of the IscR regulon. However, in the case of *cka*, decreased IscR levels in cells entering late exponential phase and early stationary phase are important for derepression of colicin C production. This is apparently because when IscR levels are sufficient, such as in nonstressed, exponentially growing cells, IscR stabilizes LexA at the *cka* promoter to prevent transcription [44]. Given that IscR levels in aerobic cells are very abundant (~10,300 molecules per cell [23], compared with ~5,000 and ~4,100 molecules per cell for the global regulators Fur [64] and FNR [65], respectively), it is possible that post-translational regulation may be important for decreasing the cellular concentration of IscR. Furthermore, stabilization of LexA by IscR at the *cka* promoter raises the question as to whether IscR employs a similar mechanism with other transcription factors to alter gene expression. Thus, future studies will likely provide exciting new insight on additional mechanisms that regulate IscR.

## 13.11 Summary

Here, we have reviewed the structural and functional analyses of IscR that have advanced our understanding of how *E. coli* maintains Fe-S cluster homeostasis. Critical to this process is the ability of IscR to sense the demand for Fe-S synthesis through its [2Fe-2S] cluster and, accordingly, adjust transcription of relevant target genes through binding two different DNA sites. The capacity of IscR to distinguish between type 1 and type 2 sites based on its cluster ligation state is an extremely versatile trait, providing an efficient means to alter gene expression in response to environmental changes in $O_2$ availability. As our phylogenetic analysis of IscR suggests, this mechanism of differential gene expression by a single transcription factor appears to be conserved in numerous bacteria, further demonstrating the significance of IscR versatility.

## Acknowledgments

The authors are grateful to past and current members of Patricia Kiley's laboratory who have contributed much of the work discussed in this chapter. We specifically thank Jennifer Giel for providing the data shown in Fig. 12.2. This work was supported by NIH grant GM45844 to P.J.K.

## References


[1] Beinert H. Iron-sulfur proteins: ancient structures, still full of surprises. J Biol Inorg Chem 2000;5:2–15.

[2] Fontecave M. Iron-sulfur clusters: ever-expanding roles. Nat Chem Biol 2006;2:171–4.

[3] Kiley PJ, Beinert H. The role of Fe-S proteins in sensing and regulation in bacteria. Curr Opin Microbiol 2003;6:181–5.

[4] Fleischhacker AS, Kiley PJ. Iron-containing transcription factors and their roles as sensors. Curr Opin Chem Biol 2011;15:335–41.

[5] Unden G, Achebach S, Holighaus G, Tran HG, Wackwitz B, Zeuner Y. Control of FNR function of Escherichia coli by $O_2$ and reducing conditions. J Mol Microbiol Biotechnol 2002;4:263–8.

[6] Kiley PJ, Beinert H. Oxygen sensing by the global regulator, FNR: the role of the iron-sulfur cluster. FEMS Microbiol Rev 1998;22:341–52.

[7] Green J, Crack JC, Thomson AJ, LeBrun NE. Bacterial sensors of oxygen. Curr Opin Microbiol 2009;12:145–51.

[8] Sawers G. The aerobic/anaerobic interface. Curr Opin Microbiol 1999;2:181–7.

[9] Imlay JA. Iron-sulphur clusters and the problem with oxygen. Mol Microbiol 2006;59:1073–82.

[10] Imlay JA. Cellular defenses against superoxide and hydrogen peroxide. Annu Rev Biochem 2008;77:755–76.

[11] Imlay JA. The molecular mechanisms and physiological consequences of oxidative stress: lessons from a model bacterium. Nat Rev Microbiol 2013;11:443–54.
[12] Schwartz CJ, Giel JL, Patschkowski T, et al. IscR, an Fe-S cluster-containing transcription factor, represses expression of Escherichia coli genes encoding Fe-S cluster assembly proteins. Proc Natl Acad Sci USA 2001;98:14895–900.
[13] Shepard W, Soutourina O, Courtois E, England P, Haouz A, Martin-Verstraete I. Insights into the Rrf2 repressor family – the structure of CymR, the global cysteine regulator of Bacillus subtilis. FEBS J 2011;278:2689–701.
[14] Bodenmiller DM, Spiro S. The yjeB (nsrR) gene of Escherichia coli encodes a nitric oxide-sensitive transcriptional regulator. J Bacteriol 2006;188:874–81.
[15] Todd JD, Wexler M, Sawers G, Yeoman KH, Poole PS, Johnston AW. RirA, an iron-responsive regulator in the symbiotic bacterium Rhizobium leguminosarum. Microbiology 2002;148:4059–71.
[16] Rajagopalan S, Teter SJ, Zwart PH, Brennan RG, Phillips KJ, Kiley PJ. Studies of IscR reveal a unique mechanism for metal-dependent regulation of DNA binding specificity. Nat Struc Mol Biol 2013;20:740–7.
[17] Berman HM, Westbrook J, Feng Z, et al. The Protein Data Bank. Nucleic Acids Res 2000;28: 235–42.
[18] Nesbit AD, Giel JL, Rose JC, Kiley PJ. Sequence-specific binding to a subset of IscR-regulated promoters does not require IscR Fe-S cluster ligation. J Mol Biol 2009;387:28–41.
[19] Fleischhacker AS, Stubna A, Hsueh KL, et al. Characterization of the [2Fe-2S] Cluster of Escherichia coli Transcription Factor IscR. Biochemistry 2012;51:4453–62.
[20] Li H, Mapolelo DT, Dingra NN, et al. The yeast iron regulatory proteins Grx3/4 and Fra2 form heterodimeric complexes containing a [2Fe-2S] cluster with cysteinyl and histidyl ligation. Biochemistry 2009;48:9569–81.
[21] Paddock ML, Wiley SE, Axelrod HL, et al. MitoNEET is a uniquely folded 2Fe 2S outer mitochondrial membrane protein stabilized by pioglitazone. Proc Natl Acad Sci USA 2007;104:14342–7.
[22] Giel JL, Rodionov D, Liu M, Blattner FR, Kiley PJ. IscR-dependent gene expression links iron-sulphur cluster assembly to the control of $O_2$-regulated genes in Escherichia coli. Mol Microbiol 2006;60:1058–75.
[23] Giel JL, Nesbit AD, Mettert EL, Fleischhacker AS, Wanta BT, Kiley PJ. Regulation of iron-sulphur cluster homeostasis through transcriptional control of the Isc pathway by [2Fe-2S]-IscR in Escherichia coli. Mol Microbiol 2013;87:478–92.
[24] Giel JL. Role of IscR in regulation of iron-sulfur biogenesis in Escherichia coli: Identification of the IscR regulon and mechanisms of autoregulation. PhD dissertation. University of Wisconsin-Madison; 2007.
[25] Vinella D, Loiseau L, de Choudens SO, Fontecave M, Barras F. In vivo [Fe-S] cluster acquisition by IscR and NsrR, two stress regulators in Escherichia coli. Mol Microbiol 2013;87:493–508.
[26] Vickery LE, Silberg JJ, Ta DT. Hsc66 and Hsc20, a new heat shock cognate molecular chaperone system from Escherichia coli. Protein Sci 1997;6:1047–56.
[27] Hoff KG, Silberg JJ, Vickery LE. Interaction of the iron-sulfur cluster assembly protein IscU with the Hsc66/Hsc20 molecular chaperone system of Escherichia coli. Proc Natl Acad Sci USA 2000;97:7790–5.
[28] Py B, Barras F. Building Fe-S proteins: bacterial strategies. Nat Rev Microbiol 2010;8:436–46.
[29] Jordan PA, Thomson AJ, Ralph ET, Guest JR, Green J. FNR is a direct oxygen sensor having a biphasic response curve. FEBS Lett 1997;416:349–52.
[30] Khoroshilova N, Popescu C, Münck E, Beinert H, Kiley PJ. Iron-sulfur cluster disassembly in the FNR protein of Escherichia coli by $O_2$: [4Fe-4S] to [2Fe-2S] conversion with loss of biological activity. Proc Natl Acad Sci USA 1997;94:6087–92.

[31] Popescu CV, Bates DM, Beinert H, Münck E, Kiley PJ. Mössbauer spectroscopy as a tool for the study of activation/inactivation of the transcription regulator FNR in whole cells of Escherichia coli. Proc Natl Acad Sci USA 1998;95:13431–5.
[32] Gu M, Imlay JA. The SoxRS response of Escherichia coli is directly activated by redox-cycling drugs rather than by superoxide. Mol Microbiol 2011;79:1136–50.
[33] Outten FW, Djaman O, Storz G. A suf operon requirement for Fe-S cluster assembly during iron starvation in Escherichia coli. Mol Microbiol 2004;52:861–72.
[34] Yeo WS, Lee JH, Lee KC, Roe JH. IscR acts as an activator in response to oxidative stress for the suf operon encoding Fe-S assembly proteins. Mol Microbiol 2006;61:206–18.
[35] Zheng M, Wang X, Templeton LJ, Smulski DR, LaRossa RA, Storz G. DNA microarray-mediated transcriptional profiling of the Escherichia coli response to hydrogen peroxide. J Bacteriol 2001;183:4562–70.
[36] Blanchard JL, Wholey WY, Conlon EM, Pomposiello PJ. Rapid changes in gene expression dynamics in response to superoxide reveal SoxRS-dependent and independent transcriptional networks. PLoS One 2007;2:e1186.
[37] Fantino JR, Py B, Fontecave M, Barras F. A genetic analysis of the response of Escherichia coli to cobalt stress. Environ Microbiol 2010;12:2846–57.
[38] Spiro S. Regulators of bacterial responses to nitric oxide. FEMS Microbiol Rev 2007;31:193–211.
[39] Tucker NP, Le Brun NE, Dixon R, Hutchings MI. There's NO stopping NsrR, a global regulator of the bacterial NO stress response. Trends Microbiol 2010;18:149–56.
[40] Lee KC, Yeo WS, Roe JH. Oxidant-responsive induction of the suf operon, encoding a Fe-S assembly system, through Fur and IscR in Escherichia coli. J Bacteriol 2008;190:8244–7.
[41] Desnoyers G, Morissette A, Prévost K, Massé E. Small RNA-induced differential degradation of the polycistronic mRNA iscRSUA. EMBO J 2009;28:1551–61.
[42] Wu Y, Outten FW. IscR controls iron-dependent biofilm formation in Escherichia coli by regulating type I fimbria expression. J Bacteriol 2009;191:1248–57.
[43] Otsuka Y, Miki K, Koga M, et al. IscR regulates RNase LS activity by repressing rnlA transcription. Genetics 2010;185:823–30.
[44] Butala M, Sonjak S, Kamensek S, et al. Double locking of an Escherichia coli promoter by two repressors prevents premature colicin expression and cell lysis. Mol Microbiol 2012;86:129–39.
[45] Martin JE, Imlay JA. The alternative aerobic ribonucleotide reductase of Escherichia coli, NrdEF, is a manganese-dependent enzyme that enables cell replication during periods of iron starvation. Mol Microbiol 2011;80:319–34.
[46] Zheng L, Cash VL, Flint DH, Dean DR. Assembly of iron-sulfur clusters. Identification of an iscSUA-hscBA-fdx gene cluster from Azotobacter vinelandii. J Biol Chem 1998;273:13264–72.
[47] Justino MC, Vicente JB, Teixeira M, Saraiva LM. New genes implicated in the protection of anaerobically grown Escherichia coli against nitric oxide. J Biol Chem 2005;280:2636–43.
[48] Mukhopadhyay P, Zheng M, Bedzyk LA, LaRossa RA, Storz G. Prominent roles of the NorR and Fur regulators in the Escherichia coli transcriptional response to reactive nitrogen species. Proc Natl Acad Sci USA 2004;101:745–50.
[49] Hidalgo E, Demple B. An iron-sulfur center essential for transcriptional activation by the redox-sensing SoxR protein. EMBO J 1994;13:138–46.
[50] Butcher J, Sarvan S, Brunzelle JS, Couture JF, Stintzi A. Structure and regulon of Campylobacter jejuni ferric uptake regulator Fur define apo-Fur regulation. Proc Natl Acad Sci USA 2012;109:10047–52.
[51] Ouyang Z, Deka RK, Norgard MV. BosR (BB0647) controls the RpoN-RpoS regulatory pathway and virulence expression in Borrelia burgdorferi by a novel DNA-binding mechanism. PLoS Pathog 2011;7:e1001272.
[52] Friedman YE, O'Brian MR. A novel DNA-binding site for the ferric uptake regulator (Fur) protein from Bradyrhizobium japonicum. J Biol Chem 2003;278:38395–401.

[53] Miles S, Carpenter BM, Gancz H, Merrell DS. Helicobacter pylori apo-Fur regulation appears unconserved across species. J Microbiol 2010;48:378–86.
[54] Rodionov DA, Gelfand MS, Todd JD, Curson AR, Johnston AW. Computational reconstruction of iron- and manganese-responsive transcriptional networks in α-Proteobacteria. PLoS Comput Biol 2006;2:e163.
[55] Rincon-Enriquez G, Crete P, Barras F, Py B. Biogenesis of Fe/S proteins and pathogenicity: IscR plays a key role in allowing Erwinia chrysanthemi to adapt to hostile conditions. Mol Microbiol 2008;67:1257–73.
[56] Kim SH, Lee BY, Lau GW, Cho YH. IscR modulates catalase A (KatA) activity, peroxide resistance and full virulence of Pseudomonas aeruginosa PA14. J Microbiol and Biotechn 2009;19:1520–6.
[57] da Silva Neto JF, Lourenco RF, Marques MV. Global transcriptional response of Caulobacter crescentus to iron availability. BMC Genomics 2013;14:549.
[58] Partridge JD, Bodenmiller DM, Humphrys MS, Spiro S. NsrR targets in the Escherichia coli genome: new insights into DNA sequence requirements for binding and a role for NsrR in the regulation of motility. Mol Microbiol 2009;73:680–94.
[59] Tardat B, Touati D. Iron and oxygen regulation of Escherichia coli MnSOD expression: competition between the global regulators Fur and ArcA for binding to DNA. Mol Microbiol 1993;9:53–63.
[60] Presutti DG, Hassan HM. Binding of integration host factor (IHF) to the Escherichia coli sodA gene and its role in the regulation of a sodA-lacZ fusion gene. Mol Genet 1995;246:228–35.
[61] Fawcett WP, Wolf RE, Jr. Purification of a MalE-SoxS fusion protein and identification of the control sites of Escherichia coli superoxide-inducible genes. Mol Microbiol 1994;14:669–79.
[62] Nesbit AD, Fleischhacker AS, Teter SJ, Kiley PJ. ArcA and AppY antagonize IscR repression of hydrogenase-1 expression under anaerobic conditions, revealing a novel mode of $O_2$ regulation of gene expression in Escherichia coli. J Bacteriol 2012;194:6892–9.
[63] Flynn JM, Neher SB, Kim YI, Sauer RT, Baker TA. Proteomic discovery of cellular substrates of the ClpXP protease reveals five classes of ClpX-recognition signals. Mol Cell 2003;11:671–83.
[64] Zheng M, Doan B, Schneider TD, Storz G. OxyR and SoxRS regulation of fur. J Bacteriol 1999;181:4639–43.
[65] Sutton VR, Mettert EL, Beinert H, Kiley PJ. Kinetic analysis of the oxidative conversion of the $[4Fe\text{-}4S]^{2+}$ cluster of FNR to a $[2Fe\text{-}2S]^{2+}$ cluster. J Bacteriol 2004;186:8018–25.

# 14 Fe-S assembly in Gram-positive bacteria

Patricia C. Dos Santos

## 14.1 Introduction

Biosynthetic pathways involved in the assembly and trafficking of Fe-S clusters in Gram-positive bacteria are distinct from those found in Gram-negative bacteria and Eukaryotes. Inspection of Gram-positive bacterial genome sequences showed that only one type of biosynthetic pathway is represented in each species, which suggests that each pathway will be vital for function. These single pathways varied across taxonomic classes of Gram-positive bacteria where three types of genetically distinct pathways were identified: Clostridia-ISC, Actinobacteria-SUF, and Bacilli-SUF encoded by *iscRSU, sufRCDBSUT*, and *sufCDSUB* gene regions, respectively. Close inspection of these systems show that they differ in the number of genes contained in each operon, genomic organization, and composition from previously studied systems. Although components of the Clostridia-ISC are more similar to the components of the Gram-negative ISC system, the components of the Actinobacteria- and Bacilli-SUF system share high levels of sequence similarities to each other and have more in common with SUF systems. The assembly of Fe-S clusters in Gram-positive bacteria containing the SUF systems involves SufS, which catalyzes the PLP-dependent Cys:SufU sulfurtransferase reaction. The subsequent assembly of Fe-S clusters is proposed to be mediated by SufU and/or the SufBDC complex. The distinct biosynthetic makeup of these systems combined with the essential role of their individual components render this pathway a potential target for specific pharmacological interventions in infections caused by Gram-positive pathogens.

## 14.2 Fe-S proteins in Gram-positive bacteria

The requirement for Fe-S clusters in promoting life-sustaining reactions discussed in previous chapters for nitrogen-fixing and Gram-negative bacteria is also an important metabolic feature in Gram-positive bacteria. As in other forms of life, Fe-S clusters, when associated with proteins, facilitate numerous reactions ranging from electron transfer, to substrate activation, to sulfur mobilization, to molecular signaling, to transcriptional and post-translational regulation of gene expression [1, 2]. The inherent oxygen sensitivity of Fe-S proteins and the existence of species-specific Fe-S cluster assembly machineries pose technical challenges for identification of their functionalities and their protein partners.

With the development of affordable technologies for genome sequencing, computational tools have aided in the identification of putative Fe-S proteins and gene products involved in the biosynthesis and cellular trafficking of Fe-S clusters. Although bioinformatic predictions are *in silico* estimations of biological processes, they have provided direction for experimental validation of the involvement of Fe-S enzymes in metabolism. Sequence analysis of Fe-S cluster binding motifs using manually curated genomes showed that some microbial genomes encode a large number of Fe-S proteins (Tab. 14.1). The genome of the well-studied *Escherichia coli*, for example, encodes 143 known and putative Fe-S proteins, one-third of which are of unidentified function [1, 3]. Similarly, the genome of other Gram-negative bacteria, including *Azotobacter vinelandii* and *Salmonella enterica*, encode over 100 proteins containing Fe-S binding motifs. In contrast, the genomes of selected Gram-positive bacteria encode fewer proteins that contain Fe-S binding motifs (Tab. 14.1), with the exception of the microaerophilic *Clostridium difficile*, which is predicted to encode nearly 100 proteins that have Fe-S binding motifs [3].

The reduced number of Fe-S proteins in Gram-positive bacteria may be attributable to multiple causes. First, these selected genomes often lack the duplicated genes that encode multiple versions of Fe-S enzymes that perform similar and overlapping functions. For example, the *E. coli* and *A. vinelandii* genomes encode two copies of

**Tab. 14.1:** Number of bacterial proteins known or suspected to contain Fe-S cluster(s).

| | **Fe-S proteins[a] (total proteins)** |
|---|---|
| Gram-negative bacteria | |
| *A. vinelandii* DJ | 129 (5,050) |
| *E. coli* MG1655 | 143 (4,519) |
| *Pseudomonas aeruginosa* DK2 | 92 (5,884) |
| *S. enterica enterica* serovar Heidelberg str. B182 | 113 (4,333) |
| Gram-positive bacteria | |
| *B. subtilis subtilis 168* | 63 (4,229) |
| *C. botulinum* B1 str. Okra | 101 (3,862) |
| *C. difficile* CD196 | 109 (3,444) |
| *En. faecalis* OG1RF | 25 (2,579) |
| *Lactobacillus casei* BL23 | 8 (3,015) |
| *Listeria monocytogenes* 07PF0776 | 29 (2,797) |
| *M. tuberculosis* H37Ra | 63 (4,034) |
| *St. aureus* 04-02981 | 35 (2,650) |
| *Streptomyces avermitilis* MA-4680 | 83 (7,681) |

[a] Number of gene products identified through gene ontology classes: GO: 0051536, Fe-S cluster binding; GO: 0051537, 2Fe-2S cluster binding; GO: 0051538, 3Fe-4S cluster binding; GO: 0051539, 4Fe-4S cluster binding.

aconitase, whereas *Bacillus subtilis* only expresses one. Second, some of the reactions catalyzed by Fe-S enzymes in *E. coli* are performed by Fe-S independent enzymes in *B. subtilis*, as in the case of fumarase and ribonucleotide reductase. Third, in some species, pathways that contain Fe-S enzymes are not present at all, such as the isoprenoid and quinolinic acid biosynthetic pathways in *Staphylococcus aureus* [3]. These modifications may be seen as evolutionary adaptation events in response to nutritional and environmental challenges faced by certain bacteria. Consequently, pathways involved in the assembly of Fe-S cofactors differ in their biosynthetic components and are subjected to distinct types of regulation.

## 14.3 Fe-S cluster assembly orthologous proteins

The Fe-S cluster biosynthetic pathways found in Gram-positive bacteria are different from those of well-studied Gram-negative bacteria. Initial identification of gene products suspected to be involved in Fe-S cluster formation was based on amino acid sequence similarity to proteins known to participate in the metabolism of Fe-S clusters in other organisms (Tab. 14.2). Prior to experimental validation, their functions are proposed to be analogous to those identified in other bacterial model systems. Genomes of Gram-positive bacteria, despite a few exceptions, contain one single pathway for the assembly and trafficking of Fe-S clusters. The apparent lack of redundancy suggests that gene-encoding components of this pathway are essential. In fact, high-throughput genomic analysis established that *suf* genes are essential in *B. subtilis* [4]; likewise, inactivation of the *Mycobacterium tuberculosis sufB* demonstrated that *SufB* was essential [5, 6]. Disruption of the *Str. thermophilus sufD* gene resulted in a slow-growth phenotype under iron-limiting conditions [7]. Identification of main Fe-S biosynthetic components in Gram-positive species, exemplified for selected species in Tab. 14.2, showed the occurrence of three types of genomic arrangements found in Actinobacteria, Bacilli, and Clostridia taxonomic classes (Fig. 14.1). When compared with the systems found in Gram-negative bacteria, they vary in composition and gene organization.

### 14.3.1 Clostridia-ISC system

A shorter version of the ISC system from *E. coli* is found in Clostridium species, a group of mainly rod-shaped anaerobes that includes *C. difficile*, *C. botulinum*, *C. beijerinckii*, *C. cellulolyticum*, *C. perfringens*, *C. thermocellum*, and *Carboxydothermus hydrogenoformans* (Fig. 14.1). The first gene in the operon encodes a transcriptional regulator similar to IscR. The clostridial IscR consensus sequence retains the three conserved cysteine residues proposed to coordinate a Fe-S cluster in the IscR homologue of *E. coli* but lacks the homologous histidine cluster ligand (His107 in *E. coli*). In

**Tab. 14.2:** Proteins known or suspected to be involved in Fe-S cluster biogenesis in Gram-positive bacteria.

| | *B. subtilis* | *St. aureus* | *En. faecalis* | *L. casei* | *Li. monocytogenes* | *M. tuberculosis* | *M. smegmatis* | *Str. avermitilis* | *C. botulinum* | *C. difficile* |
|---|---|---|---|---|---|---|---|---|---|---|
| Cysteine desulfurase | | | | | | | | | | |
| CD-Class I | BSU27150 (YrvO)[a] | SA2981_1580[a] | OG1RF_11765[a] | LCABL_15200[a] | MUO_07795[a] | MRA_3056[a] | MSMEI_2297[a] | SAV_2933 | CLD_1997[d] | CD196_1141[d] |
| | BSU29590 (NifZ)[b] | SA2981_1674 | OG1RF_10258 | LCABL_14780[b] | MUO_08195[b] | | MSMEI_1207 | SAV_2756 | CLD_3387 | CD196_0700 |
| | BSU27800 (NifS)[c] | | OG1RF_11949 | | MUO-10340[c] | | MSMEI_6085 | SAV_6046 | | |
| | | | | | | | MSMEI_6196 | | | |
| Class II | BSU32690 (SufS)[e] | SA2981_0799[e] | OG1RF_11827[e] | LCABL_14040[e] | MUO-12040[e] | MRA_1473[e] | MSMEI_3046[e] | SAV_6329[e] | CLD_0389 | D196_3484 |
| | | | | | | MRA_3818 | MSMEI_4425 | SAV_6057 | CLD_0841 | CD196_2820 |
| | | | | | | | | SAV_7203 | | |
| | | | | | | | | SAV_1061 | | |
| Sulfur transfer protein | | | | | | | | | | |
| E-type (SufE) | – | – | – | – | – | – | – | – | – | – |
| U-type (SufU) | BSU32680 (SufU) | SA2981_0800 | OG1RF_11826 | LCABL_14050 | MUO_12035 | MRA_1474 | MSMEI_3047 | SAV_6330 | – | – |
| Fe-S scaffold | | | | | | | | | | |
| *IscU* | – | – | – | – | – | – | – | – | CLD_1998 | CD196_1142 |
| *SufBCD* | BSU32710 (SufC) | SA2981_0797 | OG1RF_11829 | LCABL_14020 | MUO_12050 | MRA_1472 | MSMEI_3045 | SAV_6328 | – | – |
| | BSU32700 (SufD) | SA2981_0798 | OG1RF_11828 | LCABL_14030 | MUO_12045 | MRA_1471 | MSMEI_3044 | SAV_6326 | – | – |
| | BSU32670 (SufB) | SA2981_0801 | OG1RF_11825 | LCABL_14060 | MUO_12030 | MRA_1470 | MSMEI_3043 | SAV_6325 | – | – |

| | | | | | | | | | | |
|---|---|---|---|---|---|---|---|---|---|---|
| Fe-S carrier | | | | | | | | | | |
| A-type | BSU32160 (SufA) | SA2981_0895 | – | – | – | MRA_2220 | MSMEI_4171 | SAV6042 | – | – |
| Nfu-type | BSU32220 (Nfu) | SA2981_0891 | – | – | MUO_11975 | – | MSMEI_2651 | – | – | CD196_0799 |
| Iron sensor/donor | | | | | | | | | | |
| Frataxin | BSU05750 (Fra) | – | – | LCABL_24650 | MUO_02090 | – | – | – | – | – |
| Unknown function | | | | | | | | | | |
| SufT | BSU11160 (SufT) | SA2981_0927 | OG1RF_11226 | LCABL_17190 | MUO_11595 | MRA_1475 | MSMEI_3048 | SAV6331 | – | – |

*B. subtilis*, *St. aureus*, *En. faecalis*, *L. casei*, and *Li. monocytogenes* contain the Bacilli-SUF system; *M. tuberculosis*, *M. smegmatis*, and *Str. avermitilis* contain the Actinobacteria-SUF system; *C. botulinum* and *C. difficile* contain the Clostridia-ISC system.

Relevant genomic location coding sequence adjacent to [a] *mnmA*, [b] *thiI*, [c] *nadBCA* operon, [d] *iscU*, and [e] *sufU*.

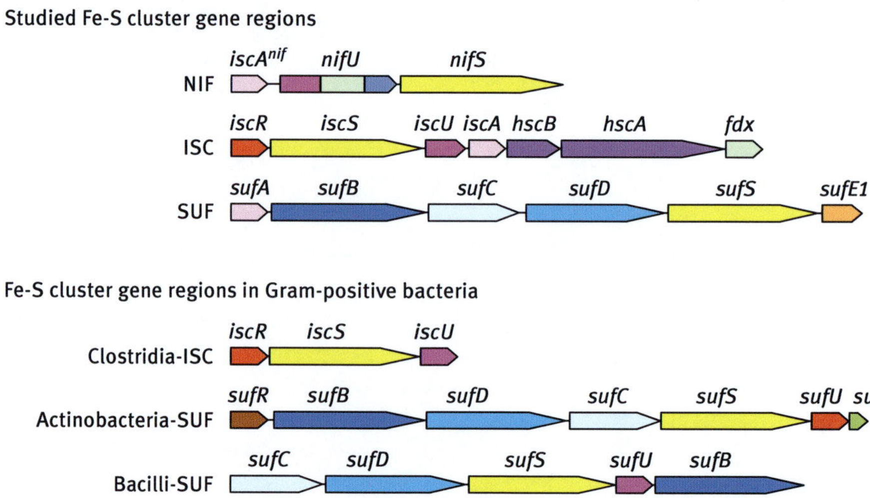


**Fig. 14.1:** Bacterial Fe-S cluster biosynthetic gene regions. Diagram representation shows the gene regions containing genes coding for the main Fe-S cluster biosynthetic components. Gene products displaying similar amino acid sequences are color-coded accordingly. Although not shown, some species of Actinobacteria class contain a ferredoxin-like coding sequence between *sufD* and *sufC*.

addition, clostridial IscR sequences also contain the conserved residues Tyr41, Glu43, Gln44, Arg59, which are involved in binding of IscR to DNA regulatory sequences [8]. Sequence similarities suggest that the mechanism by which clostridial IscR transcription regulates transcription is likely similar to the one described for *E. coli* (Mettert *et al.*, Chapter 13). Downstream of *iscR*, the *iscS* and *iscU* orthologous genes are located. Clostridial IscS is a class I cysteine desulfurase, similar to *E. coli* IscS and *A. vinelandii* IscS and NifS. It contains the conserved lysine involved in PLP binding and a conserved cysteine involved in the persulfide formation and sulfur transfer. The IscU protein sequences share more than 50% identity with the well-studied Fe-S cluster scaffold protein, IscU of *E. coli*, including (i) three conserved cysteine residues along with an aspartate located two residues away from the first cysteine, (ii) the LPP(V/A)K chaperone binding motif, and (iii) a histidine that precedes the third cysteine (Fig. 14.2). Although clostridial species lack the *hscA* and *hscB* chaperone and co-chaperone genes, their genomes encode one to two copies of *dnaK*, *grpE*, and *dnaJ* genes, which are always remotely located but may function as chaperones. It has previously been shown that the *Thermatoga maritima* DnaK is able to bind to *T. maritima* and *Homo sapiens* IscU and that complex formation affects the rate of cluster transfer [9]. Based on these observations, it seems plausible to propose that the Clostridia-ISC system uses the generic DnaKJ pair as chaperones during Fe-S cluster assembly.

In most clostridial species, the *mnmA* gene, which encodes tRNA-specific 2-thiouridylate, is located downstream of the *iscU* coding sequence. The proximity of the gene coding for the 2-thiouridylase MnmA to IscS suggests that IscS is involved in 2-thiouridine tRNA formation in clostridial species. Among Gram-positive bacteria

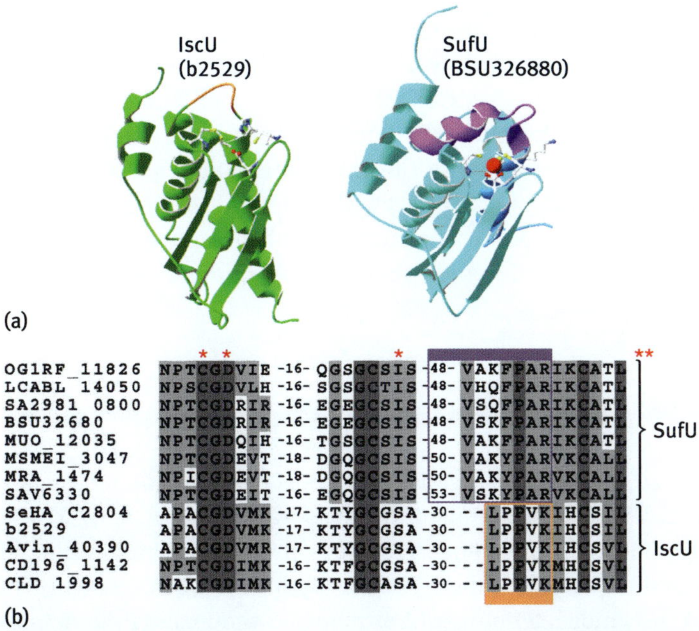


**Fig. 14.2:** Structural and amino acid sequence comparison among SufU and IscU sequences. (a) Ribbon representation of *E. coli IscU* (3LVIL) and *B. subtilis* SufU (1XJS). (b) Amino acid sequence alignment of IscU and SufU sequences from Gram-positive bacteria included in Tab. 14.1 and IscU sequences from *E. coli* (b2529) and *S. enterica* (SeHA_C2804). The alignment shows the three most conserved regions among IscU and SufU. The number of residues between regions are indicated. Strictly conserved residues in both SufU and IscU are highlighted in dark gray, whereas residues partially conserved are highlighted in light gray. The chaperone binding motif within IscU is colored in orange in the structure and the alignment. SufU sequences do not contain a standard LPPVK motif, but residues occupying equivalent positions show conservation within the last three residues (PAR), including the proline, followed by a small hydrophobic residue, and further followed by a positively charged residue. The 18–23 residue insertion unique to SufU sequences is colored in purple in the structure and the alignment. Although not shown in the figure, the consensus sequence of this insertion is FsemvqgkexvedxxxxilgD. Residues known to be important for the function of IscU and SufU are displayed in the structure as a stick-and-ball model and indicated with a star mark in the alignment. The zinc present in the SufU structure is displayed in space filling representation in red.

with sequenced genomes, components of the Clostridia-ISC are phylogenetically closer to components of the ISC system found in Gram-negative bacteria but are very different from systems found in other Gram-positive bacteria (Actinobacteria-SUF and Bacilli-SUF). This observation is not only supported by sequence identity of pathway components but also by potential involvement of the cysteine desulfurase IscS in the biosynthesis of Fe-S clusters and other thio-cofactors such as tRNA thio-modification, which is a feature of other Isc systems.

Notable exceptions to the generalization that clostridial species rely mainly on Isc systems were found in the nitrogen fixing *C. beijerinckii, C. lentocellum,* and *C. acetobutylicum*. These species contain one copy of the Bacilli-SUF gene region (Fig. 14.1 and Tab. 14.2). Interestingly, *C. beijerinckii* contains two proposed Fe-S biosynthetic gene regions: one Clostridia-ISC and one Bacilli-SUF. Among 22 (known or proposed) diazotrophic (nitrogen-fixing) Gram-positive bacteria with sequenced genomes, 20 are species of the Clostridia taxonomic class. None of nitrogen fixing Clostridia contain components of the nitrogen fixing (NIF) Fe-S cluster biosynthetic pathway, suggesting that the general Fe-S cluster biosynthetic machinery is able to provide Fe-S clusters for both housekeeping and nitrogen fixation purposes [10].

### 14.3.2 Actinobacteria-SUF

In the species of Actinobacteria, aerobic bacteria that sometimes form branching filaments, the main Fe-S cluster gene region includes genes with sequence similarity to the *E. coli suf* genes (Fig. 14.1). The first gene in the operon encodes a transcriptional regulator SufR, which is similar to the transcriptional regulator of *sufBCDS* operon in Synechococcus, a marine cyanobacterium. SufR orthologous sequences from Actinobacteria also contain the three conserved cysteine residues thought to be the ligands of a 4Fe-4S cluster in Synechococcal IscR [11]. Despite utilizing a similar mechanism of transcription regulation, SufR is not homologous to IscR; their amino acid sequences show less than 20% identity and the cysteine residues proposed to coordinate a Fe-S cluster do not occupy equivalent positions within the primary sequence.

The Actinobacteria-SUF system also includes the SufB, SufC, and SufD proteins similar to the SufBCD complex found in Gram-negative bacteria. In *E. coli*, SufC is a soluble ATPase that forms a complex with SufD and/or SufB [12]. The latter is thought to serve as a platform for the synthesis of Fe-S clusters [13]. In *M. tuberculosis*, SufB is thought to be essential because a viable deletion strain could not be obtained [5]. Interestingly, in this organism, the *sufB* gene contains an intein domain, and its post-translational splicing of the peptide is crucial for the ability of SufB to interact with other SUF components, namely SufC and SufD [6]. Amino acid sequence alignment of SufB sequences indicated that there are only two cysteine residues, both of which are conserved and present in the *E. coli* SufB sequence (Cys254 and Cys405).

The sulfur mobilization step of Actinobacteria species is thought to involve the class II cysteine desulfurase, SufS. Its amino acid sequence is 64% and 60% similar to the *B. subtilis* and *E. coli* SufS enzymes, respectively. In this reaction, the sulfur is mobilized from the free amino acid cysteine to the proposed sulfur acceptor SufU *via* a PLP-dependent mechanism mediated by SufS. SufU shows 25% and 37% amino acid sequence identity to the Fe-S cluster scaffold IscU from *E. coli* and SufU from *B. subtilis*, respectively (Fig. 14.2). In some Actinobacteria species such as *Arthrobacter aurescens* and *Rothia dentocariosa*, *sufSU* genes are located in a separate gene region away from *sufRBDCT* operon. As discussed later in this chapter, the co-occurrence of a cysteine desulfurase SufS and a U-type protein SufU is not a unique feature of Actinobacteria-SUF because Bacilli species also contain the *sufSU* gene pair.

Downstream of *sufU* is located *sufT*, whose function remains unidentified. SufT is a member of the Domain of Unknown Function 59 family (DUF59), which also includes components of eukaryotic Fe-S cluster biosynthetic pathways, such as AE7 [14], CIA2 [15], and the N-terminus of HC101 [16]. SufT orthologous sequences are also found in other Gram-positive species containing the Bacilli-SUF system but not in Clostridia-ISC or Gram-negative bacteria containing the ISC system (Tab. 14.2). The Cys62 residue in the *M. tuberculosis* sequence is conserved throughout SufT sequences from both Actinobacteria and Bacilli species.

### 14.3.3 Bacilli-SUF

The third and best studied Fe-S cluster biosynthetic gene region in Gram-positive bacteria is the Bacilli-SUF. Genes encoding the SUF components are located in the *sufCDSUB* operon, which displays an alternate order of genes of the Actinobacteria *suf* gene region operon and lacks the regulator *sufR*. In this system, the cysteine desulfurase SufS along with the sulfur acceptor protein SufU catalyzes the sulfur mobilization step [17]. The Bacilli-SUF also includes the SufC, SufD, and SufB proteins similar to the SufBCD from Actinobacteria and Gram-negative bacteria. In *B. subtilis*, SufC is an ATPase that forms a transient complex with SufD, and the binding of SufD to SufC enhances the rate of ATP hydrolysis by nearly 30-fold (unpublished data). Neither the mechanisms of SufC activation nor the step(s) in Fe-S cluster biogenesis and trafficking that involves ATP hydrolysis have been determined in any arrangement of SUF system found in either Gram-positive or Gram-negative bacteria. In *E. coli*, the $SufBC_2D$ complex is proposed to be the scaffold for the synthesis of Fe-S clusters during iron starvation and oxidative stress conditions [18]. Although a similar role is anticipated for Gram-positive SufBCD proteins, this proposal remains to be verified in Gram-positive bacteria.

Transcriptomic and proteomic analyses indicated that patterns of expression of genes involved in Fe-S biogenesis change in species containing the Bacilli-SUF system according to changes in environmental conditions. In a genome-wide expression study, Nicolas *et al.* quantified expression levels of 5,874 unique transcript sequences of *B. subtilis* samples [19]. Levels of whole genome transcript expression varied under 144 growth conditions, growth stages, and stress conditions such as temperature (16°C–51°C) and exposure to salt, $H_2O_2$, and paraquat. The *suf* genes were included in a subset of genes (<3% of coding sequences) that were always highly expressed, regardless of the condition tested. Additionally, a two-dimensional proteomics approach showed that SufC and SufB proteins accumulated upon challenge with salt and paraquat [20]. Expression of genes involved in Fe-S metabolism including biosynthetic SUF and Fe-S enzymes were also sensitive to copper stress through the copper-responsive transcriptional repressor, CsoR

[21]. In *B. anthracis*, proteomic and transcriptomic analysis after challenge with $H_2O_2$ and paraquat also showed increased expression of *suf* along with genes involved in Fe acquisition [22]. Likewise, *Str. thermophilus suf* genes have been associated with the defense against oxidative stress, as inactivation of *sufD* and *sufU* genes caused higher sensitivity to superoxide [7]. In group A Streptococcus, the expression of these genes was enhanced 2- to 4-fold upon challenge with $H_2O_2$ in both wild-type and *perR* deletion strains [23], indicating that upregulation of the Fe-S cluster biosynthetic system in this organism is not regulated by the peroxide stress response transcriptional regulator, PerR. Despite mounting evidence showing the expression of the *suf* genes, regulatory elements directly controlling expression of Bacilli-SUF have not been identified to date.

#### 14.3.3.1 Sulfurtransferase reaction of SufS

The kinetic mechanism of the *B. subtilis* cysteine desulfurase SufS has been determined [17]. Although the overall mechanism of all cysteine desulfurases is thought to follow the same general path [24, 25], their physiological roles and catalytic reactivities are very distinct *in vitro*. Like other cysteine desulfurases, the first half of the *B. subtilis* SufS reaction involves the binding of free cysteine to the active site pyridoxal-5′-phosphate (PLP), leading to the formation of a Cys-PLP internal aldimine intermediate. This event triggers the activation of the substrate for the nucleophilic attack of the active site cysteine (Cys361 in *B. subtilis* SufS), which is the first committed step of the SufS reaction. The first half of this reaction is marked by the formation of a covalent enzyme S-intermediate (Cys361-persulfide) and release of the first product, alanine [17]. *In vitro*, the *B. subtilis* SufS can be partially regenerated through release of free sulfide by artificial reducing agents, such as DTT. *In vivo*, the second half of the reaction involves transfer of the terminal persulfide sulfur to the acceptor molecule SufU, which regenerates the SufS enzyme for the next catalytic cycle.

The *B. subtilis* SufS displayed very modest activity *in vitro* (~5 nmol alanine/min/mg) [17, 26]. These levels are comparable to the activities found for the *E. coli* SufS enzyme [27, 28]. Kinetic assays following the profile of alanine release showed that a kinetic burst was associated with the first turnover of the reaction. Interestingly, the amplitude of the burst phase (nanomoles of alanine product/nanomole of enzyme) was close to 0.5, suggesting that only one active site of the SufS dimer would be active at a time. This behavior, also observed in dimers and tetramers of other PLP-dependent enzymes, is characteristic of a flip-flop mechanism [17]. Analysis of the crystal structure of Synechocystis SufS enzyme (50% identical to *B. subtilis* SufS) shows that the protein environment surrounding the access of the active site is partially occupied by 13-residue loop from the other subunit of the dimer (residues 259–272 of the of Synechocystis SufS) [29]. This structural element could provide communication between subunits and serve as basis for a flip-flop mechanism. This segment is unique to class II cysteine desulfurases including enzymes from both Gram-positive and Gram-negative bacteria.

In the presence of its physiological substrate SufU, the catalytic rate of alanine and sulfide formation by SufS is enhanced nearly 200-fold. The kinetic scheme of this reaction has been characterized as a double displacement mechanism (ping-pong), wherein the release of the first product alanine precedes the binding of the second substrate, SufU. Besides the occurrence of the aforementioned kinetic burst, the characteristic activity profile of substrate saturation curves along with the essential role of the SufS Cys361 residue provided evidence for the proposed kinetic scheme (Fig. 14.3). The affinity of *B. subtilis* SufS for cysteine ($K_{mCys}$ of 86 μM) was found to be near the physiological concentration of reduced cysteine. The $K_m$ of the SufS for SufU was much lower (3 μM), indicating the high affinity of SufS for its sulfur acceptor molecule, SufU [17]. The cysteine:SufU sulfurtransferase reaction of the *B. subtilis* enzyme led to the proposal that other cysteine desulfurases would follow the same scheme. In fact, a recent study showed a similar kinetic profile for the cysteine:SufE sulfurtransferase reaction of the *E. coli* SufS enzyme [28]. However, in this case, the sulfur transfer reaction was resistant to reductive cleavage by DTT [27].

SufS is proposed to be the major S-donor for the biosynthesis of Fe-S clusters in Bacilli species. The reaction is thought to involve the direct sulfur transfer to SufU. However, it is possible that SufS also forms partnerships with yet-unidentified sulfur acceptor molecules and participates in additional sulfur transfer reactions for the biosynthesis of other thio-cofactors that do not contain metals.

#### 14.3.3.2 Sulfur acceptor SufU

The second half of the cysteine desulfurase reaction is controlled by the nucleophilic attack of a thiolate group of SufU onto the persulfide sulfur of SufS Cys361. SufU has three conserved cysteine residues (Cys41, Cys63, and Cys128 in *B. subtilis*), which are mandatory for its sulfurtransferase activity [26]. Alkylation of SufU

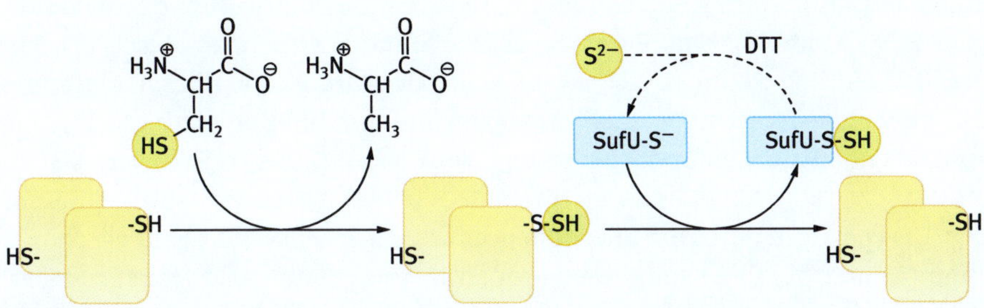


**Fig. 14.3:** Kinetic scheme of the cysteine:SufU sulfur transferase reaction of SufS. The reaction proceeds *via* a ping-pong mechanism where the formation of a covalent S-enzyme intermediate at Cys361 of SufS and release of alanine precede the binding of the second substrate, SufU. In *in vitro* reactions in the presence of DTT, the second product, SufU-SSH, is recycled through release of sulfide (dashed lines).

with iodoacetamide ($SufU_{alk}$) eliminates its ability to participate in the sulfurtransferase reaction of SufS; $SufU_{alk}$ acts as a noncompetitive inhibitor of Suf in competition assays, although $SufU_{alk}$ is still able to interact with SufS. The strong inhibition of $SufU_{alk}$ ($IC_{50}$ of 0.37 μM) displays positive cooperative behavior, suggesting that one molecule of $SufU_{alk}$ inhibits two active sites of SufS. This pattern of inhibition is consistent with the flip-flop mechanism of SufS where the binding of one molecule of $SufU_{alk}$ to one subunit of SufS inhibits both subunits of the SufS dimer [17]. Site-directed mutagenesis revealed that substitution of Cys41 to Ala uniquely elicited an inhibitory effect similar to that of $SufU_{alk}$. Interestingly, the tight binding of SufS to the SufUC41A variant occurs only in the presence of cysteine, indicating that the SufUC41A does not interfere with the free form of the cysteine desulfurase [26].

In *Enterococcus faecalis*, another member of the Bacillus genus, SufU also enhances the activity of SufS [30]. Similar to *B. subtilis*, the three conserved cysteine residues (Cys35, Cys41, and Cys128) are also critical for the sulfurtransferase activity of the *En. faecalis* SufU. Mass spectrometry analysis of a trypsin digest of a reaction containing *En. faecalis* SufU, SufS, and cysteine showed that there was a +64 mass shift (i.e. two additional sulfur atoms) associated with the peptide containing the SufS active site Cys365. This analysis did not show the presence of any modifications in peptides containing the three conserved cysteine residues, but a +32 mass shift was identified in the peptide containing the Cys153 of SufU. Unlike other SufU sequences, the *En. faecalis* protein contains a fourth nonconserved cysteine residue (Cys153) near the C-terminal end [30].

SufU displays primary and tertiary structural similarities to IscU, with some notable differences. The structural alignment of the *B. subtilis* SufU NMR and IscU crystal structures shows on average 3-Å root mean square deviation, a measure of tertiary alignment according to a network service that compare three-dimensional protein structures, the DALI server (Fig. 14.2). The three conserved cysteines along with the aspartate (Asp43) are located in close proximity to both IscU and SufU structures. In the *B. subtilis* SufU structure, these residues provide the tetracoordination to a zinc atom [31]. Notably, zinc was also found in structures of IscU [32, 33]. However, the binding of zinc to IscU seems to be adventitious and does not interfere with the Fe-S cluster assembly reaction. Meanwhile, the binding of zinc to SufU is tight and it is not displaced under conditions known to retain a Fe-S cluster on scaffold protein [62], indicating important differences between these two subfamilies for proteins. Diagnostic features of SufU sequences are the absence of a complete chaperone binding motif (LPPVK) and presence of an 18- to 21-amino acid sequence inserted before the third cysteine (Fig. 14.2) [34]. Interestingly, all SufU sequences also contain a conserved lysine residue (Lys127) in place of the essential histidine immediately preceding the last cysteine (His105 in *E. coli* IscU) [35]. Neither the function of the histidine on IscU nor the function of the lysine on SufU has been determined.

In *B. subtilis*, inactivation of *sufU* has been suggested to be lethal [4], whereas a conditional knockout strain showed pronounced growth defects [36]. Cell extracts depleted of SufU have impaired Fe-S metabolism as evidenced by more than 50% reduction in the levels of aconitase and succinate dehydrogenase activity. Additionally, SufU was able to accelerate the rate of *in vitro* reconstitution of the yeast 4Fe-4S cluster-containing Leu1 enzyme upon incubation with iron and sulfide, leading to the proposal that this protein could serve as a platform for the synthesis of Fe-S clusters [36]. However, unlike the standard scaffold IscU, Fe-S cluster-loaded SufU has not been isolated, an observation that justifies additional investigation on the role of this protein.

#### 14.3.3.3 Cysteine desulfurases involved in the biosynthesis of other thio-cofactors

The SufU-dependent activity of SufS suggests the existence of alternate routes for sulfur transfer. In fact, the *B. subtilis* genome contains four additional putative cysteine desulfurases genes: *yrvO, nifS, nifZ,* and *ycbU.* Amino acid sequence comparison to known cysteine desulfurases shows that all four are members of group I, which includes the studied NifS and IscS from Gram-negative bacteria [37]. Initial analyses of the kinetic properties of these enzymes showed that, with the exception of YcbU, all enzymes displayed reactivity toward cysteine [38, 39]. In *E. coli* and other organisms containing the ISC system, the biosynthesis of Fe-S clusters is coupled to the formation of other sulfur-containing cofactors [40]. First, IscS catalyzes the first step of sulfur mobilization for the synthesis other thio-cofactors. Second, Fe-S enzymes are essential participants in the biosynthesis of sulfur-containing cofactors such as thiamine (ThiH [41] and ThiC [42]), thio-tRNA (MiaB [43, 44] and TtcA [45]), biotin (BioB [43]), Mo-cofactor (MoaA [46]), and lipoic acid (LipA [47]). In Bacilli and Actinobacteria, the presence of additional cysteine desulfurases suggests that sulfur mobilization for the synthesis of thio-cofactors is not coupled to Fe-S cluster biogenesis, and their genomic locations gives hints into their physiological functions.

NifZ is a cysteine desulfurase devoted to the biosynthesis of 4-thiouridine tRNA [38]. In *B. subtilis* and most Gram-positive bacteria, a cysteine desulfurase gene (*nifZ*) is located adjacent to *thiI*. The co-occurrence of the *nifZ-thiI* gene pair in several Gram-positive bacterial species suggests that they are partners in mediating sulfur transfer reactions. In *E. coli* and *S. enterica*, IscS and ThiI participate in the biosynthesis of thiamine and 4-thiouridine ($s^4U8$) [48]. Gene inactivation studies in *B. subtilis* indicated that both NifZ and ThiI are essential for the synthesis of $s^4U8$ but not thiamine. Interestingly, complementation studies in *ΔiscS* or *ΔthiI* strains of *E. coli* showed that $s^4U8$ synthesis was restored only when both NifZ and ThiI were present [38], indicating their mutual dependence. It is possible that Gram-positive ThiI sequences, which lack the rhodanase sulfurtransferase domain, required a dedicated cysteine desulfurase, as they do not compete well with other sulfur acceptors for the general cysteine desulfurase, IscS.

A similar pattern of specific partnership between cysteine desulfurase and sulfur acceptor might also involve the cysteine desulfurase, YrvO from *B. subtilis*. Following a similar biosynthetic scheme as $s^4U8$, post-transcriptional modification of the wobble base on $tRNA^{Lys/Gln/Glu}$ involves the insertion of a sulfur atom at the C2 position of a uridine base ($s^2U34$) [49]. In *E. coli*, this pathway also includes IscS, which transfers the sulfur indirectly to the activating thiouridylase, MnmA, *via* a five-component sulfur-relay system, TusABCDE [50]. In *B. subtilis*, the *mnmA* (*trmU*) gene coding for tRNA (5-methylaminomethyl-2-thiouridylate)-methyltransferase is adjacent to the *yrvO* coding sequence, and most Gram-positive bacteria lack the *tusABCDE* genes, leading to the hypothesis that YrvO provides sulfur directly to MnmA. Gene inactivation studies in *B. subtilis* indicate that both *yrvO* and *mnmA* are essential genes [4], suggesting that the sulfur transfer reaction from YrvO to MnmA is indispensible.

Located upstream of *yrvO* is *cymR*, which encodes the global regulator of cysteine metabolism. In *St. aureus* [51], this transcriptional regulator mediates changes in gene expression upon response to oxidative stress. It has been proposed that oxidative challenge with $H_2O_2$ induces oxidation through formation of sulfenic acid on the single cysteine residue (Cys25) of CymR. This modification decreases its DNA binding affinity by nearly 20-fold [52]. In *B. subtilis*, deletion of *cymR* causes several metabolic defects manifested by (i) slow growth rate, (ii) sensitivity to $H_2O_2$ and paraquat, (iii) depletion of branched chain amino acids, and (iv) elevated cysteine levels and hydrogen sulfide production [53]. CymR is a member of the Rrf2 family of transcription regulators along with IscR and NsrR, but it differs in that it lacks residues involved in Fe-S cluster coordination [54]. In fact, the *B. subtilis* CymR sequence does not contain any cysteine residues, indicating that it employs a distinctive mode of regulation. Unlike *St. aureus*, CymR affinity for DNA in *B. subtilis* is modulated upon complex formation with CysK [55]. Nevertheless, the *cymR-yrvO-mnmA* operon connects distinct aspects of sulfur acquisition and metabolism to oxidative stress, conditions known to affect Fe-S metabolism.

NifS was the first cysteine desulfurase to be studied at the genetic level in *B. subtilis*. In 1993, Sun and Setlow [56] identified two genes upstream of the *B. subtilis nadBCA* operon associated with the biosynthesis of nicotinamide adenine dinucleotide (NAD). The first gene, *nifS*, encodes a cysteine desulfurase, and *nadR* encodes a transcriptional repressor of both the *nadBCA* and *nifS-nadR* operons (Fig. 14.4) [57]. The *B. subtilis* Δ*nifS* strain is unable to synthesize quinolinic acid, an intermediate of the NAD pathway [56]. This metabolic defect can be circumvented by activating the salvage pathway through provision of nicotinic acid in the growth media. Neither nicotinic acid nor NAD are thio-cofactors, but the enzyme that catalyzes the second step in the *de novo* pathway, NadA, is a [4Fe-4S] cluster-containing enzyme [58]. Therefore, it has been suggested that NifS participation in NAD synthesis is indirect, as it facilitates the synthesis of the Fe-S cofactor of NadA, although a potential scaffold partner for this NifS has not been identified.

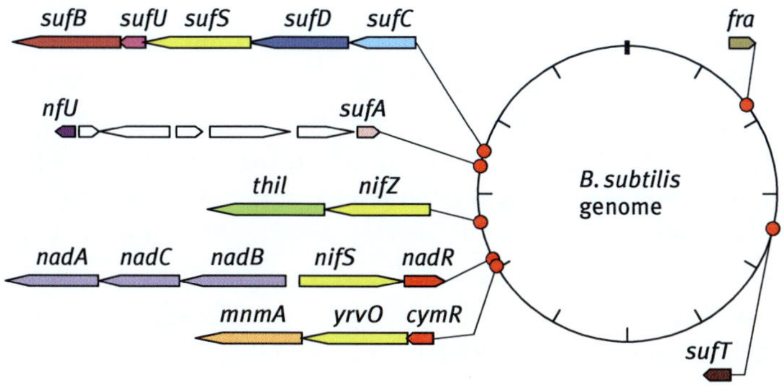


**Fig. 14.4:** Genomic location of genes known or suspect to be involved in Fe-S metabolism in *B. subtilis*. Diagram representation shows the relative chromosomal location, size, and orientation of genes included in Tab. 14.1 and the relevant genomic neighborhood.

#### 14.3.3.4 Additional proteins involved in Fe-S metabolism

Besides the main Fe-S biosynthetic system, additional coding sequences similar to proteins involved in Fe-S metabolism were identified in Gram-positive bacteria (Tab. 14.2). Single copies of A-type Fe-S carrier similar to ErpA, IscA, and SufA are present in some species [59]. Amino acid sequence alignments showed that these sequences are more similar to ErpA than IscA and SufA. In *E. coli*, the three A-type carriers exert partial redundant functions in Fe-S metabolism and ErpA has been shown to be an essential respiratory protein required for activation of Fe-S enzymes IspH and IspG [60]. Although IspH and IspG are essential enzymes in *B. subtilis*, inactivation of the single A-type carrier gene *sufA* (*yutM*) showed no growth phenotype or any defects in Fe-S enzymes [36]. These observations indicate that in Gram-positive bacteria, A-type carriers may serve alternative cellular purposes or that the route of Fe-S delivery may also utilize a distinct subset of cluster carriers. Approximately 5 kb downstream of *sufA*, an *nfu* gene that encodes a second type of Fe-S cluster scaffold/carrier is found in Gram-positive bacteria. Similarly to A-type carriers, genes encoding Nfu domain proteins are present as single copies, within small proteins that are devoid of additional domains. Nfu sequences from Gram-positive bacteria also contain a CxxC motif, suggesting that these Nfu proteins may coordinate transient Fe-S clusters. Transcriptomic analysis in *B. subtilis* cultured under a variety of growth conditions and stress challenges showed that *sufA* and *nfu* were always expressed regardless of growth stage or condition similar to the expression pattern of the *sufCDSUB* genes. Interestingly, these seven genes, proposed to promote the assembly and trafficking of Fe-S clusters, were expressed on average five times higher than other *B. subtilis* genes [19].

Lastly, frataxin orthologous genes (related to Yfh of *Saccharomyces cerevisisiae* and human frataxin, which causes Friedreich ataxia when deficient) are scattered across Gram-positive bacterial genomes. In *B. subtilis*, Fra (YdhG) is important for maintaining iron homeostasis. When cultured in minimum medium, *B. subtilis* strains lacking Fra displayed severe growth defects. In rich medium, *B. subtilis fra* knockout strains lacked the bacillibactin siderophore and showed increased levels of iron and decreased activity of the Fe-S enzyme, aconitase [61]. *In vitro* assays showed that iron-bound Fra could serve as the iron donor for the SufU-mediated activation of aconitase. These observations led to a model of which Fra could serve as a cellular iron donor for the synthesis of Fe-S clusters by the SUF system [61]. Although Fra is an important component of iron-metabolism in *B. subtilis*, its function(s) are not limited to its role in Gram-positive SUF systems because phenotypes associated with its deletion are distinct from those observed upon *sufU* deletion. In addition, *fra* is absent in most Gram-positive species that contain the SUF system.

## 14.4 Concluding remarks and remaining questions

In the last few years, the availability of genomic information has resulted in the identification of numerous proteins involved in Fe-S biogenesis, which has propelled initial biochemical investigation of their associated functions. However, many biochemical steps involved in the synthesis of Fe-S clusters remain to be explored. Apart from the involvement of SufS in sulfur mobilization, the subsequent reactions involving the synthesis and delivery of the Fe-S clusters have not been well defined. The identity and roles of proposed scaffold protein(s), the involvement of an ATP-dependent reaction during synthesis and/or delivery of clusters, and the participation of protein carriers constitute some of the key questions to be addressed in future studies. The Actinobacteria- and Bacilli-SUF systems contain SufU, a sulfur acceptor of SufS that is proposed to be a Fe-S cluster scaffold based on its sequence similarity to IscU and its ability to enhance the activation of Fe-S enzymes. This system also includes SufBCD, which is orthologous to the Fe-S cluster scaffold of the *E. coli* SUF system. The occurrence of two potential scaffolds within the main pathway seems paradoxical [62]. Whether these components have evolved to serve different cellular purposes or to assist Fe-S cluster biogenesis in a single concerted pathway remains to be identified. Interestingly, like the Gram-negative-SUF system, the Gram-positive SUF systems also include the SufC ATPase, whose activity is crucial for function of the pathway. The mechanisms by which ATPase activity of SufC enables cluster synthesis and/or delivery is not understood. Furthermore, the involvement of additional cellular components such as SufA and Nfu remains to be identified. The lack of phenotypes associated with the inactivation of SufA raises questions about its function: When are additional biosynthetic components recruited? Do they serve a specific clientele of Fe-S targets? Are they recruited during specific physiological conditions?

Although the essential role of SUF components in Gram-positive bacteria provides relevance for the study of Fe-S cluster biogenesis, it also poses a technical challenge for *in vivo* validation of proposed biosynthetic schemes. The dependency of Fe-S biosynthetic components for cell survival can be seen as a consequence of the requirement for participation of certain Fe-S enzymes catalyzing indispensable reactions. The identification of Fe-S proteins performing critical reactions not only opens an avenue for the discovery of new metabolic targets to treat infections caused by Gram-positive pathogens but also enables the development of viable genetic approaches to study the functions of their biosynthetic components. It appears that species that contain fewer Fe-S enzymes (Tab. 14.1) may offer good model systems for dissecting the specific roles of Fe-S biogenesis enzymes.

## References


[1] Py B, Barras F. Building Fe-S proteins: bacterial strategies. Nat Rev Microbiol 2010;8:436–46.

[2] Saini V, Farhana A, Glasgow JN, Steyn AJ. Iron sulfur cluster proteins and microbial regulation: implications for understanding tuberculosis. Curr Opin Chem Biol 2012;16:45–53.

[3] Caspi R, Altman T, Dreher K, et al. The MetaCyc database of metabolic pathways and enzymes and the BioCyc collection of pathway/genome databases. Nucleic Acids Res 2012;40:D742–53.

[4] Kobayashi K, Ehrlich SD, Albertini A, et al. Essential Bacillus subtilis genes. Proc Natl Acad Sci USA 2003;100:4678–83.

[5] Huet G, Daffe M, Saves I. Identification of the Mycobacterium tuberculosis SUF machinery as the exclusive mycobacterial system of [Fe-S] cluster assembly: evidence for its implication in the pathogen's survival. J Bacteriol 2005;187:6137–46.

[6] Huet G, Castaing JP, Fournier D, Daffe M, Saves I. Protein splicing of SufB is crucial for the functionality of the Mycobacterium tuberculosis SUF machinery. J Bacteriol 2006;188:3412–4.

[7] Thibessard A, Borges F, Fernandez A, Gintz B, Decaris B, Leblond-Bourget N. Identification of Streptococcus thermophilus CNRZ368 genes involved in defense against superoxide stress. Appl Environ Microbiol 2004;70:2220–9.

[8] Rajagopalan S, Teter SJ, Zwart PH, Brennan RG, Phillips KJ, Kiley PJ. Studies of IscR reveal a unique mechanism for metal-dependent regulation of DNA binding specificity. Nat Struct Mol Biol 2013;20:740–7.

[9] Wu SP, Mansy SS, Cowan JA. Iron-sulfur cluster biosynthesis. Molecular chaperone DnaK promotes IscU-bound [2Fe-2S] cluster stability and inhibits cluster transfer activity. Biochemistry 2005;44:4284–93.

[10] Dos Santos PC, Fang Z, Mason SW, Setubal JC, Dixon R. Distribution of nitrogen fixation and nitrogenase-like sequences amongst microbial genomes. BMC Genomics 2012;13:162.

[11] Shen G, Balasubramanian R, Wang T, et al. SufR coordinates two [4Fe-4S]2+,1+ clusters and functions as a transcriptional repressor of the sufBCDS operon and an autoregulator of sufR in cyanobacteria. J Biol Chem 2007.

[12] Outten FW, Wood MJ, Munoz FM, Storz G. The SufE protein and the SufBCD complex enhance SufS cysteine desulfurase activity as part of a sulfur transfer pathway for Fe-S cluster assembly in Escherichia coli. J Biol Chem 2003;278:45713–9.

[13] Layer G, Gaddam SA, Ayala-Castro CN, et al. SufE Transfers Sulfur from SufS to SufB for Iron-Sulfur Cluster Assembly. J Biol Chem 2007;282:13342–50.

[14] Luo D, Bernard DG, Balk J, Hai H, Cui X. The DUF59 family gene AE7 acts in the cytosolic iron-sulfur cluster assembly pathway to maintain nuclear genome integrity in Arabidopsis. Plant Cell 2012;24:4135–48.

[15] Stehling O, Mascarenhas J, Vashisht AA, et al. Human CIA2A-FAM96A and CIA2B-FAM96B integrate iron homeostasis and maturation of different subsets of cytosolic-nuclear iron-sulfur proteins. Cell Metab 2013;18:187–98.

[16] Schwenkert S, Netz DJ, Frazzon J, et al. Chloroplast HCF101 is a scaffold protein for [4Fe-4S] cluster assembly. Biochem J 2009;425:207–14.
[17] Selbach B, Earles E, Dos Santos PC. Kinetic analysis of the bisubstrate cysteine desulfurase SufS from Bacillus subtilis. Biochemistry 2010;49:8794–802.
[18] Saini A, Mapolelo DT, Chahal HK, Johnson MK, Outten FW. SufD and SufC ATPase activity are required for iron acquisition during in vivo Fe-S cluster formation on SufB. Biochemistry 2010;49:9402–12.
[19] Nicolas P, Mader U, Dervyn E, et al. Condition-dependent transcriptome reveals high-level regulatory architecture in Bacillus subtilis. Science 2012;335:1103–6.
[20] Tam le T, Antelmann H, Eymann C, Albrecht D, Bernhardt J, Hecker M. Proteome signatures for stress and starvation in Bacillus subtilis as revealed by a 2-D gel image color coding approach. Proteomics 2006;6:4565–85.
[21] Chillappagari S, Miethke M, Trip H, Kuipers OP, Marahiel MA. Copper acquisition is mediated by YcnJ and regulated by YcnK and CsoR in Bacillus subtilis. J Bacteriol 2009;191:2362–70.
[22] Pohl S, Tu WY, Aldridge PD, et al. Combined proteomic and transcriptomic analysis of the response of Bacillus anthracis to oxidative stress. Proteomics 2011;11:3036–55.
[23] Le Breton Y, Mistry P, Valdes KM, et al. Genome-wide identification of genes required for fitness of group A Streptococcus in human blood. Infect Immun 2013;81:862–75.
[24] Zheng L, White RH, Cash VL, Dean DR. Mechanism for the desulfurization of L-cysteine catalyzed by the nifS gene product. Biochemistry 1994;33:4714–20.
[25] Behshad E, Bollinger JM Jr. Kinetic analysis of cysteine desulfurase CD0387 from Synechocystis sp. PCC 6803: formation of the persulfide intermediate. Biochemistry 2009;48:12014–23.
[26] Albrecht AG, Peuckert F, Landmann H, Miethke M, Seubert A, Marahiel MA. Mechanistic characterization of sulfur transfer from cysteine desulfurase SufS to the iron-sulfur scaffold SufU in Bacillus subtilis. FEBS Lett 2011;585:465–70.
[27] Selbach BP, Pradhan PK, Dos Santos PC. Protected sulfur transfer reactions by the Escherichia coli Suf system. Biochemistry 2013;52:4089–96.
[28] Dai Y, Outten FW. The E. coli SufS-SufE sulfur transfer system is more resistant to oxidative stress than IscS-IscU. FEBS Lett 2012;586:4016–22.
[29] Tirupati B, Vey JL, Drennan CL, Bollinger JM Jr. Kinetic and structural characterization of Slr0077/SufS, the essential cysteine desulfurase from Synechocystis sp. PCC 6803. Biochemistry 2004;43:12210–9.
[30] Riboldi GP, de Oliveira JS, Frazzon J. Enterococcus faecalis SufU scaffold protein enhances SufS desulfurase activity by acquiring sulfur from its cysteine-153. Biochim Biophys Acta 2011;1814:1910–8.
[31] Kornhaber GJ, Snyder D, Moseley HN, Montelione GT. Identification of zinc-ligated cysteine residues based on 13Calpha and 13Cbeta chemical shift data. J Biomol NMR 2006;34:259–69.
[32] Liu J, Oganesyan N, Shin DH, et al. Structural characterization of an iron-sulfur cluster assembly protein IscU in a zinc-bound form. Proteins 2005;59:875–81.
[33] Ramelot TA, Cort JR, Goldsmith-Fischman S, et al. Solution NMR structure of the iron-sulfur cluster assembly protein U (IscU) with zinc bound at the active site. J Mol Biol 2004;344:567–83.
[34] Johnson DC, Dean DR, Smith AD, Johnson MK. Structure, function, and formation of biological iron-sulfur clusters. Annu Rev Biochem 2005;74:247–81.
[35] Johnson DC, Unciuleac MC, Dean DR. Controlled expression and functional analysis of iron-sulfur cluster biosynthetic components within Azotobacter vinelandii. J Bacteriol 2006;188:7551–61.
[36] Albrecht AG, Netz DJ, Miethke M, et al. SufU is an essential iron-sulfur cluster scaffold protein in Bacillus subtilis. J Bacteriol 2010.

[37] Mihara H, Esaki N. Bacterial cysteine desulfurases: their function and mechanisms. Appl Microbiol Biotechnol 2002;60:12–23.
[38] Rajakovich LJ, Tomlinson J, Dos Santos PC. Functional analysis of Bacillus subtilis genes involved in the biosynthesis of 4-thiouridine in tRNA. J Bacteriol 2012;194:4933–40.
[39] Park JH, Dorrestein PC, Zhai H, Kinsland C, McLafferty FW, Begley TP. Biosynthesis of the thiazole moiety of thiamin pyrophosphate (vitamin B1). Biochemistry 2003;42:12430–8.
[40] Mueller EG. Trafficking in persulfides: delivering sulfur in biosynthetic pathways. Nat Chem Biol 2006;2:185–94.
[41] Leonardi R, Fairhurst SA, Kriek M, Lowe DJ, Roach PL. Thiamine biosynthesis in Escherichia coli: isolation and initial characterisation of the ThiGH complex. FEBS Lett 2003;539:95–9.
[42] Raschke M, Burkle L, Muller N, et al. Vitamin B1 biosynthesis in plants requires the essential iron sulfur cluster protein, THIC. Proc Natl Acad Sci USA 2007;104:19637–42.
[43] Booker SJ, Cicchillo RM, Grove TL. Self-sacrifice in radical S-adenosylmethionine proteins. Curr Opin Chem Biol 2007;11:543–52.
[44] Anton BP, Russell SP, Vertrees J, et al. Functional characterization of the YmcB and YqeV tRNA methylthiotransferases of Bacillus subtilis. Nucleic Acids Res 2010.
[45] Jager G, Leipuviene R, Pollard MG, Qian Q, Bjork GR. The conserved Cys-X1-X2-Cys motif present in the TtcA protein is required for the thiolation of cytidine in position 32 of tRNA from Salmonella enterica serovar Typhimurium. J Bacteriol 2004;186:750–7.
[46] Hanzelmann P, Schindelin H. Crystal structure of the S-adenosylmethionine-dependent enzyme MoaA and its implications for molybdenum cofactor deficiency in humans. Proc Natl Acad Sci USA 2004;101:12870–5.
[47] Cicchillo RM, Lee KH, Baleanu-Gogonea C, Nesbitt NM, Krebs C, Booker SJ. Escherichia coli lipoyl synthase binds two distinct [4Fe-4S] clusters per polypeptide. Biochemistry 2004;43:11770–81.
[48] Martinez-Gomez NC, Palmer LD, Vivas E, Roach PL, Downs DM. The rhodanese domain of ThiI is both necessary and sufficient for synthesis of the thiazole moiety of thiamine in Salmonella enterica. J Bacteriol 2011;193:4582–7.
[49] Numata T, Ikeuchi Y, Fukai S, Suzuki T, Nureki O. Snapshots of tRNA sulphuration via an adenylated intermediate. Nature 2006;442:419–24.
[50] Ikeuchi Y, Shigi N, Kato J, Nishimura A, Suzuki T. Mechanistic insights into sulfur relay by multiple sulfur mediators involved in thiouridine biosynthesis at tRNA wobble positions. Mol Cell 2006;21:97–108.
[51] Soutourina O, Dubrac S, Poupel O, Msadek T, Martin-Verstraete I. The pleiotropic CymR regulator of Staphylococcus aureus plays an important role in virulence and stress response. PLoS Pathog 2010;6:e1000894.
[52] Ji Q, Zhang L, Sun F, et al. Staphylococcus aureus CymR is a new thiol-based oxidation-sensing regulator of stress resistance and oxidative response. J Biol Chem 2012;287:21102–9.
[53] Hullo MF, Martin-Verstraete I, Soutourina O. Complex phenotypes of a mutant inactivated for CymR, the global regulator of cysteine metabolism in Bacillus subtilis. FEMS Microbiol Lett 2010;309:201–7.
[54] Shepard W, Soutourina O, Courtois E, England P, Haouz A, Martin-Verstraete I. Insights into the Rrf2 repressor family – the structure of CymR, the global cysteine regulator of Bacillus subtilis. FEBS J 2011;278:2689–701.
[55] Tanous C, Soutourina O, Raynal B, et al. The CymR regulator in complex with the enzyme CysK controls cysteine metabolism in Bacillus subtilis. J Biol Chem 2008;283:35551–60.
[56] Sun D, Setlow P. Cloning, nucleotide sequence, and regulation of the Bacillus subtilis nadB gene and a nifS-like gene, both of which are essential for NAD biosynthesis. J Bacteriol 1993;175:1423–32.

[57] Rossolillo P, Marinoni I, Galli E, Colosimo A, Albertini AM. YrxA is the transcriptional regulator that represses de novo NAD biosynthesis in Bacillus subtilis. J Bacteriol 2005;187:7155–60.
[58] Marinoni I, Nonnis S, Monteferrante C, et al. Characterization of L-aspartate oxidase and quinolinate synthase from Bacillus subtilis. FEBS J 2008;275:5090–107.
[59] Vinella D, Brochier-Armanet C, Loiseau L, Talla E, Barras F. Iron-sulfur (Fe/S) protein biogenesis: phylogenomic and genetic studies of A-type carriers. PLoS Genet 2009;5:e1000497.
[60] Loiseau L, Gerez C, Bekker M, et al. ErpA, an iron sulfur (Fe S) protein of the A-type essential for respiratory metabolism in Escherichia coli. Proc Natl Acad Sci USA 2007;104:13626–31.
[61] Albrecht AG, Landmann H, Nette D, et al. The frataxin homologue Fra plays a key role in intracellular iron channeling in Bacillus subtilis. Chembiochem 2011;12:2052–61.
[62] Selbach BP, Chung AH, Scott AD, George SJ, Cramer SP, Dos Santos PC. Fe-S cluster biogenesis in Gram-positive bacteria: SufU is a zinc-dependent sulfur transfer protein. Biochemistry 2014;53:152–60.

# 15 Fe-S cluster assembly and regulation in yeast

Debkumar Pain and Andrew Dancis

## 15.1 Introduction

The general scheme by which iron-sulfur (Fe-S) clusters are assembled has come into clearer view recently through work in the model eukaryote *Saccharomyces cerevisiae*. Fe-S cluster assembly components have been discovered by a variety of genetic screens based on iron homeostasis, amino acid biosynthesis, ribosome biogenesis, DNA repair, tRNA thiolation, and oxidant stress protection, all linking to Fe-S cluster assembly components. The mitochondrial system, termed ISC for Fe-S cluster assembly, is a complete machinery and is able to synthesize Fe-S clusters on its own. The central feature is the Isu scaffold, and many components are involved in making the Fe-S cluster intermediate on the Isu scaffold. These include Nfs1/Isd11, which provides cysteine desulfurase activity and contributes sulfur from the amino acid cysteine. Several regulatory conformational changes in Nfs1 are associated with the cysteine desulfurase activity, including stimulation by frataxin and Isd11. The Fe-S cluster intermediate on the Isu scaffold is transferred to recipients in a process involving ATP-dependent chaperones, Ssq1 and Jac1, and Grx5, a glutaredoxin able to coordinate Fe-S clusters in the presence of glutathione. Glutathione is involved in multiple stages of Fe-S cluster assembly *via* interactions with monothiol glutaredoxins inside and outside mitochondria. A parallel cytoplasmic Fe-S cluster machinery, termed CIA for cytoplasmic Fe-S protein assembly, has never been reconstituted, but its constituents, including scaffold, reductase, and transfer components, are required for formation of cytoplasmic and nuclear Fe-S cluster proteins. The Atm1 transporter is an exporter of the mitochondrial inner membrane with its substrate-binding site oriented toward the mitochondrial interior, and thus, it may export a substrate or a signal from mitochondria to the cytoplasm. Defects of ISC components, but not the CIA components, lead to a special iron homeostatic phenotype, characterized by activation of the cellular iron uptake system and accumulation of iron in mitochondria.

## 15.2 Yeast and Fe-S cluster assembly – evolutionary considerations

All kingdoms of life contain Fe-S cluster proteins, and all kingdoms possess at least one Fe-S cluster assembly machinery. Prokaryotes use ISC, plants and Archaea use SUF (for sulfur mobilization), and high-output systems such as *Azotobacter vinelandii* with abundant nitrogenase use NIF (for nitrogen fixation) [1]. The systems have common features such as the use of cysteine desulfurase to provide sulfur, the central role of scaffold proteins for Fe-S cluster synthesis, and the chaperone functions to transfer Fe-S cluster intermediates to recipient proteins [2]. In eukaryotes, such as the yeast *S. cerevisiae*, the ISC machinery was acquired from purple bacteria as part of the endosymbiotic events that gave rise to mitochondria. Thus, many of the components of the yeast mitochondrial Fe-S cluster system are conserved and are recognizable by orthology or homology with the bacterial counterparts [3].

### 15.2.1 Nfs1 and the surprise of Isd11

The sulfur for Fe-S cluster synthesis is derived exclusively from the amino acid cysteine *via* the action of enzymes called cysteine desulfurases (Fig. 15.1). The role of these enzymes in Fe-S cluster assembly was discovered by the pioneering work of Dennis Dean. The *A. vinelandii* NIF operon carries NifS, the founding member of this enzyme group [4]. Key features of the enzyme include a substrate-binding site containing the cofactor pyridoxal phosphate that interacts with the substrate cysteine, and a cysteine residue in the active site that performs a nucleophilic attack on the enzyme-bound substrate for sulfur abstraction [5]. The bacterial ISC operon carries a cysteine desulfurase with these features [2], and the yeast nuclear genome contains a single essential gene, *NFS1*, encoding the cysteine desulfurase [6]. In yeast, the Nfs1 protein is initially synthesized on cytoplasmic ribosomes with an N-terminal mitochondrial targeting signal. Upon import into mitochondria, the targeting signal is removed by two distinct mitochondrial peptidases, generating the mature form of the enzyme [7]. Thus, Nfs1 is present primarily in mitochondria, although a small amount of the enzyme is found in the cytoplasm and nucleus [7]. The protein is essential for viability and functions to provide an activated form of sulfur for Fe-S cluster assembly and other vital cellular processes. The bacterial and mature yeast proteins exhibit 48.4% identity and 62.8% amino acid similarity. The *E. coli* protein has a characteristic biochemical activity – it acts on the amino acid cysteine, releasing sulfide that can be detected by a colorimetric assay [2]. A surprising result was that the mature yeast enzyme, even in the presence of the pyridoxal phosphate cofactor, was inactive [8]. This result was later explained by the discovery of Isd11 [9, 10], a small mitochondrial protein that forms a tight complex with Nfs1 and is required for the Nfs1 cysteine desulfurase activity [8]. *In vivo*, Isd11 is required for cysteine desulfurase activity [8], Nfs1 protein stability, and cellular Fe-S cluster assembly [9, 10]. Isd11 has no prokaryotic homologue and seems to be an entirely eukaryotic invention. Homologous proteins, however, are found in all eukaryotic subgroups, including plants, animals and lineages with reduced organelle contents such as hydrogenosomes of *Trichomonas*, and mitosomes of *Cryptosporidium* species [11]. In hydrogenosomes or mitosomes, the oxidative phosphorylation function has been lost, and only rudimentary components of Fe-S cluster assembly remain as organelle contents. The retention of Nfs1/Isd11 in these organelles suggests that the Fe-S cluster synthesis may be the most essential and irreplaceable function. The evolutionary origin of Isd11 is unclear; it might have been originated at the time of the key endosymbiotic event shared by all eukaryotes [12]. The way in which this special eukaryotic component influences the process of Fe-S cluster assembly is still being worked out, and distinct regulatory features are being uncovered.

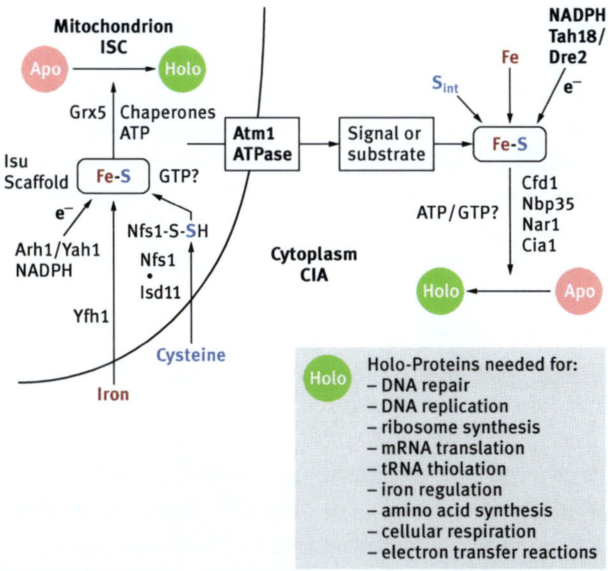


**Fig. 15.1:** General scheme for Fe-S cluster assembly in yeast. A mitochondrial system termed ISC for Fe-S cluster assembly is complete and able to synthesize Fe-S clusters on its own [49]. The central feature is the Isu scaffold, and many components are involved in making an Fe-S cluster intermediate on the Isu scaffold [50]. These include Nfs1/Isd11, which contribute sulfur from cysteine [8], Arh1/Yah1, which provide electrons [51, 52], and Yfh1, which provides iron and/or regulates the sulfur donation [22, 23]. The intermediate on the Isu scaffold is transferred to recipients in a process involving ATP-dependent chaperones, Ssq1 and Jac1 [29], and Grx5, a glutaredoxin able to coordinate Fe-S clusters in the presence of glutathione [53]. A parallel cytoplasmic Fe-S cluster system termed CIA for cytoplasmic Fe-S protein assembly, has never been reconstituted and may be incomplete. It is made up of scaffold components (Cfd1, Npb35), reductase components (Dre2, Tah18), and transfer components (Nar1, Cia1) [3]. The iron source is not well defined. The sulfur intermediate, $S_{int}$, presumably originates from cysteine *via* cysteine desulfurase activity of Nfs1/Isd11, but it is also not well defined in terms of its chemical nature or its cellular source. The Atm1 transporter is an exporter of the mitochondrial inner membrane [54], and this may export a substrate or a signal from mitochondria to the cytoplasm. The output of the Fe-S cluster assembly systems is made up of critical Fe-S cluster holoproteins that perform many functions [1], some of which are listed.

### 15.2.2 Scaffold proteins in yeast mitochondria

The basic scheme for Fe-S cluster assembly in cells involves transient assembly of a cluster on a scaffold protein and subsequent transfer of the intermediate to recipients (Fig. 15.1) [2]. In yeast mitochondria, Isu1 and Isu2 are redundant scaffold proteins encoded by separate genes [13]. These proteins are synthesized on cytoplasmic ribosomes and are imported into mitochondria. Inside mitochondria, the targeting sequence is removed by the processing peptidase, and the proteins interact with other components of the Fe-S cluster assembly machinery. The yeast proteins are highly homologous to the *E. coli* IscU scaffold, exhibiting 62.4% identity and 74.4% similarity at the amino acid level. Accordingly, the bacterial protein targeted to yeast mitochondria by addition of a matrix targeting sequence, is able to complement the essential function of the yeast Isu scaffolds [14]. The yeast protein is also highly conserved with metazoan mitochondrial scaffold proteins exhibiting 64.9% identity and 73.9% similarity, and the human protein likewise can complement the yeast mutant lacking Isu1/2. The retention of high sequence similarity and orthologous relationships likely reflect constraints that result from multiple conserved protein-protein interactions with the scaffold that are required during the process of Fe-S cluster assembly [15].

### 15.2.3 Frataxin's roles throughout evolution

Frataxin is a highly conserved protein, found in prokaryotes such as *E. coli*, in primitive eukaryotes such as yeast, and in metazoans such as humans. The human frataxin was cloned in 1996 after identifying the disease locus of patients with the inherited neurodegenerative disease, Friedreich ataxia [16]. The homologous yeast gene (*YFH1*) was knocked out, and the phenotypic characterization of the deletion strain led to realization of its role in iron homeostasis, and Fe-S cluster assembly [17]. Recently frataxin was found to be a constituent of an Fe-S cluster assembly complex in mitochondria (Fig. 15.1), associated with Nfs1/Isd11, and Isu, and cells lacking components of the complex were found to be deficient in Fe-S cluster proteins [18, 19]. A regulatory function of frataxin was hinted at early on because yeast mutants completely lacking the frataxin homologue (Yfh1) were still viable and exhibited a small amount of basal Fe-S cluster assembly [20]. The Δ*yfh1* deletion strain accumulated extragenic suppressor mutants [21]. By contrast, knockouts of integral components (e.g. Nfs1) of the Fe-S cluster assembly machinery were completely nonviable with no detectable escape mutants. Frataxin might be involved in promoting and coordinating sulfur [22] and iron [23] delivery to Isu for the synthesis of Fe-S cluster intermediates, although the mechanistic details are still being clarified.

The *E. coli* genome carries a frataxin protein called CyaY, and it is not present on the ISC operon. CyaY was shown to function in yeast if properly targeted to mitochondria; it was able to complement the Fe-S cluster assembly defect of the Δ*yfh1* mutant, presumably by interacting with yeast Nfs1 and the Fe-S cluster assembly complex [24]. Whereas the eukaryotic Nfs1/Isd11 exhibited low cysteine desulfurase activity that was stimulated by frataxin, the prokaryotic Nfs1 homologue, IscS, showed constitutively high activity that was inhibited by frataxin [25]. Thus, the regulatory effects of frataxin were different for prokaryotes and eukaryotes. These differences were further underscored by the finding of a genetic suppressor of the yeast Δ*yfh1* deletion that promoted growth and Fe-S cluster assembly independent of frataxin. Specifically a single point mutation in Isu1 was identified with frataxin-bypassing activity. The substitution changed methionine 107, an amino acid present in all eukaryotic Isu proteins, to isoleucine, an amino acid present in *E. coli*, and many other prokaryotic Isu homologues [26]. The basal cysteine desulfurase of the yeast Δ*yfh1* mitochondria was increased in the presence of Isu1 (M107I), thus mimicking the situation in prokaryotes [27]. In summary, there appear to be fundamental differences between prokaryotic and mitochondrial Fe-S cluster assembly apparatus in the way the components are wired. A critical function of frataxin in stimulating cysteine desulfurase in yeast can be bypassed by a mutation that renders the Isu scaffold more "prokaryote-like."

### 15.2.4 Ssq1 is a specialized Hsp70 chaperone arising by convergent evolution

Hsp70 chaperones and co-chaperones in mitochondria are required for the transfer of Fe-S cluster intermediates formed on Isu to recipient proteins (Fig. 15.1) [28]. In yeast, there are two mitochondrial Hsp70 (mtHsp70) proteins. The major one, Ssc1, is involved in protein import and protein folding in the organelle, and thus, it has innumerable substrates. The less abundant one, Ssq1, apparently arose by a gene duplication event about 300 million years ago, and over evolutionary time it became adapted to interacting with a single client protein, Isu. The Isu scaffold contains the LPPVK motif that binds to the Ssq1 substrate-binding site, and this interaction is highly specific, as Ssq1 does not interact with more general mitochondrial targeting peptides. The Ssq1 ATPase activity is optimally stimulated by the LPPVK peptide in conjunction with the specialized co-chaperone, Jac1, a J protein that is not stimulated by other peptides [29]. In yeast mitochondria, Mge1 is required to release ADP from both Ssc1 and Ssq1, allowing initiation of another catalytic cycle. For the human and *E. coli* systems, the evolutionary aspects of the chaperones are somewhat different. In humans, the gene duplication giving rise to Ssq1 did not occur. Therefore, there is a single mtHsp70 that must handle all clients, including protein import substrates and Isu. The existence of a specific J protein, orthologous to the yeast Jac1 and adapted to interacting with Isu, however, may provide some degree of specificity to the system and allow efficient handling of Fe-S cluster protein substrates [29]. In *E. coli*, a specialized Hsp70 and J protein, called HscA and HscB, respectively, are found on the ISC operon. However, these proteins arose from the major cellular Hsp70, DnaK, by an independent gene duplication. They are not considered to be orthologues of the yeast proteins but rather the products of independent convergent evolution. Remarkably, the substrate binding specificity of the HscA is identical to that of Ssq1, and the cognate scaffold protein has the identical LPPVK motif [30]. Other subtle differences in the systems have been found; for example, the *E. coli* chaperones do not need a recycling factor because of their lower binding affinities for nucleotides [31].

### 15.2.5 Atm1 and CIA components

Atm1 is an ABC transporter, with its substrate-binding site oriented toward the mitochondrial matrix, suggesting that it exports its substrate(s) from the matrix to other compartments of the cell (Fig. 15.1). Atm1 in yeast has been implicated in Fe-S cluster assembly and iron homeostasis. Strong homologues exist in proteobacteria, suggesting that it could have come along during the original endosymbiotic event. The importance of a transporter that would move its substrate from the inside to the outside of the bacterial cell is unclear, although one could imagine that some key activities for Fe-S cluster assembly take place in the periplasm. The substrate for the bacterial transporter is unknown, and phenotypes of the knockout have not been characterized. Therefore, this evolutionary link does not shed much light on the function of these proteins, but rather adds to the mystery.

CIA components are involved in cytoplasmic Fe-S cluster assembly (Fig. 15.1), and in yeast there are now eight constituents that have been identified [3]. They are thought to perform reductase (Dre2 and Tah18), scaffold (Nbp35 and Cfd1), and transfer functions (Cia1, Cia2, Nar1, and Mms18), somewhat analogous to counterparts in the mitochondrial system. In terms of evolution, however, these genes are found in the last common eukaryotic ancestor before the diversification of eukaryotic lineages, and thus, they are present in protists and primitive eukaryotes [32]. Their prior origins in prokaryotes are harder to discern. In fact, only the scaffold components, Nbp35 and Cfd1, have a clear prokaryotic homologue. This protein, called ApbC, was studied in *Salmonella enterica* and shown to function in Fe-S cluster biogenesis. Furthermore, *in vitro* biochemical analysis showed that the purified ApbC could function as a scaffold in binding and transferring Fe-S cluster intermediates to Leu1, the yeast cytoplasmic $Fe_4S_4$ enzyme [33]. Interestingly, Archaea also have ApbC homologues that were able to functionally complement the *S. enterica* mutant, indicating that at least for this case, the CIA machinery seems to have origins in Archaea [34].

### 15.2.6 Yeast components are conserved with their human counterparts

Moving up the evolutionary ladder, the components of the human Fe-S cluster assembly machineries (ISC and CIA) are conserved with yeast, with a few exceptions. Many human proteins have yeast orthologues, and the human proteins expressed in yeast often correct the Fe-S cluster assembly defects of the corresponding yeast mutants, demonstrating real functional equivalence. This is true for Isu1/2 (our unpublished data), frataxin [35], Atm1 [36], Dre2 [37], Tah18 [38], and others. Grx5 from zebrafish rescued the corresponding yeast mutant, and so this mitochondrial glutaredoxin is also conserved between yeast and metazoans [39]. The story for the yeast ferredoxin Yah1 is complicated by existence of two human ferredoxins encoded by separate genes, Fdx1 and Fdx2. In one study, Fdx2 but not Fdx1 was able to complement the phenotypes of the Yah1-depleted yeast, including Fe-S cluster assembly, overall heme levels, and heme O formation [40]. In another study, knockdown of either Fdx1 or Fdx2 in various human cell lines impaired Fe-S cluster assembly [41]. The basis for the specificity of the Fdx2 yeast complementation is unclear as both isoforms are very similar. As mentioned above (Section 15.2.4), human mitochondria make use of a single Hsp70 chaperone for handling protein import, protein folding, and Fe-S cluster assembly. Thus, the human Hsp70 has multiple client proteins, including Isu1, as contrasted with the yeast specialized Hsp70, Ssq1, which interacts with Isu as its sole client [29]. The human Hsc20, which encodes a J protein with similarity to yeast Jac1, interacts with the human Isu1, although other substrates have not been ruled out. The presence of an insertion sequence between the aspartic acid of the HPD motif and an invariant serine in the helix 3 apparently enhances the interaction of the Hsc20 with the generic mtHsp70 (called HspA9). In any case, the human Hsc20 is able to complement the yeast Jac1 deletion strain [42].

### 15.2.7 Yeast Fe-S cluster assembly mutants modeling aspects of human diseases

The yeast frataxin homologue (Yfh1), the scaffold proteins of mitochondria (Isu1/2), and the mitochondrial exporter (Atm1) all have human homologues. Human diseases may result from deficiency of each of these conserved Fe-S cluster components [43]. Because Fe-S cluster assembly is essential for viability of all cells, the disease-causing mutations are usually hypomorphic rather than complete loss-of-function mutations. The phenotype-genotype correlations for Fe-S cluster diseases are very complicated, especially as regards to tissue specificity. It remains unclear why one tissue is targeted in some cases and another is affected in other cases. Friedreich ataxia results from a GAA expansion in the first intron of the frataxin gene, resulting in decreased expression of frataxin [16]. The disease primarily affects certain nerves, especially dorsal root ganglion cells. Cardiomyocytes are also sensitive to lack of frataxin, and cardiomyopathy is a major feature of the disease [44]. Blood cells of various lineages are spared, even though frataxin is deficient in these cells and Fe-S cluster assembly is necessary for blood cell development [45]. In ISCU myopathy, skeletal muscles are particularly affected due to low levels of ISCU (homologous to yeast Isu1/2), with characteristic episodes of muscle fatigue and lactic acidosis. More severe stress leads to rhabdomyolysis and myoglobinuria as a result of muscle cell death. A tissue specific splicing defect interfering with expression of the ISCU gene may partly explain the predilection for skeletal muscle [46]. In a disease called XLSA/A, X-linked sideroblastic anemia with ataxia, a defect in ABCB7 (homologous to yeast Atm1) has been implicated. A nonprogressive neurological defect causing ataxia and anemia are the principal manifestations. The anemia is characterized by ringed sideroblasts, indicating red cell precursors with mitochondrial iron accumulations that stain with Prussian blue. The ring pattern derives from the propensity of iron-loaded mitochondria to encircle the nucleus in these cells [47]. Although these diseases are very different, each one is caused by deficiency of a protein involved in Fe-S cluster assembly [43]. In addition, in each case the diseased tissues show deficiency of Fe-S cluster proteins and perturbed iron metabolism, thereby resembling the corresponding yeast mutants. The perturbed iron metabolism, also similar to the yeast mutants, is characterized by increased cellular iron and increased mitochondrial iron, primarily in a nonproteinaceous form [48].

## 15.3 Yeast genetic screens pointing to the Fe-S cluster assembly apparatus

The constituents of the Fe-S cluster machinery were initially discovered by homology with components from other species, especially *E. coli*, or by genetic screens. The discovery of NifS in *A. vinelandii* by Dennis Dean opened the field and showed that Fe-S cluster assembly in cells was a catalyzed process [4]. This finding provided a fulcrum for uncovering other components of the Fe-S cluster assembly machinery, initially by association with operon groupings of genes in prokaryotes. The identification of the NIF operon for nitrogenase production and the ISC operon for housekeeping Fe-S cluster assembly in *A. vinelandii* led to the identification of cysteine desulfurase, reductase, scaffold, and chaperone components [55]. These were found to have homologues in eukaryotes that localized primarily to mitochondria. Studies in yeast allowed for phenotypic characterization of the corresponding genes and processes. Some components were not found on the operons, either because they were elsewhere in the prokaryotic genome (e.g. Arh1, frataxin) or because they were not present at all in prokaryotes (e.g. Isd11) [56]. The discovery of the role of frataxin was made possible following cloning of the human disease gene and identification of the yeast homologue. The yeast frataxin homologue was deleted, with striking phenotypic effects on iron homeostasis and Fe-S clusters [17].

The yeast *S. cerevisiae* is conducive to genetic screens, and many informative screens were performed. In general, mutants were identified with easy-to-score phenotypes: some mutants exhibited altered iron homeostasis [57], amino acid auxotrophies [58, 59], ribosome assembly abnormalities [60], DNA replication defects [61], or absent tRNA modifications [62]. The phenotypes were traced back to one or more Fe-S cluster proteins (in some cases, the relevant Fe-S cluster protein has not yet been identified) and to an underlying problem with Fe-S cluster synthesis. In this way, hypomorphic mutants in many of the constituents of the Fe-S cluster machinery were found (Tab. 15.1).

### 15.3.1 Misregulation of iron uptake

In yeast, the high affinity cellular iron uptake system consists of an externally directed surface reductase encoded by *FRE1* and *FRE2*, and an oxidase-permease complex encoded by *FET3/FTR1* [63]. The entirety is homeostatically regulated by iron availability *via* the Aft1 transcriptional regulator and the Cth1 and Cth2 RNA stability regulators, driving increased expression of the iron uptake system in response to iron deprivation [64]. For the screen, the *FRE1* promoter was fused to the *HIS3* open reading frame, and mutants were selected under iron replete conditions that constitutively expressed His3. The selected mutants showed constitutively high iron uptake and included components of the mitochondrial Fe-S cluster assembly machinery,

**Tab. 15.1:** Yeast genetic screens leading to Fe-S cluster assembly components.

| | Genetic screen | Mutant phenotype | Fe-S cluster(s) targeted | Assembly components found | References |
|---|---|---|---|---|---|
| 1 | FRE1-HIS3 selection | Misregulated iron uptake | Aft1, Fra1, Grx3/4 | Ssq1, Nfs1, Jac1, Atm1 | [57] |
| 2 | Sod1 suppression | Lysine, methionine auxotrophy | Unknown | Nfs1, Ssq1, Jac1 | [59] |
| 3 | IRP1 in yeast | Glutamate auxotrophy | IRP1 in cytoplasm | Cfd1 | [58] |
| 4 | Pol3 interaction | *pol3-13* synthetic lethality | Rad3, Met10, Dna2, Ntg2 | Dre2, Tah18, Nbp35, Mms19 | [61] |
| 5 | Mrs3/4 interaction | Δ*mrs3* Δ*mrs4* synthetic lethality | Unknown | Dre2 | [37] |
| 6 | Ribosome processing | Rpl25-GFP in nucleus | Rli1 | Cfd1, Nbp35, Nar1 | [60] |
| 7 | Cytosolic tRNA thiolation | Sensitive to *K. lactis* killer toxin | Unknown | Cfd1, Npb35, Cia1 | [62, 65] |
| 8 | Yap5 high iron response | No Ccc1 induction | ?Yap5 cluster | Ssq1 | [66] |

including a hypomorphic allele of Nfs1 [6], loss-of-function mutants of Ssq1 [57], Jac1 [67], and Atm1 [68]. Cytoplasmic Fe-S cluster assembly components such as Dre2 and others were not selected. The iron homeostatic effects of perturbing mitochondrial Fe-S cluster assembly led to activation of *FRE1-HIS3* expression, allowing selection of the mutants.

### 15.3.2 Suppression of Δ*sod1* amino acid auxotrophies

The Cu-Zn superoxide dismutase, Sod1, is responsible for defense against oxidative stress by dismutation of superoxide to peroxide and water. The yeast Δ*sod1* mutant is sensitive to oxygen exposure, and in air, it develops lysine and methionine auxotrophies [59]. The most likely explanation for the auxotrophies is that a key step in the biosynthesis of these amino acids is sensitive to oxidants that are ordinarily inactivated by Sod1. In a genetic screen, mutants of Δ*sod1* were selected for the ability to grow in the absence of lysine and methionine. These mutants included partial loss-of-function alleles of Nfs1, Ssq1, and Jac1 i.e. mitochondrial Fe-S cluster assembly components [59]. The manner in which partial loss-of-function of Fe-S cluster assembly rescues Sod1 deficiency is not immediately clear. Interestingly, all of these mutants were associated with mitochondrial iron accumulation, and mutations minimizing this effect also abrogated the Δ*sod1* suppression. Thus, the mitochondrial iron might cause a stress response in the mutants that enhanced specific pathways of amino acid synthesis [59].

### 15.3.3 tRNA modification and the *SPL1-1* allele

Nfs1 was first identified in a complicated genetic screen involving splicing of certain nuclear-cytoplasmic tRNAs [69]. Briefly, an opal stop codon was inserted into the *HIS4* and *LEU2* genes, leading to auxotrophy for histidine and leucine. A cytoplasmic/nuclear tRNA with a spliced intron introduced into this strain was able to relieve the auxotrophy by stop codon suppression. A modified and nonfunctional version of the suppressor tRNA with an altered splice site was introduced instead, and this was the starting point for the screen. A randomly selected mutant that conferred splicing activity and stop codon suppression was identified, and the activity was found to depend on a dominant allele of Nfs1 called *SPL1-1* for its newly acquired splicing activity [69]. It remains to be explained how the altered Nfs1 enhanced splicing of the inactive opal tRNA suppressor. However, subsequent studies have shown that some tRNA modifications such as tRNA thiolation are dependent on Fe-S cluster synthesis in general and cytoplasmic Fe-S cluster synthesis in particular [65]. One can imagine that an Nfs1 mutant could result in an altered tRNA, maybe incompletely thiolated, which might be a more efficient splicing substrate.

### 15.3.4 tRNA thiolation and resistance to killer toxin

Certain cytoplasmic tRNAs such as cy-tRNA-Lys2$_{UUU}$ and cy-tRNA-Glu3$_{UUC}$ carry a thiol modified uridine in the 5′ wobble position, the so-called mcm$^5$s$^2$U (5-methoxycarbonylmethyl-2-thiouridine). The modification is essential for correct base pairing and for viability of the organism. A role for Fe-S clusters in cytoplasmic tRNA modifications was indirectly shown by cellular depletion experiments. The depletion of CIA components (Cfd1, Nbp35, or Cia1) was associated with decreased tRNA modification [65]. The presumption is that a critical Fe-S cluster protein is involved in uridine thiolation, although the identity and role of the presumed Fe-S cluster protein has not been clarified. A global genetic screen was undertaken to find the genes required for modification of the 5′ wobble nucleoside [70]. The idea for the screen comes from the observation that the γ subunit of the *Kluyveromyces lactis* killer toxin is a tRNA endonuclease that cleaves modified tRNAs 3′ of the wobble nucleoside, utilizing the modified base as a recognition motif. Mutants that fail to modify the uridine were identified based on their resistance to the killer toxin. Various resistant mutants including Urm1, Uba4, Ncs2, Ncs6, and Tum1 were found in this screen, and it is possible that one of these proteins carries an Fe-S cluster or requires an Fe-S cluster for its activation [70].

### 15.3.5 Cytoplasmic aconitase maturation

Aconitase, a [$Fe_4S_4$] cluster protein, is responsible for isomerizing citrate to isocitrate, which in turn gives rise to α-ketoglutarate as part of the tricarboxylic acid cycle. The yeast Δ*aco1* deletion strain is viable but develops auxotrophy for glutamate because of the inadequate supply of the α-ketoglutarate precursor. Interestingly, although yeast aconitase is primarily found in mitochondria, expression of aconitase (in this case, the human IRP1 [71]) in the cytoplasm of a Δ*aco1* strain corrects the glutamate auxotrophy, probably *via* exchange of intermediates in and out of mitochondria [72]. The yeast Δ*aco1* strain expressing IRP1 in the cytoplasm was the starting point for a genetic screen. Mutants with glutamate auxotrophy included a hypomorphic allele of Cfd1 in which the AUG start codon was changed to a less efficient AUA initiator. The auxotrophy for glutamate developed because the cytoplasmic IRP1 was no longer properly loaded with its [$Fe_4S_4$] cluster, and as a consequence, the enzyme was less active [58]. The results demonstrated that Cfd1 was required for cytoplasmic Fe-S cluster assembly, making it the first member of the CIA machinery to be discovered. The Cfd1-depletion phenotype was instructive and novel because the defect was restricted to cytoplasmic Fe-S clusters, with preservation of the mitochondrial Fe-S cluster enzyme activities [58].

### 15.3.6 Ribosome assembly

The biogenesis of eukaryotic ribosomes is a highly coordinated process, beginning in the nucleus and ending with the assembly of 60S and 40S subunits in the cytoplasm. The nuclear export of various ribosomal components is subjected to quality control, and improper processing of ribosomal proteins leads to their retention in the nucleus [73]. For the screen, a fusion of the ribosomal protein Rpl25 (or Rps2) to GFP was inserted into a library of temperature-sensitive mutants [60]. After shifting to non-permissive temperature, mutants were identified with enhanced nuclear GFP fluorescence due to nuclear retention of the ribosomal precursor protein. The mutants identified in this manner, so called rix or ribosomal protein export mutants, included CIA components Cfd1, Nbp35, and Nar1 [60]. The explanation for the ribosomal phenotype of CIA mutants was provided by the protein Rli1. Rli1 is required for rRNA maturation and nuclear export of 60S and 40S subunits, and Rli1 possesses two [$Fe_4S_4$] clusters that are critical to its function [74]. Thus, mutations interfering with Fe-S cluster assembly resulted in defects in Rli1 function and indirectly led to a block in nuclear-cytoplasmic trafficking of ribosomal proteins. The mutants could only have been discovered with such a library of temperature-sensitive alleles because the corresponding genes are essential and complete loss-of-function mutants are nonviable.

### 15.3.7 Synthetic lethality with the *pol3-13* allele

DNA polymerases are multi-subunit enzyme complexes that are responsible for priming, initiating, and extending the lagging and leading strands of the double stranded genome during the process of DNA replication. The catalytic subunits, Pol1,

Pol2, and Pol3, contain conserved sets of cysteine residues in an essential carboxyl terminal domain [75]. A change of $^{104}$Cys to Ser in the carboxyl terminal domain of Pol3 conferred a temperature-sensitive growth phenotype, presumably because of compromised efficiency of replication [61]. This finding was the basis for a synthetic lethal sectoring screen, starting with the *pol3-13* allele covered by a plasmid with the wild-type *POL3*. Synthetic lethal screens are commonly used in model organisms to identify other genes that function in the same genetic pathway. They identify other genes that are needed for survival in a compromised strain. The strain with the $^{104}$Cys to Ser in the carboxyl terminal domain of Pol3 was mutagenized, and mutants were recognized that were unable to survive without the plasmid-borne copy of wild-type *POL3*. The procedure identified second site mutations (mutations in other genes) that worsened the phenotype of the *pol3-13* allele, and these included alleles of CIA components such as Dre2, Tah18, Nbp35, and Mms19 [61]. These genes have been implicated in assembly of cytoplasmic or nuclear Fe-S clusters. The observation linking Pol3 and Fe-S cluster assembly was subsequently explained when Pol3 itself was found to coordinate an Fe-S cluster [76]. The cluster plays a role in stabilizing the protein and mediating critical protein-protein interactions among the DNA polymerase subunits. Thus, the synthetic lethal genetic relationship occurred because the *pol3-13* allele with $^{104}$Cys replaced by Ser was less effective in coordinating its Fe-S cluster, and second site mutations interfering with Fe-S cluster assembly made matters even worse. Another genetic screen was based on the putative mitochondrial iron carriers, Mrs3 and Mrs4. Cells lacking both of these carrier proteins grow robustly on rich media, and a synthetic lethal screen identified Dre2, an essential component of the CIA [37]. The *dre2-137* mutant allele was synthetically lethal with the double mutant Δ*mrs3*Δ*mrs4*. Other Fe-S cluster assembly components including Cfd1 and Atm1 showed a similar synthetic lethality when combined with Δ*mrs3*Δ*mrs4*, and the synthetic lethality was made worse by iron deprivation [37]. A possible explanation for these findings is that mitochondrial iron (involving Mrs3/4) and cytoplasmic Fe-S cluster assembly (involving Dre2, Cfd1, and Atm1) are dependent on each other.

### 15.3.8 Factors needed for Yap5 response to high iron

The cellular responses to iron exposure involve downregulation of the cell surface iron uptake systems (Aft1 and Cth1/2 mediated) [64] and upregulation of vacuolar iron transport *via* increased Ccc1 expression (Yap5 mediated) [77]. The signals for the Yap5 effects were unknown, and so a genetic screen was undertaken [66]. The Yap5 regulatory domain responsible for conveying the iron signal was fused to Gal4, thereby conferring an iron-dependent response to the *GAL1* promoter. The *GAL1* promoter was fused to *HIS3*, and the hybrid gene was used to identify mutants that are unresponsive to the iron signal. In a second construct, the *CCC1* promoter was fused to *lacZ*. Selection of mutants that were histidine auxotrophs and were white identified genes that were required for the Yap5-dependent iron induction. Interestingly Ssq1 mutants were identified in this screen, and mutants of Yfh1 and Isu1 were also shown to prevent the high iron response. The implication is that both low-iron (Aft1) and high-iron (Yap5) responses involve Fe-S cluster intermediates in the sensing pathways. An attractive possibility is that Yap5 itself might form an Fe-S cluster in its cysteine-rich domain [66].

### 15.3.9 Screening of essential genes coding for mitochondrial proteins

A genome-wide screen in yeast was performed by deletion and sporulation to identify essential genes. Isd11 was shown to be essential by this approach [78]. A separate screen for protein localization using GFP fusions identified Isd11 as a mitochondrial protein [79]. The essentiality and mitochondrial localization led to the conjecture that Isd11 must be involved in a critical function of mitochondria. Mitochondrial protein import has been known for a long time to be an essential process. However, Isd11 mutants or conditional alleles exhibited normal protein import activity [9, 80]. Fe-S cluster synthesis is also an essential function of mitochondria. Therefore, Fe-S cluster assembly was examined in Isd11 mutants, and shown to be defective. No prokaryotic homologue exists, and so Isd11 could not have been found by comparison with prokaryotes [9, 80]. The Isd11 protein was found to form a tight complex with the cysteine desulfurase Nfs1 and to participate in stabilization of the protein in mitochondria [9, 80]. Isd11 was also found to activate Nfs1, enabling its cysteine desulfurase activity and promoting persulfide formation [8].

## 15.4 Mitochondrial Fe-S cluster assembly

Mitochondria possess a complete and functional Fe-S cluster assembly machinery. The basic pieces were acquired during evolution from purple bacteria, although important tweaking and major modifications occurred over evolutionary time [11]. Intact mitochondria isolated from yeast and "fed" metabolic precursors are capable of synthesizing new [$Fe_2S_2$] or [$Fe_4S_4$] clusters, and inserting them into endogenous or freshly imported apo-proteins [49]. For mitochondrial Fe-S cluster assembly to proceed, iron, sulfur, and various nucleotides are required.

1. Iron supply: Iron is already present inside mitochondria in a form that can be utilized for Fe-S cluster synthesis. The functionally important mitochondrial iron pool may correspond to the nonheme high-spin (NHHS) $Fe^{2+}$ iron pool described by Lindahl and colleagues [81] in Mössbauer spectroscopic studies of isolated mitochondria. In some settings, iron in this pool is present at concentrations of 100–200 μM, and the abundance is subjected to regulatory shifts depending on growth conditions. The precise chemical form and its ligands have not been ascertained, but the iron pool is sensitive to membrane permeable ferrous iron

chelators such as *o*-phenanthroline [81]. The iron in this pool can be "chased" into new Fe-S clusters as shown by experiments with isolated mitochondria, suggesting that it functions as a true intermediate [82]. The source of this mitochondrial iron pool remains poorly defined. It is presumed to originate from cytoplasmic iron that is transported into mitochondria. The transporters responsible have not been clearly identified either, so this entire issue remains somewhat nebulous. Mrs3 and Mrs4 are redundant mitochondrial carrier proteins, and extensive work has shown that they play a role in providing iron to mitochondria for Fe-S cluster and heme synthesis [83, 84]. However, these proteins have never been reconstituted into liposomes and shown to transport iron or any iron ligand, and so their transport substrate remains unknown. Another mitochondrial carrier protein, the pyrimidine exchanger Rim2, also plays a role in iron delivery into mitochondria, and overexpression is able to completely rescue the phenotype of a Δ*mrs3*Δ*mrs4* deletion [85]. Once inside mitochondria, iron must be delivered to the scaffold Isu, and frataxin may play a role here [18, 23].

2. Sulfur supply: The sulfur for Fe-S cluster assembly is derived from cysteine by the action of cysteine desulfurase [4]. Although Fe-S clusters can be made *in vitro* using sulfide and iron under defined conditions, only the cysteine desulfurase can supply sulfur for Fe-S cluster synthesis in intact cells or in isolated mitochondria, and sulfur in other chemical forms cannot substitute [2]. The yeast cysteine desulfurase, similar to the bacterial enzymes, interacts with cysteine and forms a persulfide on the enzyme [82]. The persulfide sulfur is the real precursor, and is used in Fe-S cluster assembly following transfer to Isu. In the eukaryotic setting, the cysteine desulfurase Nfs1 requires Isd11 for its activity [8].
3. Nucleotide requirements: Mitochondria depleted of nucleotides are unable to synthesize Fe-S clusters, and thus, *in vitro* assays with isolated mitochondria must be supplemented with nucleotides for supporting efficient Fe-S cluster assembly [49, 86]. This is because key steps depend on NAD(P)H, ATP, or GTP. More specifically, NAD(P)H is required for reductases, such as the ferredoxin reductase (Arh1) that provides electrons to ferredoxin (Yah1) and then probably to an Fe-S cluster assembly complex. The reducing equivalents are needed for converting the persulfide sulfur ($S^0$) on Nfs1 to $S^{2-}$ for incorporation into the nascent cluster on Isu [87]. Likewise, only ferrous iron is useful for the assembly process, and the Arh1/Yah1 reductase chain may also be involved in maintaining iron in reduced ($Fe^{2+}$) form, as this may have a tendency to reoxidize. Furthermore, the conversion of $[Fe_2S_2]$ to $[Fe_4S_4]$ clusters requires additional electrons in a process called reductive coupling, and this might also depend on the mitochondrial NAD(P)H reductase system [88]. Like NAD(P)H, ATP is also required for one or more steps in the assembly process. The role of Hsp70 chaperones in Fe-S cluster assembly was discovered at the time of the discovery of the ISC operon because the Hsp70 HscA was found on the operon and linked to IscS and the other key elements for Fe-S cluster assembly [55]. The yeast

mitochondrial Ssq1, although not orthologous to HscA, is an Hsp70 that interacts with Isu as its sole client. The requirement of ATP for the Hsp70 reaction cycle likely explains the requirement of ATP for mitochondrial Fe-S cluster assembly [29, 89]. Finally, GTP is required, although the target for this nucleotide requirement has not yet been identified. The role of GTP was discovered because mutant mitochondria lacking the GTP/GDP carrier, Ggc1, were deficient in both mitochondrial GTP and in Fe-S cluster assembly activity [90]. Presumably, there exists a GTPase that interacts with the assembly machinery and conveys a signal or performs a key catalytic function. Thus, mitochondria provided with cysteine, iron, NAD(P)H, ATP, and GTP are able to catalyze efficient synthesis of new Fe-S clusters (Fig. 15.1).

4. Synthesis of Fe-S cluster intermediates on Isu *vs* transfer of cluster intermediates to apo-proteins: The scaffold proteins, Isu1 and Isu2, in yeast mitochondria are redundant proteins, but at least one of them is absolutely essential [13]. The Fe-S cluster biogenesis can be understood in terms of two steps: formation of an Fe-S cluster intermediate on Isu and transfer of the intermediate from Isu to apo-protein recipients (Fig. 15.1) [50]. A number of independent components must interact in a complex choreography to perform each of these steps and to switch between them. To classify different components in terms of their importance for the synthesis or transfer step, an iron-labeling approach was undertaken. The yeast deletion mutant (or depleted strain if the corresponding gene was found to be essential) was labeled with radioactive iron ($^{55}Fe$), and Isu was recovered by immunoprecipitation. If the mutant was defective in formation of the Isu intermediate, $^{55}Fe$ label on Isu was decreased. Conversely, if the mutant was capable of generating the Isu intermediate but defective in transfer of the intermediate to apo-proteins, $^{55}Fe$ label on Isu was increased. In this manner, Fe-S cluster synthesis/assembly components (Nfs1, Isd11, Yfh1, and Yah1) were distinguished from Fe-S cluster transfer components (Ssq1, Jac1, and Grx5) [50].

### 15.4.1 Mitochondrial cysteine desulfurase

Cysteine desulfurases are conserved enzymes that act on the substrate cysteine, removing sulfur and forming a covalent persulfide on the active site cysteine. It is this form of sulfur that is exclusively utilized in the synthesis of cofactors such as Fe-S clusters, biotin, lipoic acid, and thiamine [91, 92] and also for the thio modification of tRNAs [65]. In *S. cerevisiae*, Nfs1 is the only known cysteine desulfurase, and the enzyme is required for activities of the Fe-S cluster proteins throughout the cell, including aconitase (Aco1) in mitochondria and isopropylmalate isomerase (Leu1) in the cytoplasm [6, 93]. These enzymes require a $[Fe_4S_4]$ cluster for activity. In the absence of adequate levels of Nfs1, cellular Fe-S cluster biogenesis does not occur, leading to global deficiency of these cofactors. Likewise, a hypomorphic mutant allele, *nfs1-14*, was associated with low activities of Fe-S proteins in mitochondria and constitutively induced cellular and mitochondrial iron uptake activities [6]. Lipoic acid proteins and enzyme activities were also deficient in the *nfs1* mutant because of deficiency of lipoic acid synthase, which requires Fe-S cluster cofactors [94].

Nfs1 forms a persulfide intermediate in mitochondria. Cysteine desulfurase activity can be assessed by incubating the enzyme with cysteine and DTT, and then assessing the amount of sulfide released utilizing methylene blue and ferric chloride [95]. However, this colorimetric assay is too insensitive and is not suitable for detecting the cysteine desulfurase activity in mitochondria. The assay requires a minimum of 2–5 μg of the purified Nfs1/Isd11 complex or *A. vinelandii* NifS. Such an insensitive assay therefore cannot be reliably used for measuring the cysteine desulfurase activity in mitochondria in which Nfs1 is present in small amounts. Furthermore, mitochondria with no detectable Nfs1 exhibit strong background sulfide-generating activity [8], likely due to cystathionine β-synthase or similar enzymes that release $H_2S$ and perform transulfuration reactions. A highly sensitive radioactive assay has therefore been developed based on the fact that the enzymatic cycle of cysteine desulfurase involves formation of a covalent persulfide on the active site. If the active enzyme is incubated with $^{35}$S-cysteine, the covalently attached $^{35}$S can be visualized on a nonreducing SDS gel. The radioactivity on the active enzyme can be removed by reduction of the persulfide with DTT. By contrast with the spectrophotometric assays, the radioactive assay using $^{35}$S-cysteine is at least 50–100 times more sensitive, and $^{35}$S-persulfide formation can be easily detected with as little as 5–50 ng of the purified Nfs1/Isd11 complex or 50–200 μg of total mitochondrial proteins [8]. Nfs1 persulfide formation (a necessary step in cysteine desulfurase activity) requires Isd11. Nfs1-S-$^{35}$SH, the Nfs1 persulfide, cannot be detected in Nfs1-depleted or *nfs1-14* mutant mitochondria. In these assays, several other radiolabeled proteins are also detected regardless of the presence or absence of Nfs1 in mitochondria. However, the authentic Nfs1-bound persulfide (Nfs1-S-$^{35}$SH) can be easily separated and distinguished from the Nfs1-independent radiolabeled background bands by nonreducing SDS-PAGE [8].

Isd11 is required for Nfs1 to form a persulfide. Nfs1 cannot form persulfide in mitochondria lacking Isd11, and consequently, Fe-S cluster synthesis does not occur in these mitochondria. Cells depleted of Isd11 or carrying a temperature-sensitive allele that has been shifted to the nonpermissive temperature fail to make Fe-S clusters [9, 10]. Bacterially expressed and purified Isd11 imported into these isolated mitochondria restores both of these processes [8]. Importantly, Isd11-independent and Isd11-dependent steps involved in the Nfs1-mediated cysteine desulfuration reaction can be distinguished. Nfs1 is able to bind the substrate cysteine but is unable to form the persulfide. Then Nfs1 with prebound cysteine, after interacting with Isd11, is able to form the persulfide bound to its active site. An interaction of Isd11 with Nfs1 may induce a conformational change in the enzyme that brings the active site cysteine and the bound substrate cysteine together, thereby allowing persulfide formation [8].

Nfs1 persulfide formation is stimulated by frataxin. Another level of Nfs1 regulation involves a unique role of frataxin/Yfh1. A stimulatory effect of frataxin on cysteine desulfurase was first noted for the human enzyme [22]. In studies of the yeast proteins, frataxin directly interacts with Nfs1, exposing substrate-binding sites and enhancing the binding of cysteine. This function of frataxin does not require iron, Isu1, or Isd11. Once bound to Nfs1, the thiol of the substrate cysteine can be used to form the persulfide on the active site cysteine of the enzyme. Persulfide formation depends on the interaction with Isd11. Thus, for full activation of cysteine desulfurase activity, the mitochondrial enzyme must undergo at least two conformational changes. One change is mediated by frataxin/Yfh1 interaction that promotes substrate binding. A second change is induced by Isd11 binding that brings the bound substrate in proximity to the active site cysteine for persulfide formation [27] (Fig. 15.2).

The Nfs1-bound persulfide functions as an intermediate in mitochondrial Fe-S cluster assembly. Upon incubation of isolated mitochondria with $^{35}$S-cysteine, a radiolabeled persulfide was detected on Nfs1 (Nfs1-S-$^{35}$SH). The $^{35}$S label detected on Nfs1 depended on the cysteine desulfurase activity of the enzyme, and the radioactive signal varied reciprocally with conditions that blocked or allowed Fe-S cluster

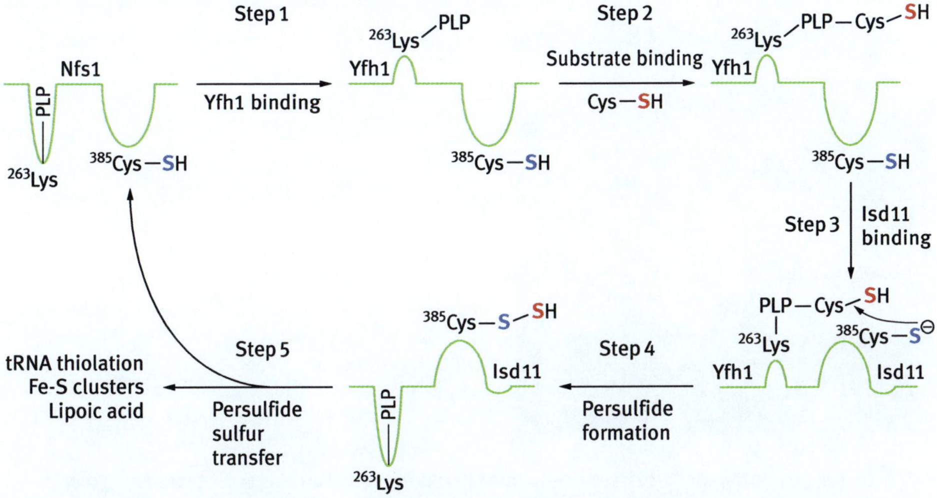


**Fig. 15.2:** Scheme for steps involved in cysteine desulfurase activation [27]. Yeast Nfs1 (green) contains pyridoxal phosphate (PLP) bound to lysine ($^{263}$Lys) in the substrate-binding site. A free thiol is present on a cysteine of the active site loop ($^{385}$Cys-SH, blue S). Step 1: Yfh1 binds to Nfs1-exposing substrate-binding sites. Step 2: The substrate cysteine (Cys-SH, red S) binds to substrate-binding sites on Nfs1, including sites exposed by the Yfh1 interaction. Step 3: Isd11 binds to Nfs1, producing a conformational change that brings the active site close to the substrate site with substrate already bound. Step 4: The persulfide (blue S and red S) is formed on the active site $^{385}$Cys of Nfs1. Step 5: Persulfide sulfur (red S) on Nfs1 is transferred from Nfs1 to recipients for Fe-S cluster synthesis and other uses including tRNA thiolation and lipoic acid synthesis.

synthesis in mitochondria. Nucleotide and/or iron depletion of mitochondria block Fe-S cluster synthesis, and the Nfs1-bound persulfide persisted under these conditions. When nucleotides, and iron were added, the radiolabeled Nfs1 persulfide was greatly reduced and radiolabeled aconitase was formed. Thus, the persulfide sulfur was productively chased from Nfs1 and utilized for Fe-S cluster synthesis under optimal metabolic conditions of mitochondria [82].

In summary, the mitochondrial cysteine desulfurase activity is highly regulated, with important controls exerted by interactions with frataxin and Isd11. The optimal control of cysteine desulfurase activity is important for normal cellular functions, and it must be tightly regulated. Too little cysteine desulfurase activity leads to Fe-S cluster deficiency in mitochondria (e.g. *nfs11-14* or Isd11-depleted mitochondria). Conversely, too much activity is toxic. Overexpression of the active Nfs1/Isd11 cysteine desulfurase complex causes the cells to become nonviable, whereas overexpression of Nfs1 alone (inactive without Isd11) is not toxic and does not confer a growth/viability phenotype (Fig. 15.3). Regulation of cysteine desulfurase is thus critical for cell viability. The stimulatory effects of frataxin and Isd11 on cysteine desulfurase activity are apparently eukaryotic features, and the prokaryotic enzymes behave differently in terms of their regulation [25]. The mitochondrial Nfs1 is different from the bacterial cysteine desulfurases that are active in the absence of accessory proteins [4, 5].

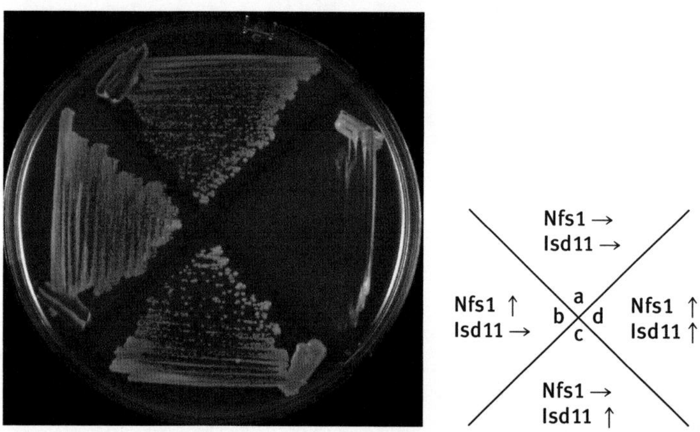


**Fig. 15.3:** Toxicity of too much cysteine desulfurase. The following strains were tested for growth and viability on galactose plates. (a) Wild type. (b) Promoter swap strain overexpressing Nfs1 from the Gal1 promoter. Isd11 was expressed from the genomic locus [6]. (c) Promoter swap strain overexpressing Isd11 from the Gal1 promoter. Nfs1 was expressed from the genomic locus. (d) Promoter swap strain overexpressing Isd11 from the Gal1 promoter [8] and overexpressing Nfs1 from a high-copy-number plasmid YEp352-Nfs1. Note that the slow growth of (d) indicating toxicity of too much cysteine desulfurase was observed only when both Nfs1 and Isd11 were overexpressed together. Nfs1 overexpression alone (b) was not toxic because it was inactive as a cysteine desulfurase without Isd11.

### 15.4.2 Formation of the Isu Fe-S cluster intermediate in mitochondria

Isu physically interacts with Nfs1/Isd11 and frataxin [19]. The multi-protein complex forms a core machinery for Fe-S cluster biosynthesis, facilitating and accelerating the formation of cluster intermediates on Isu [22]. The protein complex was isolated from yeast [19] and human [96] mitochondria, although no crystal structures are available and there is little overall structural information. Two particularly informative prokaryotic complexes have been structurally characterized. In one case, the Nfs1 (bacterial IscS) dimer was viewed in head to tail configuration [97]. Two Isu (bacterial IscU) monomers were seen binding at the ends of each Nfs1, maximally separated from each other and precluded from interacting with each other. Frataxin (bacterial CyaY) monomers were found in a groove between each Isu and the Nfs1 surface. The molecular stoichiometry (2 Nfs1:2 Isu:2 frataxin) and orientation of the components suggest that Fe-S cluster intermediates might be formed separately on each Isu monomer (Fig. 15.4). Isd11 was not present in this structure because it is not present in prokaryotes [97]. In another structural study, the *A. vinelandii* NifS (Nfs1 homologue) with a D39A substitution was co-crystallized with the IscU scaffold (Isu homologue), and the Fe-S cluster intermediate was captured in the crystal [98]. The cluster was liganded by three cysteine residues from IscU, and surprisingly, the fourth ligand was provided by the active site cysteine from NifS. The presence of a complete $Fe_2S_2$ cluster within the NifS/IscU complex showed that the complex did not have to be dissociated for each cycle of cysteine desulfuration; two sulfur atoms were contained in the cluster intermediate and two cycles would be required to form it. Importantly, the fact that the cluster intermediate on IscU was captured in the crystal showed that synthesis could occur separately from transfer [98].

The protein-protein interactions in the Fe-S cluster assembly complex of mitochondria are important for formation of the Fe-S cluster intermediate on Isu (Fig. 15.4). The sulfur for the intermediate is probably directly transferred from the Nfs1 active site to Isu thiol, thus involving direct interaction between Nfs1 and Isu proteins [99]. The flow of sulfur to Isu must be regulated because too little sulfur is deleterious and too much is toxic (Fig. 15.3). The components of the Fe-S cluster assembly complex, frataxin and Isd11, both exert regulatory effects on the cysteine desulfurase for controlling sulfur flow, but these effects are mediated differently. Frataxin stimulates cysteine binding to Nfs1 by exposing new substrate-binding sites. By contrast Isd11, induces a conformational change of Nfs1 that brings the protein loop with the active site cysteine into proximity with the substrate-binding site, so that the persulfides can be formed [27]. At this point, sulfur transfer to Isu can proceed, even without dissociation of the protein complex. Fe-S cluster assembly on Isu involves iron insertion as well as sulfur donation, and this process too is stimulated in the setting of this protein complex (Nfs1/Isd11/Isu/frataxin), as assessed by *in vitro* assays using cysteine, ferrous iron, and dithiothreitol as a reductant [100].

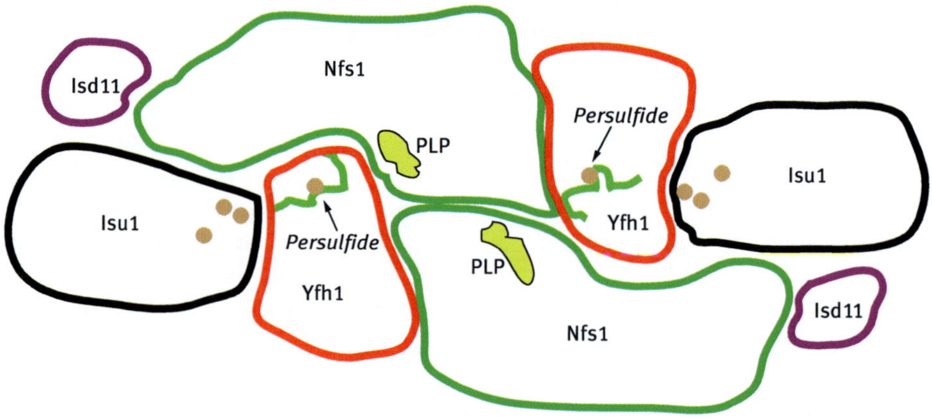


**Fig. 15.4:** Fe-S cluster assembly complex of the mitochondria. No structure for a eukaryotic or mitochondrial assembly complex has been obtained. The possible locations of key components were outlined based on the positions of the *E. coli* counterparts in the SASX structure [97]. Nfs1 is outlined in green, and it is shown as a head to tail dimer, with substrate-binding sites indicated as yellow "lakes" of PLP in each monomer. The Nfs1 protein loop with the active site cysteine is shown as a green "squiggle," and the active site cysteine is shown as a brown circle labeled "persulfide" because this is the site of persulfide formation. The Isu1 scaffold proteins are shown as black polygons with three brown circles inside, indicating the cysteine residues that coordinate the Fe-S cluster intermediate [97]. The position of the active site loop is such that it could toggle between the Nfs1 substrate-binding site and the Isu1-liganding cysteines without dissociating the Nfs1-Isu1 complex. The Yfh1 proteins are shown in red, and they fit neatly in a space between Isu1 and Nfs1. Isd11 is not present in prokaryotes [11], and so the binding sites in the complex are unknown. Isd11 is shown modeled in purple interacting with Nfs1 and participating in the Nfs1/Isd11/Isu1/Yfh1 Fe-S cluster assembly complex. The manner in which Fe-S cluster assembly on each Isu1 monomer is coordinated (if at all) is unknown.

### 15.4.3 Roles of frataxin

Frataxin is the protein implicated in the inherited neurodegenerative disease Friedreich ataxia [16]. The finding that the yeast homologue associates with Nfs1/Isd11 and Isu in mitochondria underscores its important role in Fe-S cluster assembly [19]. Two functions ascribed to frataxin relate to sulfur and iron effects, and these are not mutually exclusive. Frataxin stimulates cysteine desulfurase activity of Nfs1 as noted above (Section 15.4.1; Fig. 15.2). Frataxin also plays a role in iron donation for forming the Isu intermediate. Evidence for this function is that iron was found to interact with an acidic surface of purified frataxin with a stoichiometry of 2 atoms per molecule, albeit with a low micromolar (3 μM) affinity [23]. In addition, the interaction of frataxin with the other core assembly components in mitochondria was shown to be iron dependent because it was enhanced by the addition of iron and inhibited by iron chelation [18]. Using purified proteins, Yfh1-Isu1 interaction was found to take place in 1:1 stoichiometry as assessed by ITC measurements, and the interaction occurred only if iron was added to the reaction [23]. Further evidence of frataxin's role was provided by detailed studies using EXAFS, in which iron was tracked during transfer from frataxin to Isu. Oxygen and nitrogen liganding was detected for iron intermediates bound to frataxin or bound to Isu1, and only when a sulfur source was provided did sulfur ligation supersede the oxygen environment [23]. A picture can be imagined according to which the iron cofactor for Fe-S cluster assembly is bound to frataxin, an iron-dependent interaction with Isu occurs, and iron is then transferred to Isu. When sulfur becomes available in the Fe-S cluster-binding site on Isu, iron migrates to that location and assembles to form the Fe-S cluster intermediate. Of course, iron and sulfur delivery for Fe-S cluster intermediate formation need to be coordinated because it does no good to have one without the other. To address this issue, an interesting set of experiments was performed with protein complexes reconstituted from the mammalian components [100]. In the three-component complex (Nfs1/Isd11/Isu), iron supplied as ferrous ammonium sulfate could be loaded on Isu, up to a stoichiometry of two atoms per complex. In the four-component complex (Nfs1/Isd11/Isu/frataxin), however, iron could not be loaded at all. Only when iron and cysteine were added together to the four-component complex was iron loading achieved, and in that case, a complete Fe-S cluster intermediate was formed on Isu [100]. The implication is that frataxin was blocking iron entry into the Fe-S cluster assembly site on Isu until sulfur was provided *via* the cysteine desulfurase, and only then did it allow iron donation and cluster assembly. How such an iron-gating function would operate is unknown, but multiple conformational changes have been ascribed to Isu [15], and such a change induced by frataxin might close or open a pathway for iron entry into Isu.

### 15.4.4 Bypass mutation in Isu

A spontaneously occurring mutation in Isu1 was found to confer improved growth and improved Fe-S cluster assembly to a Δ*yfh1* mutant [26]. The mutant Isu1 was able to stimulate cysteine desulfurase of purified Nfs1/Isd11 or of Nfs1 in isolated mitochondria lacking frataxin [27]. These effects explain the frataxin bypass activity because the mutant Isu1 was performing frataxin's function. Interestingly, the bypass substitution mutation altered an amino acid M107 that is conserved with all eukaryotic Isu scaffold proteins. By contrast, the substituted amino acid, isoleucine, occurs in many prokaryotic Isu scaffolds, including *E. coli*. The change can be envisaged as rendering the eukaryotic Fe-S cluster machinery into a more prokaryotic form. The stimulatory effect of frataxin on cysteine desulfurase is a feature of eukaryotic enzyme and contrasts with the prokaryotic enzymes, which are constitutively active. Thus, by changing M107 of Isu1 to I, the need for the stimulatory effect of frataxin was lessened. A detailed examination of the bypass mutant phenotype, however, showed a more nuanced picture. Under usual laboratory conditions, the growth and Fe-S cluster proteins were restored to normal levels. However, under stress conditions exerted by peroxide exposure, the bypass suppressor grew more slowly than the wild-type strain expressing frataxin. The reason for this phenotype is that the bypass Isu mutant corrected the cysteine desulfurase deficit but did not restore the rate of Fe-S cluster assembly to normal [101].

These data suggest that frataxin possesses an additional function (other than cysteine desulfurase stimulation) that is not replaced by the Isu suppressor substitution. A possibility is that iron donation or regulation of iron donation represents the function of frataxin that is not corrected in the bypass mutant.

### 15.4.5 Transfer of the mitochondrial Isu Fe-S cluster intermediate

The Hsp70s are a large group of chaperones that perform essential functions in protein folding and protein import [101]. They contain two domains – a substrate-binding domain and an enzymatic domain with ATP-binding and hydrolysis activity. The two domains talk to each other, such that substrate binding in one domain is influenced by the nucleotide binding state in the other. The higher affinity substrate binding state is conferred by ADP interaction in the nucleotide-binding site. Co-chaperones or special J proteins are important for conferring functional specificity, bringing the substrate to the substrate-binding site and enhancing nucleotide hydrolysis. Finally, recycling factors are needed to facilitate release of ADP and replacement by ATP [102].

The remarkable story of the mechanism of Isu cluster transfer in mitochondria relates to the special adaptation of a mitochondrial Hsp70 chaperone for this purpose. Ssq1 is an Hsp70 chaperone that has become specialized for handling a single client protein, Isu. The manner in which this specialized reaction cycle functions for accelerating Fe-S cluster transfer is as follows. The Fe-S cluster intermediate is formed on the Isu scaffold, and the holo-Isu thus formed interacts with Jac1, a specialized co-chaperone [103]. Jac1 targets the holo-Isu1 to the ATP-bound Hsp70 chaperone, Ssq1, where it binds in the substrate-binding cleft *via* the conserved LPPVK motif of Isu [104]. Substrate binding to the Hsp70, especially in the context of Jac1 interaction, triggers a burst of ATP hydrolysis and a conformational change in the chaperone [29]. The conformational change compels release of the Isu-bound cluster intermediates with concomitant transfer to the recipient apo-protein. Finally, recycling of ADP and binding of ATP to the nucleotide-binding domain of Ssq1 is facilitated by Mge1, the recycling factor, and this step induces release of apo-Isu1 and initiation of another reaction cycle [31]. In this way, the transfer of Fe-S cluster intermediates is coupled to the activity of the Hsp70, which has been especially adapted for this purpose. The design of the chaperone machinery also adds another level of regulatory control because cluster transfer depends on the ATPase activity of Ssq1, and the latter is activated only when presented with the Isu1 client in the context of Jac1 [29].

### 15.4.6 Role of Grx5

Grx5, the monothiol glutaredoxin, is a conserved protein required in the later stages of mitochondrial Fe-S cluster assembly [105]. The Δ*grx5* mutant accumulates Fe-S cluster intermediates on Isu, showing that it is not required for the formation of the intermediates but that it is required for transfer of the intermediates [50]. In addition, the Grx5 protein is able to coordinate an Fe-S cluster *via* thiols of two monomers and two glutathione molecules [106]. A missing link in the larger scheme then was to show that Grx5 interacts with the chaperones in some fashion and that the Grx5-bound cluster functions as an intermediate in Fe-S cluster synthesis. Two recent papers provide some new information on these points. In experiments with the bacterial constituents, Johnson and colleagues [106] were able to show that Fe-S clusters assembled on IscU were transferred to Grx5-gluathione much more efficiently in the presence of chaperones HscA and HscB. In experiments with yeast proteins and yeast mutants, Uzarska *et al.* showed that holo-Isu1 and apo-Grx5 were able to bind Ssq1 simultaneously, with the Grx5 binding outside of the substrate-binding pocket [107]. Such interactions would be well poised to facilitate transfer of the Fe-S cluster intermediate from Isu to Grx5, while all the constituents were still bound on the surface of the Hsp70 chaperone. After forming on Grx5, Fe-S cluster intermediates are probably further modified and distributed to specialized branch pools. Some [$Fe_4S_4$] proteins such as mitochondrial aconitase apparently need special components (e.g. Iba57) for maturation of its cofactor [108] and others such as lipoic acid synthase use different maturation factors (e.g. Nfu1) [109, 110].

### 15.4.7 The switch between cluster synthesis and cluster transfer

Fe-S cluster synthesis in mitochondria involves assembly protein complexes (Nfs1 containing) and transfer protein complexes (Jac1 containing) that are set to function sequentially. Recently, it was found that Nfs1 and Jac1 interact with Isu1 at the same site, implying that there must be a switch such that Nfs1 comes off Isu and Jac1 comes on [111]. The physiological signal for the switch has not been identified with certainty, but a logical scheme would be for the switch to depend on the presence or absence of the Isu Fe-S cluster. A proposal for how this might work is as follows: if Isu is in the apo form, it assumes an unstructured conformation, as described for the disordered form of the bacterial IscU [15]. In this form, it is more adapted to Nfs1 binding and Fe-S cluster formation. Meanwhile, if Isu is in the holo form, it assumes a compact conformation, as described for the structured form of the bacterial IscU [15]. In this form, it is more adapted to Jac1 binding and Fe-S cluster transfer. Thus, the formation of the Fe-S cluster on Isu may generate the structural features suited to interacting with the chaperone system, initiating the switch from cluster forming to cluster transferring protein complexes [111].

In summary, Fe-S cluster assembly in mitochondria involves the formation of intermediates on the Isu scaffold, as performed by closely coordinated cysteine desulfurase activation and iron donation steps. More than one cycle of cysteine desulfurase can occur without disassembling the protein complex, and the cluster assembly can occur independent from transfer. The iron source and means of coordinating iron with sulfur delivery are still poorly characterized, although frataxin plays a role. A switch probably initiated by a conformational change in Isu promotes formation of the Fe-S cluster transfer complexes with Hsp70 chaperones and co-chaperones. The Hsp70 reaction cycle, specially adapted to Isu, facilitates Fe-S cluster transfer. Following this, glutathione and Grx5 receive and distribute the clusters. Insights into the molecular details of this process have been achieved over the last few years, and genetic and biochemical experiments using *S. cerevisiae* have played an important part in providing these insights.

## 15.5 Role of glutathione

Glutathione is a conserved tripepide consisting of glutamate, cysteine, and glycine [112]. It is synthesized in sequential steps mediated by Gsh1 and Gsh2, and the process is independent of ribosomal protein synthesis [113]. Although these enzymes reside in the cytoplasm, glutathione is found in millimolar amounts in all cellular compartments (cytoplasm, nucleus, endoplasmic reticulum, and mitochondria), implying that specific transporters exist for distributing the compound within the cell, although these putative transporters have not been molecularly identified [114]. Glutathione serves two different roles: on the one hand, it performs a redox buffering function, and on the other hand, it interacts with iron and Fe-S clusters. For the redox buffering function, the effect is mediated by interconversions between reduced (GSH) and oxidized (GSSG) forms, controlled by oxidative stresses and enzymes such as glutathione reductases [114]. The balance between the two forms defines a redox state characteristic of each cellular subcompartment, and this may be important for protein folding within that compartment [115]. For the iron-related function, it is significant that glutathione can interact directly with $Fe^{2+}$ *via* its free thiol, reducing and maintaining solubility of the metal. According to one hypothesis $Fe^{2+}$ glutathione may serve as a labile pool in cells, facilitating trafficking and transfer steps before being incorporated into Fe-S clusters and other cofactors [116]. A new insight into the functions of glutathione came with the discovery of a unique property of monothiol glutaredoxins. These proteins were found to be capable of coordinating Fe-S clusters using the thiols from their CGFS peptide motif and thiols from glutathione [117]. The $[Fe_2S_2]$ ligation was achieved by the contribution of two glutaredoxin molecules and two molecules of reduced glutathione. The lability of these glutaredoxin-glutathione clusters under some *in vitro* conditions led to the suggestion that these molecules might function as biosynthetic intermediates, binding and transferring cofactors or cofactor constituents for downstream utilization. Molecular genetic approaches in yeast have highlighted the roles of monothiol glutaredoxins in mitochondrial Fe-S cluster assembly, both inside mitochondria (Grx5) [118] and outside mitochondria (Grx3 and Grx4) [119]. These proteins appear to act by forming Fe-S cluster intermediates with glutathione.

The yeast Δ*gsh1* mutant was instrumental in the discovery of a role for glutathione in Fe-S cluster assembly. The deletion mutant was found to be auxotrophic for glutathione [120]. Growth of Δ*gsh1* yeast slowed progressively as the intracellular glutathione level declined, reaching about 60 μM after three divisions (down from 3 mM), and growth stopped completely after about ten divisions [121]. Examination of the phenotypes as the fuel ran down, so to speak, revealed striking changes. Genes involved in iron uptake were strongly induced, and genes involved in iron utilization were repressed, consistent with Aft1 activation and an iron starvation response [121]. Tests of iron homeostasis showed induction of cellular iron uptake and simultaneous mitochondrial iron accumulation as nanoparticles, similar to some mutants with Fe-S cluster deficiency [122]. In terms of Fe-S cluster proteins and activities, the observed effects depended on the degree of glutathione depletion achieved. In the early stages, cytoplasmic Fe-S clusters were more strongly affected [122], whereas in later stages, mitochondrial Fe-S cluster proteins were also deficient [121]. Leu1, a cytoplasmic Fe-S cluster protein, was deficient and activity was almost undetectable after seven divisions, whereas mitochondrial aconitase, Aco1, was relatively preserved but also deficient at this point. Interestingly, during the time course of depletion, oxidant stress genes were not induced. Yap1, a redox sensor transcriptional activator, remained in the reduced state and so was not turned on. Aft1, an iron sensor and transcriptional activator, was turned on, indicating that the iron uptake response was occurring earlier (at a higher glutathione threshold) than the oxidative stress response [121]. The experiment of progressive cellular glutathione depletion may be pointing to the existence of different thresholds that titrate different targets of glutathione action: one for cytoplasmic Fe-S cluster synthesis, another one for mitochondrial Fe-S cluster synthesis, and a third one for oxidative stress responses [121].

### 15.5.1 Glutathione and monothiol glutaredoxins in mitochondria

In yeast, Grx5 is the only monothiol glutaredoxin found in mitochondria. The deletion mutant phenotype was associated with deficiencies of Fe-S cluster proteins and iron homeostasis abnormalities [118]. This phenotypic analysis suggested that it was acting in the same pathway as other mitochondrial ISC components. Additional observations placed Grx5 as a late acting component. In an iron-labeling epistasis experiment, the deletion strain was found to accumulate $^{55}$Fe on Isu, suggesting that synthesis of the Fe-S cluster intermediate was able to proceed, whereas transfer of the Fe-S cluster intermediate was blocked [50]. The phenotype was similar to the Ssq1 Hsp70 chaperone mutant, and Ssq1 and Grx5 were recently shown to physically interact. Interestingly, the Ssq1 interaction with Grx5 could occur simultaneously with Isu1 interaction i.e. outside the substrate-binding site and without stimulation of ATP hydrolysis [107]. Thus, Isu1 with bound scaffold intermediate was probably interacting with the Ssq1 chaperone and Grx5, facilitating transfer of the Fe-S cluster intermediate to Grx5. Glutathione was presumed to be an essential ligand of the cluster intermediate on the glutaredoxin Grx5 and therefore required for the transfer step. Change of the cysteine residue of the CGFS motif in Grx5 to alanine or serine abrogated or decreased the transfer activity [107].

A direct demonstration of a role for glutathione in Fe-S cluster transfer to Grx5 was achieved by elegant experiments with *A. vinelandii* constituents [106]. Fe-S cluster transfer between IscU and Grx5 was monitored by circular dichroism. There was very little basal transfer even in the presence of glutathione; however, a tremendous rate enhancement was achieved by the addition of the chaperones HscA, HscB (homologues of the yeast Ssq1 and Jac1), and MgATP in the presence of glutathione. The second-order rate constant of the transfer of the $[Fe_2S_2]$ clusters increased from 30 to 20,000 $M^{-1}$ $min^{-1}$ [106]. Such a tremendous *in vitro* effect reflecting synergistic interaction among these components almost certainly reflects a physiological role in Fe-S cluster transfer. This experiment also directly demonstrated a role for glutathione because glutathione had to be included in the assays for cluster transfer to occur. Identification of the downstream acceptors for Grx5-glutathione intermediates has not been accomplished as of yet, but candidates include cluster apo-proteins, both $[Fe_2S_2]$ and $[Fe_4S_4]$ proteins, and carrier proteins of branch pathways such as A-type and Nfu1-type carriers. Although these experiments were performed with bacterial proteins, the eukaryotic homologues, Isu1/2, Ssq1, Jac1, and Grx5, will very likely interact with glutathione and behave in a fashion similar to their prokaryotic counterparts during Fe-S cluster transfers.

### 15.5.2 Glutathione and monothiol glutaredoxins Grx3 and Grx4 outside of mitochondria

In yeast, two redundant monothiol glutaredoxins, Grx3 and Grx4, reside outside of mitochondria, primarily in the cytoplasm and nucleus [53]. In addition to a glutaredoxin domain, these proteins also possess an amino terminal thioredoxin domain and a carboxy terminal Aft1-binding domain [53]. The Grx3 and Grx4 proteins, like Grx5, were shown to coordinate Fe-S clusters [123]. This ability required the presence of the critical cysteine of the CGFS motif and glutathione, as demonstrated by *in vivo* $^{55}Fe$ immunoprecipitation studies [119] as well as by *in vitro* reconstitution [123]. The loss-of-function Δ*grx3*/Δ*grx4* mutant or the C-to-A (AGFS) allele, exhibited a complicated iron regulatory phenotype, in part due to Aft1 activation and in part due to other downstream effects [119]. In these mutants, cellular iron uptake was increased, reflecting Aft1 activation, and a direct interaction of the C-terminal domain of Grx3 or Grx4 with Aft1 was shown [124, 125]. The manner in which the iron regulatory signal was conveyed to Aft1 is thought to involve the Fe-S clusters on Grx3/4. When the Fe-S clusters on Grx3/4 were present, there was a direct physical interaction with Aft1 protein and other intermediate regulators, e.g. Fra1, causing Aft1 to be retained in the cytoplasm and preventing the translocation of Aft1 to the nucleus. Conversely, in the absence of the Fe-S clusters on Grx3/4, Aft1 did not interact and nuclear translocation occurred concurrent with transcriptional activation of iron uptake and other effects [124].

The iron-related phenotypes, however, were more complicated and not entirely explained by the Grx3/4-Aft1 axis. In the Δ*grx3*Δ*grx4* mutant, iron uptake was activated but mitochondrial iron accumulation did not occur, distinguishing this phenotype from other mutants with Fe-S cluster assembly deficits [119]. Another unique aspect of the phenotype was a more global iron deficiency, affecting iron proteins in various cellular compartments and with various types of cofactors. Heme and di-iron proteins were also found to be deficient in these mutants, suggesting that a global problem with biological iron was affecting all types of cellular iron proteins. The problem affected both cytoplasmic and mitochondrial proteins. For example, nuclear-cytoplasmic ribonucleotide reductase, which requires a di-iron cofactor for generating a critical sulfur radical was deficient in *grx3/4* mutants [126]. Similarly, Coq7, a protein located on the matrix face of the mitochondrial inner membrane, utilizes a di-iron cofactor for synthesis of ubiquinone in mitochondria. This enzyme was also deficient in the *grx3/4* mutant as shown by accumulation of the DMQ6 precursor metabolite in the matrix fraction [119]. The implication is that Grx3/4-glutathione Fe-S clusters may be involved, either directly or indirectly, in providing iron for synthesis of cellular iron proteins. It is not yet clear how this works. Possibly, iron atoms of the Grx3/4-glutathione Fe-S clusters are cannibalized in a process that generates iron intermediates for biosynthesis of iron cofactors. In one scheme that has been proposed, an iron-glutathione conjugate with iron bound to the free thiol of a GSH molecule acts as the low molecular weight labile iron pool [116]. Measurements of this pool suggest it may measure about 1 μM in the cytoplasm of most cell types [116]. The cytoplasmic pool could then be tapped for cytoplasmic cofactor synthesis and transfer into mitochondria for mitochondrial Fe-S cluster and heme synthesis.

## 15.6 Role of Atm1, an ABC transporter of the mitochondrial inner membrane

Atm1 is an ABC transporter of the mitochondrial inner membrane [54]. A key feature has to do with its topology, with its predicted ATP- and substrate-binding sites oriented toward the interior or matrix side of mitochondria [54]. The implication of this finding is that Atm1 likely exports its substrate(s) from inside to outside of mitochondria. A major missing piece of the puzzle is the unknown identity of the substrate(s).

The *ATM1* gene was disrupted, and the mutant phenotypes were determined. The mutant grew slowly on rich medium and did not grow at all on minimal medium. The mitochondrial genome was destabilized, and the isolated mitochondria were completely white, indicating a lack of cytochromes [54]. Fe-S cluster proteins were also deficient in the mutant. In some studies, both mitochondrial and cytoplasmic Fe-S clusters were deficient [127], whereas in others, mitochondrial Fe-S clusters were relatively preserved [93]. The iron accumulation phenotype characteristic of Fe-S cluster deficiency was prominently displayed by the Δ*atm1* mutant, with loss of homeostasis and iron accumulation in mitochondria [68, 128]. The biophysical characteristics of the mitochondrial iron were investigated [127]. Similar to the situation for mitochondrial Fe-S cluster assembly mutants (e.g. Yah1- or Yfh1-deficient cells), mitochondrial iron in Atm1-depleted cells accumulated as ferric phosphate nanoparticles. Fe-S clusters and heme detected by Mössbauer and spectroscopic methods were decreased, but an NHHS $Fe^{2+}$ (nonheme high-spin ferrous) species was noted [127]. Evidence has been presented that this mitochondrial iron species, abundant in *atm1* mutant and also present in wild-type cells, could represent an iron intermediate for Fe-S cluster and heme synthesis [129].

Microarray analysis of Atm1-depleted cells revealed a large number of induced and repressed transcripts [130]. The effects reflected a general iron starvation response. The Aft1 regulon was strongly induced, causing activation of the cellular iron uptake systems. Transcripts for Isu1/2 scaffold proteins were increased, whereas Grx5 was repressed. The genes for citric acid cycle enzymes and for proteins involved in oxidative phosphorylation and ergosterol biosynthesis were repressed, consistent with an effort to conserve iron by minimizing utilization. The effects were virtually identical to the phenotypes of Fe-S cluster deficiency associated with depletion of the mitochondrial ferredoxin, Yah1 [130]. However, there was a set of genes that did not match. The retrograde response genes encoding citrate synthase (*CIT2*), dicarboxylic amino acid permease (*DIP5*), and isocitrate dehydrogenase (*IDH2*) were induced by Yah1 depletion and not by Atm1 depletion [130]. The retrograde response refers to the effects of an unknown signal, produced by dysfunctional mitochondria and acting on rest of the cell, likely mediating different types of compensatory responses [131]. The inability of *atm1* mutant mitochondria to induce a retrograde response could mean that Atm1 in some fashion mediates this signaling pathway.

### 15.6.1 Cells lacking Atm1 lose mtDNA

Iron accumulates in Atm1-deficient cells, causing oxidative damage to cellular constituents, and iron proteins are deficient in mitochondria and cytoplasm. Thus, the pleiomorphic phenotypes of these cells make it very difficult to determine the primary function of the exporter. Two types of experiments have aided in further parsing these phenotypes. If Gal-Atm1 cells were grown under repressing conditions (glucose) and at the same time were kept anaerobic, Atm1 was depleted from the cells, but the characteristic iron accumulation was abrogated. The lack of iron accumulation was associated with improved heme and improved Fe-S cluster levels. Oxidative damage to mitochondrial proteins as assessed by carbonyl adducts was also prevented [127]. By contrast, cytoplasmic isopropylmalate isomerase (Leu1), which requires a $[Fe_4S_4]$ cluster for catalyzing a critical step in the leucine biosynthetic pathway, was diminished in both anaerobic and aerobic conditions [127]. However, in another work [132], both wild-type and Gal-Atm1 cells grown in glucose were shown to be Leu1 deficient, raising the question of whether the enzyme deficiency was a carbon source effect rather than an Atm1 effect. In another type of experiment, the Gal-Atm1 cells were shifted from the inducing carbon source, galactose, to a repressing carbon source, glucose, and Atm1 was progressively depleted. Cytoplasmic Leu1 activity declined, whereas mitochondrial aconitase and succinate dehydrogenase were relatively preserved [93]. However, the differences very much depended on experimental conditions, such that with shorter depletion times only minor effects were noted, whereas with extremely long depletion times, mtDNA effects ensued and both mitochondrial and cytoplasmic Fe-S clusters were compromised. In the human disease resulting from missense mutations in the gene for the homologous protein ABCB7, the principal manifestation is anemia. In studies of blood cells from such individuals, protoporphyrin was found to accumulate, indicating that these mitochondria were abnormal and exhibited problems with iron metabolism [133]. Both mitochondrial and cytoplasmic aconitases were deficient in HeLa cell knockdowns of the homologous ABCB7, indicating an effect of ABCB7 on Fe-S cluster biogenesis in both compartments [133]. Thus, the precise roles of Atm1 in mitochondrial *vs* cytoplasmic iron metabolism and Fe-S cluster assembly remain to be elucidated.

The answers to Atm1 function may ultimately come from better understanding of the protein itself and by the identification of its transport substrate(s). Atm1 is encoded by a nuclear gene with an N-terminal mitochondrial targeting signal [54]. The protein is synthesized on cytoplasmic ribosomes and is post-translationally imported and matured by cleavage of the targeting sequence within mitochondria. Atm1 is an integral membrane protein of the inner membrane with both N- and C-termini oriented toward the matrix side, suggesting that six transmembrane domains span the inner membrane. The finding that the C-terminal domain of the protein is in the matrix is especially important because this domain includes the nucleotide-binding site as well as the putative substrate-binding site [54]. These topology data led to the prediction that the Atm1 substrate originates within mitochondria and that Atm1 must function to export its substrate from the mitochondrial matrix to the intermembrane space. The protein self-assembles to form homodimers *in vivo*, and this has been experimentally demonstrated [134]. Other mitochondrial ABC half transporters with similar topology include Mdl1 and Mdl2. These proteins are called half transporters because of their requirement for dimerization. The former may be a peptide exporter, and as for the latter, the substrate is undefined [135]. ABC transporters often show stimulation of their ATPase activity when challenged with the correct transport substrate. Experiments have been performed with Atm1-containing liposomes, seeking to identify the substrate, and stimulation was observed with micromolar concentrations of thiol rich peptides [136]. Further identification of the substrate has not been achieved, however. Thus, it will be important to identify the Atm1 substrate(s) to finally ascertain its function. The phenotypes of Atm1-depleted cells suggest that its transport substrate has something to do with Fe, S, Fe-S, and/or heme. The iron accumulation phenotype and gene regulation phenotype coincide closely with the phenotypes of other Fe-S cluster assembly mutants [135]. A role of Atm1 in Fe-S cluster assembly therefore seems likely. It is possible that the Atm1 transport substrate is a constituent needed for Fe-S cluster assembly such as iron, an activated form of sulfur, or glutathione. Alternatively, a molecule mediating a form of retrograde signaling might be exported by Atm1, and this could be a peptide, nucleotide, siderophore, or other specialized molecule.

## 15.7 Relationship between Fe-S cluster biogenesis and iron homeostasis

A distinctive iron homeostatic phenotype was first observed in yeast mutants with defects in frataxin [17] or Ssq1 [57]. It was subsequently recognized that these mutants were also deficient in Fe-S cluster assembly and that there was a connection between Fe-S cluster assembly and iron homeostasis [56, 137]. Yeast deploys cellular iron uptake systems that homeostatically respond to iron availability, increasing cellular uptake in response to iron starvation and repressing cellular uptake in response to iron repletion. In various mutants (e.g. *nfs1-14*) with Fe-S cluster defects, however, the iron uptake systems were constitutively induced and unresponsive to iron availability. If more iron was added to the growth medium, more was taken up into the cells. Furthermore, the excess cellular iron was diverted to mitochondria, reaching levels of 100 times or more the normal range [6]. Mitochondrial iron accumulating in these mutants was also different from that of the wild-type in terms of the physical properties. When mitochondria isolated from the mutants were treated with a nonionic detergent such as Triton X-100, most of the proteins were solubilized, leaving the majority of the iron in the insoluble fraction. This iron insolubility correlated with a lack of iron availability for cofactor synthesis. In Yfh1-depleted cells, for example, heme synthesis was deficient because iron was unavailable to the ferrochelatase enzyme in mitochondria; instead, zinc was inserted in place of iron, generating zinc protoporphyrin [21].

A more sophisticated and quantitative approach to characterizing iron in mitochondria has been undertaken by Lindahl and colleagues. Briefly, a combination of methods including Mössbauer, EPR, and UV-Vis spectroscopy combined with ICP-MS for metal quantification was applied to studies of isolated mitochondria. A packing efficiency parameter was measured for preparations of isolated mitochondria so that concentrations of different iron species could be calculated [138]. In wild-type mitochondria, the total iron was found to be present at very high concentrations of about 700–800 μM, and this could be divided into different pools. In respiring cells, 70% of the mitochondrial iron was present in Fe-S cluster and heme cofactors of respiratory complexes. An iron pool within mitochondria was recognized by a distinctive signature in Mössbauer studies, the nonheme high-spin ferrous pool (NHHS $Fe^{2+}$). This was proposed to act as feedstock for biosynthetic activities by supplying iron for heme, Fe-S cluster, and probably di-iron-containing proteins. Accordingly, shifting of growth conditions from fermentable to non-fermentable carbon source induced a shift in the mitochondrial iron distribution, with a marked decrease in the NHHS $Fe^{II}$, and an increase in the protein-bound iron pools, consistent with utilization of iron intermediates and a metabolic shift to oxidative phosphorylation [81]. Iron nanoparticles were also detected in mitochondria. These particles, with characteristic Mössbauer features, are 2- to 4-nm-diameter spheroids containing magnetically interacting $Fe^{3+}$ ions coordinated by oxygen donor ligands, including phosphate and polyphosphate. A small quantity of iron nanoparticles was detected in wild-type mitochondria grown under respiring conditions and more were detected under fermenting conditions. In the Fe-S cluster assembly mutants the total mitochondrial iron could be as high as 10 mM, mostly in the form of nanoparticles [139]. Thus, the nanoparticles represent the pool of insoluble iron noted in biochemical studies of the same mutants. In various Fe-S cluster assembly mutant cells examined by electron microscopy, iron nanoparticles were visible as electron dense bodies in the mitochondrial matrix (Fig. 15.5). Aggregates or nanoparticles were found sporadically in different parts of the mitochondria. Some areas were spared and others were heavily involved, consistent with the importance of local conditions; iron aggregation may depend on local ferroxidation and nucleation events (Fig. 15.5).

The question has been posed whether mitochondrial iron overload produces the Fe-S cluster assembly defect or vice versa. Experiments performed with Yfh1 (or Nfs1) under control of a regulated promoter allowed the protein to be depleted from cells and then reintroduced. During these time course experiments, the Fe-S cluster assembly defect appeared prior to mitochondrial iron accumulation [6, 21]. Conversely, recovery of the Fe-S cluster defect occurred rapidly upon re-expression of the protein and prior to clearing of mitochondrial iron [21]. Thus, the Fe-S cluster defect was considered to be primary, and the iron homeostatic effect was considered to be secondary. However, iron accumulation can worsen the Fe-S cluster defect in a situation with some residual activity, such as in a deletion of Yfh1 [140]. Under anaerobic conditions, no accumulation occurred during Gal-Atm1 depletion, perhaps because iron uptake was blocked in the absence of ferroxidase requiring Fet3 or because iron oxidation in mitochondria was a driving force and did not occur in the absence of oxygen. Under these conditions, iron proteins and enzyme activities were much improved because iron toxicity was lessened [127]. Thus, iron toxicity does contribute to the Fe-S cluster defect, although the major causal arrows seem to point the other way, from Fe-S cluster defect to the iron homeostatic defect.

The iron homeostatic phenotype can be viewed as Fe-S cluster deficiency and increased cellular uptake associated with mitochondrial iron accumulation in the form of nanoparticles. Various Fe-S cluster assembly mutants have been associated

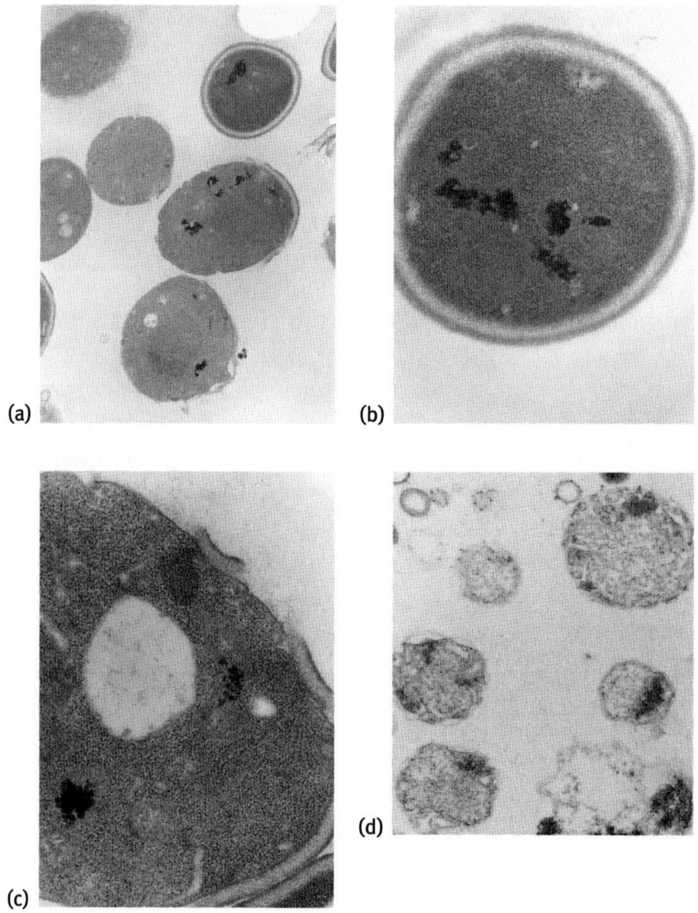

**Fig. 15.5:** Iron accumulation phenotype. A mutant carrying a truncated form of Ssq1 was identified in a genetic screen [57]. The *ssq1* mutant cells were grown in defined medium with 10 μM ferric chloride, fixed with glutaraldehyde, lightly stained with osmium tetroxide and potassium ferrocyanide, and visualized using a Jeol 100CX model electron microscope as described [57]. Shown are (a) multiple cells, most of which contain electron dense iron aggregates in mitochondria, (b) and (c) higher magnification views of affected cells, (d) isolated intact mitochondria with iron accumulations. Aggregates or nanoparticles occur sporadically in different parts of the mitochondria, which may contain larger or smaller deposits of iron.

with the homeostatic phenotype. These include mutations in components affecting formation of the mitochondrial Isu intermediate (Nfs1 [6], Isd11 [9, 10], Yfh1 [17], Yah1 [51], Arh1 [52]) (Tab. 15.2, item 1) and components involved in transfer of the Isu intermediate to recipients (Ssq1 [57], Jac1 [67, 141], Grx5 [117]) (Tab. 15.2, item 2). The Fe-S cluster scaffold Isu1 is redundant with Isu2, and so the single mutants exhibited a mild phenotype, although some iron accumulation was noted [13]. Depletion of Isu1 in an Δ*isu2* mutant [14] or hypomorphic alleles of Isu1 (M107E) in the Δ*isu2* mutant produced a strong iron accumulation phenotype (Tab. 15.2, item 3). The existence of branch pathways required for distribution of Fe-S cluster intermediates to specialized pools has recently been discovered in yeast and *E. coli*. The purview and substrates for these pathways are subjects of active research [1]. The Nfu1 mutant, originally isolated by its synthetic lethality with Ssq1 [89], was recently found to be associated with deficiencies of maturation of lipoate-containing 2-oxoacid dehydrogenases and assembly of some respiratory chain complexes [109, 110]. No iron homeostatic changes were noted in this mutant. Similarly, other proteins associated with subsets of Fe-S cluster protein maturation, such as Iba57 [108] and Mms19 [141], were not associated with perturbed iron homeostasis. By contrast, mutants of Isa1 and Isa2 did show the phenotype [142]. These proteins were originally proposed to act as alternative scaffold proteins in mitochondria but they did not show redundancy with Isu proteins, and later they were shown to be involved in iron donation for [$Fe_4S_4$] cluster assembly in mitochondria [143]. Isa1 and Isa2 proteins are not redundant with each other, and the corresponding deletion strains were independently associated with activation of the iron uptake system and iron homeostatic abnormalities [142] (Tab. 15.2, item 4).

Mutation or deletion of the ABC transporter Atm1 also gave a strong iron homeostatic phenotype, with iron accumulation in the cell and mitochondria [68, 128]. The role of Atm1 in Fe-S cluster assembly is still unresolved and quite controversial (see Section 15.6). However, the iron homeostatic effects are unambiguous, and they strongly resemble the iron homeostatic effects caused by defects in the core mitochondrial Fe-S cluster assembly components (Tab. 15.2, item 5). Erv1 is a sulfhydryl oxidase of the mitochondrial intermembrane space and is required for protein import and trapping of precursor proteins in that compartment. Interestingly, it is also involved in some aspects of Fe-S cluster assembly, and temperature-sensitive mutations have been associated with mitochondrial iron accumulation [144]. Glutathione appears to participate in multiple steps of Fe-S cluster assembly *via* interactions with monothiol glutaredoxins (Grx5 and Grx3/Grx4) and perhaps by other mechanisms. Depletion of cellular glutathione has been experimentally achieved by depriving the Δ*gsh1* mutant of exogenous glutathione, and in this setting, iron homeostatic abnormalities ensued, including increased cellular uptake and iron accumulation in mitochondria [121, 122]. It is unknown which of the cellular targets primarily mediates the glutathione iron homeostasis effects, but targets may include Grx3 and Grx4 (Tab. 15.2, item 6). A machinery dedicated to cytoplasmic and nuclear Fe-S cluster assembly has been characterized in yeast [3]. The components are encoded by essential genes, and they perform reductase, scaffold, ATPase, and transfer functions. These include Dre2, Tah18, Nbp35, Cfd1, Nar1, Cia1, Cia2, and Mms19 (Tab. 15.2, item 7). Depletion or mutation of these genes has been associated with deficiencies of many cytoplasmic and nuclear Fe-S cluster proteins, but cellular iron uptake was unaffected, and mitochondrial iron accumulation was not observed. Fe-S cluster assembly mutants of the CIA components did not perturb iron regulation [3].

**Tab. 15.2:** Yeast mutants that do or do not accumulate iron in mitochondria.

| Gene(s) | Function | Aft1 activation | Mitochondrial iron accumulation | References |
|---|---|---|---|---|
| 1. Core assembly component involved in formation of the Isu intermediate in mitochondria | | | | |
| Nfs1 | Cysteine desulfurase | + | + | [6] |
| Isd11 | Accessory protein | + | + | [9, 10] |
| Yah1 | Ferredoxin | + | + | [51] |
| Arh1 | Ferredoxin reductase | + | + | [52] |
| Yfh1 | Yeast frataxin homologue | + | + | [17] |
| 2. Core assembly component involved in transfer of the Isu intermediate to recipient proteins in mitochondria | | | | |
| Ssq1 | Hsp70 chaperone | + | + | [57] |
| Jac1 | J protein co-chaperone | + | + | [67, 145] |
| Grx5 | Monothiol glutaredoxin | + | + | [118] |
| 3. Mitochondrial scaffold protein | | | | |
| Isu1 | Mitochondrial scaffold | + | + | [13, 14] |
| Isu1 (M107E) | | | | |
| 4. Branch pathway proteins mediating Fe-S cluster assembly of a subset of proteins | | | | |
| Nfu1 | Lipoic acid synthase requires | – | – | [109, 110] |
| Iba57 | Aconitase requires | – | – | [108] |
| Isa1, Isa2 | $[Fe_4S_4]$ clusters require | + | + | [142, 143] |
| 5. Mitochondrial export and associated proteins | | | | |
| Atm1 | ABC transporter | + | + | [68, 128] |
| Erv1 | Sulfhydryl oxidase of intermembrane space | + | + | [144] |
| 6. Glutathione | | | | |
| Gsh1 | Glutathione synthesis | + | + | [121, 122] |
| Grx3, Grx4 | Monothiol glutaredoxins | + | – | [119] |
| 7. Cystosolic Fe-S cluster assembly components | | | | |
| Dre2, Tah18 | Reductase components | – | – | [37, 38] |
| Cfd1, Nbp35 | Scaffold components | – | – | [76] |
| Nar1, Cia1, Cia2, Mms19 | Transfer components | – | – | [3] |
| 8. Mitochondrial carrier proteins | | | | |
| Mtm1 | Unknown substrate | + | + | [146] |
| Ggc1 | GTP/GDP exchanger | + | + | [147, 148] |
| Mrs3, Mrs4 | Putative iron importers | + | – | [84] |
| 9. Iron regulatory | | | | |
| Aft1-$1^{up}$ | Iron transcriptional regulator | + | – | [17] |
| Fra1, Fra2 | Co-regulatory factors | + | – | [149] |
| 10. Heme biosynthesis | | | | |
| Hem1, Hem15 | Porphyrin synthesis | – | – | [150] |

Mitochondrial carrier proteins Mtm1 [146] and Ggc1 [147] were also required for iron homeostasis (Tab. 15.2, item 8). The carrier proteins belong to a family of proteins with uncertain evolutionary origin and are responsible for moving varied substrates in and out of mitochondria by transport across the mitochondrial inner membrane [151]. Mtm1 exhibits a network of cytoplasmic and matrix salt bridges and therefore is expected to function as an obligate exchanger, with substrate binding in the intermembrane space followed by exchange for a matrix substrate [152]. The identity of the substrates, either imported from the cytoplasm or exported from the mitochondria, is unknown. A guess is that they may resemble amino acids, as indicated by an evolutionary analysis that groups Mtm1 with known amino acid transporters [152]. The phenotype of the Δ*mtm1* deletion strain is similar to that of other mitochondrial iron accumulators. Iron uptake is induced, mitochondrial iron builds up in the form of nanoparticles, and Fe-S cluster proteins are deficient [153, 154]. Thus, it is possible that Mtm1 transports a critical but unknown substrate that is required for mitochondrial Fe-S cluster assembly. The deficiency of this unknown molecule might impair mitochondrial Fe-S cluster assembly, triggering a secondary signal that disrupts iron homeostasis. A role for Mtm1 in heme synthesis has also been proposed [155]. Because both heme synthesis and Fe-S cluster synthesis are affected in the mutant, an explanation could be that the Mtm1 substrate mediates iron delivery for cofactor synthesis in the mitochondrial matrix.

Ggc1 is another mitochondrial carrier protein (Tab. 15.2, item 8). In this case, the substrate is known and transport has been demonstrated by vesicle studies [148]. The yeast protein was expressed in *E. coli* and reconstituted into liposomes. Based on the transport studies showing exchange activity for GTP and GDP, the function of the transporter was proposed to deliver GTP into the mitochondrial matrix for important processes such as nucleic acid and protein synthesis [148]. An additional function was revealed by studies with the Δ*ggc1* mutant showing defective mitochondrial Fe-S cluster assembly and iron homeostasis [147]. The GTP dependence of the Fe-S cluster assembly defect was demonstrated by a bypass experiment. In this experiment, the mammalian GTP-generating enzyme, Nm23-H4, was targeted to the mitochondrial matrix of the yeast Δ*ggc1* mutant. The Δ*ggc1* mitochondria expressing the bypass enzyme, although they could still not transport GTP from the cytoplasm, were now able to synthesize GTP inside mitochondria. The mutant phenotypes were thereby reverted, including recovery of normal Fe-S cluster assembly and normal iron regulation [147]. These studies showed yet again that Fe-S cluster assembly and iron homeostasis are linked; GTP depletion of the mitochondria was associated with both defective Fe-S cluster assembly and perturbed iron homeostasis, and GTP repletion was associated with recovery of Fe-S cluster assembly and normal iron homeostasis [90]. The results also suggest that there is a GTP-dependent enzyme in mitochondria, perhaps a GTPase, involved in Fe-S cluster assembly. The identification of this enzyme is the objective of an ongoing search.

At the cellular level, the uptake systems are regulated by the Aft1 sensor/regulator with important inputs *via* the Cth1/2 mRNA stability modulators [156]. Recent discoveries have shown that Aft1 is retained in the cytoplasm by interactions with Fe-S cluster proteins Grx3/4-Fra1. In the absence of the regulatory clusters such as occurs in Fe-S cluster assembly mutants, Aft1 is translocated to the nucleus, and there it activates a set of target genes (including Fet3 and Ftr1, the iron transporters of the plasma membrane) by direct interaction with sequences in the target promoters [157]. Thus, lack of Fe-S clusters turns on the cellular iron uptake system. Aft1-1$^{up}$, an activating mutation in Aft1, is associated with constitutive cellular iron uptake but not mitochondrial iron accumulation [17] (Tab. 15.2, item 9). Similarly, *fra1* or *fra2* mutants activate cellular but not mitochondrial uptake [149] (Tab. 15.2, item 9). The means by which lack of Fe-S clusters augments mitochondrial iron levels is less clear. The lack of heme synthesis in some mammalian cells (e.g. from mutant alleles of erythroid 5-aminolevulinic acid synthase) is associated with mitochondrial iron accumulation and sideroblastic anemia [158]. However, the regulatory circuits are different for yeast, and *hem1* or *hem15* mutants do not accumulate iron in mitochondria [150] (Tab. 15.2, item 10).

A hypothetical scheme for explaining the iron accumulation phenotype of Fe-S cluster assembly mutants is as follows. Iron in mitochondria is regulated, and the functional iron pool, perhaps the NHHS $Fe^{II}$ pool [138] or the soluble iron pool, is maintained in a narrow range by homeostatic controls. If the level declines, more is taken up by transport from the cytoplasm, and if the level increases, less is taken up. In the pathological situation of the iron accumulation mutants, the functional iron pool in mitochondria becomes non-utilizable because of precipitation and aggregation. Precipitated iron is inert to the sensing/feedback system in mitochondria, and so more iron enters, continuously feeding the insoluble pool, while the functional pool remains diminished. The defect in Fe-S cluster assembly leads to a decrease in the soluble iron pool, which in turn leads to more mitochondrial iron uptake and accumulation of the nanoparticles. Unfortunately, many of the key variables for this scheme are undefined. The chemical form of the soluble iron pool in cytoplasm is unknown. The transport of iron into mitochondria is thought to be mediated by the carrier proteins, Mrs3 and Mrs4 [80, 84], Rim2 [85], and Mtm1 [146]. Mmt1/2 [159] may also be involved, but the iron-related substrates for these transporters are not well defined, and feedback controls for iron transport are unknown. Finally, the key Fe-S cluster protein(s) responsible for the complex iron regulatory phenotype has not been identified. Recently a number of regulatory Fe-S clusters have been discovered. Aft1, Fra1, Grx3, Grx4, and maybe Yap5 are extra-mitochondrial proteins shown to interact with clusters that perform regulatory functions in iron homeostasis [157]. However, none of these (when interrupted or mutated) fully account for the mitochondrial iron accumulation phenotype. The precise manner in which Fe-S clusters control mitochondrial iron remains to be clarified.

## 15.8 Conclusion and missing pieces

A variety of biochemical and genetic approaches undertaken through studies in yeast have yielded the outlines of a picture of the Fe-S cluster assembly process in eukaryotes. However, key pieces are still obscured. In mitochondria, the Fe-S cluster assembly complex and Fe-S cluster transfer complexes are better defined, but how the switch between them occurs is not known. How is the assembly process calibrated so that enough Fe-S clusters are made to satisfy the demands of new protein synthesis and to keep up with turnover occurring during the course of cellular stresses? The regulatory controls that occur within mitochondria are just beginning to be defined. Cysteine desulfurase in mitochondria appears to be highly regulated, but this is probably just the first glimpse of the types of extensive regulatory controls involved. The iron source for Fe-S cluster synthesis is still mysterious, and the manner in which iron and sulfur availability are coordinated is not known. The existence of mitochondrial and cytoplasmic Fe-S cluster proteins necessitates assembly systems that function inside and outside of mitochondria. However, the signaling between these compartments and the coordination of the Fe-S cluster synthesis in mitochondria and cytoplasm needs to be better understood. Finally, the iron homeostatic abnormality resulting from Fe-S cluster synthesis defects represents the failure of a regulatory signal, and the nature of the signal needs to be better defined.

## Acknowledgments

A.D. and D.P. are supported by National Institutes of Health Grants R37DK053953 and RO1AG030504, respectively. We thank Alok Pandey for help with some of the graphics.

## References

[1] Roche B, Aussel L, Ezraty B, Mandin P, Py B, Barras F. Iron/sulfur proteins biogenesis in prokaryotes: formation, regulation and diversity. Biochim Biophys Acta 2013;1827:455–69.
[2] Johnson DC, Dean DR, Smith AD, Johnson MK. Structure, function, and formation of biological iron-sulfur clusters. Annu Rev Biochem 2005;74:247–81.
[3] Stehling O, Lill R. The role of mitochondria in cellular iron-sulfur protein biogenesis: mechanisms, connected processes, and diseases. Cold Spring Harb Perspect Med 2013;3:1–17.
[4] Zheng L, White RH, Cash VL, Jack RF, Dean DR. Cysteine desulfurase activity indicates a role for NIFS in metallocluster biosynthesis. Proc Natl Acad Sci USA 1993;90:2754–8.
[5] Zheng L, White RH, Cash VL, Dean DR. Mechanism for the desulfurization of L-cysteine catalyzed by the nifS gene product. Biochemistry 1994;33:4714–20.
[6] Li J, Kogan M, Knight SA, Pain D, Dancis A. Yeast mitochondrial protein, Nfs1p, coordinately regulates iron-sulfur cluster proteins, cellular iron uptake, and iron distribution. J Biol Chem 1999;274:33025–34.
[7] Naamati A, Regev-Rudzki N, Galperin S, Lill R, Pines O. Dual targeting of Nfs1 and discovery of its novel processing enzyme, Icp55. J Biol Chem 2009;284:30200–8.
[8] Pandey A, Golla R, Yoon H, Dancis A, Pain D. Persulfide formation on mitochondrial cysteine desulfurase: enzyme activation by a eukaryote-specific interacting protein and Fe-S cluster synthesis. Biochem J 2012;448:171–87.
[9] Adam AC, Bornhovd C, Prokisch H, Neupert W, Hell K. The Nfs1 interacting protein Isd11 has an essential role in Fe/S cluster biogenesis in mitochondria. EMBO J 2006;25:174–83.

[10] Wiedemann N, Urzica E, Guiard B, et al. Essential role of Isd11 in mitochondrial iron-sulfur cluster synthesis on Isu scaffold proteins. EMBO J 2006;25:184–95.
[11] Richards TA, van der Giezen M. Evolution of the Isd11-IscS complex reveals a single alpha-proteobacterial endosymbiosis for all eukaryotes. Mol Biol Evol 2006;23:1341–4.
[12] Goldberg AV, Molik S, Tsaousis AD, et al. Localization and functionality of microsporidian iron-sulphur cluster assembly proteins. Nature 2008;452:624–8.
[13] Garland SA, Hoff K, Vickery LE, Culotta VC. Saccharomyces cerevisiae ISU1 and ISU2: members of a well-conserved gene family for iron-sulfur cluster assembly. J Mol Biol 1999;294:897–907.
[14] Gerber J, Neumann K, Prohl C, Muhlenhoff U, Lill R. The yeast scaffold proteins Isu1p and Isu2p are required inside mitochondria for maturation of cytosolic Fe/S proteins. Mol Cell Biol 2004;24:4848–57.
[15] Markley JL, Kim JH, Dai Z, et al. Metamorphic protein IscU alternates conformations in the course of its role as the scaffold protein for iron-sulfur cluster biosynthesis and delivery. FEBS Lett 2013;587:1172–9.
[16] Campuzano V, Montermini L, Molto MD, et al. Friedreich's ataxia: autosomal recessive disease caused by an intronic GAA triplet repeat expansion. Science 1996;271:1423–7.
[17] Babcock M, de Silva D, Oaks R, et al. Regulation of mitochondrial iron accumulation by Yfh1p, a putative homolog of frataxin. Science 1997;276:1709–12.
[18] Gerber J, Muhlenhoff U, Lill R. An interaction between frataxin and Isu1/Nfs1 that is crucial for Fe/S cluster synthesis on Isu1. EMBO Rep 2003;4:906–11.
[19] Wang T, Craig EA. Binding of yeast frataxin to the scaffold for Fe-S cluster biogenesis, Isu. J Biol Chem 2008;283:12674–9.
[20] Duby G, Foury F, Ramazzotti A, Herrmann J, Lutz T. A non-essential function for yeast frataxin in iron-sulfur cluster assembly. Hum Mol Genet 2002;11:2635–43.
[21] Lesuisse E, Santos R, Matzanke BF, Knight SA, Camadro JM, Dancis A. Iron use for haeme synthesis is under control of the yeast frataxin homologue (Yfh1). Hum Mol Genet 2003;12: 879–89.
[22] Tsai CL, Barondeau DP. Human frataxin is an allosteric switch that activates the Fe-S cluster biosynthetic complex. Biochemistry 2010;49:9132–9.
[23] Cook JD, Kondapalli KC, Rawat S, et al. Molecular details of the yeast frataxin-Isu1 interaction during mitochondrial Fe-S cluster assembly. Biochemistry 2010;49:8756–65.
[24] Bedekovics T, Gajdos GB, Kispal G, Isaya G. Partial conservation of functions between eukaryotic frataxin and the Escherichia coli frataxin homolog CyaY. FEMS Yeast Res 2007;7:1276–84.
[25] Bridwell-Rabb J, Iannuzzi C, Pastore A, Barondeau DP. Effector role reversal during evolution: the case of frataxin in Fe-S cluster biosynthesis. Biochemistry 2012;51:2506–14.
[26] Yoon H, Golla R, Lesuisse E, et al. Mutation in the Fe-S scaffold protein Isu bypasses frataxin deletion. Biochem J 2012;441:473–80.
[27] Pandey A, Gordon DM, Pain J, Stemmler TL, Dancis A, Pain D. Frataxin directly stimulates mitochondrial cysteine desulfurase by exposing substrate-binding sites and a mutant Fe-S cluster scaffold protein with frataxin-bypassing ability acts similarly. J Biol Chem 2013;288:36773–86.
[28] Hoff KG, Silberg JJ, Vickery LE. Interaction of the iron-sulfur cluster assembly protein IscU with the Hsc66/Hsc20 molecular chaperone system of Escherichia coli. Proc Natl Acad Sci USA 2000;97:7790–5.
[29] Schilke B, Williams B, Knieszner H, et al. Evolution of mitochondrial chaperones utilized in Fe-S cluster biogenesis. Curr Biol 2006;16:1660–5.
[30] Tapley TL, Cupp-Vickery JR, Vickery LE. Structural determinants of HscA peptide-binding specificity. Biochemistry 2006;45:8058–66.

[31] Dutkiewicz R, Schilke B, Knieszner H, Walter W, Craig EA, Marszalek J. Ssq1, a mitochondrial Hsp70 involved in iron-sulfur (Fe/S) center biogenesis. Similarities to and differences from its bacterial counterpart. J Biol Chem 2003;278:29719–27.
[32] Tsaousis AD, Gentekaki E, Eme L, Gaston D, Roger AJ. Evolution of the cytosolic Iron/sulfur cluster assembly machinery in Blastocystis sp. and other microbial eukaryotes. Eukaryot Cell 2013;13:143–53.
[33] Boyd JM, Pierik AJ, Netz DJ, Lill R, Downs DM. Bacterial ApbC can bind and effectively transfer iron-sulfur clusters. Biochemistry 2008;47:8195–202.
[34] Boyd JM, Drevland RM, Downs DM, Graham DE. Archaeal ApbC/Nbp35 homologs function as iron-sulfur cluster carrier proteins. J Bacteriol 2009;191:1490–7.
[35] Wilson RB, Roof DM. Respiratory deficiency due to loss of mitochondrial DNA in yeast lacking the frataxin homologue. Nat Genet 1997;16:352–7.
[36] Csere P, Lill R, Kispal G. Identification of a human mitochondrial ABC transporter, the functional orthologue of yeast Atm1p. FEBS Lett 1998;441:266–70.
[37] Zhang Y, Lyver ER, Nakamaru-Ogiso E, et al. Dre2, a conserved eukaryotic Fe/S cluster protein, functions in cytosolic Fe/S protein biogenesis. Mol Cell Biol 2008;28:5569–82.
[38] Netz DJ, Stumpfig M, Dore C, Muhlenhoff U, Pierik AJ, Lill R. Tah18 transfers electrons to Dre2 in cytosolic iron-sulfur protein biogenesis. Nat Chem Biol 2010;6:758–65.
[39] Wingert RA, Galloway JL, Barut B, et al. Deficiency of glutaredoxin 5 reveals Fe-S clusters are required for vertebrate haem synthesis. Nature 2005;436:1035–9.
[40] Sheftel AD, Stehling O, Pierik AJ, et al. Humans possess two mitochondrial ferredoxins, Fdx1 and Fdx2, with distinct roles in steroidogenesis, heme, and Fe/S cluster biosynthesis. Proc Natl Acad Sci USA 2010;107:11775–80.
[41] Shi Y, Ghosh M, Kovtunovych G, Crooks DR, Rouault TA. Both human ferredoxins 1 and 2 and ferredoxin reductase are important for iron-sulfur cluster biogenesis. Biochim Biophys Acta 2012;1823:484–92.
[42] Uhrigshardt H, Singh A, Kovtunovych G, Ghosh M, Rouault TA. Characterization of the human HSC20, an unusual DnaJ type III protein, involved in iron-sulfur cluster biogenesis. Hum Mol Genet 2010;19:3816–34.
[43] Rouault TA. Biogenesis of iron-sulfur clusters in mammalian cells: new insights and relevance to human disease. Dis Model Mech 2012;5:155–64.
[44] Payne RM, Pride PM, Babbey CM. Cardiomyopathy of Friedreich's ataxia: use of mouse models to understand human disease and guide therapeutic development. Pediatr Cardiol 2011;32:366–78.
[45] Selak MA, Lyver E, Micklow E, et al. Blood cells from Friedreich ataxia patients harbor frataxin deficiency without a loss of mitochondrial function. Mitochondrion 2011;11:342–50.
[46] Crooks DR, Jeong SY, Tong WH, et al. Tissue specificity of a human mitochondrial disease: differentiation-enhanced mis-splicing of the Fe-S scaffold gene ISCU renders patient cells more sensitive to oxidative stress in ISCU myopathy. J Biol Chem 2012;287:40119–30.
[47] Allikmets R, Raskind WH, Hutchinson A, Schueck ND, Dean M, Koeller DM. Mutation of a putative mitochondrial iron transporter gene (ABC7) in X-linked sideroblastic anemia and ataxia (XLSA/A). Hum Mol Genet 1999;8:743–9.
[48] Whitnall M, Suryo Rahmanto Y, Huang ML, et al. Identification of nonferritin mitochondrial iron deposits in a mouse model of Friedreich ataxia. Proc Natl Acad Sci USA 2012;109:20590–95.
[49] Amutha B, Gordon DM, Dancis A, Pain D. Chapter 14 Nucleotide-dependent iron-sulfur cluster biogenesis of endogenous and imported apoproteins in isolated intact mitochondria. Methods Enzymol 2009;456:247–66.
[50] Muhlenhoff U, Gerber J, Richhardt N, Lill R. Components involved in assembly and dislocation of iron-sulfur clusters on the scaffold protein Isu1p. EMBO J 2003;22:4815–25.

[51] Lange H, Kaut A, Kispal G, Lill R. A mitochondrial ferredoxin is essential for biogenesis of cellular iron-sulfur proteins. Proc Natl Acad Sci USA 2000;97:1050–5.
[52] Li J, Saxena S, Pain D, Dancis A. Adrenodoxin reductase homolog (Arh1p) of yeast mitochondria required for iron homeostasis. J Biol Chem 2001;276:1503–9.
[53] Herrero E, Belli G, Casa C. Structural and functional diversity of glutaredoxins in yeast. Curr Protein Peptide Sci 2010;11:659–68.
[54] Leighton J, Schatz G. An ABC transporter in the mitochondrial inner membrane is required for normal growth of yeast. EMBO J 1995;14:188–95.
[55] Zheng L, Cash VL, Flint DH, Dean DR. Assembly of iron-sulfur clusters. Identification of an iscSUA-hscBA-fdx gene cluster from Azotobacter vinelandii. J Biol Chem 1998;273:13264–72.
[56] Lill R, Hoffmann B, Molik S, et al. The role of mitochondria in cellular iron-sulfur protein biogenesis and iron metabolism. Biochim Biophys Acta 2012;1823:1491–508.
[57] Knight SA, Sepuri NB, Pain D, Dancis A. Mt-Hsp70 homolog, Ssc2p, required for maturation of yeast frataxin and mitochondrial iron homeostasis. J Biol Chem 1998;273:18389–93.
[58] Roy A, Solodovnikova N, Nicholson T, Antholine W, Walden WE. A novel eukaryotic factor for cytosolic Fe-S cluster assembly. EMBO J 2003;22:4826–35.
[59] Strain J, Lorenz CR, Bode J, et al. Suppressors of superoxide dismutase (SOD1) deficiency in Saccharomyces cerevisiae. Identification of proteins predicted to mediate iron-sulfur cluster assembly. J Biol Chem 1998;273:31138–44.
[60] Yarunin A, Panse VG, Petfalski E, Dez C, Tollervey D, Hurt EC. Functional link between ribosome formation and biogenesis of iron-sulfur proteins. EMBO J 2005;24:580–8.
[61] Chanet R, Heude M. Characterization of mutations that are synthetic lethal with pol3–13, a mutated allele of DNA polymerase delta in Saccharomyces cerevisiae. Curr Genet 2003;43:337–50.
[62] Huang B, Lu J, Bystrom AS. A genome-wide screen identifies genes required for formation of the wobble nucleoside 5-methoxycarbonylmethyl-2-thiouridine in Saccharomyces cerevisiae. RNA 2008;14:2183–94.
[63] Stearman R, Yuan DS, Yamaguchi-Iwai Y, Klausner RD, Dancis A. A permease-oxidase complex involved in high-affinity iron uptake in yeast. Science 1996;271:1552–7.
[64] Puig S, Vergara SV, Thiele DJ. Cooperation of two mRNA-binding proteins drives metabolic adaptation to iron deficiency. Cell Metab 2008;7:555–64.
[65] Nakai Y, Nakai M, Lill R, Suzuki T, Hayashi H. Thio modification of yeast cytosolic tRNA is an iron-sulfur protein-dependent pathway. Mol Cell Biol 2007;27:2841–7.
[66] Li L, Miao R, Bertram S, Jia X, Ward DM, Kaplan J. A role for iron-sulfur clusters in the regulation of transcription factor Yap5-dependent high iron transcriptional responses in yeast. J Biol Chem 2012;287:35709–21.
[67] Kim R, Saxena S, Gordon DM, Pain D, Dancis A. J-domain protein, Jac1p, of yeast mitochondria required for iron homeostasis and activity of Fe-S cluster proteins. J Biol Chem 2001;276:17524–32.
[68] Chen S, Sanchez-Fernandez R, Lyver ER, Dancis A, Rea PA. Functional characterization of AtATM1, AtATM2, and AtATM3, a subfamily of Arabidopsis half-molecule ATP-binding cassette transporters implicated in iron homeostasis. J Biol Chem 2007;282:21561–71.
[69] Kolman C, Soll D. SPL1–1, a Saccharomyces cerevisiae mutation affecting tRNA splicing. J Bacteriol 1993;175:1433–42.
[70] Bjork GR, Huang B, Persson OP, Bystrom AS. A conserved modified wobble nucleoside (mcm5s2U) in lysyl-tRNA is required for viability in yeast. RNA 2007;13:1245–55.
[71] Rouault TA, Stout CD, Kaptain S, Harford JB, Klausner RD. Structural relationship between an iron-regulated RNA-binding protein (IRE-BP) and aconitase: functional implications. Cell 1991;64:881–3.
[72] Brown NM, Kennedy MC, Antholine WE, Eisenstein RS, Walden WE. Detection of a [3Fe-4S] cluster intermediate of cytosolic aconitase in yeast expressing iron regulatory protein 1. Insights into the mechanism of Fe-S cluster cycling. J Biol Chem 2002;277:7246–54.

[73] Tschochner H, Hurt E. Pre-ribosomes on the road from the nucleolus to the cytoplasm. Trends Cell Biol 2003;13:255–63.
[74] Kispal G, Sipos K, Lange H, et al. Biogenesis of cytosolic ribosomes requires the essential iron-sulphur protein Rli1p and mitochondria. EMBO J 2005;24:589–98.
[75] Johansson E, Macneill SA. The eukaryotic replicative DNA polymerases take shape. Trends Biochem Sci 2010;35:339–47.
[76] Netz DJ, Pierik AJ, Stumpfig M, et al. A bridging [4Fe-4S] cluster and nucleotide binding are essential for function of the Cfd1-Nbp35 complex as a scaffold in iron-sulfur protein maturation. J Biol Chem 2012;287:12365–78.
[77] Li L, Bagley D, Ward DM, Kaplan J. Yap5 is an iron-responsive transcriptional activator that regulates vacuolar iron storage in yeast. Mol Cell Biol 2008;28:1326–37.
[78] Winzeler EA, Shoemaker DD, Astromoff A, et al. Functional characterization of the S. cerevisiae genome by gene deletion and parallel analysis. Science 1999;285:901–6.
[79] Huh WK, Falvo JV, Gerke LC, et al. Global analysis of protein localization in budding yeast. Nature 2003;425:686–91.
[80] Wiesenberger G, Link TA, von Ahsen U, Waldherr M, Schweyen RJ. MRS3 and MRS4, two suppressors of mtRNA splicing defects in yeast, are new members of the mitochondrial carrier family. J Mol Biol 1991;217:23–37.
[81] Holmes-Hampton GP, Miao R, Garber Morales J, Guo Y, Munck E, Lindahl PA. A nonheme high-spin ferrous pool in mitochondria isolated from fermenting Saccharomyces cerevisiae. Biochemistry 2010;49:4227–34.
[82] Pandey A, Yoon H, Lyver ER, Dancis A, Pain D. Identification of a Nfs1p-bound persulfide intermediate in Fe-S cluster synthesis by intact mitochondria. Mitochondrion 2012;12:539–49.
[83] Muhlenhoff U, Stadler JA, Richhardt N, et al. A specific role of the yeast mitochondrial carriers MRS3/4p in mitochondrial iron acquisition under iron-limiting conditions. J Biol Chem 2003;278:40612–20.
[84] Zhang Y, Lyver ER, Knight SA, Pain D, Lesuisse E, Dancis A. Mrs3p, Mrs4p, and frataxin provide iron for Fe-S cluster synthesis in mitochondria. J Biol Chem 2006;281:22493–502.
[85] Yoon H, Zhang Y, Pain J, et al. Rim2, a pyrimidine nucleotide exchanger, is needed for iron utilization in mitochondria. Biochem J 2011;440:137–46.
[86] Pain J, Balamurali MM, Dancis A, Pain D. Mitochondrial NADH kinase, Pos5p, is required for efficient iron-sulfur cluster biogenesis in Saccharomyces cerevisiae. J Biol Chem 2010;285:39409–24.
[87] Lill R. Function and biogenesis of iron-sulphur proteins. Nature 2009;460:831–8.
[88] Agar JN, Krebs C, Frazzon J, Huynh BH, Dean DR, Johnson MK. IscU as a scaffold for iron-sulfur cluster biosynthesis: sequential assembly of [2Fe-2S] and [4Fe-4S] clusters in IscU. Biochemistry 2000;39:7856–62.
[89] Schilke B, Voisine C, Beinert H, Craig E. Evidence for a conserved system for iron metabolism in the mitochondria of Saccharomyces cerevisiae. Proc Natl Acad Sci USA 1999;96:10206–11.
[90] Amutha B, Gordon DM, Gu Y, Lyver ER, Dancis A, Pain D. GTP is required for iron-sulfur cluster biogenesis in mitochondria. J Biol Chem 2008;283:1362–71.
[91] Kessler D. Enzymatic activation of sulfur for incorporation into biomolecules in prokaryotes. FEMS Microbiol Rev 2006;30:825–40.
[92] Mueller EG. Trafficking in persulfides: delivering sulfur in biosynthetic pathways. Nat Chem Biol 2006;2:185–94.
[93] Kispal G, Csere P, Prohl C, Lill R. The mitochondrial proteins Atm1p and Nfs1p are essential for biogenesis of cytosolic Fe/S proteins. EMBO J 1999;18:3981–89.
[94] Onder O, Yoon H, Naumann B, Hippler M, Dancis A, Daldal F. Modifications of the lipoamide-containing mitochondrial subproteome in a yeast mutant defective in cysteine desulfurase. Mol Cell Proteomics 2006;5:1426–36.

[95] Jaschkowitz K, Seidler A. Role of a NifS-like protein from the cyanobacterium Synechocystis PCC 6803 in the maturation of FeS proteins. Biochemistry 2000;39:3416–23.
[96] Schmucker S, Martelli A, Collin F, et al. Mammalian frataxin: an essential function for cellular viability through an interaction with a preformed ISCU/NFS1/ISD11 iron-sulfur assembly complex. PLoS One 2011;26:e16199.
[97] Prischi F, Konarev PV, Iannuzzi C, et al. Structural bases for the interaction of frataxin with the central components of iron-sulphur cluster assembly. Nat Commun 2010;1:95.
[98] Marinoni EN, de Oliveira JS, Nicolet Y, et al. (IscS-IscU)2 complex structures provide insights into $Fe_2S_2$ biogenesis and transfer. Angewandte Chemie 2012;51:5439–42.
[99] Bonomi F, Iametti S, Morleo A, Ta D, Vickery LE. Facilitated transfer of IscU-[2Fe2S] clusters by chaperone-mediated ligand exchange. Biochemistry 2011;50:9641–50.
[100] Colin F, Martelli A, Clemancey M, et al. Mammalian frataxin controls sulfur production and iron entry during de novo Fe4S4 cluster assembly. J Am Chem Soc 2013;135:733–40.
[101] Yoon H, Knight AB, Pandey A, Pain J, Zhang Y, Pain D, Dancis A. Frataxin-bypassing Isu1: characterization of the bypass activity in cells and mitochondria. Biochem J 2014;459:71–81.
[102] Kampinga HH, Craig EA. The HSP70 chaperone machinery: J proteins as drivers of functional specificity. Nat Rev Mol Cell Biol 2010;11:579–92.
[103] Andrew AJ, Dutkiewicz R, Knieszner H, Craig EA, Marszalek J. Characterization of the interaction between the J-protein Jac1p and the scaffold for Fe-S cluster biogenesis, Isu1p. J Biol Chem 2006;281:14580–7.
[104] Dutkiewicz R, Schilke B, Cheng S, Knieszner H, Craig EA, Marszalek J. Sequence-specific interaction between mitochondrial Fe-S scaffold protein Isu and Hsp70 Ssq1 is essential for their in vivo function. J Biol Chem 2004;279:29167–74.
[105] Molina-Navarro MM, Casas C, Piedrafita L, Belli G, Herrero E. Prokaryotic and eukaryotic monothiol glutaredoxins are able to perform the functions of Grx5 in the biogenesis of Fe/S clusters in yeast mitochondria. FEBS Lett 2006;580:2273–80.
[106] Shakamuri P, Zhang B, Johnson MK. Monothiol glutaredoxins function in storing and transporting [Fe(2)S(2)] clusters assembled on IscU scaffold proteins. J Am Chem Soc 2012;134:15213–6.
[107] Uzarska MA, Dutkiewicz R, Freibert SA, Lill R, Muhlenhoff U. The mitochondrial Hsp70 chaperone Ssq1 facilitates Fe/S cluster transfer from Isu1 to Grx5 by complex formation. Mol Biol Cell 2013;24:1830–41.
[108] Gelling C, Dawes IW, Richhardt N, Lill R, Muhlenhoff U. Mitochondrial Iba57p is required for Fe/S cluster formation on aconitase and activation of radical SAM enzymes. Mol Cell Biol 2008;28:1851–61.
[109] Cameron JM, Janer A, Levandovskiy V, et al. Mutations in iron-sulfur cluster scaffold genes NFU1 and BOLA3 cause a fatal deficiency of multiple respiratory chain and 2-oxoacid dehydrogenase enzymes. Am J Hum Genet 2011;89:486–95.
[110] Navarro-Sastre A, Tort F, Stehling O, et al. A fatal mitochondrial disease is associated with defective NFU1 function in the maturation of a subset of mitochondrial Fe-S proteins. Am J Hum Genet 2011;89:656–67.
[111] Majewska J, Ciesielski SJ, Schilke B, et al. Binding of the chaperone Jac1 protein and cysteine desulfurase Nfs1 to the iron-sulfur cluster scaffold Isu protein is mutually exclusive. J Biol Chem 2013;288:29134–42.
[112] Meister A, Anderson ME. Glutathione. Annu Rev Biochem 1983;52:711–60.
[113] Grant CM, MacIver FH, Dawes IW. Glutathione synthetase is dispensable for growth under both normal and oxidative stress conditions in the yeast Saccharomyces cerevisiae due to an accumulation of the dipeptide gamma-glutamylcysteine. Mol Biol Cell 1997;8:1699–707.
[114] Toledano MB, Delaunay-Moisan A, Outten CE, Igbaria A. Functions and cellular compartmentation of the thioredoxin and glutathione pathways in yeast. Antioxid Redox Signal 2013;18:1699–711.
[115] Hu J, Dong L, Outten CE. The redox environment in the mitochondrial intermembrane space is maintained separately from the cytosol and matrix. J Biol Chem 2008;283:29126–34.

[116] Hider RC, Kong XL. Glutathione: a key component of the cytoplasmic labile iron pool. Biometals 2011;24:1179–87.
[117] Picciocchi A, Saguez C, Boussac A, Cassier-Chauvat C, Chauvat F. CGFS-type monothiol glutaredoxins from the cyanobacterium Synechocystis PCC6803 and other evolutionary distant model organisms possess a glutathione-ligated [2Fe-2S] cluster. Biochemistry 2007;46:15018–26.
[118] Rodriguez-Manzaneque MT, Tamarit J, Belli G, Ros J, Herrero E. Grx5 is a mitochondrial glutaredoxin required for the activity of iron/sulfur enzymes. Mol Biol Cell 2002;13:1109–21.
[119] Muhlenhoff U, Molik S, Godoy JR, et al. Cytosolic monothiol glutaredoxins function in intracellular iron sensing and trafficking via their bound iron-sulfur cluster. Cell Metab 2010;12:373–85.
[120] Wu AL, Moye-Rowley WS. GSH1, which encodes gamma-glutamylcysteine synthetase, is a target gene for yAP-1 transcriptional regulation. Mol Cell Biol 1994;14:5832–9.
[121] Kumar C, Igbaria A, D'Autreaux B, et al. Glutathione revisited: a vital function in iron metabolism and ancillary role in thiol-redox control. EMBO J 2011;30:2044–56.
[122] Sipos K, Lange H, Fekete Z, Ullmann P, Lill R, Kispal G. Maturation of cytosolic iron-sulfur proteins requires glutathione. J Biol Chem 2002;277:26944–9.
[123] Li H, Mapolelo DT, Dingra NN, et al. The yeast iron regulatory proteins Grx3/4 and Fra2 form heterodimeric complexes containing a [2Fe-2S] cluster with cysteinyl and histidyl ligation. Biochemistry 2009;48:9569–81.
[124] Hoffmann B, Uzarska MA, Berndt C, et al. The multidomain thioredoxin-monothiol glutaredoxins represent a distinct functional group. Antioxid Redox Signal 2011;15:19–30.
[125] Ojeda L, Keller G, Muhlenhoff U, Rutherford JC, Lill R, Winge DR. Role of glutaredoxin-3 and glutaredoxin-4 in the iron regulation of the Aft1 transcriptional activator in Saccharomyces cerevisiae. J Biol Chem 2006;281:17661–9.
[126] Zhang Y, Liu L, Wu X, An X, Stubbe J, Huang M. Investigation of in vivo diferric tyrosyl radical formation in Saccharomyces cerevisiae Rnr2 protein: requirement of Rnr4 and contribution of Grx3/4 AND Dre2 proteins. J Biol Chem 2011;286:41499–509.
[127] Miao R, Kim H, Koppolu UM, Ellis EA, Scott RA, Lindahl PA. Biophysical characterization of the iron in mitochondria from Atm1p-depleted Saccharomyces cerevisiae. Biochemistry 2009;48:9556–68.
[128] Kispal G, Csere P, Guiard B, Lill R. The ABC transporter Atm1p is required for mitochondrial iron homeostasis. FEBS Lett 1997;418:346–50.
[129] Garber Morales J, Holmes-Hampton GP, Miao R, Guo Y, Munck E, Lindahl PA. Biophysical characterization of iron in mitochondria isolated from respiring and fermenting yeast. Biochemistry 2010;49:5436–44.
[130] Hausmann A, Samans B, Lill R, Muhlenhoff U. Cellular and mitochondrial remodeling upon defects in iron-sulfur protein biogenesis. J Biol Chem 2008;283:8318–30.
[131] Liu Z, Butow RA. Mitochondrial retrograde signaling. Annu Rev Genet 2006;40:159–85.
[132] Bedekovics T, Li H, Gajdos GB, Isaya G. Leucine biosynthesis regulates cytoplasmic iron-sulfur enzyme biogenesis in an Atm1p-independent manner. J Biol Chem 2011;286:40878–88.
[133] Cavadini P, Biasiotto G, Poli M, et al. RNA silencing of the mitochondrial ABCB7 transporter in HeLa cells causes an iron-deficient phenotype with mitochondrial iron overload. Blood 2007;109:3552–9.
[134] Chloupkova M, Reaves SK, LeBard LM, Koeller DM. The mitochondrial ABC transporter Atm1p functions as a homodimer. FEBS Lett 2004;569:65–69.
[135] Zutz A, Gompf S, Schagger H, Tampe R. Mitochondrial ABC proteins in health and disease. Biochim Biophys Acta 2009;1787:681–90.
[136] Kuhnke G, Neumann K, Muhlenhoff U, Lill R. Stimulation of the ATPase activity of the yeast mitochondrial ABC transporter Atm1p by thiol compounds. Mol Membr Biol 2006;23:173–84.
[137] Rouault TA, Tong WH. Iron-sulfur cluster biogenesis and human disease. Trends Genet 2008;24:398–407.

[138] Lindahl PA, Holmes-Hampton GP. Biophysical probes of iron metabolism in cells and organelles. Curr Opin Chem Biol 2011;15:342–6.
[139] Miao R, Martinho M, Morales JG, et al. EPR and Mossbauer spectroscopy of intact mitochondria isolated from Yah1p-depleted Saccharomyces cerevisiae. Biochemistry 2008;47:9888–99.
[140] Bulteau AL, Dancis A, Gareil M, Montagne JJ, Camadro JM, Lesuisse E. Oxidative stress and protease dysfunction in the yeast model of Friedreich ataxia. Free Radic Biol Med 2007;42:1561–70.
[141] Stehling O, Vashisht AA, Mascarenhas J, et al. MMS19 assembles iron-sulfur proteins required for DNA metabolism and genomic integrity. Science 2012;337:195–9.
[142] Jensen LT, Culotta VC. Role of Saccharomyces cerevisiae ISA1 and ISA2 in iron homeostasis. Mol Cell Biol 2000;20:3918–27.
[143] Muhlenhoff U, Richter N, Pines O, Pierik AJ, Lill R. Specialized function of yeast Isa1 and Isa2 proteins in the maturation of mitochondrial [4Fe-4S] proteins. J Biol Chem 2011;286:41205–16.
[144] Lange H, Lisowsky T, Gerber J, Muhlenhoff U, Kispal G, Lill R. An essential function of the mitochondrial sulfhydryl oxidase Erv1p/ALR in the maturation of cytosolic Fe/S proteins. EMBO Rep 2001;2:715–20.
[145] Voisine C, Cheng YC, Ohlson M, et al. Jac1, a mitochondrial J-type chaperone, is involved in the biogenesis of Fe/S clusters in Saccharomyces cerevisiae. Proc Natl Acad Sci USA 2001;98:1483–8.
[146] Yang M, Cobine PA, Molik S, et al. The effects of mitochondrial iron homeostasis on cofactor specificity of superoxide dismutase 2. EMBO J 2006;25:1775–83.
[147] Gordon DM, Lyver ER, Lesuisse E, Dancis A, Pain D. GTP in the mitochondrial matrix plays a crucial role in organellar iron homoeostasis. Biochem J 2006;400:163–8.
[148] Vozza A, Blanco E, Palmieri L, Palmieri F. Identification of the mitochondrial GTP/GDP transporter in Saccharomyces cerevisiae. J Biol Chem 2004;279:20850–57.
[149] Lesuisse E, Knight SA, Courel M, Santos R, Camadro JM, Dancis A. Genome-wide screen for genes with effects on distinct iron uptake activities in Saccharomyces cerevisiae. Genetics 2005;169:107–22.
[150] Crisp RJ, Pollington A, Galea C, Jaron S, Yamaguchi-Iwai Y, Kaplan J. Inhibition of heme biosynthesis prevents transcription of iron uptake genes in yeast. J Biol Chem 2003;278:45499–506.
[151] Palmieri F, Pierri CL, De Grassi A, Nunes-Nesi A, Fernie AR. Evolution, structure and function of mitochondrial carriers: a review with new insights. Plant J Cell Mol Biol 2011; 66:161–81.
[152] Robinson AJ, Overy C, Kunji ER. The mechanism of transport by mitochondrial carriers based on analysis of symmetry. Proc Natl Acad Sci USA 2008;105:17766–71.
[153] Park J, McCormick SP, Chakrabarti M, Lindahl PA. Insights into the iron-ome and manganese-ome of Δmtm1 Saccharomyces cerevisiae mitochondria. Metallomics 2013;5:656–72.
[154] Yang Y, Yamashita T, Nakamaru-Ogiso E, et al. Reaction mechanism of single subunit NADH-ubiquinone oxidoreductase (Ndi1) from Saccharomyces cerevisiae: evidence for a ternary complex mechanism. J Biol Chem 2011;286:9287–97.
[155] Nilsson R, Schultz IJ, Pierce EL, et al. Discovery of genes essential for heme biosynthesis through large-scale gene expression analysis. Cell Metab 2009;10:119–30.
[156] Puig S, Askeland E, Thiele DJ. Coordinated remodeling of cellular metabolism during iron deficiency through targeted mRNA degradation. Cell 2005;120:99–110.
[157] Outten CE, Albetel AN. Iron sensing and regulation in Saccharomyces cerevisiae: ironing out the mechanistic details. Curr Opin Microbiol 2013;16:662–8.
[158] Bottomley SS. Congenital sideroblastic anemias. Curr Hematol Rep 2006;5:41–9.
[159] Li L, Kaplan J. Characterization of two homologous yeast genes that encode mitochondrial iron transporters. J Biol Chem 1997;272:28485–93.

# 16 The role of Fe-S clusters in regulation of yeast iron homeostasis

Caryn E. Outten

## 16.1 Introduction

The intracellular concentration and subcellular speciation of iron in eukaryotic cells is tightly regulated to maintain sufficient levels of this essential micronutrient while avoiding the toxic effects of iron overload. Iron is present in all subcellular compartments in a variety of chemical forms, including as protein-bound cofactors in Fe-S clusters, heme, and mononuclear and dinuclear iron-binding sites, as ferric oxide biominerals in iron storage proteins, as ferric oxyhydroxide nanoparticles, and as chelatable or "free" ferrous and/or ferric iron bound to low-molecular-weight ligands. Which of these forms is used as a gauge for intracellular iron concentrations and what are the molecular mechanisms for sensing and regulating iron bioavailability within eukaryotic cells? An emerging theme from recent studies in yeast model systems is that Fe-S clusters play key roles in iron regulation, which parallels previous findings in mammalian cells. This chapter highlights studies in the budding yeast *Saccharomyces cerevisiae*, and the fission yeast *Schizosaccharomyces pombe* that demonstrate the importance of Fe-S clusters as signals of intracellular iron bioavailability. A brief overview of yeast iron uptake and trafficking is provided, followed by detailed descriptions of the transcriptional regulators in *S. cerevisiae* and *S. pombe* that coordinate the cell's response to both high- and low-iron-growth conditions. Current knowledge on the role of Fe-S clusters in modulating the activity of yeast transcriptional regulators is discussed for each model eukaryote. The chapter concludes with a section comparing and contrasting the mechanisms of iron regulation in *S. cerevisiae* and *S. pombe*.

## 16.2 Iron acquisition and trafficking in yeast

Microbial iron acquisition is complicated by the low bioavailability of iron in oxygenated environments at neutral pH. Thus, eukaryotic microbes such as *S. cerevisiae* and *S. pombe* have devised multiple strategies for scavenging this essential resource (Fig. 16.1). One strategy involves acquisition and transport of ionic iron at the cell surface. Transmembrane proton pumps help acidify the extracellular environment to increase $Fe^{3+}$ solubility. Environmental $Fe^{3+}$ is converted to $Fe^{2+}$ by cell surface ferrireductases (Fre1 and Fre2 in *S. cerevisiae*, Frp1 in *S. pombe*) [1]. $Fe^{2+}$ is the substrate for the high-affinity oxidase-permease complexes (*S. cerevisiae* Fet3 and Ftr1, *S. pombe* Fio1 and Fip1) that first oxidize $Fe^{2+}$ to $Fe^{3+}$ prior to transporting $Fe^{3+}$ into the cell. Another strategy is to scavenge iron from the environment using highly specific small-molecule $Fe^{3+}$ chelators

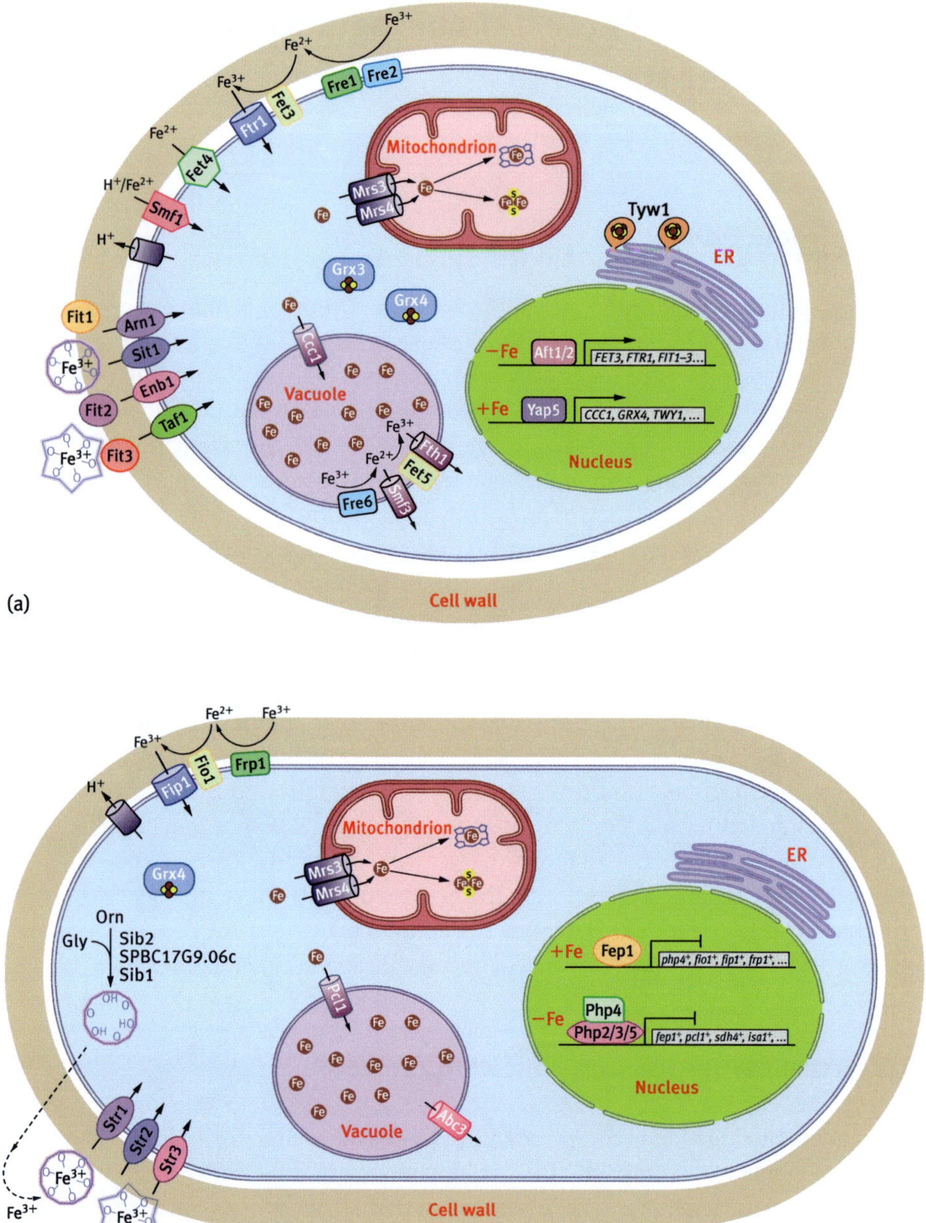


**Fig. 16.1:** Iron uptake, trafficking, and regulation systems in (a) *S. cerevisiae* and (b) *S. pombe*. (a) *S. cerevisiae* Fit1, Fit2, and Fit3 are siderophore-binding mannoproteins anchored in the cell wall. Arn1, Sit1, Enb1, and Taf1 are cell surface transporters for $Fe^{3+}$-siderophore complexes. The ferrireductases Fre1 and Fre2 reduce environmental $Fe^{3+}$ to $Fe^{2+}$. Fet3 and Ftr1 form the cell surface high-affinity iron uptake system in which Fet3 oxidizes $Fe^{2+}$ to $Fe^{3+}$ and Ftr1 transports $Fe^{3+}$ across the plasma membrane. Fet4 is a low-affinity transporter responsible for cell surface uptake under iron-replete

conditions. Smf1 is a $H^+/M^+$ symporter for transition metals such as $Fe^{2+}$, $Mn^{2+}$, and $Zn^{2+}$. Vacuolar iron transport systems include the importer Ccc1, the ferrireductase Fre6, and the exporters Smf3 and Fet5/Fth1. Grx3 and Grx4 are cytosolic [2Fe-2S]-binding proteins implicated in intracellular iron trafficking and Aft1/2 regulation. Mrs3 and Mrs4 import iron into the mitochondria, which is used for both heme and Fe-S cluster biogenesis, whereas Tyw1 is an ER-anchored radical SAM enzyme that sequesters Fe-S clusters during iron overload. Iron homeostasis in *S. cerevisiae* is coordinated by the low-iron-responsive activators, Aft1 and Aft2, which activate the expression of iron acquisition pathways, and the high-iron-responsive activator Yap5, which activates expression of proteins involved in iron sequestration. (b) *S. pombe* synthesizes the siderophore ferrichrome from ornithine and glycine using the ornithine $N^5$ monooxygenase Sib2, the gene product of *SPBC17G9.06c*, and the non-ribosomal peptide synthetase Sib1. Str1, Str2, and Str3 are plasma membrane transporters that import iron-siderophore complexes. Iron is imported into vacuoles *via* Pcl1 and exported *via* Abc3. *S. pombe* also has orthologues to *S. cerevisiae* Mrs3 and Mrs4 that likely import iron into mitochondria for Fe-S cluster and heme biosynthesis. The CGFS monothiol glutaredoxin Grx4 signals iron availability to the two transcriptional repressors Fep1 and Php4 that regulate iron homeostasis in *S. pombe*.

known as siderophores. *S. pombe* synthesizes and excretes ferrichrome, a siderophore with hydroxamate groups for high-affinity $Fe^{3+}$ binding [2]. Ferrichrome is synthesized in the cytosol using three glycines and three modified ornithines. Sib2 and the product of the gene *SPBC17G9.06c* are predicted to catalyze $N^5$-hydroxylation and acetylation of ornithine, respectively [3]. A third unidentified enzyme may catalyze formation of ornithine-glycine dipeptides, whereas Sib1 is a non-ribosomal peptide synthetase that polymerizes and circularizes the ornithine-glycine dipeptides to form the finished ferrichrome siderophore [4]. $Fe^{3+}$-loaded ferrichrome is taken up by the transporters Str1 and Str2 [5]. In addition to iron acquisition, ferrichrome may also play a role in intracellular iron storage because *S. pombe* accumulates intracellular ferrichrome under both high- and low-iron conditions [2].

Unlike *S. pombe*, *S. cerevisiae* does not synthesize siderophores; nevertheless, it expresses proteins at the cell surface for capturing (Fit1, Fit2, and Fit3) and internalizing (Arn1, Sit1, Enb1, and Taf1) a variety of siderophores produced by other microorganisms [6]. Likewise, *S. pombe* is capable of importing exogenous $Fe^{3+}$-bound siderophores (*via* Str1, Str2, and Str3) in addition to ferrichrome as sources of iron [5]. In addition to these high-affinity systems, *S. cerevisiae* possesses nonspecific low-affinity transporters (Fet4 and the NRAMP homologue Smf1) that import $Fe^{2+}$ as well as $Cu^{2+}$, $Mn^{2+}$, and $Zn^{2+}$ (Fig. 16.1) [7].

A number of pathways for trafficking iron from the cytosol to other subcellular compartments have been identified. The two main hubs for iron trafficking in yeast are the mitochondria and vacuoles. Iron import into mitochondria in *S. cerevisiae* is primarily mediated by the paralogous mitochondrial carrier proteins Mrs3 and Mrs4 [8, 9], whereas the vacuolar transporter Ccc1 facilitates storage and sequestration of iron in this organelle [10]. Orthologues for these mitochondrial and vacuolar iron importers are found in *S. pombe*, although only the function of the vacuolar importer (Pcl1) has been verified [11, 12]. *S. cerevisiae* additionally expresses a low-affinity NRAMP vacuolar transporter, Smf3, that is homologous to Smf1 [13]. Under conditions of iron depletion, vacuolar iron stores are mobilized by Smf3 and the oxidase-permease complex Fet5/Fth1, which is analogous to the Fet3/Ftr1 complex at the plasma membrane [7]. Orthologues for Smf3 and Fet5/Fth1 have not been identified in *S. pombe*. However, the *S. pombe* ABC-type transporter Abc3 carries out a similar function to Smf3 and Fet5/Fth1 by transporting vacuolar iron or iron conjugates to the cytosol when iron is scarce (Fig. 16.1) [12].

## 16.3 Regulation of iron homeostasis in *S. cerevisiae*

### 16.3.1 Aft1/Aft2 low-iron transcriptional regulators and target genes

*S. cerevisiae* responds to iron deficiency by activating the high-affinity ionic iron and iron-siderophore uptake systems as well as the vacuolar iron export systems to increase cytosolic iron levels. Activation of these systems is primarily controlled at the transcriptional level by the DNA binding regulator Aft1 and its paralogue Aft2 [6, 14–17]. Both Aft1 and Aft2 have an N-terminal DNA-binding domain and a C-terminal activation domain (Fig. 16.2). The N-terminal DNA-binding domains are homologous to the WRKY-GCM1 superfamily of eukaryotic transcriptional factors and have conserved Cys/His residues that are zinc finger (ZF) ligands in some WRKY-GCM1 family members [18]. A crystal structure for Aft2 bound to DNA confirms that Aft2 binds a structural ZF with these residues, and the same is likely true for Aft1 (C. Outten

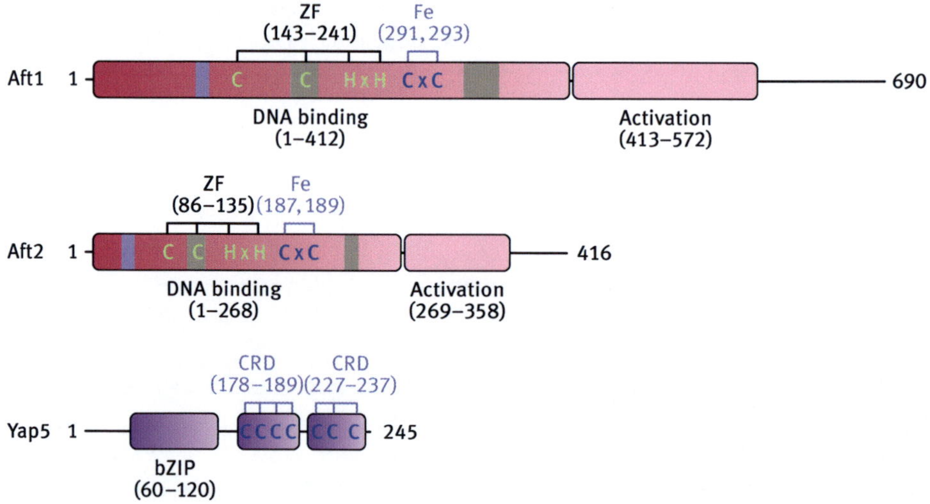


**Fig. 16.2:** Domain structure of iron-responsive transcriptional regulators Aft1, Aft2, and Yap5 from *S. cerevisiae*. Conserved ZF ligands in Aft1 and Aft2 are shown in green, and iron-responsive cysteine residues are shown in blue. The dark blue boxes indicate the positions of nuclear export signals, whereas the gray boxes show the locations of nuclear import signals. The conserved cysteines in the Yap5 cysteine-rich domains (CRDs) are shown in blue.

and C. He, unpublished data). Downstream of this ZF in the DNA-binding domain is a CDC motif that is essential to regulate Aft1 and Aft2 activity in response to iron, as discussed in Section 16.3.3 (Fig. 16.2). In total, Aft1 and Aft2 activate expression of 27 different genes that together comprise the iron regulon (Tab. 16.1) [1]. In addition to the plasma membrane and vacuolar iron transport systems, the iron regulon includes genes encoding the mitochondrial iron importer Mrs4, the Fe-S cluster assembly

**Tab. 16.1:** *S. cerevisiae* genes regulated by Aft1 and Aft2.

| Gene name | Protein function and location |
|---|---|
| **Cell surface siderophore iron uptake** | |
| *ARN1* | Cell surface iron-siderophore transporter |
| *ARN2/TAF1* | Cell surface iron-siderophore transporter |
| *ARN3/SIT1* | Cell surface iron-siderophore transporter |
| *ARN4/ENB1* | Cell surface iron-siderophore transporter |
| *FIT1* | Cell wall iron-siderophore-binding protein |
| *FIT2* | Cell wall iron-siderophore-binding protein |
| *FIT3* | Cell wall iron-siderophore-binding protein |
| **Cell surface ionic iron uptake** | |
| *FRE1* | Cell surface ferrireductase |
| *FRE2* | Cell surface ferrireductase |
| *FRE3* | Cell surface ferrireductase |
| *FET3* | Cell surface multicopper oxidase |
| *FTR1* | Cell surface $Fe^{2+}$ permease |
| *CCC2* | Golgi $Cu^{2+}$-transporting ATPase required for Fet3/Fet5 assembly |
| *ATX1* | Cytosolic Cu chaperone required for Cu delivery to Ccc2 |
| **Vacuolar metal transport** | |
| *FRE6* | Vacuolar ferrireductase |
| *FET5* | Vacuolar multicopper oxidase |
| *FTH1* | Vacuolar $Fe^{2+}$ permease |
| *SMF3* | Vacuolar $Fe^{2+}$ exporter |
| *COT1* | Vacuolar $Co^{2+}/Zn^{2+}$ importer |
| **Mitochondrial iron metabolism** | |
| *ISU1* | Mitochondrial Fe-S cluster assembly scaffold protein |
| *MRS4* | Mitochondrial $Fe^{2+}$ importer |
| **Post-transcriptional iron regulation** | |
| *CTH1* | Nuclear/cytosolic mRNA-binding protein |
| *CTH2* | Nuclear/cytosolic mRNA-binding protein |
| **Other functions** | |
| *VHT1* | Cell surface biotin importer |
| *AKR1* | Golgi palmitoyl transferase |
| *HMX1* | ER heme oxygenase |
| *BNA2* | Cytosolic tryptophan 2,3 dioxygenase required for *de novo* NAD biosynthesis (predicted) |
| *ECM4* | Cytosolic glutathione S-transferase |

scaffold protein Isu1, and the mRNA-binding proteins Cth1 and Cth2 [1]. Cth1 and Cth2 play roles in post-transcriptional control of iron homeostasis by facilitating targeted turnover of transcripts encoding proteins in nonessential, Fe-rich metabolic pathways. The net result is to redirect diminishing iron pools to essential Fe-dependent functions [19]. For example, Cth1 and Cth2 downregulate mRNA transcripts for TCA cycle Fe-S enzymes such as succinate dehydrogenase and aconitase, as well as components of the respiratory chain including the Rieske Fe-S protein, and cytochromes *c* and $c_1$. Consequently, cellular metabolism shifts from Fe-dependent respiration to Fe-independent fermentation to conserve iron [19–21].

Aft1 and Aft2 have overlapping as well as independent DNA targets. Aft1 preferentially binds Fe response elements with the consensus sequence TGCACCC, whereas Aft2 binds the related but shorter consensus sequence G/ACACCC [22, 23]. Aft1 has a more dominant role in iron homeostasis because *aft1Δ* mutants exhibit a stronger iron-deficiency phenotype than *aft2Δ* mutants. However, an *aft1Δaft2Δ* double mutant is more sensitive to iron-deficient conditions than an *aft1Δ* single mutant, indicating that Aft2 can partially compensate for loss of Aft1 in iron regulation [16, 17, 23]. Transcriptional analysis of Aft1 and Aft2 target genes suggest that Aft1 is primarily involved in cellular iron uptake, whereas Aft2 specifically regulates intracellular trafficking to vacuoles and mitochondria [22, 23].

### 16.3.2 Yap5 high-iron transcriptional regulator and target genes

In addition to the low-iron-sensing transcriptional activators Aft1 and Aft2, *S. cerevisiae* possesses a high-iron-sensing transcriptional activator named Yap5. Yap5 is a member of the basic leucine zipper (bZIP) transcriptional activator family (Fig. 16.2), which includes eight homologues in the *S. cerevisiae* genome. The list of genes regulated by Yap5 is much shorter than the Aft1/Aft2 regulon, limited to genes encoding the vacuolar iron importer Ccc1, Fe-S cluster-binding proteins Twy1 and Grx4, and the metallothionein Cup1 (Tab. 16.2) [24, 25]. Each of these Yap5-regulated genes has a known or potential role in sequestering excess iron, albeit *via* different mechanisms.

**Tab. 16.2:** *S. cerevisiae* genes regulated by Yap5.

| Gene name | Protein function and location |
|---|---|
| **Vacuolar iron import** | |
| *CCC1* | Vacuolar iron importer |
| **Metal-binding proteins** | |
| *GRX4* | Nuclear/cytosolic multidomain [2Fe-2S]-binding glutaredoxin |
| *TYW1* | ER-tethered radical SAM [4Fe-4S] enzyme required for synthesis of wybutosine-modified tRNA |
| *CUP1* | Cytosolic $Cu^{2+}$-binding metallothionein |

Activation of *CCC1* allows iron import into the vacuole to decrease the cytosolic iron pool (Fig. 16.1) [24]. Tyw1 is an ER-anchored, [4Fe-4S] cluster-binding radical SAM enzyme involved in modification of tRNA bases; however, the catalytic function of Tyw1 is not implicated in protection against iron toxicity. Instead, the data suggest that Yap5 protects the cell from excess iron by increasing Tyw1 levels, which sequesters iron as protein-bound Fe-S clusters [24]. Upregulation of Grx4 expression may serve a similar role because this [2Fe-2S]-binding protein is implicated in iron trafficking [26]. In addition, Grx4 plays a key role in inhibiting Aft1 and Aft2 during iron sufficiency (see Section 16.3.3), and increased levels of this protein may thereby enhance Aft1 and Aft2 inactivation. *CUP1* encodes a Cu-binding metallothionein that is important for resistance to copper toxicity [27]. Interestingly, biophysical studies indicate that Cup1 also binds four $Fe^{2+}$ atoms/monomer *in vitro* [28] and thus could play a role in iron sequestration under toxic iron conditions. Taken together, these studies suggest that Yap5 responds to high cellular iron levels by decreasing free cytosolic iron through sequestration in vacuoles or incorporation into Fe- or Fe-S cluster-binding proteins [24].

### 16.3.3 Links between mitochondrial Fe-S cluster biogenesis, the Grx3/Grx4/Fra2/Fra1 signaling pathway, and Aft1/Aft2 regulation

A number of studies have started to shed light on the specific molecular mechanism by which Aft1 and Aft2 sense intracellular iron levels, revealing the critical role of Fe-S clusters. The first clues to the Aft1/2 regulation mechanism were provided by *in vivo* DNA footprinting studies demonstrating that Aft1 binds its genetic targets under iron-deplete, but not iron-replete conditions [29]. In addition, Aft1 nuclear localization is controlled by the cellular iron status: in iron-starved cells, Aft1 accumulates in the nucleus, whereas under iron-replete conditions, Aft1 dissociates from its DNA targets and is shuttled to the cytosol by the nuclear exportin Msn5 [30, 31]. Interaction with Msn5 and export of Aft1/2 to the cytosol requires multimerization of Aft1 or Aft2, which is dependent on a conserved CDC located in the DNA-binding domain (Fig. 16.2). Mutation of either Cys residue in the CDC motif (termed Aft1/$2^{up}$ mutations) leads to constitutive nuclear localization of Aft1/2 and activation of the iron regulon [14, 30, 32]. Interestingly, Aft1/2 localization and DNA binding are independent of changes in cytosolic iron levels. Instead, the activity of Aft1 and Aft2 is dependent one the ability of mitochondria to assemble Fe-S clusters [33]. When Fe-S cluster biogenesis is disrupted, Aft1 and Aft2 are constitutively active regardless of cytosolic iron levels. In addition, iron signaling to Aft1/2 also requires the mitochondrial ABC transporter Atm1 [34]. Atm1 is proposed to export an unknown sulfur-containing substrate from mitochondria that is used to build and/or insert Fe-S clusters into cytosolic proteins [35]. However, deletion of proteins in the CIA (cytosolic iron-sulfur assembly) pathway has no effect on Aft1 and Aft2 activity, indicating that iron sensing is only linked to the mitochondrial pathway [32].

Aft1/2 inhibition in response to mitochondrial Fe-S cluster (ISC) biogenesis is mediated by a signaling pathway involving four cytosolic proteins: Grx3, Grx4, Fra2, and Fra1, all of which have homologues in mammalian cells [36–38]. Components of this signaling pathway were primarily identified by gene deletion and *in vivo* protein-protein interaction studies. The two cytosolic *S. cerevisiae* CGFS Grxs perform essential but redundant functions in iron metabolism because single *grx3Δ* and *grx4Δ* deletion strains have no discernable growth phenotypes, whereas *grx3grx4* double mutants are severely growth impaired or inviable, depending on the strain background [26, 36, 37]. Intracellular iron overaccumulates in *grx3grx4* strains due to constitutive activation of the iron regulon, although mitochondrial iron levels are decreased. In addition, the enzymatic activity and *in vivo* iron binding for a variety of Fe-S cluster, heme, and nonheme iron-binding proteins in both the cytosol and mitochondria are dramatically reduced in a *grx3grx4* strain, indicating that intracellular iron is much less bioavailable [26]. Thus, in addition to regulation of Aft1/2 activity, the cytosolic, multidomain CGFS Grxs in *S. cerevisiae* are suggested to play an essential role in intracellular iron trafficking. In *fra2Δ* and *fra1Δ* strains, the iron regulon is also constitutively activated, although disruption of iron homeostasis is less severe in these strains [38, 39]. Iron signaling to Aft1 and Aft2 is controlled by specific protein-protein interactions between components of this signaling pathway. Physical interactions between Aft1 and Grx3/4, Grx3/4 and Fra2, and Fra1 and Fra2 have been demonstrated *in vivo* by high-throughput and targeted yeast two-hybrid, co-immunoprecipitation, and affinity capture studies [36–38, 40].

### 16.3.4 Fe-S cluster binding by Grx3/4 and Fra2 is important for their function in *S. cerevisiae* iron regulation

*S. cerevisiae* Grx3 and Grx4 are paralogous members of the monothiol glutaredoxin (Grx) family found a wide variety of prokaryotes and eukaryotes. Members of this family possess a signature CGFS active site sequence that differs from the classical CPY/FC active site found in dithiol Grxs. Dithiol Grxs are glutathione (GSH)-dependent thiol-disulfide oxidoreductases, whereas monothiol CGFS Grxs have little or no oxidoreductase activity when tested with standard Grx model substrates [41–46]. CGFS-type monothiol Grxs can be further classified into two groups: single-domain Grxs that have a single Grx-like domain and multidomain Grxs that possess an N-terminal thioredoxin (TRX)-like domain and one or more Grx-like domains. Single-domain CGFS Grxs are found in prokaryotes and in the mitochondria and chloroplasts of eukaryotes, whereas multidomain CGFS Grxs are exclusively eukaryotic, exhibiting cytosolic/nuclear localization. Both single and multidomain CGFS Grxs form $[2Fe\text{-}2S]^{2+}$-bridged homodimers with all-cysteinyl ligation provided by the two CGFS active site cysteines in the Grx-like domain and two GSH molecules (Fig. 16.3a) [46–53]. Members of the single-domain subfamily are proposed to facilitate maturation of Fe-S cluster proteins (see Chapter 15 by Debkumar Pain and

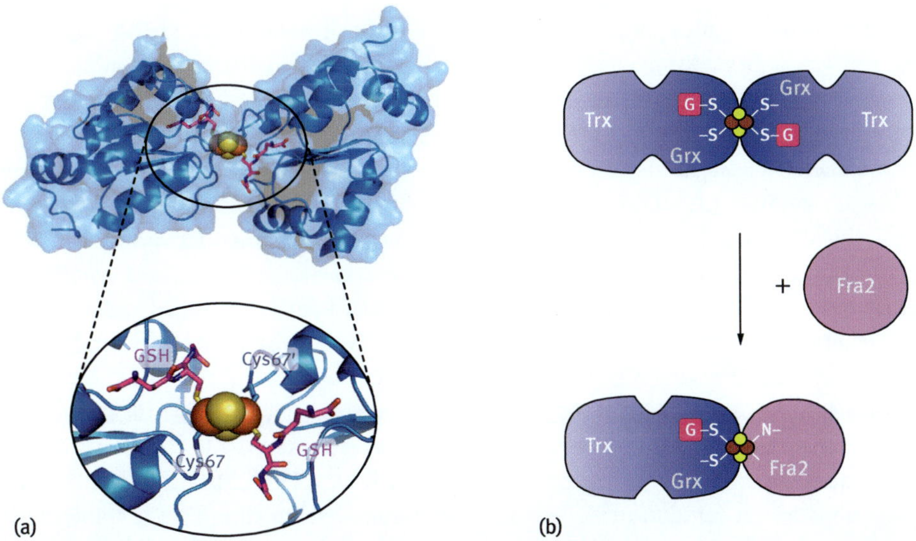


**Fig. 16.3:** Structures of [2Fe-2S]-bridged CGFS Grx complexes. (a) X-ray crystal structure of *H. sapiens* GLRX5 (PDB 2WUL) [47] with close-up view of the GSH-ligated [2Fe-2S] cluster. Cys67 is the conserved cysteine in the CGFS active site of human Glrx5. Glrx5 crystallizes as two interacting homodimers but only one homodimer is shown in the figure. (b) Model for $[2Fe\text{-}2S]^{2+}$ Grx3/4 homodimer (top) and $[2Fe\text{-}2S]^{2+}$ Fra2-Grx3/4 heterodimer (bottom) characterized from *S. cerevisiae*. The $[2Fe\text{-}2S]^{2+}$ Fra2-Grx3/4 heterodimer is formed by titration of $[2Fe\text{-}2S]^{2+}$ Grx3/4 homodimers with apo Fra2. For the Grx3/4 homodimer, the active site cysteines in the GRX-like domains and 2 GSH [G] molecules ligate the Fe-S cluster. For the Fra2-Grx3/4 heterodimer, the Fe-S cluster is ligated by one GRX domain active site cysteine, one GSH, a histidine from Fra2 (N), and an unidentified fourth ligand.

Andrew Dancis and Chapter 18 by Wing-Hang Tong), whereas multidomain CGFS Grxs are proposed to play dual roles in cytosolic iron trafficking (see Section 16.3.3) and iron regulation [26, 36–38, 54]. The Fe-S cluster biogenesis function of single-domain CGFS Grxs and the trafficking and regulatory functions of multidomain CGFS Grxs in yeast are all dependent on the presence of the conserved Cys in the CGFS active site, suggesting that coordination of the [2Fe-2S] cluster is essential to these functions [26, 36, 54–57]. Formation of this Fe-S complex *in vivo* is supported by studies in *S. cerevisiae* demonstrating that iron binding to CGFS Grxs requires the GRX domain CGFS motif, sufficient intracellular GSH levels, and functional mitochondrial Fe-S cluster biogenesis [26]. The role of the less well-conserved TRX domain in multidomain Grxs has also been studied, although its specific function remains elusive. Mutation of a conserved Cys in the TRX domain of Grx3/4 disrupts neither Fe-S binding or homodimer formation *in vitro* [51] nor iron binding, iron trafficking, and Aft1/2 iron regulation *in vivo* [26]. Complete removal of this domain also does not impact Fe-S binding to the GRX domain *in vivo* and *in vitro* [51, 58], although the TRX domain is essential for Grx3/4 trafficking and regulation functions *in vivo* [58]. Based on these studies, the current hypothesis is that the TRX domain mediates specific protein-protein interactions with Grx3/4 binding partners.

*In vitro* biochemical and biophysical studies have demonstrated that *S. cerevisiae* Grx3 and Grx4 form $[2Fe\text{-}2S]^{2+}$-bridged heterodimers with Fra2 in addition to $[2Fe\text{-}2S]^{2+}$-bridged homodimers (Fig. 16.3b) [51, 59]. Fra2-Grx3/4 heterodimers exhibit significant differences in cluster stability and coordination environment in comparison to Grx3/4 homodimers. The Fe-S cluster in Fra2-Grx3/4 heterodimers is more stable to oxidation/reduction than the Fe-S cluster in Grx3/4 homodimers. Furthermore, the [2Fe-2S] Grx3 homodimer can be rapidly and efficiently converted to a [2Fe-2S] Grx3-Fra2 heterodimer by titration with apo-Fra2, demonstrating that formation of the heterodimer is thermodynamically and kinetically favored. In Fra2-Grx3/4 heterodimers, Fra2 replaces one Grx3/4 monomer and one GSH, providing a histidine ligand (His103) and presumably a fourth ligand that has not yet been identified. Mutation of Fra2 His103 to a Cys or Ala does not abolish formation of a [2Fe-2S]-bridged heterodimer with Grx3/4 *in vitro*; however, EPR and EXAFS data indicate that the stability of the mutant complex is compromised [59]. The H103A Fra2 mutant is also unable to rescue the iron regulation defects in *fra2Δ* strains, suggesting that a stable and/or redox-active cluster may be important for Fra2-Grx3/4 function [59]. Fra2 His103 is highly conserved in both prokaryotic and eukaryotic BolA homologues; therefore, it may play an important structural or functional role for other members of this protein family [60].

### 16.3.5 Working model for Fe-dependent regulation of Aft1/2 *via* the Fra1/Fra2/Grx3/Grx4 signaling pathway

The available data lend itself to a working model for Fe-dependent regulation of Aft1/2 *via* the Fra1/Fra2/Grx3/Grx4 signaling pathway (Fig. 16.4). According to the model, the mitochondrial inner membrane transporter Atm1 exports a sulfur-containing substrate produced by the mitochondrial ISC assembly machinery during conditions of iron sufficiency. This substrate and GSH are both essential for Fe-S cluster binding to Grx3/4 homodimers. The mechanism by which the [2Fe-2S] cluster is assembled and/or delivered to Grx3 and Grx4 is unknown. However, the cytosolic iron-sulfur assembly (CIA) system is not required for *in vivo* iron incorporation into Grx3/4 [26] or iron-dependent inhibition of Aft1/2 activity [32]. Thus, an unidentified parallel pathway must exist for Fe-S loading into Grx3/4 complexes. Formation of [2Fe-2S] Fra2-Grx3/4 heterodimers may proceed *via* a simple substitution reaction whereby apo-Fra2 displaces GSH and one of the Grx3/4 monomers. This mechanism is supported by *in vitro* results demonstrating that apo-Fra2 binds tightly and stoichiometrically to Grx3/4 homodimers leading to [2Fe-2S] Fra2-Grx3/4 heterodimer formation [59]. The [2Fe-2S] Fra2-Grx3/4 complex is then proposed to facilitate oligomerization of Aft1/2. The specific compartment in which the Fra-Grx inhibitory complex interacts with Aft1/2 is unclear. An initial study using overexpressed, GFP-tagged Grx3 suggested that this protein is primarily localized to the nucleus [55]. However, natively expressed, untagged Grx3 was found to exhibit mainly cytosolic

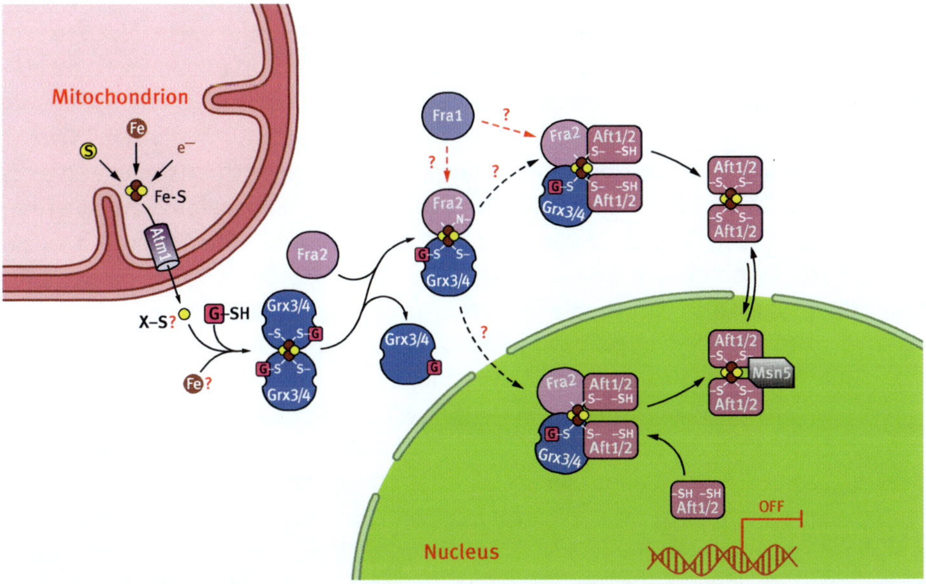


**Fig. 16.4:** Proposed model for *S. cerevisiae* iron regulation *via* Aft1 and Aft2 under iron-replete conditions. During conditions of iron sufficiency, Fe-S clusters are synthesized in mitochondria *via* integration of iron, sulfur, and redox control pathways. An unknown substrate produced by the mitochondrial ISC machinery is exported out to the cytosol by the transporter Atm1. GSH is also required for export of this signal. Grx3 and Grx4, which form GSH-ligated, Fe-S-bridged homodimers, are proposed to form heterodimers with Fra2 to relay this signal to Aft1 and Aft2. Fra1 interacts with Fra2; however, its specific role is not known. In addition, the specific compartment (nucleus, cytosol, or both) in which the Fra-Grx inhibitory complex interacts with Aft1 and Aft2 is unclear. Interaction of Grx3/4 with Aft1 promotes dissociation of the transcriptional activator from its target DNA and export to the cytosol, leading to deactivation of Aft1/2-regulated genes. The exportin Msn5 facilitates iron-dependent export of both Aft1 and Aft2.

localization. In addition, restriction of Grx4 to the cytosol *via* tethering to the mitochondrial outer membrane was shown to have little effect on inhibition of Aft1 activity in response to iron, suggesting that the Fra-Grx inhibitory complex acts in the cytosol [38]. In either case, oligomerization of Aft1 and Aft2 requires the conserved CDC motif. This motif in Aft1/2 is key to the regulation mechanism because substitution of these Cys residues disrupts the *in vivo* interaction between Grx3/4 and Aft1/2 and prevents Aft1/2 oligomerization. The CDC motif is also required for Aft1/2 interaction with the exportin Msn5- and Fe-dependent nuclear export, although recent evidence suggests that Msn5-mediated export of Aft1/2 is not essential to the inhibition mechanism because Aft1 functions normally in an *msn5Δ* strain despite constitutive localization to the nucleus [34]. Nevertheless, the essential role of the CDC motif in iron-dependent Aft1/2 inhibition, taken together with the dependence of Grx3/4 and Fra2 function on [2Fe-2S] cluster binding, strongly suggests that thiol redox chemistry and/or Fe-S or Fe binding to Aft1/2 drives multimerization, DNA dissociation, and translocation of Aft1/2 to the cytosol under iron-replete conditions (Fig. 16.4).

Despite considerable progress in defining the genetic and molecular interactions among Aft1/2, Grx3/4, Fra2, and Fra1, some key aspects of the iron sensing and regulation mechanism still remain unresolved. Most importantly, what is the specific molecular mechanism for inhibiting Aft1/2 activity in response to iron? Recent work by our group suggests that dimerization of Aft2 is driven by transfer of a [2Fe-2S] cluster from Fra2-Grx3 to Aft2, with two Cys from each Aft2 monomer ligating the [2Fe-2S] cluster (Fig. 16.4) (C. Outten and C. He, unpublished data). Furthermore, we find that Fe-S cluster binding to Aft2 lowers its DNA-binding affinity, which agrees with *in vivo* data suggesting that Grx3/4 induces Aft1 dissociation from DNA [34]. Experiments to further test this regulation model are ongoing in our laboratory. Fra1 also plays a role in Aft1-mediated iron signaling and interacts with Fra2 *in vivo*, but how is it specifically involved in the signaling pathway? Additionally, the Fra-Grx signaling pathway may not be the only method for inhibiting Aft1/2 activity under iron-replete conditions. Transcriptional reporter assays of Aft1/2-regulated genes indicate that the iron regulon is not fully activated in iron-sufficient medium in *fra1Δ* or *fra2Δ* mutants or upon disruption of mitochondrial ISC biogenesis pathways, suggesting that a separate signal may partially inhibit Aft1/2 activity in these mutants [38]. Finally, it is unknown how Fe-S dependent inhibition of Aft1 and Aft2 influences their interaction with transcriptional co-activators. Ssn6/Tup1 and Nhp6 associate with Aft1 *in vivo* and *in vitro* and are required for transcriptional activation of certain Aft1-regulated genes [61, 62]. It is possible that Aft1/2 metal binding and/or dimerization decreases these interactions, which in turn facilitates deactivation of Aft1/2 target genes.

### 16.3.6 Yap5 regulation and mitochondrial Fe-S cluster biogenesis

Unlike Aft1 and Aft2, Yap5 is constitutively localized to the nucleus and bound to its DNA targets under both high- and low-iron conditions. Thus, the iron-dependent switch that controls Yap5 activity does not involve changes in DNA-binding affinity as suggested for Aft1 and Aft2. However, similar to Aft1/2, cysteine residues are involved in controlling Yap5 function. In the case of Yap5, changes in iron availability are proposed to modify sulfhydryls located in two cysteine-rich domains (CRDs) downstream of the DNA-binding domain (Fig. 16.2). Single mutations in these Cys residues decrease Yap5 transcriptional activity to varying degrees, with mutations in the N-terminal CRD having the strongest effect. Mutation of all seven Cys residues leads to complete loss of Yap5 transcriptional activity without affecting DNA binding [63]. Furthermore, thiol modification assays on cellular extracts suggested that high iron induces disulfide formation on Yap5, implicating thiol redox chemistry in the regulation mechanism (Fig. 16.5). Interestingly, mitochondrial Fe-S cluster biogenesis also plays a key role in the response to high iron mediated by the *S. cerevisiae* transcription factor Yap5. As previously demonstrated for Aft1 and Aft2, iron-dependent control of Yap5-regulated genes is independent of changes in cytosolic iron levels or vacuolar/mitochondrial iron

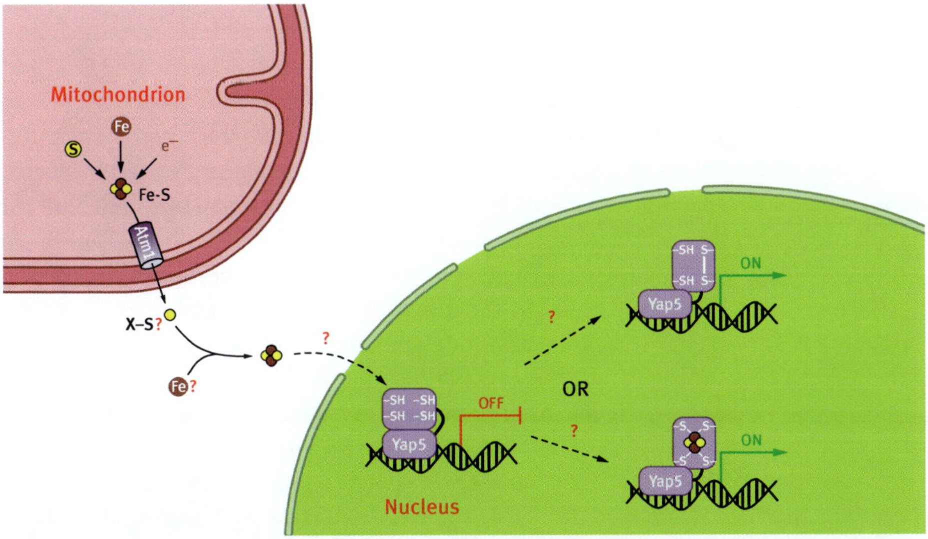


**Fig. 16.5:** Proposed model for *S. cerevisiae* iron regulation *via* Yap5 under iron-replete conditions. Under iron-replete conditions, the mitochondria ISC-dependent signal exported by Atm1 is relayed to Yap5 *via* as-yet-unidentified factor(s). This mitochondrial Fe-S-dependent signal stimulates Yap5 activation of its target genes *via* thiol modification and/or Fe-S cluster binding to conserved cysteine residues located in the C-terminal domain. Yap5 is bound to DNA in both high- and low-iron conditions.

sequestration. Instead, the transcriptional response is tuned to the availability of Fe-S clusters assembled in the mitochondria [64]. This intriguing result raises the alternate possibility that Yap5 directly binds an Fe-S cluster as suggested for Aft1/2 (Fig. 16.5); however, at present there is no direct evidence for this hypothesis.

Although both Aft1/2 and Yap5 are connected to mitochondrial Fe-S cluster assembly, one significant difference is the pathway by which the Fe-S cluster-dependent signal is relayed to the low- *vs* high-iron transcriptional regulators. Both are dependent on Atm1 function, but iron signaling to Yap5 is independent of the cytosolic Fe-S cluster-binding proteins Grx3 and Grx4, which play an essential role in Aft1/2 regulation [64]. The CIA assembly pathway is also not required for iron signaling to Yap5 [64]. Thus, identifying the molecular pathway for activating Yap5 in response to Fe-S cluster availability requires further study.

## 16.4 Regulation of iron homeostasis in *S. pombe*

### 16.4.1 Fep1 and Php4 transcriptional repressors and target genes

Although the pathways for iron uptake and trafficking are similar between *S. pombe* and *S. cerevisiae*, the mechanisms for regulation of iron homeostasis in these two fungi are significantly different, reflecting their evolutionary divergence. The genome of

*S. pombe* does not encode homologues of *S. cerevisiae* Aft1 and Aft2. Instead, iron homeostasis in *S. pombe* is primarily modulated by two repressors designated Fep1 and Php4, which are responsible for regulating iron uptake and iron utilization, respectively [1, 3, 65]. Fep1 is a GATA-type transcription factor that represses the expression of iron uptake and transport genes. Fep1-regulated genes encode proteins involved in siderophore iron uptake (*str1*$^+$, *str2*$^+$, and *str3*$^+$) and ionic iron uptake (*fio1*$^+$, *frp1*$^+$, and *fip1*$^+$) at the cell surface, as well as vacuolar iron export (*abc3*$^+$) (Tab. 16.3) [5, 12, 66]. Fep1 binds to GATA-containing sequences in promoters of these genes in iron-replete conditions, thereby repressing their expression to avoid iron overload. The co-repressors Tup11 and Tup12 bind to DNA-bound Fep1 in an iron-independent manner, facilitating repression of the target genes. These co-repressors downregulate numerous genes involved in a variety of different pathways [67]. When iron levels are low, Fep1 is released from the promoter region and the Fep1 regulon is turned on to promote iron uptake [68].

As mentioned in Section 16.3.1, metabolic remodeling in response to low iron availability in *S. cerevisiae* is controlled at the post-transcriptional level by the mRNA-binding proteins Cth1 and Cth2, whose expression is activated by Aft1 and Aft2 when iron is limiting. These regulators facilitate degradation of mRNA-encoding enzymes that function in nonessential Fe-rich metabolic pathways to preserve iron pools for essential functions. In contrast, iron-dependent metabolic remodeling in *S. pombe* is primarily controlled at the transcriptional level by the repressor Php4. Php4 binds to a heterotrimeric CCAAT-binding complex composed of Php2, Php3, and Php5. Under iron-replete conditions, the Php2/Php3/Php5 complex activates expression of its target genes by binding to CCAAT sequences in their promoters. Php2/Php3/Php5-regulated

**Tab. 16.3:** *S. pombe* genes regulated by Fep1.

| Gene name | Protein function and location |
|---|---|
| **Ferrichrome synthesis** | |
| *sib1*$^+$ | Cytosolic ferrichrome synthetase |
| *sib2*$^+$ | Cytosolic ornithine $N^5$ monooxygenase (predicted) |
| **Cell surface siderophore iron uptake** | |
| *str1*$^+$ | Cell surface iron-siderophore transporter |
| *str2*$^+$ | Cell surface iron-siderophore transporter |
| *str3*$^+$ | Cell surface iron-siderophore transporter |
| **Cell surface ionic iron uptake** | |
| *fio1*$^+$ | Cell surface multicopper oxidase |
| *fip1*$^+$ | Cell surface $Fe^{2+}$ permease |
| *frp1*$^+$ | Cell surface ferrireductase |
| **Vacuolar iron export** | |
| *abc3*$^+$ | Vacuolar iron exporter |
| **Transcriptional iron regulation** | |
| *php4*$^+$ | Cytosolic/nuclear Fe-sensing transcriptional repressor, subunit of CCAAT-binding factor complex |

genes encode proteins involved in iron-rich metabolic pathways such as the mitochondrial electron transport chain (*cyt1*$^+$, *cyc1*$^+$), the tricarboxylic acid cycle (*sdh2*$^+$, *sdh3*$^+$, *sdh4*$^+$), amino acid biosynthesis (*leu2*$^+$, *glt1*$^+$), and iron-sulfur cluster (*isa1*$^+$) and heme biosynthesis (*hem3*$^+$) (Tab. 16.4) [11]. Under low-iron conditions, Php4 binds to the Php2/Php3/Php5 complex, causing it to switch from an activator to a repressor. Thus, nonessential iron-utilizing pathways are downregulated as an iron-sparing response.

Php4 and Fep1 each regulate expression of the other iron-responsive transcription factor creating a reciprocal regulatory loop. The *php4*$^+$ gene contains GATA elements

**Tab. 16.4:** *S. pombe* genes regulated by Php4.[a]

| Gene name | Protein function and location |
|---|---|
| **Vacuolar iron import** | |
| *pcl1*$^+$ | Vacuolar iron importer |
| **Mitochondrial energy metabolism** | |
| *sdh2*$^+$ | Mitochondrial Fe-S subunit of succinate dehydrogenase |
| *sdh3*$^+$ | Mitochondrial cytochrome *b* subunit of succinate dehydrogenase |
| *sdh4*$^+$ | Mitochondrial inner membrane anchor subunit of succinate dehydrogenase |
| *cyc1*$^+$ | Mitochondrial heme-binding cytochrome *c* (predicted) |
| *cyt1*$^+$ | Mitochondrial heme-binding cytochrome $c_1$ (predicted) |
| SPAC20G8.04C | Mitochondrial electron-transfer flavoprotein-ubiquinone oxidoreductase (predicted) |
| SPAC24C9.06C | Mitochondrial Fe-S binding aconitase (predicted) |
| **Amino acid biosynthesis** | |
| *leu2*$^+$ | Cytosolic Fe-S binding isopropylmalate dehydratase (predicted) |
| *glt1*$^+$ | Mitochondrial Fe-S binding glutamate synthase (predicted) |
| *sir1*$^+$ | Cytosolic Fe-S binding sulfite reductase, beta subunit |
| **Cofactor biosynthesis** | |
| *isa1*$^+$ | Mitochondrial Fe-S cluster trafficking protein (Fe-S cluster assembly) |
| *bio2*$^+$ | Mitochondrial Fe-S binding biotin synthase |
| *hem3*$^+$ | Cytosolic/nuclear hydroxymethylbilane synthase (heme biosynthesis) |
| **Transcriptional iron regulation** | |
| *fep1*$^+$ | Nuclear Fe-sensing GATA-type transcriptional repressor |
| **Other functions** | |
| *aif1*$^+$ | Mitochondrial Fe-S binding apoptosis-inducing factor homologue (predicted) |
| *ctt1*$^+$ | Cytosolic/nuclear heme-binding catalase |
| *erg11*$^+$ | ER heme-binding sterol 14-demethylase, cytochrome P450 member (predicted) |

[a] A recent microarray study identified 86 potential Php4-regulated genes [69]. Only the target genes confirmed by RNase protection assays are shown here.

within its promoter allowing Fep1 repression [11]. Likewise, the *fep1*$^+$ gene contains CCAAT *cis*-acting elements in its promoter region and is downregulated in a Php4-dependent manner. Php2, Php3, and Php5 are constitutively synthesized so the Php2/Php3/Php5 complex itself is not directly responsive to iron levels [11]. Thus, when iron levels are high, Fep1 is expressed and binds to the promoter of *php4*$^+$, thereby inactivating its transcription (Fig. 16.6, left). Without the Php4 repressor, the Php2/Php3/Php5 complex activates expression of *fep1*$^+$. Conversely, when iron levels are low, *php4*$^+$ is expressed enabling repression of *fep1*$^+$ *via* Php4 binding to the Php2/Php3/Php5 complex (Fig. 16.6, right). Reduced Fep1 levels, in turn, relieve the Fep1-dependent repression of the *php4*$^+$ gene. This reciprocal regulatory loop thus allows direct crosstalk between iron acquisition and iron utilization pathways to fine tune iron homeostasis in *S. pombe*.

### 16.4.2 Roles for Grx4 in regulation of Fep1 and Php4 activity

In addition to iron-dependent cross-regulation at the transcriptional level, the activities of both Fep1 and Php4 are controlled at the post-translational level by *S. pombe* Grx4, a member of the multidomain CGFS Grx subfamily. Thus, as demonstrated for *S. cerevisiae* Aft1 and Aft2 iron regulators, a CGFS Grx plays a key role in *S. pombe* iron regulation, even though Fep1 and Php4 do not share significant sequence identity with Aft1 or Aft2 and use different regulation mechanisms. Although *S. cerevisiae* has two paralogous cytosolic CGFS Grxs (Grx3/4), the *S. pombe*

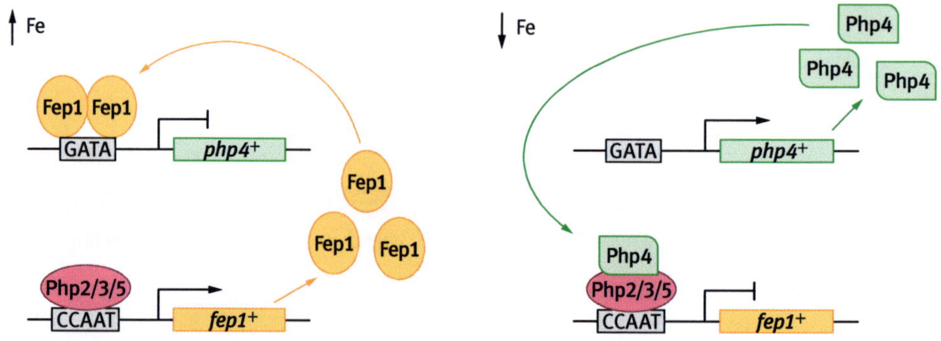


**Fig. 16.6:** Mutual transcriptional regulation by Php4 and Fep1 in *S. pombe*. When iron levels are high, *php4*$^+$ expression is repressed by Fep1 binding to GATA-containing sequences in the promoter region of this gene. Without Php4, the heterotrimeric Php2/Php3/Php5 CCAAT-binding factor complex that is bound to the CCAAT *cis*-activating elements in the promoter region of the *fep1*$^+$ gene is able to activate expression of *fep1*$^+$. In iron-deficient conditions, Fep1 dissociates from the promoter of *php4*$^+$, allowing expression of this gene. Increased Php4 levels, in turn, facilitate repression of *fep1*$^+$ *via* binding of Php4 to the CCAAT-binding factor complex, switching this complex from an activator to a repressor.

genome only encodes one, designated Grx4 [60, 70]. Grx4 impacts both Fep1 and Php4 function because Fep1- and Php4-regulated genes are constitutively repressed in *grx4* deletion strains [54, 57, 71]. Control of Fep1 and Php4 activity by Grx4 is governed by specific protein-protein interactions. Grx4 physically interacts with Fep1 and Php4 *in vivo* in an iron-independent manner as demonstrated by yeast two-hybrid and bimolecular fluorescence complementation experiments [54, 57, 71, 72]. Under iron-replete conditions, Grx4 promotes Php4 export to the cytosol by facilitating interaction with the nuclear exportin Crm1 (Fig. 16.7b). However, as demonstrated for Aft1/2, nucleocytoplasmic shuttling is not the primary mechanism for control of Php4 activity because Php4-regulated genes are activated in iron-replete conditions upon inhibition of Crm1 nuclear export activity [71]. Thus, in addition to promoting nuclear export of Php4 when iron is abundant, the interaction of Grx4 with Php4 also prevents Php4 from repressing the Php2/Php3/Php5 complex [71]. In

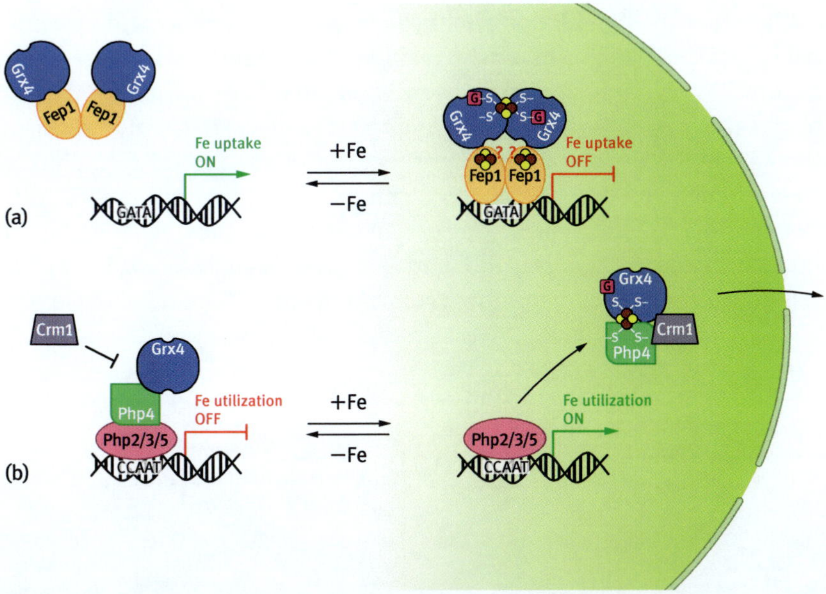


**Fig. 16.7:** Proposed model for *S. pombe* iron regulation under iron-replete and iron-depleted conditions for (a) Fep1 and (b) Php4. (a) (left) When iron is limiting, Fep1 dissociates from DNA but maintains an interaction with both the TRX and GRX domains of Grx4. (right) Under iron-replete conditions when Grx4 forms [2Fe-2S]-bridged homodimers, Fep1 binds DNA and interacts only with the TRX domain of Grx4. Fep1 itself may also bind iron or an Fe-S cluster. The Fep1 co-repressors Tup11 and Tup12 are not included in the model for the sake of simplicity. (b) (left) When iron levels are low, Php4 interacts with the TRX domain of Grx4 while binding the Php2/3/5 heterotrimeric complex, thereby repressing expression of iron utilization genes. Under these conditions, the exportin Crm1 is unable to interact with Php4. When iron is abundant, Php4 interacts with both the GRX domain and the TRX domain of Grx4, likely forming an Fe-S cluster-bound heterocomplex. Php4 dissociates from the Php2/3/4 heterotrimer and the Php4-Grx4 complex is exported by Crm1.

contrast, the interaction between Fep1 and Grx4 does not involve iron-dependent nucleocytoplasmic shuttling. Grx4 constitutively interacts with Fep1 in the nucleus regardless of intracellular iron status. When iron is limited, Grx4 inhibits Fep1 function, leading to dissociation of Fep1 from DNA and derepression of the Fep1 regulon. When iron is abundant, Grx4 is unable to inhibit Fep1 repressor function, although it still physically interacts with Fep1 (Fig. 16.7a) [54, 57].

### 16.4.3 Molecular basis of iron-dependent control of Fep1 activity

The specific domains and residues that mediate protein-DNA and protein-protein interactions for Fep1 have been mapped out by a variety of techniques. The N-terminal DNA-binding domain of Fep1 (and other iron-dependent GATA-type transcription factors found in similar fungi) includes two ZF domains (ZF1 and ZF2) that are required for high-affinity DNA binding *in vivo* and *in vitro* (Fig. 16.8) [68, 73]. Sandwiched between the two ZF domains is a conserved 27 amino acid cysteine-rich region [65, 74]. The four conserved Cys residues in this region are also required for high-affinity DNA binding and have been implicated in $Fe^{3+}$ binding [11, 74]. Whether or not this cysteine-rich region binds an Fe-S cluster has not yet been tested, although the UV-visible absorption spectrum of as-purified recombinant SRE [75], a Fep1 homologue from *Neurospora crassa*, is reminiscent of a $[2Fe\text{-}2S]^{2+}$-binding protein [76]. Substitution of these conserved cysteines in recombinant Fep1 (and in its orthologues *N. crassa* SRE and *Histoplasma capsulatum* Sre1) leads to loss of this characteristic reddish-brown color, confirming the role of these resi-

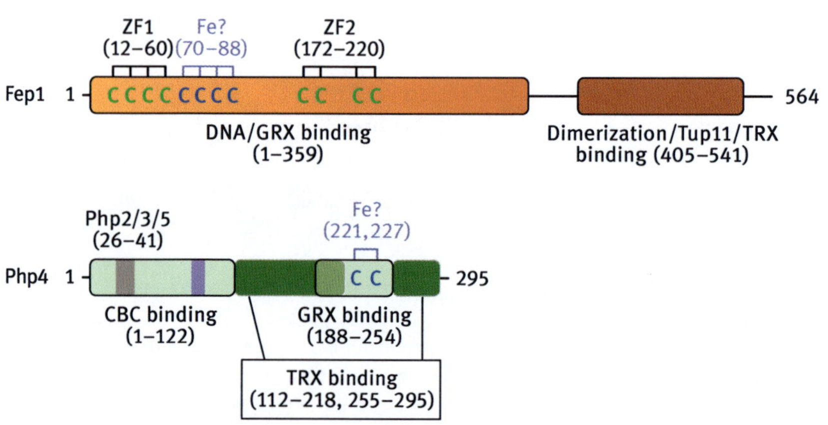


**Fig. 16.8:** Domain structure of iron-responsive transcriptional regulators Fep1 and Php4 from *S. pombe*. Conserved ZF ligands are shown in green and putative iron-responsive cysteine residues are shown in blue. The dark blue box in Php4 indicates the position of the nuclear export signal, whereas the gray box in the Php4 CCAAT-binding complex (CBC)-binding domain depicts the specific location of the Php2/3/5-binding residues.

dues in metal binding [65, 74, 75]. Nevertheless, Fep1 remains bound to DNA and represses its target genes under both low- and high-iron conditions when *grx4* is deleted, suggesting that iron or Fe-S binding is not absolutely necessary for Fep1 DNA binding *in vivo* [54, 57].

The C-terminus of Fep1 includes a predicted leucine zipper coiled-coil region that favors Fep1-Fep1 dimer formation and is required for efficient repression of Fep1 target genes [73]. The Tup11 interaction domain has been mapped to the C-terminal domain as well (Fig. 16.8) [77]. The interaction between Fep1 and Grx4 involves both the N- and the C-terminal regions of both proteins. The N-terminal TRX domain of Grx4 invariably and strongly interacts with the C-terminus of Fep1, whereas the C-terminal GRX domain of Grx4 weakly interacts with the N-terminus of Fep1 only in iron-depleted conditions. The association between the N-terminus of Fep1 and the GRX domain of Grx4 requires Grx4 Cys172 located in the CGFS motif, whereas the interaction between the C-terminus of Fep1 and the TRX domain of Grx4 is dependent on Cys35 located in the putative active site of the TRX domain [54]. Analysis of Fep1 function upon expression of C35A or C35S Grx4 in *grx4* mutant strains indicated that Cys35 in the TRX domain is dispensable for Grx4-dependent inhibition of Fep1 activity. In contrast, mutation of Cys172 in the GRX domain of Grx4 led to constitutive repression by Fep1, demonstrating that the weak interaction between the GRX domain of Grx4 and the N-terminal domain of Fep1 is critical to the inhibition mechanism [54, 57]. A key difference between the role of CGFS Grxs in iron regulation for *S. pombe* in comparison to *S. cerevisiae* is that inhibition of Fep1 activity *via* interaction with Grx4 occurs under low-iron conditions in *S. pombe*, whereas Grx3/4-mediated inhibition of Aft1/2 occurs under iron-replete conditions in *S. cerevisiae*. This suggests that the CGFS Grx-dependent inhibition mechanisms for *S. cerevisiae* Aft1/2 and *S. pombe* Fep1 are distinctly different.

### 16.4.4 Molecular basis of iron-dependent control of Php4 activity

In contrast to Fep1, *S. pombe* Php4 is inhibited by Grx4 in iron-replete conditions and undergoes iron-dependent nucleocytoplasmic shuttling similar to *S. cerevisiae* Aft1/2. The Php4 nuclear export signal is located in the N-terminal domain along with the putative interaction site for the Php2/3/5 complex (Fig. 16.8) [11, 71]. Interestingly, Php4 and its orthologues in other fungi have conserved cysteine-rich regions in the C-terminal domain that are proposed to bind iron [78]. *S. pombe* Php4 has two conserved cysteines in this region, Cys221 and Cys227 (Fig. 16.8), that are essential for the iron-dependent interaction between the Php4 C-terminal domain (residues 188–254) and the GRX domain of Grx4 [72]. Grx4 Cys172 located in the CGFS Fe-S binding site is also required for this interaction with Php4. In addition, transcriptional analysis of Php4-regulated genes upon expression of C172S/A Grx4 mutants demonstrates that Cys172 is absolutely required for Grx4-dependent inhibition of Php4 activity [57,

72]. Based on these findings, Php4 and Grx4 are proposed to form an Fe-S binding complex that inhibits Php4 activity in iron-replete conditions (Fig. 16.7b) [72].

The interaction between Php4 and Grx4 is also partially mediated by the TRX domain of Grx4. Yet, in contrast to the GRX domain, the TRX domain of Grx4 binds to Php4 in an iron-independent manner. The TRX domain interaction site on Php4 has been mapped to two regions between residues 112–218 and 255–295 (Fig. 16.8). Grx4 Cys35 is required for the strong interaction between the TRX domain and Php4 [72]; however, unlike the conserved Cys in the GRX domain, mutation of the conserved Cys in the TRX domain has no effect on Php4 activity *in vivo* [57]. This interaction pattern between Grx4 and Php4 parallels the Grx4-Fep1 interaction: the TRX domain Cys35 facilitates a strong, iron-independent interaction with the repressor that is dispensable for repressor inhibition, whereas the GRX domain Cys172 maintains a weak, iron-dependent interaction that is essential for repressor inhibition. However, one key difference is that the GRX domain of Grx4 binds to Php4 under iron-replete conditions *in vivo* but interacts with Fep1 exclusively when iron is depleted.

## 16.5 Summary

The budding yeast *S. cerevisiae* and the fission yeast *S. pombe* utilize different mechanisms for regulation of iron homeostasis because *S. cerevisiae* regulators (Aft1, Aft2, Yap5) are transcriptional activators, whereas *S. pombe* regulators (Fep1, Php4) are transcriptional repressors. However, a common theme in yeast iron regulation is the central role of Fe-S clusters. In both model eukaryotes, the transcriptional response to changing iron bioavailability is specifically tuned to the ability of the cell to assemble Fe-S clusters. These versatile and ancient cofactors are a logical choice for this role due to their involvement in a wide variety of cellular functions, including mitochondrial energy metabolism, cofactor biosynthesis (e.g. heme, lipoate), amino acid biosynthesis, nucleotide metabolism, ribosome assembly, and DNA replication and repair. Tapping into the available pool of Fe-S clusters likely provides a sensitive gauge of intracellular iron availability.

An additional similarity between iron regulation in *S. cerevisiae* and *S. pombe* is the important regulatory role of cytosolic CGFS Grxs. Formation of GSH-ligated Fe-S cluster-binding complexes has been demonstrated for a wide variety of CGFS Grxs [60], and in most cases, the Fe-S cluster has proven essential for Grx function. This observation holds true for the two model yeasts described in this chapter because the available evidence suggests that Fe-S binding *via S. cerevisiae* Grx3/4 and *S. pombe* Grx4 is required for inhibition of *S. cerevisiae* Aft1/2 as well as *S. pombe* Php4 and Fep1. In each case, mutation of the CGFS motif in Grx3/Grx4 disrupts Fe-dependent inhibition of the transcriptional regulators, leading to constitutive activation by Aft1/Aft2 or constitutive repression by Php4 and Fep1. In addition, GSH depletion in both *S. cerevisiae* and *S. pombe* leads to a similar loss of iron-dependent inhibition of Aft1/

Aft2 and Php4. This phenomenon presumably stems from the requirement for GSH as a [2Fe-2S] cluster ligand in CGFS Grx complexes and/or for export of the Atm1 substrate [33, 69, 79]. The discovery of GSH-ligated Fe-S clusters thus uncovers a novel function for this ubiquitous thiol-containing tripeptide and highlights the intimate connection between thiol redox homeostasis and iron metabolism.

Despite the common themes in iron regulation in *S. cerevisiae* and *S. pombe*, there are some key differences between the regulation mechanisms. For instance, Aft1/Aft2 regulation involves two additional proteins to some extent, the BolA homologue Fra2 and the aminopeptidase P-like protein Fra1. Roles for either of these proteins in *S. pombe* iron homeostasis have not been reported. Another important distinction is that Aft1/Aft2 and Php4 activities are inhibited by cytosolic CGFS Grxs under iron-replete conditions, whereas Fep1 activity is inhibited under low-iron conditions. Thus, if ligation of a [2Fe-2S] cluster by Grx3/Grx4 is a key factor, Fe-S cluster binding must promote the inhibitive interaction between Aft1/Aft2/Php4 and Grx3/Grx4 while it prevents the inhibitive interaction between Fep1 and Grx4. Finally, *S. cerevisiae* Yap5 is unique compared with the other regulators because deletion of the genes encoding Grx3 and Grx4 does not affect the transcriptional activity of this regulator. Nevertheless, the available information on Yap5 confirms that both Fe-S cluster assembly and thiol redox biochemistry impact the function of this regulator as demonstrated for the other iron-responsive regulators in *S. cerevisiae* and *S. pombe*.

## Acknowledgments

Work in the laboratory of Dr. Caryn E. Outten has been supported by National Institutes of Health grants K22 ES103780, R01 GM086619, and R01 GM100069.

## References


[1] Kaplan CD, Kaplan J. Iron acquisition and transcriptional regulation. Chem Rev 2009; 109:4536–52.

[2] Schrettl M, Winkelmann G, Haas H. Ferrichrome in Schizosaccharomyces pombe – an iron transport and iron storage compound. Biometals 2004;17:647–54.

[3] Labbé S, Khan MG, Jacques JF. Iron uptake and regulation in Schizosaccharomyces pombe. Curr Opin Microbiol 2013;16:669–76.

[4] Pluskal T, Nakamura T, Villar-Briones A, Yanagida M. Metabolic profiling of the fission yeast S. pombe: quantification of compounds under different temperatures and genetic perturbation. Mol Biosyst 2010;6:182–98.

[5] Pelletier B, Beaudoin J, Philpott CC, Labbé S. Fep1 represses expression of the fission yeast Schizosaccharomyces pombe siderophore-iron transport system. Nucleic Acids Res 2003;31:4332–44.

[6] Protchenko O, Ferea T, Rashford J, et al. Three cell wall mannoproteins facilitate the uptake of iron in Saccharomyces cerevisiae. J Biol Chem 2001;276:49244–50.

[7] Kosman DJ. Molecular mechanisms of iron uptake in fungi. Mol Microbiol 2003;47:1185–97.
[8] Li L, Kaplan J. A mitochondrial-vacuolar signaling pathway in yeast that affects iron and copper metabolism. J Biol Chem 2004;279:33653–61.
[9] Mühlenhoff U, Stadler JA, Richhardt N, et al. A specific role of the yeast mitochondrial carriers MRS3/4p in mitochondrial iron acquisition under iron-limiting conditions. J Biol Chem 2003;278:40612–20.
[10] Li L, Chen OS, McVey Ward D, Kaplan J. CCC1 is a transporter that mediates vacuolar iron storage in yeast. J Biol Chem 2001;276:29515–9.
[11] Mercier A, Pelletier B, Labbé S. A transcription factor cascade involving Fep1 and the CCAAT-binding factor Php4 regulates gene expression in response to iron deficiency in the fission yeast Schizosaccharomyces pombe. Eukaryot Cell 2006;5:1866–81.
[12] Pouliot B, Jbel M, Mercier A, Labbé S. *abc3*$^{+}$ encodes an iron-regulated vacuolar ABC-type transporter in Schizosaccharomyces pombe. Eukaryot Cell 2010;9:59–73.
[13] Portnoy ME, Liu XF, Culotta VC. Saccharomyces cerevisiae expresses three functionally distinct homologues of the Nramp family of metal transporters. Mol Cell Biol 2000;20:7893–902.
[14] Yamaguchi-Iwai Y, Dancis A, Klausner RD. AFT1: a mediator of iron regulated transcriptional control in Saccharomyces cerevisiae. EMBO J 1995;14:1231–9.
[15] Casas C, Aldea M, Espinet C, Gallego C, Gil R, Herrero E. The AFT1 transcriptional factor is differentially required for expression of high-affinity iron uptake genes in Saccharomyces cerevisiae. Yeast 1997;13:621–37.
[16] Blaiseau PL, Lesuisse E, Camadro JM. Aft2p, a novel iron-regulated transcription activator that modulates, with Aft1p, intracellular iron use and resistance to oxidative stress in yeast. J Biol Chem 2001;276:34221–6.
[17] Rutherford JC, Jaron S, Ray E, Brown PO, Winge DR. A second iron-regulatory system in yeast independent of Aft1p. Proc Natl Acad Sci USA 2001;98:14322–7.
[18] Babu MM, Iyer LM, Balaji S, Aravind L. The natural history of the WRKY-GCM1 zinc fingers and the relationship between transcription factors and transposons. Nucleic Acids Res 2006;34:6505–20.
[19] Puig S, Vergara SV, Thiele DJ. Cooperation of two mRNA-binding proteins drives metabolic adaptation to iron deficiency. Cell Metab 2008;7:555–64.
[20] Puig S, Askeland E, Thiele DJ. Coordinated remodeling of cellular metabolism during iron deficiency through targeted mRNA degradation. Cell 2005;120:99–110.
[21] Philpott CC, Leidgens S, Frey AG. Metabolic remodeling in iron-deficient fungi. Biochim Biophys Acta 2012;1823:1509–20.
[22] Courel M, Lallet S, Camadro JM, Blaiseau PL. Direct activation of genes involved in intracellular iron use by the yeast iron-responsive transcription factor Aft2 without its paralog Aft1. Mol Cell Biol 2005;25:6760–71.
[23] Rutherford JC, Jaron S, Winge DR. Aft1p and Aft2p mediate iron-responsive gene expression in yeast through related promoter elements. J Biol Chem 2003;278:27636–43.
[24] Li L, Jia X, Ward DM, Kaplan J. Yap5 protein-regulated transcription of the TYW1 gene protects yeast from high iron toxicity. J Biol Chem 2011;286:38488–97.
[25] Pimentel C, Vicente C, Menezes RA, Caetano S, Carreto L, Rodrigues-Pousada C. The role of the Yap5 transcription factor in remodeling gene expression in response to Fe bioavailability. PLoS One 2012;7:e37434.
[26] Mühlenhoff U, Molik S, Godoy JR, et al. Cytosolic monothiol glutaredoxins function in intracellular iron sensing and trafficking via their bound iron-sulfur cluster. Cell Metab 2010;12:373–85.
[27] Winge DR, Nielson KB, Gray WR, Hamer DH. Yeast metallothionein. Sequence and metal-binding properties. J Biol Chem 1985;260:14464–70.

[28] Ding XQ, Bill E, Trautwein AX, Hartmann HJ, Weser U. Mossbauer studies on iron(II)-substituted yeast metallothionein. Eur J Biochem 1994;223:841–5.
[29] Yamaguchi-Iwai Y, Stearman R, Dancis A, Klausner RD. Iron-regulated DNA binding by the AFT1 protein controls the iron regulon in yeast. EMBO J 1996;15:3377–84.
[30] Ueta R, Fujiwara N, Iwai K, Yamaguchi-Iwai Y. Mechanism underlying the iron-dependent nuclear export of the iron-responsive transcription factor Aft1p in Saccharomyces cerevisiae. Mol Biol Cell 2007;18:2980–90.
[31] Yamaguchi-Iwai Y, Ueta R, Fukunaka A, Sasaki R. Subcellular localization of Aft1 transcription factor responds to iron status in Saccharomyces cerevisiae. J Biol Chem 2002;277:18914–8.
[32] Rutherford JC, Ojeda L, Balk J, Mühlenhoff U, Lill R, Winge DR. Activation of the iron regulon by the yeast Aft1/Aft2 transcription factors depends on mitochondrial but not cytosolic iron-sulfur protein biogenesis. J Biol Chem 2005;280:10135–40.
[33] Chen OS, Crisp RJ, Valachovic M, Bard M, Winge DR, Kaplan J. Transcription of the yeast iron regulon does not respond directly to iron but rather to iron-sulfur cluster biosynthesis. J Biol Chem 2004;279:29513–8.
[34] Ueta R, Fujiwara N, Iwai K, Yamaguchi-Iwai Y. Iron-induced dissociation of the Aft1p transcriptional regulator from target gene promoters is an initial event in iron-dependent gene suppression. Mol Cell Biol 2012;32:4998–5008.
[35] Lill R, Hoffmann B, Molik S, et al. The role of mitochondria in cellular iron-sulfur protein biogenesis and iron metabolism. Biochim Biophys Acta 2012;1823:1491–508.
[36] Ojeda L, Keller G, Mühlenhoff U, Rutherford JC, Lill R, Winge DR. Role of glutaredoxin-3 and glutaredoxin-4 in the iron regulation of the Aft1 transcriptional activator in Saccharomyces cerevisiae. J Biol Chem 2006;281:17661–9.
[37] Pujol-Carrion N, Bellí G, Herrero E, Nogues A, de la Torre-Ruiz MA. Glutaredoxins Grx3 and Grx4 regulate nuclear localisation of Aft1 and the oxidative stress response in Saccharomyces cerevisiae. J Cell Sci 2006;119:4554–64.
[38] Kumánovics A, Chen OS, Li L, et al. Identification of FRA1 and FRA2 as genes involved in regulating the yeast iron regulon in response to decreased mitochondrial iron-sulfur cluster synthesis. J Biol Chem 2008;283:10276–86.
[39] Lesuisse E, Knight SA, Courel M, Santos R, Camadro JM, Dancis A. Genome-wide screen for genes with effects on distinct iron uptake activities in Saccharomyces cerevisiae. Genetics 2005;169:107–22.
[40] Huynen MA, Spronk CA, Gabaldon T, Snel B. Combining data from genomes, Y2H and 3D structure indicates that BolA is a reductase interacting with a glutaredoxin. FEBS Lett 2005;579:591–6.
[41] Ken CF, Chen IJ, Lin CT, Liu SM, Wen L. Monothiol glutaredoxin cDNA from Taiwanofungus camphorata: a novel CGFS-type glutaredoxin possessing glutathione reductase activity. J Agric Food Chem 2011;59:3828–35.
[42] Mesecke N, Mittler S, Eckers E, Herrmann JM, Deponte M. Two novel monothiol glutaredoxins from Saccharomyces cerevisiae provide further insight into iron-sulfur cluster binding, oligomerization, and enzymatic activity of glutaredoxins. Biochemistry 2008;47:1452–63.
[43] Tamarit J, Bellí G, Cabiscol E, Herrero E, Ros J. Biochemical characterization of yeast mitochondrial Grx5 monothiol glutaredoxin. J Biol Chem 2003;278:25745–51.
[44] Zaffagnini M, Michelet L, Massot V, Trost P, Lemaire SD. Biochemical characterization of glutaredoxins from Chlamydomonas reinhardtii reveals the unique properties of a chloroplastic CGFS-type glutaredoxin. J Biol Chem 2008;283:8868–76.
[45] Fernandes AP, Fladvad M, Berndt C, et al. A novel monothiol glutaredoxin (Grx4) from Escherichia coli can serve as a substrate for thioredoxin reductase. J Biol Chem 2005;280:24544–52.

[46] Johansson C, Roos AK, Montano SJ, et al. The crystal structure of human GLRX5: iron-sulfur cluster co-ordination, tetrameric assembly and monomer activity. Biochem J 2010;433:303–11.
[47] Picciocchi A, Saguez C, Boussac A, Cassier-Chauvat C, Chauvat F. CGFS-type monothiol glutaredoxins from the cyanobacterium Synechocystis PCC6803 and other evolutionary distant model organisms possess a glutathione-ligated [2Fe-2S] cluster. Biochemistry 2007;46:15018–26.
[48] Bandyopadhyay S, Gama F, Molina-Navarro MM, et al. Chloroplast monothiol glutaredoxins as scaffold proteins for the assembly and delivery of [2Fe-2S] clusters. EMBO J 2008;27:1122–33.
[49] Iwema T, Picciocchi A, Traore DA, Ferrer JL, Chauvat F, Jacquamet L. Structural basis for delivery of the intact [Fe2S2] cluster by monothiol glutaredoxin. Biochemistry 2009;48:6041–3.
[50] Haunhorst P, Berndt C, Eitner S, Godoy JR, Lillig CH. Characterization of the human monothiol glutaredoxin 3 (PICOT) as iron-sulfur protein. Biochem Biophys Res Commun 2010;394:372–6.
[51] Li H, Mapolelo DT, Dingra NN, et al. The yeast iron regulatory proteins Grx3/4 and Fra2 form heterodimeric complexes containing a [2Fe-2S] cluster with cysteinyl and histidyl ligation. Biochemistry 2009;48:9569–81.
[52] Yeung N, Gold B, Liu NL, et al. The E. coli monothiol glutaredoxin GrxD forms homodimeric and heterodimeric FeS cluster containing complexes. Biochemistry 2011;50:8957–69.
[53] Ye H, Jeong SY, Ghosh MC, et al. Glutaredoxin 5 deficiency causes sideroblastic anemia by specifically impairing heme biosynthesis and depleting cytosolic iron in human erythroblasts. J Clin Invest 2010;120:1749–61.
[54] Jbel M, Mercier A, Labbé S. Grx4 monothiol glutaredoxin is required for iron limitation-dependent inhibition of Fep1. Eukaryot Cell 2011;10:629–45.
[55] Molina MM, Bellí G, de la Torre MA, Rodriguez-Manzaneque MT, Herrero E. Nuclear monothiol glutaredoxins of Saccharomyces cerevisiae can function as mitochondrial glutaredoxins. J Biol Chem 2004;279:51923–30.
[56] Bellí G, Polaina J, Tamarit J, et al. Structure-function analysis of yeast Grx5 monothiol glutaredoxin defines essential amino acids for the function of the protein. J Biol Chem 2002;277:37590–6.
[57] Kim KD, Kim HJ, Lee KC, Roe JH. Multi-domain CGFS-type glutaredoxin Grx4 regulates iron homeostasis via direct interaction with a repressor Fep1 in fission yeast. Biochem Biophys Res Commun 2011;408:609–14.
[58] Hoffmann B, Uzarska MA, Berndt C, et al. The multidomain thioredoxin-monothiol glutaredoxins represent a distinct functional group. Antioxid Redox Signal 2011;15:19–30.
[59] Li H, Mapolelo DT, Dingra NN, et al. Histidine 103 in Fra2 is an iron-sulfur cluster ligand in the [2Fe-2S] Fra2-Grx3 complex and is required for in vivo iron signaling in yeast. J Biol Chem 2011;286:867–76.
[60] Li H, Outten CE. Monothiol CGFS glutaredoxins and BolA-like proteins: [2Fe-2S] binding partners in iron homeostasis. Biochemistry 2012;51:4377–89.
[61] Lesuisse E, Blaiseau PL, Dancis A, Camadro JM. Siderophore uptake and use by the yeast Saccharomyces cerevisiae. Microbiology 2001;147:289–98.
[62] Fragiadakis GS, Tzamarias D, Alexandraki D. Nhp6 facilitates Aft1 binding and Ssn6 recruitment, both essential for FRE2 transcriptional activation. EMBO J 2004;23:333–42.
[63] Li L, Bagley D, Ward DM, Kaplan J. Yap5 is an iron-responsive transcriptional activator that regulates vacuolar iron storage in yeast. Mol Cell Biol 2008;28:1326–37.
[64] Li L, Miao R, Bertram S, Jia X, Ward DM, Kaplan J. A role for iron-sulfur clusters in the regulation of transcription factor Yap5-dependent high iron transcriptional responses in yeast. J Biol Chem 2012;287:35709–21.
[65] Labbé S, Pelletier B, Mercier A. Iron homeostasis in the fission yeast Schizosaccharomyces pombe. Biometals 2007;20:523–37.

[66] Pelletier B, Beaudoin J, Mukai Y, Labbé S. Fep1, an iron sensor regulating iron transporter gene expression in Schizosaccharomyces pombe. J Biol Chem 2002;277:22950–8.
[67] Fagerstrom-Billai F, Wright AP. Functional comparison of the Tup11 and Tup12 transcriptional corepressors in fission yeast. Mol Cell Biol 2005;25:716–27.
[68] Jbel M, Mercier A, Pelletier B, Beaudoin J, Labbé S. Iron activates in vivo DNA binding of Schizosaccharomyces pombe transcription factor Fep1 through its amino-terminal region. Eukaryot Cell 2009, 8;649–64.
[69] Mercier A, Watt S, Bahler J, Labbé S. Key function for the CCAAT-binding factor Php4 to regulate gene expression in response to iron deficiency in fission yeast. Eukaryot Cell 2008;7:493–508.
[70] Chung WH, Kim KD, Roe JH. Localization and function of three monothiol glutaredoxins in Schizosaccharomyces pombe. Biochem Biophys Res Commun 2005;330:604–10.
[71] Mercier A, Labbé S. Both Php4 function and subcellular localization are regulated by iron via a multistep mechanism involving the glutaredoxin Grx4 and the exportin Crm1. J Biol Chem 2009;284:20249–62.
[72] Vachon P, Mercier A, Jbel M, Labbé S. The monothiol glutaredoxin Grx4 exerts an iron-dependent inhibitory effect on Php4 function. Eukaryot Cell 2012;11:806–19.
[73] Pelletier B, Trott A, Morano KA, Labbé S. Functional characterization of the iron-regulatory transcription factor Fep1 from Schizosaccharomyces pombe. J Biol Chem 2005;280:25146–61.
[74] Chao LY, Marletta MA, Rine J. Sre1, an iron-modulated GATA DNA-binding protein of iron-uptake genes in the fungal pathogen Histoplasma capsulatum. Biochemistry 2008;47:7274–83.
[75] Harrison KA, Marzluf GA. Characterization of DNA binding and the cysteine rich region of SRE, a GATA factor in Neurospora crassa involved in siderophore synthesis. Biochemistry 2002;41:15288–95.
[76] Dailey HA, Finnegan MG, Johnson MK. Human ferrochelatase is an iron-sulfur protein. Biochemistry 1994;33:403–7.
[77] Znaidi S, Pelletier B, Mukai Y, Labbé S. The Schizosaccharomyces pombe corepressor Tup11 interacts with the iron-responsive transcription factor Fep1. J Biol Chem 2004;279:9462–74.
[78] Hortschansky P, Eisendle M, Al-Abdallah Q, et al. Interaction of HapX with the CCAAT-binding complex – a novel mechanism of gene regulation by iron. EMBO J 2007;26:3157–68.
[79] Kumar C, Igbaria A, D'Autreaux B, et al. Glutathione revisited: a vital function in iron metabolism and ancillary role in thiol-redox control. EMBO J 2011;30:2044–56.

# 17 Biogenesis of Fe-S proteins in mammals

Tracey A. Rouault

## 17.1 Introduction

Although mammalian cells were known to contain Fe-S proteins as early as in 1960 [1], little was known about how Fe-S clusters were assembled. Extensive work on mammalian mitochondrial aconitase had proven that mitochondrial aconitase contained a cubane [4Fe-4S] cluster to which the aconitase substrates citrate or isocitrate were ligated during the enzymatic reaction. The Fe-S cluster of mitochondrial aconitase could be assembled from sulfide and iron *in vitro* under anaerobic conditions using high concentrations of sulfide, iron, and reductant, but the assembly process was arduous and inefficient. In 1990, when the bifunctional protein now known as iron regulatory protein 1 (IRP1) was purified and cloned, it became apparent that this protein functioned as a cytosolic aconitase in iron-replete cells. In iron-deficient cells, the Fe-S cluster was absent, and binding of apo-IRP1 to RNA stem-loop structures in transcripts repressed the translation of some transcripts such as ferritin and extended the half-life of others. The presence or absence of the Fe-S cluster determined whether the protein functioned as an enzyme or in post-transcriptional regulation, and this kindled interest in how the Fe-S cluster was assembled and degraded. Building on the progress that had been made in identifying Fe-S biogenesis proteins in bacteria [2], human homologues of these proteins were identified through sequence homology, and when the human genome was fully sequenced, it was possible to identify a full complement of Fe-S biogenesis proteins in the genome based on homology and function. What follows is an account of how the field of mammalian Fe-S biogenesis progressed upon the discovery of Fe-S biogenesis proteins and unexpectedly proved to be key to understanding a group of rare human diseases.

## 17.2 The Fe-S regulatory switch of IRP1

In the 1980s, several genes important in mammalian iron metabolism were cloned and sequenced, including genes that encoded ferritin H and L chains [3], which form heteropolymers that sequester iron [4] and transferrin receptor 1 (TFR1), which internalizes iron from circulating diferric transferrin [5]. Direct measurements of protein expression had shown that cells expressed high amounts of TFR1 and low ferritin when they were iron-starved, whereas they synthesized low amounts of TFR1 and more ferritin when they were rich in iron. To determine how cells achieved these desirable changes in ferritin and TFR1 expression and maintained iron homeostasis, the sequences of the newly cloned genes were used to quantify the expression levels of the mRNA that encoded each gene in cell lysates. There was very little change in ferritin mRNA levels when cells were iron-starved or loaded with iron [6], consistent with earlier indirect observations [7]. The fact that protein levels varied up to 50-fold, whereas mRNA levels did not change, led to focus on the translation of the mRNA itself as the point of regulation. A deletional analysis focusing on the 5′ end of the mRNA led to the conclusion that an RNA stem-loop in the untranslated region (UTR) near the 5′ end of the mRNA was both necessary and sufficient to confer translational regulation upon the mRNA [8]. This RNA stem-loop became known as the iron-responsive element (IRE), and it was soon apparent in gel-shift experiments that the IRE could tightly bind a cellular protein, which slowed the migration of negatively charged radiolabeled IRE in an acrylamide gel [9, 10]. Five IREs were found in the 3′ UTR of the TFR1 mRNA [11] that were later shown to flank and protect the mRNA from endonucleolytic cleavage in iron-depleted cells [12] (Fig. 17.1).

Cellular lysates contained a protein(s) that binds to the IRE in low or high amounts, depending on whether the cells from which the lysates were made were rich or deficient in iron [9, 10]. Using the gel-shift assay, the IRE-binding protein (IRE-BP) was purified and cloned, and the peptide sequences were used to predict the mRNA sequence [13–16]. A remarkable sequence homology to the primary sequence of mitochondrial aconitase was noted, which included the identity of all of the residues involved in the formation of the enzymatic active site [17]. The presence of a cytosolic aconitase had been previously identified in mammalian cells [18, 19], raising the possibility that the IRE-BP could also function as an aconitase. Indeed, the IRE-BP readily acquired aconitase function in iron-rich cells [20–22], and EPR and sulfide measurements verified that a [4Fe-4S] cluster was present [23] in specifically purified cytosolic aconitase.

The novel discovery that the presence or absence of an Fe-S cluster determined how a protein functioned raised questions about how clusters were synthesized and degraded. To cover its two functions, the proteins was renamed IRP1, and much work indicated that the Fe-S cluster of both aconitases was relatively solvent exposed and could be readily disassembled by oxidants ranging from oxygen and superoxide to nitric oxide [24, 25], whereas measurements of the half-life of IRP1 protein indicated that the protein itself was relatively stable [26]. Thus, the regulatory switch might depend on the ability of the cell to rebuild an Fe-S cluster that was readily disassembled because of its exposed position in the aconitase active site cleft [27]. Moreover, the two activities of IRP1 appeared to be mutually exclusive because each depended on portions of the enzymatic active site for function [28–31], and opening of the active site cleft was predicted to be required for the IRE to fit into the active site cleft. Indeed, years later, the structure was solved for the cytosolic aconitase form of IRP1 [32] and for the apo-IRP1-IRE complex [33], and conformational changes were identified that permitted access of the IRE to otherwise inaccessible internal sites

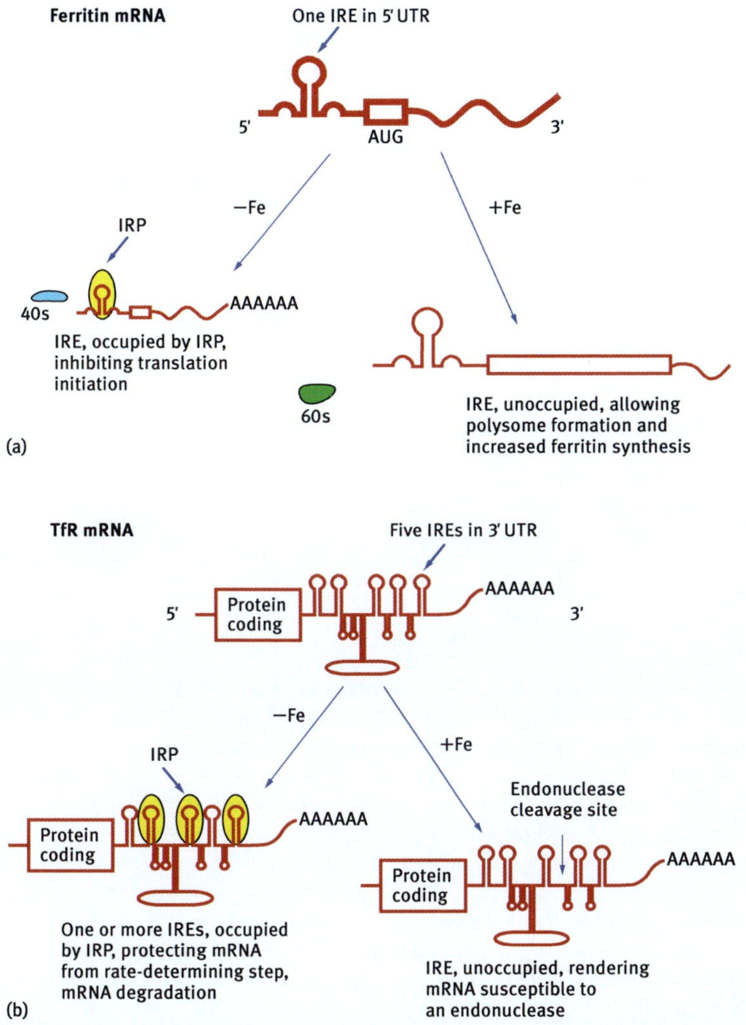


**Fig. 17.1:** Model of how binding of IRPs (IRP1 and IRP2) to RNA stem-loops in the transcripts of ferritin (a) and transferrin receptor (TfR mRNA) regulates synthesis of each protein. (Reproduced from Tracey A. Rouault, "If the RNA fits, use it", Science 314:1886 (22 December 2006) Image: N. Kevitiyagala/Science [34].) In iron-depleted cells, when IRP1 lacks an Fe-S cluster, it binds to IRE stem-loops in many different mRNAs. Binding to an IRE near the 5′ end of the transcript prevents ribosomal subunits from binding to the 5′ end and then synthesizing the protein encoded by the open reading frame (box). New translation of ferritin (and other proteins encoded by transcripts that contain an IRE near the 5′ end of the transcript) is repressed; thus, there is little synthesis of the iron sequestration protein, ferritin, when cells are depleted of iron. In the TfR transcript, binding of IRPs protects the mRNA from degradation, which leads to more synthesis of TfR. Conversely, when cells are iron-replete, ferritin is actively synthesized, whereas TfR synthesis is reduced because the TfR mRNA is degraded. Thus, expression of the iron uptake protein, TfR, increases when cells are depleted of iron. By regulating the expression of ferritin, the major iron sequestration protein, and TfR, a major iron uptake protein, the IRP regulatory system adjusts iron homeostasis to maintain optimal intracellular iron levels in mammalian cells.

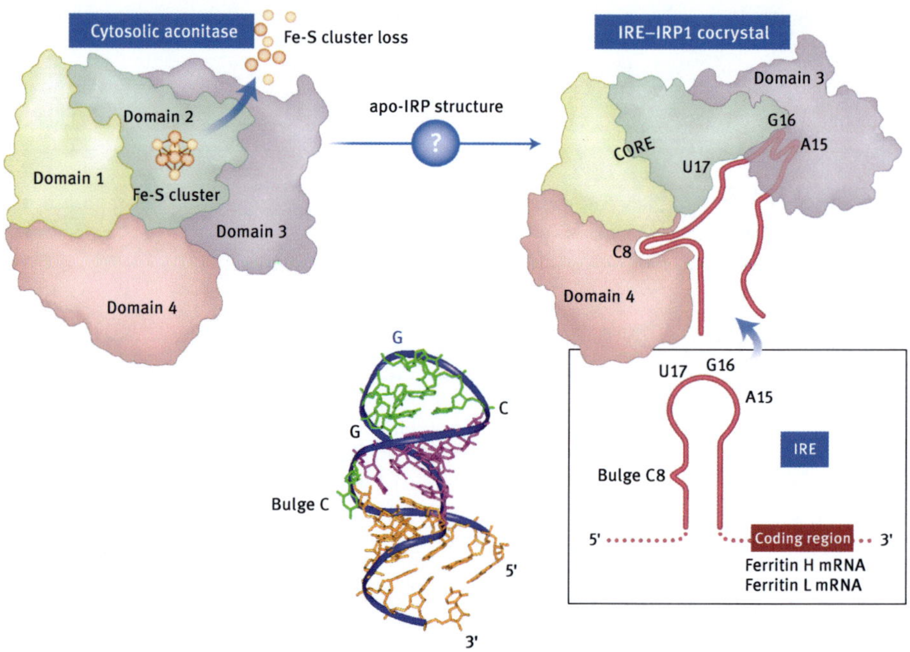


**Fig. 17.2:** The presence or absence of an Fe-S cluster determines whether IRP1 will function as an active aconitase or as an IRE-binding protein. Upon loss of the Fe-S cluster, which is readily degraded by oxidants, IRP1 undergoes a conformational shift in several of its four domains, creating a binding pocket that contains interstices into which residues C8, G16, A15, and U17 of the IRE can insert. Multiple hydrogen bonds form between IRP1 and these highly conserved residues of the IRE, which are optimally spatially separated from one another by the length of the upper stem and the configuration of the loop structure of the IRE. Conserved structural features of the IRE function as molecular rulers that spatially separate specific contact points [35] and permit formation of multiple intermolecular bonds. Multiple contacts between residues of the IRE with peptide sequences that are accessible in the open conformation of IRP1 contribute to high affinity binding of IRP1 to the IRE. (From Rouault TA, If the RNA fits, use it, *Science*, 314, 1886–7, 2006 [36].)

where multiple high-affinity contact points between apo-IRP1 and the IRE formed. Phosphorylation also was shown to influence the reversibility of IRP functions [37] (Fig. 17.2).

Therefore, attention increasingly centered on identifying the mechanisms involved in the synthesis of Fe-S clusters. The large body of work by Richard Holm and colleagues had shown that Fe-S clusters could spontaneously assemble [38], but reassembly of the Fe-S cluster on apo-IRP1 was difficult to achieve using inorganic materials alone. When the importance of a cysteine desulfurase in Fe-S assembly in bacteria became known, it appeared that a mechanism for enzymatic mobilization of sulfur might be the answer to the challenge of how to generate the sulfur needed for Fe-S biogenesis [39]. Indeed, the addition of a bacterial cysteine desulfurase to Fe-S reconstitutions of aconitase revealed that the efficiency of reconstitution was greatly increased (M.C. Kennedy, personal communication).

## 17.3 IRP2, a highly homologous gene, also post-transcriptionally regulates iron metabolism, but iron sensing occurs through the regulation of its degradation rather than through an Fe-S switch mechanism

When IRP1 was cloned, a second highly homologous gene was cloned [14], which proved to encode a second IRP, IRP2 [40, 41], that functioned equally well as IRP1 in translational regulation [42]. IRP2 lacked an Fe-S cluster but rather was degraded by the proteasome in iron-replete cells [43, 44]. Ultimately, iron sensing was attributed to the protein FBXL5, which contained a hemerythin-type iron-binding site and orchestrated ubiquitination and proteasomal degradation of IRP2 in iron-replete cells [45, 46]. *IRP1* and *IRP2* likely represent a duplicated gene pair, a common feature of mammalian genomes [47] in which gene function is largely conserved, but redundancy makes it possible for a system to evolve, and in this case, perhaps to enable a different mechanism of iron sensing to prevail. From one perspective, degradation of IRP2 in iron-replete cells represents a more robust solution to the problem of sensing and regulation because the protein is rendered physically absent in iron-replete cells, whereas IRP1 is still present, although its Fe-S cluster eliminates its IRE-binding activity. The lability of the Fe-S cluster allows the IRE-binding activity of IRP1 to be activated even when cells are otherwise iron-replete, potentially leading to iron overload when cells are oxidatively stressed (Fig. 17.3).

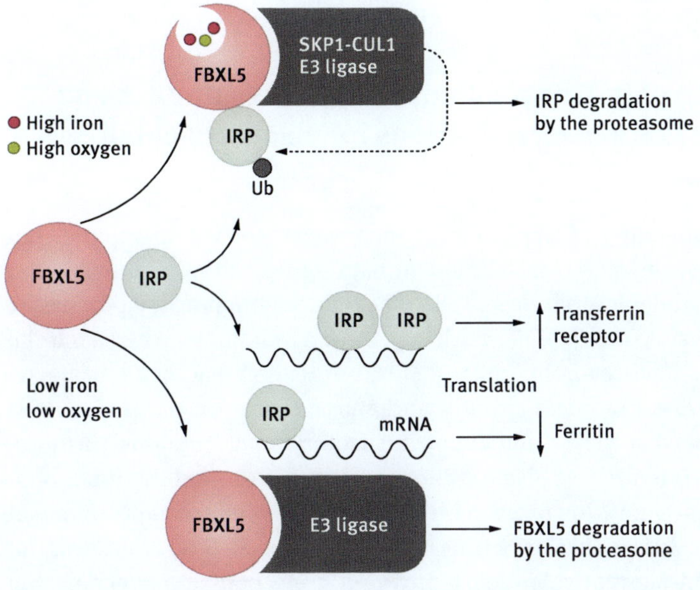


**Fig. 17.3:** Iron-dependent regulation of IRPs through regulation of their degradation. In addition to the Fe-S switch of IRP1, IRP2, which is highly homologous to IRP1, does not contain an Fe-S cluster but rather is degraded by the proteasome when cells are iron-rich because the protein FBXL5 contains an iron-binding site and targets IRP2 for degradation when cells are iron-rich [45, 46]. (From Tracey A. Rouault, "An ancient gauge for iron", Science 326: 67 (30 October 2009). Image: N. Kevitiyagala/Science [48].)

Although the two IRPs appear to be expressed in most or all cell types, expression levels of IRP1 are highest in the kidney and brown fat, whereas IRP2 expression is highest in the central nervous system [49, 50]. The differences in relative distribution among tissues may account for the fact that animals engineered to lack *IRP2* develop neurodegeneration [51–53] and anemia [54, 55], whereas animals that lack *IRP1* develop polycythemia [56–58] and pulmonary hypertension [56]. Animals cannot develop beyond the blastocyst stage when both *IRP*s are deleted, illustrating the physiological significance and partial redundancy of this important post-transcriptional regulatory system [59].

Since the IRE was initially characterized in transcripts of ferritin H and L chains and TFR1, numerous transcripts that contain IREs at the 5′ end have been discovered that mediate translational repression, including ALAS2, which encodes a heme biosynthetic enzyme [60, 61], mitochondrial aconitase [62, 63], the iron exporter, ferroportin [64–66], and HIF2α [67]. Interestingly, cells that need to evade suppression of ferroportin synthesis by IRPs for physiological reasons synthesize a ferroportin transcript that lacks an IRE, using a different promoter [68, 69]. Several other transcripts that contain IREs of unknown significance have been identified using a search program for IREs [70], including several transcripts that contain IREs in the 3′ UTR [71], and by sequencing mRNAs that bind to IRPs [72]. The physiological relevance of the IRE in regulation of HIF2α expression was recently demonstrated when animals that lacked IRP1 developed high erythropoietin levels, pulmonary hypertension, and polycythemia [56–58].

## 17.4 Identification of the mammalian cysteine desulfurase and two scaffold proteins: implications for compartmentalization of the process

In the 1990s, the sequencing of the human genome was not yet completed, but a database of short sequences from cloned human genes, known as expressed sequence tags, was available and was used to search for human homologues to bacterial Fe-S biogenesis genes [73]. Remarkably, a human homologue of bacterial NifS was identified as having approximately 60% sequence homology to a bacterial cysteine desulfurase. Complete cloning and characterization of human NFS1 revealed that it encoded a major mitochondrial isoform that contained a mitochondrial targeting sequence and also a minor isoform that lacked the mitochondrial targeting sequence and localized to the cytosolic/nuclear compartment of mammalian cells [74]. Using similar techniques, the major Fe-S scaffold protein, ISCU, was cloned, and alternative splicing provided a mechanism for generating both mitochondrial and cytosolic/nuclear isoforms [75]. Human NFU1, which is homologous to a portion of bacterial IscU, also encoded two isoforms through alternative splicing and was shown to bind a [4Fe-4S] cluster [76]. The function of the cytosolic form of ISCU was demonstrated by reducing the expression of the cytosolic isoform but not the mitochondrial isoform [77]. These experiments suggested that a parallel machinery for nascent Fe-S assembly existed in the cytosolic/nuclear compartments of mammalian cells, but researchers working on *Saccharomyces cerevisiae* disputed this point because they found no evidence for cytosolic or nuclear isoforms of these proteins being involved in early Fe-S assembly [78]. However, a nuclear isoform of NFS1 is present and necessary in *S. cerevisiae* [79, 80], and its required partner, ISD11, is also present in the nucleus and cytosol of mammalian cells [81]. Moreover, the co-chaperone, HSC20, is readily found in mammalian cytosolic lysates [82], again raising the question of whether Fe-S synthesis can be initiated in the cytosol and nucleus of mammalian cells.

There are also numerous mammalian homologues of proteins identified in *S. cerevisiae* as important in cytosolic Fe-S synthesis, referred to as the CIA machinery, for cytosolic iron-sulfur assembly (see Chapter 15, by Pain and Dancis, and Chapter 16, by Outten). Recently, these proteins were recognized as important for Fe-S biogenesis in enzymes involved in mammalian DNA metabolism and maintenance [83, 84] (discussed further in Chapter 20, by Gari, and Chapter 21, by Lill *et al.*) and for Fe-S acquisition by IRP1 [85].

## 17.5 Sequential steps in Fe-S biogenesis – an initial Fe-S assembly process on a scaffold, followed by Fe-S transfer to recipient proteins, aided by a chaperone-co-chaperone system

Based on biochemical and structural characterization of bacterial proteins [86, 87], a model for synthesis of nascent Fe-S clusters emerged in which Nfs1 dimerizes and the scaffold protein IscU binds to each end of the complex. In eukaryotes, Isd11 is an obligate binding partner of Nfs1, and a ternary complex has been identified in mammalian cells composed of NFS1, ISD11, and ISCU [88, 89] that works in concert with frataxin (FTX) to synthesize nascent clusters on ISCU. In 1996 [90], FTX was initially identified as the diseased gene in Friedreich ataxia, but its function was completely unknown until 1997, when experiments in *S. cerevisiae* tied it to mitochondrial iron metabolism [91]. Subsequently, experiments have demonstrated that FTX is important in early Fe-S synthesis, where it may function as an iron sequestration protein [92], iron carrier [93], or as an allosteric modifier [94] of the complex. Recent studies suggest that this tetrameric complex can generate [4Fe-4S] clusters that can ligate directly to mitochondrial aconitase [95] (Fig. 17.4).

After the initial cluster is synthesized on ISCU, a co-chaperone known as HSC20 binds holo-ISCU and forms a complex with the chaperone protein HSPA9 (also known as HSC70, HSCA, or mortalin), an ATPase that is activated by the HPD

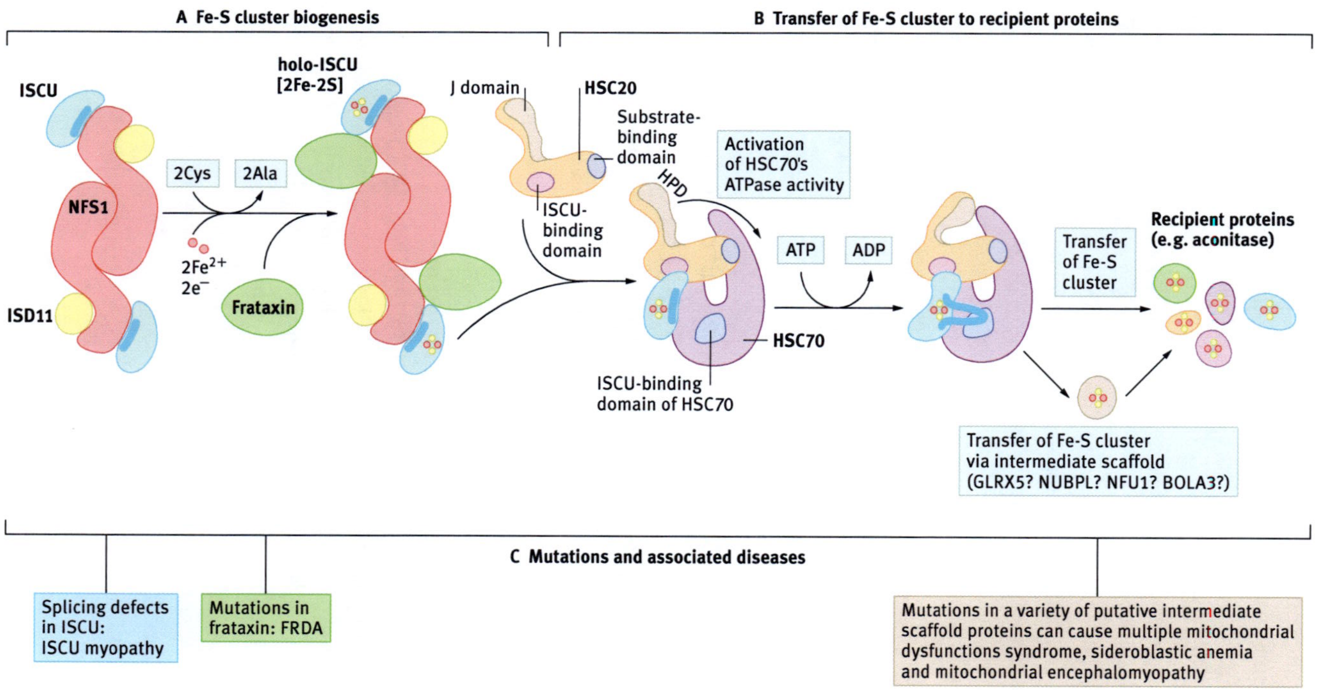


**Fig. 17.4:** A model for mammalian Fe-S biogenesis based on the crystal structure of the Nfs1-IscU complex in bacteria [86, 87] and proposed roles of FTX [95, 96]. The first stage (a) of Fe-S synthesis depends on the head-to-tail dimerization of the cysteine desulfurase, NFS1; in eukaryotes, each NFS1 monomer binds in an obligate manner to a small associated protein, ISD11, and to the Fe-S scaffold protein, ISCU. Nascent [2Fe-2S] clusters form on ISCU, aided in an unknown manner by transient FTX binding. (b) A co-chaperone, HSC20, binds holo-ISCU and forms a complex with the chaperone HSC70 (as labeled in the figure), also known as HSPA9 or mortalin, an ATPase that is activated by the HPD tripeptide of HSC20. Activation may generate a conformational change that facilitates release of the Fe-S cluster from ISCU, allowing it to bind to recipient proteins or to intermediate scaffold-type carrier proteins (see text). Mutations of various Fe-S biogenesis proteins cause specific diseases. (From Rouault TA, *Dis Model Mech*, 5, 155–64, 2012. With permission [97].)

tripeptide of HSC20 [98, 99]. Activation of HSPA9 may generate a conformational change that releases the Fe-S cluster from ISCU, allowing it to bind to recipient proteins or to intermediate scaffold-type carrier proteins. The sequence LPPVK of ISCU interacts with HSPA9, binding ISCU to the chaperone in bacterial [98] and *S. cerevisiae* model systems [100], where the critical motif of Isu was identified as the tripeptide PVK.

Once ISCU, with its nascent cluster, binds to the chaperone-co-chaperone complex through independent contacts with HSC20 and HSPA9, the cluster is transferred to recipient proteins, but few details are known about how Fe-S clusters are acquired by recipient proteins. Notably, some proteins such as zinc fingers could readily accommodate the binding of an Fe-S cluster through their available cysteine ligands, but mammalian cells appear to be able to specifically transfer Fe-S clusters to proteins that need Fe-S clusters for function. Secondary scaffold proteins are likely to be involved in ferrying Fe-S clusters from the basic Fe-S synthesis complexes to specific subsets of recipient proteins. Monothiol glutaredoxins have been proposed to function as secondary scaffolds in bacteria [101], and mutations in human glutaredoxin 5 disrupt Fe-S biogenesis and cause sideroblastic anemia [102].

Several recently described human diseases underscore the importance of various steps of Fe-S biogenesis. Mutations of ISD11, also known as LYRM4, cause severe impairments of mitochondrial oxidative phosphorylation in infants, as might be expected when initial Fe-S complex formation is compromised [103]. Abnormal splicing of ISCU causes skeletal myopathy [104, 105], and a missense mutation at this locus in combination with the abnormal splicing allele causes cardiac and skeletal disease [106]. Reduced expression of FTX due to heterochromatin formation on an expanded trinucleotide repeat leads to abnormal Fe-S protein functions in neurons and cardiomyocytes of patients and in animal models [107]. Mutations in NFU1 and BOLA3 cause defective function of proteins that require lipoic acid modification for function, including the E2 component of the pyruvate dehydrogenase complex, α-ketoglutarate dehydrogenase, branched-chain keto acid dehydrogenase), and the H protein of the glycine cleavage system [108, 109]. Accordingly, infants with mutations in these genes had lactic acidosis along with elevated serum levels of glycine and the branched-chain amino acids leucine, valine, and isoleucine. The clinical symptoms imply that NFU1 and BOLA3 are important for the delivery of the Fe-S cluster to lipoic acid synthase (see Chapter 9, by Lanz and Booker), perhaps by functioning as secondary scaffolds. Other candidates for secondary scaffolds include IBA57 [110] and glutaredoxin 5 [102]. Absence of glutaredoxin 5 was identified as a cause of sideroblastic anemia [111], a disease in which the mitochondria of immature red cells become markedly iron-loaded and too few mature red cells are produced. Thus, human diseases can highlight specific roles of Fe-S assembly proteins at various points in the pathway.

## 17.6 Mitochondrial iron overload in response to defects in Fe-S biogenesis raises important questions about how mitochondrial iron homeostasis is regulated

In Friedreich ataxia [107], ISCU myopathy [104], and glutaredoxin 5 deficiency [111], abnormal mitochondrial iron accumulation occurs in affected tissues. The same phenotype has been observed in *S. cerevisiae* resulting from loss of Fe-S biogenesis proteins such as Nfs1 [112], the yeast FTX homologue Yfh1 [91], and others (see Chapter 15, by Pain and Dancis). The fact that mitochondrial iron overload occurs (Fig. 17.5), sometimes at the expense of cytosolic iron [113], suggests that intact Fe-S biogenesis is required for the maintenance of mitochondrial iron homeostasis. In skeletal muscle from humans with ISCU deficiency, the expression of the mitochondrial iron importer, mitoferrin, increases [114]. Fe-S clusters have previously been shown to have regulatory functions in bacteria [115], and the regulatory role of the Fe-S cluster of IRP1 was discussed earlier in this chapter. The occurrence of mitochondrial iron overload as a consequence of problems with Fe-S biogenesis hints that an Fe-S protein is involved in regulating mitochondrial iron homeostasis, perhaps by transcriptionally remodeling the cell to increase mitochondrial iron uptake and decrease mitochondrial iron export. In the past, researchers have postulated that iron may have no dedicated mitochondrial exporter and may exit the mitochondrial matrix in the form of heme or an Fe-S cluster. Thus far, neither a

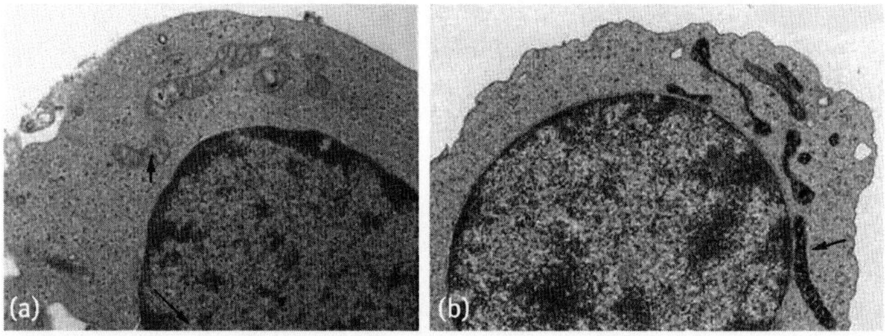


**Fig. 17.5:** Mitochondrial iron overload occurs in human and animal models when functions of some proteins involved in Fe-S biogenesis are impaired. Electron micrographs of mitochondrial iron overload in developing erythroblasts from the early stages (a) to advanced stages (b) are shown from patients with sideroblastic anemia. Sideroblastic anemia is a term for anemias in which mitochondria of the developing red blood cells show severe iron overload, which may be observed with the Prussian blue stain for iron or by electron microscopy, where iron appears as black deposits. (a) A patient with mild disease has a small black iron deposit in a mitochondrion (small arrow) located near the nucleus (large arrow). (b) Perinuclear mitochondria are loaded with black iron precipitates in a patient with severe disease. Mitochondrial cristae are still visible as white band-like lines in the mitochondria (an arrow points to one iron-laden rod-shaped mitochondrion that is loaded with black iron deposits). (Courtesy of Dr M. Djaldetti.)

mitochondrial iron exporter nor a heme exporter has yet been discovered. However, mutations in an inner mitochondrial membrane transporter of unidentified substrate known as ABCB7 in humans, or Atm1 in the *S. cerevisiae* homologue, lead to mitochondrial iron accumulation. In humans, these ABCB7 mutations cause a disease known as X-linked sideroblastic anemia with ataxia because the mutations are on the X chromosome, and males are typically affected [116]. Molecular dissection of other rare diseases in humans may lead to specific insights into the Fe-S biogenesis pathway, particularly in the downstream steps through which recipient proteins acquire their Fe-S clusters.

## 17.7 Perspectives and future directions

Much of the progress made in identifying the main genes involved in mammalian Fe-S biogenesis exploited the homology of mammalian to bacterial genes and built on biochemical and spectroscopic studies of bacterial processes. However, there are several issues in Fe-S biogenesis that have not yet been resolved by studies in bacterial model systems or cannot be studied in bacteria because of their lack of compartmentalizaion. The problem of the role of FTX remains unresolved, even though bacteria express FTX. In addition, it is not yet known how the Fe-S biogenesis machinery specifically targets recipient proteins. Moreover, studies in bacteria cannot tackle the question of how eukaryotic cells coordinate Fe-S biogenesis in multiple compartments. If bacteria regulate Fe-S biogenesis by regulating transcription of the ISC operon (see Chapter 13, by Mettert *et al.*), what regulates eukaryotic Fe-S biogenesis? Some progress was made when it was recognized that ISCU expression was repressed during hypoxia by micro-RNA-mediated repression [117], which suppressed mitochondrial respiration. Other regulatory inputs must also govern the Fe-S biogenesis pathway, and mechanisms that coordinate mitochondrial and cytosolic/nuclear Fe-S biogenesis need to be elucidated. It is not clear whether mammalian cytosolic/nuclear FeS biogenesis requires a product of mitochondrial Fe-S biogenesis, as has been asserted for *S. cerevisiae*, or whether *de novo* Fe-S biogenesis can occur in the cytosol and/or nucleus using cytosolic isoforms of NFS1, ISCU, NFU1, HSC20, and HSPA9. In addition, regulatory mechanisms may have evolved that do not exist in simple eukaryotes such as *S. cerevisiae*. As an illustration, it is noteworthy that the IRE-IRP regulatory system does not exist in *S. cerevisiae*. Studies performed in mammalian cells will likely be important in resolving numerous questions because the strong protein-protein interactions characteristic of mammalian cells may facilitate biochemical identification of critical protein complexes. In addition, the excellent phenotyping associated with diagnosis and characterization of rare human diseases may lead to other unexpected insights into the Fe-S biogenesis pathway.

## References


[1] Beinert H, Sands RH. Biochem Biophys Res Commun 1960;3:41–6.

[2] Zheng L, Cash VL, Flint DH, Dean DR. Assembly of iron-sulfur clusters. Identification of an iscSUA-hscBA-fdx gene cluster from Azotobacter vinelandii. J Biol Chem 1998;273:13264–72.

[3] Leibold EA, Aziz N, Brown AJ, Munro HN. Conservation in rat liver of light and heavy subunit sequences of mammalian ferritin. Presence of unique octopeptide in the light subunit. J Biol Chem 1984;259:4327–34.

[4] Theil EC. Ferritin: at the crossroads of iron and oxygen metabolism. J Nutr 2003;133:1549S–53S.

[5] Kuhn LC, Mcclelland A, Ruddle FH. Gene transfer, expression, and molecular cloning of the human transferrin receptor gene. Cell 1984;37:95–103.

[6] Rouault TA, Hentze MW, Dancis A, et al. Influence of altered transcription on the translational control of human ferritin expression. Proc Natl Acad Sci U S A 1987;84:6335–9.

[7] Zahringer J, Baliga BS, Munro HN. Relative abundance of specific messenger-RNA species in the free mRNP fraction of rat liver. FEBS Lett 1979;108:317–20.

[8] Hentze MW, Caughman SW, Rouault TA, et al. Identification of the iron-responsive element for the translational regulation of human ferritin mRNA. Science 1987;238:1570–3.

[9] Rouault TA, Hentze MW, Caughman SW, Harford JB, Klausner RD. Binding of a cytosolic protein to the iron-responsive element of human ferritin messenger RNA. Science 1988;241:1207–10.

[10] Leibold EA, Munro HN. Cytoplasmic protein binds in vitro to a highly conserved sequence in the 5′ untranslated region of ferritin heavy- and light-subunit mRNAs. Proc Natl Acad Sci U S A 1988;85:2171–5.

[11] Casey JL, Hentze MW, Koeller DM, et al. Iron-responsive elements: regulatory RNA sequences that control mRNA levels and translation. Science 1988;240:924–8.

[12] Binder R, Horowitz JA, Basilion JP, et al. Evidence that the pathway of transferrin receptor mRNA degradation involves an endonucleolytic cleavage within the 3' UTR and does not involve poly(A) tail shortening. EMBO J 1994;13:1969–80.

[13] Rouault TA, Tang CK, Kaptain S, et al. Cloning of the cDNA encoding an RNA regulatory protein--the human iron-responsive element-binding protein. Proc Natl Acad Sci U S A 1990;87:7958–62.

[14] Yu Y, Radisky E, Leibold EA. The iron-responsive element binding protein. Purification, cloning, and regulation in rat liver. J Biol Chem 1992;267:19005–10.

[15] Patino MM, Walden WE. Cloning of a functional cDNA for the rabbit ferritin mRNA repressor protein. Demonstration of a tissue-specific pattern of expression. J Biol Chem 1992;267:19011–6.

[16] Mullner EW, Rothenberger S, Muller AM, Kuhn LC. In vivo and in vitro modulation of the mRNA-binding activity of iron-regulatory factor. Tissue distribution and effects of cell proliferation, iron levels and redox state. Eur J Biochem 1992;208:597–605.

[17] Rouault TA, Stout CD, Kaptain S, Harford JB, Klausner RD. Structural relationship between an iron-regulated RNA-binding protein (IRE-BP) and aconitase: functional implications. Cell 1991;64:881–3.

[18] Eanes RZ, Kun E. Separation and characterization of aconitate hydratase isoenzymes from pig tissues. Biochim Biophys Acta 1971;227:204–10.

[19] Shows TB, Brown JA. Mapping AK1, ACONs, and AK3 to chromosome 9 in man employing and X/9 translocation and somatic cell hybrids. Cytogenet Cell Genet 1977;19:26–37.

[20] Kaptain S, Downey WE, Tang C, et al. A regulated RNA binding protein also possesses aconitase activity. Proc Natl Acad Sci U S A 1991;88:10109–13.

[21] Haile DJ, Rouault TA, Harford JB, et al. Cellular regulation of the iron-responsive element binding protein: disassembly of the cubane iron-sulfur cluster results in high-affinity RNA binding. Proc Natl Acad Sci U S A 1992;89:11735–9.

[22] Beinert H, Kennedy MC. Aconitase, a two-faced protein: enzyme and iron regulatory factor. FASEB J 1993;7:1442–9.
[23] Kennedy MC, Mende-Mueller L, Blondin GA, Beinert H. Purification and characterization of cytosolic aconitase from beef liver and its relationship to the iron-responsive element binding protein. Proc Natl Acad Sci U S A 1992;89:11730–4.
[24] Weiss G, Goossen B, Doppler W, et al. Translational regulation via iron-responsive elements by the nitric oxide/NO-synthase pathway. EMBO J 1993;12:3651–7.
[25] Drapier JC, Hirling H, Wietzerbin J, Kaldy P, Kuhn LC. Reciprocal modulation of aconitase activity and RNA-binding activity of iron regulatory factor by nitric oxide. Adv Exp Med Biol 1994;356:141–8.
[26] Tang CK, Chin J, Harford JB, Klausner RD, Rouault TA. Iron regulates the activity of the iron-responsive element binding protein without changing its rate of synthesis or degradation. J Biol Chem 1992;267:24466–70.
[27] Robbins AH, Stout CD. Structure of activated aconitase: formation of the [4Fe-4S] cluster in the crystal. Proc Natl Acad Sci U S A 1989;86:3639–43.
[28] Philpott CC, Haile D, Rouault TA, Klausner RD. Modification of a free Fe-S cluster cysteine residue in the active iron-responsive element-binding protein prevents RNA binding. J Biol Chem 1993;268:17655–8.
[29] Philpott CC, Klausner RD, Rouault TA. The bifunctional iron-responsive element binding protein/cytosolic aconitase: the role of active-site residues in ligand binding and regulation. Proc Natl Acad Sci U S A 1994;91:7321–5.
[30] Basilion JP, Rouault TA, Massinople CM, Klausner RD, Burgess WH. The iron-responsive element-binding protein: localization of the RNA-binding site to the aconitase active-site cleft. Proc Natl Acad Sci U S A 1994;91:574–8.
[31] Hirling H, Henderson BR, Kuhn LC. Mutational analysis of the [4Fe-4S]-cluster converting iron regulatory factor from its RNA-binding form to cytoplasmic aconitase. EMBO J 1994;13:453–61.
[32] Dupuy J, Volbeda A, Carpentier P, et al. Crystal structure of human iron regulatory protein 1 as cytosolic aconitase. Structure 2006;14:129–39.
[33] Walden WE, Selezneva AI, Dupuy J, et al. Structure of dual function iron regulatory protein 1 complexed with ferritin IRE-RNA. Science 2006;314:1903–8.
[34] Rouault TA. The role of iron regulatory proteins in mammalian iron homeostasis and disease. Nat Chem Biol 2006;2:406–14.
[35] Addess KJ, Basilion JP, Klausner RD, Rouault TA, Pardi A. Structure and dynamics of the iron responsive element RNA: implications for binding of the RNA by iron regulatory binding proteins. J Mol Biol 1997;274:72–83.
[36] Rouault TA. Biochemistry. If the RNA fits, use it. Science 2006;314:1886–7.
[37] Eisenstein RS. Iron regulatory proteins and the molecular control of mammalian iron metabolism. Annu Rev Nutr 2000;20:627–62.
[38] Venkateswara Rao P, Holm RH. Synthetic analogues of the active sites of iron-sulfur proteins. Chem Rev 2004;104:527–59.
[39] Zheng L, White RH, Cash VL, Jack RF, Dean DR. Cysteine desulfurase activity indicates a role for NIFS in metallocluster biosynthesis. Proc Natl Acad Sci U S A 1993;90:2754–8.
[40] Samaniego F, Chin J, Iwai K, Rouault TA, Klausner RD. Molecular characterization of a second iron-responsive element binding protein, iron regulatory protein 2. Structure, function, and post-translational regulation. J Biol Chem 1994;269:30904–10.
[41] Guo B, Yu Y, Leibold EA. Iron regulates cytoplasmic levels of a novel iron-responsive element-binding protein without aconitase activity. J Biol Chem 1994;269:24252–60.
[42] Kim HY, Klausner RD, Rouault TA. Translational repressor activity is equivalent and is quantitatively predicted by in vitro RNA binding for two iron-responsive element-binding proteins, IRP1 and IRP2. J Biol Chem 1995;270:4983–6.

[43] Guo B, Phillips JD, Yu Y, Leibold EA. Iron regulates the intracellular degradation of iron regulatory protein 2 by the proteasome. J Biol Chem 1995;270:21645–51.

[44] Iwai K, Klausner RD, Rouault TA. Requirements for iron-regulated degradation of the RNA binding protein, iron regulatory protein 2. EMBO J 1995;14:5350–7.

[45] Salahudeen AA, Thompson JW, Ruiz JC, et al. An E3 ligase possessing an iron responsive hemerythrin domain is a regulator of iron homeostasis. Science 2009;326:722–6.

[46] Vashisht AA, Zumbrennen KB, Huang X, et al. Control of iron homeostasis by an iron-regulated ubiquitin ligase. Science 2009;326:718–21.

[47] Zhang Z. Evolution by gene duplication: an update. Trends Ecol Evol 2003;18:292–8.

[48] Rouault TA. Cell biology. An ancient gauge for iron. Science 2009;326:676–7.

[49] Meyron-Holtz EG, Ghosh MC, Iwai K, et al. Genetic ablations of iron regulatory proteins 1 and 2 reveal why iron regulatory protein 2 dominates iron homeostasis. EMBO J 2004;23:386–95.

[50] Meyron-Holtz EG, Ghosh MC, Rouault TA. Mammalian tissue oxygen levels modulate iron-regulatory protein activities in vivo. Science 2004;306:2087–90.

[51] Lavaute T, Smith S, Cooperman S, et al. Targeted deletion of the gene encoding iron regulatory protein-2 causes misregulation of iron metabolism and neurodegenerative disease in mice. Nat Genet 2001;27:209–14.

[52] Galy B, Holter SM, Klopstock T, et al. Iron homeostasis in the brain: complete iron regulatory protein 2 deficiency without symptomatic neurodegeneration in the mouse. Nat Genet 2006;38:967–9; discussion 969.

[53] Jeong SY, Crooks DR, Wilson-Ollivierre H, et al. Iron insufficiency compromises motor neurons and their mitochondrial function in Irp2-null mice. PLoS One 2011;6:e25404.

[54] Cooperman SS, Meyron-Holtz EG, Olivierre-Wilson H, et al. Microcytic anemia, erythropoietic protoporphyria, and neurodegeneration in mice with targeted deletion of iron-regulatory protein 2. Blood 2005;106:1084–91.

[55] Galy B, Ferring D, Minana B, et al. Altered body iron distribution and microcytosis in mice deficient in iron regulatory protein 2 (IRP2). Blood 2005;106:2580–9.

[56] Ghosh MC, Zhang DL, Jeong SY, et al. Deletion of iron regulatory protein 1 causes polycythemia and pulmonary hypertension in mice through translational derepression of HIF2alpha. Cell Metab 2013;17:271–81.

[57] Anderson SA, Nizzi CP, Chang YI, et al. The IRP1-HIF-2alpha axis coordinates iron and oxygen sensing with erythropoiesis and iron absorption. Cell Metab 2013;17:282–90.

[58] Wilkinson N, Pantopoulos K. IRP1 regulates erythropoiesis and systemic iron homeostasis by controlling HIF2alpha mRNA translation. Blood 2013;122:1658–68.

[59] Smith SR, Ghosh MC, Ollivierre-Wilson H, Hang Tong W, Rouault TA. Complete loss of iron regulatory proteins 1 and 2 prevents viability of murine zygotes beyond the blastocyst stage of embryonic development. Blood Cells Mol Dis 2006;36:283–7.

[60] Cox TC, Bawden MJ, Martin A, May BK. Human erythroid 5-aminolevulinate synthase: promoter analysis and identification of an iron-responsive element in the mRNA. EMBO J 1991;10: 1891–902.

[61] Gray NK, Hentze MW. Iron regulatory protein prevents binding of the 43S translation pre-initiation complex to ferritin and eALAS mRNAs. EMBO J 1994;13:3882–91.

[62] Kim HY, Lavaute T, Iwai K, Klausner RD, Rouault TA. Identification of a conserved and functional iron-responsive element in the 5′-untranslated region of mammalian mitochondrial aconitase. J Biol Chem 1996;271:24226–30.

[63] Schalinske KL, Chen OS, Eisenstein RS. Iron differentially stimulates translation of mitochondrial aconitase and ferritin mRNAs in mammalian cells. Implications for iron regulatory proteins as regulators of mitochondrial citrate utilization. J Biol Chem 1998;273:3740–6.

[64] Abboud S, Haile DJ. A novel mammalian iron-regulated protein involved in intracellular iron metabolism. J Biol Chem 2000;275:19906–12.

[65] Mckie AT, Marciani P, Rolfs A, et al. A novel duodenal iron-regulated transporter, IREG1, implicated in the basolateral transfer of iron to the circulation. Mol Cell 2000;5:299–309.
[66] Donovan A, Brownlie A, Zhou Y, et al. Positional cloning of zebrafish ferroportin1 identifies a conserved vertebrate iron exporter. Nature 2000;403:776–81.
[67] Sanchez M, Galy B, Muckenthaler MU, Hentze MW. Iron-regulatory proteins limit hypoxia-inducible factor-2alpha expression in iron deficiency. Nat Struct Mol Biol 2007;14:420–6.
[68] Zhang DL, Hughes RM, Ollivierre-Wilson H, Ghosh MC, Rouault TA. A ferroportin transcript that lacks an iron-responsive element enables duodenal and erythroid precursor cells to evade translational repression. Cell Metab 2009;9:461–73.
[69] Zhang DL, Senecal T, Ghosh MC, et al. Hepcidin regulates ferroportin expression and intracellular iron homeostasis of erythroblasts. Blood 2011;118:2868–77.
[70] Campillos M, Cases I, Hentze MW, Sanchez M. SIREs: searching for iron-responsive elements. Nucleic Acids Res 2010;38:W360–7.
[71] Hentze MW, Muckenthaler MU, Galy B, Camaschella C. Two to tango: regulation of Mammalian iron metabolism. Cell 2010;142:24–38.
[72] Sanchez M, Galy B, Schwanhaeusser B, et al. Iron regulatory protein-1 and -2: transcriptome-wide definition of binding mRNAs and shaping of the cellular proteome by iron regulatory proteins. Blood 2011;118:e168–79.
[73] Adams MD, Soares MB, Kerlavage AR, Fields C, Venter JC. Rapid cDNA sequencing (expressed sequence tags) from a directionally cloned human infant brain cDNA library. Nat Genet 1993;4:373–80.
[74] Land T, Rouault TA. Targeting of a human iron-sulfur cluster assembly enzyme, nifs, to different subcellular compartments is regulated through alternative AUG utilization. Mol Cell 1998;2:807–15.
[75] Tong WH, Rouault T. Distinct iron-sulfur cluster assembly complexes exist in the cytosol and mitochondria of human cells. EMBO J 2000;19:5692–700.
[76] Tong WH, Jameson GN, Huynh BH, Rouault TA. Subcellular compartmentalization of human Nfu, an iron-sulfur cluster scaffold protein, and its ability to assemble a [4Fe-4S] cluster. Proc Natl Acad Sci U S A 2003;100:9762–7.
[77] Tong WH, Rouault TA. Functions of mitochondrial ISCU and cytosolic ISCU in mammalian iron-sulfur cluster biogenesis and iron homeostasis. Cell Metab 2006;3:199–210.
[78] Lill R, Muhlenhoff U. Maturation of iron-sulfur proteins in eukaryotes: mechanisms, connected processes, and diseases. Annu Rev Biochem 2008;77:669–700.
[79] Nakai Y, Nakai M, Hayashi H, Kagamiyama H. Nuclear localization of yeast Nfs1p is required for cell survival. J Biol Chem 2001;276:8314–20.
[80] Nakai Y, Nakai M, Lill R, Suzuki T, Hayashi H. Thio modification of yeast cytosolic tRNA is an iron-sulfur protein-dependent pathway. Mol Cell Biol 2007;27:2841–7.
[81] Shi Y, Ghosh MC, Tong WH, Rouault TA. Human ISD11 is essential for both iron-sulfur cluster assembly and maintenance of normal cellular iron homeostasis. Hum Mol Genet 2009;18:3014–25.
[82] Uhrigshardt H, Singh A, Kovtunovych G, Ghosh M, Rouault TA. Characterization of the human HSC20, an unusual DnaJ type III protein, involved in iron-sulfur cluster biogenesis. Hum Mol Genet 2010;19:3816–34.
[83] Stehling O, Vashisht AA, Mascarenhas J, et al. MMS19 assembles iron-sulfur proteins required for DNA metabolism and genomic integrity. Science 2012;337:195–9.
[84] Gari K, Leon Ortiz AM, Borel V, et al. MMS19 links cytoplasmic iron-sulfur cluster assembly to DNA metabolism. Science 2012;337:243–5.
[85] Stehling O, Mascarenhas J, Vashisht AA, et al. Human CIA2A-FAM96A and CIA2B-FAM96B integrate iron homeostasis and maturation of different subsets of cytosolic-nuclear iron-sulfur proteins. Cell Metab 2013;18:187–98.
[86] Shi R, Proteau A, Villarroya M, et al. Structural basis for Fe-S cluster assembly and tRNA thiolation mediated by IscS protein-protein interactions. PLoS Biol 2010;8:e1000354.

[87] Marinoni EN, De Oliveira JS, Nicolet Y, et al. (IscS-IscU)2 complex structures provide insights into Fe2S2 biogenesis and transfer. Angew Chem Int Ed Engl 2012;51:5439–42.

[88] Schmucker S, Martelli A, Colin F, et al. Mammalian frataxin: an essential function for cellular viability through an interaction with a preformed ISCU/NFS1/ISD11 iron-sulfur assembly complex. PLoS One 2011;6:e16199.

[89] Bridwell-Rabb J, Winn AM, Barondeau DP. Structure-function analysis of Friedreich's ataxia mutants reveals determinants for frataxin binding and activation of the Fe-S assembly complex. Biochemistry 2011;50:7265–74.

[90] Campuzano V, Montermini L, Molto MD, et al. Friedreich's ataxia: autosomal recessive disease caused by an intronic GAA triplet repeat expansion. Science 1996;271:1423–7.

[91] Babcock M, De Silva D, Oaks R, et al. Regulation of mitochondrial iron accumulation by Yfh1p, a putative homolog of frataxin. Science 1997;276:1709–12.

[92] Vaubel RA, Isaya G. Iron-sulfur cluster synthesis, iron homeostasis and oxidative stress in Friedreich ataxia. Mol Cell Neurosci 2013;55:50–61.

[93] Cook JD, Kondapalli KC, Rawat S, et al. Molecular details of the yeast frataxin-Isu1 interaction during mitochondrial Fe-S cluster assembly. Biochemistry 2010;49:8756–65.

[94] Tsai CL, Barondeau DP. Human frataxin is an allosteric switch that activates the Fe-S cluster biosynthetic complex. Biochemistry 2010;49:9132–9.

[95] Colin F, Martelli A, Clemancey M, et al. Mammalian frataxin controls sulfur production and iron entry during de novo Fe4S4 cluster assembly. J Am Chem Soc 2013;135:733–40.

[96] Pastore A, Puccio H. Frataxin: a protein in search for a function. J Neurochem 2013;126:43–52.

[97] Rouault TA. Biogenesis of iron-sulfur clusters in mammalian cells: new insights and relevance to human disease. Dis Model Mech 2012;5:155–64.

[98] Vickery LE, Cupp-Vickery JR. Molecular chaperones HscA/Ssq1 and HscB/Jac1 and their roles in iron-sulfur protein maturation. Crit Rev Biochem Mol Biol 2007;42:95–111.

[99] Kampinga HH, Craig EA. The HSP70 chaperone machinery: J proteins as drivers of functional specificity. Nat Rev Mol Cell Biol 2010;11:579–92.

[100] Dutkiewicz R, Schilke B, Cheng S, et al. Sequence-specific interaction between mitochondrial Fe-S scaffold protein Isu and Hsp70 Ssq1 is essential for their in vivo function. J Biol Chem 2004;279:29167–74.

[101] Shakamuri P, Zhang B, Johnson MK. Monothiol glutaredoxins function in storing and transporting [Fe2S2] clusters assembled on IscU scaffold proteins. J Am Chem Soc 2012;134:15213–6.

[102] Ye H, Jeong SY, Ghosh MC, et al. Glutaredoxin 5 deficiency causes sideroblastic anemia by specifically impairing heme biosynthesis and depleting cytosolic iron in human erythroblasts. J Clin Invest 2010;120:1749–61.

[103] Lim SC, Friemel M, Marum JE, et al. Mutations in LYRM4, encoding iron-sulfur cluster biogenesis factor ISD11, cause deficiency of multiple respiratory chain complexes. Hum Mol Genet 2013;22:4460–73.

[104] Mochel F, Knight MA, Tong WH, et al. Splice mutation in the iron-sulfur cluster scaffold protein ISCU causes myopathy with exercise intolerance. Am J Hum Genet 2008;82:652–60.

[105] Olsson A, Lind L, Thornell LE, Holmberg M. Myopathy with lactic acidosis is linked to chromosome 12q23.3-24.11 and caused by an intron mutation in the ISCU gene resulting in a splicing defect. Hum Mol Genet 2008;17:1666–72.

[106] Kollberg G, Tulinius M, Melberg A, et al. Clinical manifestation and a new ISCU mutation in iron-sulphur cluster deficiency myopathy. Brain 2009;132:2170–9.

[107] Perdomini M, Hick A, Puccio H, Pook MA. Animal and cellular models of Friedreich ataxia. J Neurochem 2013;126:65–79.

[108] Cameron JM, Janer A, Levandovskiy V, et al. Mutations in Iron-Sulfur Cluster Scaffold Genes NFU1 and BOLA3 Cause a Fatal Deficiency of Multiple Respiratory Chain and 2-Oxoacid Dehydrogenase Enzymes. Am J Hum Genet 2011;89:486–95.

[109] Navarro-Sastre A, Tort F, Stehling O, et al. A fatal mitochondrial disease is associated with defective NFU1 function in the maturation of a subset of mitochondrial Fe-S proteins. Am J Hum Genet 2011;89:656–67.

[110] Ajit Bolar N, Vanlander AV, Wilbrecht C, et al. Mutation of the iron-sulfur cluster assembly gene IBA57 causes severe myopathy and encephalopathy. Hum Mol Genet 2013;22:2590–602.

[111] Ye H, Rouault TA. Human iron-sulfur cluster assembly, cellular iron homeostasis, and disease. Biochemistry 2010;49:4945–56.

[112] Li J, Kogan M, Knight SA, Pain D, Dancis A. Yeast mitochondrial protein, Nfs1p, coordinately regulates iron-sulfur cluster proteins, cellular iron uptake, and iron distribution. J Biol Chem 1999;274:33025–34.

[113] Li K, Besse EK, Ha D, Kovtunovych G, Rouault TA. Iron-dependent regulation of frataxin expression: implications for treatment of Friedreich ataxia. Hum Mol Genet 2008;17:2265–73.

[114] Crooks DR, Natarajan TG, Jeong SY, et al. Elevated FGF21 secretion, PGC-1alpha and ketogenic enzyme expression are hallmarks of iron-sulfur cluster depletion in human skeletal muscle. Hum Mol Genet 2013;23:24–39.

[115] Kiley PJ, Beinert H. The role of Fe-S proteins in sensing and regulation in bacteria. Curr Opin Microbiol 2003;6:181–5.

[116] Allikmets R, Raskind WH, Hutchinson A, et al. Mutation of a putative mitochondrial iron transporter gene (ABC7) in X-linked sideroblastic anemia and ataxia (XLSA/A). Hum Mol Genet 1999;8:743–9.

[117] Chan SY, Zhang YY, Hemann C, et al. MicroRNA-210 controls mitochondrial metabolism during hypoxia by repressing the iron-sulfur cluster assembly proteins ISCU1/2. Cell Metab 2009;10:273–84.

# 18 Iron-sulfur proteins and human diseases

Wing-Hang Tong

## 18.1 Introduction

Iron-sulfur (Fe-S) clusters are essential for numerous biological processes, including electron transfer in the mitochondrial respiratory chain, enzymatic catalysis in metabolic pathways, performing structural roles in DNA repair proteins, and sensing of intracellular iron levels and reactive oxygen species (ROS). Disruptions of Fe-S proteins and Fe-S cluster biogenesis proteins by genetic defects or oxidative stress have been shown to cause human diseases with a variety of symptoms and distinctive tissue-specific manifestations. This chapter presents an overview of the clinical, biochemical, and cell physiological aspects of the diseases associated with defects in Fe-S proteins or Fe-S cluster biogenesis and discusses current therapeutic approaches.

Fe-S proteins constitute one of the most ubiquitous and functionally versatile classes of metalloproteins [1, 2]. More than 120 distinct types of proteins contain Fe-S clusters, and Fe-S proteins are involved in fundamental processes such as respiration, photosynthesis, intermediary metabolism and nitrogen fixation. Fe-S clusters are essential for electron transfer processes in energy metabolism. Nicotinamide adenine dinucleotide (NADH):ubiquinone oxidoreductase (respiratory complex I) contains eight Fe-S clusters, and four additional Fe-S clusters are involved in electron transport in complexes II and III. A solvent-exposed Fe-S cluster directly participates in substrate binding and acid-base catalysis in a family of dehydratases, which includes the mitochondrial aconitase (ACO2) in the tricarboxylic acid (TCA) cycle [3]. Fe-S clusters are essential components of diverse nucleic acid processing machineries including glycosylases, helicases, and primases [4]. Fanconi anemia group J protein (FANCJ) and xeroderma pigmentosum (XP) group D protein (XPD) are Fe-S proteins involved in DNA repair and transcription [5, 6], whereas DNA polymerases have an essential Fe-S cluster-binding domain that is important for accessory subunit recruitment and replisome stability [7]. Fe-S clusters are also critical for the function of *S*-adenosyl-L-methionine (SAM)-dependent lipoic acid synthase (LIAS), both as the sulfur donor and in the generation of a catalytic 5′-deoxyadenosyl 5′-radical [8]. In addition, the sensitivities of Fe-S clusters to ROS, reactive nitrogen species (RNS), and iron bioavailability provide cells with a sensitive mechanism to register oxidative stress and intracellular iron status and to regulate gene expressions accordingly. For instance, the bifunctional iron regulatory protein 1/cytosolic aconitase (IRP1/ACO1) registers cytosolic iron levels through its labile Fe-S cluster. Reduced iron availability or oxidative stress limits Fe-S cluster assembly in IRP1, and IRP1 loses aconitase activity and binds to its mRNA targets and thereby regulates iron trafficking, intermediary metabolism, and heme biosynthesis (reviewed in [9–11]). Given the critical roles of Fe-S proteins in a wide range of cellular activities, disruption of Fe-S clusters or the Fe-S cluster assembly and repair processes, either as a result of genetic defects or oxidative stress, can affect basic cellular processes and result in human diseases.

## 18.2 Oxidative susceptibility of Fe-S proteins

Oxidative stress can cause damage to an array of cellular components and is linked to serious diseases in humans, including amyotrophic lateral sclerosis, Parkinson disease, Huntington disease, cardiomyopathy, and cancer [12, 13]. Studies from bacterial to human model systems have shown that Fe-S proteins are primary targets of oxidative damage [14, 15]. Early studies revealed that high oxygen levels block the ability of *Escherichia coli* to synthesize branched-chain amino acids [16], and studies from the Fridovich and Flint laboratories showed that superoxide ($O_2^-$) can inactivate the [4Fe-4S] family of dehydratases, including dihydroxyacid dehydratase, aconitase, and fumarase, which are key enzymes of the branched-chain amino acid metabolisms and TCA cycle in *E. coli* [17–19]. Studies in *Saccharomyces cerevisiae* showed that yeast deficient in cytosolic superoxide dismutase (CuZnSOD) and mitochondrial superoxide dismutase (MnSOD) exhibit deficiencies in [4Fe-4S]-dependent homoaconitase (which participates in lysine biosynthesis) and ACO2 [20, 21]. Similarly, flies depleted of SOD exhibit loss of aconitase activities [22]. Rli1 in yeast (human ABCE1) is an essential and highly conserved protein that requires an Fe-S cluster for its functions in ribosome biogenesis and maturation [23], and Rli1 was shown to be an important target for inhibition of cell growth by ROS [24].

In mammals, Fe-S clusters in the respiratory complexes and ACO2 are targets of oxidative damage during aging [15, 25–28]. C57BL/6J mice lacking MnSOD have reduced aconitase activity, suffer from dilated cardiomyopathy, and die within 10 days of birth [29, 30]. It has been suggested that the lipid accumulation observed in the livers of MnSOD-deficient mice resulted from oxidative damage to the Fe-S clusters in ACO2 and succinate dehydrogenase (SDH), which inhibited TCA cycle activity, allowing citrate to be diverted to lipid biosynthesis [31]. In addition to ROS, nitric oxide (NO) has been shown to inactivate respiratory complexes II and III through its reaction with the Fe-S clusters [32, 33], and electron paramagnetic resonance (EPR) studies of rat gastric tissue and human biopsy specimens revealed that dinitrosyl iron complexes were formed, indicating that Fe-S clusters are intracellular targets for NO that is generated from dietary nitrate [34]. Oxidative damages to ACO2 and respiratory complexes II and III have also been reported in the most affected regions in the brains of patients with Huntington disease [35, 36], and defects in mitochondrial energy metabolism leading to increased susceptibility to excitotoxic injury is a potential mechanism contributing to the pathogenesis of Huntington disease [37, 38]. A genetic modifier screen in *Drosophila* identified aconitase as a dominant suppressor of PTEN-induced kinase 1 (PINK1), a mitochondrial kinase linked to early-onset Parkinson disease in humans, and it was suggested that increased $O_2^-$ in *pink1* mutants led to the release of iron from the [4Fe-4S] cluster in ACO2, resulting in more oxidative stress and mitochondrial swelling [39]. In addition to damaging energy metabolism through the inactivation of aconitase and respiratory complexes, ROS and RNS can damage the Fe-S cluster in the heme biosynthetic enzyme ferrochelatase (FECH), leading to rapid degradation of the protein, and resulting in decreased heme biosynthesis [40, 41].

The molecular mechanisms of Fe-S cluster disassembly have been examined in a number of proteins, including dihydroxy-acid dehydratase, isopropylmalate isomerase, and fumarate and nitrate reduction (FNR) regulatory proteins from *E. coli* and HydE from *Thermotoga maritima* [42–44]. $O_2^-$-mediated inactivation of Fe-S-dependent dehydratases is thought to be initiated by oxidation of the solvent-exposed cluster, resulting in the formation of $H_2O_2$ and the release of $Fe^{2+}$ from an unstable $[4Fe\text{-}4S]^{3+}$ intermediate [19] (Fig. 18.1a). The resultant inactive enzyme containing a $[3Fe\text{-}4S]^+$ cluster can undergo further cluster disassembly or be reactivated *in vitro* and *in vivo* by reduction and remetallation. Similar to $O_2^-$, $H_2O_2$ can oxidize the catalytic

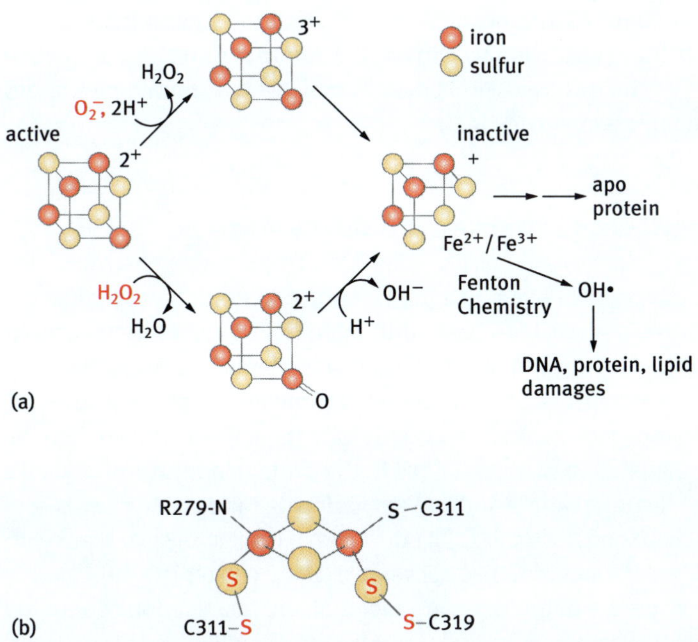


**Fig. 18.1:** Oxidative susceptibility of Fe-S clusters. (a) Superoxide ($O_2^-$) and hydrogen peroxide ($H_2O_2$) can directly oxidize the solvent-exposed catalytic iron atom of dehydratase Fe-S clusters, precipitating Fe loss and enzyme inactivation. $O_2^-$ oxidizes the $[4Fe\text{-}4S]^{2+}$ cluster, generating $H_2O_2$ and an unstable $[4Fe\text{-}4S]^{3+}$ species. Subsequent release of $Fe^{2+}$ generates a $[3Fe\text{-}4S]^+$ cluster [19]. Oxidation of the $[4Fe\text{-}4S]^{2+}$ cluster by $H_2O_2$ presumably generates a transient ferryl species, followed by release of $Fe^{3+}$ [45]. (b) X-ray crystallography studies suggested that oxygen-induced degradation of the $[4Fe\text{-}4S]^{2+}$ cluster in the hydrogenase maturase HydE from *T. maritime* proceeds with retention of the bridging sulfides to generate $[2Fe\text{-}2S]^{2+}$ cluster-bound cysteine persulfides [44].

iron atom of the dehydratase Fe-S cluster, generating a transient ferryl species that abstracts a second electron from the cluster, followed by the release of a $Fe^{3+}$ to form a catalytically inactive $[3Fe\text{-}4S]^{+}$ form [45].

In the case of the $O_2^{-}$-induced $[4Fe\text{-}4S]^{2+}$ to $[2Fe\text{-}2S]^{2+}$ cluster conversion in FNR, a metal-based oxidation mechanism was initially proposed that involved 2 one-electron oxidations of the cluster, followed by the release of one $Fe^{3+}$, one $Fe^{2+}$, and two $S^{2-}$ to yield a $[2Fe\text{-}2S]^{2+}$ cluster [46]. Subsequently, a sulfur-based mechanism was proposed based on the data from Raman and UV-vis absorption/circular dichroism spectroscopies and mass spectrometry, which suggested that the cluster conversion proceeded with the retention of two sulfides as cysteine persulfide ligands [43]. X-ray crystallography studies of HydE from *T. maritima* also identified an [2Fe-2S] cluster bound by two cysteine persulfide residues [44] (Fig. 18.1b). The study suggested that cluster degradation starts with iron oxidation that leads to an intermediate [3Fe-4S] cluster with the formation of $O_2^{-}$, which reacts with one of the cysteine ligands to form a cysteine persulfenate, which is then converted to cysteine persulfide with production of $H_2O_2$ and release of a second $Fe^{2+}$ ion. Subsequent oxidation of a second cysteine ligand generates a second cysteine persulfide. The differences in oxygen tolerance and mechanism of oxidative disassembly of the different Fe-S clusters may be determined by protein environments and cofactor geometries [47].

### 18.2.1 Aconitases: targets of oxidative stress in disease and aging

Aconitase catalyzes the reversible isomerization of citrate, *cis*-aconitate, and isocitrate [3, 48]. Two aconitase isozymes are present in mammalian cells: the mitochondrial enzyme (ACO2) that functions in the TCA cycle and the bifunctional cytosolic enzyme (ACO1/IRP1) that also plays a role in the regulation of iron metabolism. Aconitase contains a $[4Fe\text{-}4S]^{2+}$ cluster that is essential for its activity; three of the Fe atoms in the cluster are ligated to three cysteines in the protein; a fourth catalytic Fe atom is uniquely free to ligate substrate and solvent molecules and act as a Lewis acid during the enzymatic reaction [3]. $O_2^{-}$, $H_2O_2$, and NO have been shown to react with the exposed catalytic Fe atom and initiate disassembly of the cluster [18, 49].

Given the key role of citrate in intermediary metabolism, inactivation of aconitases can have profound effects on many aspects of physiology [49]. Citrate is an intermediate in major pathways of energy and intermediary metabolism. In the mitochondria, citrate is an intermediate in the TCA cycle, which is important for the generation of NADH and $FADH_2$ needed for oxidative phosphorylation. In the cytosol, citrate can be converted to acetyl-CoA, the building block for cholesterol and fatty acid biosynthesis in liver and adipose tissue. Citrate also has regulatory roles in glycolysis, fatty acid synthesis, and oxidation. Citrate is a negative regulator of the glycolytic enzyme phosphofructokinase [50, 51] and an allosteric activator of acetyl-CoA carboxylase, the enzyme that generates malonyl-CoA [52], a potent regulator of the partition between fatty acid biosynthesis and oxidation [53]. Decreased expression of a plasma membrane citrate transporter, Indy, in *Drosophila* and ceNAC-2 in *Caenorhabditis elegans* resulted in decreased lipid content and increased life span [54–56]. Thus, through its effects on malonyl-CoA production, glucose utilization, fatty acid synthesis, and oxidation, changes in aconitase activity can have profound impacts on obesity, insulin resistance and diabetes [57, 58].

In addition to intermediary metabolism, citrate homeostasis is also important for many other biological processes. In the central nervous system and the retina, ACO1 is thought to be important for generating glutamate, the major excitatory neurotransmitter [59]. In cholinergic neurons, citrate can be used to generate acetyl-CoA for acetylcholine synthesis [60]. In the kidney, citrate is an inhibitor of kidney stone formation [61]. In prostate cells, zinc has been shown to be an inhibitor of ACO2 activity, and the downregulation of zinc transporters is thought to be an important factor in the metabolic transformation of normal citrate-secreting epithelial cells to malignant citrate-oxidizing cells in prostate cancer [62].

In addition, isocitate has been identified as a mediator of the erythroid response to iron restriction [63]. Erythropoiesis (the production of red blood cells) is regulated by the cytokine erythropoietin (EPO). In iron deficiency, erythroid progenitors lose responsiveness to EPO, resulting in hypoplastic anemia. *In vitro*, the aconitase inhibitor fluorocitrate blocked erythroid differentiation in a manner similar to iron deprivation [63] and infusion of an aconitase inhibitor sodium fluoroacetate in mice caused a rapid-onset hypoplastic anemia [64]. Administration of isocitrate abrogated the eythroid iron restriction response *in vitro* and reversed anemia progression in iron-deprived mice [63]. Taken together, these studies indicated that isocitrate production by ACO2 plays a key role in the regulation of erythropoiesis.

The importance of aconitases was illustrated by studies of animal models and human patients with aconitase deficiency. Oxidative inactivation of aconitase has been associated with decreased life span in flies [25]. Deletion of ACO2 in *Drosophila* was lethal, whereas RNA interference-generated ACO2-knockdown flies showed reduced locomotor activity, a shortened life span, and increased cell death in the developing brain [65]. In humans, aconitase deficiency causes infantile cerebellar-retinal disorder (MIM614559) [66]. Homozygosity mapping with exome sequencing in eight patients with severe early-onset neurodegenerative degeneration identified a homozygous S112A mutation in ACO2 (Fig. 18.2). These patients presented at 2–6 months of age with truncal hypotonia (low muscle tone), athetosis (repetitive involuntary, slow movements in the extremities), seizure disorder, retinal abnormalities, culminating in profound psychomotor retardation. MRI showed progressive cerebral and cerebellar degeneration. One patient presented with peripheral demyelinating neuropathy. The mutant human ACO2 failed to

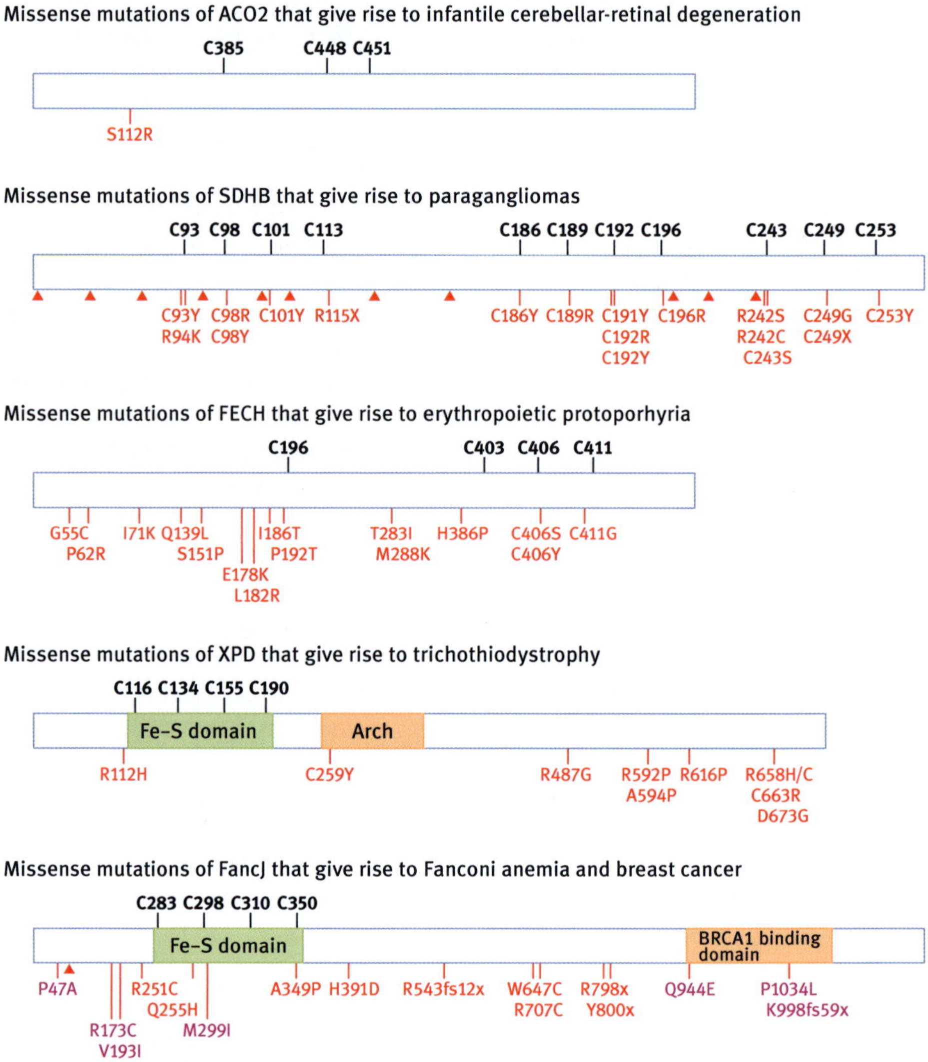


**Fig. 18.2:** Clinically relevant missense mutations in ACO2 [66], SDHB [67–74], FECH [75–77], and XPD [78, 79] are indicated in red. The positions of frameshift mutations in SDHB are indicated by red arrowheads [68]. Mutations in FANCJ associated with Fanconi anemia are indicated in red, whereas those associated with breast cancers are indicated in pink [80]. Cysteine ligands of the Fe-S clusters are indicated in black.

complement a yeast aconitase deletion strain, indicating that this mutation inactivates the enzyme activity.

Impairment of aconitase has also been identified in a cancer cell line derived from a patient with hereditary leiomyomatosis and renal cell cancer (HLRCC) [81], which is caused by germline mutations of the TCA cycle enzyme fumarate hydratase

(FH) [82–84]. A FH-deficient HLRCC cell line, UOK262, exhibited decreased ACO1 activity, which may be a result of cellular iron limitation (that is caused by decreased expression of the divalent metal transporter 1 DMT1 [81]) and/or increased oxidative stress [85]. In addition, a proteomic-based screen in renal cysts from FH-deficient mice revealed that the three cysteine residues in ACO2 crucial for Fe-S cluster binding were succinated, suggesting that loss of FH activity resulted in accumulation of intracellular fumarate, which modified cysteine residues to form 2-succinocysteine [86]. *In vitro* experiments confirmed that increased succination by fumarate correlated with loss of ACO2 activity. Loss of aconitase activity may contribute to the dysregulation of cell metabolism that promotes oncogenesis [87].

Oxidative stress can also disrupt intracellular iron homeostasis *via* the disassembly of the Fe-S cluster in IRP1/ACO1 [88]. Given that iron is both essential for basic cellular processes and toxic through its ability to generate ROS, systemic as well as intracellular iron status are tightly regulated by transcriptional and post-transcriptional mechanisms [89]. IRP1/ACO1 and IRP2 are RNA-binding proteins that play important roles in the post-transcriptional regulation of intracellular iron metabolism (reviewed in [9, 11]). IRP2 is regulated by iron- and oxygen-dependent protein degradation, whereas IRP1 regulation largely depends on the assembly and disassembly of its Fe-S cluster. When IRP1 contains an intact Fe-S cluster, it functions as a cytosolic aconitase; when it lacks the Fe-S cluster, it binds RNA stem-loop structures known as iron-responsive elements (IREs) that are present in a number of genes involved in iron metabolism. The IRE-binding activity of ACO1/IRP1 can be induced by $O_2^-$, $H_2O_2$, and NO [90–92] through oxidative disassembly of the Fe-S cluster and activation of IRP1 by oxidative stress can disrupt iron homeostasis by stimulating iron uptake *via* transferrin receptor 1 (TfR1) and DMT1, reducing iron export *via* ferroportin, and suppressing the iron storage protein ferritin.

## 18.3 Diseases associated with genetic defects in Fe-S proteins

### 18.3.1 Mitochondrial respiratory complexes and human diseases

More than 70 polypeptides are organized in complexes I–IV of the mitochondrial respiratory chain. Each complex consists of multiple protein components that are associated with a variety of redox-active prosthetic groups, including flavins, Fe-S clusters, ubiquinone, cytochromes, and copper centers that make up the electron transport chain. Genetic defects or oxidative damages of respiratory complexes are associated with a range of clinical conditions (Tab. 18.1) [93–98], some of which have been shown to involve the Fe-S proteins in the respiratory complexes I–III. This section will focus on the diseases associated with the Fe-S protein SDHB in respiratory complex II.

**Tab. 18.1:** Diseases associated with Fe-S proteins.

| Human protein | Function | Pathology |
|---|---|---|
| Complex I (NADH:ubiquinone oxidoreductase) | Respiration | Leigh syndrome, cardiomyopathy, encephalopathy, skeletal muscle myopathy [93, 94] |
| Complex II (SDH) | Respiration | Encephalomyopathy, cardiomyopathy, skeletal muscle myopathy, Kearns-Sayre syndrome, optic defects, leukoencephalopathy, paraganglioma-pheochromocytoma syndrome, renal cell carcinoma, Carney-Stratakis syndrome [95] |
| Complex III (ubiquinone:cytochrome *c* oxidoreductase) | Respiration | Encephalopathy, hypoglycemia, Leigh syndrome, Leber hereditary optic neuropathy, Bjornstad syndrome, lactic acidosis, skeletal muscle myopathy, hepatic failure [96] |
| Electron transfer flavoprotein-ubiquinone oxidoreductase (ETFDH) | Fatty acid and amino acid oxidation | Multiple acyl-CoA dehydrogenation deficiency/type II glutaric acidemia [99] |
| Ferrochelatase (FECH) | Last step in heme biosynthesis | Erythropoietic protoporphyria: light-sensitive dermatitis [100] |
| Iron regulatory protein 1/cytosolic aconitase (IRP1/ACO1) | Sensing and regulation of cellular iron homeostasis | ACO1−/− mice develop polycythemia and pulmonary hypertension [101, 102] |
| Mitochondrial aconitase (ACO2) | TCA cycle | Infantile cerebellar-retinal disorder: hypotonia, athetosis, seizures, retinal abnormalities, psychomotor retardation [66] |
| Dihydropyrimidine dehydrogenase (DPYD) | Pyrimidine degradation | Thymine-uraciluria; convulsive disorder with motor and mental retardation [103] |
| Lipoic acid synthase (LIAS) | Biosynthesis of lipoic acid cofactor for PDH, OGDH, and glycine cleavage system | Neonatal-onset encephalopathy, epilepsy, psychomotor retardation hypotonia, lactic acidosis, elevated glycine concentrations in plasma and urine [104] |
| Molybdenum cofactor synthesis 1 (MOCS1) | Biosynthesis of cofactor for aldehyde oxidase, xanthine dehydrogenase and sulfite oxidase | Moco deficiency: neonatal seizures, fatal neurological abnormalities [105] |
| Xanthine dehydrogenase/ xanthine oxidase (XDH) | Oxidative metabolism of purines | Xanthinuria: renal failure and xanthine kidney stones [106] |
| Xeroderma pigmentosum group D protein (XPD) | DNA repair helicase | XP, TTD, Cockayne syndrome [107] |
| Fanconi anemia group J protein (FANCJ) (BRIP1) | DNA repair helicase | Fanconi anemia, breast cancer [107] |
| MutY homologue (MUTYH) | DNA repair glycosylase | Colorectal cancer [108] |

#### 18.3.1.1 Diseases associated with complex II deficiencies

As complex II of the respiratory chain, SDH (succinate:ubiquinone oxidoreductase, EC 1.3.5.1) [109] is situated at the intersection of the TCA cycle and the mitochondrial respiratory chain, two pathways essential for energy production and anabolic processes. SDHA and SDHB encode the two catalytic subunits that project into the mitochondrial matrix, whereas SDHC and SDHD encode the membrane anchors in the mitochondrial inner membrane. The substrate succinate binds to the flavoprotein SDHA, and electrons from the oxidation of succinate to fumarate are channeled through three Fe-S clusters in SDHB to the ubiquinone site associated with SDHC and SDHD. Major insights into the workings of human complex II have been provided by crystal structures, mutagenesis studies, and biophysical studies [EPR and magnetic circular dichroism (MCD) spectroscopies] of closely related enzymes in several species [110–115].

Loss of complex II function leads to a range of clinical conditions [95, 116–118]. Generally speaking, enzyme depletion and mutations in SDHA compromise TCA cycle activities and energy production, resulting in early-onset encephalopathy, whereas heterozygous mutations in SDHB, SDHC, and SDHD induce later-onset paraganglioma syndromes. Several biallelic SDHA mutations have been shown to cause Leigh syndrome, a rare severe neonatal neurological disorder [119]. Congenital complex II deficiencies due to inherited homozygous mutations of SDHA, SDHB, and the SDHAF1 assembly factor also lead to cardiomyopathy and infantile leukodystrophies (inflammation in the white matter of the brain). Heterozygous mutations/deletions of SDHB, SDHC, or SDHD result in increased risk for autosomal dominant tumor syndromes. These patients have a loss-of-function mutation in one of the germline copies of an SDH subunit gene. When the second, intact copy is somatically lost or mutated in particular tissues, tumors develop. SDHD mutations are associated with mostly benign paragangliomas and phaeochromocytomas, whereas SDHB mutations are associated with a higher proportion of malignant lesions and a greater risk of phaeochromocytoma [120]. Renal cell carcinoma (in patients with SDHB mutations) and gastrointestinal stromal tumors (GISTs in patients with SDHB, SDHC, and SDHD mutations) affect a minority of patients. Loss of complex II can promote tumorigenesis through several processes, including the activation of hypoxia inducible factor (HIF) as a result of succinate accumulation [121], the generation of ROS [122], the inhibition of histone demethylases [123], and the inhibition of EglN3-mediated apoptosis [124].

Cellular and animal models have provided valuable insights into the diseases associated with mutations in SDH. Homozygous deletion of SDHD in mice is embryonic lethal, underscoring the essential role of SDH in energy production [125, 126]. Several models of complex II mutations have increased ROS production and increased sensitivity to oxidative stress [127, 128]. Mutations in the ubiquinone-binding site in *S. cerevisiae* SDH promote $O_2^-$ production [129]. Transgenic flies expressing a dominant-negative form of SDHC (I71E) [130] and a P-element insertion mutant with reduced expression of SDHB [131] exhibit hypersensitivity to oxygen, reduced life span, and age-related behavioral decay. Transgenic mice with a V69E mutation in SDHC have respiratory dysfunction, increased $O_2^-$ levels in the mitochondria of their hearts and muscles, abnormal mitochondrial structures in muscles, decreased body weight and locomotion activity, and infertility [128]. Knockdown of SDHB in mammalian cells produce features of tumor phenotypes, with alterations in cellular proliferation and respiration, a metabolic shift to glycolysis, and changes in the expression levels of many genes involved in proliferation, adhesion, and the hypoxia pathway [132].

In addition to increasing oxidative stress, there is mounting evidence that points to the hyperactivation of the hypoxia-response pathway as a major cause of tumorigenesis in individuals with SDH mutations. Mammalian cells respond to changes in oxygen availability through a conserved pathway that is regulated by the HIF [133]. HIF is a heterodimer composed of HIF-1α (or HIF-2α) and HIF-1β. In normoxia, HIF-1α and HIF-2α are hydroxylated at a proline residue by HIF prolyl hydroxylases (PHD) 1–3 and targeted for degradation by the proteasome [134, 135]. Pathological stabilization of HIFα can initiate the transcription of many genes including those that encode vascular endothelial growth factor (VEGF), transforming growth factor, phosphoglycerate kinase, lactate dehydrogenase A, and EPO, thereby fueling tumorigenesis by promoting angiogenesis, changes in energy production, and growth stimulation [133, 136]. Mutations in *SDH* have been shown to induce the hypoxia-response pathway in tumors [137]. In particular, paragangliomas and extra-adrenal phaeochromocytoma from patients with the SDHD R22X mutation or R46Q SDHB mutation showed increased expression of VEGF [138, 139], whereas cells depleted of SDH using siRNA have increased succinate levels, overexpression of HIF-1α, and activation of HIF-1α target VEGF [84, 121]. Mass spectrometric analysis and *in vitro* assays confirmed that succinate can bind and inhibit PHD [140]. These studies indicate that accumulation of succinate due to SDH deficiency results in stabilization of HIF and that pseudo-hypoxia promotes tumorigenesis in patients with SDH deficiency.

#### 18.3.1.2 Mutations in SDHB, an Fe-S protein, in paraganglioma-pheochromocytoma syndrome, renal cell carcinoma, Carney-Stratakis syndrome/GIST, and infantile leukoencephalopathy

**Paraganglioma-pheochromocytoma syndrome**

*SDHB* mutation carriers are prone to developing extra-adrenal paragangliomas, adrenal pheochromocytomas, or head and neck paragangliomas (Fig. 18.2) [141]. Paragangliomas are neuroendocrine neoplasms that originate from both the sympathetic and parasympathetic branches of the autonomic nervous system and may develop at various body sites, including the head, neck, thorax, and abdomen. Mutations of *SDHA, SDHB, SDHC, SDHD*, and SDH assembly factor 2 (*SDHAF2*) have been identified in familial head and neck paragangliomas [142–147]. A type of paraganglioma known as a pheochromocytoma develops in the adrenal glands, and mutations of *VHL, RET, NF1* (gene 17 neurofibromatosis type 1), *SDHB*, and *SDHD* are associated with familial pheochromocytoma. Around 20% of *SDHB* mutation carriers will develop metastatic disease [67, 148]. To date, more than 100 pathological genetic defects in *SDHB* have been reported in paraganglioma-pheochromocytoma syndrome (TCA cycle gene mutation database [68]), including mutations of 10 of the 11 cysteine ligands of the three Fe-S clusters in SDHB [67, 69–74]. In addition to mutations of the cysteine ligands, nonsense and frameshift mutations, deletions, or splicing defects can result in the loss of one or more Fe-S cluster domains. Missense mutations close to the Fe-S binding sites can also potentially affect the binding and stability of the Fe-S clusters. For instance, paraganglioma patients have been reported to carry a c.281G>A mutation, which results in the substitution of a highly conserved R94 that is adjacent to a putative cysteine ligand for an Fe-S cluster in SDHB [149].

#### Renal cell carcinoma

Kidney cancer is a heterogeneous disorder, currently associated with mutations of 12 different genes (*VHL, MET, FLCN, TSC1, TSC2, TFE3, TFEB, MITF, FH, SDHB, SDHD*, and *PTEN*). A hereditary kidney cancer syndrome caused by *SDHB* mutation was initially reported by Vanharanta *et al.* [145] and was subsequently identified in individuals with or without a personal or family history of paraganglioma and/or pheochromocytoma [120, 150–154].

#### Carney-Stratakis syndrome/GIST

GISTs are rare mesenchymal tumors of the gastrointestinal tract, mostly associated with activating mutations in KIT (CD117) or platelet derived growth factor A (*PDFRA*) [155]. The Carney-Stratakis syndrome is a very rare subset of pediatric GIST, with a co-occurrence of GIST and familial paraganglioma and characteristic decreased SDHB immunostaining [156]. Germline mutations in the *SDHB, SDHC*, and *SDHD* genes have been found in a number of these patients [157], although reduced SDHB protein levels have also been observed in GIST tumors without identifiable *SDH* mutations [158].

#### Infantile leukoencephalopathy

Leukoencephalopathies are a group of disorders characterized by degeneration of the white matter of the brain. The disorders arise during infancy or childhood, with progressive loss in body tone, movements, gait, speech, vision, hearing, and behavior, leading to premature death. A novel homozygous mutation in the *SDHB* gene (D48V) was recently observed in a specific infantile leukoencephalopathy [159]. The patient presented with hypotonia and leukodystrophy (degeneration of myelin) with elevated brain succinate levels. Patient fibroblasts showed decreased levels of fully assembled complex II and near-complete absence of the SDHB subunit. A muscle biopsy showed markedly reduced SDHB protein levels and complex II activity. This represents the first description of a homozygous SDHB mutation in any patient. Studies of the equivalent mutation in yeast showed only 50% reduction in SDH activity, consistent with the idea that only mutations with relatively mild effects on the catalytic functions of SDH can occur in a homozygous state.

### 18.3.2 FECH deficiency causes erythropoietic protoporhyria (MIM 177000)

FECH (EC 4.99.1.1) is the terminal enzyme of the heme biosynthetic pathway, catalyzing the insertion of $Fe^{2+}$ into protoporphyrin IX (PPIX) to form heme [160]. Mammalian FECH contains a labile [2Fe-2S] cluster, similar to those found in 2Fe ferredoxins (FDXs), as judged by variable-temperature MCD, Mössbauer, and EPR studies [161, 162]. However, no redox role has been found for the cluster in FECH. The cluster is essential for the function of human FECH [163], but the cluster does not participate directly in enzymatic catalysis because FECHs in many organisms do not possess the cluster. The purified protein exhibited reduced stability under oxidative conditions [161, 164], and in cultured cells, FECH is degraded during conditions of iron deficiency or oxidative stress [41, 165]. A set of pulse/chase experiments following the levels of the transcripts, newly synthesized and steady-state levels of FECH protein demonstrated that iron limitation diminishes FECH levels by decreasing the stability of newly formed FECH protein [41]. FECH was also severely depleted in muscle biopsies and cultured myoblasts from patients with ISCU myopathy, a disease caused by deficiency of the scaffold protein ISCU that is essential for Fe-S cluster assembly [41]. Taken together, these data suggest that oxidative degradation of the Fe-S cluster or impaired Fe-S cluster assembly causes reduced maturation and destabilization of apo-FECH. Crystallographic studies revealed a stabilizing bridge formed by the Fe-S cluster between the three C-terminal cysteines and a fourth internal cysteine, suggesting that the cluster-ligating C-terminal region of the protein is important for folding of the mature enzyme and its protection from proteolysis [166, 167].

Erythropoietic protoporhyria (EPP) is an autosomal recessive disorder resulting from mutations in the *FECH* gene, which reduces the activity of FECH, leading to the accumulation of the enzyme substrate PPIX. The disorder is characterized clinically by painful photosensitivity to visible light [168] and biochemically by the accumulation of PPIX in bone marrow reticulocytes, liver, bile, and skin [169–171]. In ~5% of the patients, protoporphyrin deposits cause liver disease that may progress to liver failure caused by biliary occlusions of crystalline protoporphyrin [172, 173]. Hemolysis and anemia are usually absent or mild. A mouse model for EPP harboring homozygous FECH M98K mutation exhibited low FECH activity (3%–6%), hemolytic anemia, photosensitivity, cholestasis, and severe hepatic dysfunction [174, 175].

EPP patients often show a 70%–95% loss in FECH activity [169, 173], which usually results from the inheritance of a nonfunctional mutated *FECH* allele together with a low-expressing wild type *FECH* allele [176]. Exon deletions are a common cause of the nonfunctional *FECH* alleles [75, 76]. Individual deletions of exons 3 through 11 all resulted in proteins that failed to acquire the Fe-S cluster and lost enzyme activity [177]. It was suggested that with cluster ligands spanning the entire length of the protein, exon skipping may affect the ability of the protein to properly assume its native conformation with an intact Fe-S cluster. Subsequent studies revealed that three EPP patients had point mutations of the [2Fe-2S] cluster ligands, firmly establishing the importance of the Fe-S cluster in the stability and function of FECH (Fig. 18.2) [77].

### 18.3.3 DNA repair Fe-S proteins and human disorders

Many studies have indicated that the postmitotic neurons are particularly prone to accumulation of unrepaired DNA lesions, and deficiency in the repair of nuclear and mitochondrial DNA (mtDNA) damage has been linked to several neurodegenerative disorders [178, 179]. Nucleotide excision repair removes helix-distorting DNA damage and deficiency in such repair is found in XP and Cockayne syndrome [107]. Mismatch repair corrects base mispairs generated during replication and oxidative DNA damage and is linked to the trinucleotide repeat expansion in Huntington disease [180]. Single-strand DNA breaks are associated with the neurodegenerative diseases, ataxia-oculomotor apraxia 1 [181], and spinocerebellar ataxia with axonal neuropathy [182]. Defects in homologous recombination and nonhomologous end-joining for repairing DNA double-strand breaks are associated with ataxia telangiectasia [183]. Accumulation of oxidative mtDNA damage has also been linked to the age-associated neurodegenerative disorders Alzheimer disease [184] and amyotrophic lateral sclerosis [185].

Fe-S clusters are present in DNA repair glycosylases of the EndoIII/MutY family [186], in the family 4 uracil DNA glycosylases [187], and in DNA helicases XPD and FANCJ [5, 6]. In MutY DNA glycosylase, the Fe-S cluster is thought to position a loop that is important for DNA binding [188], but it has also been suggested that these DNA repair proteins detect DNA lesions through electron transfer *via* their Fe-S clusters [189]. Fe-S cluster DNA helicases, including XPD and FANCJ, unwind DNA and allow access to single-stranded DNA during DNA replication, repair and recombination, and transcription of RNA. The observations that the Fe-S domain is found in a variety of helicases with different functions suggest that the role of the Fe-S cluster is not to detect specific forms of DNA damages. Structural studies of XPD and XPD homologues, and characterization of Fe-S cluster site-directed mutants suggested that the integrity of the Fe-S domain is required for the proper folding and structural stability of the enzymes [5, 6].

Mutations in Fe-S cluster DNA repair enzymes have been shown to cause diseases in human patients and in animal models. MutY homologue (MUTYH) glycosylase is a component of a base excision repair system that protects the genome from oxidative damage. Bi-allelic germline mutations of *MUTYH* are linked to the mutation of cancer-related genes *APC* and *KRAS* in an autosomal-recessive colorectal cancer syndrome [108]. Mutations that affect the formation or stability of the Fe-S cluster in XPD and FANCJ have been implicated in trichothiodystrophy (TTD) and Fanconi anemia, respectively [5]. Other members of the XPD helicase family, RTEL (regular of telomere length) and Chl1, also possess four conserved cysteine residues analogous to the cysteinyl ligands of the Fe-S clusters in XPD and FANCJ. RTEL-knockout mice died during gestation with defects in the nervous system, heart, vasculature, and extra-embryonic tissues as a result of telomere loss and genomic instability [190]. Human ChlR1 interacts with the cohesin complex, and deletion of the mouse gene causes lethality due to defects in chromosome segregation, chromosome cohesion, and placental malformation [191].

#### 18.3.3.1 Mutations in XPD in XP, TTD, and combined XP with Cockayne syndrome

The XPD/Rad3 helicase family comprises a group of related superfamily 2 DNA helicases with a 5′-to-3′ directionality [192]. In eukaryotes, XPD functions as part of the transcription factor IIH (TFIIH) complex, which has a dual role in transcription initiation and nucleotide excision repair. Mutations in the *XPD* gene give rise to three different genetic conditions in humans – XP, TTD, and combined XP with Cockayne syndrome (XP/CS) – with a wide spectrum of symptoms [193]. XP is characterized by extreme light sensitivity and highly elevated rates of skin cancer as a result of reduced repair of UV photoproducts in DNA. Mutations causing XP are thought to disrupt XPD helicase activity while preserving the role of TFIIH in transcription initiation. In contrast, TTD mutations cause developmental and growth abnormalities, but typically do not result in elevated cancer rates. These symptoms are thought to arise due to defects in both nucleotide excision repair and DNA transcription initiation. Defective transcription probably prevents cells from becoming cancerous, explaining the distinction between XP and TTD. TTD mice exhibit many symptoms of premature aging, including osteoporosis and kyphosis, osteosclerosis, early graying, cachexia, infertility, and reduced life span [194]. XP/CS is a rare disease characterized by segmental progeria (premature aging), in which both transcription and repair are defective. Four XPD mutations that give rise to XP/CS are known: G47R, G602D, R666W, and G675R [195]. Mouse models for XP/CS harboring the XPD G602D mutation exhibited cancer predisposition and symptoms of segmental progeria, including cachexia (loss of body mass) and progressive loss of germinal epithelium [196].

The presence of an Fe-S cluster in a DNA helicase was first demonstrated when the *XPD* homologue from *Sulfolobus acidocaldarius* was cloned and overexpressed in *E. coli* [5]. The Fe-S cluster-binding domain is not important for the stability, binding of single-stranded DNA substrate, or ATPase activity of the enzyme but is essential for collaborating with the Arch domain for DNA strand displacement during helicase action [5, 197, 198]. The abolition of Fe-S cluster binding by mutagenesis of a conserved cysteine in yeast Rad3 resulted in loss of Rad3 activity and a severe UV-sensitive phenotype [5]. Structural, mutational, and biophysical studies suggested that the TTD mutation in XPD, R112H [199, 200], inactivates the helicase activity by disrupting the Fe-S cluster [5] (Fig. 18.2).

#### 18.3.3.2 Mutations in FANCJ in Fanconi anemia

Fanconi anemia is a clinically and genetically heterogeneous disorder that causes genomic instability in about 1 in every 100,000 births [201, 202]. Biallelic mutations in Fanconi anemia genes lead to developmental abnormalities in major organs, early-onset bone-marrow failure, predisposition to acute myeloid leukemia and solid tumors, congenital abnormalities, and infertility. Thus far, 15 genes have been identified as mutated in patients, and many more interacting genes have been discovered [202]. These proteins work together in the repair of deleterious interstrand DNA cross-links, and are important in maintaining genomic stability during DNA replication. On detection of a DNA cross-link, the core complex, which comprises FANCA, FANCB, FANCC, FANCE, FANCF, FANCG, FANCL, and FANCM, and accessory proteins, including FAAP20, FAAP24, and FAAP100 is activated and ubiquitinates the FANCI-FANCD2 complex, which then coordinates the action of downstream repair factors. SLX4 (also known as FANCP) interacts with multiple nucleases that make incisions at the site of DNA damage. FAN1 and SNM1A have a role in processing the cross-link after incision, whereas TLS polymerases (TLS Pol) are recruited to bypass the unhooked cross-link. The break is then repaired through homologous recombination by the Fanconi anemia proteins BRCA2, BRIP1, PALB2, and RAD51C. BRIP1 (also known as FANCJ) is implicated in homologous recombination but is also required at an earlier step for pathway activation.

Several clinically relevant mutations map close to the Fe-S cluster domain of FANCJ (Fig. 18.2) [5, 203]. FANCJ not only plays a role in DNA cross-link repair pathway [203–206] but also interacts with the breast cancer susceptibility protein BRCA1 and has a role in double-strand break repair [207]. A Fanconi anemia-associated mutation, A349P, reduced FANCJ enzyme activity, and A349 is positioned next to one of cysteine ligands (C350) of the Fe-S cluster in human FANCJ [203, 208]. Mutation of the equivalent residue in archaeal XPD (F136P) appeared to disrupt the hydrogen bond between the main-chain nitrogen and Fe ion ligand C137, resulting in the destabilization of the Fe-S cluster and a loss of helicase activity, suggesting that the A349P mutation of human FANCJ probably causes Fanconi anemia by disrupting the Fe-S cluster-binding domain [5]. The breast cancer-associated M299I mutant has enhanced ATPase, helicase and translocase activities [209], and M299 is positioned next to another cysteine ligand (C298) of the Fe-S cluster in human FANCJ (Fig. 18.2).

## 18.4 Diseases associated with genetic defects in Fe-S cluster biogenesis

Fe-S cluster biogenesis is a complex biological process, involving more than 20 different proteins in eukaryotes [210–214]. The basic components of this process were originally identified in bacterial *nif*, *isc*, and *suf* operons [2, 214], and analogous processes and protein homologues have been identified in yeast, plants, and animals. The process involves the assembly of nascent clusters from iron and sulfur atoms on scaffold proteins, followed by transfer of these nascent clusters to apo-target proteins directly or *via* intermediate carriers. In non-plant eukaryotes, Fe-S cluster biogenesis factors, including NFS1/ISCS, LYRM4/ISD11, ISCU (yeast Isu1 and Isu2), FXN (yeast Yfh1), FDX1 (yeast Yah1), FDX1L and FDXR (Arh1), are thought to be important in the early steps of cluster assembly, whereas HSPA9 (yeast Ssq1), HSC20 (yeast Jac1), GLRX5 (yeast Grx5), NFU1, BOLA3, ISCA (yeast Isa1 and Isa2), IBA57, ABCB7 (yeast Atm1), GFER (yeast Erv1), NUBPL/IND1, NUBP1 (yeast Nbp35), NUBP2 (yeast Cfd1), NARFL/IOP1 (yeast Nar1), CIAO1 (yeast Cia1), NDOR1 (yeast Tah18), CIAPIN1 (yeast Dre2), MMS19, FAM96A/CIA2A, and FAM96B/CIA2B are involved in subsequent steps in which Fe-S clusters from the scaffold protein(s) are transferred and assembled on target apo-proteins in different subcellular compartments.

The early steps of cluster assembly involve the abstraction of sulfur atoms from cysteine molecules by the cysteine desulfurase NFS1 [215–217], iron acquisition, and sulfur transfer to the scaffold protein ISCU [218–220], leading to the formation of [2Fe-2S] and [4Fe-4S] clusters [221–223]. LYRM4 (also known as ISD11) appears to be important for the stability and activity of NFS1 [224–227]. Frataxin (FXN) is critical in this early stage of Fe-S cluster biogenesis, although the exact function of FXN has been elusive. Various roles have been proposed for FXN, including iron storage, iron chaperone activity, and functioning as an allosteric factor [223, 228–231]. Redox proteins such as FDX and FDX reductase (FDXR) are thought to provide electrons for cluster assembly on scaffold proteins [232, 233], whereas the chaperone system comprising the heat-shock 70-kDa protein HSPA9 (also known as GRP75 or mortalin), the DnaJ-like co-chaperone HSC20, and the nucleotide exchange factor SIL1 (also known as BAP; yeast Mge1) utilizes energy derived from ATP hydrolysis to drive conformational changes in scaffold proteins to facilitate cluster transfer to intermediate carrier or final recipient proteins [234–241].

Additional Fe-S cluster biogenesis components are involved in guaranteeing the accurate and specific transfer of Fe-S clusters from the scaffold protein to target apo-proteins. Homologues of the mitochondrial monothiol glutaredoxin GLRX5 have been implicated in Fe-S cluster biogenesis in yeast [242, 243], zebrafish [244], and humans [245], but its precise function is unclear. Chloroplast monothiol glutaredoxins were suggested to be scaffolds for the formation and delivery of [2Fe-2S] clusters [246], whereas yeast and mammalian GLRX5 were suggested to function in Fe-S cluster transfer [243, 247]. Yeast Iba57 physically interacts with Isa1 and Isa2, and it was suggested that the complex of these three proteins functions to promote cluster transfer/assembly on a subset of Fe-S proteins including aconitase and radical SAM proteins [248, 249]. NFU1 was initially proposed to be an alternative scaffold protein [250–253]. However, more recent studies suggested that Nfu-type proteins can accept an Fe-S cluster from holo ISCU and serve as intermediate Fe-S cluster carriers to deliver Fe-S cluster to apo-targets [254–256]. The mitochondrial P-loop NTPase NUBPL/IND1 is thought to serve as a specific scaffold or transfer protein for the assembly of the eight Fe-S clusters into complex I [257, 258], but more recent studies suggested that IND1 has a primary role in mitochondrial translation that indirectly affects the assembly of complex I [429]. The ABC transporter ABCB7 (yeast Atm1) in the inner mitochondrial membrane, sulfydryl oxidase GFER (yeast Erv1) in the intermembrane space [259] together with glutathione have been described as an "export machinery" for Fe-S cluster biogenesis, although the identity of the transported compound(s) has remained unresolved [213].

Assembly and repair of Fe-S proteins in the cytosol and nucleus involve the extramitochondrial isoforms of NFS1, ISCU, and NFU1 [216, 251, 260] as well as a number of additional proteins in the cytosolic iron-sulfur protein assembly (CIA) machinery. In yeast, the P-loop NTPases Cfd1 and Nbp35 form a heterotetrameric complex and serve as a scaffold or transfer proteins for cytosolic Fe-S cluster biogenesis [261–263]. Cfd1 and Nbp35 facilitate the assembly of two Fe-S clusters on the hydrogenase-like protein Nar1 [264, 265], which then assists the transfer of Fe-S clusters to target apo-proteins by interacting with Cia1, a WD40 repeat protein [266, 267]. Electrons are transferred from NADPH *via* the FAD- and FMN-containing Tah18 to the Fe-S clusters of Dre2, a process required for the assembly of cytosolic target proteins [268]. MMS19 functions as part of the CIA machinery that facilitates Fe-S cluster insertion into a specific subset of apo-proteins involved in methionine biosynthesis, DNA replication, DNA repair, and telomere maintenance [269]. In mammalian cells, the complex composed of FAM96B/CIA2B, CIA1, and MMS19 facilitates cluster assembly on cytosolic-nuclear Fe-S proteins including phosphoribosylpyrophosphate aminotranferase (GPAT), dihydropyrimidine dehydrogenase (DYPD) and DNA polymerases, whereas the complex composed of FAM96A/CIA2A, CIA1, and MMS19 facilitates Fe-S cluster assembly of IRP1 [270]. The functions of many of these cytosolic Fe-S cluster biogenesis factors appear to be conserved in plants and animals [268–276].

Given the essential roles of Fe-S clusters in electron transfer, enzyme catalysis, and sensing functions in many proteins, defects in Fe-S cluster biogenesis can disrupt many cellular process and cause human diseases [210, 277]. In *S. cerevisiae,* deletions of many genes involved in Fe-S cluster assembly, including *nfs1, isd11, jac1, yah1, arh1, cfd1, erv1, nbp35, nar1,* and *cia1,* are lethal. In addition, synthetic lethality was observed in pairwise combinations of several other Fe-S cluster assembly genes including *isu1, isu2, nfu1, ssq1,* and *grx5* [278–280]. In vertebrates, deletions of the GLRX5 homologue in zebrafish and deletion of FXN, ABCB7, ISCU, NARFL, and MMS19 in mice are embryonic lethal [244, 273, 275, 281–283]. Depletion of mammalian Fe-S cluster assembly factors in cellular and animal models showed inactivation of many important proteins including SDH, ACO1, ACO2, xanthine oxidoreductase (XOR), and GPAT, and misregulation of proteins involved in iron metabolism [232, 260, 270, 284–287]. The following sections provide an overview of the human genetic defects that have now been reported for FXN, ABCB7, GLRX5, ISCU, NUBPL, NFU1, BOLA3, IBA57, and ISD11 (Tab. 18.2 and Fig. 18.3).

**Tab. 18.2:** Diseases associated with Fe-S cluster biogenesis.

| Human protein | Yeast homologue(s) | Proposed function(s) | Pathology |
|---|---|---|---|
| FXN | Yfh1 | Facilitate Fe-S cluster assembly in ISCU scaffold protein | FRDA |
| ABCB7 | Atm1 | ABC transporter, mitochondrial export machinery component | XLSA and cerebellar ataxia |
| GLRX5 | Grx5 | Monothiol glutaredoxins; transfer of Fe-S clusters from ISCU to apo-target | Autosomal recessive pyridoxine-refractory sideroblastic anemia |
| ISCU | Isu1, Isu2 | Scaffold for Fe-S cluster assembly | Myopathy with lactic acidosis |
| NUBPL | Ind1 | A primary role in mitochondrial translation that indirectly affects the assembly of respiratory complex I | Childhood-onset mitochondrial encephalomyopathy and respiratory complex I deficiency |
| NFU1 | Nfu1 | Fe-S cluster transfer protein | Multiple mitochondrial dysfunctions syndrome 1 |
| BOLA3 | Bola3 | Fe-S cluster biogenesis in a subset of Fe-S proteins including LIAS | Multiple mitochondrial dysfunctions syndrome 2 |
| IBA57 | Iba57 | Fe-S cluster biogenesis in a subset of Fe-S proteins including LIAS | Severe myopathy and encephalopathy |
| ISD11 | Isd11 | Important for cysteine desulfurase activity | Deficiencies of respiratory complexes |

### 18.4.1 A GAA trinucleotide repeat expansion in FXN is the major cause of the neurodegenerative disorder Friedreich ataxia

Friedreich ataxia (FRDA) (MIM 229300), an autosomal recessive neurodegenerative disorder caused by deficiency of FXN [288], is the most prevalent form of hereditary ataxia in Caucasians, occurring in about 1 in 50,000 individuals [289]. FXN deficiency leads to progressive spinocerebellar neurodegeneration associated with gait and limb ataxia, muscle weakness, as well as cardiomyopathy and diabetes [290, 291]. Most of neurological manifestations result from the degeneration of the dorsal root ganglia and the posterior columns, followed by degeneration in the spinocerebellar tracts and the corticospinal tracts of the spinal cord. Although cognitive functions remain largely intact during disease progression, patients develop communication

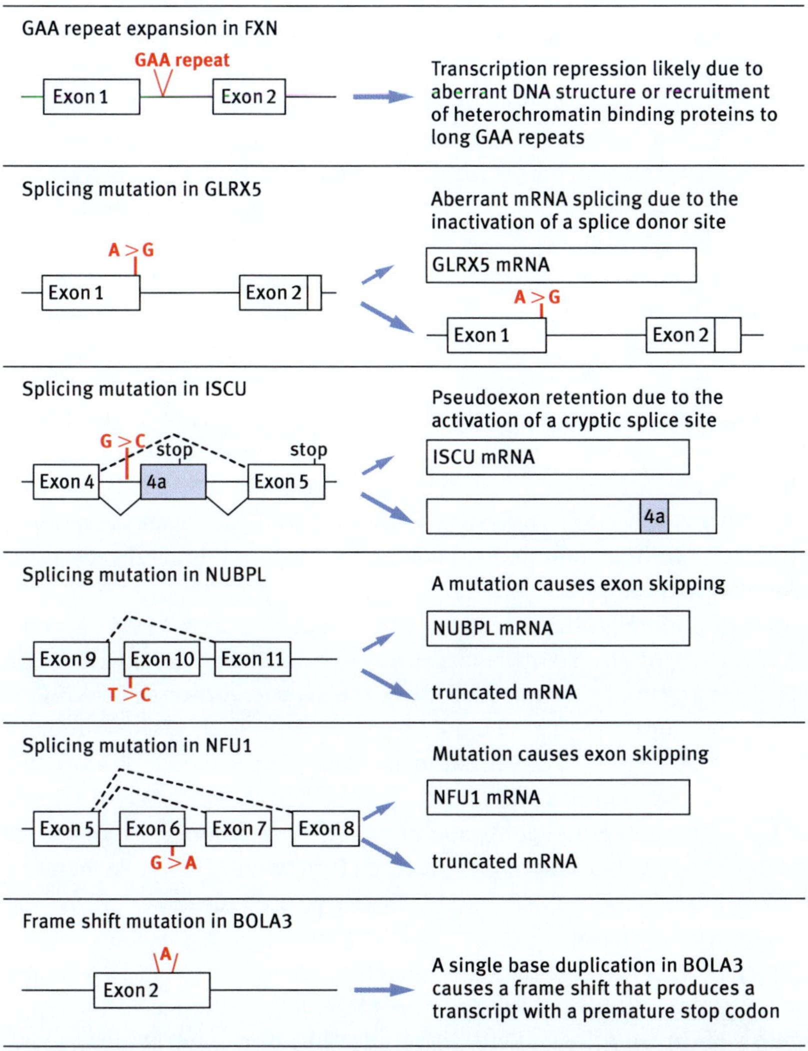


**Fig. 18.3:** In addition to missense mutations in FXN, ABCB7, GLRX5, ISCU, NUBPL, BOL3, and ISD11, dysfunction of Fe-S cluster biogenesis can also be the result of a variety of genetic defects, including trinucleotide repeat expansion, intron retention, pseudo-exon retention, exon skipping, and frameshift, that lead to reduced production of the mRNA and/or protein products. In more than 98% of FRDA cases, the defect is a GAA repeat expansion in the first intron of the FXN gene. In one patient with GLRX5 deficiency, a homozygous silent mutation in exon 1 interferes with splicing and reduces the level of fully processed mature GLRX5 mRNA. In ISCU myopathy, a single G-to-C mutation in the fourth intron of ISCU activates a weak splice acceptor site, resulting in the increased production of an alternative splice form with a premature stop codon. In patients with NUBPL deficiency, a mutation in exon 10 of NUBPL results in exon skipping and production of a truncated transcript. In multiple mitochondrial dysfunctions syndrome 1, a homozygous missense mutation in exon 6 of NFU1 results in exon skipping and production of several truncated transcripts. In one of the patients with multiple mitochondrial dysfunctions syndrome 2, a single nucleotide duplication in exon 2 of BOLA3 causes a frameshift that produces a transcript with a premature stop codon.

difficulties due to dysarthria, vision, and hearing loss. Cardiac failure is a frequent cause of death at a young age.

Expansion of an unstable GAA trinucleotide repeat in intron 1 of *FXN* is the most common causal mutation of FRDA [288]. Most FRDA individuals are homozygous for this mutation, but 4% of the patients are compound heterozygous for the GAA expansion and a second inactivating mutation (nonsense, missense, deletions, insertions) [288, 292, 293]. Normal alleles have 5 to 30 repeats, whereas FRDA alleles have 70 to more than 1,000 repeats. This GAA expansion leads to transcriptional silencing of *FXN* by inducing the formation of repressive heterochromatin of the locus, resulting in reduced expression of FXN (~5%–30% of normal) (reviewed in [294]). It has been suggested that the progressive pathology in the dorsal root ganglia may result from an age-dependent, tissue-specific, increased expansion of the GAA triplet-repeat sequence [295]. Other disease mutations in *FXN* lead to the production of nonfunctional or partially functional proteins [296, 297]. In addition, rare *FXN* isoforms that are specifically expressed in affected tissues have been shown to decrease more in patients with FRDA than in healthy individuals, suggesting additional mechanisms for the tissue-specific pathology [298].

Early studies of endomyocardial biopsies of two patients with FRDA revealed that activities of ACO2 and respiratory complexes I, II, and III were markedly decreased, providing the first hint of a role of FXN in the synthesis or stability of Fe-S proteins [299]. Disruption of the *FXN* homologue in *S. cerevisiae, yfh1*, resulted in a severe defect in respiration, deficiencies in multiple Fe-S dependent enzymes, mitochondrial iron accumulation, and loss of mtDNA [299–301]. A role of FXN in Fe-S cluster biogenesis was further confirmed in yeast, *Drosophila*, and human cell lines depleted of FXN [302–308]. However, the exact function of FXN in Fe-S cluster biogenesis has remained a subject of debate. Extensive structural, complementation, and biochemical studies on FXN in the last 15 years have yielded complex and contradictory results. *In vitro* studies of bacterial, yeast, or human FXN homologues showed a modest affinity of the proteins for iron (micromolar range) [309, 310]. Observation of an iron-dependent oligomerization of the recombinant Yfh1 led to a proposed role of FXN oligomers in mitochondrial iron storage [311], and *in vitro* reconstitution studies showed that the oligomeric forms of yeast and human FXN can provide iron for Fe-S cluster formation on the scaffold protein ISCU [312, 313]. However, the importance of FXN oligomerization in Fe-S cluster biogenesis is unclear because mutations of the acidic residues on the surface of Yfh1 abrogated iron-dependent oligomerization but showed no deleterious effects on Fe-S protein activities and iron homeostasis *in vivo* [314]. Other studies have suggested that FXN serves as an iron-binding chaperone protein that delivers Fe for Fe-S cluster assembly in the scaffold protein ISCU [228, 230, 315, 316]. It has also been suggested that FXN provides iron through direct protein-protein interactions with target Fe-S proteins, including ACO2 [317], FECH [318, 319], and SDH [320], although the relevance of these interactions has been questioned [321]. Other *in vitro* studies suggested that mammalian FXN interacts with a complex composed of NFS1, ISCU, and ISD11 and acts as an allosteric factor that activates sulfur production from cysteine [223, 231] and controls iron entry during *de novo* Fe-S cluster assembly [223]. Interestingly, a point mutation in the yeast *isu1* gene (M107I) was found to restore many deficient functions in Yfh1-depleted yeast cells [322]: iron homeostasis and Fe-S cluster enzyme activities were improved and cytochrome levels and heme synthesis were restored. It was suggested that the Isu1 mutation bypasses the putative allosteric activator Yfh1 by increasing exposure of an Fe-S cluster binding cysteine ligand and improving sulfur transfer from the cysteine desulfurase complex Nfs1-Isd11 to Isu1.

The effect of FXN depletion has been modeled in diverse systems, including *E. coli* [323], yeast [300, 324, 325], *C. elegans* [279, 326], *Drosophila* [305, 327], mice [328–330], and human cell lines [302, 304, 306, 307], and these cellular and animal models of FRDA have provided many insights into the pathophysiology of the disease (reviewed in [331]). Depletion of FXN homologues in yeast and *Drosophila* leads to Fe-S cluster deficiencies, mitochondrial iron accumulation and increased sensitivity to oxidative stress [305, 324]. The complete knockout of *FXN* in mice leads to early embryonic lethality [281], whereas cardiac-specific and neuronal models of *FXN* deletion [328, 332] are viable and recapitulate most of the characteristic features of the disease, including progressive spinocerebellar and sensory ataxia and hypertrophic cardiomyopathy. More specifically, the cardiac-specific FRDA model revealed that a deficit in Fe-S clusters precedes cardiac dysfunction and mitochondrial iron accumulation [328], whereas the neuronal model showed that abnormal autophagy, with formation of lipofuscin and large vacuoles within large sensory neurons, might be involved in the neurodegeneration of the dorsal root ganglia [332]. Conditional models lacking FXN primarily in the pancreas were generated to model the pathophysiology of diabetes mellitus associated with FRDA [333], and these mice showed a progressive reduction in the number of pancreatic islets, resulting in an impaired insulin response to glucose and subsequent diabetes. In addition to the models of *FXN* deletion, several mouse models with GAA expansions in *FXN* have been generated to better mimic the human disease [329, 330]. These animal models are invaluable in revealing the molecular and cellular mechanisms associated with GAA-mediated silencing of the *FXN* gene *in vivo* and for the development of drugs to target FRDA pathology.

Notably, mitochondrial iron overload was observed in the hearts and brains of patients with FRDA [334, 335] and heart-specific *FXN* deletion mouse models [328] and Yfh1-deficient yeast strains also develop profound mitochondrial iron overload [324]. These results suggest that Fe-S cluster biogenesis might be required for signaling information about the status of mitochondrial iron stores to the nucleus [336, 337]. In the absence of an appropriate signal, mitochondrial iron uptake might be upregulated [338], efflux might be diminished, and iron might accumulate as insoluble ferric phosphate nanoparticles in the mitochondrial matrix [339, 340] or within the iron sequestration protein, mitochondrial ferritin [307, 335]. Whether the mitochondrial iron overload might further cause mitochondrial damage *via* iron-catalyzed oxidation reactions and thereby contribute to FRDA pathology remains a subject of debate [327, 341–346].

### 18.4.2 Mutations in ABCB7 cause X-linked sideroblastic anemia with ataxia

The sideroblastic anemias are a heterogeneous group of acquired and heritable disorders, characterized by bone marrow ringed sideroblasts due to pathological iron overload in the mitochondria of erythroid precursors [347–350]. The most common congenital sideroblastic anemia is the X-linked sideroblastic anemia (XLSA) caused by mutations in the erythroid-specific 5-aminolevulinate synthase gene (ALAS2), the first and rate-limiting enzyme in the mammalian heme biosynthetic pathway. Patients develop hepatic and systemic iron overload, but not ataxia. Other genetic defects found in congenital sideroblastic anemia include mutations in mitochondrial ATP-binding cassette transporter ABCB7 [351], high-affinity thiamine transporter SLC19A2 [352], RNA-modifying enzyme pseudo-uridine synthase 1 (PUS1) [353], glutaredoxin 5 (GLRX5) [245], erythroid specific mitochondrial transporter SLC25A38 [354], and deletions, duplications, and rearrangements of mtDNA [355].

Mutations in ABCB7 are associated with XLSA with ataxia, a rare form of congenital sideroblastic anemia with early-onset nonprogressive spinocerebellar ataxia, cerebellar hypoplasia, dysarthria, mild anemia, and elevated levels of protoporphyrin [351, 356–358]. Systemic iron overload is not detected. *ABCB7* has high sequence similarity to the yeast ABC transporter gene *ATM1*, whose protein product localizes to the inner mitochondrial membrane. Depletion of ABCB7 homologues in plants, mice, and human cell lines resulted in loss of ACO1 activity in the cytosol, but no significant changes in the activities of ACO2 and SDH in the mitochondria [282, 359, 360]. Moreover, deletion of yeast *ATM1* leads to mitochondrial iron accumulation [361, 362]. In mice, tissue-specific deletions of *ABCB7* in brain and bone marrow were lethal [282], whereas liver-specific deletion of *ABCB7* revealed changes in several proteins involved in cytosolic iron homeostasis, including activations of IRP1 and IRP2, increased IRP2 levels, and decreased expression of ferritin. Iron is essential for heme biosynthesis and red blood cell development [363, 364]. Thus, it appears that the anemia and the ataxia observed in XLSA-A may be related to disruption of intracellular iron homeostasis and damages caused by excess iron in the mitochondria in developing red blood cells and neural cells [365].

The substrate transported by ABCB7 remains unclear. Given its association with a form of anemia characterized by mitochondrial iron accumulation in erythroid precursors, ABCB7 was initially thought to be involved in transporting heme that is generated in the mitochondria to the cytosol for incorporation into hemoglobin, but *in vitro* studies in *S. cerevisiae* argued against heme export and iron import as possible functions of Atm1 [361, 366]. Instead, Kispal *et al.* observed that yeast depleted of *ATM1* contained no detectable activity of the cytosolic Fe-S enzyme isopropyl malate isomerase Leu1 and proposed that Atm1 was involved in the export of Fe-S clusters generated by the mitochondrial Fe-S cluster assembly machinery into the cytosol for incorporation into extra-mitochondrial proteins [367, 368]. However, subsequent studies indicated that Atm1 is not required for exporting Fe-S clusters for incorporation into Leu1 [369], and the overexpression of an Atm1 homologue, Mdl1, which exports peptides out of mitochondrial matrix [370], can partially compensate for *ATM1* loss. Furthermore, *Arabidopsis* ATM3 can functionally complement a yeast *Δatm1* mutant [371], but, unlike yeast *atm1* mutants, mitochondria from *atm3* mutant plants do not accumulate significant amounts of iron [360]. These results have led to the suggestion that a sulfur-containing compound of as-yet unknown nature is exported to the cytosol by Atm1 [360, 372]. It is also interesting to note that Atm3 in *Arabidopsis* is required for the export of the cyclic pyranopterin monophosphate intermediate from the mitochondria into the cytosol for the biogenesis of molybdenum cofactor (Moco) [373]. Thus, *Arabidopsis* ATM3 may transport either two distinct compounds or a single compound required for both Fe-S,and Moco assembly machineries in the cytosol [360]. Neither deletion of *ATM1* in yeast nor *ATM3* in *Arabidopsis* is lethal, suggesting that either there is some low level diffusion of the transported compound(s) through the mitochondrial inner membrane or that the transported compound(s) can be transported with low efficiency by other transporter(s).

### 18.4.3 Mutations in glutaredoxin 5 cause an autosomal recessive pyridoxine-refractory sideroblastic anemia

A point mutation in *GLRX5* gene has been shown to be the genetic cause of an autosomal recessive pyridoxine-refractory sideroblastic anemia (MIM205950) [245]. The patient had mild anemia until midlife, when anemia worsened, and diabetes, splenomegaly, and cirrhosis were diagnosed. The anemia was worsened by blood transfusions but partially reversed by iron chelation therapy. DNA sequencing and RT-PCR experiments showed that a homozygous mutation in the penultimate nucleotide of exon 1 of GLRX5 interfered with splicing, resulting in drastically reduced mRNA levels. More recently, two heterozygous missense mutations in *GLRX5* were identified in a Chinese patient affected with sideroblastic anemia [374].

Glutaredoxins were initially defined as thiol disulfide oxidoreductases that catalyze thiol-disulfide exchange reactions using reduced glutathione as the electron donor, but genome sequencing in recent years has shown that these proteins constitute a complex family of proteins with diverse structural and functional properties [375, 376]. In yeast, three members of the subfamily of monothiol glutaredoxins – Grx3, Grx4, and Grx5 – have a conserved CGFS active site and have roles in iron homeostasis, facilitating Fe-S cluster biogenesis in the mitochondria (Grx5) or signaling intracellular iron status for the regulation of iron trafficking (Grx3 and Grx4). The CGFS active site is required for the coordination of a [2Fe-2S] cluster; CGFS glutaredoxins form $[2Fe\text{-}2S]^{2+}$-bridged homodimers with cysteine ligands provided by the two CGFS active sites and two glutathione molecules [246, 377–381].

The effects of GLRX5 depletion have been modeled in a number of studies. Yeast that lack Grx5 exhibit respiratory deficiency, iron accumulation, reduction in ACO2 and SDH activities, a loss of mtDNA, and the absence of Rip1, the Fe-S Rieske protein in complex III [242, 382, 383]. Silencing of GLRX5 in human cells also resulted in reduced aconitase activities [384]. Depletion of Grx5 in yeast resulted in accumulation of $^{55}Fe$ on the scaffold protein Isu1, suggesting that Grx5 has a role in a step after the assembly of an Fe-S cluster on Isu1 [243]. Grx5 deficiency has also been shown to increase oxidative stress in yeast cells [385] and human osteoblasts [386]. In zebrafish, Grx5 mutations cause anemia as a result of IRP1-mediated translational repression of erythroid aminolevulinate synthase (ALAS2), the first enzyme in the heme biosynthetic pathway [244]. Human and zebrafish genomes contain two *ALAS* genes; the *ALAS2* transcript contains an IRE in its 5′ untranslated region (UTR) and is highly expressed in erythroid cells, whereas the *ALAS1* transcript is expressed in other tissues and lacks an IRE. In the GLRX5-deficient patient and in the zebrafish model, IRP1 becomes an active IRE-binding protein that inhibits *ALAS2* translation and thereby blocks heme biosynthesis. It appeared that loss of mitochondrial GLRX5 resulted in mitochondrial iron overload and concomitant cytosolic iron depletion, which prevents *de novo* cytosolic cluster assembly in IRP1, and activates IRP1-mediated translational repression of *ALAS2*. These alterations may explain the sideroblastic anemia-associated phenotype resulting from the aberrant splicing of human *GLRX5* mRNA [245, 384].

Several roles have been proposed for GLRX5. One proposal is that the deglutathionylation activity of glutaredoxins might be important for repairing mixed disulfides between glutathione and Fe-S cluster assembly factors in the oxidizing environment of the mitochondrial matrix [280]. Other studies in plants have shown that a GLRX5 homologue is able to bind a [2Fe-2S] cluster and transfer the cluster to apo FDX *in vitro* and suggested that Grx5 may be a scaffold protein [246]. In yeast, depletion of Grx5 resulted in accumulation of $^{55}Fe$ on the scaffold protein Isu1, suggesting that Grx5 function as an intermediate carrier in transferring Fe-S clusters from the scaffold proteins to dedicated apo-proteins [243]. Spectroscopic evidence provided by circular dichroism spectroscopic studies of recombinant proteins from *Azotobacter vinelandii* showed a rapid, ATP-driven, [2Fe-2S] cluster transfer from [2Fe-2S]-IscU to apo-Grx5 in the presence of chaperone proteins, HscA and HscB [247]. Phenotypic defects associated with the absence of Grx5 in yeast were suppressed by overexpression of the *HSPA9* homologue *SSQ1* [242], and yeast two-hybrid studies have indicated that Grx5 interacts with Ssq1 [387]. Defects in yeast lacking Grx5 were also suppressed by overexpression of *ISA2* [242]. This observation, together with yeast two-hybrid studies and bimolecular fluorescence complementation experiments showing physical interactions between Grx5 with Isa1 [382, 388] and Isa2 [382], led to the proposal that Grx5 functions with Isa type proteins in Fe-S cluster biogenesis.

### 18.4.4 Mutations in ISCU cause myopathy with lactic acidosis (MIM 255125)

ISCU is a scaffold protein on which nascent Fe-S clusters are assembled and from which the clusters are delivered to intermediate carriers or directly to target apo-proteins [221, 223]. In mammalian cells, two splice isoforms of *ISCU* exist [260, 389].

These transcripts have different 5′ UTRs and generate proteins that are localized to different subcellular compartments: ISCU1 is located in the cytosol and nucleus, whereas ISCU2 is located in the mitochondrial matrix. Human ISCU works in a complex with NFS1, ISD11, and FXN [231, 321]. Sulfur transfer from NFS1 and iron acquisition leads to assembly of [2Fe-2S] and/or [4Fe-4S] clusters in ISCU [221–223].

Because of its essential role in the very early steps in Fe-S cluster biogenesis, complete loss of ISCU homologues in *S. cerevisiae* and in mice is lethal [278, 283] and depletion of human ISCU has been shown to affect all Fe-S proteins that have been examined, including SDH, ACO1/IRP1, ACO2, FECH, GPAT, and LIAS (as indicated by the decrease in lipolyated proteins) [41, 255, 260, 390]. Notably, ISCU has now been shown to be one of the mediators of the Pasteur effect, a metabolic shift in hypoxic cells resulting from the repression of TCA cycle, mitochondrial electron transport, and oxidative phosphorylation in favor of glycolysis [391–393]. This metabolic versatility of mammalian cells is essential for the maintenance of energy production and cell survival throughout a range of oxygen concentrations. Both ISCU1 and ISCU2 were found to be targets for repression by the microRNA-210 (miR-210) [394–397], which is specifically induced by HIF-1α during hypoxia [398, 399]. MiR-210 is upregulated and ISCU is downregulated in several clinical settings, including pregnancies complicated by preeclampsia, clear cell renal cancer, and head and neck paragangliomas [400–403]. By downregulating the expression of ISCU during hypoxia or in pseudohypoxia conditions, miR-210 decreases the activities of respiratory complex I and aconitases. These findings indicate that both miR-210 and ISCU are important factors in the regulation of mitochondrial respiration and metabolism during hypoxic stress and mitochondrial dysfunction in a variety of diseases [404, 405].

Deficiency in ISCU causes myopathy with lactic acidosis, a rare autosomal recessive hereditary disease found mainly in individuals of northern Swedish descent [406, 407]. Patients develop muscle weakness and experience severe activity-related muscle pain, associated with rhabdomyolysis (breakdown of skeletal muscle fibers with leakage of muscle contents into the circulation) and myoglobinuria (reddish urine caused by excretion of myoglobin), followed by muscle regeneration and temporary resolution of these symptoms [406, 408]. Even minor exertion causes markedly increased heart rate and palpitations, dyspnea, muscle fatigue, and lactic acidosis [406, 409]. Physiological investigations of these patients during exercise showed impaired muscle oxidative phosphorylation and low maximal muscle oxygen extraction associated with exaggerated circulatory responses. Biochemical studies indicated a deficiency in SDH [410] and aconitase activities [390, 411] and the presence of iron-rich mitochondrial inclusions [390, 411] (Fig. 18.4). Modest deficiencies in respiratory complex I and the Rieske protein in complex III were also reported [411].

Two independent studies revealed that the ISCU myopathy is caused by a homozygous intronic point mutation (g.7044 G>C) in intron 4 of *ISCU* gene, which activates a cryptic splice site, resulting in the retention of a ~100-base pair fragment of intronic sequence [390, 412]. A third study reported a more severe progressive

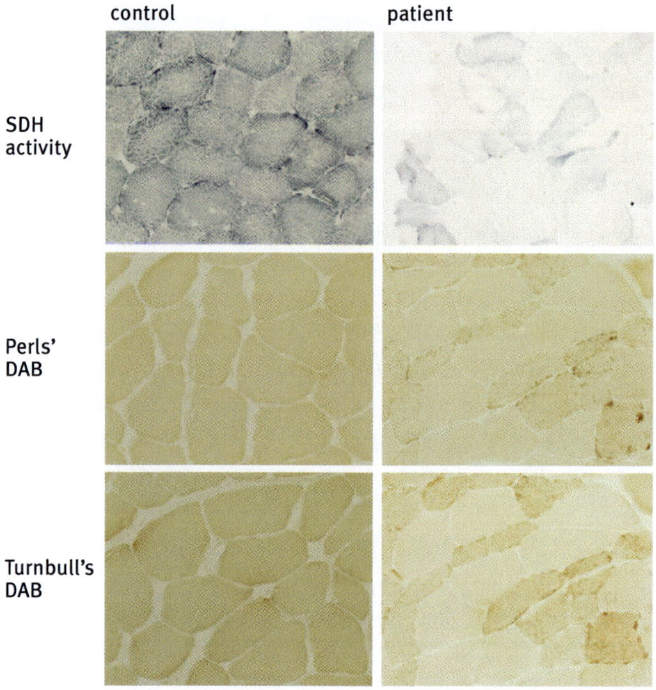


**Fig. 18.4:** Histochemistry indicated the loss of SDH activity and an increase in mitochondrial iron levels in the skeletal muscle of patients with ISCU myopathy. (a) Skeletal muscles from healthy individuals show robust SDH activity staining (blue), whereas patient muscles show very little SDH activity. (b) Perls' Prussian blue staining enhanced with 3,3'-diaminobenzidine (Perls/DAB), a histochemical test for nonheme ferric iron, indicated iron overload (brown punctate stains) in the muscle fibers of patients with ISUC myopathy. The punctate distribution of the iron staining was consistent with the mitochondrial iron overload previously detected in ultrastructural studies of these patients. (c) Turnbull staining enhanced with DAB, a stain for nonheme ferrous iron, indicated an increase in ferrous iron in the skeletal muscle of ISCU myopathy patients. (Courtesy of Karen Ayyad and Ron Haller, University of Texas Southwestern Medical Center and Veterans Administration North Texas Medical Center.)

myopathy associated with hypertrophic cardiomyopathy that is caused by the g.7044 G>C mutation on one allele in combination with a heterozygous missense allele (c.149G>A) in *ISCU* exon 3 [413]. The incorporation of a premature translational stop codon in the aberrantly spliced ISCU mRNA gave rise to a truncated ISCU protein, and pulse-chase experiments in patient myoblasts demonstrated that the truncated ISCU protein is rapidly degraded [414]. Gene expression analysis on muscle biopsies from patients with the homozygous g.7044 G>C mutation revealed expression of the mitochondrial iron importer mitoferrin 2 [415], suggesting that increased mitochondrial iron uptake may contribute to the mitochondrial iron overload seen in these patients [390, 411]. Gene expression analysis and biochemistry study also showed alterations of several key pathways involved in muscle fiber composition, fatty acid metabolism, and ketogenesis and induced expression of the starvation response hormone, FGF-21 [415], indicating a metabolic response to energy deprivation in the muscle tissue on the organismal level [416].

The g.7044 G>C mutation that underlies ISCU myopathy has the potential to cause aberrant splicing of ISCU in all cells and cause catastrophic loss of Fe-S cluster biogenesis in all tissues. Yet the patients with homozygous g.7044 G>C mutation all present with muscle-specific pathology. Several studies have attempted to identify the molecular features that may contribute to the tissue specificity and localized clinical phenotypes of ISCU myopathy. The ratio between aberrantly spliced to normal ISCU transcripts in patient muscle biopsies was markedly higher than those in patient myoblasts, fibroblasts, and heart tissue [283, 417]. Furthermore, Nordin *et al.* identified several muscle-specific cellular RNA-binding factors that facilitate the splicing of the mutant ISCU mRNA [418], suggesting that tissue-specific differences in RNA processing give rise to muscle-specific pathology and that disease will only manifest in tissues where the level of normal transcripts is below a certain threshold level. Interestingly, Crooks *et al.* showed that heterologous expression of the muscle-specific transcription factor MyoD1 decreased the amount of normal ISCU mRNA expressed in patient myoblasts, further suggesting that altered expression of RNA splicing factors during the process of terminal muscle differentiation enhances aberrant ISCU mRNA splicing [414]. In addition, they observed a decrease in ISCU protein levels in cells under oxidative stress and proposed that increased production of ROS during exercise may further contribute to a decrease in ISCU protein levels in the muscle tissue and to the phenotype of exercise-induced rhabdomyolysis in ISCU myopathy patients [414].

### 18.4.5 NUBPL mutations cause childhood-onset mitochondrial encephalomyopathy and respiratory complex I deficiency (MIM252010)

Mitochondrial respiratory complex I (NADH:ubiquinone oxidoreductase; EC 1.6.5.3) is the main entry point to the mitochondrial respiratory chain and catalyzes the transfer of electrons from NADH to ubiquinone. Complex I is a large ~1-MDa macromolecule composed of 45 protein subunits encoded by both the nuclear and the mitochondrial genomes. Defects in complex I activity are the most common type of mitochondrial disease [419], can present at a young age, and often result in multisystem disorders with a fatal outcome [93, 94]. The wide range of clinical manifestations include Leigh syndrome (an early-onset, fatal neurodegenerative disorder), dystonia, developmental delay, seizures, respiratory irregularities, cardiomyopathy, skeletal muscle myopathy, hypotonia, stroke, ataxia, lactic acidosis, and hepatopathy with renal tubulopathy. Mutations underlying human complex I deficiency have been identified in 19 of the subunits of the complex and 7 nuclear-encoded accessory factors that are required for the assembly, maturation, or stability of complex I [420, 421].

A high-throughput sequencing project identified several mutations in NUBPL (also known as IND1) in an individual who presented at 2 years of age with developmental delay, leukodystrophy, elevated CSF lactate, and complex I deficiency [422]. The patient carried one *NUBPL* allele harboring a deletion that spans exons 1–4 and a second allele that harbors both a c.815-27T>C mutation that probably causes exon 10 skipping and a p.Gly56Arg missense mutation [422, 423]. Functional studies showed that expression of wild-type NUBPL rescued complex I activity in fibroblasts from the patients, establishing NUBPL as the causal gene [422]. The c.815-27T>C branch-site mutation was subsequently found in six other cases, usually in combination with a null allele [424, 425]. RT-PCR and protein analysis indicated that the branch-site mutation led to a decrease in NUBPL1 mRNA and/or protein levels [423, 425, 426].

Deletion of the *IND1* in the respiratory yeast *Yarrowia lipolytica* carrying an alternative NADH dehydrogenase resulted in slower growth and strongly decreased complex I activity, whereas the activities of aconitase, SDH, and cytochrome bc1 complex were not affected, leading to the suggestion that Ind1 is specifically involved in the assembly/stability of complex I [257]. Knockdown of NUBPL in human HeLa cells caused decreases in several complex I subunits (NDUFS1, NDUFV1, NDUFS3, and NDUFA13), improper assembly of the peripheral arm of complex I, decreased complex I activity, and abnormal mitochondrial morphology [258]. Complex I contains eight Fe-S clusters that are associated with five different subunits (NDUFS1, NDUFS7, NDUFS8, NDUFV1, and NDUFV2) [427]. NDUFS7 and NDUFS8 assemble relatively early in the assembly pathway, and defects in these subunits usually do not result in accumulation of assembly intermediates. Kevelam *et al.* observed no accumulation of assembly intermediates of the peripheral arm in the patients and concluded that NUBPL is involved in early assembly of the Fe-S clusters [425]. NUBPL shows sequence similarity to Nbp35 and Cfd1 in yeast, which are P-loop NTPases that are involved in cytosolic Fe-S protein maturation [428]. Recombinant NUBPL/IND1 binds a labile [4Fe-4S] cluster, as judged by UV-vis spectroscopies, and the Fe-S cluster could be transferred to an Fe-S apo-protein *in vitro* [258, 428]. These studies led to the proposal that Ind1 serves as a specific scaffold or transfer protein for the assembly of the Fe-S clusters in respiratory complex I. However, more recent studies designed to unravel primary from secondary phenotypes in *Arabidopsis IND1* mutants suggested that IND1 has a primary role in mitochondrial translation that indirectly but specifically affects the assembly of complex I [429].

### 18.4.6 Mutations in NFU1 cause multiple mitochondrial dysfunctions syndrome 1 (MIM 605711)

Multiple mitochondrial dysfunctions syndrome, a fatal neonatal autosomal recessive disorder characterized by weakness, respiratory failure, developmental delay, and metabolic acidosis with elevated blood lactate levels, was first described in two families in 2001 [430]. Biochemical analysis indicated severe deficiencies of pyruvate dehydrogenase (PDH) complex, oxoglutarate dehydrogenase (OGDH) complex (also known as α-ketoglutarate dehydrogenase complex), and complexes I, II, and III of the respiratory chain. Notably, ACO2 activity was normal. Metabolite analysis performed on the patient urine samples revealed high levels of glycine, leucine, valine, and isoleucine, indicating abnormalities of the glycine cleavage system and the branched-chain α-keto-acid dehydrogenase [430]. Using the technique of microcell-mediated chromosome transfer, the genetic defect in three patients in one family was mapped to chromosome 2p14-p13 [430] and subsequently mapped to NFU1 [251, 431], a gene with sequence homology to the C-terminal domain of *A. vinelandii* Fe-S cluster scaffold protein NifU. A homozygous missense mutation c.545G>A near the splice donor site of exon 6 in NFU1 causes abnormal mRNA splicing, leading to the production of several truncated transcripts, and no detectable NFU1 protein product [431]. Transduction of fibroblast lines with retroviral vectors expressing the mitochondrial isoform of NFU1 restored respiratory chain function and oxoacid dehydrogenase complexes, confirming the pathogenicity of the NFU1 mutation. Immunostaining using lipoate antibodies indicated a severe reduction in lipoylated E2 proteins of the PDH and OGDH complexes [255, 431], suggesting that the loss of NFU1 might be specifically affecting the Fe-S enzyme LIAS [432].

A link between NFU1 defects and severe mitochondrial disorders was also independently identified in ten individuals with fatal infantile encephalopathy, pulmonary hypertension, hyperglycemia, lactic acidosis, and decreased glycine cleavage system, and PDH complex functions [255]. Nine of these individuals were homozygous for a G208C mutation in NFU1, and the tenth was compound heterozygous for a G208C mutation and a splice-site (c.545fl5G>A) mutation. G208 is highly conserved and is only one residue away from the Fe-S cluster-binding motif. *In vitro* studies confirmed that the G208C mutation of yeast Nfu1 leads to its functional impairment. Depletion or inactivation of NFU1 resulted in decreased SDHA and SDHB subunits of complex II and an assembly defect in complex II [255, 431, 433]. Interestingly, fibroblasts from NFU1 patients showed heterogeneous patterns of respiratory chain alterations, with isolated complex II deficiency in some patients [255] or combined deficiencies of complexes I, II, and III in others [431].

Several of the biochemical phenotypes observed in the NFU1 patients suggested that loss of NFU1 impaired LIAS function [255, 431]. Although the PDH complex, OGDH complex, and glycine cleavage system do not contain Fe-S clusters, these enzyme complexes all require a covalently linked lipoic acid moiety for their enzyme functions, and mutations in LIAS have been reported to impair PDH complex and glycine cleavage system activities [104]. Quantitative analysis for iron and sulfide on the *E. coli* LIAS, LipA, have shown that the active form of LipA contains two [4Fe-4S] clusters per polypeptide [434]. One of the Fe-S clusters binds S-adenosyl-L-methionine (SAM) and is essential for the reductive cleavage of SAM to generate a methionine and a 5′-deoxyadenosyl 5′-radical (5′-dA•). The second Fe-S cluster is sacrificed during catalytic turnover as the source of the two sulfur atoms inserted into the substrate octanoic acid to form lipoic acid [8]. Thus, the Fe-S cluster biogenesis pathway is important for the activities of lipoate-dependent enzymes, not only for the generation of 5′-dA• for the catalytic activity of LIAS but also as sulfur donors for the formation of lipoic acid (see Chapter 9, by Lanz and Booker, for further discussion of LIAS).

*A. vinelandii* NifU consists of three distinct domains: the N-terminal ISCU-like domain, the central FDX-like domain, and the C-terminal NFU-like domain [435]. The NFU domain contains a CXXC motif, and proteins containing the NFU domain are present in bacteria, yeast, *Arabidopsis thaliana* and humans [251, 253, 254, 256, 436, 437]. Depletion of NfuA in *E. coli* and in *A. vinelandii* resulted in increased sensitivity to oxidative stress [253]. In cyanobacteria, the NFU homologue is essential for viability [254, 438]. In *A. thaliana, Nfu2* gene disruption resulted in a dwarf phenotype, with impairment in the assembly of [4Fe-4S] and [2Fe-2S] FDX clusters, whereas Rieske and [3Fe-4S] glutamate synthase clusters were not affected [436, 437].

Early studies of Nfu homologues suggested that these proteins may serve as a scaffold for Fe-S cluster assembly. The protein containing Nfu domain in cyanobacterium *Synechocystis* PCC6803 was found by UV-vis spectroscopy to bind a [2Fe-2S] cluster and transfer it to apo-FDX [250, 436]. Mössbauer spectroscopic studies showed that nfuA from *Synechococcus* PCC7002 assembles and transfers [4Fe-Fs] clusters to apo-PsaC in photosystem I [252], whereas *E. coli* and *A. vinelandii* NfuA bind a [4Fe-4S] cluster, which is transferred to apo-aconitase [253, 254]. UV-visible spectral analysis suggested that *A. thaliana* NFU-like proteins are capable of binding a labile [2Fe-2S] cluster *in vitro* and transferring them to apo-FDX [439]. Mössbauer spectroscopy analysis indicated that human NFU1 is able to assemble a [4Fe-4S] cluster [251]. Liu and Cowan [440] reported that human NFU forms a complex with the cysteine desulfurase NifS and mediates persulfide bond cleavage of sulfur-loaded NifS persulfide and suggested that human NFU1 mediates sulfide delivery to ISCU in the final step of [2Fe-2S] cluster assembly. However, a different model has been proposed, which suggests that Fe-S clusters initially formed on the IscU type of scaffolds are subsequently transferred to Nfu-type proteins. In this model, NfuA-type proteins serve as intermediate Fe-S cluster carriers to deliver Fe-S cluster to apo-targets [254]. In support of this proposal, siRNA-mediated depletion of ISCU resulted in marked decreases in the activity of SDH, mitochondrial and cytosolic aconitases, and cytosolic GPAT and decreases in the levels of lipoylated a-KGDH, PDH, and GCS, whereas depletion of NFU1 affected a subset of these proteins (lipoylated proteins and SDH) [255]. Furthermore, apo-NfuA in *E. coli* is able to accept an Fe-S cluster from holo ISCU or holo-SufBC2D, whereas no cluster transfer was observed between holo-NfuA and apo-ISCU or apo-SufBC2D, suggesting that NfuA is not a scaffold but rather an Fe-S cluster carrier [256]. Taken together, these studies support the idea that NFU1 acts downstream of ISCU and may be involved in transferring newly synthesized Fe-S clusters from ISCU to specific apo-target proteins.

### 18.4.7 Mutations in BOLA3 cause multiple mitochondrial dysfunctions syndrome 2 (MIM 614299)

A mutation in *BOLA3* was first described in an infant who developed epileptic seizures, dilated cardiomyopathy, encephalopathy, respiratory distress, hepatomegaly, acidosis, with elevated levels of glycine in his serum and cerebrospinal fluid, and subsequently died at 11 months of age [431]. Studies of patient skin fibroblasts showed elevated lactate-to-pyruvate ratios and decreased activities of the PDH complex, branched chain α-keto-acid dehydrogenase, and mitochondrial respiratory chain complexes I and II. In contrast, the activity of ACO2 was normal. A single base-pair duplication c.123dupA was identified in exon 2 of *BOLA3*, which causes a frameshift that produces a premature stop codon in both isoforms of *BOLA3*. A second study soon followed, reporting the case of two siblings who presented with severe neonatal lactic acidosis, hypotonia, respiratory insufficiency, and intractable cardiomyopathy [441]. Both patients died within the first months of life due to multiorgan failure. Organic acid analyses of urine revealed elevated metabolites of the TCA cycle (succinate, fumarate, malate, aconitic acid, citric acid) together with increased excretion of lactate and pyruvate. Assessment of skeletal muscle biopsies and fibroblasts showed combined deficiency of respiratory chain complexes I and II accompanied by a defect of the PDH complex. A decrease of complex IV activity was also observed in the muscle biopsy of one of the patients. Brain MRI indicated lesions in several areas in the brain. The identified homozygous missense mutation, c.200T>A I67N, is located in exon 3 and solely affects isoform I of *BOLA3*.

Little is known about the exact function of human BOLA3. Mammalian cells have three BOLA homologues, denoted as BOLA1, BOLA2, and BOLA3. On the basis of functional genomics data, affinity purification, yeast two-hybrid studies, and structural characterizations, BOLA family members were postulated to act as reductases, interacting with the monothiol glutaredoxin family that includes the cytosolic GLRX3 and GLRX4 and the mitochondrial GLRX5 [442]. Genetic studies in yeast demonstrate a crucial role for complexes between Grx3 and Grx4 with the BolA-like protein Fra2 in iron homeostasis and that the Grx-BolA interaction is required for efficient iron-dependent inhibition of the Aft1 and Aft2 transcriptional activators [443] (see Chapter 16, by Outten, for an in-depth discussion). Studies in zebrafish also indicated a crucial function of vertebrate Grx3 (PICOT) in iron homeostasis and hemoglobin maturation [444]. UV-vis absorption, CD, resonance Raman, EPR, ENDOR, Mössbauer, and EXAFS studies showed that yeast Fra2 forms a [2Fe-2S]-bridged heterodimeric complex with Grx3 or Grx4, with iron ligands provided by a cysteine from Grx3 or Grx4, a histidine from Fra2, and glutathione [381]. This complex plays a key role in iron regulation in *S. cerevisiae* [443, 445]. Based on analogy to the cytosolic Fra2-Grx3/4 complex, it has been suggested that BOLA3 possibly functions by interacting with GLRX5 in the mitochondria.

### 18.4.8 IBA57 deficiency causes severe myopathy and encephalopathy

IBA57 was first identified as an Fe-S protein biogenesis factor in a genome-wide screen for yeast mutants that are auxotrophic for glutamate and lysine due to defects in ACO2 and homoaconitase [248]. Depletion of yeast Iba57 results in diminished *de novo* Fe-S cluster formation on aconitase and homoaconitase, but not in FDX (Yah1) or isopropylmalate dehydratase (Leu1p). Depletion of Iba57 also resulted in loss of catalytic function of radical SAM Fe-S proteins biotin synthase and LIAS. In *E. coli*, suppression of an IBA57-related protein, YgfZ, resulted in deficiency in Fe-S proteins SDH and tRNA modification enzyme MiaB, but not aconitase [446]. In IBA57-depleted HeLa cells, catalytic activities of complexes I and II were compromised, and steady-state levels of several Fe-S proteins, including subunits of complex I (NDUFA9, NDUFA13, NDUFB4, and NDUFS3), complex II (SDH), and complex IV (MT-CO2) were affected [447]. Notably, the activity of the [2Fe-2S] cluster-containing complex III was not affected in these individuals, consistent with the suggestion that IBA57 is specific for [4Fe-4S] cluster formation [248, 447]. Yeast Iba57 physically interacts with the Fe-S cluster assembly proteins Isa1 and Isa2 [248], and *E. coli* YgfZ occurs in complexes with IscA-type proteins *in vivo* [448], suggesting that the complex of these three proteins forms the functional unit [248, 249].

Mutations in IBA57 were found in two siblings who became critically ill shortly after birth [449]. Patient 1 presented with severe hypotonia, absent primitive reflexes, microcephaly, dysmorphic features, hyperglycemia, elevated lactate and glycine levels in serum and cerebrospinal fluid, but normal blood cell counts and hemoglobin and hematocrit measurements. Cerebral MRI showed cerebral atrophy. Patient 2 died shortly after birth due to hypoventilation and low cardiac activity. Biochemical studies revealed decreased catalytic activities of complexes I and II, whereas the activity of complex III was normal. Immunoblot analysis showed severely decreased levels of complexes I, II, and IV subunits in skeletal muscle, whereas levels of complexes III and V were unchanged. In addition, defects in ACO2 and the lipoate-containing enzymes, OGDH and PDH, were also found in these patients. Sequence analysis revealed a homozygous mutation, c.941A>C, resulting in a change of the residue G314 to a Pro. Analysis of mitochondrial extracts from skeletal muscle and cultured skin fibroblasts, and IBA57-depleted HeLa cells indicated that the mutation resulted in a severe decrease in IBA57 protein due to proteolytic degradation. The lipoylated groups in PDH and OGDH were significantly decreased in cultured skin fibroblasts and skeletal muscle tissue of the patients. These findings are consistent with the proposal that IBA57 is important for [4Fe-4S] cluster assembly in ACO2 and the radical SAM enzyme, LIAS.

### 18.4.9 A mutation in ISD11 causes deficiencies of respiratory complexes

LYRM4 encodes the ISD11 protein that is essential for cell viability in yeast [224]. Yeast strains lacking Isd11 are deficient in aconitase and SDH activities [224, 225], and knockdown of ISD11 results in aconitase deficiency in human cell lines [226, 450], implicating an essential role of ISD11 in Fe-S cluster assembly. Pull-down and co-immunoprecipitation experiments showed that ISD11 forms a complex with the cysteine desulfurase NFS1 in yeast and human cells [224, 225, 450]. In yeast without Isd11, it was initially reported that Nfs1 had normal desulfurase activity but was prone to aggregation and proteolytic degradation [224]. However, other studies suggested that Nfs1 by itself was inactive and that Isd11 is important for inducing an activating conformational change in Nfs1 [227]. In addition, Shi *et al.* [226] showed that ISD11 depletion in human cell lines activated the iron regulatory proteins IRP1 and IRP2, which led to upregulation of TfR1 and mitochondrial iron accumulation. These results suggest that ISD11 is important in the biogenesis of Fe-S clusters, and loss of ISD11 disrupts normal mitochondrial and cytosolic iron homeostasis. In addition to sequestering the sulfur for Fe-S cluster biogenesis, the Nfs1•ISD11 complex might be important for other biological pathways that require sulfur transfer, including the biogenesis pathways of biotin, thiamine, lipoic acid, molybdopterin, and sulfur containing bases in tRNA [451–453].

A homozygous mutation in LYRM4 was recently identified *via* massive parallel sequencing of >1,000 mitochondrial genes in two patients with deficiencies of the respiratory complexes containing Fe-S clusters [454]. Patient 1 had stridor (a high-pitched wheezing sound resulting from turbulent air flow in the upper airway), hypotonia, and lactic acidosis shortly after birth, but improved later. Patient 2 had stridor, hepatomegaly, metabolic acidosis, severe ketosis, and died at 2 months of age. Postmortem muscle histology of patient 2 revealed reduced glycogen levels and increased lipid levels. Liver analysis showed steatosis (indicative of impairment of lipid metabolism), markedly reduced glycogen, and mildly increased iron levels. The mutation in *LYRM4* was predicted to cause a missense change affecting a highly conserved amino acid residue in ISD11, and ISD11 protein was undetectable in patient muscle and liver biopsies. Levels of complex I subunits NDUFB8 and NDUFS3, complex II subunit SDHB, complex III subunit UQCRFS1, and complex IV subunits COX2 and COX1 were reduced in both patients. Other Fe-S proteins, including ACO1, ACO2, and FECH were also reduced in the patients.

## 18.5 Fe-S cluster biogenesis and iron homeostasis

Another important finding from the studies of genes involved in Fe-S cluster biogenesis is that Fe-S cluster biogenesis is important for the regulation of mitochondrial iron homeostasis [213, 336]. Many yeast strains that were depleted of Fe-S cluster biogenesis proteins (e.g. *yfh1, nfs1, isu1, isu2, isa1, isa2, nfu1, ssq1, jac1, yah1*) showed marked iron accumulation in their mitochondria [235, 278, 324, 455–457]. Disruption of intracellular iron homeostasis is also a prominent feature in human patients depleted of FXN, ABCB7, GLRX5, and ISCU [245, 335, 390] (Fig. 18.4) and in human cell lines depleted of ISCU, ABCB7, GLRX5, ISD11, and FDX *via* RNA silencing [226, 232, 260, 359, 384]. The increased mitochondrial iron uptake is facilitated by the mitochondrial iron transporters, mitoferrin 1 (in erythroid cells) and mitoferrin 2 (in non-erythroid cells) [338, 458]. In yeast, depletion of components of the mitochondrial Fe-S assembly systems induced strong transcriptional responses of many genes, and these alterations were very similar to the transcriptional profiles developed upon iron starvation, including many that are involved in cellular iron uptake [383, 459, 460]. In human cells, disruptions of Fe-S cluster biogenesis and mitochondrial iron homeostasis also activate iron regulatory proteins IRP1 and IRP2 [260, 270, 271, 287, 341], which triggers an increase in cellular iron uptake *via* TfR1 and a decrease in iron sequestration by ferritin. Together, these results suggest that an Fe-S protein might function as a sensor for mitochondrial iron status or in signaling the status of mitochondrial iron stores to the nucleus. In this scenario, dysfunction in Fe-S cluster biogenesis would be registered by the cell as mitochondrial iron deficiency, which triggers an increase in mitochondrial iron uptake *via* mitoferrin [338, 415], and iron accumulates in amorphous nanoparticles of ferric phosphate in the mitochondrial matrix [339, 340] or within mitochondrial ferritin [335]. In addition, cytosolic iron pools might become functionally depleted because of mitochondrial iron sequestration, which triggers the activation of IRP1 and IRP2, resulting in upregulation of iron import through TfR1, decrease in ferroportin-mediated iron export, and decreased iron sequestration by iron storage protein ferritin [338]. The aberrant upregulation of TfR1 and increase in mitochondrial iron uptake has the potential to engage the cell in a vicious cycle in which increased cellular iron uptake further exacerbates mitochondrial iron overload.

Interestingly, mitochondrial iron overload in cells with defects in Fe-S cluster biogenesis can cause increased oxidative stress not only directly *via* iron-catalyzed oxidation reactions [324, 461] but also by inactivating MnSOD [462]. Although mitochondrial iron does not normally bind MnSOD, iron will misincorporate into *S. cerevisiae* MnSOD when mitochondrial iron homeostasis is disrupted in yeast mutants that have defects in the late stages of Fe-S cluster biogenesis (e.g. *grx5, ssq1, atm1*). Iron binding inactivates the MnSOD enzyme, presumably by causing changes in the redox potential at the active site or by blocking substrate access.

## 18.6 Therapeutic strategies

A greater understanding of the physiological roles of Fe-S proteins and the process of Fe-S cluster biogenesis, together with technological advances in genetic diagnosis in recent years, have led to the discovery of disease-causing mutations in a number of Fe-S proteins and Fe-S cluster biogenesis factors and provided the opportunities to design more effective diagnostics and potential treatments for several devastating diseases [294, 463]. Some therapeutic approaches target the specific disease mutation and disease symptoms. For instance, palliative treatments of patients with FRDA typically consist of physical therapy and β-blockers, ACE inhibitors, and surgery for cardiac symptoms [464]. In the case of patients with mutations in LYRM4/ISD11, the realization that the disease gene encodes a protein that is important for the enzymatic activity of a cysteine desulfurase led to the suggestion that providing a sulfur donor such as cysteine or *N*-acetyl cysteine may constitute a potential therapy [454]. Additional therapeutic approaches involve agents that aim to boost the expression levels of the disease genes. In the case of FRDA, EPO has been shown to promote the translation of FXN [465, 466], and histone deacetylase inhibitors have been shown to reverse FXN silencing in FRDA cells [467, 468]. In the case of ISCU myopathy, *in vitro* studies have demonstrated that RNA modulating therapy with an antisense phosphorodiamidate morpholino oligonucleotide that specifically targets the aberrantly activated splice site in ISCU is able to restore the normal splicing pattern of ISCU [469]. An upregulation of normally spliced ISCU mRNA produced by blocking the cryptic splice site may thus be a therapeutic possibility for these patients.

On the one hand, because mitochondrial iron overload and mitochondrial failure are common features in diseases associated with defects in Fe-S proteins and Fe-S cluster biogenesis, several common therapeutic strategies have emerged that involve administering antioxidants to reduce oxidative damages and iron chelators to decrease iron-mediated oxidative stress and attenuate iron overload. For instance, adenoviral delivery of the human SOD2 gene was shown to suppress the optic nerve degeneration in a mouse model of complex I deficiency [470], suggesting that antioxidant therapy may attenuate the disease process in patients with defective respiratory complexes. Patients with FRDA also have increased DNA damage [343] and lipid peroxidation [471], and much work has been done to evaluate the potential of antioxidants in preventing mitochondrial damage and preserving aerobic respiration. Studies by the Schoumacher laboratory suggested that small-molecule glutathione peroxidase mimetics have the potential to treat FRDA [472]. On the other hand, neither the administration of a MnSOD mimetic (MnTBAP) nor the overexpression of CuZnSOD improved the cardiomyopathy symptoms in a murine FRDA model [341]. Idebenone, a CoQ10 analogue that can reduce intracellular ROS as well as shuttle electrons between damaged respiratory complex proteins, showed promise in animal models [346, 473] and early clinical studies [342–344] but failed the phase III clinical trial because it did not significantly improve cardiac outcomes in patients over a 6-month treatment period [345]. It has been suggested that antioxidants may prove to be more effective if given at an early stage of disease progression before significant neuronal and cardiomyocyte damage accumulates.

The consistent finding that dysregulation of iron homeostasis is a pathological hallmark of FRDA has led to efforts in testing iron chelators for use in the treatment of FRDA [463, 474]. Iron chelators, including deferoxamine and deferiprone, have been evaluated in *in vitro* models and in clinical trials, with mixed results. Although iron chelators reduced ROS damage to mitochondrial proteins, and reduced iron buildup in the brain with improvements in neurological function in a number of studies [346, 475, 476], other studies indicated that iron chelation has the undesirable effects of reducing the mRNA levels of FXN and aconitase and in impairing aconitase function [287, 477].

In most of the diseases described in this chapter, pathology resulted from loss of function of the disease gene. In the future, it may be feasible to overcome the disease by expressing a nonmutated gene using viral gene therapy, a field that is advancing rapidly [478].

## Acknowledgments

I thank Drs Karen Ayyad and Ron Haller for generously providing the SDH activity and iron stainings of skeletal muscle from patients with ISCU deficiency and Dr Daniel R. Crooks for critical reading of the manuscript.

## References

[1] Beinert H, Holm RH, Munck E. Iron-sulfur clusters: nature's modular, multipurpose structures. Science 1997;277:653–9.
[2] Johnson DC, Dean DR, Smith AD, et al. Structure, function, and formation of biological iron-sulfur clusters. Annu Rev Biochem 2005;74:247–81.
[3] Beinert H, Kennedy MC, Stout CD. Aconitase as Iron-Protein, Enzyme, and Iron-Regulatory Protein. Chem Rev 1996;96:2335–74.
[4] White MF, Dillingham MS. Iron-sulphur clusters in nucleic acid processing enzymes. Curr Opin Struct Biol 2012;22:94–100.
[5] Rudolf J, Makrantoni V, Ingledew WJ, et al. The DNA repair helicases XPD and FancJ have essential iron-sulfur domains. Mol Cell 2006;23:801–8.
[6] Wu Y, Brosh RM. DNA helicase and helicase-nuclease enzymes with a conserved iron-sulfur cluster. Nucleic Acids Res 2012;40:4247–60.
[7] Netz DJ, Stith CM, Stumpfig M, et al. Eukaryotic DNA polymerases require an iron-sulfur cluster for the formation of active complexes. Nat Chem Biol 2012;8:125–32.
[8] Lanz ND, Booker SJ. Identification and function of auxiliary iron-sulfur clusters in radical SAM enzymes. Biochim Biophys Acta 2012;1824:1196–212.
[9] Rouault TA. The role of iron regulatory proteins in mammalian iron homeostasis and disease. Nat Chem Biol 2006;2:406–14.
[10] Hentze MW, Muckenthaler MU, Galy B, et al. Two to tango: regulation of Mammalian iron metabolism. Cell 2010;142:24–38.
[11] Anderson CP, Shen M, Eisenstein RS, et al. Mammalian iron metabolism and its control by iron regulatory proteins. Biochim Biophys Acta 2012;1823:1468–83.
[12] Roberts RA, Smith RA, Safe S, et al. Toxicological and pathophysiological roles of reactive oxygen and nitrogen species. Toxicology 2010;276:85–94.
[13] Federico A, Cardaioli E, Da Pozzo P, et al. Mitochondria, oxidative stress and neurodegeneration. J Neurol Sci 2012;322:254–62.
[14] Imlay JA. Iron-sulphur clusters and the problem with oxygen. Mol Microbiol 2006;59:1073–82.
[15] Yarian CS, Toroser D, Sohal RS. Aconitase is the main functional target of aging in the citric acid cycle of kidney mitochondria from mice. Mech Ageing Dev 2006;127:79–84.

[16] Boehm DE, Vincent K, Brown OR. Oxygen and toxicity inhibition of amino acid biosynthesis. Nature 1976;262:418–20.
[17] Kuo CF, Mashino T, Fridovich I. Alpha, beta-Dihydroxyisovalerate dehydratase. A superoxide-sensitive enzyme. J Biol Chem 1987;262:4724–7.
[18] Gardner PR, Fridovich I. Superoxide sensitivity of the Escherichia coli aconitase. J Biol Chem 1991;266:19328–33.
[19] Flint DH, Tuminello JF, Emptage MH. The inactivation of Fe-S cluster containing hydro-lyases by superoxide. J Biol Chem 1993;268:22369–76.
[20] Longo VD, Liou LL, Valentine JS, et al. Mitochondrial superoxide decreases yeast survival in stationary phase. Arch Biochem Biophys 1999;365:131–42.
[21] Wallace MA, Liou LL, Martins J, et al. Superoxide inhibits 4Fe-4S cluster enzymes involved in amino acid biosynthesis.Cross-compartment protection by CuZn-superoxide dismutase. J Biol Chem 2004;279:32055–62.
[22] Missirlis F, Hu J, Kirby K, et al. Compartment-specific protection of iron-sulfur proteins by superoxide dismutase. J Biol Chem 2003;278:47365–9.
[23] Kispal G, Sipos K, Lange H, et al. Biogenesis of cytosolic ribosomes requires the essential iron-sulphur protein Rli1p and mitochondria. EMBO J 2005;24:589–98.
[24] Alhebshi A, Sideri TC, Holland SL, et al. The essential iron-sulfur protein Rli1 is an important target accounting for inhibition of cell growth by reactive oxygen species. Mol Biol Cell 2012;23:3582–90.
[25] Yan LJ, Levine RL, Sohal RS. Oxidative damage during aging targets mitochondrial aconitase. Proc Natl Acad Sci USA 1997;94:11168–72.
[26] Van Remmen H, Richardson A. Oxidative damage to mitochondria and aging. Exp Gerontol 2001;36:957–68.
[27] Bulteau AL, Ikeda-Saito M, Szweda LI. Redox-dependent modulation of aconitase activity in intact mitochondria. Biochemistry 2003;42:14846–55.
[28] Delaval E, Perichon M, Friguet B. Age-related impairment of mitochondrial matrix aconitase and ATP-stimulated protease in rat liver and heart. Eur J Biochem 2004;271:4559–64.
[29] Li Y, Huang TT, Carlson EJ, et al. Dilated cardiomyopathy and neonatal lethality in mutant mice lacking manganese superoxide dismutase. Nat Genet 1995;11:376–81.
[30] Huang TT, Carlson EJ, Kozy HM, et al. Genetic modification of prenatal lethality and dilated cardiomyopathy in Mn superoxide dismutase mutant mice. Free Radic Biol Med 2001;31:1101–10.
[31] Armstrong JS, Whiteman M, Yang H, et al. The redox regulation of intermediary metabolism by a superoxide-aconitase rheostat. Bioessays 2004;26:894–900.
[32] Welter R, Yu L, Yu CA. The effects of nitric oxide on electron transport complexes. Arch Biochem Biophys 1996;331:9–14.
[33] Tinberg CE, Tonzetich ZJ, Wang H, et al. Characterization of iron dinitrosyl species formed in the reaction of nitric oxide with a biological Rieske center. J Am Chem Soc 2010;132:18168–76.
[34] Asanuma K, Iijima K, Ara N, et al. Fe-S cluster proteins are intracellular targets for nitric oxide generated luminally at the gastro-oesophageal junction. Nitric Oxide 2007;16:395–402.
[35] Tabrizi SJ, Cleeter MW, Xuereb J, et al. Biochemical abnormalities and excitotoxicity in Huntington's disease brain. Ann Neurol 1999;45:25–32.
[36] Sorolla MA, Reverter-Branchat G, Tamarit J, et al. Proteomic and oxidative stress analysis in human brain samples of Huntington disease. Free Radic Biol Med 2008;45:667–78.
[37] Martin WR, Clark C, Ammann W, et al. Cortical glucose metabolism in Huntington's disease. Neurology 1992;42:223–9.
[38] Antonini A, Leenders KL, Spiegel R, et al. Striatal glucose metabolism and dopamine D2 receptor binding in asymptomatic gene carriers and patients with Huntington's disease. Brain 1996;119:2085–95.

[39] Esposito G, Vos M, Vilain S, et al. Aconitase causes iron toxicity in Drosophila pink1 mutants. PLoS Genet 2013;9:e1003478.
[40] Sellers VM, Johnson MK, Dailey HA. Function of the [2FE-2S] cluster in mammalian ferrochelatase: a possible role as a nitric oxide sensor. Biochemistry 1996;35:2699–704.
[41] Crooks DR, Ghosh MC, Haller RG, et al. Posttranslational stability of the heme biosynthetic enzyme ferrochelatase is dependent on iron availability and intact iron-sulfur cluster assembly machinery. Blood 2010;115:860–9.
[42] Imlay JA. The molecular mechanisms and physiological consequences of oxidative stress: lessons from a model bacterium. Nat Rev Microbiol 2013;11:443–54.
[43] Zhang B, Crack JC, Subramanian S, et al. Reversible cycling between cysteine persulfide-ligated [2Fe-2S] and cysteine-ligated [4Fe-4S] clusters in the FNR regulatory protein. Proc Natl Acad Sci USA 2012;109:15734–9.
[44] Nicolet Y, Rohac R, Martin L, et al. X-ray snapshots of possible intermediates in the time course of synthesis and degradation of protein-bound Fe4S4 clusters. Proc Natl Acad Sci USA 2013;110:7188–92.
[45] Jang S, Imlay JA. Micromolar intracellular hydrogen peroxide disrupts metabolism by damaging iron-sulfur enzymes. J Biol Chem 2007;282:929–37.
[46] Crack JC, Jervis AJ, Gaskell AA, et al. Signal perception by FNR: the role of the iron-sulfur cluster. Biochem Soc Trans 2008;36:1144–8.
[47] Bruska MK, Stiebritz MT, Reiher M. Analysis of differences in oxygen sensitivity of Fe-S clusters. Dalton Trans 2013;42:8729–35.
[48] Breusch FL. Citric acid in tissue metabolism. Physiol Chem 1937;250:262–80.
[49] Tong WH, Rouault TA. Metabolic regulation of citrate and iron by aconitases: role of iron-sulfur cluster biogenesis. Biometals 2007;20:549–64.
[50] Denton RM, Randle PJ. Citrate and the regulation of adipose-tissue phosphofructokinase. Biochem J 1966;100:420–3.
[51] Randle PJ. Regulatory interactions between lipids and carbohydrates: the glucose fatty acid cycle after 35 years. Diabetes Metab Rev 1998;14:263–83.
[52] Munday MR. Regulation of mammalian acetyl-CoA carboxylase. Biochem Soc Trans 2002;30:1059–64.
[53] Saha AK, Ruderman NB. Malonyl-CoA and AMP-activated protein kinase: an expanding partnership. Mol Cell Biochem 2003;253:65–70.
[54] Rogina B, Reenan RA, Nilsen SP, et al. Extended life-span conferred by cotransporter gene mutations in Drosophila. Science 2000;290:2137–40.
[55] Knauf F, Mohebbi N, Teichert C, et al. The life-extending gene Indy encodes an exchanger for Krebs-cycle intermediates. Biochem J 2006;397:25–9.
[56] Fei YJ, Liu JC, Inoue K, et al. Relevance of NAC-2, an Na+-coupled citrate transporter, to life span, body size and fat content in Caenorhabditis elegans. Biochem J 2004;379:191–8.
[57] Belfiore F, Iannello S. Insulin resistance in obesity: metabolic mechanisms and measurement methods. Mol Gen Metab 1998;65:121–8.
[58] Wolfgang MJ, Lane MD. Control of energy homeostasis: role of enzymes and intermediates of fatty acid metabolism in the central nervous system. Annu Rev Nutr 2006;26:23–44.
[59] McGahan MC, Harned J, Mukunnemkeril M, et al. Iron alters glutamate secretion by regulating cytosolic aconitase activity. Am J Physiol Cell Physiol 2005;288:C1117–24.
[60] Sterling GH, O'Neill JJ. Citrate as the precursor of the acetyl moiety of acetylcholine. J Neurochem 1978;31:525–30.
[61] Caudarella R, Vescini F, Buffa A, et al. Citrate and mineral metabolism: kidney stones and bone disease. Front Biosci 2003;8:s1084–106.

[62] Costello LC, Franklin RB. The clinical relevance of the metabolism of prostate cancer; zinc and tumor suppression: connecting the dots. Mol Cancer 2006;5:17.
[63] Bullock GC, Delehanty LL, Talbot AL, et al. Iron control of erythroid development by a novel aconitase-associated regulatory pathway. Blood 2010;116:97–108.
[64] Talbot AL, Bullock GC, Delehanty LL, et al. Aconitase regulation of erythropoiesis correlates with a novel licensing function in erythropoietin-induced ERK signaling. PLoS One 2011;6:e23850.
[65] Cheng Z, Tsuda M, Kishita Y, et al. Impaired energy metabolism in a Drosophila model of mitochondrial aconitase deficiency. Biochem Biophys Res Commun 2013;433:145–50.
[66] Spiegel R, Pines O, Ta-Shma A, et al. Infantile cerebellar-retinal degeneration associated with a mutation in mitochondrial aconitase, ACO2. Am J Hum Genet 2012;90:518–23.
[67] Benn DE, Gimenez-Roqueplo AP, Reilly JR, et al. Clinical presentation and penetrance of pheochromocytoma/paraganglioma syndromes. J Clin Endocrinol Metab 2006;91:827–36.
[68] Bayley JP, Devilee P, Taschner PE. The SDH mutation database: an online resource for succinate dehydrogenase sequence variants involved in pheochromocytoma, paraganglioma and mitochondrial complex II deficiency. BMC Med Genet 2005;6:39.
[69] Cascon A, Pita G, Burnichon N, et al. Genetics of pheochromocytoma and paraganglioma in Spanish patients. J Clin Endocrinol Metab 2009;94:1701–5.
[70] Neumann HP, Bausch B, McWhinney SR, et al. Germ-line mutations in nonsyndromic pheochromocytoma. N Engl J Med 2002;346:1459–66.
[71] Lima J, Feijao T, Ferreira da Silva A, et al. High frequency of germline succinate dehydrogenase mutations in sporadic cervical paragangliomas in northern Spain: mitochondrial succinate dehydrogenase structure-function relationships and clinical-pathological correlations. J Clin Endocrinol Metab 2007;92:4853–64.
[72] Burnichon N, Rohmer V, Amar L, et al. The succinate dehydrogenase genetic testing in a large prospective series of patients with paragangliomas. J Clin Endocrinol Metab 2009;94:2817–27.
[73] Korpershoek E, Petri BJ, van Nederveen FH, et al. Candidate gene mutation analysis in bilateral adrenal pheochromocytoma and sympathetic paraganglioma. Endocr Relat Cancer 2007;14:453–62.
[74] Amar L , Baudin E, Burnichon N, et al. Succinate dehydrogenase B gene mutations predict survival in patients with malignant pheochromocytomas or paragangliomas. J Clin Endocrinol Metab 2007;92:3822–8.
[75] Nakahashi Y, Fujita H, Taketani S, et al. The molecular defect of ferrochelatase in a patient with erythropoietic protoporphyria. Proc Natl Acad Sci USA 1992;89:281–5.
[76] Sarkany RP, Alexander GJ, Cox TM. Recessive inheritance of erythropoietic protoporphyria with liver failure. Lancet 1994;343:1394–6.
[77] Schneider-Yin X, Gouya L, Dorsey M, et al. Mutations in the iron-sulfur cluster ligands of the human ferrochelatase lead to erythropoietic protoporphyria. Blood 2000;96:1545–9.
[78] Broughton BC, Steingrimsdottir H, Weber CA, et al. Mutations in the xeroderma pigmentosum group D DNA repair/transcription gene in patients with trichothiodystrophy. Nat Genet 1994;7:189–94.
[79] Botta E, Nardo T, Broughton BC, et al. Analysis of mutations in the XPD gene in Italian patients with trichothiodystrophy: site of mutation correlates with repair deficiency, but gene dosage appears to determine clinical severity. Am J Hum Genet 1998;63:1036–48.
[80] Cantor SB, Guillemette S. Hereditary breast cancer and the BRCA1-associated FANCJ/BACH1/BRIP1. Future Oncol 2011;7:253–61.
[81] Tong WH, Sourbier C, Kovtunovych G, et al. The glycolytic shift in fumarate-hydratase-deficient kidney cancer lowers AMPK levels, increases anabolic propensities and lowers cellular iron levels. Cancer Cell 2011;20:315–27.

[82] Tomlinson IP, Alam NA, Rowan AJ, et al. Germline mutations in FH predispose to dominantly inherited uterine fibroids, skin leiomyomata and papillary renal cell cancer. Nat Genet 2002;30:406–10.
[83] Isaacs JS, Jung, YJ, Mole DR, et al. HIF overexpression correlates with biallelic loss of fumarate hydratase in renal cancer: novel role of fumarate in regulation of HIF stability. Cancer Cell 2005;8:143–53.
[84] Pollard PJ, Briere JJ, Alam NA, et al. Accumulation of Krebs cycle intermediates and over-expression of HIF1alpha in tumours which result from germline FH and SDH mutations. Hum Mol Genet 2005;14:2231–9.
[85] Sudarshan S, Sourbier C, Kong HS, et al. Fumarate hydratase deficiency in renal cancer induces glycolytic addiction and hypoxia-inducible transcription factor 1alpha stabilization by glucose-dependent generation of reactive oxygen species. Mol Cell Biol 2009;29:4080–90.
[86] Ternette N, Yang M, Laroyia M, et al. Inhibition of mitochondrial aconitase by succination in fumarate hydratase deficiency. Cell Rep 2013;3:689–700.
[87] DeBerardinis RJ, Lum JJ, Hatzivassiliou G, et al. The biology of cancer: metabolic reprogramming fuels cell growth and proliferation. Cell Metab 2008;7:11–20.
[88] Galaris D, Pantopoulos K. Oxidative stress and iron homeostasis: mechanistic and health aspects. Crit Rev Clin Lab Sci 2008;45:1–23.
[89] Andrews NC. Forging a field: the golden age of iron biology. Blood 2008;112:219–30.
[90] Pantopoulos K, Hentze MW. Rapid responses to oxidative stress mediated by iron regulatory protein. EMBO J 1995;14:2917–24.
[91] Drapier JC, Hirling H, Wietzerbin J, et al. Biosynthesis of nitric oxide activates iron regulatory factor in macrophages. EMBO J 1993;12:3643–9.
[92] Weiss G, Goossen B, Doppler W, et al. Translational regulation via iron-responsive elements by the nitric oxide/NO-synthase pathway. EMBO J 1993;12:3651–7.
[93] Smeitink J, van den Heuvel L, DiMauro S. The genetics and pathology of oxidative phosphorylation. Nat Rev Genet 2001;2:342–52.
[94] Papa S, De Rasmo D. Complex I deficiencies in neurological disorders. Trends Mol Med 2013;19:61–9.
[95] Hoekstra AS, Bayley JP. The role of complex II in disease. Biochim Biophys Acta 2013;1827:543–51.
[96] Benit P, Lebon S, Rustin P. Respiratory-chain diseases related to complex III deficiency. Biochim Biophys Acta 2009;1793:181–5.
[97] Pestronk A. Immune and inflammatory myopathies: pathology. Neuromuscular Disease Center. (Accessed at http://neuromuscular.wustl.edu/mitosyn.html-leigh); 2011.
[98] Koopman WJ, Distelmaier F, Smeitink JA, et al. OXPHOS mutations and neurodegeneration. EMBO J 2013;32:9–29.
[99] Olsen RK, Olpin SE, Andresen BS, et al. ETFDH mutations as a major cause of riboflavin-responsive multiple acyl-CoA dehydrogenation deficiency. Brain 2007;130:2045–54.
[100] Balwani M, Desnick RJ. The porphyrias: advances in diagnosis and treatment. Blood 2012;120:4496–504.
[101] Ghosh MC, Zhang DL, Jeong SY, et al. Deletion of iron regulatory protein 1 causes polycythemia and pulmonary hypertension in mice through translational derepression of HIF2alpha. Cell Metab 2013;17:271–81.
[102] Anderson SA, Nizzi CP, Chang YI, et al. The IRP1-HIF-2alpha axis coordinates iron and oxygen sensing with erythropoiesis and iron absorption. Cell Metab 2013;17:282–90.
[103] Van Kuilenburg AB, Vreken P, Abeling NG, et al. Genotype and phenotype in patients with dihydropyrimidine dehydrogenase deficiency. Hum Genet 1999;104:1–9.
[104] Mayr JA, Zimmermann FA, Fauth C, et al. Lipoic acid synthetase deficiency causes neonatal-onset epilepsy, defective mitochondrial energy metabolism, and glycine elevation. Am J Hum Genet 2011;89:792–7.

[105] Reiss J, Hahnewald R. Molybdenum cofactor deficiency: mutations in GPHN, MOCS1, and MOCS2. Hum Mutat 2011;32:10–8.
[106] Ichida K, Amaya Y, Okamoto K, et al. Mutations associated with functional disorder of xanthine oxidoreductase and hereditary xanthinuria in humans. Int J Mol Sci 2012;13:15475–95.
[107] Suhasini AN, Brosh RM Jr. Disease-causing missense mutations in human DNA helicase disorders. Mutat Res 2013;752:138–52.
[108] Mazzei F, Viel A, Bignami M. Role of MUTYH in human cancer. Mutat Res 2013;743–744:33–43.
[109] Scheffler IE, Molecular genetics of succinate:quinone oxidoreductase in eukaryotes. Prog Nucleic Acid Res Mol Biol 1998;60:267–315.
[110] Kok JD, Muller JL, Slater EC. EPR studies on the respiratory chain of wild-type Saccharomyces cerevisiae and mutants with a deficiency in succinate dehydrogenase. Biochim Biophys Acta 1975;387:441–50.
[111] Ackrell BA, Kearney EB, Singer TP. Mammalian succinate dehydrogenase. Methods Enzymol 1978;53:466–83.
[112] Singer TP, Johnson MK. The prosthetic groups of succinate dehydrogenase: 30 years from discovery to identification. FEBS Lett 1985;190:189–98.
[113] Lemire BD, Oyedotun KS. The Saccharomyces cerevisiae mitochondrial succinate:ubiquinone oxidoreductase. Biochim Biophys Acta 2002;1553:102–16.
[114] Huang LS, Sun G, Cobessi D, et al. 3-Nitropropionic acid is a suicide inhibitor of mitochondrial respiration that, upon oxidation by complex II, forms a covalent adduct with a catalytic base arginine in the active site of the enzyme. J Biol Chem 2006;281:5965–72.
[115] Zhou Q, Zhai Y, Lou J, et al. Thiabendazole inhibits ubiquinone reduction activity of mitochondrial respiratory complex II via a water molecule mediated binding feature. Protein Cell 2011;2:531–42.
[116] Ackrell BA. Cytopathies involving mitochondrial complex II. Mol Aspects Med 2002;23:369–84.
[117] Briere JJ, Favier J, El Ghouzzi V, et al. Succinate dehydrogenase deficiency in human. Cell Mol Life Sci 2005;62:2317–24.
[118] Gill AJ. Succinate dehydrogenase (SDH) and mitochondrial driven neoplasia. Pathology 2012;44:285–92.
[119] Bourgeron T, Rustin P, Chretien D, et al. Mutation of a nuclear succinate dehydrogenase gene results in mitochondrial respiratory chain deficiency. Nat Genet 1995;11:144–9.
[120] Neumann HP, Pawlu C, Peczkowska M, et al. Distinct clinical features of paraganglioma syndromes associated with SDHB and SDHD gene mutations. JAMA 2004;292:943–51.
[121] Selak MA, Armour SM, MacKenzie ED, et al. Succinate links TCA cycle dysfunction to oncogenesis by inhibiting HIF-alpha prolyl hydroxylase. Cancer Cell 2005;7:77–85.
[122] Ishii T, Yasuda K, Akatsuka A, et al. A mutation in the SDHC gene of complex II increases oxidative stress, resulting in apoptosis and tumorigenesis. Cancer Res 2005;65:203–9.
[123] Cervera AM, Bayley JP, Devilee P, et al. Inhibition of succinate dehydrogenase dysregulates histone modification in mammalian cells. Mol Cancer 2009;8:89.
[124] Lee S, Nakamura E, Yang H, et al. Neuronal apoptosis linked to EglN3 prolyl hydroxylase and familial pheochromocytoma genes: developmental culling and cancer. Cancer Cell 2005;8:155–67.
[125] Piruat JI, Pintado CO, Ortega-Saenz P, et al. The mitochondrial SDHD gene is required for early embryogenesis, and its partial deficiency results in persistent carotid body glomus cell activation with full responsiveness to hypoxia. Mol Cell Biol 2004;24:10933–40.
[126] Bayley JP, van Minderhout I, Hogendoorn P, et al. Sdhd and SDHD/H19 knockout mice do not develop paraganglioma or pheochromocytoma. PLoS One 2009;4:e7987.
[127] Turrens JF, Mitochondrial formation of reactive oxygen species. J Physiol 2003;552:335–44.
[128] Ishii T, Miyazawa M, Onouchi H, et al. Model animals for the study of oxidative stress from complex II. Biochim Biophys Acta 2013;1827:588–97.

[129] Szeto SS, Reinke SN, Sykes BD, et al. Ubiquinone-binding site mutations in the Saccharomyces cerevisiae succinate dehydrogenase generate superoxide and lead to the accumulation of succinate. J Biol Chem 2007;282:27518–26.
[130] Tsuda M, Sugiura T, Ishii T, et al. A mev-1-like dominant-negative SdhC increases oxidative stress and reduces lifespan in Drosophila. Biochem Biophys Res Commun 2007;363:342–6.
[131] Walker DW, Hajek P, Muffat J, et al. Hypersensitivity to oxygen and shortened lifespan in a Drosophila mitochondrial complex II mutant. Proc Natl Acad Sci USA 2006;103:16382–7.
[132] Cervera AM, Apostolova N, Crespo L, et al. Cells silenced for SDHB expression display characteristic features of the tumor phenotype. Cancer Res 2008;68:4058–67.
[133] Semenza GL, Hypoxia-inducible factors in physiology and medicine. Cell 2012;148:399–408.
[134] Bruick RK, McKnight SL. A conserved family of prolyl-4-hydroxylases that modify HIF. Science 2001;294:1337–40.
[135] Epstein AC, Gleadle JM, McNeill A, et al. C. elegans EGL-9 and mammalian homologs define a family of dioxygenases that regulate HIF by prolyl hydroxylation. Cell 2001;107:43–54.
[136] Gordan JD, Simon MC. Hypoxia-inducible factors: central regulators of the tumor phenotype. Curr Opin Genet Dev 2007;17:71–7.
[137] Pollard PJ, Wortham NC, Tomlinson IP. The TCA cycle and tumorigenesis: the examples of fumarate hydratase and succinate dehydrogenase. Ann Med 2003;35:632–9.
[138] Gimenez-Roqueplo AP, Favier J, Rustin P, et al. The R22X mutation of the SDHD gene in hereditary paraganglioma abolishes the enzymatic activity of complex II in the mitochondrial respiratory chain and activates the hypoxia pathway. Am J Hum Genet 2001;69:1186–97.
[139] Gimenez-Roqueplo AP, Favier J, Rustin P, et al. Functional consequences of a SDHB gene mutation in an apparently sporadic pheochromocytoma. J Clin Endocrinol Metab 2002;87:4771–4.
[140] Hewitson KS, Lienard BM, McDonough MA, et al. Structural and mechanistic studies on the inhibition of the hypoxia-inducible transcription factor hydroxylases by tricarboxylic acid cycle intermediates.
J Biol Chem 2007;282:3293–301.
[141] Astuti D, Latif F, Dallol A, et al. Gene mutations in the succinate dehydrogenase subunit SDHB cause susceptibility to familial pheochromocytoma and to familial paraganglioma. Am J Hum Genet 2001;69:49–54.
[142] Baysal BE, Ferrell RE, Willett-Brozick JE, et al. Mutations in SDHD, a mitochondrial complex II gene, in hereditary paraganglioma. Science 2000;287:848–51.
[143] Niemann S, Muller U. Mutations in SDHC cause autosomal dominant paraganglioma, type 3. Nat Genet 2000;26:268–70.
[144] Astuti D, Douglas F, Lennard TW, et al. Germline SDHD mutation in familial phaeochromocytoma. Lancet 2001;357:1181–2.
[145] Vanharanta S, Buchta M, McWhinney SR, et al. Early-onset renal cell carcinoma as a novel extraparaganglial component of SDHB-associated heritable paraganglioma. Am J Hum Genet 2004;74:153–9.
[146] Burnichon N, Briere JJ, Libe R, et al. SDHA is a tumor suppressor gene causing paraganglioma. Hum Mol Genet 2010;19:3011–20.
[147] Hao HX, Khalimonchuk O, Schraders M, et al. SDH5, a gene required for flavination of succinate dehydrogenase, is mutated in paraganglioma. Science 2009;325:1139–42.
[148] Mannelli M, Castellano M, Schiavi F, et al. Clinically guided genetic screening in a large cohort of Italian patients with pheochromocytomas and/or functional or nonfunctional paragangliomas. J Clin Endocrinol Metab 2009;94:1541–7.

[149] Bayley JP, van Minderhout I, Weiss MM, et al. Mutation analysis of SDHB and SDHC: novel germline mutations in sporadic head and neck paraganglioma and familial paraganglioma and/or pheochromocytoma. BMC Med Genet 2006;7:1.
[150] Henderson A, Douglas F, Perros P, et al. SDHB-associated renal oncocytoma suggests a broadening of the renal phenotype in hereditary paragangliomatosis. Fam Cancer 2009;8:257–60.
[151] Housley SL, Lindsay RS, Young B, et al. Renal carcinoma with giant mitochondria associated with germ-line mutation and somatic loss of the succinate dehydrogenase B gene. Histopathology 2010;56:405–8.
[152] Gill AJ, Pachter NS, Clarkson A, et al. Renal tumors and hereditary pheochromocytoma-paraganglioma syndrome type 4. N Engl J Med 2011;364:885–6.
[153] Ricketts CJ, Shuch B, Vocke CD, et al. Succinate dehydrogenase kidney cancer: an aggressive example of the Warburg effect in cancer. J Urol 2012;188:2063–71.
[154] Ricketts C, Woodward ER, Killick P, et al. Germline SDHB mutations and familial renal cell carcinoma. J Natl Cancer Inst 2008;100:1260–2.
[155] Miettinen M, Lasota J. Gastrointestinal stromal tumors – definition, clinical, histological, immunohistochemical, and molecular genetic features and differential diagnosis. Virchows Arch 2001;438:1–12.
[156] Gill AJ, Chou A, Vilain R, et al. Immunohistochemistry for SDHB divides gastrointestinal stromal tumors (GISTs) into 2 distinct types. Am J Surg Pathol 2010;34:636–44.
[157] Pasini B, McWhinney SR, Bei T, et al. Clinical and molecular genetics of patients with the Carney-Stratakis syndrome and germline mutations of the genes coding for the succinate dehydrogenase subunits SDHB, SDHC, and SDHD. Eur J Hum Genet 2008;16:79–88.
[158] Janeway KA, Kim SY, Lodish M, et al. Defects in succinate dehydrogenase in gastrointestinal stromal tumors lacking KIT and PDGFRA mutations. Proc Natl Acad Sci USA 2011;108:314–8.
[159] Alston CL, Davison JE, Meloni F, et al. Recessive germline SDHA and SDHB mutations causing leukodystrophy and isolated mitochondrial complex II deficiency. J Med Genet 2012;49: 569–77.
[160] Dailey HA, Terminal steps of haem biosynthesis. Biochem Soc Trans 2002;30:590–5.
[161] Dailey HA, Finnegan MG, Johnson MK. Human ferrochelatase is an iron-sulfur protein. Biochemistry 1994;33:403–7.
[162] Ferreira GC, Franco R, Lloyd SG, et al. Mammalian ferrochelatase, a new addition to the metalloenzyme family.
J Biol Chem 1994;269:7062–5.
[163] Medlock AE, Dailey HA, Examination of the activity of carboxyl-terminal chimeric constructs of human and yeast ferrochelatases. Biochemistry 2000;39:7461–7.
[164] Furukawa T, Kohno H, Tokunaga R, et al. Nitric oxide-mediated inactivation of mammalian ferrochelatase in vivo and in vitro: possible involvement of the iron-sulphur cluster of the enzyme. Biochem J 1995;310:533–8.
[165] Taketani S, Adachi Y, Nakahashi Y. Regulation of the expression of human ferrochelatase by intracellular iron levels. Eur J Biochem 2000;267:4685–92.
[166] Crouse BR, Sellers VM, Finnegan MG, et al. Site-directed mutagenesis and spectroscopic characterization of human ferrochelatase: identification of residues coordinating the [2Fe-2S] cluster. Biochemistry 1996;35:16222–9.
[167] Wu CK, Dailey HA, Rose JP, et al. The 2.0 A structure of human ferrochelatase, the terminal enzyme of heme biosynthesis. Nat Struct Biol 2001;8:156–60.
[168] Peterka ES, Fusaro RM, Runge WJ, et al. Erythropoietic Protoporphyria. I. Clinical and Laborabory Features in Seven New Cases. JAMA 1965;193:1036–42.

[169] Bonkowsky HL, Bloomer JR, Ebert PS, et al. Heme synthetase deficiency in human protoporphyria. Demonstration of the defect in liver and cultured skin fibroblasts. J Clin Invest 1975;56:1139–48.
[170] Brenner DA, Didier JM, Frasier F, et al. A molecular defect in human protoporphyria. Am J Hum Genet 1992;50:1203–10.
[171] Balwani M, Desnick RJ. The porphyrias: advances in diagnosis and treatment. Hematology Am Soc Hematol Educ Program 2012;2012:19–27.
[172] Bloomer JR, Pathogenesis and therapy of liver disease in protoporphyria. Yale J Biol Med 1979;52:39–48.
[173] Bloomer J, Bruzzone C, Zhu L, et al. Molecular defects in ferrochelatase in patients with protoporphyria requiring liver transplantation. J Clin Invest 1998;102:107–14.
[174] Tutois S, Montagutelli X, Da Silva V, et al. Erythropoietic protoporphyria in the house mouse. A recessive inherited ferrochelatase deficiency with anemia, photosensitivity, and liver disease. J Clin Invest 1991;88:1730–6.
[175] Boulechfar S, Lamoril J, Montagutelli X, et al. Ferrochelatase structural mutant (Fechm1Pas) in the house mouse. Genomics 1993;16:645–8.
[176] Gouya L, Puy H, Lamoril J, et al. Inheritance in erythropoietic protoporphyria: a common wild-type ferrochelatase allelic variant with low expression accounts for clinical manifestation. Blood 1999;93:2105–10.
[177] Sellers VM, Dailey HA. Expression, purification, and characterization of recombinant mammalian ferrochelatase. Methods Enzymol 1997;281:378–87.
[178] McKinnon PJ. DNA repair deficiency and neurological disease. Nat Rev Neurosci 2009;10: 100–12.
[179] Jeppesen DK, Bohr VA, Stevnsner T. DNA repair deficiency in neurodegeneration. Prog Neurobiol 2011;94:166–200.
[180] Kovtun IV, McMurray CT. Trinucleotide expansion in haploid germ cells by gap repair. Nat Genet 2001;27:407–11.
[181] Moreira MC, Barbot C, Tachi N, et al. The gene mutated in ataxia-ocular apraxia 1 encodes the new HIT/Zn-finger protein aprataxin. Nat Genet 2001;29:189–93.
[182] Takashima H, Boerkoel CF, John J, et al. Mutation of TDP1, encoding a topoisomerase I-dependent DNA damage repair enzyme, in spinocerebellar ataxia with axonal neuropathy. Nat Genet 2002;32:267–72.
[183] Gatti RA, Becker-Catania S, Chun HH, et al. The pathogenesis of ataxia-telangiectasia. Learning from a Rosetta Stone. Clin Rev Allergy Immunol 2001;20:87–108.
[184] Wang J, Xiong S, Xie C, et al. Increased oxidative damage in nuclear and mitochondrial DNA in Alzheimer's disease. J Neurochem 2005;93:953–62.
[185] Kikuchi H, Furuta A, Nishioka K, et al. Impairment of mitochondrial DNA repair enzymes against accumulation of 8-oxo-guanine in the spinal motor neurons of amyotrophic lateral sclerosis. Acta Neuropathol 2002;103:408–14.
[186] Kuo CF, McRee DE, Fisher CL, et al. Atomic structure of the DNA repair [4Fe-4S] enzyme endonuclease III. Science 1992;258:434–40.
[187] Hinks JA, Evans MC, De Miguel Y, et al. An iron-sulfur cluster in the family 4 uracil-DNA glycosylases. J Biol Chem 2002;277:16936–40.
[188] Thayer MM, Ahern H, Xing D, et al. Novel DNA binding motifs in the DNA repair enzyme endonuclease III crystal structure. EMBO J 1995;14:4108–20.
[189] Merino EJ, Boal AK, Barton JK. Biological contexts for DNA charge transport chemistry. Curr Opin Chem Biol 2008;12:229–37.
[190] Ding H, Schertzer M, Wu X, et al. Regulation of murine telomere length by Rtel: an essential gene encoding a helicase-like protein. Cell 2004;117:873–86.

[191] Inoue A, Li T, Roby SK, et al. Loss of ChlR1 helicase in mouse causes lethality due to the accumulation of aneuploid cells generated by cohesion defects and placental malformation. Cell Cycle 2007;6:1646–54.
[192] White MF, Structure, function and evolution of the XPD family of iron-sulfur-containing 5′->3′ DNA helicases. Biochem Soc Trans 2009;37:547–51.
[193] Lehmann AR, DNA repair-deficient diseases, xeroderma pigmentosum, Cockayne syndrome and trichothiodystrophy. Biochimie 2003;85:1101–11.
[194] de Boer J, Andressoo JO, de Wit J, et al. Premature aging in mice deficient in DNA repair and transcription. Science 2002;296:1276–9.
[195] Theron T, Fousteri MI, Volke M, et al. Transcription-associated breaks in xeroderma pigmentosum group D cells from patients with combined features of xeroderma pigmentosum and Cockayne syndrome. Mol Cell Biol 2005;25:8368–78.
[196] Andressoo JO, Mitchell JR, de Wit J, et al. An Xpd mouse model for the combined xeroderma pigmentosum/Cockayne syndrome exhibiting both cancer predisposition and segmental progeria. Cancer Cell 2006;10:121–32.
[197] Pugh RA, Honda M, Leesley H, et al. The iron-containing domain is essential in Rad3 helicases for coupling of ATP hydrolysis to DNA translocation and for targeting the helicase to the single-stranded DNA-double-stranded DNA junction. J Biol Chem 2008;283:1732–43.
[198] Fan L, Fuss JO, Cheng QJ, et al. XPD helicase structures and activities: insights into the cancer and aging phenotypes from XPD mutations. Cell 2008;133:789–800.
[199] Dubaele S, Proietti De Santis L, Bienstock RJ, et al. Basal transcription defect discriminates between xeroderma pigmentosum and trichothiodystrophy in XPD patients. Mol Cell 2003;11:1635–46.
[200] Botta E, Nardo T, Lehmann AR, et al. Reduced level of the repair/transcription factor TFIIH in trichothiodystrophy. Hum Mol Genet 2002;11:2919–28.
[201] Andreassen PR, Ren K. Fanconi anemia proteins, DNA interstrand crosslink repair pathways, and cancer therapy. Curr Cancer Drug Targets 2009;9:101–17.
[202] Kottemann MC, Smogorzewska A. Fanconi anaemia and the repair of Watson and Crick DNA crosslinks. Nature 2013;493:356–63.
[203] Wu Y, Sommers JA, Suhasini A, et al. Fanconi anemia group J mutation abolishes its DNA repair function by uncoupling DNA translocation from helicase activity or disruption of protein-DNA complexes. Blood 2010;116:3780–91.
[204] Levitus M, Waisfisz Q, Godthelp BC, et al. The DNA helicase BRIP1 is defective in Fanconi anemia complementation group J. Nat Genet 2005;37:934–5.
[205] Litman R, Peng M, Jin Z, et al. BACH1 is critical for homologous recombination and appears to be the Fanconi anemia gene product FANCJ. Cancer Cell 2005;8:255–65.
[206] Schwab RA, Nieminuszczy J, Shin-ya K, et al. FANCJ couples replication past natural fork barriers with maintenance of chromatin structure. J Cell Biol 2013;201:33–48.
[207] Cantor SB, Bell DW, Ganesan S, et al. BACH1, a novel helicase-like protein, interacts directly with BRCA1 and contributes to its DNA repair function. Cell 2001;105:149–60.
[208] Levran O, Diotti R, Pujara K, et al. Spectrum of sequence variations in the FANCA gene: an International Fanconi Anemia Registry (IFAR) study. Hum Mutat 2005;25:142–9.
[209] Gupta R, Sharma S, Doherty KM, et al. Inhibition of BACH1 (FANCJ) helicase by backbone discontinuity is overcome by increased motor ATPase or length of loading strand. Nucleic Acids Res 2006;34:6673–83.
[210] Rouault TA, Tong WH. Iron-sulfur cluster biogenesis and human disease. Trends Genet 2008;24:398–407.
[211] Balk J, Pilon M. Ancient and essential: the assembly of iron-sulfur clusters in plants. Trends Plant Sci 2011;16:218–26.
[212] Py B, Barras F. Building Fe-S proteins: bacterial strategies. Nat Rev Microbiol 2010;8:436–46.

[213] Lill R, Hoffmann B, Molik S, et al. The role of mitochondria in cellular iron-sulfur protein biogenesis and iron metabolism. Biochim Biophys Acta 2012;1823:1491–508.

[214] Peters JW, Broderick JB. Emerging paradigms for complex iron-sulfur cofactor assembly and insertion. Annu Rev Biochem 2012;81:429–50.

[215] Zheng L, White RH, Cash VL, et al. Cysteine desulfurase activity indicates a role for NIFS in metallocluster biosynthesis. Proc Natl Acad Sci USA 1993;90:2754–8.

[216] Land T, Rouault TA. Targeting of a human iron-sulfur cluster assembly enzyme, nifs, to different subcellular compartments is regulated through alternative AUG utilization. Molecular Cell 1998;2:807–15.

[217] Biederbick A, Stehling O, Rosser R, et al. Role of human mitochondrial Nfs1 in cytosolic iron-sulfur protein biogenesis and iron regulation. Mol Cell Biol 2006;26:5675–87.

[218] Urbina HD, Silberg JJ, Hoff KG, et al. Transfer of sulfur from IscS to IscU during Fe/S cluster assembly. J Biol Chem 2001;276:44521–26.

[219] Smith AD, Agar JN, Johnson KA, et al. Sulfur transfer from IscS to IscU: the first step in iron-sulfur cluster biosynthesis. J Am Chem Soc 2001;123:11103–4.

[220] Shi R, Proteau A, Villarroya M, et al. Structural basis for Fe-S cluster assembly and tRNA thiolation mediated by IscS protein-protein interactions. PLoS Biol 2010;8:e1000354.

[221] Agar JN, Krebs C, Frazzon J, et al. IscU as a scaffold for iron-sulfur cluster biosynthesis: sequential assembly of [2Fe-2S] and [4Fe-4S] clusters in IscU. Biochemistry 2000;39: 7856–62.

[222] Marinoni EN, de Oliveira JS, Nicolet Y, et al. (IscS-IscU)2 complex structures provide insights into Fe2S2 biogenesis and transfer. Angew Chem Int Ed Engl 2012;51:5439–42.

[223] Colin F, Martelli A, Clemancey M, et al. Mammalian frataxin controls sulfur production and iron entry during de novo Fe4S4 cluster assembly. J Am Chem Soc 2013;135:733–40.

[224] Adam AC, Bornhovd C, Prokisch H, et al. The Nfs1 interacting protein Isd11 has an essential role in Fe/S cluster biogenesis in mitochondria. EMBO J 2006;25:174–83.

[225] Wiedemann N, et al. Essential role of Isd11 in mitochondrial iron-sulfur cluster synthesis on Isu scaffold proteins. EMBO J 2006;25:184–95.

[226] Shi Y, Urzica E, Guiard B, et al. Human ISD11 is essential for both iron-sulfur cluster assembly and maintenance of normal cellular iron homeostasis. Hum Mol Genet 2009;18:3014–25.

[227] Pandey A, Golla R, Yoon H, et al. Persulfide formation on mitochondrial cysteine desulfurase: enzyme activation by a eukaryote-specific interacting protein and Fe-S cluster synthesis. Biochem J 2012;448:171–87.

[228] Gerber J, Muhlenhoff U, Lill R. An interaction between frataxin and Isu1/Nfs1 that is crucial for Fe/S cluster synthesis on Isu1. EMBO Rep 2003;4:906–11.

[229] Gakh O, Park S, Liu G, et al. Mitochondrial iron detoxification is a primary function of frataxin that limits oxidative damage and preserves cell longevity. Hum Mol Genet 2006;15:467–79.

[230] Cook JD, Kondapalli KC, Rawat S, et al. Molecular details of the yeast frataxin-Isu1 interaction during mitochondrial Fe-S cluster assembly. Biochemistry 2010;49:8756–65.

[231] Tsai CL, Barondeau DP. Human frataxin is an allosteric switch that activates the Fe-S cluster biosynthetic complex. Biochemistry 2010;49:9132–9.

[232] Sheftel AD, Stehling O, Pierik AJ, et al. Humans possess two mitochondrial ferredoxins, Fdx1 and Fdx2, with distinct roles in steroidogenesis, heme, and Fe/S cluster biosynthesis. Proc Natl Acad Sci USA 2010;107:11775–80.

[233] Shi Y, Ghosh M, Kovtunovych G, et al. Both human ferredoxins 1 and 2 and ferredoxin reductase are important for iron-sulfur cluster biogenesis. Biochim Biophys Acta 2012;1823:484–92.

[234] Hoff KG, Silberg JJ, Vickery LE. Interaction of the iron-sulfur cluster assembly protein IscU with the Hsc66/Hsc20 molecular chaperone system of Escherichia coli. Proc Natl Acad Sci USA 2000;97:7790–5.

[235] Voisine C, Cheng YC, Ohlson M, et al. Jac1, a mitochondrial J-type chaperone, is involved in the biogenesis of Fe/S clusters in Saccharomyces cerevisiae. Proc Natl Acad Sci USA 2001;98:1483–8.
[236] Chandramouli K, Johnson MK. HscA and HscB stimulate [2Fe-2S] cluster transfer from IscU to apoferredoxin in an ATP-dependent reaction. Biochemistry 2006;45:11087–95.
[237] Andrew AJ, Dutkiewicz R, Knieszner H, et al. Characterization of the interaction between the J-protein Jac1p and the scaffold for Fe-S cluster biogenesis, Isu1p. J Biol Chem 2006;281:14580–7.
[238] Schilke B, Williams B, Knieszner H, et al. Evolution of mitochondrial chaperones utilized in Fe-S cluster biogenesis. Curr Biol 2006;16:1660–5.
[239] Bonomi F, Iametti S, Morleo A, et al. Studies on the mechanism of catalysis of iron-sulfur cluster transfer from IscU[2Fe2S] by HscA/HscB chaperones. Biochemistry 2008;47: 12795–801.
[240] Uhrigshardt H, Singh A, Kovtunovych G, et al. Characterization of the human HSC20, an unusual DnaJ type III protein, involved in iron-sulfur cluster biogenesis. Hum Mol Genet 2010;19:3816–34.
[241] Kampinga HH, Craig EA. The HSP70 chaperone machinery: J proteins as drivers of functional specificity. Nat Rev Mol Cell Biol 2010;11:579–92.
[242] Rodriguez-Manzaneque MT, Tamarit J, Belli G, et al. Grx5 is a mitochondrial glutaredoxin required for the activity of iron/sulfur enzymes. Mol Biol Cell 2002;13:1109–21.
[243] Muhlenhoff U, Gerber J, Richhardt N, et al. Components involved in assembly and dislocation of iron-sulfur clusters on the scaffold protein Isu1p. EMBO J 2003;22:4815–25.
[244] Wingert RA, Galloway JL, Barut B, et al. Deficiency of glutaredoxin 5 reveals Fe-S clusters are required for vertebrate haem synthesis. Nature 2005;436:1035–39.
[245] Camaschella C, Campanella A, De Falco L, et al. The human counterpart of zebrafish shiraz shows sideroblastic-like microcytic anemia and iron overload. Blood 2007;110:1353–8.
[246] Bandyopadhyay S, Gama F, Molina-Navarro MM, et al. Chloroplast monothiol glutaredoxins as scaffold proteins for the assembly and delivery of [2Fe-2S] clusters. EMBO J 2008;27:1122–33.
[247] Shakamuri P, Zhang B, Johnson MK. Monothiol glutaredoxins function in storing and transporting [Fe2S2] clusters assembled on IscU scaffold proteins. J Am Chem Soc 2012;134:15213–6.
[248] Gelling C, Dawes IW, Richhardt N, et al. Mitochondrial Iba57p is required for Fe/S cluster formation on aconitase and activation of radical SAM enzymes. Mol Cell Biol 2008;28:1851–61.
[249] Muhlenhoff U, Richter N, Pines O, et al. Specialized function of yeast Isa1 and Isa2 proteins in the maturation of mitochondrial [4Fe-4S] proteins. J Biol Chem 2011;286:41205–16.
[250] Nishio K, Nakai M. Transfer of iron-sulfur cluster from NifU to apoferredoxin. J Biol Chem 2000;275:22615–8.
[251] Tong WH, Jameson GN, Huynh BH, et al. Subcellular compartmentalization of human Nfu, an iron-sulfur cluster scaffold protein, and its ability to assemble a [4Fe-4S] cluster. Proc Natl Acad Sci USA 2003;100:9762–7.
[252] Jin Z, Heinnickel M, Krebs C, et al. Biogenesis of iron-sulfur clusters in photosystem I: holo-NfuA from the cyanobacterium Synechococcus sp. PCC 7002 rapidly and efficiently transfers [4Fe-4S] clusters to apo-PsaC in vitro. J Biol Chem 2008;283:28426–35.
[253] Angelini S, Gerez C, Ollagnier-de Choudens S, et al. NfuA, a new factor required for maturing Fe/S proteins in Escherichia coli under oxidative stress and iron starvation conditions. J Biol Chem 2008;283:14084–91.
[254] Bandyopadhyay S, Naik SG, O'Carroll IP, et al. A proposed role for the Azotobacter vinelandii NfuA protein as an intermediate iron-sulfur cluster carrier. J Biol Chem 2008;283:14092–9.

[255] Navarro-Sastre A, Tort F, Stehling O, et al. A fatal mitochondrial disease is associated with defective NFU1 function in the maturation of a subset of mitochondrial Fe-S proteins. Am J Hum Genet 2011;89:656–67.
[256] Py B, Gerez C, Angelini S, et al. Molecular organization, biochemical function, cellular role and evolution of NfuA, an atypical Fe-S carrier. Mol Microbiol 2012;86:155–71.
[257] Bych K, Kerscher S, Netz DJ, et al. The iron-sulphur protein Ind1 is required for effective complex I assembly. EMBO J 2008;27:1736–46.
[258] Sheftel AD, Stehling O, Pierik AJ, et al. Human ind1, an iron-sulfur cluster assembly factor for respiratory complex I. Mol Cell Biol 2009;29:6059–73.
[259] Mesecke N, Terziyska N, Kozany C, et al. A disulfide relay system in the intermembrane space of mitochondria that mediates protein import. Cell 2005;121:1059–69.
[260] Tong WH, Rouault TA. Functions of mitochondrial ISCU and cytosolic ISCU in mammalian iron-sulfur cluster biogenesis and iron homeostasis. Cell Metab 2006;3:199–210.
[261] Roy A, Solodovnikova N, Nicholson T, et al. A novel eukaryotic factor for cytosolic Fe-S cluster assembly. EMBO J 2003;22:4826–35.
[262] Netz DJ, Pierik AJ, Stumpfig M, et al. The Cfd1-Nbp35 complex acts as a scaffold for iron-sulfur protein assembly in the yeast cytosol. Nat Chem Biol 2007;3:278–86.
[263] Hausmann A, Aguilar Netz DJ, Balk J, et al. The eukaryotic P loop NTPase Nbp35: an essential component of the cytosolic and nuclear iron-sulfur protein assembly machinery. Proc Natl Acad Sci USA 2005;102:3266–71.
[264] Balk J, Pierik AJ, Netz DJ, et al. The hydrogenase-like Nar1p is essential for maturation of cytosolic and nuclear iron-sulphur proteins. EMBO J 2004;23:2105–15.
[265] Urzica E, Pierik AJ, Muhlenhoff U, et al. Crucial role of conserved cysteine residues in the assembly of two iron-sulfur clusters on the CIA protein Nar1. Biochemistry 2009;48:4946–58.
[266] Balk J, Aguilar Netz DJ, Tepper K, et al. The essential WD40 protein Cia1 is involved in a late step of cytosolic and nuclear iron-sulfur protein assembly. Mol Cell Biol 2005;25:10833–41.
[267] Srinivasan V, Netz DJ, Webert H, et al. Structure of the yeast WD40 domain protein Cia1, a component acting late in iron-sulfur protein biogenesis. Structure 2007;15:1246–57.
[268] Netz DJ, Stumpfig M, Dore C, et al. Tah18 transfers electrons to Dre2 in cytosolic iron-sulfur protein biogenesis. Nat Chem Biol 2010;6:758–65.
[269] Stehling O, Vashisht AA, Mascarenhas J, et al. MMS19 assembles iron-sulfur proteins required for DNA metabolism and genomic integrity. Science 2012;337:195–9.
[270] Stehling O, Mascarenhas J, Vashisht AA, et al. Human CIA2A-FAM96A and CIA2B-FAM96B integrate iron homeostasis and maturation of different subsets of cytosolic-nuclear iron-sulfur proteins. Cell Metab 2013;18:187–98.
[271] Stehling O, Netz DJ, Niggemeyer B, et al. Human Nbp35 is essential for both cytosolic iron-sulfur protein assembly and iron homeostasis. Mol Cell Biol 2008;28:5517–28.
[272] Seki M, Takeda Y, Iwai K, et al. IOP1 protein is an external component of the human cytosolic iron-sulfur cluster assembly (CIA) machinery and functions in the MMS19 protein-dependent CIA pathway. J Biol Chem 2013;288:16680–9.
[273] Song D, Lee FS. Mouse knock-out of IOP1 protein reveals its essential role in mammalian cytosolic iron-sulfur protein biogenesis. J Biol Chem 2011;286:15797–805.
[274] Nakamura M, Buzas DM, Kato A, et al. The role of Arabidopsis thaliana NAR1, a cytosolic iron-sulfur cluster assembly component, in gametophytic gene expression and oxidative stress responses in vegetative tissue. New Phytol 2013;199:925–35.
[275] Gari K, Leon Ortiz AM, Borel V, et al. MMS19 links cytoplasmic iron-sulfur cluster assembly to DNA metabolism. Science 2012;337:243–5.
[276] van Wietmarschen N, Moradian A, Morin GB, et al. The mammalian proteins MMS19, MIP18, and ANT2 are involved in cytoplasmic iron-sulfur cluster protein assembly. J Biol Chem 2012;287:43351–8.

[277] Sheftel A, Stehling O, Lill R. Iron-sulfur proteins in health and disease. Trends Endocrinol Metab 2010;21:302–14.
[278] Schilke B, Voisine C, Beinert H, et al. Evidence for a conserved system for iron metabolism in the mitochondria of Saccharomyces cerevisiae. Proc Natl Acad Sci USA 1999;96:10206–11.
[279] Vazquez-Manrique RP, Gonzalez-Cabo P, Ros S, et al. Reduction of Caenorhabditis elegans frataxin increases sensitivity to oxidative stress, reduces lifespan, and causes lethality in a mitochondrial complex II mutant. FASEB J 2006;20:172–4.
[280] Herrero E, de la Torre-Ruiz MA. Monothiol glutaredoxins: a common domain for multiple functions. Cell Mol Life Sci 2007;64:1518–30.
[281] Cossee M, Puccio H, Gansmuller A, et al. Inactivation of the Friedreich ataxia mouse gene leads to early embryonic lethality without iron accumulation. Hum Mol Genet 2000;9:1219–26.
[282] Pondarre C, Antiochos BB, Campagna DR, et al. The mitochondrial ATP-binding cassette transporter Abcb7 is essential in mice and participates in cytosolic iron-sulfur cluster biogenesis. Hum Mol Genet 2006;15:953–64.
[283] Nordin A, Larsson E, Thornell LE, et al. Tissue-specific splicing of ISCU results in a skeletal muscle phenotype in myopathy with lactic acidosis, while complete loss of ISCU results in early embryonic death in mice. Hum Genet 2011;129:371–8.
[284] Fosset C, Chauveau MJ, Guillon B, et al. RNA silencing of mitochondrial m-Nfs1 reduces Fe-S enzyme activity both in mitochondria and cytosol of mammalian cells. J Biol Chem 2006;281:25398–406.
[285] Martelli A, Wattenhofer-Donze M, Schmucker S, et al. Frataxin is essential for extramitochondrial Fe-S cluster proteins in mammalian tissues. Hum Mol Genet 2007;16:2651–8.
[286] Song D, Lee FS. A role for IOP1 in mammalian cytosolic iron-sulfur protein biogenesis. J Biol Chem 2008;283:9231–8.
[287] Li K, Besse EK, Ha D, et al. Iron-dependent regulation of frataxin expression: implications for treatment of Friedreich ataxia. Hum Mol Genet 2008;17:2265–73.
[288] Campuzano V, Montermini L, Molto MD, et al. Friedreich's ataxia: autosomal recessive disease caused by an intronic GAA triplet repeat expansion. Science 1996;271:1423–7.
[289] Cossee M, Schmitt M, Campuzano V, et al. Evolution of the Friedreich's ataxia trinucleotide repeat expansion: founder effect and premutations. Proc Natl Acad Sci USA 1997;94:7452–7.
[290] Harding AE, Friedreich's ataxia: a clinical and genetic study of 90 families with an analysis of early diagnostic criteria and intrafamilial clustering of clinical features. Brain 1981;104: 589–620.
[291] Pandolfo M, Pastore A. The pathogenesis of Friedreich ataxia and the structure and function of frataxin. J Neurol 2009;256:9–17.
[292] Cossee M, et al. Friedreich's ataxia: point mutations and clinical presentation of compound heterozygotes. Ann Neurol 1999;45:200–6.
[293] Gellera C, et al. Frataxin gene point mutations in Italian Friedreich ataxia patients. Neurogenetics 2007;8:289–99.
[294] Gottesfeld JM. Small molecules affecting transcription in Friedreich ataxia. Pharmacol Ther 2007;116:236–48.
[295] De Biase I, Rasmussen A, Endres D, et al. Progressive GAA expansions in dorsal root ganglia of Friedreich's ataxia patients. Ann Neurol 2007;61:55–60.
[296] Correia AR, Pastore C, Adinolfi S, et al. Dynamics, stability and iron-binding activity of frataxin clinical mutants. FEBS J 2008;275:3680–90.
[297] Bridwell-Rabb J, Winn AM, Barondeau DP. Structure-function analysis of Friedreich's ataxia mutants reveals determinants of frataxin binding and activation of the Fe-S assembly complex. Biochemistry 2011;50:7265–74.

[298] Xia H, Cao Y, Dai X, et al. Novel frataxin isoforms may contribute to the pathological mechanism of Friedreich ataxia. PLoS One 2012;7:e47847.
[299] Rotig A, de Lonlay P, Chretien D, et al. Aconitase and mitochondrial iron-sulphur protein deficiency in Friedreich ataxia. Nat Genet 1997;17:215–7.
[300] Babcock M, de Silva D, Oaks R, et al. Regulation of mitochondrial iron accumulation by Yfh1p, a putative homolog of frataxin. Science 1997;276:1709–12.
[301] Koutnikova H, Campuzano V, Foury F, et al. Studies of human, mouse and yeast homologues indicate a mitochondrial function for frataxin. Nat Genet 1997;16:345–51.
[302] Santos MM, Ohshima K, Pandolfo M. Frataxin deficiency enhances apoptosis in cells differentiating into neuroectoderm. Hum Mol Genet 2001;10:1935–44.
[303] Muhlenhoff U, Richhardt N, Ristow M, et al. The yeast frataxin homolog Yfh1p plays a specific role in the maturation of cellular Fe/S proteins. Hum Mol Genet 2002;11:2025–36.
[304] Stehling O, Elsasser HP, Bruckel B, et al. Iron-sulfur protein maturation in human cells: evidence for a function of frataxin. Hum Mol Genet 2004;13:3007–15.
[305] Anderson PR, Kirby K, Hilliker AJ, et al. RNAi-mediated suppression of the mitochondrial iron chaperone, frataxin, in Drosophila. Hum Mol Genet 2005;14:3397–405.
[306] Napoli E, Morin D, Bernhardt R, et al. Hemin rescues adrenodoxin, heme a and cytochrome oxidase activity in frataxin-deficient oligodendroglioma cells. Biochim Biophys Acta 2007;1772:773–80.
[307] Zanella I, Derosas M, Corrado M, et al. The effects of frataxin silencing in HeLa cells are rescued by the expression of human mitochondrial ferritin. Biochim Biophys Acta 2008;1782:90–8.
[308] Lu C, Schoenfeld R, Shan Y, et al. Frataxin deficiency induces Schwann cell inflammation and death. Biochim Biophys Acta 2009;1792:1052–61.
[309] Nair M, Adinolfi S, Pastore C, et al. Solution structure of the bacterial frataxin ortholog, CyaY: mapping the iron binding sites. Structure 2004;12:2037–48.
[310] Cook JD, Bencze KZ, Jankovic AD, et al. Monomeric yeast frataxin is an iron-binding protein. Biochemistry 2006;45:7767–77.
[311] Adamec J, Rusnak F, Owen WG, et al. Iron-dependent self-assembly of recombinant yeast frataxin: implications for Friedreich ataxia. Am J Hum Genet 2000;67:549–62.
[312] Li H, Gakh O, Smith DY, et al. Oligomeric yeast frataxin drives assembly of core machinery for mitochondrial iron-sulfur cluster synthesis. J Biol Chem 2009;284:21971–80.
[313] Gakh O, Bedekovics T, Duncan SF, et al. Normal and Friedreich ataxia cells express different isoforms of frataxin with complementary roles in iron-sulfur cluster assembly. J Biol Chem 2010;285:38486–501.
[314] Aloria K, Schilke B, Andrew A, et al. Iron-induced oligomerization of yeast frataxin homologue Yfh1 is dispensable in vivo. EMBO Rep 2004;5:1096–101.
[315] Layer G, Ollagnier-de Choudens S, Sanakis Y, et al. Iron-sulfur cluster biosynthesis: characterization of Escherichia coli CYaY as an iron donor for the assembly of [2Fe-2S] clusters in the scaffold IscU. J Biol Chem 2006;281:16256–63.
[316] Bencze KZ, Kondapalli KC, Cook JD, et al. The structure and function of frataxin. Crit Rev Biochem Mol Biol 2006;41:269–91.
[317] Bulteau AL, O'Neill HA, Kennedy MC, et al. Frataxin acts as an iron chaperone protein to modulate mitochondrial aconitase activity. Science 2004;305:242–5.
[318] He Y, Alam SL, Proteasa SV, et al. Yeast frataxin solution structure, iron binding, and ferrochelatase interaction. Biochemistry 2004;43:16254–62.
[319] Yoon T, Cowan JA. Frataxin-mediated iron delivery to ferrochelatase in the final step of heme biosynthesis. J Biol Chem 2004;279:25943–6.

[320] Gonzalez-Cabo P, Vazquez-Manrique RP, Garcia-Gimeno MA, et al. Frataxin interacts functionally with mitochondrial electron transport chain proteins. Hum Mol Genet 2005;14:2091–8.
[321] Schmucker S, Martelli A, Colin F, et al. Mammalian frataxin: an essential function for cellular viability through an interaction with a preformed ISCU/NFS1/ISD11 iron-sulfur assembly complex. PLoS One 2011;6:e16199.
[322] Yoon H, Golla R, Lesuisse E, et al. Mutation in the Fe-S scaffold protein Isu bypasses frataxin deletion. Biochem J 2012;441:473–80.
[323] Li DS, Ohshima K, Jiralerspong S, et al. Knock-out of the cyaY gene in Escherichia coli does not affect cellular iron content and sensitivity to oxidants. FEBS Lett 1999;456:13–6.
[324] Foury F, Cazzalini O. Deletion of the yeast homologue of the human gene associated with Friedreich's ataxia elicits iron accumulation in mitochondria. FEBS Lett 1997;411:373–7.
[325] Wilson RB, Roof DM. Respiratory deficiency due to loss of mitochondrial DNA in yeast lacking the frataxin homologue. Nat Genet 1997;16:352–7.
[326] Zarse K, Schulz TJ, Birringer M, et al. Impaired respiration is positively correlated with decreased life span in Caenorhabditis elegans models of Friedreich Ataxia. FASEB J 2007;21:1271–5.
[327] Llorens JV, Navarro JA, Martinez-Sebastian MJ, et al. Causative role of oxidative stress in a Drosophila model of Friedreich ataxia. FASEB J 2007;21:333–44.
[328] Puccio H, Simon D, Cossee M, Criqui-Filipe P, et al. Mouse models for Friedreich ataxia exhibit cardiomyopathy, sensory nerve defect and Fe-S enzyme deficiency followed by intramitochondrial iron deposits. Nat Genet 2001;27:181–6.
[329] Miranda CJ, Santos MM, Ohshima K, et al. Frataxin knockin mouse. FEBS Lett 2002;512:291–7.
[330] Al-Mahdawi S, Pinto RM, Varshney D, et al. GAA repeat expansion mutation mouse models of Friedreich ataxia exhibit oxidative stress leading to progressive neuronal and cardiac pathology. Genomics 2006;88:580–90.
[331] Martelli A, Napierala M, Puccio H. Understanding the genetic and molecular pathogenesis of Friedreich's ataxia through animal and cellular models. Dis Model Mech 2012;5:165–76.
[332] Simon D, Seznec H, Gansmuller A, et al. Friedreich ataxia mouse models with progressive cerebellar and sensory ataxia reveal autophagic neurodegeneration in dorsal root ganglia. J Neurosci 2004;24:1987–95.
[333] Ristow M, Mulder H, Pomplun D, et al. Frataxin deficiency in pancreatic islets causes diabetes due to loss of beta cell mass. J Clin Invest 2003;112:527–34.
[334] Lamarche JB, Cote M, Lemieux B. The cardiomyopathy of Friedreich's ataxia morphological observations in 3 cases. Can J Neurol Sci 1980;7:389–96.
[335] Michael S, Petrocine SV, Qian J, et al. Iron and iron-responsive proteins in the cardiomyopathy of Friedreich's ataxia. Cerebellum 2006;5:257–67.
[336] Rouault TA, Tong WH. Iron-sulphur cluster biogenesis and mitochondrial iron homeostasis. Nat Rev Mol Cell Biol 2005;6:345–51.
[337] Richardson DR, Huang ML, Whitnall M, et al. The ins and outs of mitochondrial iron-loading: the metabolic defect in Friedreich's ataxia. J Mol Med (Berl) 2010;88:323–9.
[338] Huang ML, Becker EM, Whitnall M, et al. Elucidation of the mechanism of mitochondrial iron loading in Friedreich's ataxia by analysis of a mouse mutant. Proc Natl Acad Sci USA 2009;106:16381–6.
[339] Lesuisse E, Santos R, Matzanke BF, et al. Iron use for haeme synthesis is under control of the yeast frataxin homologue (Yfh1). Hum Mol Genet 2003;12:879–89.
[340] Miao R, Martinho M, Morales JG, et al. EPR and Mossbauer spectroscopy of intact mitochondria isolated from Yah1p-depleted Saccharomyces cerevisiae. Biochemistry 2008;47:9888–99.

[341] Seznec H, Simon D, Bouton C, et al. Friedreich ataxia: the oxidative stress paradox. Hum Mol Genet 2005;14:463–74.
[342] Rustin P, von Kleist-Retzow JC, Chantrel-Groussard K, et al. Effect of idebenone on cardiomyopathy in Friedreich's ataxia: a preliminary study. Lancet 1999;354:477–9.
[343] Schulz JB, Dehmer T, Schols L, et al. Oxidative stress in patients with Friedreich ataxia. Neurology 2000;55:1719–21.
[344] Di Prospero NA, Baker A, Jeffries N, et al. Neurological effects of high-dose idebenone in patients with Friedreich's ataxia: a randomised, placebo-controlled trial. Lancet Neurol 2007;6:878–86.
[345] Lagedrost SJ, Sutton MS, Cohen MS, et al. Idebenone in Friedreich ataxia cardiomyopathy-results from a 6-month phase III study (IONIA). Am Heart J 2011;161:639–45 e1.
[346] Soriano S, Llorens JV, Blanco-Sobero L, et al. Deferiprone and idebenone rescue frataxin depletion phenotypes in a Drosophila model of Friedreich's ataxia. Gene 2013;521:274–81.
[347] Fleming MD. The genetics of inherited sideroblastic anemias. Semin Hematol 2002;39:270–81.
[348] Camaschella C. Recent advances in the understanding of inherited sideroblastic anaemia. Br J Haematol 2008;143:27–38.
[349] Sheftel AD, Richardson DR, Prchal J, et al. Mitochondrial iron metabolism and sideroblastic anemia. Acta Haematol 2009;122:120–33.
[350] Harigae H, Furuyama K. Hereditary sideroblastic anemia: pathophysiology and gene mutations. Int J Hematol 2010;92:425–31.
[351] Allikmets R, Raskind WH, Hutchinson A, et al. Mutation of a putative mitochondrial iron transporter gene (ABC7) in X-linked sideroblastic anemia and ataxia (XLSA/A). Hum Mol Genet 1999;8:743–9.
[352] Neufeld EJ, Fleming JC, Tartaglini E, et al. Thiamine-responsive megaloblastic anemia syndrome: a disorder of high-affinity thiamine transport. Blood Cells Mol Dis 2001;27:135–8.
[353] Bykhovskaya Y, Casas K, Mengesha E, et al. Missense mutation in pseudouridine synthase 1 (PUS1) causes mitochondrial myopathy and sideroblastic anemia (MLASA). Am J Hum Genet 2004;74:1303–8.
[354] Guernsey DL, Jiang H, Campagna DR, et al. Mutations in mitochondrial carrier family gene SLC25A38 cause nonsyndromic autosomal recessive congenital sideroblastic anemia. Nat Genet 2009;41:651–3.
[355] Camaschella C, Hereditary sideroblastic anemias: pathophysiology, diagnosis, and treatment. Semin Hematol 2009;46:371–7.
[356] Bekri S, Kispal G, Lange H, et al. Human ABC7 transporter: gene structure and mutation causing X-linked sideroblastic anemia with ataxia with disruption of cytosolic iron-sulfur protein maturation. Blood 2000;96:3256–64.
[357] Hellier KD, Hatchwell E, Duncombe AS, et al. X-linked sideroblastic anaemia with ataxia: another mitochondrial disease? J Neurol Neurosurg Psychiatry 2001;70:65–9.
[358] D'Hooghe M, Selleslag D, Mortier G, et al. X-linked sideroblastic anemia and ataxia: a new family with identification of a fourth ABCB7 gene mutation. Eur J Paediatr Neurol 2012;16:730–5.
[359] Cavadini P, Biasiotto G, Poli M, et al. RNA silencing of the mitochondrial ABCB7 transporter in HeLa cells causes an iron-deficient phenotype with mitochondrial iron overload. Blood 2007;109:3552–9.
[360] Bernard DG, Cheng Y, Zhao Y, et al. An allelic mutant series of ATM3 reveals its key role in the biogenesis of cytosolic iron-sulfur proteins in Arabidopsis. Plant Physiol 2009;151:590–602.

[361] Kispal G, Csere P, Guiard B, et al. The ABC transporter Atm1p is required for mitochondrial iron homeostasis. FEBS Lett 1997;418:346–50.
[362] Miao R, Kim H, Koppolu UM, et al. Biophysical characterization of the iron in mitochondria from Atm1p-depleted Saccharomyces cerevisiae. Biochemistry 2009;48:9556–68.
[363] Fleming RE, Feng Q, Britton RS. Knockout mouse models of iron homeostasis. Annu Rev Nutr 2011;31:117–37.
[364] Chung J, Chen C, Paw BH. Heme metabolism and erythropoiesis. Curr Opin Hematol 2012;19:156–62.
[365] Napier I, Ponka P, Richardson DR. Iron trafficking in the mitochondrion: novel pathways revealed by disease. Blood 2005;105:1867–74.
[366] Leighton J, Schatz G. An ABC transporter in the mitochondrial inner membrane is required for normal growth of yeast. EMBO J 1995;14:188–95.
[367] Kispal G, Csere P, Prohl C, et al. The mitochondrial proteins Atm1p and Nfs1p are essential for biogenesis of cytosolic Fe/S proteins. EMBO J 1999;18:3981–89.
[368] Lill R, Muhlenhoff U. Iron-sulfur-protein biogenesis in eukaryotes. Trends Biochem Sci 2005;30:133–41.
[369] Bedekovics T, Li H, Gajdos GB, et al. Leucine biosynthesis regulates cytoplasmic iron-sulfur enzyme biogenesis in an Atm1p-independent manner. J Biol Chem 2011;286:40878–88.
[370] Chloupkova M, LeBard LS, Koeller DM. MDL1 is a high copy suppressor of ATM1: evidence for a role in resistance to oxidative stress. J Mol Biol 2003;331:155–65.
[371] Chen S, Sanchez-Fernandez R, Lyver ER, et al. Functional characterization of AtATM1, AtATM2, and AtATM3, a subfamily of Arabidopsis half-molecule ATP-binding cassette transporters implicated in iron homeostasis. J Biol Chem 2007;282:21561–71.
[372] Lill R, Function and biogenesis of iron-sulphur proteins. Nature 2009;460:831–8.
[373] Teschner J, Lachmann N, Schulze J, et al. A novel role for Arabidopsis mitochondrial ABC transporter ATM3 in molybdenum cofactor biosynthesis. Plant Cell 2010;22:468–80.
[374] Liu G, Guo S, Han B, et al. Novel heterozygous missense mutations in GLRX5 gene of a chinese patient affected by sideroblastic anemia. In: Fifth Congress of the International BioIron Society, Binennial World Meeting, London, UK; 2013.
[375] Herrero E, Belli G, Casa C. Structural and functional diversity of glutaredoxins in yeast. Curr Protein Pept Sci 2010;11:659–68.
[376] Li H, Outten CE. Monothiol CGFS glutaredoxins and BolA-like proteins: [2Fe-2S] binding partners in iron homeostasis. Biochemistry 2012;51:4377–89.
[377] Johansson C, Roos AK, Montano SJ, et al. The crystal structure of human GLRX5: iron-sulfur cluster co-ordination, tetrameric assembly and monomer activity. Biochem J 2011;433:303–11.
[378] Picciocchi A, Saguez C, Boussac A, et al. CGFS-type monothiol glutaredoxins from the cyanobacterium Synechocystis PCC6803 and other evolutionary distant model organisms possess a glutathione-ligated [2Fe-2S] cluster. Biochemistry 2007;46:15018–26.
[379] Iwema T, Picciocchi A, Traore DA, et al. Structural basis for delivery of the intact [Fe2S2] cluster by monothiol glutaredoxin. Biochemistry 2009;48:6041–3.
[380] Haunhorst P, Berndt C, Eitner S, et al. Characterization of the human monothiol glutaredoxin 3 (PICOT) as iron-sulfur protein. Biochem Biophys Res Commun 2010;394:372–6.
[381] Li H, Mapolelo DT, Dingra NN, et al. The yeast iron regulatory proteins Grx3/4 and Fra2 form heterodimeric complexes containing a [2Fe-2S] cluster with cysteinyl and histidyl ligation. Biochemistry 2009;48:9569–81.
[382] Kim KD, Chung WH, Kim HJ, et al. Monothiol glutaredoxin Grx5 interacts with Fe-S scaffold proteins Isa1 and Isa2 and supports Fe-S assembly and DNA integrity in mitochondria of fission yeast. Biochem Biophys Res Commun 2010;392:467–72.

[383] Belli G, Molina MM, Garcia-Martinez J, et al. Saccharomyces cerevisiae glutaredoxin 5-deficient cells subjected to continuous oxidizing conditions are affected in the expression of specific sets of genes. J Biol Chem 2004;279:12386–95.
[384] Ye H, Jeong SY, Ghosh MC, et al. Glutaredoxin 5 deficiency causes sideroblastic anemia by specifically impairing heme biosynthesis and depleting cytosolic iron in human erythroblasts. J Clin Invest 2010;120:1749–61.
[385] Rodriguez-Manzaneque MT, Ros J, Cabiscol E, et al. Grx5 glutaredoxin plays a central role in protection against protein oxidative damage in Saccharomyces cerevisiae. Mol Cell Biol 1999;19:8180–90.
[386] Linares GR, Xing W, Govoni KE, et al. Glutaredoxin 5 regulates osteoblast apoptosis by protecting against oxidative stress. Bone 2009;44:795–804.
[387] Uzarska MA, Dutkiewicz R, Freibert SA, et al. The mitochondrial Hsp70 chaperone Ssq1 facilitates Fe/S cluster transfer from Isu1 to Grx5 by complex formation. Mol Biol Cell 2013;24:1830–41.
[388] Vilella F, Alves R, Rodriguez-Manzaneque MT, et al. Evolution and cellular function of monothiol glutaredoxins: involvement in iron-sulphur cluster assembly. Comp Funct Genomics 2004;5:328–41.
[389] Tong W.-H, Rouault T. Distinct iron-sulfur cluster assembly complexes exist in the cytosol and mitochondria of human cells. EMBO J 2000;19:5692–700.
[390] Mochel F, Knight MA, Tong WH, et al. Splice mutation in the iron-sulfur cluster scaffold protein ISCU causes myopathy with exercise intolerance. Am J Hum Genet 2008;82:652–60.
[391] Racker E, History of the Pasteur effect and its pathobiology. Mol Cell Biochem 1974;5:17–23.
[392] Aisenberg AC, Potter VR. Studies on the Pasteur effect. II. Specific mechanisms. J Biol Chem 1957;224:1115–27.
[393] Aisenberg AC, Reinafarje B, Potter VR. Studies on the Pasteur effect. I. General observations. J Biol Chem 1957;224:1099–113.
[394] Chan SY, Zhang YY, Hemann C, et al. MicroRNA-210 controls mitochondrial metabolism during hypoxia by repressing the iron-sulfur cluster assembly proteins ISCU1/2. Cell Metab 2009;10:273–84.
[395] Fasanaro P, Greco S, Lorenzi M, et al. An integrated approach for experimental target identification of hypoxia-induced miR-210. J Biol Chem 2009;284:35134–43.
[396] Favaro E, Ramachandran A, McCormick R, et al. MicroRNA-210 regulates mitochondrial free radical response to hypoxia and Krebs cycle in cancer cells by targeting iron sulfur cluster protein ISCU. PLoS One 2010;5:e10345.
[397] Chen Z, Li Y, Zhang H, et al. Hypoxia-regulated microRNA-210 modulates mitochondrial function and decreases ISCU and COX10 expression. Oncogene 2010;29:4362–8.
[398] Kulshreshtha R, Ferracin M, Wojcik SE, et al. A microRNA signature of hypoxia. Mol Cell Biol 2007;27:1859–67.
[399] Camps C, Buffa FM, Colella S, et al. hsa-miR-210 Is induced by hypoxia and is an independent prognostic factor in breast cancer. Clin Cancer Res 2008;14:1340–8.
[400] Lee DC, Romero R, Kim JS, et al. miR-210 targets iron-sulfur cluster scaffold homologue in human trophoblast cell lines: siderosis of interstitial trophoblasts as a novel pathology of preterm preeclampsia and small-for-gestational-age pregnancies. Am J Pathol 2011;179: 590–602.
[401] Muralimanoharan S, Maloyan A, Mele J, et al. MIR-210 modulates mitochondrial respiration in placenta with preeclampsia. Placenta 2012;33:816–23.
[402] Merlo A, de Quiros SB, Secades P, et al. Identification of a signaling axis HIF-1alpha/ microRNA-210/ISCU independent of SDH mutation that defines a subgroup of head and neck paragangliomas. J Clin Endocrinol Metab 2012;97:E2194–200.

[403] McCormick RI, Blick C, Ragoussis J, et al. miR-210 is a target of hypoxia-inducible factors 1 and 2 in renal cancer, regulates ISCU and correlates with good prognosis. Br J Cancer 2013;108:1133–42.
[404] Huang X, Le QT, Giaccia AJ. MiR-210 – micromanager of the hypoxia pathway. Trends Mol Med 2010;16:230–7.
[405] Devlin C, Greco S, Martelli F, et al. miR-210: more than a silent player in hypoxia. IUBMB Life 2011;63:94–100.
[406] Larsson LE, Linderholm H, Mueller R, et al. Hereditary metabolic myopathy with paroxysmal myoglobinuria due to abnormal glycolysis. J Neurol Neurosurg Psychiatry 1964;27:361–80.
[407] Linderholm H, Muller R, Ringqvist T, et al. Hereditary abnormal muscle metabolism with hyperkinetic circulation during exercise. Acta Med Scand 1969;185:153–66.
[408] Kollberg G, Melberg A, Holme E, et al. Transient restoration of succinate dehydrogenase activity after rhabdomyolysis in iron-sulphur cluster deficiency myopathy. Neuromuscul Disord 2011;21:115–20.
[409] Drugge U, Holmberg M, Holmgren G, et al. Hereditary myopathy with lactic acidosis, succinate dehydrogenase and aconitase deficiency in northern Sweden: a genealogical study. J Med Genet 1995;32:344–7.
[410] Linderholm H, Essen-Gustavsson B, Thornell LE. Low succinate dehydrogenase (SDH) activity in a patient with a hereditary myopathy with paroxysmal myoglobinuria. J Intern Med 1990;228:43–52.
[411] Haller RG, Henriksson KG, Jorfeldt L, et al. Deficiency of skeletal muscle succinate dehydrogenase and aconitase. Pathophysiology of exercise in a novel human muscle oxidative defect. J Clin Invest 1991;88:1197–206.
[412] Olsson A, Lind L, Thornell LE, et al. Myopathy with lactic acidosis is linked to chromosome 12q23.3-24.11 and caused by an intron mutation in the ISCU gene resulting in a splicing defect. Hum Mol Genet 2008;17:1666–72.
[413] Kollberg G, Tulinius M, Melberg A, et al. Clinical manifestation and a new ISCU mutation in iron-sulphur cluster deficiency myopathy. Brain 2009;132:2170–9.
[414] Crooks DR, Jeong SY, Tong WH, et al. Tissue specificity of a human mitochondrial disease: differentiation-enhanced mis-splicing of the Fe-S scaffold gene ISCU renders patient cells more sensitive to oxidative stress in ISCU myopathy. J Biol Chem 2012;287:40119–30.
[415] Crooks DR, Natarajan TG, Jeong SY, et al. Elevated FGF21 secretion, PGC-1alpha and ketogenic enzyme expression are hallmarks of iron-sulfur cluster depletion in human skeletal muscle. Hum Mol Genet 2013;23:24–39.
[416] Reitman ML. FGF21: a missing link in the biology of fasting. Cell Metab 2007;5:405–7.
[417] Sanaker PS, Toompuu M, Hogan VE, et al. Differences in RNA processing underlie the tissue specific phenotype of ISCU myopathy. Biochim Biophys Acta 2010;1802:539–44.
[418] Nordin A, Larsson E, Holmberg M. The defective splicing caused by the ISCU intron mutation in patients with myopathy with lactic acidosis is repressed by PTBP1 but can be derepressed by IGF2BP1. Hum Mutat 2012;33:467–70.
[419] Skladal D, Halliday J, Thorburn DR. Minimum birth prevalence of mitochondrial respiratory chain disorders in children. Brain 2003;126:1905–12.
[420] Distelmaier F, Koopman WJ, van den Heuvel LP, et al. Mitochondrial complex I deficiency: from organelle dysfunction to clinical disease. Brain 2009;132:833–42.
[421] Lazarou M, Thorburn DR, Ryan MT, et al. Assembly of mitochondrial complex I and defects in disease. Biochim Biophys Acta 2009;1793:78–88.
[422] Calvo SE, Tucker EJ, Compton AG, et al. High-throughput, pooled sequencing identifies mutations in NUBPL and FOXRED1 in human complex I deficiency. Nat Genet 2010;42:851–8.

[423] Tucker EJ, Mimaki M, Compton AG, et al. Next-generation sequencing in molecular diagnosis: NUBPL mutations highlight the challenges of variant detection and interpretation. Hum Mutat 2012;33:411–8.
[424] Tenisch EV, Lebre AS, Grevent D, et al. Massive and exclusive pontocerebellar damage in mitochondrial disease and NUBPL mutations. Neurology 2012;79:391.
[425] Kevelam SH, Rodenburg RJ, Wolf NI, et al. NUBPL mutations in patients with complex I deficiency and a distinct MRI pattern. Neurology 2013;80:1577–83.
[426] Wydro MM, Balk J. Insights into the pathogenic character of a common NUBPL branch-site mutation associated with mitochondrial disease and complex I deficiency using a yeast model. Dis Model Mech 2013;6:1279–84.
[427] Sazanov LA, Hinchliffe P. Structure of the hydrophilic domain of respiratory complex I from Thermus thermophilus. Science 2006;311:1430–6.
[428] Bych K, Netz DJ, Vigani G, et al. The essential cytosolic iron-sulfur protein Nbp35 acts without Cfd1 partner in the green lineage. J Biol Chem 2008;283:35797–804.
[429] Wydro MM, Sharma P, Foster JM, et al. The evolutionarily conserved iron-sulfur protein INDH is required for complex I assembly and mitochondrial translation in Arabidopsis. Plant Cell 2013; 25:4014–27.
[430] Seyda A, Newbold RF, Hudson TJ, et al. A novel syndrome affecting multiple mitochondrial functions, located by microcell-mediated transfer to chromosome 2p14-2p13. Am J Hum Genet 2001;68:386–96.
[431] Cameron JM, Janer A, Levandovskiy V, et al. Mutations in Iron-Sulfur Cluster Scaffold Genes NFU1 and BOLA3 Cause a Fatal Deficiency of Multiple Respiratory Chain and 2-Oxoacid Dehydrogenase Enzymes. Am J Hum Genet 2011;89:486–95.
[432] Schonauer MS, Kastaniotis AJ, Kursu VA, et al. Lipoic acid synthesis and attachment in yeast mitochondria. J Biol Chem 2009;284:23234–42.
[433] Ferrer-Cortes X, Font A, Bujan N, et al. Protein expression profiles in patients carrying NFU1 mutations. Contribution to the pathophysiology of the disease. J Inherit Metab Dis 2013;36:841–7.
[434] Cicchillo RM, Lee KH, Baleanu-Gogonea C, et al. Escherichia coli lipoyl synthase binds two distinct [4Fe-4S] clusters per polypeptide. Biochemistry 2004;43:11770–81.
[435] Dos Santos PC, Smith AD, Frazzon J, et al. Iron-sulfur cluster assembly: NifU-directed activation of the nitrogenase Fe protein. J Biol Chem 2004;279:19705–11.
[436] Yabe T, Morimoto K, Kikuchi S, et al. The Arabidopsis chloroplastic NifU-like protein CnfU, which can act as an iron-sulfur cluster scaffold protein, is required for biogenesis of ferredoxin and photosystem I. Plant Cell 2004;16:993–1007.
[437] Touraine B, Boutin JP, Marion-Poll A, et al. Nfu2: a scaffold protein required for [4Fe-4S] and ferredoxin iron-sulphur cluster assembly in Arabidopsis chloroplasts. Plant J 2004;40:101–11.
[438] Balasubramanian R, Shen G, Bryant DA, et al. Regulatory roles for IscA and SufA in iron homeostasis and redox stress responses in the cyanobacterium Synechococcus sp. strain PCC 7002. J Bacteriol 2006;188:3182–91.
[439] Leon S, Touraine B, Ribot C, et al. Iron-sulphur cluster assembly in plants: distinct NFU proteins in mitochondria and plastids from Arabidopsis thaliana. Biochem J 2003;371:823–30.
[440] Liu Y, Cowan JA. Iron sulfur cluster biosynthesis. Human NFU mediates sulfide delivery to ISU in the final step of [2Fe-2S] cluster assembly. Chem Commun (Camb) 2007:3192–4.
[441] Haack TB, Rolinski B, Haberberger B, et al. Homozygous missense mutation in BOLA3 causes multiple mitochondrial dysfunctions syndrome in two siblings. J Inherit Metab Dis 2013;36: 55–62.
[442] Huynen MA, Spronk CA, Gabaldon T, et al. Combining data from genomes, Y2H and 3D structure indicates that BolA is a reductase interacting with a glutaredoxin. FEBS Lett 2005;579:591–6.

[443] Kumanovics A, Chen OS, Li L, et al. Identification of FRA1 and FRA2 as genes involved in regulating the yeast iron regulon in response to decreased mitochondrial iron-sulfur cluster synthesis. J Biol Chem 2008;283:10276–86.
[444] Haunhorst P, Hanschmann EM, Brautigam L, et al. Crucial function of vertebrate glutaredoxin 3 (PICOT) in iron homeostasis and hemoglobin maturation. Mol Biol Cell 2013;24:1895–903.
[445] Li H, Mapolelo DT, Dingra NN, et al. Histidine 103 in Fra2 is an iron-sulfur cluster ligand in the [2Fe-2S] Fra2-Grx3 complex and is required for in vivo iron signaling in yeast. J Biol Chem 2011;286:867–76.
[446] Waller JC, Alvarez S, Naponelli V, et al. A role for tetrahydrofolates in the metabolism of iron-sulfur clusters in all domains of life. Proc Natl Acad Sci USA 2010;107:10412–7.
[447] Sheftel AD, Wilbrecht C, Stehling O, et al. The human mitochondrial ISCA1, ISCA2, and IBA57 proteins are required for [4Fe-4S] protein maturation. Mol Biol Cell 2012;23:1157–66.
[448] Hu P, Janga SC, Babu M, et al. Global functional atlas of Escherichia coli encompassing previously uncharacterized proteins. PLoS Biol 2009;7:e96.
[449] Ajit Bolar N, Vanlander AV, Wilbrecht C, et al. Mutation of the iron-sulfur cluster assembly gene IBA57 causes severe myopathy and encephalopathy. Hum Mol Genet 2013;22:2590–602.
[450] Shan Y, Napoli E, Cortopassi G. Mitochondrial frataxin interacts with ISD11 of the NFS1/ISCU complex and multiple mitochondrial chaperones. Hum Mol Genet 2007;16:929–41.
[451] Marquet A, Enzymology of carbon-sulfur bond formation. Curr Opin Chem Biol 2001;5:541–9.
[452] Mueller EG. Trafficking in persulfides: delivering sulfur in biosynthetic pathways. Nat Chem Biol 2006;2:185–94.
[453] Marelja Z, Stocklein W, Nimtz M, et al. A novel role for human Nfs1 in the cytoplasm: Nfs1 acts as a sulfur donor for MOCS3, a protein involved in molybdenum cofactor biosynthesis. J Biol Chem 2008;283:25178–85.
[454] Lim SC, Friemel M, Marum JE, et al. Mutations in LYRM4, encoding iron-sulfur cluster biogenesis factor ISD11, cause deficiency of multiple respiratory chain complexes. Hum Mol Genet 2013;22:4460–73.
[455] Li J, Kogan M, Knight SAB, et al. Yeast mitochondrial protein, Nfs1p, coordinately regulates iron-sulfur cluster proteins, cellular iron uptake, and iron distribution. J Biol Chem 1999;274:33025–34.
[456] Jensen LT, Culotta VS. Role of Saccharomyces cerevisiae ISA1 and ISA2 in iron homeostasis. Mol Cell Biol 2000;20:3918–27.
[457] Lange H, Kaut A, Kispal G, et al. A mitochondrial ferredoxin is essential for biogenesis of cellular iron-sulfur proteins. Proc Natl Acad Sci USA 2000;97:1050–5.
[458] Foury F, Roganti T. Deletion of the mitochondrial carrier genes MRS3 and MRS4 suppresses mitochondrial iron accumulation in a yeast frataxin-deficient strain. J Biol Chem 2002;277:24475–83.
[459] Foury F, Talibi D. Mitochondrial control of iron homeostasis. A genome wide analysis of gene expression in a yeast frataxin-deficient strain. J Biol Chem 2001;276:7762–8.
[460] Hausmann A, Samans B, Lill R, et al. Cellular and mitochondrial remodeling upon defects in iron-sulfur protein biogenesis. J Biol Chem 2008;283:8318–30.
[461] Anderson PR, Kirby K, Orr W, et al. Hydrogen peroxide scavenging rescues frataxin deficiency in a Drosophila model of Friedreich's ataxia. Proc Natl Acad Sci USA 2008;105:611–6.
[462] Naranuntarat A, Jensen LT, Pazicni S, et al. The interaction of mitochondrial iron with manganese superoxide dismutase. J Biol Chem 2009;284:22633–40.
[463] Richardson TE, Kelly HN, Yu AE, et al. Therapeutic strategies in Friedreich's ataxia. Brain Res 2013;1514:91–7.
[464] Pandolfo M, Friedreich ataxia: the clinical picture. J Neurol 2009;256:3–8.
[465] Sturm B, Stupphann D, Kaun C, et al. Recombinant human erythropoietin: effects on frataxin expression in vitro. Eur J Clin Invest 2005;35:711–7.

[466] Acquaviva F, Castaldo I, Filla A, et al. Recombinant human erythropoietin increases frataxin protein expression without increasing mRNA expression. Cerebellum 2008;7:360–5.
[467] Herman D, Jenssen K, Burnett R, et al. Histone deacetylase inhibitors reverse gene silencing in Friedreich's ataxia. Nat Chem Biol 2006;2:551–8.
[468] Rai M, Soragni E, Jenssen K, et al. HDAC inhibitors correct frataxin deficiency in a Friedreich ataxia mouse model. PLoS One 2008;3:e1958.
[469] Kollberg G, Holme E. Antisense oligonucleotide therapeutics for iron-sulphur cluster deficiency myopathy. Neuromuscul Disord 2009;19:833–6.
[470] Qi X, Lewin AS, Sun L, et al. SOD2 gene transfer protects against optic neuropathy induced by deficiency of complex I. Ann Neurol 2004;56:182–91.
[471] Emond M, Lepage G, Vanasse M, et al. Increased levels of plasma malondialdehyde in Friedreich ataxia. Neurology 2000;55:1752–3.
[472] Jauslin ML, Wirth T, Meier T, et al. A cellular model for Friedreich Ataxia reveals small-molecule glutathione peroxidase mimetics as novel treatment strategy. Hum Mol Genet 2002;11: 3055–63.
[473] Seznec H, Simon D, Monassier L, et al. Idebenone delays the onset of cardiac functional alteration without correction of Fe-S enzymes deficit in a mouse model for Friedreich ataxia. Hum Mol Genet 2004;13:1017–24.
[474] Kakhlon O, Breuer W, Munnich A, et al. Iron redistribution as a therapeutic strategy for treating diseases of localized iron accumulation. Can J Physiol Pharmacol 2010;88:187–96.
[475] Boddaert N, Le Quan Sang KH, Rotig A, et al. Selective iron chelation in Friedreich ataxia: biologic and clinical implications. Blood 2007;110:401–8.
[476] Kakhlon O, Manning H, Breuer W, et al. Cell functions impaired by frataxin deficiency are restored by drug-mediated iron relocation. Blood 2008;112:5219–27.
[477] Goncalves S, Paupe V, Dassa EP, et al. Deferiprone targets aconitase: implication for Friedreich's ataxia treatment. BMC Neurol 2008;8:20.
[478] O'Reilly M, Kohn DB, Bartlett J, et al. Gene therapy for rare diseases: summary of a National Institutes of Health workshop, September 13, 2012. Hum Gene Ther 2013;24, 355–362.

# 19 Connecting the biosynthesis of the molybdenum cofactor, Fe-S clusters, and tRNA thiolation in humans

Silke Leimkühler

## 19.1 Introduction

The thiolation of biomolecules is a complex process that involves the activation of sulfur. The L-cysteine desulfurase NFS1 is the main sulfur-mobilizing protein that provides the sulfur from L-cysteine to several important biomolecules in the cell such as iron-sulfur (Fe-S) clusters, molybdopterin (MPT), and thionucleosides of tRNA. Various biomolecules mediate the transfer of sulfur from NFS1 to various biomolecules using different interaction partners. A direct connection among the sulfur-containing molecules, Fe-S clusters, thiolated tRNA, and the molybdenum cofactor (Moco) has been identified. The biosynthesis of Moco in humans originates in mitochondria, which is also the main compartment for Fe-S cluster biosynthesis. The first step of Moco biosynthesis involves the conversion of 5′GTP to cyclic pyranopterin monophosphate (cPMP), a reaction catalyzed by an Fe-S cluster containing protein. After transfer of cPMP to the cytosol, cPMP is converted to MPT by insertion of two sulfur atoms. The sulfur for this reaction is provided by the L-cysteine desulfurase NFS1 in the cytosol. Further sulfur transfer is mediated by the rhodanese-like protein, MOCS3, which forms a persulfide group on its active site cysteine. MOCS3 in humans directly connects Moco biosynthesis and tRNA thiolation in the cytosol by interacting with two proteins, URM1 and MOCS2A. A thiocarboxylate intermediate is formed on URM1 and MOCS2A, from which sulfur is further transferred to the target molecules, tRNA and cPMP, respectively. This review will dissect the sulfur transfer pathway in humans for both tRNA and Moco and their connection to Fe-S cluster biosynthesis and will explore how these reactions occur in the different compartments in the cell.

Sulfur is an essential element to all living organisms, and the presence of sulfur in cofactors was discovered more than a century ago [1]. The trafficking and delivery of sulfur to cofactors and nucleosides is a highly regulated process that occurs by complex sulfur relay systems involving numerous proteins in reactions that remain incompletely understood [2, 3]. In the last few years, several studies of sulfur transfer pathways in the cell concluded that the major enzymes involved in mobilization of sulfur are L-cysteine desulfurases and rhodanese homology domain proteins [3, 4]. These enzymes catalyze the formation of a persulfide group ($R\text{-}S\text{-}S^-$)] on specific conserved cysteine residues, which serves as a sulfur donor for the biosynthesis of sulfur-containing cofactors such as Fe-S clusters, thiamine, biotin, lipoic acid, and MPT and for thiomodification of tRNA [3]. In eukaryotes, only Fe-S clusters, the Moco, and thionucleosides in tRNA can be synthesized *de novo* (Fig. 19.1) [5–7]. The chemistry

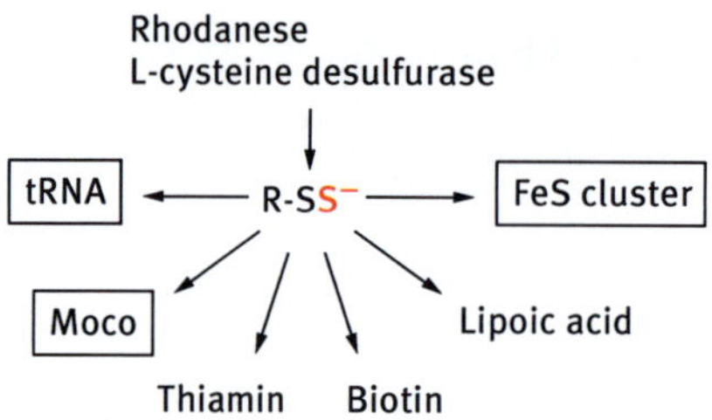


**Fig. 19.1:** Sulfur transfer to sulfur-containing biomolecules *via* protein-bound persulfide. A protein-bound persulfide group can be either formed on a rhodanese-like protein or on an L-cysteine desulfurase. The persulfide sulfur is further transferred by involvement of different proteins for the formation of sulfur-containing cofactors such as Moco, thiamine, biotin, lipoic acid, Fe-S clusters and for the thiolation of tRNA. Sulfur-containing molecules synthesized in humans are boxed.

of persulfide groups involves three oxidation states: $S^0$ (sulfane), $S^{1-}$ (persulfide), or $S^{2-}$ (sulfide) [1]. The terminal sulfur of the persulfide group can serve as a nucleophile to form a disulfide bond (R-S-S-R) with an electrophile [8]. The persulfide is usually generated by L-cysteine desulfurases that use L-cysteine as their sulfur source [3, 4].

L-cysteine desulfurases are pyridoxal-phosphate-containing homodimers, which decompose L-cysteine to L-alanine and sulfane sulfur *via* the formation of an enzyme-bound persulfide intermediate [9]. Further, the persulfide sulfur can be transferred in a relay mechanism to other proteins containing conserved cysteine residues, including the family of rhodanese-like proteins or other sulfur-transferring proteins [4]. The persulfide sulfur from the L-cysteine desulfurase is further incorporated either directly or *via* sulfur relay systems into the biosynthetic pathways of several sulfur-containing biofactors, thus providing an elegant mechanism for making sulfur atoms available without releasing them in solution [3, 8]. Many of the sulfur-relay proteins are highly conserved across species, e.g. the L-cysteine desulfurase that initiates the sulfur transfer step, the enzyme IscS from *Escherichia coli*, shares 60% amino acid sequence identity with its human homologue, NFS1 [3]. In this chapter, we focus on the connections between the pathways for the biosynthesis of three sulfur-containing biofactors in humans: Fe-S clusters, Moco, and thiolated tRNAs.

Fe-S clusters, which are thought to be among the earliest catalysts in the evolution of biomolecules, serve as electron carriers in redox reactions, regulatory sensors, stabilizers of protein structure, and chemical catalysts [1]. The mitochondria constitute the main compartment for the biosynthesis of Fe-S clusters in humans. However, Fe-S cluster biosynthesis also occurs in the cytosol, and some protein components of the pathway have also been identified in the nucleus [5]. The main proteins required for Fe-S cluster biosynthesis in mitochondria are NFS1, ISD11, ISCU, and FXN, which form the quaternary core complex [10–12]. After the synthesis of the Fe-S clusters on ISCU, the cluster can be transferred to acceptor proteins within the mitochondria [13]. Among these proteins is the MOCS1A protein, which is involved in the first step of Moco biosynthesis [14]. MOCS1A contains two [4Fe-4S] clusters.

Moco is a tricyclic pyranopterin containing a unique dithiolene group to which the molybdenum atom is coordinated [15]. Moco biosynthesis first starts with the conversion from 5′-GTP to cPMP, a reaction that is catalyzed by MOCS1A and MOCS1B in mitochondria (Fig. 19.2) [14]. All further steps for the formation of Moco from cPMP are localized in the cytosol and do not require Fe-S-containing proteins [6, 16]. However, it is believed that the sulfur for the dithiolene group originates from L-cysteine, which is mobilized by NFS1 in the cytosol [17, 18]. Thus, NFS1 would mobilize the sulfur for both pathways, Moco biosynthesis and Fe-S cluster biosynthesis, although mobilization of sulfur for Moco synthesis occurs in the cytosolic compartment. The direct insertion of two sulfur atoms into cPMP for the formation of MPT is catalyzed by the proteins MOCS2A, MOCS2B, and MOCS3 [19, 20]. The dithiolene group acts as a backbone for the ligation of the molybdenum atom in Moco. MOCS3 belongs to the class of rhodanese-like proteins, which accepts the sulfur from NFS1 and transfers the activated sulfur in a relay system to various acceptor proteins [17]. MOCS3 is not only involved in Moco biosynthesis because besides interacting with MOCS2A, it also interacts with URM1, which acts as a sulfur acceptor protein that is involved in the thiolation of some tRNAs [21]. Thus, Moco biosynthesis and tRNA thiolation are connected and share the same sulfur delivery pathway composed of NFS1 and MOCS3.

The wobble bases of tRNAs contain two thiouridines, 5-methoxycarbonylmethyl-2-thiouridine ($mcm^5s^2U_{34}$) in cytoplasmic tRNAs and 5-taurinomethyl-2-thiouridine ($\tau m^5s^2U_{34}$) in mitochondrial tRNAs [22]. More than five proteins are involved in the insertion of the sulfur group in $mcm^5s^2U$ in humans in the cytosol: NFS1, URM1, MOCS3, CTU1, and CTU2 [7, 21]. For the thiolation of tRNA, thiomodification of uridine in the 2 position is known to ensure accurate deciphering of the genetic code and stabilization of tRNA structure [22, 23].

In this chapter, we will describe the biosynthesis of each biomolecule with respect to the different compartments with a focus on mitochondria, cytosol, and the nucleus. This chapter will mainly focus on Moco biosynthesis and tRNA thiomodification because mammalian Fe-S cluster biosynthesis is described in detail in other chapters.

## 19.2 Pathways for the formation of Moco and thiolated tRNAs in humans

### 19.2.1 Moco biosynthesis in mammals

In all organisms including humans, Moco is synthesized by a conserved biosynthetic pathway that can be divided into three steps, according to the stable biosynthetic intermediates that can be isolated and were first studied in *E. coli* (Fig. 19.2) [24]: the synthesis of cPMP [25], conversion of cPMP into MPT by introduction of two sulfur atoms [26], and insertion of molybdate to form Moco [27]. Moco is present in five enzymes in humans, sulfite oxidase (SUOX), xanthine oxidoreductase (XOR),

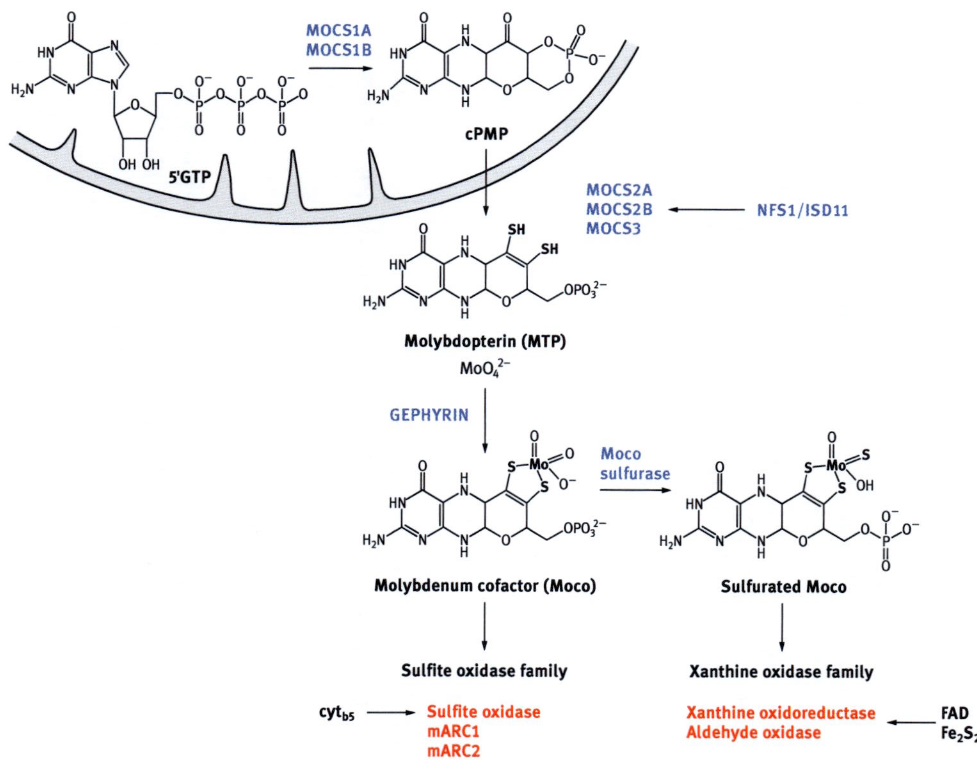


**Fig. 19.2:** The biosynthesis of Moco. Shown is a scheme of the biosynthetic pathway for Moco biosynthesis in humans. The proteins involved in the reactions are colored in blue. Moco can be modified by the replacement of one oxo ligand by a sulfido ligand, forming the mono-oxo Moco. The molybdenum-containing enzymes are divided into different families (xanthine oxidase and sulfite oxidase families in humans) according to their active-site structures.

aldehyde oxidase (AOX1), and two mitochondrial amidoxime-reducing components, mARC1 and mARC2 [6, 28]. SUOX is the only one of the five molybdoenzymes that is essential for humans [29, 30].

Six proteins that directly catalyze Moco biosynthesis have been identified in humans: MOCS1A, MOCS1B, MOCS2A, MOCS2B, MOCS3, and GPHN (gephyrin) [6]. Genes and the encoded proteins were named in humans as MOCS (molybdenum cofactor synthesis) [31–33].

#### 19.2.1.1 Conversion of 5′-GTP to cPMP

The biosynthesis of Moco starts from 5′-GTP, which results in the formation of cPMP, the first stable intermediate of Moco biosynthesis [25] (Fig. 19.3). cPMP is an oxygen-sensitive 6-alkyl pterin with a cyclic phosphate group at the C2′ and C4′ atoms [34, 35]. The *MOCS1* locus encodes two proteins, MOCS1A and MOCS1B, which are involved in the conversion of 5′-GTP into cPMP [36]. MOCS1A belongs to the superfamily of *S*-adenosylmethionine (SAM)-dependent radical enzymes [14]. Members of this family catalyze the formation of protein and/or substrate radicals by reductive cleavage of SAM by a [4Fe-4S] cluster [37]. MOCS1A is a protein containing two oxygen-sensitive Fe-S clusters, each of which is coordinated by only three cysteine residues. The N-terminal [4Fe-4S] cluster, present in all radical SAM proteins, binds SAM and carries out the reductive cleavage of SAM to generate the 5′-deoxyadenosyl radical, which subsequently initiates the transformation of 5′-GTP bound through the C-terminal [4Fe-4S] cluster [14].

The mechanism of conversion of 5′-GTP to cPMP has been best studied using the bacterial proteins MoaA and MoaC [38, 39], but it has been shown that the human proteins catalyze the reaction in the same manner [40]. As shown in Fig. 19.3, formation of cPMP from GTP involves several steps including H-abstraction by the 5′-deoxyadenosyl radical of MOCS1A (MoaA) from the 3′ position of the ribose, attack of the C8 in the guanine ring by the C3′ radical, and formation of a (8*S*)-3′,8-cyclo-7,8-dihydroguanosine 5′-triphosphate (3′,8-c$H_2$GTP) intermediate, which further serves as a substrate for MOCS1B (MoaC) [40]. The additional reducing equivalents required for this step might be provided by the C-terminal [4Fe-4S] cluster in MOCS1A (MoaA). The reaction of MOCS1A (MoaC) is suggested to occur *via* general acid/base catalysis, converting 3′,8-c$H_2$GTP to cPMP, including pyrophosphate cleavage and formation of the cyclic phosphate group [40].

#### 19.2.1.2 Conversion of cPMP to MPT

The next step involves the conversion of cPMP to MPT in which two sulfur atoms are incorporated in the C1′ and C2′ positions of cPMP [34] (Fig. 19.4). This reaction is catalyzed by MPT synthase, a protein consisting of two small (~10 kDa) and two large subunits (~21 kDa), encoded by *MOCS2A* and *MOCS2B*, respectively [41]. It was shown

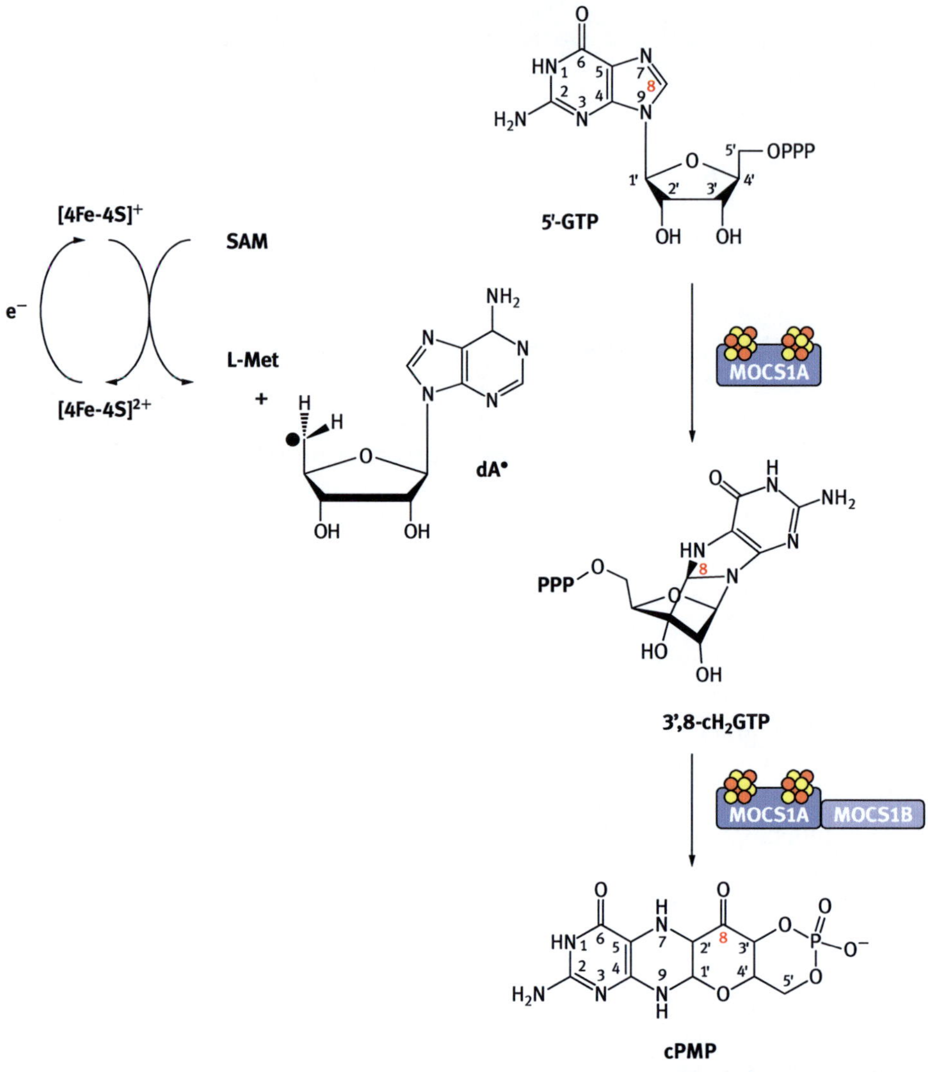


**Fig. 19.3:** Synthesis of cPMP from 5′-GTP. All carbon atoms of the 5′-GTP are found within cPMP. The C8 atom from the guanine ring is inserted between the C2′ and the C3′ atoms of the ribose. This reaction is catalyzed by the MOCS1A protein, a SAM-dependent enzyme. The MOCS1A/MOCS1B fusion protein cleaves the pyrophosphate group of the cyclic intermediate. cPMP is shown in the tetrahydropyrano form with a keto group at the C1′ position.

that MPT synthase carries the sulfur in form of a thiocarboxylate at the C-terminal glycine of MOCS2A [20, 42]. The central dimer is formed by two MOCS2B subunits containing one MOCS2A at each end, as revealed by the crystal structure of the bacterial homologues [43]. It was shown for the *E. coli* proteins that the two MOCS2A/MOCS2B dimers act independently. Thus, for the insertion of two sulfurs into cPMP,

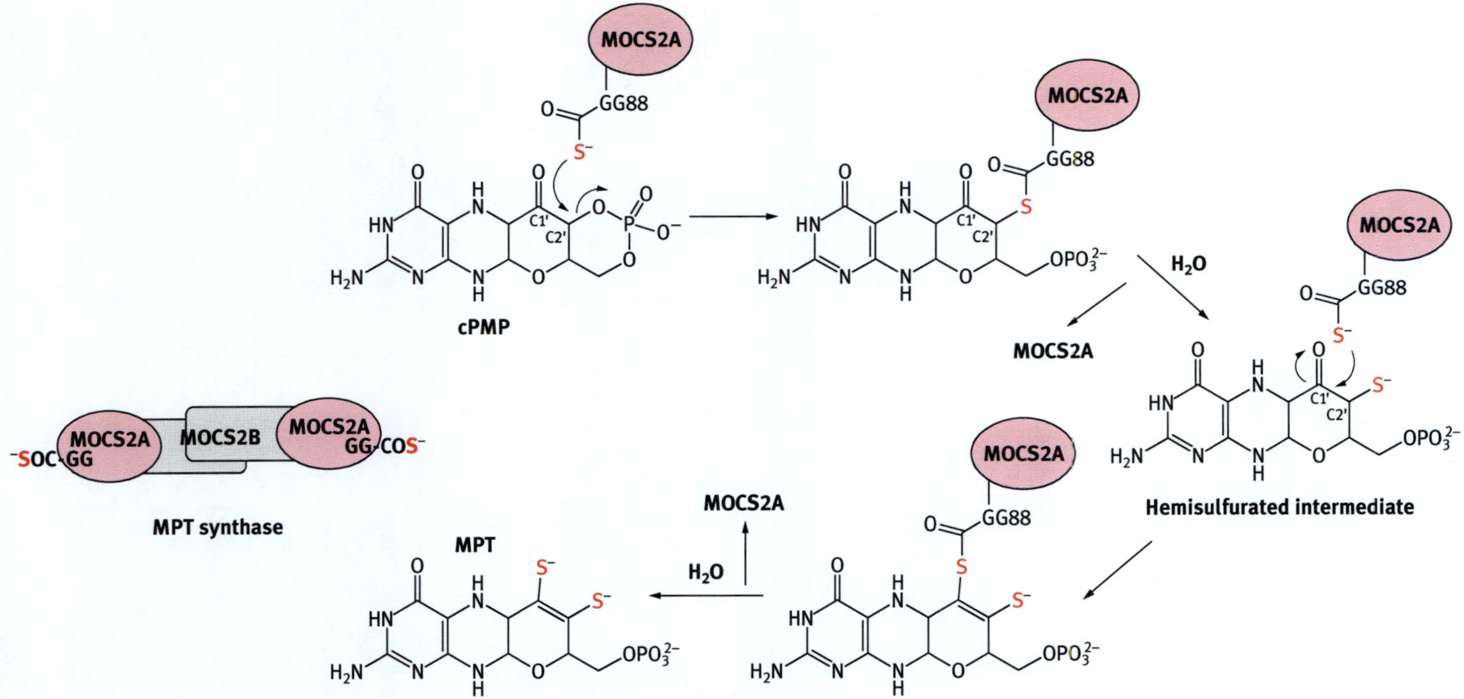


**Fig. 19.4:** The biosynthesis of MPT from cPMP. In the MPT synthase mechanism, the initial attack, and transfer of the first thiocarboxylated MOCS2A sulfur atom occurs at the C2′ position, coupled to the hydrolysis of the cPMP cyclic phosphate. An intermediate is formed in which the MOCS2A C-terminus is covalently linked to the substrate *via* a thioester linkage that is subsequently hydrolyzed by a water molecule to generate a hemi-sulfurated intermediate at C2′. Opening of the cyclic phosphate shifts the location of the intermediate within the complex to a position where C1′ becomes more accessible. A new MOCS2A thiocarboxylate attacks the C1,′ resulting again in a second covalent intermediate that is converted to MPT *via* the elimination of a water molecule and hydrolysis of the thioester intermediate. On the left side, the MPT synthase tetramer is shown. During the reaction, cPMP remains bound to the MOCS2B molecule. The mechanism was adapted from the one proposed for the *E. coli* proteins [44].

two MOCS2A proteins are required at each end of the MPT synthase tetramer [45]. The first sulfur is added by one MOCS2A molecule at the C2′ position of cPMP (Fig. 19.4), a reaction that is coupled to the hydrolysis of the cPMP cyclic phosphate [44]. During the course of this reaction, a hemi-sulfurated intermediate is formed in which the MOCS2A C-terminus is covalently linked to the substrate *via* a thioester linkage, which subsequently is hydrolyzed by a water molecule. After transfer of its thiocarboxylate sulfur to cPMP, the first MOCS2A subunit dissociates from the MPT synthase complex [44, 45]. During the reaction of the first sulfur transfer, the opening of the cyclic phosphate is proposed to shift the location of the intermediate within the protein so that the C1′ position now becomes more accessible to attack by the second MOCS2A thiocarboxylate (Fig. 19.4). This results in a second covalent intermediate that is converted to MPT *via* the elimination of a water molecule and hydrolysis of the thioester intermediate. During the reaction, cPMP and the hemisulfurated intermediate remain bound to one MOCS2B subunit [31]. This reaction has been mainly revealed by studies with the homologous proteins MoaD and MoaE from *E. coli*; however, the same reaction can be catalyzed by hybrid complexes formed between the bacterial and the mammalian proteins, indicating that conservation of the reaction components is high among species [20].

The regeneration of sulfur at the C-terminal glycine of MOCS2A is catalyzed by MOCS3 [46] and resembles the first step of the ubiquitin-dependent protein degradation system [47] (Fig. 19.5). It was shown that the N-terminal domain of MOCS3 has homologies to E1-like proteins and activates the C-terminus of MOCS2A (which has a ubiquitin-like β-grasp fold) by addition of AMP forming an activated acyl-adenylate group on Gly88, the last amino acid of a highly conserved double glycine motif [19]. In the second reaction, the MOCS2A acyl-adenylate is converted to a thiocarboxylate by sulfur transfer from a persulfide present at the C-terminal rhodanese-like domain (RLD) of MOCS3 [31, 48]. In this reaction, the deprotonated persulfide group of RLD-C412 serves as a nucleophile at the activated MOCS2A-adenylate to form a disulfide intermediate (Fig. 19.5). Reductive cleavage of the disulfide bond could then occur by attack of another thiol group, e.g. the conserved MOCS3-C239 to form a disulfide bond with C412, and in turn, thiocarboxylated MOCS2A is generated and released [48]. The disulfide bond could be reduced by a thioredoxin system *in vivo*. The sulfur donor for the generation of the persulfide on C412 of MOCS3 was shown to be the L-cysteine desulfurase NFS1 in the cytosol [17, 18]. Here, the sulfur is transferred in a sulfur relay system from the NFS1 persulfide *via* the MOCS3 persulfide to MOCS2A and further to cPMP, resulting in the formation of MPT in the cytosol.

#### 19.2.1.3 Insertion of molybdate into MPT and further modification of Moco

After synthesis of the dithiolene moiety in MPT, the chemical backbone is built for binding and coordination of the molybdenum atom. In humans, the *GPHN* gene encodes a two-domain protein known as gephyrin, which consists of an N-terminal

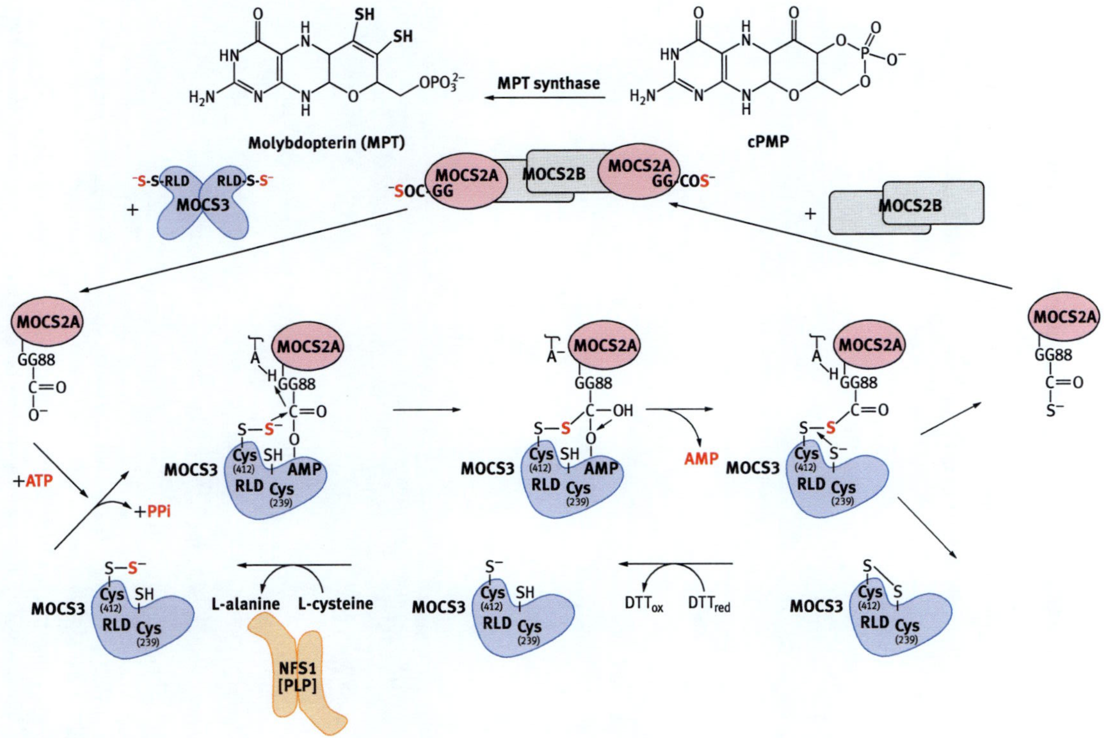


**Fig. 19.5:** Formation of the thiocarboxylate group on MOCS2A. cPMP is converted to MPT by the transfer of two sulfur groups from the C-terminal thiocarboxylate of the MOCS2A subunit of MPT synthase. Regeneration of the MOCS2A-SH sulfur occurs in a MOCS2A/MOCS3 complex. Here, adenylated MOCS2A is formed by attachment of an AMP moiety at the N-terminal E1-like domain of MOCS3. MOCS2A-AMP is then sulfurated by a protein-bound persulfide on Cys412 from the C-terminal RLD. Likely, an MOCS3-MOCS2A disulfide intermediate is formed during the reaction, which is further cleaved by reductive cleavage (e.g. MOCS3-Cys239). The disulfide bond formed between MOCS3-Cys412 and MOCS3-Cys239 is likely reduced by the thioredoxin system *in vivo* (*in vitro* reduction by DTT is shown). The sulfur donor for the persulfide group on MOCS3-Cys412 is NFS1, which acquires the sulfur from L-cysteine. After formation of the thiocarboxylate group, MOCS2A-SH dissociates from the MOCS3 dimer and reassociates with MOCS2B, forming the active MPT synthase.

domain that binds MPT and forms an MPT-AMP intermediate and a C-terminal domain that inserts the molybdate anion and splits the AMP from the activated MPT-AMP intermediate in a molybdate and $Mg^{2+}$-dependent manner [49, 50]. This reaction leads to the formation of Moco [31] (Fig. 19.6). Apart from the Moco biosynthesis, GPHN is also involved in synaptic anchoring of inhibitory ligand-gated ion channels [51]. Alternative splicing has been proposed to contribute to functional diversity of gephyrin proteins within the cell [52].

The completed Moco can be directly inserted into SUOX or mARC1 and mARC2 [28]. For XOR (also known as XDH) and AOX1, Moco is further modified by an exchange of the equatorial oxygen ligand by sulfur, forming the sulfurated or mono-oxo form of Moco [53, 54] (Fig. 19.2). This reaction is carried out by a Moco sulfurase, MOCOS (also known as HMCS) in humans, a two-domain protein with a N-terminal L-cysteine desulfurase domain and a C-terminal Moco binding (MOSC) domain [28, 55, 56]. L-Cysteine is the sulfur source in this reaction [57]. The Moco likely gets sulfurated while bound to the MOCOS protein and is then inserted into XOR and AOX1 [58, 59].

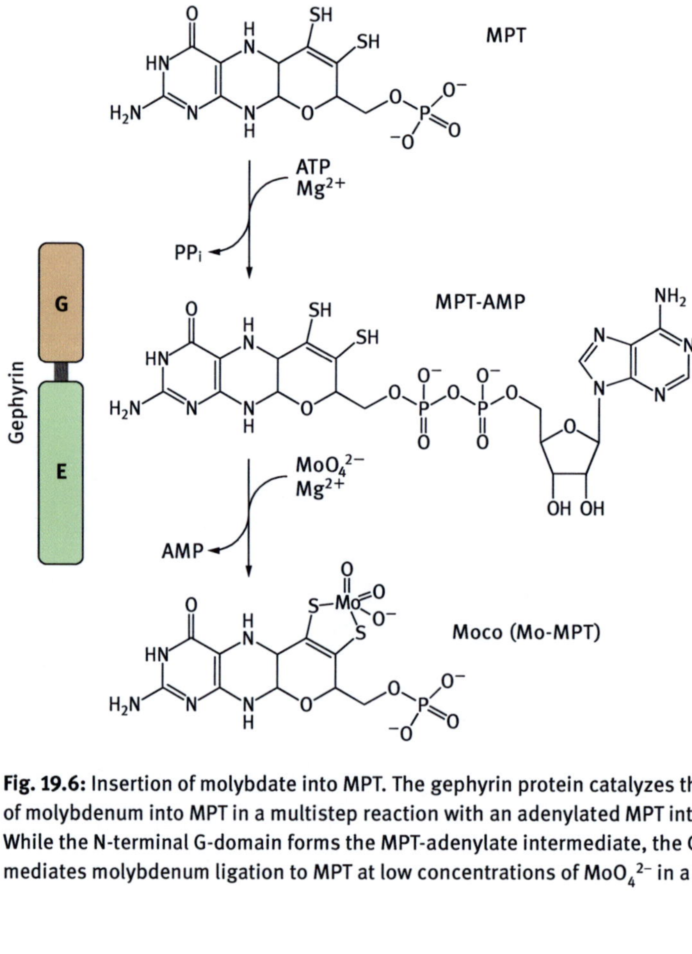


**Fig. 19.6:** Insertion of molybdate into MPT. The gephyrin protein catalyzes the specific incorporation of molybdenum into MPT in a multistep reaction with an adenylated MPT intermediate (MPT-AMP). While the N-terminal G-domain forms the MPT-adenylate intermediate, the C-terminal E-domain mediates molybdenum ligation to MPT at low concentrations of $MoO_4^{2-}$ in a $Mg^{2+}$-dependent manner.

#### 19.2.1.4 Finishing Moco biosynthesis: maturation and insertion into complex molybdoenzymes

After insertion of the molybdenum atom into MPT, the completed Moco can be either directly inserted into SUOX or mARC1 and mARC2 or is further modified by an exchange of the equatorial oxygen ligand by sulfur, forming the sulfurated or mono-oxo form of Moco present in AOX1 or XOR [28]. The basis of the structures of their molybdenum centers divides the molybdoenzymes into three distinct families, two of which are present in eukaryotes (Fig. 19.2): the xanthine oxidase family, the sulfite oxidase family, and the dimethylsulfoxide (DMSO) reductase family (not shown) [54]. The xanthine oxidase family is characterized by an MPT-$Mo^{VI}$OS(OH) core in the oxidized state, with one MPT equivalent coordinated to the metal and no additional ligand from the polypeptide chain. The sulfido-group is cyanide labile. Enzymes of the sulfite oxidase family coordinate a single equivalent of the pterin cofactor with an MPT-$Mo^{VI}O_2$ core in its oxidized state and usually an additional cysteine ligand, which is provided by the polypeptide chain of the enzyme. The DMSO-reductase family is exclusively found in Bacteria and Archaea, and all members have two equivalents of the pterin cofactor bound to the metal. The molybdenum coordination sphere is usually characterized by an $MPT_2$-$Mo^{VI}$S(X) core. The sixth ligand, X, can be a serine, a cysteine, a selenocysteine, or a hydroxide and/or water molecule (not shown).

All molybdoenzymes in eukaryotes are oxidoreductases that catalyze substrate hydroxylation or transfer an oxo group and two electrons from the substrate. The electrons are further transferred from the MoIV, which accepted two electrons from the substrate, and these electrons are donated in single one-electron steps to an electron acceptor such as FAD or cytochromes through an intramolecular pathway that can additionally be mediated by Fe-S clusters [28, 54]. The mechanism of Moco insertion into SUOX is not well understood [60], and little is known so far about how the cytochrome is inserted into the protein in the intermembrane space in humans. Insertion of the mature Moco for members of the xanthine oxidase family is now fairly well understood and has been studied in prokaryotes and eukaryotes [61, 62]. The final step in maturation of the molybdenum center is the incorporation of the catalytically essential sulfur as ligand to the molybdenum atom. The protein involved in the sulfuration binds the Moco and inserts the mature cofactor into the XOR or AOX1 apo-protein after its sulfuration [58, 63, 64]. It was shown that insertion of the molybdenum center occurs after incorporation of both the two [2FeS] clusters and FAD in a protein that is already largely folded into its final three-dimensional structure and assembled as a dimer [65].

SUOX is the only molybdoenzyme that is essential for humans. SUOX is a protein that is located in the intermembrane space of mitochondria in mammals [66]. It catalyzes the terminal step in the catabolism of the sulfur-containing amino acids cysteine and methionine. Electrons derived from sulfite oxidation to sulfate are transferred from Moco *via* the $cyt_{b5}$ cofactor in SUOX to the final electron acceptor cytochrome *c* [67]. A mutation in the gene for SUOX leads to isolated SUOX deficiency, an inherited sulfur metabolic disorder in humans that results in profound birth defects, severe neonatal neurological problems, and early death, and no effective therapies are known [31]. It has been suggested that the pathology associated with SUOX deficiency is due to the toxic buildup of sulfite in the developing brain [68–71]. Excess sulfite can react with disulfide bonds to form sulfonated Cys residues that distort protein structures [30]. In addition, the lack of sulfate in the brain may interfere with the production of sulfatides, lipid sulfate esters found in the white matter of the brain that are a component of myelin. SUOX is synthesized as a pre-protein with a mitochondrial targeting sequence that localizes to the mitochondrial intermembrane space after processing [60]. Subsequently, insertion of Moco, insertion of the $cyt_{b5}$ and dimerization occurs in the intermembrane space [60].

Recent investigations of the aerobic reduction of amidoxime structures led to the discovery of a previously unknown molybdenum-containing enzyme system [28, 72]. It was named "mitochondrial amidoxime-reducing component" (mARC) because, initially, *N*-reduction of amidoxime structures was studied with this enzyme purified from mammalian liver mitochondria. The human genome harbors two mARC genes, referred to as *hmARC1* and *hmARC2*, which are organized in a tandem arrangement on chromosome 1 (designated as *MOSC1* and *MOSC2* in the databases) [73]. The two enzymes form the catalytic part of a three-component enzyme system, consisting of mARC, heme/cytochrome $b_5$, and NADH/FAD-dependent cytochrome $b_5$ reductase. It was shown that the mARC proteins are associated with the outer mitochondrial membrane [74]. However, nothing is known about the targeting, assembly, and cofactor insertion of the protein complex. Additionally, although many N-hydroxylated compounds have been found to serve as substrates for native and recombinant mARC proteins, the physiological substrates and physiological functions of mARC proteins remain unknown [28].

In contrast to SUOX and mARC1 and mARC2, XOR and AOX1 are cytosolic enzymes. XOR generally is involved in the later stages of purine catabolism, catalyzing the oxidation of hypoxanthine to xanthine and of xanthine to uric acid, which is finally excreted in the urine [75]. XOR and AOX1 belong to the class of molybdoflavoenzymes, containing in addition to Moco, two distinct [2Fe-2S] centers and FAD as catalytically acting units [76, 77]. XOR in mammals is synthesized as the dehydrogenase form (XDH, EC 1.17.1.4) but can be converted to the oxidase form (XO, EC 1.1.3.22) either reversibly by oxidation of sulfhydryl residues of the protein molecule or irreversibly by proteolysis [78]. XDH shows a preference for $NAD^+$ reduction at the FAD reaction site, whereas XO exclusively uses dioxygen as its terminal electron acceptor, leading to the formation of superoxide and hydrogen peroxide [79]. The enzyme has been implicated in diseases characterized by oxygen radical-induced tissue damage, such as post-ischemic reperfusion injury [80]. The oxidation of xanthine takes place at the molybdenum center and the electrons thus introduced are rapidly distributed in one-electron transfer reactions to the Fe-S centers and finally to FAD according to their relative redox potentials [54]. The reoxidation of the reduced enzyme by either $NAD^+$ or molecular oxygen occurs through the FAD cofactor [81].

Although the biochemical function of XOR is well established, the biochemical and physiological functions of AOX1 are still largely obscure. Only limited information is available about the physiological substrates of AOX1 or about the role of this enzyme in the mammalian organism [77]. It has been shown that the enzyme is located in the cytosol and is involved in the metabolism of drugs and xenobiotics of toxicological importance, and AOX1 metabolizes N-heterocyclic compounds and aldehydes of pharmacological relevance [77]. Single monogenic AOX1 deficits have not been described yet in mammalia; absence of a phenotype is perhaps not surprising because genetic deficiencies in the downstream Moco sulfurase (*MCSF*) gene are not associated with pathophysiological consequences [28]. It is likely that AOX1 serves to detoxify exogenously derived unphysiological compounds of wide structural diversity in animals, and it is believed that the absence of AOX1 produces symptoms in animals after high intake of such xenobiotics.

Although Moco is inserted into XOR and AOX1 in the cytosol by the MCSF after the insertion of the terminal sulfido ligand, nothing is known so far about which proteins are involved in Fe-S cluster insertion or FAD insertion in the cytosol. However, the presence of two [2Fe-2S] clusters in XOR and AOX1 directly link the activity of these two enzymes to the cytosolic iron-sulfur cluster assembly (CIA) machinery for Fe-S cluster biosynthesis. It still remains to be elucidated which proteins are directly involved in [2Fe-2S] cluster biosynthesis and insertion for cytosolic molybdo-flavoenzymes in mammals.

### 19.2.2 The role of tRNA thiolation in the cell

Post-transcriptional RNA modifications are a characteristic structural feature of RNA molecules. The diverse modifications have been shown to play critical roles in biogenesis, metabolism, structural stability, and function of RNA molecules [23]. More than 100 different RNA modifications have been reported to date [22, 82]. These post-transcriptional modifications are required for several functions in translation including codon recognition, maintenance of reading frame, stabilization of tertiary structure and directing the specificity of the codons recognition by the ribosome. Base modifications at the wobble positions in anticodon loops of tRNA have been shown to play important roles in deciphering genetic codes, which include the precise codon-anticodon interactions at the ribosomal A-site. In particular, the wobble bases of tRNAs for Glu, Gln and Lys are modified, and sulfurated to form 5-methyl-2-thiouridine derivatives ($xm^5s^2U$), such as 5-taurinomethyl-2-thiouridine ($\tau m^5s^2U$) in mammalian mitochondrial tRNAs, and 5-methoxycarbonylmethyl-2-thiouridine ($mcm^5s^2U$) in eukaryotic cytoplasmic tRNAs [22]. The result of these thio modifications is that the conformation of $xm^5s^2U$ is trapped in the C3′-*endo* form of the ribose because the large van der Waals' radius of the 2-thio group causes a steric clash with its 2′-OH group [22, 83]. This conformational rigidity causes preferential pairing of the $xm^5s^2U$-modified bases with purines and prevents misreading of codons ending in pyrimidines [83–85].

Recently, it was shown that this 2-thio group of $mcm^5s^2U$ is required for efficient codon recognition on the ribosome [86]. Lack of the $xm^5s^2U$ modification in the mutant mitochondrial $tRNA^{Lys}$ from individuals with myoclonus epilepsy associated with ragged-red fibers (MERRFs) results in a marked defect in all mitochondrial translation [87]. For the thiolation and formation of the $mcm^5s^2U$ in the cytoplasm of eukaryotes, the biosynthesis of the 5-methoxycarbonylmethyl group of the uracil ring is required for efficient 2-thiouridine formation in the cytoplasm [7]. In humans it was shown that the proteins MOCS3, URM1, CTU1, and CTU2 are essential for the biogenesis of the 2-thiouridine group in cytoplasmic tRNAs Lys, Gln, and Glu (Fig. 19.7) [88]. The URM1 protein (ubiquitin-related modifier) was shown to have a ubiquitin-like β-grasp-fold

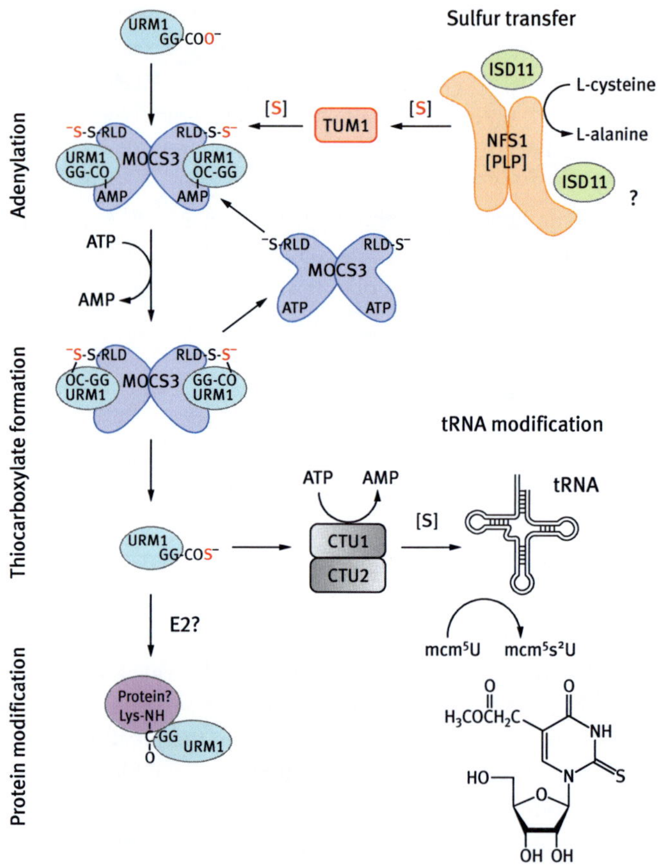


**Fig. 19.7:** Proposed mechanism of 5-methoxycarbonylmethyl-2-thiouridine ($mcm^5s^2U^{34}$) in humans. MOCS3 activates URM1 in the presence of ATP by formation of an acyl-adenylate bond. URM1 is further transferred to a persulfide group on Cys412 of MOCS3-RLD, forming a disulfide bond. URM1-thiocarboxylate is released and transfers the sulfur further onto uridine 34 on the lysine tRNA, helped by the CTU1 and CTU2 proteins under ATP consumption. URM1 was also shown to be conjugated to target proteins *via* a lysine-isopeptide bond, showing that it is a dual-function protein.

and contains a conserved C-terminal double-glycine motif on which a thiocarboxylate group is formed for direct sulfur tranfer to $mcm^5U_{34}$ in tRNA [89–92]. The sulfur is transferred to URM1 *via* MOCS3, which was originally recognized for its role in Moco biosynthesis (see Section 19.2.1.2) [21]. MOCS3 activates URM1 in the presence of ATP by formation of an acyl-adenylate bond. URM1 is further transferred to a persulfide group on Cys412 of MOCS3-RLD, forming a disulfide bond (Fig. 19.7). The disulfide bond can be cleaved by C239 of MOCS3, which releases thiocarboxylated URM1. URM1-thiocarboxylate further transfers the sulfur onto uridine 34 of the tRNA, aided by the CTU1 and CTU2 proteins and ATP consumption [21]. As described Section 19.2.1.2, NFS1 was shown to be the protein that supplies sulfur to MOCS3, which links tRNA modification to the sulfur supplied by L-cysteine desulfurases [17]. Additionally and in contrast to MOCS2A, URM1 was also shown to be conjugated to target protein *via* a lysine-isopeptide bond, revealing that it is a dual-function protein [88]. Among the targets for urmylation are the components of the tRNA thiolation machinery itself (CTU1 and CTU2) and a deubiquitinating-enzyme (USP15) and, in addition, proteins involved in nuclear transport such as the cellular apoptosis susceptibility protein, CAS, which promotes the shuttling of proteins between cytosol and nucleus [88]. However, the role of the protein urmylation is not completely clear to date and might only occur under certain conditions in the cell such as oxidative stress.

## 19.3 The connection between sulfur-containing biomolecules and their distribution in different compartments in the cell

### 19.3.1 Sulfur transfer in mitochondria

Mitochondria are the main compartment for Fe-S cluster biosynthesis [93]. The initial phase of mitochondrial Fe-S cluster biosynthesis in mammals is accomplished by a multimeric protein complex in which a dimer of the L-cysteine desulfurase NFS1 in conjunction with its stabilizing protein ISD11 builds the core for the interaction of two monomers of ISCU and two FXN (frataxin) monomers (Fig. 19.8) [11, 94]. The ISCU protein provides the scaffold for building the Fe-S clusters from sulfur, delivered by NFS1 and iron from an unknown iron donor. FXN might function as an iron donor or act as a regulator for the formation of Fe-S clusters depending on the iron availability in the cell [95]. After the formation of [2Fe-2S] clusters and [4Fe-4S] clusters, the clusters are transferred to target proteins with the help of co-chaperones and chaperones such as HSC20 and HSPA9 [5, 13].

In addition to Fe-S cluster biosynthesis, mitochondria are the compartment for the first step of Moco biosynthesis, namely the conversion of 5′-GTP to cPMP, catalyzed by MOCS1A and MOCS1B as described above (Fig. 19.8) [14]. The *MOCS1* locus encodes two proteins, MOCS1A and MOCS1B [36]. An unusual bicistronic transcript of the *MOCS1* locus was identified, with open reading frames for both *MOCS1A* and

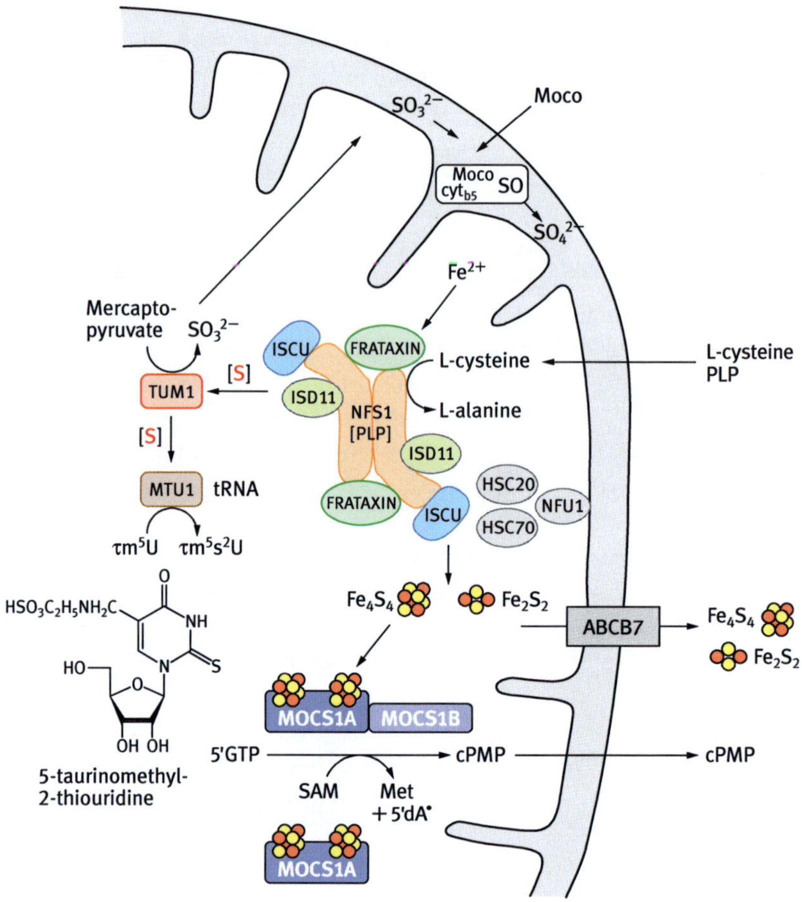


**Fig. 19.8:** Localization and role of proteins in mitochondria. Shown is the localization of proteins involved in Fe-S cluster biosynthesis, Moco biosynthesis, and tRNA thiolation and their connection in the mitochondrion. Human SUOX is localized in the intermembrane space of mitochondria. Shown are also compounds that need to be imported or exported to mitochondria.

*MOCS1B* in a single transcript, separated by a stop codon [96]. Splice variants of the *MOCS1* locus were identified that bypass the termination codon of *MOCS1A*, resulting in a new two-domain protein fusing MOCS1A and MOCS1B [97]. No evidence was found for the expression of MOCS1B from the bicistronic *MOCS1A-MOCS1B* splice-type I cDNA, indicating that MOCS1B is only expressed as a fusion with MOCS1A, whereas MOCS1A is also expressed as a separate protein [98]. Both proteins contain a mitochondrial targeting sequence for the translocation to mitochondria. The reaction catalyzed by MOCS1A and MOCS1B (in the MOCS1A-MOCS1B fusion) is described in detail Section 19.2.1.1. MOCS1A contains two [4Fe-4S] clusters that are directly involved in SAM cleavage and 5′-GTP binding and are therefore essential for the reaction [14]. The presence of these two Fe-S clusters directly links the first step of Moco biosynthesis to Fe-S cluster biosynthesis in mitochondria (Fig. 19.8). Thus, mutations in genes for Fe-S cluster biosynthesis would also result in a decrease of Moco biosynthesis in mammals and impaired cPMP production in mitochondria. After biosynthesis of cPMP from 5′-GTP in a reaction requiring reducing equivalents and SAM, cPMP has to be transported to the cytosol, where it is further modified.

The transporter involved in exporting cPMP to the cytosol was identified in plants to be ATM3 (a homologue of human ABCB7) [99]; however, cPMP has been additionally proposed to pass through the membrane without specific transport proteins due to its hydrophobic nature [25]. Thus, it remains unclear how cPMP is specifically transported to the cytosol in humans. Surprisingly, ABCB7 is also suggested to transport an essential molecule for Fe-S cluster biosynthesis in the cytosol [100]. This would present essential functions for the mitochondria in both cytosolic Fe-S cluster and Moco biosynthesis.

However, the mitochondria are also a compartment in which tRNAs are thiolated [101]. Thiolated tRNAs in mitochondria were identified to contain a 5-taurinomethyl-2-thiouridine ($\tau m^5s^2U$) modification in tRNAs for Lys, Glu, and Gln [87]. Mitochondrial tRNAs are transcribed from mitochondrial DNA, and taurine is transported into mitochondria, where the specific modification of uridine 34 at the wobble position occurs. The proteins MSS1 and MTO1 in humans have been shown to be involved in the $\tau m^5U$ modification in mitochondrial tRNA in humans. Further, the mitochondrial tRNA specific 2-thiouridylase, MTU1, is involved transferring sulfur to form the $\tau m^5s^2U$ tRNAs [101]. The sulfur donor for MTU1 is the NFS1/ISD11 complex that thus links tRNA thiolation and Fe-S cluster biosynthesis in mitochondria (Fig. 19.8) [23]. The NFS1/ISD11 complex transfers sulfur to MTU1 either directly or through an additional sulfur relay protein. This sulfur relay protein was shown in yeast to be the rhodanese-like protein, Tum1p [7]. TUM1 is also found in the mitochondria of humans, but its direct involvement in tRNA thiolation in humans still needs to be elucidated [102]. Lack of post-transcriptional modifications at the wobble positions of mitochondrial tRNAs for Leu and Lys has been associated with mitochondrial encephalomyopathy, lactic acidosis, and stroke-like episodes and myoclonic epilepsy with ragged-red fibers [103–106].

### 19.3.2 Sulfur transfer in the cytosol

Recently, several reports suggested that the mitochondria are not the sole compartment where Fe-S cluster biosynthesis is initiated [5]. Thus far, it has been proposed for yeast that the mitochondrial transporter Atm1 exports either fully formed Fe-S clusters or a special form of sulfur that is required for cytosolic Fe-S cluster biosynthesis [100, 107]. In *Saccharomyces cerevisiae*, export of a “sulfur compound” from mitochondria has been proposed to contribute to cytosolic Fe-S cluster assembly (CIA) *via* proteins known as Tah18 and Dre2 [108]. It was proposed that a cluster is transferred from Dre2 to other members of the CIA pathway [109]. The mammalian homologues of

Tah18 and Dre2 are NDOR1 and CIAPIN1, respectively, and they work with the NUBP1, NUBP2, NARFL, and CIAO1 proteins of the CIA pathway (Fig. 19.9) [110]. Because, in addition to these proteins, other proteins of the initial Fe-S cluster core complex have been identified in the cytosol (NFS1, ISCU, FXN, and HSC20), it is possible that the Fe-S clusters are also *de novo* formed in the cytosol [5]. However, although these proteins are present in the cytosol, their involvement in cytosolic Fe-S cluster biosynthesis still has to be proven.

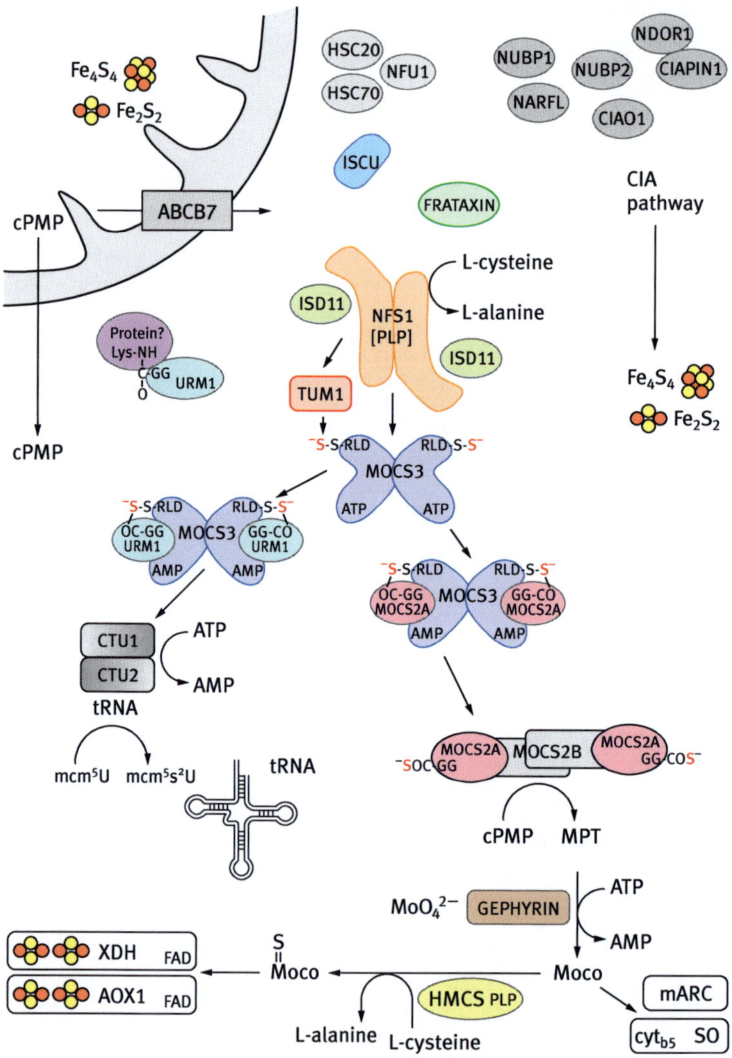


**Fig. 19.9:** Localization and role of proteins in the cytosol. Shown is the localization of proteins involved in Fe-S cluster biosynthesis, Moco biosynthesis, and tRNA thiolation and their connection in the cytosol. The enzymes in which synthesized Moco is inserted are shown schematically.

It has been reported that small amounts of NFS1 are present in the cytosol, where NFS1 interacts with MOCS3 [17]. Thus, cytosolic NFS1 is an important sulfur supplier for both Moco biosynthesis and tRNA thiolation [22]. The interaction of NFS1 and MOCS3 was revealed using Förster resonance energy transfer and a split EGFP system. The colocalization of NFS1 and MOCS3 in the cytosol was additionally confirmed by immunodetection of fractionated cells and localization studies using confocal fluorescence microscopy [17]. However, although the role of NFS1 in the cytosol for sulfur transfer to MOCS3 seems to be established, the involvement of ISD11 in this reaction still remains unclear. Thus far, ISD11 was described as a stabilizing factor of NFS1 in eukaryotes that is essential for its activity in Fe-S cluster formation in mitochondria [111]. In the absence of ISD11, NFS1 is prone to aggregation and Fe-S clusters cannot be formed [18, 112]. However, localization studies showed that ISD11 mainly is located in mitochondria and the nucleus in human cells [111]. Thus, the role of ISD11 and its involvement in the interaction of NFS1 and MOCS3 in the cytosol still needs to be investigated. It is possible that MOCS3 might replace the role of ISD11 as a stabilizing protein to NFS1 in the cytosol. This also might imply that cytosolic NFS1 is only involved in Moco biosynthesis and tRNA thiolation in conjunction with MOCS3, whereas in the absence of ISD11, it has no role in Fe-S cluster biosynthesis.

For Moco biosynthesis, cPMP is transferred from mitochondria to the cytosol, where two sulfur molecules are inserted to build the dithiolene group of MPT [99]. In this reaction, the MPT synthase composed of MOCS2A and MOCS2B is directly involved (see Section 19.2.1.2) [20]. In humans, the *MOCS2* locus encodes the two subunits of MPT synthase, MOCS2A and MOCS2B in a bicistronic transcript, from overlapping reading frames [42]. *In vitro* translation and mutagenesis experiments demonstrated that *MOCS2A* and *MOCS2B* are translated independently, from two alternative *MOCS2* splice forms, I and III, which contain different first exons in alternative transcripts [113]. The sulfur-relay system for Moco biosynthesis for the formation of the dithiolene moiety in MPT was shown to consist of the NFS1-C239 persulfide, the MOCS3-C412 persulfide, and the MOCS2A-G88 thiocarboxylate [17–20] (Fig. 19.9).

In a similar manner, the sulfur relay for tRNA thiolation in the cytosol is linked directly to Moco biosynthesis by the MOCS3 protein, which interacts with URM1 and forms the thiocarboxylate group on URM1-G101 (Fig. 19.9) [21, 92]. The interaction between MOCS3 and URM1 was revealed using Förster resonance energy transfer, in addition to functional *in vitro* studies for tRNA thiolation [21]. Here, MOCS3 is the E1-like protein of URM1 activating the C-terminal Gly88 under ATP consumption. The C-terminus of URM1 is first activated as an acyl-adenylate intermediate (-COAMP) and then thiocarboxylated (-COSH) by the persulfide group on the C-terminal RLD of MOCS3-C412, which originates from the NFS1-C239 persulfide [18]. The activated URM1-thiocarboxylate can be utilized in subsequent reactions for 2-thiouridine formation, mediated by a heterodimeric complex consisting of CTU1 and CTU2 in humans [21, 88].

In the cytosolic pathway for tRNA thiolation in yeast, the involvement of the mercaptopyruvate sulfur-tranferase Tum1p was shown [7]. Tum1p is a rhodanese-like protein with two RLDs, which was shown to be localized in the mitochondria and the cytosol [114]. An *in vitro* sulfur transfer experiment revealed that Tum1p stimulates the L-cysteine desulfurase activity of Nfs1p and accepts the sulfur from this protein [7]. Thus, Tum1p may function as a mediator between Nfs1p and its sulfur acceptor proteins, adding another protein component to the sulfur relay system. It has been suggested for yeast that the sulfur moiety in mitochondria is transferred from Nfs1 to the relay protein Tum1p. Because, in yeast, Nfs1p has not been revealed to reside in the cytosol so far, it has been suggested that Tum1 may function as a sulfur shuttle between the mitochondrial and the cytosolic compartments due to the dual localization of Tum1p. However, the proposed retrograde transfer of sulfur-loaded Tum1p after its interaction with Nfs1p from mitochondria to the cytosol occurs has not been observed and remains only speculative [22]. In humans, TUM1 was also shown to be localized in the cytosol and in mitochondria [114]. However, its involvement in tRNA thiolation and its interaction with cytosolic or mitochondrial NFS1 has not yet been reported, and its role in the sulfur-relay system between NFS1 and MOCS3 in the human cytosol remains to be elucidated. Because NFS1 is able to directly interact with MOCS3 in the cytosol, the role of TUM1 might be restricted to conditions of sulfur starvation or oxidative stress, by enhancing the activity of NFS1 or making the sulfur-tranfer process more specific.

For yeast $mcm^5s^2U$ formation (in cytoplasmic tRNAs), it was shown that components of the CIA apparatus for cytoplasmic (Fe-S) protein assembly and the scaffold proteins for the mitochondrial ISC machinery are required, which indicates that cytoplasmic 2-thiouridine formation essentially requires a protein containing an (Fe-S) cluster in yeast [22, 87]. The Fe-S containing protein has not been identified so far. Because, in humans, the proteins NFS1, MOCS3, URM1, CTU1, and CTU2 are sufficient for the formation of the $s^2U$ in $mcm^5U$-modified tRNAs, CTU1 and CTU2 would be the only candidates for being Fe-S-containing proteins. The requirement and nature of Fe-S clusters in tRNA thiolation in humans needs to be investigated in the future.

### 19.3.3 Role of NFS1, ISD11, URM1, and MOCS2A in the nucleus

In addition to the localization of NFS1 in mitochondria and the cytosol, NFS1 was also detected in the nucleus [17, 111]. However, its function of this compartment still needs to be revealed. Recent studies also showed that ISD11 is additionally detected in the nucleus, but it is not clear yet whether both proteins interact in this cellular compartment (Fig. 19.10) [17].

For yeast Nfs1p, a small but significant portion was identified to be localized to the nucleus and was shown to play an unknown essential role in cell viability that depended on its role as a L-cysteine desulfurase [115]. It has been suggested that the role of Nfs1p in the nucleus might be the synthesis Fe-S clusters for nuclear Fe-S proteins.

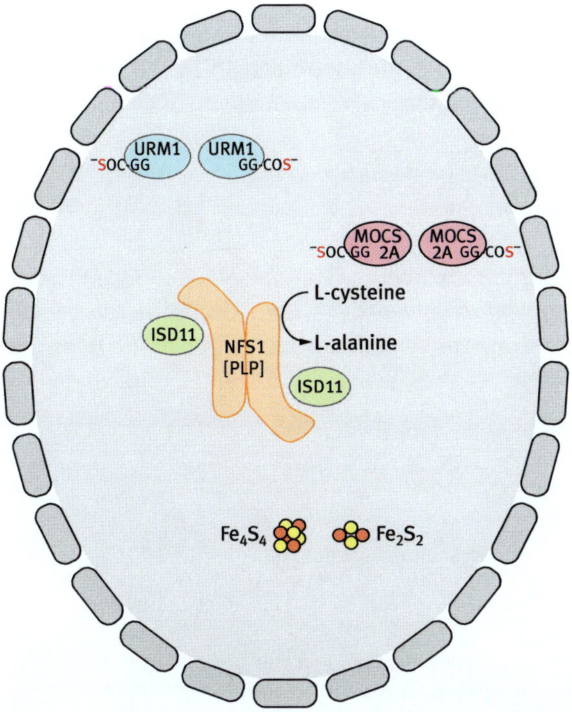


**Fig. 19.10:** Localization and role of proteins in the nucleus. Shown is the localization of proteins with a role in Fe-S cluster biosynthesis, Moco biosynthesis, and tRNA thiolation in the nucleus. The roles of NFS1, ISD11, URM1, and MOCS2A in the nucleus are not yet clear to date.

Another function might also be an involvement in the repair or maintenance of preassembled Fe-S clusters or the thiolation of nuclear tRNAs [115]. However, the role of Nfs1 in the nucleus remains unclear so far. Additionally, it is not known whether a similar phenotype exists for cell viability in human NFS1.

Additionally, URM1 and MOCS2A were identified in the nucleus (Fig. 19.10) [19, 21], where their specific functions remain unclear. For URM1, van der Veen *et al.* [88] identified targets of urmylation in response to oxidative stress in humans, among which were proteins involved in nuclear transport such as CAS, a protein involved in the shuttling between cytosol and nucleus. The role of the nucleocytoplasmic shuttling factor CAS as a target of urmylation would link the nuclear transport to oxidative stress, which is in good agreement with the observation that nuclear import is impaired upon oxidant treatment [116]. Speculation on the influence of urmylation on the arrest of nuclear transport is underlined by similar phenotypes of CAS depletion [117] and URM1 silencing [118], which both result in the arrest in G2/M phase. Thus, a contribution of URM1 to the control of global processes such as nuclear transport, cytokinesis, and cell cycle progression is possible. The role of MOCS2A in the nucleus has not been studied so far. In total, several proteins involved in cytoplasmic

Fe-S cluster formation, Moco biosynthesis and tRNA thiolation were also identified to be located in the nucleus, namely NFS1, ISD11, URM1, and MOCS2A, although their functions still remain unclear in this compartment.

## Acknowledgments

The author thanks all current and former members of their research groups in addition to collaboration partners who were involved in this work over the past years and decades. The work was mainly supported by continuous grants of the Deutsche Forschungsgemeinschaft to S.L. Special thanks goes to Mita Mullick Chowdhury for critical reading of the manuscript.

## References

[1] Beinert H. A tribute to sulfur. Eur J Biochem 2000;267:5657–64.
[2] Kessler D. Enzymatic activation of sulfur for incorporation into biomolecules in prokaryotes. FEMS Microbiol Rev 2006;30:825–40.
[3] Hidese R, Mihara H, Esaki N. Bacterial cysteine desulfurases: versatile key players in biosynthetic pathways of sulfur-containing biofactors. Applied Microbiol Biotechnol 2011;91:47–61.
[4] Bordo D, Bork P. The rhodanese/Cdc25 phosphatase superfamily. EMBO Rep 2002;3:741–746.
[5] Rouault TA. Biogenesis of iron-sulfur clusters in mammalian cells: new insights and relevance to human disease. Dis Models Mech 2012;5:155–64.
[6] Mendel RR, Kruse T. Cell biology of molybdenum in plants and humans. Biochim Biophys Acta 2012;1823:1568–79.
[7] Noma A, Sakaguchi Y, Suzuki T. Mechanistic characterization of the sulfur-relay system for eukaryotic 2-thiouridine biogenesis at tRNA wobble positions. Nucleic Acids Res 2009;37:1335–52.
[8] Mueller EG. Trafficking in persulfides: delivering sulfur in biosynthetic pathways. Nat Chem Biol 2006;2:185–94.
[9] Zheng L, White RH, Cash VL, Dean DR. Mechanism for the desulfurization of L-cysteine catalyzed by the nifS gene product. Biochemistry 1994;33:4714–20.
[10] Tsai CL, Barondeau DP. Human frataxin is an allosteric switch that activates the Fe-S cluster biosynthetic complex. Biochemistry 2010;49:9132–9.
[11] Bridwell-Rabb J, Winn AM, Barondeau DP. Structure-function analysis of Friedreich's ataxia mutants reveals determinants of frataxin binding and activation of the Fe-S assembly complex. Biochemistry 2011;50:7265–74.
[12] Schmucker S, Martelli A, Colin F, et al. Mammalian frataxin: an essential function for cellular viability through an interaction with a preformed ISCU/NFS1/ISD11 iron-sulfur assembly complex. PloS One 2011;6:e16199.
[13] Uhrigshardt H, Singh A, Kovtunovych G, Ghosh M, Rouault TA. Characterization of the human HSC20, an unusual DnaJ type III protein, involved in iron-sulfur cluster biogenesis. Hum Mol Genet 2010;19:3816–34.
[14] Hanzelmann P, Hernandez HL, Menzel C, et al. Characterization of MOCS1A, an oxygen-sensitive iron-sulfur protein involved in human molybdenum cofactor biosynthesis. J Biol Chem 2004;279:34721–32.
[15] Johnson JL, Rajagopalan KV. Structural and metabolic relationship between the molybdenum cofactor and urothione. Proc Natl Acad Sci USA 1982;79:6856–6860.

[16] Schwarz G, Mendel RR, Ribbe MW. Molybdenum cofactors, enzymes and pathways. Nature 2009;460:839–47.
[17] Marelja Z, Mullick Chowdhury M, Dosche C, et al. The L-cysteine desulfurase NFS1 is localized in the cytosol where it provides the sulfur for molybdenum cofactor biosynthesis in humans. PloS One 2013;8:e60869.
[18] Marelja Z, Stocklein W, Nimtz M, Leimkuhler S. A novel role for human Nfs1 in the cytoplasm: Nfs1 acts as a sulfur donor for MOCS3, a protein involved in molybdenum cofactor biosynthesis. J Biol Chem 2008;283:25178–85.
[19] Matthies A, Rajagopalan KV, Mendel RR, Leimkuhler S. Evidence for the physiological role of a rhodanese-like protein for the biosynthesis of the molybdenum cofactor in humans. Proc Natl Acad Sci USA 2004;101:5946–51.
[20] Leimkuhler S, Freuer A, Araujo JA, Rajagopalan KV, Mendel RR. Mechanistic studies of human molybdopterin synthase reaction and characterization of mutants identified in group B patients of molybdenum cofactor deficiency. J Biol Chem 2003;278:26127–34.
[21] Chowdhury MM, Dosche C, Lohmannsroben HG, Leimkuhler S. Dual role of the molybdenum cofactor biosynthesis protein MOCS3 in tRNA thiolation and molybdenum cofactor biosynthesis in humans. J Biol Chem 2012;287:17297–307.
[22] Noma A, Shigi N, Suzuki T. Biogenesis and functions of thio-compounds in transfer RNA: Comparison of bacterial and eukaryotic thiolation machineries. In: Grosjean H, ed. DNA and RNA modification enzymes: structure, mechanism, function and evolution. Landes Bioscience; 2009:392–405.
[23] El Yacoubi B, Bailly M, de Crecy-Lagard V. Biosynthesis and function of posttranscriptional modifications of transfer RNAs. Annu Rev Genet 2012;46:69–95.
[24] Rajagopalan KV. Biosynthesis of the molybdenum cofactor. In: Neidhardt FC, ed. Escherichia coli and Salmonella. Cellular and molecular biology. Washington (DC): ASM Press; 1996:674–9.
[25] Wuebbens MM, Rajagopalan KV. Structural characterization of a molybdopterin precursor. J Biol Chem 1993;268:13493–13498.
[26] Pitterle DM, Johnson JL, Rajagopalan KV. In vitro synthesis of molybdopterin from precursor Z using purified converting factor. Role of protein-bound sulfur in formation of the dithiolene. J Biol Chem 1993;268:13506–13509.
[27] Joshi MS, Johnson JL, Rajagopalan KV. Molybdenum cofactor biosynthesis in Escherichia coli mod and mog mutants. J Bacteriol 1996;178:4310–4312.
[28] Hille R, Nishino T, Bittner F. Molybdenum enzymes in higher organisms. Coord Chem Rev 2011;255:1179–1205.
[29] Duran M, Beemer FA, van der Heiden C, et al. Combined deficiency of xanthine oxidase and sulphite oxidase: a defect of molybdenum metabolism or transport? J Inherit Metab Dis 1978;1:175–178.
[30] Duran M, de Bree PK, de Klerk JBC, Dorland L, Berger R. Molybdenum cofactor deficiency: clinical presentation and laboratory diagnosis. Int Pediatr 1996;11:334–338.
[31] Mendel RR, Schwarz G. Molybdenum cofactor biosynthesis in plants and humans. Coord Chem Rev 2011;255:1145–1158.
[32] Reiss J, Cohen N, Dorche C, et al. Mutations in a polycistronic nuclear gene associated with molybdenum cofactor deficiency. Nat Genet 1998;20:51–53.
[33] Reiss J. Genetics of molybdenum cofactor deficiency. Hum Genet 2000;106:157–163.
[34] Leimkühler S, Wuebbens M, Rajagopalan KV. The history of the discovery of the molybdenum cofactor and novel aspects of its biosynthesis in bacteria. Coord Chem Rev 2011;255:1129–1144.
[35] Santamaria-Araujo JA, Fischer B, Otte T, et al. The tetrahydropyranopterin structure of the sulfur- and metal-free molybdenum cofactor precursor. J Biol Chem 2004;279:15994–9.
[36] Reiss J, Christensen E, Kurlemann G, Zabot M-T, Dorche C. Genomic structure and mutational spectrum of the bicistronic MOCS1 gene defective in molybdenum cofactor deficiency type A. Hum Genet 1998;103:639–644.

[37] Sofia HJ, Chen G, Hetzler BG, Reyes-Spindola JF, Miller NE. Radical SAM, a novel protein superfamily linking unresolved steps in familiar biosynthetic pathways with radical mechanisms: functional characterization using new analysis and information visualization methods. Nucleic Acids Res 2001;29:1097–1106.
[38] Hanzelmann P, Schindelin H. Crystal structure of the S-adenosylmethionine-dependent enzyme MoaA and its implications for molybdenum cofactor deficiency in humans. Proc Natl Acad Sci USA 2004;101:12870–5.
[39] Hanzelmann P, Schindelin H. Binding of 5′-GTP to the C-terminal FeS cluster of the radical S-adenosylmethionine enzyme MoaA provides insights into its mechanism. Proc Natl Acad Sci USA 2006;103:6829–34.
[40] Hover BM, Loksztejn A, Ribeiro AA, Yokoyama K. Identification of a cyclic nucleotide as a cryptic intermediate in molybdenum cofactor biosynthesis. J Am Chem Soc 2013;135:7019–32.
[41] Stallmeyer B, Drugeon G, Reiss J, Haenni AL, Mendel RR. Human molybdopterin synthase gene: identification of a bicistronic transcript with overlapping reading frames. Am J Hum Genet 1999;64:698–705.
[42] Gutzke G, Fischer B, Mendel RR, Schwarz G. Thiocarboxylation of molybdopterin synthase provides evidence for the mechanism of dithiolene formation in metal-binding pterins. J Biol Chem 2001;276:36268–36274.
[43] Rudolph MJ, Wuebbens MM, Rajagopalan KV, Schindelin H. Crystal structure of molybdopterin synthase and its evolutionary relationship to ubiquitin activation. Nat Struct Biol 2001;8:42–6.
[44] Daniels JN, Wuebbens MM, Rajagopalan KV, Schindelin H. Crystal structure of a molybdopterin synthase-precursor Z complex: insight into its sulfur transfer mechanism and its role in molybdenum cofactor deficiency. Biochemistry 2008;47:615–26.
[45] Wuebbens MM, Rajagopalan KV. Mechanistic and mutational studies of Escherichia coli molybdopterin synthase clarify the final step of molybdopterin biosynthesis. J Biol Chem 2003;278:14523–14532.
[46] Stallmeyer B, Coyne KE, Wuebbens MM, Johnson JL, Rajagopalan KV, Mendel RR. The cDNA sequence of MOCS3, human molybdopterin synthase sulfurylase. GenBank accession number AF102544; 1998.
[47] Lake MW, Wuebbens MM, Rajagopalan KV, Schindelin H. Mechanism of ubiquitin activation revealed by the structure of a bacterial MoeB-MoaD complex. Nature 2001;414:325–9.
[48] Matthies A, Nimtz M, Leimkuhler S. Molybdenum cofactor biosynthesis in humans: identification of a persulfide group in the rhodanese-like domain of MOCS3 by mass spectrometry. Biochemistry 2005;44:7912–20.
[49] Stallmeyer B, Schwarz G, Schulze J, et al. The neurotransmitter receptor-anchoring protein gephyrin reconstitutes molybdenum cofactor biosynthesis in bacteria, plants, and mammalian cells. Proc Natl Acad Sci USA 1999;96:1333–1338.
[50] Belaidi AA, Schwarz G. Metal insertion into the molybdenum cofactor: product-substrate channelling demonstrates the functional origin of domain fusion in gephyrin. Biochem J 2013;450:149–57.
[51] Schwarz G. Molybdenum cofactor biosynthesis and deficiency. Cell Mol Life Sci 2005;62:2792–810.
[52] Herweg J, Schwarz G. Splice-specific glycine receptor binding, folding, and phosphorylation of the scaffolding protein gephyrin. J Biol Chem 2012;287:12645–56.
[53] Wahl RC, Rajagopalan KV. Evidence for the inorganic nature of the cyanolyzable sulfur of molybdenum hydroxylases. J Biol Chem 1982;257:1354–1359.
[54] Hille R. The mononuclear molybdenum enzymes. Chem Rev 1996;96:2757–2816.
[55] Anantharaman V, Aravind L. MOSC domains: ancient, predicted sulfur-carrier domains, present in diverse metal-sulfur cluster biosynthesis proteins including molybdenum cofactor sulfurases. FEMS Microbiol Lett 2002;207:55–61.

[56] Peretz H, Naamati MS, Levartovsky D, et al. Identification and characterization of the first mutation (Arg776Cys) in the C-terminal domain of the Human Molybdenum Cofactor Sulfurase (HMCS) associated with type II classical xanthinuria. Mol Genet Metab 2007;91:23–9.
[57] Bittner F, Oreb M, Mendel RR. ABA3 is a molybdenum cofactor sulfurase required for activation of aldehyde oxidase and xanthine dehydrogenase in Arabidopsis thaliana. J Biol Chem 2001;276:40381–4.
[58] Wollers S, Heidenreich T, Zarepour M, et al. Binding of sulfurated molybdenum cofactor to the C-terminal domain of ABA3 from Arabidopsis thaliana provides insight into the mechanism of molybdenum cofactor sulfuration. J Biol Chem 2008;283:9642–50.
[59] Lehrke M, Rump S, Heidenreich T, Wissing J, Mendel RR, Bittner F. Identification of persulfide-binding and disulfide-forming cysteine residues in the NifS-like domain of the molybdenum cofactor sulfurase ABA3 by cysteine-scanning mutagenesis. Biochem J 2012;441:823–32.
[60] Klein JM, Schwarz G. Cofactor-dependent maturation of mammalian sulfite oxidase links two mitochondrial import pathways. J Cell Sci 2012;125:4876–85.
[61] Neumann M, Leimkuhler S. The role of system-specific molecular chaperones in the maturation of molybdoenzymes in bacteria. Biochem Res Int 2011;2011:850–924.
[62] Hille R. The molybdenum oxotransferases and related enzymes. Dalton transactions 2013;42:3029–42.
[63] Neumann M, Stocklein W, Leimkuhler S. Transfer of the molybdenum cofactor synthesized by Rhodobacter capsulatus MoeA to XdhC and MobA. J Biol Chem 2007;282:28493–500.
[64] Neumann M, Stocklein W, Walburger A, Magalon A, Leimkuhler S. Identification of a Rhodobacter capsulatus L-cysteine desulfurase that sulfurates the molybdenum cofactor when bound to XdhC and before its insertion into xanthine dehydrogenase. Biochemistry 2007;46:9586–95.
[65] Schumann S, Saggu M, Moller N, et al. The mechanism of assembly and cofactor insertion into Rhodobacter capsulatus xanthine dehydrogenase. J Biol Chem 2008;283:16602–11.
[66] Kisker C, Schindelin H, Pacheco A, et al. Molecular basis of sulfite oxidase deficiency from the structure of sulfite oxidase. Cell 1997;91:973–983.
[67] Feng C, Kedia RV, Hazzard JT, Hurley JK, Tollin G, Enemark JH. Effect of solution viscosity on intramolecular electron transfer in sulfite oxidase. Biochemistry 2002;41:5816–21.
[68] Johnson JL, Coyne KE, Rajagopalan KV, et al. Molybdopterin synthase mutations in a mild case of molybdenum cofactor deficiency. Am J Med Genet 2001;104:169–73.
[69] Johnson JL, Duran M. Molybdenum cofactor deficiency and isolated sulfite oxidase deficiency. In: Scriver CR, Beaudet AL, Sly WS, Valle D, Childs B, Vogelstein B, eds. The metabolic and molecular bases of inherited disease. 8th edn. New York: McGraw-Hill; 2001:3163–77.
[70] Johnson JL, Garrett RM, Rajagopalan KV. The biochemistry of molybdenum cofactor deficiency and isolated sulfite oxidase deficiency. Int Pediatr 1997;12:22–26.
[71] Johnson JL, Rajagopalan KV. Human sulfite oxidase deficiency. Characterization of the molecular defect in a multicomponent system. J Clin Invest 1976;58:551–556.
[72] Gruenewald S, Wahl B, Bittner F, et al. The fourth molybdenum containing enzyme mARC: cloning and involvement in the activation of N-hydroxylated prodrugs. J Med Chem 2008;51:8173–7.
[73] Wahl B, Reichmann D, Niks D, et al. Biochemical and spectroscopic characterization of the human mitochondrial amidoxime reducing components hmARC-1 and hmARC-2 suggests the existence of a new molybdenum enzyme family in eukaryotes. J Biol Chem 2010;285:37847–59.
[74] Havemeyer A, Grunewald S, Wahl B, et al. Reduction of N-hydroxy-sulfonamides, including N-hydroxy-valdecoxib, by the molybdenum-containing enzyme mARC. Drug Metab Dispos 2010;38:1917–21.
[75] Harrison R. Physiological roles of xanthine oxidoreductase. Drug Metab Rev 2004;36:363–75.

[76] Garattini E, Fratelli M, Terao M. Mammalian aldehyde oxidases: genetics, evolution and biochemistry. Cell Mol Life Sci 2008;65:1019–48.
[77] Garattini E, Mendel R, Romao MJ, Wright R, Terao M. Mammalian molybdo-flavoenzymes, an expanding family of proteins: structure, genetics, regulation, function and pathophysiology. Biochem J 2003;372:15–32.
[78] Nishino T, Okamoto K, Kawaguchi Y, et al. Mechanism of the conversion of xanthine dehydrogenase to xanthine oxidase: identification of the two cysteine disulfide bonds and crystal structure of a non-convertible rat liver xanthine dehydrogenase mutant. J Biol Chem 2005;280:24888–94.
[79] Nishino T. The conversion of xanthine dehydrogenase to xanthine oxidase and the role of the enzyme in reperfusion injury. J Biochem (Tokyo) 1994;116:1–6.
[80] McCord JM. Oxygen-derived free radicals in postischemic tissue injury. N Engl J Med 1985;312:159–163.
[81] Komai H, Massey V, Palmer G. The preparation and properties of deflavo xanthine oxidase. J Biol Chem 1969;244:1692–700.
[82] Rozenski J, Crain PF, McCloskey JA. The RNA modification database: 1999 update. Nucleic Acids Res 1999;27:196–7.
[83] Yokoyama S, Watanabe T, Murao K, et al. Molecular mechanism of codon recognition by tRNA species with modified uridine in the first position of the anticodon. Proc Natl Acad Sci USA 1985;82:4905–9.
[84] Agris PF, Soll D, Seno T. Biological function of 2-thiouridine in Escherichia coli glutamic acid transfer ribonucleic acid. Biochemistry 1973;12:4331–7.
[85] Durant PC, Bajji AC, Sundaram M, Kumar RK, Davis DR. Structural effects of hypermodified nucleosides in the Escherichia coli and human tRNALys anticodon loop the effect of nucleosides s2U, mcm5U, mcm5s2U, mnm5s2U, t6A, and ms2t6A. Biochemistry 2005;44:8078–89.
[86] Rezgui VA, Tyagi K, Ranjan N, et al. tRNA tKUUU, tQUUG, and tEUUC wobble position modifications fine-tune protein translation by promoting ribosome A-site binding. Proc Natl Acad Sci USA 2013;110:12289–94.
[87] Suzuki T. Biosynthesis and function of tRNA wobble modifications. Top Curr Genet 2005;12:23–69.
[88] Van der Veen AG, Schorpp K, Schlieker C, et al. Role of the ubiquitin-like protein Urm1 as a noncanonical lysine-directed protein modifier. Proc Natl Acad Sci USA 2011;108:1763–70.
[89] Leidel S, Pedrioli PG, Bucher T, et al. Ubiquitin-related modifier Urm1 acts as a sulphur carrier in thiolation of eukaryotic transfer RNA. Nature 2009;458:228–32.
[90] Pedrioli PG, Leidel S, Hofmann K. Urm1 at the crossroad of modifications. In: "Protein modifications: beyond the usual suspects" review series. EMBO Rep 2008;9:1196–202.
[91] Xu J, Zhang J, Wang L, et al. Solution structure of Urm1 and its implications for the origin of protein modifiers. Proc Natl Acad Sci USA 2006;103:11625–30.
[92] Schmitz J, Chowdhury MM, Hanzelmann P, et al. The sulfurtransferase activity of Uba4 presents a link between ubiquitin-like protein conjugation and activation of sulfur carrier proteins. Biochemistry 2008;47:6479–89.
[93] Lill R, Muhlenhoff U. Iron-sulfur protein biogenesis in eukaryotes: components and mechanisms. Annu Rev Cell Dev Biol 2006;22:457–86.
[94] Colin F, Martelli A, Clemancey M, et al. Mammalian frataxin controls sulfur production and iron entry during de novo Fe4S4 cluster assembly. J Am Chem Soc 2013;135:733–40.
[95] Pastore A, Puccio H. Frataxin: a protein in search for a function. J Neurochem 2013;126:43–52.
[96] Gross-Hard S, Reiss J. The bicistronic MOCS1 gene has alternative start codons on two mutually exclusive exons. Mol Genet Metab 2002;76:340–343.
[97] Gray TA, Nicholls RD. Diverse splicing mechanisms fuse the evolutionarily conserved bicistronic MOCS1A and MOCS1B open reading frames. RNA 2000;6:928–936.

[98] Hänzelmann P, Schwarz G, Mendel RR. Functionality of alternative splice forms of the first enzymes involved in human molybdenum cofactor biosynthesis. J Biol Chem 2002;277:18303–18312.
[99] Teschner J, Lachmann N, Schulze J, et al. A novel role for Arabidopsis mitochondrial ABC transporter ATM3 in molybdenum cofactor biosynthesis. Plant Cell 2010;22:468–80.
[100] Hausmann A, Samans B, Lill R, Muhlenhoff U. Cellular and mitochondrial remodeling upon defects in iron-sulfur protein biogenesis. J Biol Chem 2008;283:8318–30.
[101] Sasarman F, Antonicka H, Horvath R, Shoubridge EA. The 2-thiouridylase function of the human MTU1 (TRMU) enzyme is dispensable for mitochondrial translation. Hum Mol Genet 2011;20: 4634–43.
[102] Yadav PK, Yamada K, Chiku T, Koutmos M, Banerjee R. Structure and kinetic analysis of H2S production by human mercaptopyruvate sulfurtransferase. J Biol Chem 2013;288:20002–13.
[103] James AM, Wei YH, Pang CY, Murphy MP. Altered mitochondrial function in fibroblasts containing MELAS or MERRF mitochondrial DNA mutations. Biochem J 1996;318:401–7.
[104] Antonicka H, Floryk D, Klement P, et al. Defective kinetics of cytochrome c oxidase and alteration of mitochondrial membrane potential in fibroblasts and cytoplasmic hybrid cells with the mutation for myoclonus epilepsy with ragged-red fibres ("MERRF") at position 8344 nt. Biochem J 1999;342:537–44.
[105] Boczonadi V, Smith PM, Pyle A, et al. Altered 2-thiouridylation impairs mitochondrial translation in reversible infantile respiratory chain deficiency. Hum Mol Genet 2013;22:4602–15.
[106] Gaignard P, Gonzales E, Ackermann O, et al. Mitochondrial infantile liver disease due to TRMU gene mutations: three new cases. JIMD Rep 2013;11:117–23.
[107] Lill R, Dutkiewicz R, Elsasser HP, et al. Mechanisms of iron-sulfur protein maturation in mitochondria, cytosol and nucleus of eukaryotes. Biochim Biophys Acta 2006;1763:652–67.
[108] Netz DJ, Stumpfig M, Dore C, Muhlenhoff U, Pierik AJ, Lill R. Tah18 transfers electrons to Dre2 in cytosolic iron-sulfur protein biogenesis. Nature Chem Biol 2010;6:758–65.
[109] Sharma AK, Pallesen LJ, Spang RJ, Walden WE. Cytosolic iron-sulfur cluster assembly (CIA) system: factors, mechanism, and relevance to cellular iron regulation. J Biol Chem 2010;285:26745–51.
[110] Stehling O, Netz DJ, Niggemeyer B, et al. Human Nbp35 is essential for both cytosolic iron-sulfur protein assembly and iron homeostasis. Mol Cell Biol 2008;28:5517–28.
[111] Shi Y, Ghosh MC, Tong WH, Rouault TA. Human ISD11 is essential for both iron-sulfur cluster assembly and maintenance of normal cellular iron homeostasis. Hum Mol Genet 2009;18: 3014–25.
[112] Wiedemann N, Urzica E, Guiard B, et al. Essential role of Isd11 in mitochondrial iron-sulfur cluster synthesis on Isu scaffold proteins. EMBO J 2006;25:184–95.
[113] Hahnewald R, Leimkuhler S, Vilaseca A, Acquaviva-Bourdain C, Lenz U, Reiss J. A novel MOCS2 mutation reveals coordinated expression of the small and large subunit of molybdopterin synthase. Mol Genet Metab 2006;89:210–3.
[114] Nagahara N, Ito T, Kitamura H, Nishino T. Tissue and subcellular distribution of mercaptopyruvate sulfurtransferase in the rat: confocal laser fluorescence and immunoelectron microscopic studies combined with biochemical analysis. Histochem Cell Biol 1998;110:243–50.
[115] Nakai Y, Nakai M, Hayashi H, Kagamiyama H. Nuclear localization of yeast Nfs1p is required for cell survival. J Biol Chem 2001;276:8314–20.
[116] Kodiha M, Tran D, Morogan A, Qian C, Stochaj U. Dissecting the signaling events that impact classical nuclear import and target nuclear transport factors. PloS One 2009;4:e8420.
[117] Ogryzko VV, Brinkmann E, Howard BH, Pastan I, Brinkmann U. Antisense inhibition of CAS, the human homologue of the yeast chromosome segregation gene CSE1, interferes with mitosis in HeLa cells. Biochemistry 1997;36:9493–500.
[118] Schlieker CD, Van der Veen AG, Damon JR, Spooner E, Ploegh HL. A functional proteomics approach links the ubiquitin-related modifier Urm1 to a tRNA modification pathway. Proc Natl Acad Sci USA 2008;105:18255–60.

# 20 Iron-sulfur proteins and genome stability

Kerstin Gari

## 20.1 Introduction

The blueprint of each cell – be it a single bacterium or part of a multicellular organism, such as man – is encoded in its genome. For a cell to remain functional throughout its life cycle, its genetic program has to remain accurate and, hence, its genome has to be protected from damage. Such damage can be the result of exogenous factors, such as UV light, or endogenous factors, such as oxidative stress. Moreover, the ability of a cell to replicate its genome and divide into two identical daughter cells is a naturally complicated process that can pose a problem to genome integrity. Not surprisingly, a huge number of proteins work together to faithfully replicate DNA and to detect, signal, and repair DNA damage. Failure to do so results in genome instability, one of the hallmarks of cancer.

For a long time, it appeared that iron-sulfur (FeS) proteins are relatively rare in the processes of DNA replication and repair. Given that upon FeS cluster oxidation, free iron atoms can generate dangerous reactive oxygen species (ROS), this notion seemed rather intuitive. Over the last years, however, a surprisingly high number of proteins involved in DNA replication and repair have been identified to bind to an FeS cluster. In many cases, the function of the FeS cluster has remained elusive so far.

In this chapter, an overview of FeS proteins in DNA metabolism will be given and the potential role of the FeS cluster in these proteins will be discussed. The major focus will be on eukaryotic proteins, with some examples from the prokaryotic world.

## 20.2 The importance of genome stability

The maintenance of genome stability is essential for cellular function, and the consequences of genome instability are most dramatically exemplified in human cancer. As a common characteristic trait, tumor cells generally display gross chromosomal rearrangements and genetic mutations.

In recent years, it has become clear that genome instability is not a secondary effect of aberrant cellular behavior but rather one of the causes of tumor development. In the oncogene-induced DNA damage model (Fig. 20.1), it has been suggested that in early cancerous lesions, the activation of oncogenes leads to perturbed DNA replication due to the stalling and collapsing of DNA replication forks [1–5]. As a consequence of such DNA replication stress, the DNA damage checkpoint is activated, which involves a multitude of factors that decide to either repair the damage or (if it is beyond repair) induce cell cycle arrest, apoptosis, or senescence. This response to DNA replication stress acts as a barrier to tumorigenesis and is dependent on a functional copy of the tumor suppressor gene *p53*.

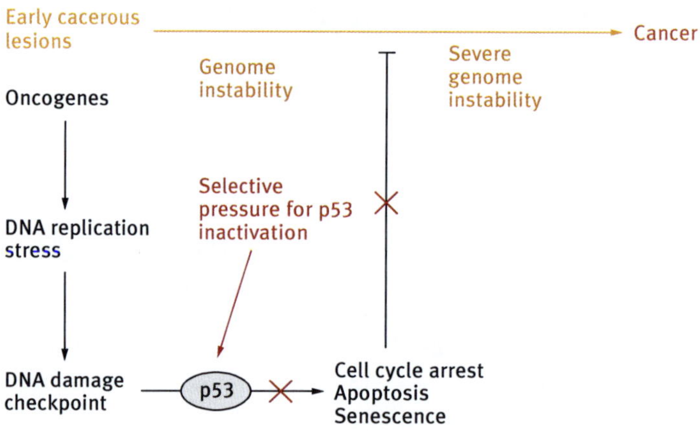


**Fig. 20.1:** Oncogene-induced DNA damage model.

This could be the end of the story of cancer if it was not for the adaptiveness of cancer cells. Instead, this initial tumorigenesis barrier creates selective pressure toward the inactivation of p53, which in turn allows full tumor development. Loss of heterozygosity (LOH), i.e. the loss of one of the two alleles of a gene, is one of the dangerous consequences of uncontrolled genome instability.

## 20.3 Link between FeS cluster biogenesis and genome stability

In 2008, Veatch *et al.* [6] showed that LOH in yeast can also be caused by the loss of mitochondrial DNA. They further demonstrated that nuclear genome instability did not occur as a consequence of dysfunctional cellular respiration but rather due to a defect in FeS cluster biogenesis. Based on their findings, they proposed that impaired FeS protein maturation might compromise the function of nuclear FeS proteins that function in DNA replication and repair, such as DNA primase [7], the helicase Rad3 [8], and the glycosylase Ntg2 [9]. Their hypothesis was based on the fact that maturation of nuclear FeS proteins – although primarily taking place in the cytoplasm – also depends on the mitochondrial iron-sulfur cluster (ISC) machinery [10].

While it is still controversial whether the ISC machinery is active both in mitochondria and the cytoplasm or whether an FeS cluster precursor is produced in mitochondria and then exported into the cytoplasm [11], it is well established that the subsequent maturation of nuclear FeS proteins is carried out in the cytoplasm by the cytoplasmic iron-sulfur assembly (CIA) machinery. In an early step, an FeS cluster is transiently assembled on a scaffold complex composed of the NTPases NBP35 and CFD1 [12, 13]. The loosely bound cluster is then transferred to FeS apo-proteins, a process that requires the function of the WD40 repeat protein CIAO1 and the iron-only hydrogenase-like protein IOP1 [14–16].

Recently, two studies have found a molecular explanation for the link between FeS cluster biogenesis and proteins involved in DNA metabolism, as had been previously proposed [6].

Coming from either the FeS field or the genome stability field, both studies identified the cytoplasmic HEAT-repeat protein MMS19 as a component of the CIA machinery that physically interacts with multiple FeS proteins involved in DNA metabolism [17, 18]. They further showed that MMS19 is required for FeS cluster transfer to FeS apo-proteins and that in the absence of MMS19, the stability of multiple FeS proteins involved in DNA replication and repair is affected. As a consequence, both yeast and human cells depleted of MMS19 showed an altered response to agents that cause DNA damage or DNA replication stress. Moreover, *Mms19* knockout mice displayed early embryonic lethality, presumably because Mms19 deficiency has pleiotropic effects on multiple proteins involved in DNA metabolism [17].

The MMS19 interacting protein MIP18 was later shown to contribute substantially to FeS protein recognition by binding to the FeS cluster-binding domain within apo-proteins [19]. Interestingly, MIP18 had originally been identified in a proteome-wide screen for factors that contain hyperreactive cysteines [20], a feature that is commonly found in FeS biogenesis proteins. Future studies will therefore have to address whether MIP18 plays a direct role in FeS cluster transfer by transiently binding to an FeS cluster and handing it over to apo-proteins.

Collectively, these data suggest that MMS19 and MIP18 physically and functionally link FeS cluster biogenesis to DNA replication and repair (Fig. 20.2).

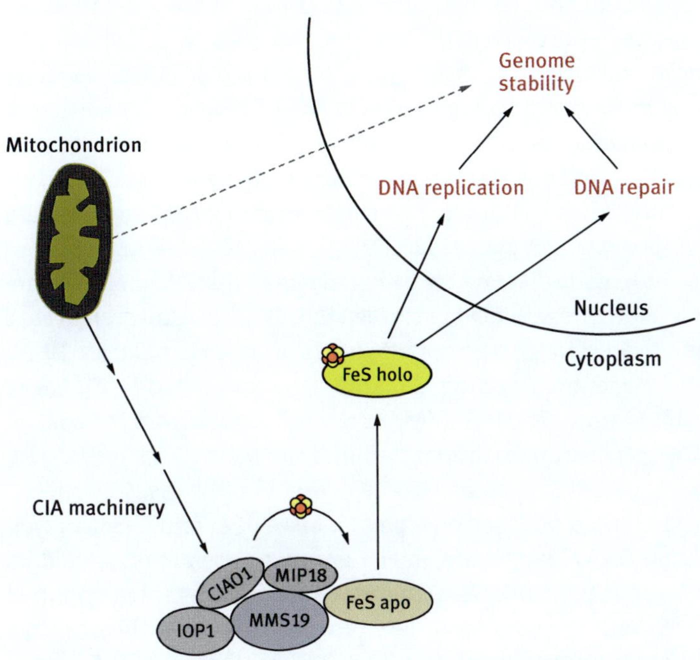


**Fig. 20.2:** MMS19 and MIP18 link FeS cluster biogenesis to DNA metabolism.

## 20.4 FeS proteins in DNA replication

DNA replication is a highly regulated process that involves the concerted action of a considerable number of proteins. Essentially, however, it requires unwinding of the parental DNA strands by a helicase activity and synthesis of the two daughter strands by the replicative DNA polymerases α, δ, and ε (Pol-α, Pol-δ, and Pol-ε, respectively) [21, 22].

Two complications arise from the intrinsic properties of DNA polymerases and the anti-parallel nature of DNA. First, all DNA polymerases have a 5′-3′ polarity, which implies that one of the two DNA strands can be elongated in a continuous manner (the leading strand), whereas the other strand has to be synthesized in a discontinuous manner (the lagging strand).

The second complication stems from the fact that DNA polymerases can only synthesize DNA by extension of an existing nucleotide primer. DNA replication therefore relies on the activity of a complex formed of DNA primase and DNA Pol-α that can carry out the *de novo* synthesis of a short chimeric RNA-DNA primer on the parental template [23]. Following primer synthesis, the leading strand template will then be primarily replicated by DNA Pol-ε, whereas lagging strand synthesis is carried out by DNA Pol-δ [24–26].

Due to its discontinuous mode of replication, initiation by DNA primase and Pol-α has to take place multiple times during synthesis of the lagging strand. The switch between initial synthesis and processive replication requires the replication clamp PCNA (proliferating cell nuclear antigen) that displaces DNA primase-Pol-α from the DNA template and loads Pol-δ onto it [22, 27].

The stretches of replicated DNA that are created due to this discontinuous mode of replication are referred to as Okazaki fragments. When Pol-δ reaches a downstream Okazaki fragment, it displaces a few nucleotides of the RNA primer, and creates a 5′-single-stranded flap structure [22]. Flap endonuclease 1 (FEN1) is required to cleave this short flap at the junction between double- and single-stranded DNA and to create a structure that can be sealed by DNA ligase I [28, 29].

It is believed that Pol-δ usually displays only limited strand displacement synthesis and that the resulting short flaps are readily cleaved by FEN1 [30]. However, in case longer flaps are created – at least in yeast – the activity of the helicase/nuclease DNA2 is required for Okazaki fragment maturation [29, 31, 32]. In contrast to FEN1, DNA2 cleaves the displaced single-stranded DNA a few nucleotides away from the base of a long flap structure, thereby creating a shorter flap that can serve as a substrate for FEN1 [33–35] (Fig. 20.3).

In recent years, it has turned out that DNA polymerases α, δ, and ε [36] as well as DNA primase [7, 37] and DNA2 [38, 39] are all FeS proteins. Although it should be kept in mind that DNA replication in this paragraph was considerably simplified and more factors are involved, the number of FeS proteins in the very heart of the DNA replication machinery is nevertheless astonishing. Given that upon FeS cluster

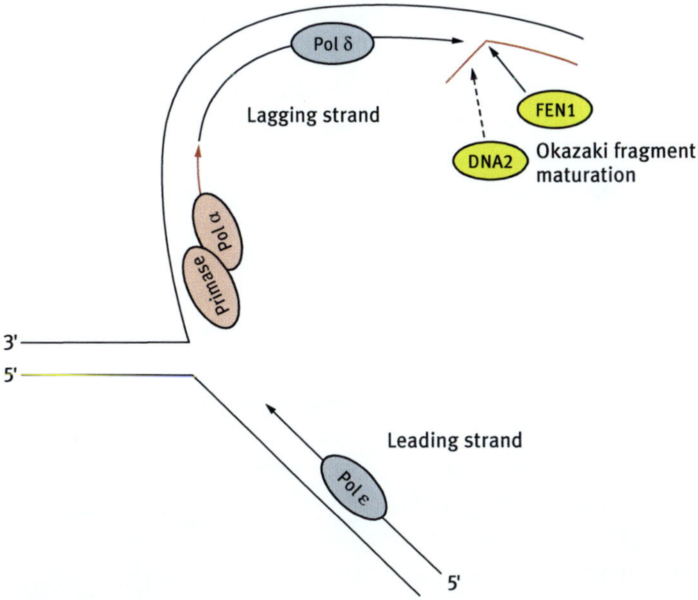


**Fig. 20.3:** DNA replication.

oxidation, free iron atoms can generate dangerous ROS *via* Fenton chemistry, having FeS clusters in DNA polymerases seems like putting the fox in charge of the henhouse. Unless, of course, FeS clusters in these enzymes carry out – as yet unclear – functions that cannot be achieved with alternative cofactors.

### 20.4.1 DNA primase and DNA polymerase α

DNA primase and Pol-α form an evolutionarily conserved complex that couples two activities crucial for replication initiation: first, the *de novo* synthesis of an 8- to 13-nucleotide-long RNA primer by the DNA-dependent RNA polymerase DNA primase, and second, the subsequent primer extension by a few dNTPs by Pol-α [23].

DNA primase is composed of a small (PRI1) and a large (PRI2) subunit, both of which are required for viability in yeast [40]. Although PRI1 carries the catalytic activity, PRI2 seems to be required for efficient primer initiation and elongation [41] and possibly handoff of the RNA primer to Pol-α [42]. In 2007, two groups reported that the C-terminal domain of PRI2 in human [37], as well as in yeast and Archaea [7], contains a $[4Fe\text{-}4S]^{2+}$ cluster. Klinge *et al.* [7] further showed that FeS cluster binding is required for efficient RNA primer synthesis in yeast.

Resolution of the crystal structure of the C-terminal part of PRI2 later showed that the FeS cluster is located at the interface of two largely independent helical folds [43].

Interestingly, the overall architecture of PRI2 has a striking similarity to the active site of DNA photolyases. Given that photolyases recognize UV-induced cyclobutane-pyrimidine dimers, an interesting possibility is that the large subunit of DNA primase might play a similar role and recognize the two initial ribonucleotides during primer synthesis [43]. Whether this is the case and which role the FeS cluster plays in this scenario – if any – needs to be further investigated.

Like DNA primase, Pol-α is composed of two subunits: the catalytic subunit (POLA1) and the regulatory B subunit (POLA2) [23]. Interaction between the two subunits is likely mediated by the C-terminal domain of the catalytic subunit [44]. For a long time, this function was assigned to a putative zinc finger motif in this domain, a notion that was supported by structural analysis that suggested the presence of a $Zn^{2+}$ ion at the binding interface of POLA1 to POLA2 [45]. As discussed in more depth in the next paragraph, however, a recent study has demonstrated that the C-terminal domain of POLA1 does not contain a zinc finger module but rather an FeS cluster-binding motif [36].

### 20.4.2 DNA polymerases δ and ε

Pol-δ and Pol-ε – like Pol-α – belong to the class B family of DNA polymerases [46]. They are multi-subunit proteins composed of a catalytic subunit, a regulatory B subunit, and (at least in human) two accessory subunits [47].

The catalytic subunits are evolutionarily conserved and contain two cysteine-rich motifs (CysA and CysB) in their C-terminal domains. For a long time, the CysA and CysB motifs were believed to be zinc finger modules [48]. The CysB motif in particular had been well studied and shown to mediate interaction with the regulatory B subunit [49].

Intrigued by the fact that *pol3-13*, a yeast mutant strain with a single point mutation (cysteine to serine) in the CysB motif of Pol-δ displays synthetic lethality with multiple members of the CIA machinery, Netz *et al.* [36] recently challenged the assumption that CysB was a zinc finger module and set out to investigate the possibility that it could be an FeS cluster-binding motif.

Using $^{55}Fe$ incorporation assays, they could indeed show that Pol-δ and Pol-α, Pol-ε, and the translesion DNA polymerase ζ are FeS proteins [36]. UV-vis and EPR spectroscopy together with site-directed mutagenesis confirmed the presence of a $[4Fe\text{-}4S]^{2+}$ cluster that is coordinated by the cysteines of the CysB motif. Upon protein purification, the FeS cluster in the Pol-δ holo-complex was mildly sensitive to oxidation, which could be exacerbated by addition of the oxidizing agent ferricyanide. In contrast, the cluster was insensitive to reduction by dithionite.

Further biochemical analysis showed that FeS cluster binding by CysB is required for Pol-δ holo-complex formation, whereas the CysA motif – which seems to represent a *bona fide* zinc finger – confers binding of Pol-δ to PCNA and allows processive DNA replication [36].

The finding that all replicative DNA polymerases in yeast and (given their evolutionary conservation) presumably in all eukaryotes are FeS proteins was surprising considering that this family had been studied rather extensively. The structural analysis of Pol-α in particular, which showed the presence of a $Zn^{2+}$ ion in the interaction domain of the catalytic subunit with the regulatory B subunit [45], seemed a rather convincing argument for the presence of a zinc finger module in this family of DNA polymerases. Meanwhile, it has to be kept in mind that in this study Pol-α had been overexpressed and purified from *Escherichia coli*. Presumably for lack of interaction with the bacterial FeS assembly system, a $Zn^{2+}$ ion was incorporated into Pol-α – an erroneous but seemingly successful attempt to produce a stable protein. Whether replacement of the FeS cluster with a $Zn^{2+}$ ion could also happen *in vivo* cannot be categorically excluded from the results presented by Netz and colleagues [36]. However, two recent studies [17, 18] showed that the protein levels of Pol-δ in the absence of the late-acting CIA component MMS19 are severely reduced, strongly reinforcing the idea that the stability of Pol-δ depends on the presence of an FeS cluster.

The question of why all replicative DNA polymerases contain an FeS cluster has remained unanswered so far. It is particularly surprising because – at least in the case of Pol-α – the FeS cluster can be replaced by a $Zn^{2+}$ ion without losing the overall structure of the protein. One would assume that there was a selective pressure against FeS clusters in favor of safer structural elements in proteins whose function it is to replicate the genome as faithfully as possible. The fact that this is visibly not the case and that the FeS cluster is a common element of all replicative polymerases points toward a specific function of FeS clusters that cannot be performed by other cofactors. Because FeS clusters are intrinsically sensitive to the surrounding redox conditions, oxidative stress-induced decomposition of the FeS cluster might be a unique way of regulating DNA polymerases in response to suboptimal conditions of DNA replication.

### 20.4.3 DNA2

DNA2 is a multifunctional protein that – apart from Okazaki fragment maturation – has been implicated in a variety of processes related to DNA metabolism: repair of DNA double-strand breaks [50, 51], telomere maintenance [52], prevention of replication fork reversal [53], and mitochondrial DNA replication and repair [54].

From a biochemical point of view, DNA2 is a particularly interesting protein because it has both helicase (5′–3′) and endonuclease activities [55, 56]. Although its primary function in Okazaki fragment maturation depends on its endonuclease activity [56, 57], its helicase activity seems to facilitate flap cleavage [58].

In 2009, Yeeles *et al.* [38] reported that a bacterial helicase/nuclease complex, *Bacillus subtilis* AddA/AddB, carries a $[4Fe\text{-}4S]^{2+}$ cluster in its nuclease domain. By sequence comparison, they suggested that this might also be true for the nuclease domain of the related eukaryotic DNA2.

Interestingly, the four cysteines required for FeS cluster binding in AddAB are not clustered together but instead flank the nuclease domain AddB on either side; the first cysteine is located on the N-terminal side of the nuclease domain, whereas the other three cysteines are found in close proximity to each other in the C-terminal portion of the nuclease domain. Due to this unusual arrangement, the FeS cluster might act as an "iron staple," which is required to pin back and stabilize the nuclease domain. In turn, impaired FeS cluster binding results in loss of structural integrity of the nuclease domain and compromised binding to DNA ends [38].

It is noteworthy that FeS cluster binding is a highly conserved feature of AddB family proteins with the exception of lactic acid bacteria that have supposedly replaced the FeS cluster with an alternative metal cofactor [38]. Because lactic acid bacteria thrive in iron-poor environments, it is not unusual for them to replace iron-containing cofactors with alternative ones [59]. Does this in turn mean that the FeS cluster in AddAB and DNA2 has an exclusively structural function? It does seem so, but more studies are required to address this question.

Recently, a study from the Campbell laboratory has formally proven that yeast DNA2 indeed is an FeS protein. As suggested, it binds to an FeS cluster *via* the proposed FeS cluster-binding motif that spans its N-terminal nuclease domain [39]. In the absence of an FeS cluster, DNA2's nuclease activity is severely compromised. In contrast to AddAB, however, the presence of an FeS cluster did not prove to be essential for DNA binding nor did it seem to influence the overall structure of DNA2, as suggested by limited protease digestion experiments with wild-type and FeS cluster-binding mutant proteins. Unexpectedly, however, loss of the FeS cluster clearly reduced the ATPase activity of DNA2, despite the fact that the ATPase domain is located far away from the FeS cluster-binding site in the C-terminal helicase domain of the protein. This might suggest that the FeS cluster functionally couples helicase and endonuclease activities [39].

In a recent biochemical study from the Cejka laboratory, it was shown for yeast DNA2 that its nuclease activity auto-inhibits its helicase activity [60]. A nuclease-dead version of DNA2 in turn displayed significantly increased helicase activity, suggesting that in a wild-type situation, DNA2's strong endonuclease activity prevails and cleaves potential helicase substrates before unwinding can occur [60]. These findings open the possibility that the dual activity of DNA2 might be highly regulated *in vivo* and – depending on the cellular context – nuclease activity or helicase activity could prevail. It is tempting to speculate whether the redox state of the FeS cluster plays a role in this regulation.

## 20.5 FeS proteins in DNA repair

In contrast to DNA replication, a role for FeS clusters in the processes of DNA repair and genome maintenance has been appreciated for many years. However, for a long time, their contribution seemed to be limited to a subset of DNA repair proteins: DNA glycosylases. More recently, the members of the Rad3 family of helicases have joined the club, underlining the fact that FeS clusters are not limited to a specific pathway but play a more general role in DNA repair.

### 20.5.1 DNA glycosylases

#### 20.5.1.1 ROS and DNA damage

Oxidative DNA damage caused by ROS is a major threat to genome stability. ROS are either generated endogenously, e.g. as by-products of mitochondrial respiration, or by exogenous factors, such as xenobiotics or ionizing radiation [61, 62]. ROS include hydroxyl radicals ($OH^\bullet$), superoxide radicals ($O_2^{\bullet-}$), and nonradical species, such as hydrogen peroxide ($H_2O_2$), all of which can cause oxidative damage to DNA [63]. A common ROS-induced lesion is the oxidation of guanine to 8-oxoguanine ($^8$OG). During DNA replication, adenine instead of cytosine is incorporated opposite of $^8$OG, which leads to the formation of $^8$OG:A mispairs that – when left unrepaired – cause G:C to T:A transversion mutations [64].

#### 20.5.1.2 Base excision repair and DNA glycosylases

DNA base oxidations or other chemical alterations, such as nonenzymatic methylations, interfere with DNA replication and transcription when left unrepaired. The evolutionary conserved base excision repair (BER) pathway has an important role in recognizing and removing altered bases [65–67].

BER is initiated when a DNA glycosylase recognizes a damaged base and cleaves the *N*-glycosylic bond that links the base to the DNA backbone. The resulting abasic site is subsequently cleaved by either the intrinsic apurinic/apyrimidinic (AP) lyase activity of the DNA glycosylase itself (in case of bifunctional glycosylases) or alternatively by an AP endonuclease (in case of monofunctional glycosylases), thereby creating a single-stranded break. Further steps include cleaning of the break-flanking region, filling of the resulting gap by a DNA polymerase, and sealing of the remaining nick by a DNA ligase [67].

In human cells, 11 DNA glycosylases are known to date, 5 of which are required for the recognition and excision of oxidized bases [67]. The specificity of DNA glycosylases for various forms of base damage depends on the size and structure of their base binding pocket; however, given that the amount of base damage variations largely exceeds the number of glycosylases, their specificities do not seem to be very stringent.

#### 20.5.1.3 Endo III and MutY

*E. coli* endonuclease III (Endo III) was the first glycosylase to be reported to bind to an FeS cluster [68]. Despite its misleading name, Endo III is not a nuclease but a bifunctional glycosylase with intrinsic AP lyase activity; it is required for the excision of a variety of oxidized pyrimidine bases [69]. Later on, the human counterpart of Endo III, NTH1, was also shown to bind to an FeS cluster [70, 71].

Another example of an FeS glycosylase is MutY [72]. Like Endo III, it is a bifunctional glycosylase that is highly conserved from *E. coli* to human (MUTYH). However, unlike Endo III, it displays specificity for $^{8}$OG:A mispairs [73].

Both Endo III and MutY were shown to bind to a $[4Fe\text{-}4S]^{2+}$ cluster that – under physiological conditions – can neither be oxidized nor reduced [68]. At first, this led to the assumption that the redox state of the FeS cluster is unlikely to have any influence on protein function [74] and that the cluster must have a purely structural role. This proved to be wrong because the FeS cluster does not contribute significantly to the overall structure of the protein, but is instead required for DNA binding and enzymatic activity [75].

Interestingly, when Endo III and MutY are bound to DNA, they become redox-active and the redox potential of their FeS clusters is shifted toward the oxidized $[4Fe\text{-}4S]^{3+}$ state [76]. At the same time, oxidized glycosylases seem to have a higher affinity for DNA [77], suggesting that DNA binding could be modulated by the redox state of the FeS cluster.

#### 20.5.1.4 Damage detection by DNA-mediated charge transfer

Double-stranded DNA has been shown to allow the transport of electric charge over distances of around 200 Å due to the $\pi$-stacking of its aromatic base pairs [78, 79]. DNA-mediated charge transfer is very sensitive to disruptions in the $\pi$-stack, and DNA molecules with mismatches or lesions, such as oxidized bases, are impaired for charge transfer [80]. Although numerous *in vitro* studies leave no doubt about the ability of DNA to mediate charge transfer, evidence that this is of importance *in vivo* is rather rare [81] and the physiological relevance of it is subject to debate.

From a theoretical point of view, however, DNA-mediated charge transfer could be an elegant explanation as to how DNA repair molecules manage to scan the enormous length of the genome for oxidative damage in a timely fashion [76, 77, 82, 83]. The fact that charge transfer can only take place through perfectly $\pi$-stacked DNA could give redox-active proteins a means to distinguish between intact and damaged DNA.

In a model put forward by the Barton laboratory (Fig. 20.4), a glycosylase (MutY or Endo III) binds to DNA, which leads to the oxidation of its FeS cluster and tighter binding to DNA [76]. The electron that is released upon FeS cluster oxidation could then travel along the DNA *via* charge transfer and reduce another glycosylase that is distantly bound. The reduced glycosylase would then have less affinity for the DNA substrate and dissociate. However, if a DNA lesion, such as an oxidized base, is present between the two proteins, charge transport would be interrupted, leaving the second protein in its oxidized state and tightly bound to DNA. As a consequence, DNA glycosylases would accumulate close to the site of DNA damage, thereby facilitating DNA damage signaling and repair.

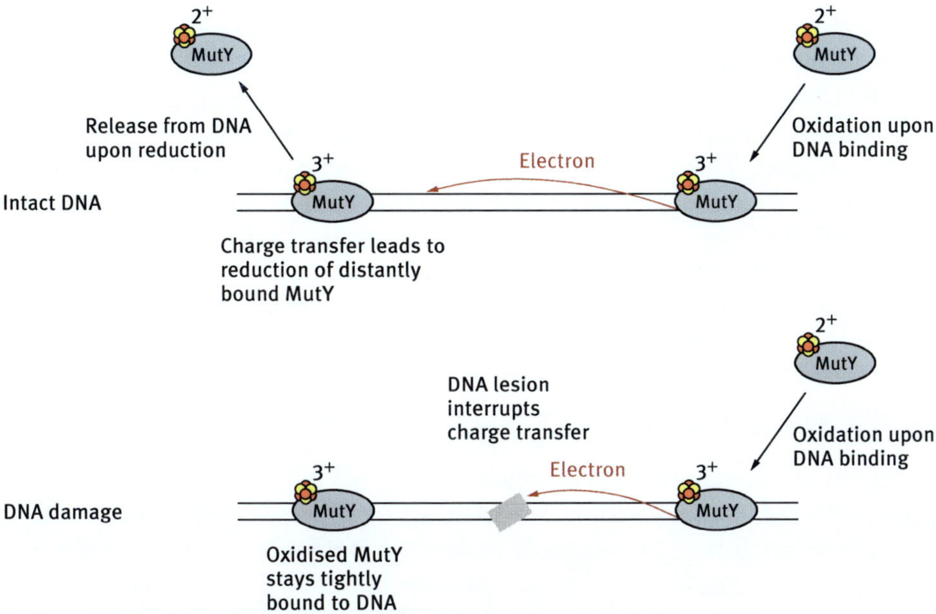


**Fig. 20.4:** DNA damage detection by charge transfer.

An overall environment of oxidative stress could further facilitate the initial binding of glycosylases to DNA because guanine radicals can directly oxidize the FeS cluster in MutY (and possibly other FeS glycosylases), which in turn increases their affinity for DNA [84]. Given that guanine radicals are generated early during physiological oxidative stress conditions [85], FeS cluster oxidation could be a means to force accumulation of MutY (and possibly other BER factors) on damaged DNA.

In favor of such a model of damage detection are two recent studies that show that DNA-mediated charge transfer is required for MutY and Endo III to locate DNA damage and to cluster in the vicinity [83, 86]. Interestingly, cooperativity between MutY and Endo III was observed in this process, which was dependent on the ability of the proteins to mediate charge transfer [86]. However attractive this model is, further studies are required to decide whether DNA-mediated charge transfer has any physiological relevance in DNA damage detection.

### 20.5.2 The Rad3 family of helicases

#### 20.5.2.1 A conserved family of helicases with links to human disease

*Saccharomyces cerevisiae* Rad3 is the founding member of a superfamily 2 of helicases that, apart from its human homologue XPD, also includes RTEL1, FANCJ, and ChlR1, all of which have been linked to human disease [87, 88]. As a common feature, they all possess 5′-3′ helicase activity but have preferences for different

DNA substrates, which presumably explains their functions in different pathways essential for the maintenance of genome stability.

RTEL1 (regulator of telomere length protein 1) does not seem to have classical DNA unwinding activity but is instead able to dismantle DNA recombination intermediates, so-called D-loop structures, in which a third strand invades the DNA duplex [89]. This anti-recombinogenic function is required to limit meiotic recombination [90], to maintain the integrity of telomeres [91], and to counteract the accumulation of toxic recombination intermediates during DNA repair [89].

SNP variations in *RTEL1* have been associated with high-grade glioma susceptibility [92], and more recently, *RTEL1* mutations have been found in several Hoyeraal-Hreidarsson syndrome patients, a severe form of dyskeratosis congenita caused by dysfunctional telomeres [93].

FANCJ is defective in a subgroup of patients with Fanconi anemia (FA), a genome instability disorder that confers an elevated cancer risk [94, 95]. FANCJ was originally identified as an interaction partner of the breast cancer susceptibility protein BRCA1 [96]; mutations in *FANCJ* have later been detected in breast and ovarian cancer [97–99], suggesting a role for FANCJ as a tumor suppressor.

From a biochemical point of view, FANCJ can bind to and unwind branched DNA molecules *in vitro* [100]. A combination of biochemical and *in vivo* data, however, suggests that FANCJ's main role is the unwinding of G-quadruplex (G4) DNA [101–103]. G4 DNA are secondary structures that are stabilized by Hoogsteen base pairing; they can form in guanine-rich regions of the genome and represent an obstacle for the DNA replication machinery [104].

ChlR1 preferentially unwinds branched DNA molecules with a 5′ single-stranded region [105, 106]. It interacts with various components of the DNA replication machinery [107] and appears to play an important role in sister chromatid cohesion during DNA replication, which is a prerequisite for normal chromosome segregation [108–112].

The gene coding for ChlR1, *DDX11*, is mutated in Warsaw breakage syndrome (WABS) [113], an extremely rare disorder characterized by congenital abnormalities and – on a cellular level – defects in sister chromatid cohesion.

XPD functions in both transcription initiation and nucleotide excision repair (NER), a DNA repair pathway specialized in the removal of damaged nucleotides, as e.g. caused by UV light [114]. XPD's 5′-3′ helicase activity has been shown to be essential for NER but dispensable for transcription initiation [115].

Surprisingly, mutations in *XPD* give rise to three symptomatically distinct human disorders: the eponymous xeroderma pigmentosum (XP), Cockayne syndrome, and trichothiodystrophy (TTD) [116]. These multiple syndromes can be explained with the fact that, depending on the mutation, transcription alone, NER alone, or both pathways are affected. Patients with XP, for example, are extremely light-sensitive and at a high risk of developing skin cancer due to a defective NER pathway [116].

### 20.5.2.2 The FeS cluster in XPD

In eukaryotes, XPD is an integral part of the multi-subunit transcription factor II H complex and is required for its stability [117, 118]. A homologue of XPD is also found in Archaea; in contrast to eukaryotes, however, archaeal XPD is active as a monomer, which makes it an interesting and amenable protein to study. In 2006, a study from the White laboratory found that *Sulfolobus acidocaldarius* XPD (sacXPD) binds to a $[4Fe\text{-}4S]^{2+}$ cluster, which is sensitive to oxidation [8]. Sequence alignment showed that the cluster-binding cysteines are highly conserved across all Rad3-like helicases and located in the HD1 helicase domain between the Walker A and B box motifs (Fig. 20.5; cysteines highlighted in yellow) [8].

Analysis of FeS cluster-binding mutants further demonstrated that the FeS cluster is dispensable for ATPase activity and the overall stability of sacXPD. In contrast, helicase activity is severely impaired in these mutants, suggesting that the FeS cluster is required to couple ATPase activity to DNA unwinding [8].

Our understanding of the FeS cluster-binding region within XPD has been significantly furthered by the structural analysis of XPD from three archaeal organisms (*S. acidocaldarius* [119], *S. tokodaii* [120], and *Thermoplasma acidophilum* [121]). All three reports are in agreement with XPD being composed of four domains: two canonical RecA-like helicase domains (HD1 and HD2) and two accessory domain inserted into HD1 (the FeS domain and an arch-shaped domain) [119–121]. Binding of an FeS cluster is required for the stability of the FeS and arch domains [119]. Together with a later study of *T. acidophilum* XPD in complex with DNA [122], these structural data suggest that the HD1, arch, and FeS domains together form a channel that can accommodate single-stranded DNA and that DNA unwinding occurs by threading single-stranded DNA through this channel.

In addition to these structural studies, much has been learned from a number of single-molecule studies that take advantage of the fact that FeS clusters can act as intrinsic quenchers of fluorescently labeled DNA substrates [123–125]. Taken together, in the current model, the FeS domain is thought to recognize the junction between single- and double-stranded DNA and to act as a wedge or ploughshare for the separation of DNA strands.

```
Rad3    102   DFRGLGLTSRKNLCLHPEVSKERKGTVVDEKCRRMTNGQAKRKLEED---PEANV---E-LCEYHENLYNIEVEDY----
XPD     103   PFLGLALSSRKNLCIHPEVTPLRFGKDVDGKCHSLTASYVRAQYQHD-----TSL---P-HCRFYEEFDAHGREVP----
RTEL1   171   ----CVLGSREQLCIHPEVKKQESNHLQIHLCRKKVAS--------------------R-SCHFYNNVEEKSLEQE----
FANCJ   309   GVPMTILSSRDHTCVHPEVVGNFNR---NEKCMELLD-----------------GKNGK-SCYFYHGVHKISDQHTLQTF
ChlR1   254   DVRLVSLGSRQNLCVNEDVKSLGSVQLINDRCVDMQRSRHEKKKGAEEEKPKRRRQEKQAACPFYNHEQMGLLRDE----

Rad3    171   ----LPKGVFSFEKLLKYCEEKTLCPYFIVRRMISLCNIIIYSYHYLLDPKIAERVSNEVSKDSIVIFDEAH
XPD     171   ----LPAGIYNLDDLKALGRRQGWCPYFLARYSILHANVVVYSYHYLLDPKIADLVSKELARKAVVVFDEAH
RTEL1   187   ----LASPILDIEDLVKSGSKHRVCPYYLSRNLKQQADIIFMPYNYLLDAKSRRAHNIDL-KGTVVIFDEAH
FANCJ   329   QGMC---KAWDIEELVSLGKKLKACPYYTARELIQDADIIFCPYNYLLDAQIRESMDLNL-KEQVVILDEAH
ChlR1   330   ----ALAEVKDMEQLLALGKEARACPYYGSRLAIPAAQLVVLPYQMLLHAATRQAAGIRL-QDQVVIIDEAH
```

**Fig. 20.5:** Rad3 family of helicases.

A possible extension to this model comes from a recent study, in which Mathieu *et al.* [126] generated rationally designed mutant versions of XPD with amino acid replacements in the FeS cluster-binding pocket. These mutants retained helicase activity but could not distinguish between damaged and undamaged nucleotides anymore, suggesting the exciting possibility that the FeS domain could play a role in the recognition of damaged nucleotides during strand unwinding.

#### 20.5.2.3 Mutations in the FeS cluster-binding region of XPD, FANCJ, and ChlR1

Interestingly, a number of clinically relevant mutations in *XPD, FANCJ*, and *DDX11* are located in the FeS cluster-binding region (Fig. 20.5; mutations marked in red).

Arginine-to-histidine replacement in XPD (R112H XPD) is common in TTD patients [127]. R112H XPD does not stably integrate an FeS cluster and displays completely abolished helicase activity [8, 127]. The aforementioned structural analysis of archaeal XPD homologues indicated that the corresponding R112 residues (R88 in *T. acidophilum* and K84 in *S. acidocaldarius*) are located in close proximity to the FeS cluster, enabling them to contribute to the stabilization of the FeS domain [119, 121].

Remarkably, the corresponding amino acid exchange in ChlR1 is also found in WABS patients, and biochemical analysis of R263H ChlR1 showed that – as in XPD – this arginine residue is essential for helicase function [128].

In the case of FANCJ, a proline-to-alanine substitution (A349P FANCJ) is found right next to one of the coordinating cysteine residues in the FeS cluster-binding domain in some patients with FA [95]. Similar to the mutations in XPD and ChlR1, this exchange abolishes FeS cluster binding and results in uncoupling of ATP hydrolysis from translocase activity [8, 129].

Taken together, these findings are in agreement with the FeS cluster being essential for helicase activity and underline the significance of FeS cluster binding *in vivo*.

#### 20.5.2.4 Possible functions of the FeS cluster in Rad3-like helicases

Future studies have to show which role exactly the FeS cluster plays in the different Rad3-like helicases. Using the FeS cluster-binding pocket for the recognition of damaged nucleotides, as suggested for XPD [126], certainly is an exciting possibility. However, taking into consideration that FANCJ, RTEL1, and ChlR1 have no specificity for damaged bases, but rather display a preference for unusual DNA structures, it seems unlikely that damage recognition is a general feature of FeS clusters in these helicases.

Another attractive hypothesis is that the redox sensitivity of FeS clusters could be used to modulate the biochemical activities of FeS repair proteins in response to altered redox conditions in the cell. For example, *in vitro* studies with DinG, the closest homologue of Rad3-like helicases in *E. coli*, have shown that reduction of its $[4Fe\text{-}4S]^{2+}$ cluster to the $[4Fe\text{-}4S]^{1+}$ state leads to a reversible switch-off of its helicase activity [130]. Meanwhile, the FeS cluster is stable in the $[4Fe\text{-}4S]^{2+}$ state under oxidizing conditions, suggesting the possibility that DinG is in an inactive ($[4Fe\text{-}4S]^{1+}$) state under normal conditions but can become activated ($[4Fe\text{-}4S]^{2+}$) under conditions of oxidative stress. However, whether this redox-regulation mechanism plays a role *in vivo* is not clear yet.

Studies from the Barton laboratory suggest that – at least in theory – a similar mechanism could regulate XPD. They could show that sacXPD has a physiologically relevant redox potential when bound to DNA (~80 mV *vs* the normal hydrogen electrode), which increases with ATP hydrolysis [131]. Importantly, the redox signal depends on the DNA substrate, raising the possibility that the ATPase and helicase activities of XPD could be regulated *via* the redox state of its FeS cluster and that regulation would depend on the DNA substrate encountered.

Taking the idea of redox signaling between XPD and DNA a step further, it is also conceivable that DNA-mediated charge transport is used by XPD to screen for DNA lesions or alterations, in a similar way as had been suggested for DNA glycosylases. Indeed, as for MutY and Endo III, location of XPD to damaged sites seems to depend on charge transfer [132]. Moreover, in the same study, charge transfer-mediated signaling occurred between XPD and other FeS proteins, suggesting a potential communication between multiple FeS repair proteins during their search for DNA damage [132]. Future studies will have to address whether long-distance screening of DNA and possibly communication between different repair factors by DNA-mediated charge transport has *in vivo* significance or whether it is merely an attractive hypothesis.

## 20.6 Summary

The significance of FeS clusters in DNA replication and repair proteins is far from being clear. In some cases, their role might be purely structural. However, given their potentially dangerous nature for the integrity of DNA, it seems likely that they have functions beyond a structural role. Their redox sensitivity certainly makes FeS clusters very interesting cofactors that could be used to modulate biochemical activities, react to conditions of oxidative stress, recognize DNA damage, or screen the DNA for lesions or abnormal structures. Further studies – both *in vitro* and *in vivo* – will be needed to address the role of FeS clusters in DNA replication and repair and decipher their importance for the maintenance of genome stability.

## References

[1] Bartkova J, Rezaei N, Liontos M, et al. Oncogene-induced senescence is part of the tumorigenesis barrier imposed by DNA damage checkpoints. Nature 2006;444:633–7.
[2] Bartkova J, Horejsí Z, Koed K, et al. DNA damage response as a candidate anti-cancer barrier in early human tumorigenesis. Nature 2005;434:864–70.

[3] Halazonetis TD, Gorgoulis VG, Bartek J. An oncogene-induced DNA damage model for cancer development. Science 2008;319:1352–5.
[4] Gorgoulis VG, Vassiliou L-VF, Karakaidos P, et al. Activation of the DNA damage checkpoint and genomic instability in human precancerous lesions. Nature 2005;434:907–13.
[5] Di Micco R, Fumagalli M, Cicalese A, et al. Oncogene-induced senescence is a DNA damage response triggered by DNA hyper-replication. Nature 2006;444:638–42.
[6] Veatch JR, McMurray MA, Nelson ZW, Gottschling DE. Mitochondrial dysfunction leads to nuclear genome instability via an iron-sulfur cluster defect. Cell 2009;137:1247–58.
[7] Klinge S, Hirst J, Maman JD, Krude T, Pellegrini L. An iron-sulfur domain of the eukaryotic primase is essential for RNA primer synthesis. Nat Struct Mol Biol 2007;14:875–7.
[8] Rudolf J, Makrantoni V, Ingledew WJ, Stark MJ, White MF. The DNA repair helicases XPD and FancJ have essential iron-sulfur domains. Mol Cell 2006;23:801–8.
[9] Alseth I, Eide L, Pirovano M, Rognes T, Seeberg E, Bjørås M. The Saccharomyces cerevisiae homologues of endonuclease III from Escherichia coli, Ntg1 and Ntg2, are both required for efficient repair of spontaneous and induced oxidative DNA damage in yeast. Mol Cell Biol 1999;19:3779–87.
[10] Lill R. Function and biogenesis of iron-sulphur proteins. Nature 2009;460:831–8.
[11] Rouault TA. Biogenesis of iron-sulfur clusters in mammalian cells: new insights and relevance to human disease. Dis Model Mech 2012;5:155–64.
[12] Netz DJA, Pierik AJ, Stümpfig M, Mühlenhoff U, Lill R. The Cfd1-Nbp35 complex acts as a scaffold for iron-sulfur protein assembly in the yeast cytosol. Nat Chem Biol 2007;3:278–86.
[13] Netz DJ, Pierik AJ, Stumpfig M, et al. A bridging [4Fe-4S] cluster and nucleotide binding are essential for function of the Cfd1-Nbp35 complex as a scaffold in iron-sulfur protein maturation. J Biol Chem 2012;287:12365–78.
[14] Balk J, Aguilar Netz DJ, Tepper K, Pierik AJ, Lill R. The essential WD40 protein Cia1 is involved in a late step of cytosolic and nuclear iron-sulfur protein assembly. Mol Cell Biol 2005;25:10833–41.
[15] Balk J, Pierik AJ, Aguilar Netz DJ, Mühlenhoff U, Lill R. The hydrogenase-like Nar1p is essential for maturation of cytosolic and nuclear iron-sulphur proteins. EMBO J 2004;23:2105–15.
[16] Song D, Lee FS. A Role for IOP1 in Mammalian Cytosolic Iron-Sulfur Protein Biogenesis. J Biol Chem 2008;283:9231–8.
[17] Gari K, León Ortiz AM, Borel V, Flynn H, Skehel JM, Boulton SJ. MMS19 links cytoplasmic iron-sulfur cluster assembly to DNA metabolism. Science 2012;337:243–5.
[18] Stehling O, Vashisht AA, Mascarenhas J, et al. MMS19 assembles iron-sulfur proteins required for DNA metabolism and genomic integrity. Science 2012;337:195–9.
[19] van Wietmarschen N, Moradian A, Morin GB, Lansdorp PM, Uringa EJ. The mammalian proteins MMS19, MIP18, and ANT2 are involved in cytoplasmic iron-sulfur cluster protein assembly. J Biol Chem 2012;287:43351–8.
[20] Weerapana E, Wang C, Simon GM, et al. Quantitative reactivity profiling predicts functional cysteines in proteomes. Nature 2010;468:790–5.
[21] Kunkel TA, Burgers PM. Dividing the workload at a eukaryotic replication fork. Trends Cell Biol 2008;18:521–7.
[22] Burgers PM. Polymerase dynamics at the eukaryotic DNA replication fork. J Biol Chem 2009;284:4041–5.
[23] Pellegrini L. The Pol alpha-primase complex. Subcell Biochem 2012;62:157–69.
[24] Larrea AA, Lujan SA, Nick McElhinny SA, et al. Genome-wide model for the normal eukaryotic DNA replication fork. Proc Natl Acad Sci USA 2010;107:17674–9.
[25] Pursell ZF, Isoz I, Lundstrom EB, Johansson E, Kunkel TA. Yeast DNA polymerase epsilon participates in leading-strand DNA replication. Science 2007;317:127–30.
[26] Miyabe I, Kunkel TA, Carr AM. The major roles of DNA polymerases epsilon and delta at the eukaryotic replication fork are evolutionarily conserved. PLoS Genet 2011;7:e1002407.

[27] Moldovan G-L, Pfander B, Jentsch S. PCNA, the maestro of the replication fork. Cell 2007;129:665–79.

[28] Liu Y, Kao HI, Bambara RA. Flap endonuclease 1: a central component of DNA metabolism. Annu Rev Biochem 2004;73:589–615.

[29] Zheng L, Shen B. Okazaki fragment maturation: nucleases take centre stage. J Mol Cell Biol 2011;3:23–30.

[30] Stith CM, Sterling J, Resnick MA, Gordenin DA, Burgers PM. Flexibility of eukaryotic Okazaki fragment maturation through regulated strand displacement synthesis. J Biol Chem 2008;283:34129–0.

[31] Budd ME, Campbell JL. A yeast gene required for DNA replication encodes a protein with homology to DNA helicases. Proc Natl Acad Sci USA 1995;92:7642–46.

[32] Bae SH, Seo YS. Characterization of the enzymatic properties of the yeast Dna2 helicase/ endonuclease suggests a new model for Okazaki fragment processing. J Biol Chem 2000;275:38022–31.

[33] Kim JH, Kim HD, Ryu GH, Kim DH, Hurwitz J, Seo YS. Isolation of human Dna2 endonuclease and characterization of its enzymatic properties. Nucleic Acids Res 2006;34:1854–64.

[34] Ayyagari R, Gomes XV, Gordenin DA, Burgers PM. Okazaki fragment maturation in yeast. I. Distribution of functions between FEN1 and DNA2. J Biol Chem 2003;278:1618–25.

[35] Masuda-Sasa T, Imamura O, Campbell JL. Biochemical analysis of human Dna2. Nucleic Acids Res 2006;34:1865–75.

[36] Netz DJA, Stith CM, Stümpfig M, et al. Eukaryotic DNA polymerases require an iron-sulfur cluster for the formation of active complexes. Nat Chem Biol 2012;8:125–32.

[37] Weiner BE, Huang H, Dattilo BM, Nilges MJ, Fanning E, Chazin WJ. An iron-sulfur cluster in the C-terminal domain of the p58 subunit of human DNA primase. J Biol Chem 2007;282:33444–51.

[38] Yeeles JTP, Cammack R, Dillingham MS. An iron-sulfur cluster is essential for the binding of broken DNA by AddAB-type helicase-nucleases. J Biol Chem 2009;284:7746–55.

[39] Pokharel S, Campbell JL. Cross talk between the nuclease and helicase activities of DNA2: role of an essential iron-sulfur cluster domain. Nucleic Acids Res 2012;40:7821–30.

[40] Foiani M, Santocanale C, Plevani P, Lucchini G. A single essential gene, PRI2, encodes the large subunit of DNA primase in Saccharomyces cerevisiae. Mol Cell Biol 1989;9:3081–7.

[41] Zerbe LK, Kuchta RD. The p58 subunit of human DNA primase is important for primer initiation, elongation, and counting. Biochemistry 2002;41:4891–900.

[42] Agarkar VB, Babayeva ND, Pavlov YI, Tahirov TH. Crystal structure of the C-terminal domain of human DNA primase large subunit: implications for the mechanism of the primase-polymerase alpha switch. Cell Cycle 2011;10:926–931.

[43] Sauguet L, Klinge S, Perera RL, Maman JD, Pellegrini L. Shared active site architecture between the large subunit of eukaryotic primase and DNA photolyase. PloS ONE 2010;5:e10083.

[44] Mizuno T, Yamagishi K, Miyazawa H, Hanaoka F. Molecular architecture of the mouse DNA polymerase alpha-primase complex. Mol Cell Biol 1999;19:7886–96.

[45] Klinge S, Nunez-Ramirez R, Llorca O, Pellegrini L. 3D architecture of DNA Pol alpha reveals the functional core of multi-subunit replicative polymerases. EMBO J 2009;28:1978–87.

[46] Burgers PM, Koonin EV, Bruford E, et al. Eukaryotic DNA polymerases: proposal for a revised nomenclature. J Biol Chem 2001;276:43487–90.

[47] Johansson E, Macneill SA. The eukaryotic replicative DNA polymerases take shape. Trends Biochem Sci 2010;35:339–47.

[48] Krishna SS, Majumdar I, Grishin NV. Structural classification of zinc fingers: survey and summary. Nucleic Acids Res 2003;31:532–50.

[49] Sanchez Garcia J, Ciufo LF, Yang X, Kearsey SE, MacNeill SA. The C-terminal zinc finger of the catalytic subunit of DNA polymerase delta is responsible for direct interaction with the B-subunit. Nucleic Acids Res 2004;32:3005–16.

[50] Zhu Z, Chung WH, Shim EY, Lee SE, Ira G. Sgs1 helicase and two nucleases Dna2 and Exo1 resect DNA double-strand break ends. Cell 2008;134:981–94.
[51] Cejka P, Cannavo E, Polaczek P, et al. DNA end resection by Dna2-Sgs1-RPA and its stimulation by Top3-Rmi1 and Mre11-Rad50-Xrs2. Nature 2010;467:112–6.
[52] Lin W, Sampathi S, Dai H, et al. Mammalian DNA2 helicase/nuclease cleaves G-quadruplex DNA and is required for telomere integrity. EMBO J 2013;32:1425–39.
[53] Hu J, Sun L, Shen F, et al. The intra-S phase checkpoint targets Dna2 to prevent stalled replication forks from reversing. Cell 2012;149:1221–32.
[54] Zheng L, Zhou M, Guo Z, et al. Human DNA2 is a mitochondrial nuclease/helicase for efficient processing of DNA replication and repair intermediates. Mol Cell 2008;32:325–36.
[55] Bae SH, Choi E, Lee KH, Park JS, Lee SH, Seo YS. Dna2 of Saccharomyces cerevisiae possesses a single-stranded DNA-specific endonuclease activity that is able to act on double-stranded DNA in the presence of ATP. J Biol Chem 1998;273:26880–90.
[56] Budd ME, Choe WC, Campbell JL. DNA2 encodes a DNA helicase essential for replication of eukaryotic chromosomes. J Biol Chem 1995;270:26766–9.
[57] Lee KH, Kim DW, Bae SH, et al. The endonuclease activity of the yeast Dna2 enzyme is essential in vivo. Nucleic Acids Res 2000;28:2873–81.
[58] Bae SH, Kim DW, Kim J, et al. Coupling of DNA helicase and endonuclease activities of yeast Dna2 facilitates Okazaki fragment processing. J Biol Chem 2002;277:26632–41.
[59] Gostick DO, Green J, Irvine AS, Gasson MJ, Guest JR. A novel regulatory switch mediated by the FNR-like protein of Lactobacillus casei. Microbiology 1998;144:705–17.
[60] Levikova M, Klaue D, Seidel R, Cejka P. Nuclease activity of Saccharomyces cerevisiae Dna2 inhibits its potent DNA helicase activity. Proc Natl Acad Sci USA 2013;110:E1992–2001.
[61] Lindahl T. Instability and decay of the primary structure of DNA. Nature 1993;362:709–15.
[62] Mitra S, Boldogh I, Izumi T, Hazra TK. Complexities of the DNA base excision repair pathway for repair of oxidative DNA damage. Environ Mol Mutagen 2001;38:180–90.
[63] Dizdaroglu M, Jaruga P, Birincioglu M, Rodriguez H. Free radical-induced damage to DNA: mechanisms and measurement. Free Radic Biol Med 2002;32:1102–15.
[64] Shibutani S, Takeshita M, Grollman AP. Insertion of specific bases during DNA synthesis past the oxidation-damaged base 8-oxodG. Nature 1991;349:431–4.
[65] Lindahl T. Repair of intrinsic DNA lesions. Mutat Res 1990;238:305–11.
[66] Dianov G, Lindahl T. Reconstitution of the DNA base excision-repair pathway. Curr Biol 1994;4:1069–76.
[67] Dianov GL, Hubscher U. Mammalian base excision repair: the forgotten archangel. Nucleic Acids Res 2013;41:3483–90.
[68] Cunningham RP, Asahara H, Bank JF, et al. Endonuclease III is an iron-sulfur protein. Biochemistry 1989;28:4450–5.
[69] Asahara H, Wistort PM, Bank JF, Bakerian RH, Cunningham RP. Purification and characterization of Escherichia coli endonuclease III from the cloned nth gene. Biochemistry 1989;28:4444–9.
[70] Aspinwall R, Rothwell DG, Roldan-Arjona T, et al. Cloning and characterization of a functional human homolog of Escherichia coli endonuclease III. Proc Natl Acad Sci USA 1997;94:109–14.
[71] Ikeda S, Biswas T, Roy R, et al. Purification and characterization of human NTH1, a homolog of Escherichia coli endonuclease III. Direct identification of Lys-212 as the active nucleophilic residue. J Biol Chem 1998;273:21585–93.
[72] Michaels ML, Pham L, Nghiem Y, Cruz C, Miller JH. MutY, an adenine glycosylase active on G-A mispairs, has homology to endonuclease III. Nucleic Acids Res 1990;18:3841–5.
[73] Tsai-Wu JJ, Liu HF, Lu AL. Escherichia coli MutY protein has both N-glycosylase and apurinic/apyrimidinic endonuclease activities on A.C and A.G mispairs. Proc Natl Acad Sci USA 1992;89:8779–83.

[74] Fu W, O'Handley S, Cunningham RP, Johnson MK. The role of the iron-sulfur cluster in Escherichia coli endonuclease III. A resonance Raman study. J Biol Chem 1992;267:16135–7.
[75] Porello SL, Cannon MJ, David SS. A substrate recognition role for the [4Fe-4S]2+ cluster of the DNA repair glycosylase MutY. Biochemistry 1998;37:6465–75.
[76] Boon EM, Livingston AL, Chmiel NH, David SS, Barton JK. DNA-mediated charge transport for DNA repair. Proc Natl Acad Sci USA 2003;100:12543–7.
[77] Boal AK, Yavin E, Lukianova OA, O'Shea VL, David SS, Barton JK. DNA-bound redox activity of DNA repair glycosylases containing [4Fe-4S] clusters. Biochemistry 2005;44:8397–407.
[78] Kelley SO, Barton JK. Electron transfer between bases in double helical DNA. Science 1999;283:375–81.
[79] Nunez ME, Hall DB, Barton JK. Long-range oxidative damage to DNA: effects of distance and sequence. Chem Biol 1999;6:85–97.
[80] Boon EM, Ceres DM, Drummond TG, Hill MG, Barton JK. Mutation detection by electrocatalysis at DNA-modified electrodes. Nat Biotechnol 2000;18:1096–100.
[81] Nunez ME, Holmquist GP, Barton JK. Evidence for DNA charge transport in the nucleus. Biochemistry 2001;40:12465–71.
[82] Boal AK, Yavin E, Barton JK. DNA repair glycosylases with a [4Fe-4S] cluster: a redox cofactor for DNA-mediated charge transport? J Inorg Biochem 2007;101:1913–21.
[83] Boal AK, Genereux JC, Sontz PA, Gralnick JA, Newman DK, Barton JK. Redox signaling between DNA repair proteins for efficient lesion detection. Proc Natl Acad Sci USA 2009;106:15237–42.
[84] Yavin E, Boal AK, Stemp ED, et al. Protein-DNA charge transport: redox activation of a DNA repair protein by guanine radical. Proc Natl Acad Sci USA 2005;102:3546–51.
[85] Cadet J, Bellon S, Berger M, et al. Recent aspects of oxidative DNA damage: guanine lesions, measurement and substrate specificity of DNA repair glycosylases. Biol Chem 2002;383:933–43.
[86] Romano CA, Sontz PA, Barton JK. Mutants of the base excision repair glycosylase, endonuclease III: DNA charge transport as a first step in lesion detection. Biochemistry 2011;50:6133–45.
[87] White MF. Structure, function and evolution of the XPD family of iron-sulfur-containing 5′-3′ DNA helicases. Biochem Soc Trans 2009;37:547–51.
[88] Wu Y, Suhasini AN, Brosh RM. Welcome the family of FANCJ-like helicases to the block of genome stability maintenance proteins. Cell Mol Life Sci 2009;66:1209–22.
[89] Barber LJ, Youds JL, Ward JD, et al. RTEL1 maintains genomic stability by suppressing homologous recombination. Cell 2008;135:261–71.
[90] Youds JL, Mets DG, McIlwraith MJ, et al. RTEL-1 enforces meiotic crossover interference and homeostasis. Science 2010;327:1254–8.
[91] Vannier JB, Pavicic-Kaltenbrunner V, Petalcorin MI, Ding H, Boulton SJ. RTEL1 dismantles T loops and counteracts telomeric G4-DNA to maintain telomere integrity. Cell 2012;149:795–806.
[92] Wrensch M, Jenkins RB, Chang JS, et al. Variants in the CDKN2B and RTEL1 regions are associated with high-grade glioma susceptibility. Nat Genet 2009;41:905–8.
[93] Le Guen T, Jullien L, Touzot F, et al. Human RTEL1 deficiency causes Hoyeraal-Hreidarsson syndrome with short telomeres and genome instability. Hum Mol Genet 2013;22:3239–49.
[94] Levitus M, Waisfisz Q, Godthelp BC, et al. The DNA helicase BRIP1 is defective in Fanconi anemia complementation group J. Nat Genet 2005;37:934–5.
[95] Levran O, Attwooll C, Henry RT, et al. The BRCA1-interacting helicase BRIP1 is deficient in Fanconi anemia. Nat Genet 2005;37:931–3.
[96] Cantor SB, Bell DW, Ganesan S, et al. BACH1, a novel helicase-like protein, interacts directly with BRCA1 and contributes to its DNA repair function. Cell 2001;105:149–60.
[97] Cantor S, Drapkin R, Zhang F, et al. The BRCA1-associated protein BACH1 is a DNA helicase targeted by clinically relevant inactivating mutations. Proc Natl Acad Sci USA 2004;101:2357–62.

[98] Seal S, Thompson D, Renwick A, et al. Truncating mutations in the Fanconi anemia J gene BRIP1 are low-penetrance breast cancer susceptibility alleles. Nat Genet 2006;38:1239–41.
[99] Rafnar T, Gudbjartsson DF, Sulem P, et al. Mutations in BRIP1 confer high risk of ovarian cancer. Nat Genet 2011;43:1104–7.
[100] Gupta R, Sharma S, Sommers JA, Jin Z, Cantor SB, Brosh RM Jr. Analysis of the DNA substrate specificity of the human BACH1 helicase associated with breast cancer. J Biol Chem 2005;280:25450–60.
[101] Kruisselbrink E, Guryev V, Brouwer K, Pontier DB, Cuppen E, Tijsterman M. Mutagenic capacity of endogenous G4 DNA underlies genome instability in FANCJ-defective C. elegans. Curr Biol 2008;18:900–5.
[102] Wu Y, Shin-ya K, Brosh RM Jr. FANCJ helicase defective in Fanconia anemia and breast cancer unwinds G-quadruplex DNA to defend genomic stability. Mol Cell Biol 2008;28:4116–28.
[103] Maizels N. Genomic stability: FANCJ-dependent G4 DNA repair. Curr Biol 2008;18:R613–4.
[104] Maizels N. Dynamic roles for G4 DNA in the biology of eukaryotic cells. Nat Struct Mol Biol 2006;13:1055–9.
[105] Hirota Y, Lahti JM. Characterization of the enzymatic activity of hChlR1, a novel human DNA helicase. Nucleic Acids Res 2000;28:917–24.
[106] Wu Y, Sommers JA, Khan I, de Winter JP, Brosh RM Jr. Biochemical characterization of Warsaw breakage syndrome helicase. J Biol Chem 2012;287:1007–21.
[107] Farina A, Shin JH, Kim DH, et al. Studies with the human cohesin establishment factor, ChlR1. Association of ChlR1 with Ctf18-RFC and Fen1. J Biol Chem 2008;283:20925–36.
[108] Petronczki M, Chwalla B, Siomos MF, et al. Sister-chromatid cohesion mediated by the alternative RF-CCtf18/Dcc1/Ctf8, the helicase Chl1 and the polymerase-alpha-associated protein Ctf4 is essential for chromatid disjunction during meiosis II. J Cell Sci 2004;117:3547–59.
[109] Skibbens RV. Chl1p, a DNA helicase-like protein in budding yeast, functions in sister-chromatid cohesion. Genetics 2004;166:33–42.
[110] Mayer ML, Pot I, Chang M, et al. Identification of protein complexes required for efficient sister chromatid cohesion. Mol Biol Cell 2004;15:1736–45.
[111] Inoue A, Li T, Roby SK, et al. Loss of ChlR1 helicase in mouse causes lethality due to the accumulation of aneuploid cells generated by cohesion defects and placental malformation. Cell Cycle 2007;6:1646–54.
[112] Parish JL, Rosa J, Wang X, Lahti JM, Doxsey SJ, Androphy EJ. The DNA helicase ChlR1 is required for sister chromatid cohesion in mammalian cells. J Cell Sci 2006;119:4857–65.
[113] van der Lelij P, Chrzanowska KH, Godthelp BC, et al. Warsaw breakage syndrome, a cohesinopathy associated with mutations in the XPD helicase family member DDX11/ChlR1. Am J Hum Genet 2010;86:262–6.
[114] Kuper J, Kisker C. Damage recognition in nucleotide excision DNA repair. Curr Opin Struct Biol 2012;22:88–93.
[115] Winkler GS, Araujo SJ, Fiedler U, et al. TFIIH with inactive XPD helicase functions in transcription initiation but is defective in DNA repair. J Biol Chem 2000;275:4258–66.
[116] Lehmann AR. The xeroderma pigmentosum group D (XPD) gene: one gene, two functions, three diseases. Genes Dev 2001;15:15–23.
[117] Laine JP, Mocquet V, Egly JM. TFIIH enzymatic activities in transcription and nucleotide excision repair. Methods Enzymol 2006;408:246–63.
[118] Egly JM, Coin F. A history of TFIIH: two decades of molecular biology on a pivotal transcription/repair factor. DNA Repair 2011;10:714–21.
[119] Fan L, Fuss JO, Cheng QJ, et al. XPD helicase structures and activities: insights into the cancer and aging phenotypes from XPD mutations. Cell 2008;133:789–800.
[120] Liu H, Rudolf J, Johnson KA, et al. Structure of the DNA repair helicase XPD. Cell 2008;133:801–12.

[121] Wolski SC, Kuper J, Hanzelmann P, et al. Crystal structure of the FeS cluster-containing nucleotide excision repair helicase XPD. PLoS Biol 2008;6:e149.
[122] Kuper J, Wolski SC, Michels G, Kisker C. Functional and structural studies of the nucleotide excision repair helicase XPD suggest a polarity for DNA translocation. EMBO J 2012;31:494–502.
[123] Pugh RA, Honda M, Leesley H, et al. The iron-containing domain is essential in Rad3 helicases for coupling of ATP hydrolysis to DNA translocation and for targeting the helicase to the single-stranded DNA-double-stranded DNA junction. J Biol Chem 2008;283:1732–43.
[124] Pugh RA, Honda M, Spies M. Ensemble and single-molecule fluorescence-based assays to monitor DNA binding, translocation, and unwinding by iron-sulfur cluster containing helicases. Methods 2010;51:313–21.
[125] Pugh RA, Wu CG, Spies M. Regulation of translocation polarity by helicase domain 1 in SF2B helicases. EMBO J 2012;31:503–14.
[126] Mathieu N, Kaczmarek N, Rüthemann P, Luch A, Naegeli H. DNA quality control by a lesion sensor pocket of the xeroderma pigmentosum group D helicase subunit of TFIIH. Curr Biol 2013;23:204–12.
[127] Dubaele S, Proietti De Santis L, Bienstock RJ, et al. Basal transcription defect discriminates between xeroderma pigmentosum and trichothiodystrophy in XPD patients. Mol Cell 2003;11:1635–46.
[128] Capo-Chichi JM, Bharti SK, Sommers JA, et al. Identification and biochemical characterization of a novel mutation in DDX11 causing Warsaw breakage syndrome. Hum Mutat 2013;34:103–7.
[129] Wu Y, Sommers JA, Suhasini AN, et al. Fanconi anemia group J mutation abolishes its DNA repair function by uncoupling DNA translocation from helicase activity or disruption of protein-DNA complexes. Blood 2010;116:3780–91.
[130] Ren B, Duan X, Ding H. Redox control of the DNA damage-inducible protein DinG helicase activity via its iron-sulfur cluster. J Biol Chem 2009;284:4829–35.
[131] Mui TP, Fuss JO, Ishida JP, Tainer JA, Barton JK. ATP-stimulated, DNA-mediated redox signaling by XPD, a DNA repair and transcription helicase. J Am Chem Soc 2011;133:16378–81.
[132] Sontz PA, Mui TP, Fuss JO, Tainer JA, Barton JK. DNA charge transport as a first step in coordinating the detection of lesions by repair proteins. Proc Natl Acad Sci USA 2012;109:1856–61.

# 21 Eukaryotic iron-sulfur protein biogenesis and its role in maintaining genomic integrity

Roland Lill, Marta A. Uzarska and James Wohlschlegel

## 21.1 Introduction

The biogenesis of iron-sulfur (Fe-S) proteins in eukaryotes is a complex biosynthetic process. Failure to assemble Fe-S clusters and insert them into apo-proteins is incompatible with life and in humans can lead to various neurological, hematological, and metabolic diseases. Fe-S protein biogenesis is initiated by the mitochondrial iron-sulfur cluster (ISC) assembly machinery which is involved in the biogenesis of all cellular Fe-S proteins including those located in the cytosol and nucleus. Maturation of the latter proteins additionally requires the cytosolic iron-sulfur protein assembly (CIA) machinery and a still unknown sulfur-containing molecule exported from mitochondria. The essential character of the biosynthetic process is explained by the function of Fe-S proteins involved in cytosolic protein translation and in numerous steps of nuclear DNA metabolism including DNA synthesis and repair, chromosome segregation, and telomere length regulation. Many of these latter Fe-S proteins have been linked to diseases such as various forms of cancer as well as ageing. Here, we first provide an overview of the components and molecular mechanisms of the ISC and CIA machineries required for Fe-S protein assembly. In a second part, we explain the intimate molecular links of this essential process to DNA maintenance and chromosome instability.

Fe-S clusters are ancient protein cofactors that function as electron carriers, catalysts in chemical reactions, regulatory sensors, or sulfur donors (see other chapters in this book). In the (non-plant) eukaryotic cell, known Fe-S proteins are localized in the mitochondria, cytosol, and nucleus where they participate in a large number of biochemical reactions (Fig. 21.1). In the model organism *Saccharomyces cerevisiae*, Fe-S proteins perform functions in energy production (respiratory complexes II and III, mitochondrial aconitase), amino acid metabolism (e.g. ketoacid hydratases Leu1 and Ilv3 or a subunit of sulfite reductase, Ecm17), cofactor biosynthesis (lipoate and biotin synthase), and in protein synthesis (Rli1 involved in translation termination) (Fig. 21.1a). In the yeast nucleus, several Fe-S proteins have been identified that perform functions in DNA synthesis (replicative DNA polymerases, primase subunit Pri2) and various aspects of DNA repair (DNA helicase Rad3 and the glycosylase Ntg2). Additionally, a conspicuous number of the Fe-S protein biogenesis components (Yah1, Grx5, Dre2, Cfd1, Nbp35, and Nar1) depend on Fe-S clusters themselves, i.e. they are both components and targets of the biogenesis systems. Mammalian cells lack some of the biosynthetic Fe-S proteins of yeast (e.g. biotin synthase or Fe-S proteins involved in amino acid synthesis) yet contain additional Fe-S proteins not present in yeast

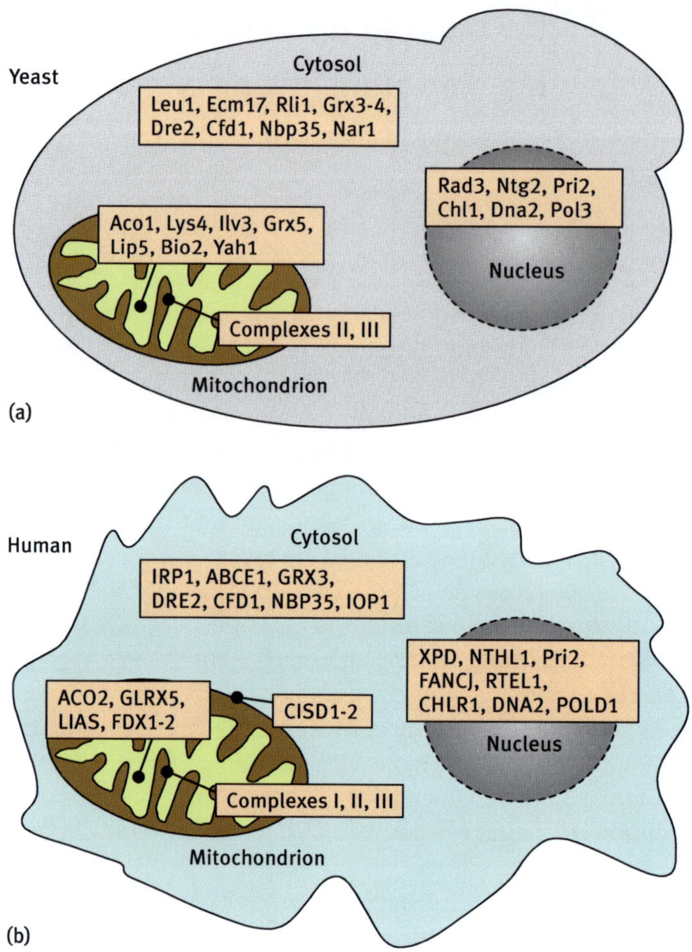


**Fig. 21.1:** Important Fe/S proteins in *S. cerevisiae* and mammalian cells. In (non-green) eukaryotes, Fe/S proteins are localized in mitochondria, cytosol, and nucleus. The figure presents a selection of the Fe/S protein inventory of (a) *S. cerevisiae* and (b) human cells. Some Fe/S proteins are present in only one of these organisms. Examples of mitochondrial Fe/S proteins (human names in parenthesis): Aco1 (ACO2), aconitase; Lys4, homoaconitase; Ilv3, dihydroxyacid dehydratase; Grx5 (GLRX5), monothiol glutaredoxin; Lip5 (LIAS), lipoate synthase; Bio2, biotin synthase; Yah1 (FDX1-FDX2), ferredoxin (adrenodoxin); respiratory complexes II and III. Cytosolic Fe/S proteins: Leu1, isopropylmalate isomerase; Ecm17, subunit of sulfite reductase; Rli1 (ABCE1), ABC protein involved in ribosome function; Grx3-Grx4 (GRX3), monothiol glutaredoxins; Dre2, Cfd1, Nbp35, and Nar1 (IOP1), CIA components. Nuclear Fe/S proteins: Rad3 (XPD), ATP-dependent DNA helicase; Ntg2 (NTHL1), DNA repair protein *N*-glycosylase, Pri2, primase subunit; Chl1 (CHLR1), DNA helicase; Dna2, ATP-dependent nuclease and helicase; Pol3 (POLD1), DNA polymerase δ subunit. Additionally, human cells possess a few Fe/S proteins that are not present in yeast, such as respiratory complex I, members of the mitoNEET (CISD) family attached to the mitochondrial outer membrane, IRP1 (cytosolic aconitase), FANCJ (DNA repair helicase mutated in Fanconi anemia), and RTEL1 (helicase involved in telomere stability).

(Fig. 21.1b). These include respiratory complex I with 8 Fe-S clusters, members of the mitoNEET (CISD) family in mitochondria and possibly the endoplasmic reticulum, and cytosolic aconitase, also known as iron regulatory protein 1 (IRP1), which uses its Fe-S cluster for sensing the intracellular iron status. The mammalian nucleus harbors, in addition to the aforementioned replicative DNA polymerases, a number of ATP-dependent DNA helicases with rather diverse functions in various aspects of DNA metabolism such as nucleotide excision repair (NER), telomere length regulation and chromosome segregation [xeroderma pigmentosum complementation group D (XPD), Fanconi anemia complement group J (FANCJ), regulator of telomere elongation 1 (RTEL1), and CHLR1]. A number of these factors have been implicated in human disease such as xeroderma pigmentosum, Fanconi anemia, and trichothiodystrophy. It is clear from this list of Fe-S proteins that knowledge of their assembly pathways in the living cell is not only of basic scientific interest but also has important medical implications.

The biogenesis of cellular Fe-S proteins is a complex process that requires the coordinated function of some 30 proteins in mitochondria and cytosol. The mitochondrial ISC assembly machinery was inherited from a similar system in bacteria yet is highly conserved from yeast to man (Fig. 21.2) [1–6]. The molecular mechanisms of the ISC assembly pathway have been worked out mainly in yeast and bacteria [7, 8] and can be divided into three main steps (Fig. 21.2a). First, a Fe-S cluster is synthesized *de novo* on a scaffold protein from iron and sulfur, which is released from cysteine by a desulfurase enzyme to form a persulfide group (-SSH) on the desulfurase. Second, the Fe-S cluster is released from the scaffold protein and transferred to proteins that transiently bind the cluster, and in a third step, insert the Fe-S cluster into different target apoproteins. Each of these steps requires the participation of several additional ISC proteins and low-molecular-mass cofactors, which will be explained in detail in Section 21.2.

Cytosolic and nuclear Fe-S proteins also require the mitochondrial ISC assembly system in both yeast and human cells [9–11]. Mitochondria export an as-yet unknown sulfur-containing molecule (X-S in Fig. 21.2) that is used by the CIA machinery for maturation. Both the ISC and CIA systems contain numerous constituents that are essential for the viability of yeast and human cells. Moreover, many diseases are linked to genetic mutations in various mitochondrial ISC components (Fig. 21.2b) [6, 12]. This demonstrates the importance of the overall biogenesis process for life. Moreover, because numerous essential Fe-S proteins with functions in protein synthesis, tRNA modification, and DNA replication and repair depend on the mitochondrial ISC system for maturation, the participation of the mitochondria in cellular Fe-S protein biosynthesis renders it essential for life [3, 4, 13]. This fact is impressively underlined by the discovery of mitosomes, i.e. mitochondria-derived organelles that, like mitochondria, possess a double membrane and import their proteins from the cytosol [14, 15]. However, during the evolution, the function of these organelles was drastically reduced by the loss of all classical mitochondrial functions including heme synthesis, citric acid cycle, oxidative phosphorylation, fatty acid oxidation, and mitochondrial

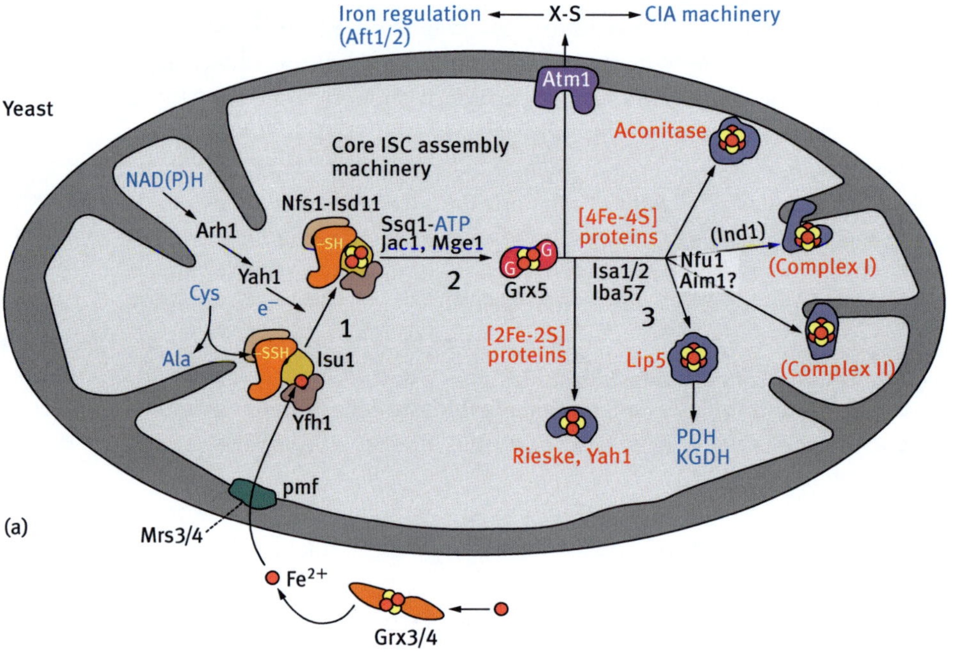


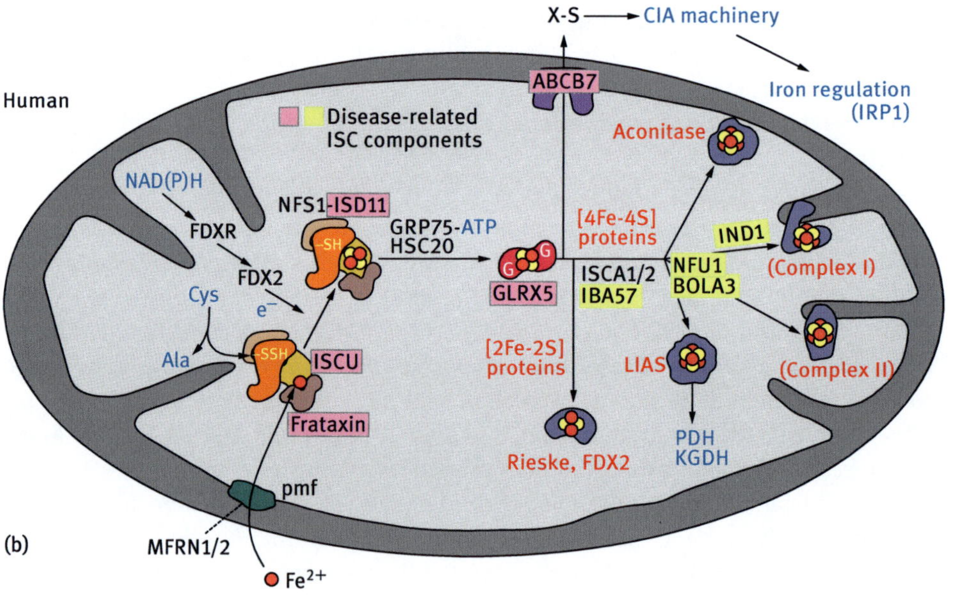


(*Figure Continued*)

(*Figure Continued*)

**Fig. 21.2:** The three steps of mitochondrial Fe/S protein assembly and diseases associated with this process. The components and mechanisms of mitochondrial Fe/S protein biogenesis are highly conserved from (a) yeast to (b) human cells. (a) In yeast, the glutaredoxins Grx3-Grx4 facilitate mitochondrial import of ferrous iron (red circle) from the cytosol *via* the inner membrane carriers Mrs3-Mrs4, which use the proton motive force (pmf) as a driving force for membrane transport. The biogenesis of mitochondrial Fe/S proteins is accomplished by the ISC assembly machinery in three major steps. First, a [2Fe-2S] cluster is synthesized on the scaffold protein Isu1, a step that requires the cysteine desulfurase complex Nfs1-Isd11 as a sulfur (yellow circle) donor and releases sulfur from cysteine *via* a Nfs1-bound persulfide intermediate (-SSH). This step further requires frataxin (yeast Yfh1), which undergoes an iron-dependent interaction with Isu1 and may serve as an iron donor and/or an allosteric regulator of the desulfurase enzyme. An electron ($e^-$) transfer chain consisting of NAD(P)H, ferredoxin reductase (Arh1) and ferredoxin (Yah1) is needed for Fe/S cluster assembly on Isu1. Second, the Isu1-bound Fe/S cluster is labilized by functional involvement of a dedicated chaperone system comprised of the ATP-dependent Hsp70 chaperone Ssq1, its co-chaperone Jac1, and the nucleotide exchange factor Mge1. Monothiol glutaredoxin Grx5 may receive the Fe/S cluster directly from Isu1 and binds it in a glutathione (G)-dependent fashion (see Fig. 21.3). The aforementioned proteins are involved in the biogenesis of all mitochondrial Fe/S proteins (including the [2Fe-2S] proteins Rieske and Yah1) and are termed core ISC assembly components. The ABC transporter Atm1 exports an unknown, sulfur-containing component (X-S) produced by the core ISC components for use in the maturation of cytosolic-nuclear Fe/S proteins by the CIA machinery (Fig. 21.4a) and for iron regulation by the Aft1-Aft2 transcription factors. In a third step, the generation of [4Fe-4S] clusters is catalyzed by Isa1-Isa2-Iba57 proteins. Further, specialized ISC targeting components (Nfu1, Aim1) assist the insertion of the Fe/S clusters into specific apo-proteins such as lipoate synthase (Lip5) and respiratory complex II (SDH), whereas Ind1 (not present in *S. cerevisiae* but in other fungi such as *Yarrowia*) is specific for complex I. The role of the BolA-like protein Aim1 is still hypothetical. Matured Lip5 produces lipoate for enzymes such as PDH and KGDH. (b) The human ISC assembly components use similar mechanisms as the yeast proteins. The function of the core ISC assembly components also impacts on cellular iron regulation *via* ABCB7- and CIA machinery-dependent maturation of IRP1 (see Fig. 21.4b). Defects in several of these mitochondrial components (highlighted in pink) lead to severe diseases with mitochondrial iron accumulation. In contrast, genetic defects in late-acting ISC components (highlighted in yellow) cause severe diseases that are not associated with alterations in cellular or mitochondrial iron levels. A variant erythropoietic protoporphyria is caused by dysfunctional mitoferrin 1 (MFRN1) in erythroid tissue. Cys, cysteine; Ala, alanine.

gene expression [16–18]. The only known process maintained in mitosomes is the Fe-S protein biosynthesis pathway, which likely has been maintained to support the maturation of extra-mitochondrial Fe-S proteins, as mitosomes do not contain any relevant Fe-S proteins themselves [15, 19, 20].

In this chapter, we will first summarize the molecular mechanisms of Fe-S protein biogenesis in eukaryotes. Then, we will provide an overview on the link between this pathway and the maintenance of genomic stability by several important Fe-S proteins. Other links of the Fe-S protein maturation process, e.g. to cellular iron regulation, and the diseases linked to Fe-S protein biogenesis will only be briefly discussed. These aspects are covered in other parts of this volume and have been comprehensively reviewed elsewhere [5, 6, 12, 21, 22].

## 21.2 Biogenesis of mitochondrial Fe-S proteins

### 21.2.1 Step 1: *De novo* Fe-S cluster assembly on the Isu1 scaffold protein

The ISC assembly machinery in yeast mitochondria consists of 17 known proteins, many of which are homologues of bacterial ISC factors [4, 5, 23] (Fig. 21.2a). Virtually all these ISC components are conserved in human cells (Fig. 21.2b), which, in addition, contains a specific maturation factor for respiratory complex I (IND1), which is present in all eukaryotes harboring this respiratory complex. In all eukaryotes, the ISC components are encoded in the nucleus and after their synthesis are imported into the mitochondrial matrix. For import, they use typical mitochondrial targeting sequences that are recognized by the TOM-TIM import machinery [24, 25].

Fe-S protein biogenesis starts with the *de novo* assembly of a Fe-S cluster on the scaffold protein Isu1 (yeast) or ISCU (human) (Fig. 21.2) [2, 26–28]. In yeast, a functionally redundant homologue termed Isu2 is present that arose by gene duplication. Isu1 scaffold proteins are highly conserved in bacteria and eukaryotes and contain three cysteine residues that are critical for Fe-S cluster synthesis. The sulfur is provided by the mitochondrial cysteine desulfurase Nfs1, which is required for all cellular Fe-S proteins [1, 9]. Nfs1 contains a conserved cysteine residue that is transiently converted to a persulfide following cysteine conversion to alanine and thus serves as a site for sulfur activation. Nfs1 tightly interacts with the small LYRM family protein Isd11. This protein is not necessary for desulfurase activity, yet essential for sulfur transfer from Nfs1 to Isu1 and Fe-S cluster formation on Isu1 [29–31].

The pathway by which iron is recruited by Isu1 is not fully understood. Iron import into the mitochondria is conducted *via* carrier proteins Mrs3 and Mrs4 (mitoferrin 1 and 2 in vertebrates) [32–35]. Recently, a minor role in mitochondrial iron import has been documented for the carrier Rim2, which co-imports iron and a pyrimidine nucleotide [36, 37]. In the matrix, the ISC protein frataxin (yeast Yfh1) has been implicated as an iron donor because it binds iron with micromolar affinities and forms an iron-stimulated complex with Isu1 and Nfs1/Isd11 [38–41]. Recent *in vitro* work on human frataxin suggests that it acts as an iron-dependent allosteric activator of Nfs1 desulfurase activity [42]. Additional experiments performed in yeast show that an Isu1 point mutation localized near one of the conserved cysteine residues is able to rescue the Fe-S protein biogenesis defect of *YFH1* deletion cells [43], showing that this function can be bypassed, at least in yeast. Although all these studies are consistent with a function of Yfh1 in Fe-S cluster synthesis on Isu1, its precise molecular function is not yet clear.

Another requirement for Fe-S cluster synthesis on Isu1 is the electron transfer from the [2Fe-2S] ferredoxin Yah1, which receives its electrons from the ferredoxin reductase Arh1 and NAD(P)H [27, 44–47] (Fig. 21.2a). It is not exactly known what this electron flow is used for. One possibility is the need for reduction of the sulfan sulfur ($S^0$) present in the cysteine to the sulfide ($S^{2-}$) present in the Fe-S cluster [4]. It has also been suggested to be needed for fusion of two [2Fe-2S] clusters into one [4Fe-4S] by reductive coupling [48, 49]. However, as pointed out later, [4Fe-4S] cluster generation requires late-acting ISC proteins. Interestingly, Yah1 is also necessary for heme A and coenzyme Q biosynthesis [50, 51]. It is the only essential Fe-S protein of yeast mitochondria (besides the scaffold protein Isu1 itself) and requires the core ISC assembly machinery for its own maturation [52]. Human cells possess two distinct mitochondrial ferredoxins, which differ in their expression pattern. The classical adrenodoxin FDX1 is found in adrenal gland and kidney cells where the protein, together with mitochondrial cytochrome P450 proteins, is involved steroid hormone and vitamin D production [46]. In contrast, the only recently characterized FDX2 is ubiquitously expressed. As expected from this tissue-specific expression pattern, FDX2 was found to be specifically involved in Fe-S protein assembly and able to replace yeast Yah1, whereas FDX1 cannot. Moreover, FDX1 does not complement FDX2-depleted HeLa cells in their defect in Fe-S protein biogenesis. In contrast, another study found effects on Fe-S protein biogenesis upon RNAi depletion of both FDX1 and FDX2 [47]. Recent *in vitro* studies reconstituting *de novo* Fe-S cluster synthesis on Isu1 confirmed the specific function of FDX2, but not of FDX1 in this process (Webert *et al.*, unpublished data). Conversely, only FDX1 but not FDX2 was active in cytochrome P450-dependent cortisol biosynthesis [46]. RNAi depletion of the human ferredoxin reductase FDXR demonstrated its requirement for the biogenesis of Fe-S proteins [47].

In conclusion, the initial step of Fe-S cluster synthesis requires the concerted action of six ISC proteins, in addition to iron, cysteine, and NADPH, and leads to a transiently bound [2Fe-2S] cluster on the Isu1 scaffold (Fig. 21.2).

### 21.2.2 Step 2: Chaperone-dependent release of the Isu1-bound Fe-S cluster

In the second major step, the Fe-S cluster is released from the Isu1 scaffold and delivered to so-called Fe-S cluster transfer proteins [27, 53] (Fig. 21.2a). Fe-S cluster release from Isu1 is facilitated by a dedicated chaperone system consisting of the Hsp70 protein Ssq1, its co-chaperone J-type protein Jac1, and nucleotide exchange factor Mge1 [54, 55]. Studies performed on these chaperones and their bacterial homologs [56, 57] gave insights into how they function within the ISC assembly pathway. The mechanistic model for the dedicated chaperone function in Fe-S protein biogenesis was derived from chaperones that function in protein folding [54] with Isu1 serving as a specific client protein of the Ssq1 chaperone. According to the current model (Fig. 21.3), the co-chaperone Jac1 recruits the holo-form of Isu1 and directs it to the ATP-bound form of Ssq1 [58, 59]. Both Jac1 and Isu1 stimulate the ATPase activity of Ssq1 thereby inducing a conformational change of the peptide-binding domain of Ssq1 to its closed state. This conformational change stabilizes the interaction between Ssq1 and the LPPVK motif of Isu1 [60, 61] and facilitates the removal of Jac1 from the complex. It is believed that Isu1 also undergoes a conformational change that results in its Fe-S cluster to be bound in a more labile fashion, thus facilitating its release from the scaffold protein [62, 63] (Fig. 21.3). To close the cycle and regenerate the

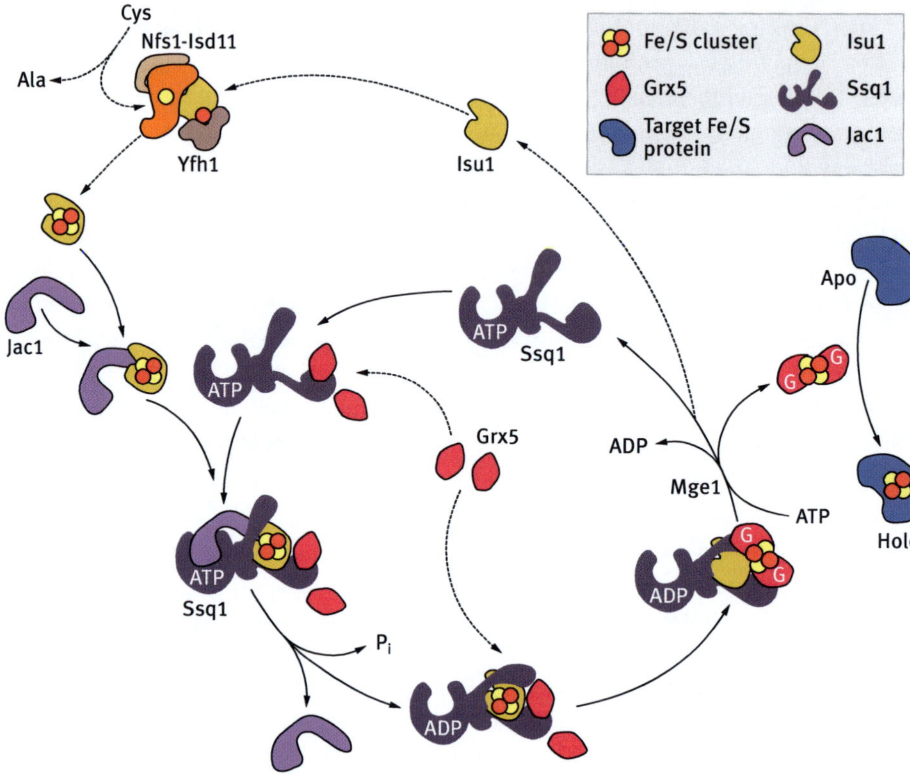


**Fig. 21.3:** The working cycle of the dedicated chaperone system in mitochondrial Fe/S protein biogenesis. The working cycle of ISC chaperone system is similar to that of the Hsp70 chaperones in protein folding [54]. After synthesis of the [2Fe-2S] cluster on the scaffold protein Isu1 (Fig. 21.2a), the co-chaperone Jac1 recruits holo-Isu1 and delivers it to the ATP-bound form of the Hsp70 chaperone Ssq1. ATP hydrolysis induced by Isu1 and Jac1 triggers a conformational change of the peptide-binding domain of Ssq1, thus tightly binding the LPPVK motif of Isu1. In turn, this is believed to induce a conformational change on Isu1 and weakens the binding of the Fe/S cluster to Isu1. Apo-Grx5 binds to Ssq1 at a site that does not overlap with that of Isu1. Eventually, this results in Fe/S cluster transfer from holo-Isu1 to Apo-Grx5. The exchange factor Mge1 facilitates ADP to ATP exchange, which triggers another conformational change in the peptide-binding domain of Ssq1 from the closed to an open state, thus leading to disassembly of the Ssq1-Isu1-holo-Grx5 complex. The Hsp70 reaction cycle can then resume with the binding of a new holo-Isu1-Jac1 complex to Ssq1-ATP. The Grx5-bound Fe/S cluster is finally transferred to target proteins (see Fig. 21.2a).

individual components, the nucleotide exchange factor Mge1 joins the complex to exchange ADP for ATP [64]. This induces another conformational change in Ssq1 that triggers disassembly of the entire protein complex and recycling of apo-Isu1 that can be used for another round of Fe-S cluster synthesis.

The Fe-S cluster released from Isu1 is then transferred toward apo-proteins, which likely involves transient binding to ISC transfer proteins. One of these proteins is the monothiol glutaredoxin Grx5 or human GLRX5 (Fig. 21.2). The protein binds to Ssq1 at a site that does not overlap with that of Isu1 [65] (Fig. 21.3). Most avid binding of Grx5 is observed for the ADP-bound form of Ssq1, which is known to tightly associate with the holo-form of Isu1. Therefore, it was suggested that the vicinity of holo-Isu1 and apo-Grx5 on Ssq1 accelerates the transfer of the labilized Fe-S cluster to Grx5. Depletion of Grx5 in yeast causes a general cellular Fe-S protein defect and leads to Fe-S cluster accumulation on Isu1, similar to what was found for Ssq1 or Jac1 depletion [27, 66]. As a result, Grx5-depleted cells display an iron overload in mitochondria and develop a severe oxidative stress. A similar phenotype, together with an impaired heme synthesis, is seen in zebrafish and human cells [67–69]. Hence, the function of Grx5 seems to be conserved throughout evolution. This is supported by the observation that *GRX5* deletion in yeast can be complemented by most monothiol glutaredoxins from both prokaryotic and eukaryotic species [70, 71]. The fact that Grx5 is involved in the biogenesis of both [2Fe-2S] and [4Fe-4S] mitochondrial Fe-S proteins and is also required for maturation of cytosolic-nuclear Fe-S proteins demonstrated that it belongs to the core ISC assembly machinery [65], thus distinguishing it from late-acting ISC factors discussed in Section 21.2.3 (Fig. 21.2a).

*In vitro* experiments on monothiol glutaredoxins from different organisms demonstrated the coordination of an unusual, glutathione (GSH)-coordinated [2Fe-2S] cluster [72]. Cysteine mutagenesis studies and the structure of the *Escherichia coli* Grx4 homodimer documented that the Grx-bound [2Fe-2S] cluster is coordinated by two GSH molecules and the active-site cysteine residues of two Grx monomers [70, 73–76]. The active-site cysteine residue is necessary for Grx5's *in vivo* function because its substitution results in the same phenotype as that seen for the null mutant [77]. The Fe-S cluster is bound to Grx5 in a rather labile fashion, making it difficult to detect *in vivo* [65, 70]. Its assembly in yeast depends on the core ISC components such as Nfs1, Isu1, and Jac1. Kinetic studies *in vitro* indicated that it is possible to transfer the [2Fe-2S] cluster to a chloroplast apo-ferredoxin [70]. These data support the view that monothiol Grx5 binds a Fe-S cluster transiently to subsequently pass it on to other Fe-S cluster-coordinating ISC proteins or to recipient target proteins including [2Fe-2S] proteins (Fig. 21.2a). However, it has to be noted that alternative Grx5 functions are not yet excluded.

In conclusion, step 2 leads to the chaperone-assisted dissociation of the Isu1-bound [2Fe-2S] cluster and its transfer to acceptor proteins including Grx5 and [2Fe-2S] target proteins (Fig. 21.2).

### 21.2.3 Step 3: Late-acting ISC assembly proteins function in [4Fe-4S] cluster synthesis and in target-specific Fe-S cluster insertion

The core ISC assembly machinery discussed so far is sufficient for maturation of mitochondrial [2Fe-2S] cluster-containing proteins. In contrast, virtually all mitochondrial

[4Fe-4S] proteins require additional ISC assembly components for their maturation. The formation of [4Fe-4S] clusters in both yeast and human cells strongly depends on the A-type ISC proteins Isa1 and Isa2 (human ISCA1 and ISCA2, respectively) [52, 78–82]. The proteins form a complex and bind either a [2Fe-2S] cluster or iron, yet the functional role of these metal cofactors is unclear. Both proteins further interact with the potential folate-binding protein Iba57 [82, 83]. Depletion of all three proteins in either *S. cerevisiae* or in human cells leads to similar and severe mitochondrial phenotypes such respiratory deficiency, loss of mitochondrial DNA, and ultrastructural alterations of mitochondria. These phenotypes can be readily explained by the maturation defect of mitochondrial [4Fe-4S] proteins prevailing in these cells. In addition, the Isa- or Iba57-depleted cells display a severe diminution of cytochrome oxidase activity. This terminal respiratory complex lacks Fe-S clusters and hence should not be affected. To date, it is unclear why its activity strictly depends on Isa-Iba57 function.

Another late-acting factor is the P-loop NTPase Ind1 (also termed NUBPL1 in humans), which binds a [4Fe-4S] cluster at two conserved cysteine residues present on its C-terminus [84, 85]. The protein is related in sequence to the two CIA proteins Cfd1-Nbp35, which act as scaffolds for [4Fe-4S] cluster generation in the cytosol (see in Section 21.4). In contrast, Ind1 is present only in organisms containing respiratory complex I, which contains eight Fe-S clusters in eukaryotes. Studies performed in *Yarrowia lipolytica* and human cells showed that deficiency of Ind1 affects the assembly of respiratory complex I, in particular of its soluble, Fe-S cluster-containing arm. The Fe-S cluster present on Ind1 is dependent on the function of the core ISC assembly machinery [84]. Therefore, it was proposed that Ind1 serves late in the biogenesis pathway and may serve as a specific Fe-S cluster-targeting factor, which delivers the cofactor to the matrix-exposed arm of complex I. In that sense, Fe-S cluster binding would serve a similar role as Fe-S cluster binding to the Cfd1-Nbp35 complex.

Proteins containing the 70-amino acid long C-terminal region of *A. vinelandii* NifU protein are termed Nfu-like proteins. Although this conserved segment possesses a CXXC motif and is able to transiently bind a [4Fe-4S] cluster [86], its location and function in the mitochondrial ISC assembly pathway remained unclear for a long period of time. Because Nfu-like proteins can assemble [2Fe-2S] or [4Fe-4S] clusters and transfer them to other Fe-S proteins, it was proposed that they might serve a scaffold function in addition to Isu1 [7, 86]. Initial evidence that Nfu1 may be involved in Fe-S protein biosynthesis was obtained in a synthetic lethal screen in yeast where the deletion of *NFU1* and *SSQ1* was synthetic lethal [2]. However, Fe-S protein activities were only slightly affected. This defect was enhanced when double mutants were analyzed in which both *NFU1* and *ISU1* were deleted. In addition to the Fe-S protein defects, mitochondria accumulated iron similar to other depletion mutants in core ISC assembly proteins [5] (Fig. 21.2).

The first insights into Nfu1 function came from studies on two groups of patients carrying mutations in the *NFU1* gene [87, 88]. In one case, a non-sense mutation resulted in abnormal mRNA splicing and complete loss of the protein, whereas in the other, a G→C point mutation led to a glycine to cysteine substitution next to the active site CXXC motif of NFU1. Affected individuals were born normally with no evident symptoms but quickly developed severe developmental retardation, brain abnormalities, and pulmonary hypertension before their deaths between 3 months and 1 year of age. The biochemical analyses of these patients showed normal aconitase activities but a massive decrease in complexes I and II [87, 88]. Additionally, strong defects in the lipoic acid-containing proteins pyruvate dehydrogenase (PDH) and α-ketoglutarate dehydrogenase (KGDH) were seen, together with hyperglycinemia and an increase in organic ketoacids. This phenotype was explained by a maturation defect in the Fe-S cluster containing-protein lipoate synthase [89] (Fig. 21.2). Because lipoate synthase activity was not routinely measured during Fe-S protein biogenesis studies, this phenotype was missed in most previous studies. RNAi depletion of NFU1 in human cell culture recapitulated the biochemical phenotype of the patients demonstrating a role of NFU1 in the assembly of complex Fe-S proteins (respiratory complexes I and II and lipoate synthase containing eight, three, and two Fe-S clusters, respectively) [5, 87, 88]. All these results indicate a function of Nfu1 not as an alternative scaffold protein but rather a role in the delivery of [4Fe-4S] clusters to dedicated target apoproteins. This idea is consistent with the finding that assembly of the Fe-S cluster transiently bound to Nfu1 requires the function of the core ISC components such as Nfs1, Isu1, and Grx5 [88]. It remains to be elucidated what the exact biochemical role of Nfu1 protein in Fe-S cluster biogenesis might be and how it coordinates its function with other components of the late part of the ISC assembly pathway.

Patients with a mutation in the *BOLA3* gene display a strikingly similar phenotype as the Canadian group of *NFU1* patients [87]. In those individuals, the *BOLA3* gene carries a frame shift mutation introducing a premature stop codon. The individuals died at a few months of age, and their cells displayed severe defects in the lipoic acid-containing proteins PDH and KGDH as well as respiratory complexes I and II, whereas mitochondrial Fe-S protein aconitase activities were unchanged. Another group of patients with a *BOLA3* mutation displayed the same phenotype, yet with an additional complex III deficiency [90]. Results of clinical and biochemical studies performed on these patients suggested that the BOLA3 protein may play an auxiliary role in the insertion of Fe-S clusters into specific target proteins such as lipoic acid synthase and the respiratory chain complexes I and II [87, 90]. Even though the participation of the BOLA3 protein in mitochondrial Fe-S protein biogenesis seems clear, its evolutionary conservation and its biochemical function remain to be unraveled.

BolA-like proteins are highly conserved throughout evolution and are, with few exceptions, present in all living organisms. BolA was first identified in bacteria that show a round or "bola" (Spanish for ball or sphere) morphology when *bolA* was overexpressed [91]. Eukaryotes from yeast to man possess three BolA-like proteins, two in the mitochondria, namely Yal044W/BOLA1 and Aim1/BOLA3, and one in the cytosol, termed Fra2/BOLA2. Not much is known about the yeast mitochondrial proteins. Deletion of *AIM1* (altered inheritance of mitochondria) in *S. cerevisiae* displayed elevated frequency of mitochondrial genome loss [92]. A bioinformatics study combining genome sequences, physical interaction data and three-dimensional structures suggested that BolA proteins are reductases that might interact with glutaredoxins [93]. Although a direct interaction of Aim1/BOLA3 with Grx5/GLRX5 in mitochondria is uncertain, yeast Fra2 is known to form a complex with Grx3/4 in the cytosol and nucleus [94, 95]. These proteins are involved in cellular iron regulation, transmitting the iron status of the yeast cell to the transcription factors Aft1-Aft2 [5, 21, 22]. *In vitro* studies showed that Grx3/4 and Fra2 can form a heterodimeric complex bridged by a [2Fe-2S] cluster. This cluster is coordinated by the active-site cysteine of Grx3/4, GSH, and a histidine residue of Fra2. A similar complex was shown for the human proteins GRX3 and BOLA2 [96], implying that Grx-BolA interaction is conserved in higher eukaryotes, even though iron regulation in higher eukaryotes is not transcriptionally controlled. Overall, a role of BOLA3 in mitochondrial Fe-S protein biogenesis seems established, but where and how the protein functions in the pathway remains to be determined.

In conclusion, step 3 is needed for proteins carrying a [4Fe-4S] cluster and involves the Isa-Iba57-mediated generation of this cluster, which subsequently is inserted into target proteins with the help of dedicated ISC targeting factors.

## 21.3 The role of the mitochondrial ABC transporter Atm1 in the biogenesis of cytosolic and nuclear Fe-S proteins and in iron regulation

Biogenesis of extra-mitochondrial Fe-S proteins in yeast strictly depends on the mitochondrial ISC assembly machinery [1, 10]. This is particularly true for the cysteine desulfurase Nfs1, which is also localized in the cytosol and/or nucleus in yeast and man [97, 98]. Only mitochondrial versions of Nfs1 and Isu1 support the biogenesis of extra-mitochondrial Fe-S proteins. Moreover, the expression of Nfs1 or Isu1 in the cytosol does not rescue the Fe-S cluster assembly defects in that compartment [1, 9, 10, 99, 100], suggesting that mitochondria produce the sulfur moiety that is utilized for cytosolic and nuclear Fe-S cluster biogenesis. Cytosolic-nuclear Nfs1 is essential for cell viability in yeast, but its function remains elusive. It was suspected that cytosolic-nuclear Nfs1 might be involved in the thio-modification of certain cytosolic tRNAs. However, this speculation was experimentally refuted [101]. The current belief why mitochondria are required for extra-mitochondrial Fe-S protein assembly is that the core components of the ISC machinery produce a sulfur-containing component termed X-S (Fig. 21.2 and Fig. 21.4), which is utilized for cytosolic and nuclear Fe-S cluster production. The nature of this component is not presently known, but the ABC transporter Atm1 may be responsible for its export to the cytosol [1, 102, 103]. Atm1 might cooperate with the FAD-dependent sulfhydryl oxidase Erv1 of the intermembrane space and GSH in this export task [44, 104]. Depletion of both Erv1 and GSH

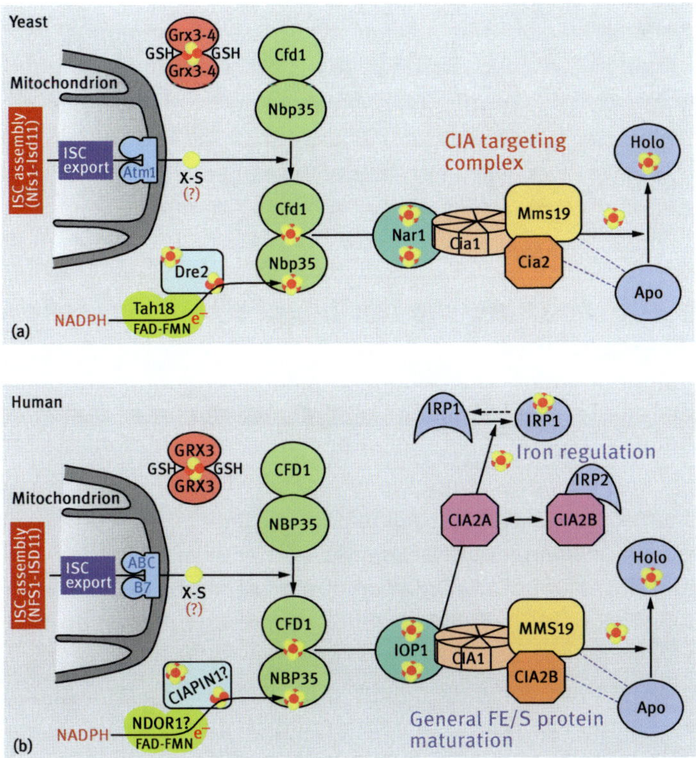


**Fig. 21.4:** The role of mitochondria and the CIA machinery in the maturation of cytosolic and nuclear Fe/S proteins. (a) In yeast, the CIA machinery encompasses eight known proteins. The assembly process can be dissected into two different steps. First, a bridging [4Fe-4S] cluster is assembled on the Cfd1-Nbp35 scaffold complex. This reaction requires a sulfur source (X-S) generated by the mitochondrial ISC assembly machinery and exported by the mitochondrial ABC transporter Atm1 (Fig. 21.2a). Generation of the functionally essential N-terminal Fe/S cluster of Nbp35 (bottom) depends on the flavoprotein Tah18 and the Fe/S protein Dre2, which serve as an NADPH-dependent electron transfer chain. Second, the bridging Fe/S cluster is released from Cfd1-Nbp35, a reaction mediated by the Fe/S protein Nar1 and the CIA-targeting complex Cia1-Cia2-Mms19. The latter three proteins interact with target (apo)proteins and assure specific Fe/S cluster insertion. Biogenesis further requires the cytosolic multidomain monothiol glutaredoxins Grx3-Grx4, which bind a GSH-coordinated, bridging [2Fe-2S] cluster, and may serve as an iron donor. (b) In humans, the components of the CIA machinery are structurally and functionally similar to those of yeast. As a major difference, humans possess two isoforms of Cia2. CIA2B is the functional orthologue of yeast Cia2 and is involved in the biogenesis of canonical cytosolic and nuclear Fe/S proteins. In contrast, CIA2A is specifically involved in the maturation of IRP1, a protein regulating cellular iron homeostasis in humans. Additionally, CIA2A tightly binds to IRP2, which does not contain a Fe/S cluster yet also plays a decisive role in cellular iron metabolism. Note that the role of some indicated (?) human CIA components has not been verified yet.

causes similar phenotypes as that of Atm1, namely cytosolic Fe-S proteins defects and an accumulation of iron in the mitochondria [1, 105]. Erv1 is known to introduce disulfide bridges into target proteins during their Mia40-dependent import into the intermembrane space [106], and hence, it is not excluded that the Erv1 function in Fe-S protein biogenesis is indirectly caused by a failure to generate disulfide bridges in this compartment.

## 21.4 The role of the CIA machinery in the biogenesis of cytosolic and nuclear Fe-S proteins

### 21.4.1 Step 1: The synthesis of a [4Fe-4S] on the scaffold complex Cfd1-Nbp35

As pointed out in Section 21.1, the cytosol and nucleus contain numerous Fe-S proteins with essential functions for life (Fig. 21.1). Their maturation depends on mitochondria and the CIA machinery, which comprises eight known yeast proteins that are conserved in eukaryotes [107, 108] (Fig. 21.4). Analogous to mitochondrial Fe-S protein biogenesis, the roles of individual CIA proteins can be attributed to functions that are formally similar to those performed by the mitochondrial ISC machinery [109–111]. First, a [4Fe-4S] cluster is transiently assembled on the P-loop NTPases Cfd1 and Nbp35, which serve as a scaffold complex. This step requires the core mitochondrial ISC assembly machinery including Nfs1-Isd11 for production of the sulfur donor (Fig. 21.4) [9, 99]. Cfd1-Nbp35 form a hetero-tetramer and bind the [4Fe-4S] cluster in a bridging manner [107, 110]. Mutation of the Walker motifs of Cfd1 or Nbp35 destroys the function of these proteins suggesting that NTP hydrolysis is required for Fe-S cluster assembly, but experimental proof for nucleotide binding or hydrolysis is still missing. In addition to its transient Fe-S cluster, Nbp35 contains another [4Fe-4S] cluster at its N-terminus that is more stably bound and essential for function [112]. Its assembly depends on electron transfer from the electron source NADPH to the flavin-containing oxidoreductase Tah18 and finally the Fe-S protein Dre2 [111, 113, 114] (Fig. 21.4a). The precise role of reduction in this early step of CIA function remains unclear. One possibility is the reduction of the sulfur moiety exported from mitochondria to sulfide, but other options are equally plausible.

In conclusion, the initial step of cytosolic-nuclear Fe-S protein maturation involves the synthesis of a [4Fe-4S] cluster on the Cfd1-Nbp35 scaffold, which requires a sulfur-containing product of mitochondria and electron input from the CIA electron transfer chain.

### 21.4.2 Step 2: Transfer of the [4Fe-4S] cluster to target apo-proteins

In the next step, the transiently bound, bridging [4Fe-4S] cluster on Cfd1-Nbp35 is transferred to apo-proteins. The CIA protein Nar1 interacts with Nbp35 and may therefore be involved in Fe-S cluster mobilization [115, 116]. Nar1 shows similarity to iron-only hydrogenases and binds two [4Fe-4S] clusters that are assembled with the help of Cfd1-Nbp35 [117]. Because later-acting CIA factors are dispensable for the assembly of the two Nar1 Fe-S clusters, the protein may act as a mediator between early and late steps of the CIA biogenesis process (Fig. 21.4a). The recently identified CIA components Cia1, Cia2, and Mms19 form the so-called CIA-targeting complex [118–122], which facilitates both Fe-S cluster transfer and target-specific cluster insertion into the various polypeptide chains. These partial reactions involve the direct physical interaction of the CIA-targeting complex components with the target Fe-S proteins, which presumably are in their apo-forms (Fig. 21.4a, dotted lines). The direct contact between late-acting CIA and Fe-S proteins became most evident from systematic affinity pull-down experiments in human cells where the various CIA targeting factors interact with a large number of cytosolic and nuclear Fe-S proteins [120–124]. The list of interacting Fe-S proteins includes DNA polymerases and primases, ATP-dependent DNA helicases, DNA glycosylases, and the ABC protein ABCE1. Possibly, there are more Fe-S proteins hidden in this collection of CIA interactors. In yeast, the CIA association with target Fe-S proteins seem to be weaker or less stable. Both dedicated and systematic approaches have identified only few such interactions including the binding of Cia2 to the helicase-nuclease Dna2, and the interaction of Mms19 with the Fe-S helicase Rad3 [120] (Mascarenhas *et al.*, unpublished data). The precise molecular function of the late-acting CIA components remains to be determined.

In addition to the aforementioned CIA proteins, the cytosolic monothiol glutaredoxins Grx3-Grx4 (in humans termed Grx3 or PICOT) were also shown to be crucial for cytosolic and nuclear Fe-S protein biogenesis [125, 126]. Because these proteins are generally involved in intracellular iron trafficking and iron uptake regulation [5, 21, 22], they appear to play a more general function and hence are not considered as CIA proteins. For instance, these glutaredoxins are involved in the maturation of yeast di-iron proteins such as ribonucleotide reductase and participate in heme biosynthesis in zebrafish erythroid cells [125–127]. Thus, Grx3-Grx4 could potentially supply iron to some step of the biosynthesis reaction, but their precise role remains to be elucidated.

In conclusion, step 2 leads to dissociation of the [4Fe-4S] from Cfd1-Nbp35 and its CIA targeting complex-dependent incorporation into specific apo-proteins.

## 21.5 Specialized functions of the human CIA-targeting complex components

### 21.5.1 Dedicated biogenesis of cytosolic and nuclear Fe-S proteins

Although yeast cells have served as a primary model organism to identify and characterize the known constituents of the CIA machinery, the process seems to be conserved in all eukaryotes including man (Fig. 21.4). All eight known yeast CIA proteins have a counterpart in human cells, and initial RNAi-mediated depletion studies have shown a function for human Nbp35 and IOP1 (yeast Nar1) in cytosolic and nuclear Fe-S protein biogenesis [116, 128]. Ongoing investigations also suggest conserved functions for CFD1 (O. Stehling, unpublished data). The *in vivo* functions of NDOR1 (human homologue of yeast Tah18) and CIAPIN1 (yeast Dre2) in human cells remain to be established. The (studied) human CIA proteins appear to be required for maturation of all analyzed cytosolic-nuclear [4Fe-4S] proteins. The members of the human equivalent of the yeast CIA-targeting complex (termed CIA1, CIA2B, and MMS19) were recently identified and functionally characterized by RNAi depletion technology and subsequent analysis of Fe-S protein function [120–122]. In contrast to yeast, the members of the human CIA-targeting complex exhibit a striking specificity for target apo-proteins. For instance, MMS19 is less important for GPAT maturation than the other CIA proteins, and CIA2B is not crucially required for DNA polymerase δ (POLD1) assembly [122]. It appears that the CIA-targeting complex members perform dedicated functions in the delivery of Fe-S clusters to specific apo-proteins. Several systematic proteomic screens identified numerous interaction partners of CIA1, CIA2B, and MMS19 proteins [120–124, 129–132]. These included a large number of cytosolic and nuclear Fe-S proteins that were mostly bound to both CIA1 and CIA2B.

Cia2 contains a C-terminal domain (CTD) of unknown function 59 (DUF59). A global study of proteins containing reactive cysteine residues identified a hyperreactive cysteine in yeast Cia2 (Cys161) and in one of the two human Cia2 homologues (Cys93 of CIA2B). Mutation of the corresponding cysteine residue was lethal in yeast and abolished the activity of the Fe-S protein Leu1 [119, 133]. Cia2 and CIA2B are general Fe-S protein maturation factors acting as part of the CIA-targeting complex in yeast and humans, respectively [122] (Mascarenhas *et al.*, unpublished data). In *Arabidopsis*, mutation of the three Cia2 homologues AE7 leads to lower activities of the [4Fe-4S] proteins cytosolic aconitase and nuclear glycosylase [134]. Hence, the function of this protein seems to be conserved in eukaryotes. The DUF59 domain is also present in the plastid Fe-S protein biogenesis factor HCF101 [135] as well as bacterial proteins, but their precise molecular role remains unresolved. Structural information has been obtained for three bacterial Cia2 (also termed SufT) homologues (see e.g. [136]). Additionally, an NMR structure and two different X-ray structures of CIA2A have been reported [137]. Thus far, the structural information has not provided any decisive functional insights.

In conclusion, the various components of the CIA-targeting complex undergo direct and specific interactions with client Fe-S proteins to assure their specific maturation.

### 21.5.2 The dual role of CIA2A in iron homeostasis

Human cells encode a second CIA2 homologue termed CIA2A, which forms a subcomplex with CIA1. Depletion of CIA2A does not elicit any defects in canonical cytosolic and nuclear Fe-S proteins [122]. However, CIA2A is exclusively required for the maturation of the [4Fe-4S] protein IRP1, which is a key regulatory element of mammalian iron metabolism. In mammals, the major impact on cellular iron supply and distribution within the cell is mediated by IRP1 and IRP2 *via* complex post-transcriptional regulatory mechanisms (for comprehensive recent reviews see [138, 139]). IRP1 is a cytosolic aconitase whose activity depends on the [4Fe-4S] cluster. Upon iron starvation, IRP1 loses its Fe-S cluster, and its apo-form can bind to iron-responsive elements (IREs) of mRNAs, which encode proteins involved in iron trafficking (e.g. transferrin receptor), storage (ferritin), and utilization (mitochondrial aconitase and eALAS). IRP1 binding differentially regulates the efficiency of translation or mRNA stability [138, 139]. Accordingly, CIA2A (but not CIA2B or MMS19) depletion increases the IRE binding activity of IRP1. In turn, ferritin levels are diminished and the expression of the transferrin receptor is increased. Thus, CIA2A depletion mimics iron deficiency. Upon iron repletion, the equilibrium between the apo- and holo-forms of IRP1 is shifted back to the Fe-S cluster form in an assembly step that is dependent on CIA2A and all earlier-acting CIA proteins, but not CIA2B or MMS19 (Fig. 21.4b).

Surprisingly, there is a second important effect of CIA2A on cellular iron regulation. The protein tightly binds to IRP2 [122], which does not contain a Fe-S cluster but is regulated by iron-dependent degradation by the proteasome (Fig. 21.4b). The precise role of CIA2A in the mechanism of this IRP2 regulatory step is currently unknown. Under iron-replete conditions, IRP2 is degraded in an iron- and oxygen-dependent fashion by the E3 ubiquitin ligase FBXL5, which responds to iron and oxygen *via* its hemerythrin domain [139]. Upon iron depletion or under low oxygen concentrations, FBXL5 is destabilized and degraded, leading to increased levels of IRP2. The stabilizing effect of CIA2A binding to IRP2 introduces another unexpected level of iron regulation *via* the CIA machinery. Thus, the CIA2A branch of the CIA machinery, through the regulation of both IRP1 and IRP2, impacts on cellular iron homeostasis in multiple ways.

In conclusion, human CIA2A has a crucial function in cellular iron homeostasis as a dedicated CIA targeting factor for IRP1 maturation and as a stabilizer of IRP2.

## 21.6 Fe-S protein assembly and the maintenance of genomic stability

The conceptual framework for Fe-S cluster assembly that been established over the past 15 years has provided novel insights into the breadth of cellular processes that are directly impacted by Fe-S proteins. The remainder of this chapter will focus specifically on how the Fe-S biogenesis pathways described above directly influence multiple aspects of genome maintenance including DNA replication and repair. Evidence for the global involvement of Fe-S cluster assembly pathways in DNA metabolism emerged from two converging discoveries: (1) proteins previously identified as DNA repair factors were actually central components of the Fe-S protein assembly machinery and (2) multiple enzymes involved in DNA metabolism require Fe-S clusters for activity. Both of these research arcs will be discussed further.

### 21.6.1 Late-acting CIA factors in DNA metabolism

The CIA-targeting complex component MMS19 was originally identified in two independent genetic screens in budding yeast as a gene required for methionine biosynthesis, NER, and RNA polymerase II transcription [140, 141]. Although its role in methionine biosynthesis remained largely unexplored, multiple follow-up studies focused on elucidating its role in transcription and DNA repair. Specifically, work in budding yeast showed that MMS19 was required for the function of the general transcription factor TFIIH and that the DNA repair and transcription defects observed in cell-free protein extracts derived from MMS19-deficient yeast strains could be complemented by the addition of purified TFIIH [142]. It was also demonstrated that while MMS19 was not itself a component of TFIIH, it was required for maintaining cellular levels of Rad3, the yeast homologue of XPD and a member of the TFIIH complex [143]. Studies of the human homologue of MMS19 were consistent with the work done in yeast and suggested an important role in DNA repair and transcription through regulation of TFIIH function [144, 145]. In both systems, however, the molecular mechanism by which MMS19 exerted its "regulatory" effects on Rad3 and TFIIH was elusive.

The first direct link between the CIA machinery and the DNA repair pathways was documented in 2012 [120, 121]. As described earlier, both articles reported the discovery that MMS19 was a late-acting factor of the CIA pathway and functioned as part of a CIA-targeting complex that physically links early CIA components to apo-protein targets (Fig. 21.4). Importantly, these studies provided multiple lines of evidence directly implicating the CIA-targeting complex in DNA repair pathways. First, the CIA-targeting complex was physically associated with a large number of DNA metabolic enzymes including DNA helicases (XPD, FANCJ, and RTEL1), the DNA glycosylase NTHL1, DNA polymerases (POLD1, POLA1, and POLE1), the nuclease DNA2, and the DNA primase PRI2 (Fig. 21.1b). Second, the depletion of MMS19 led to a variety of defects consistent with widespread deregulation of DNA metabolism including increased sensitivity to DNA damaging agents, destabilization of DNA polymerase δ, and inhibition of XPD incorporation into TFIIH. Together, these data established the central role for Fe-S protein assembly in regulating the integrity of multiple DNA metabolic pathways.

In addition to its major role in Fe-S cluster biogenesis, a recent study also implicated a protein complex consisting of MMS19, CIA2B (also known as FAM96B or MIP18),

CIA1 (also known as CIAO1), and XPD (denoted by the authors as MMXD and likely identical to the CIA targeting complex) in mitosis [146]. They demonstrated that MMXD localized to mitotic spindles and that depletion of MMS19, CIA2B, or XPD led to defects in spindle assembly and chromosome segregation as well as the accumulation of abnormal nuclei. These defects were also observed in fibroblasts derived from patients expressing XPD mutants that are known to give rise to xeroderma pigmentosum, Cockayne syndrome, and trichothiodystrophy suggesting that this mitotic role for MMXD may be of pathological relevance. Interestingly, these phenotypes cannot be readily traced to the reported function of individual Fe-S proteins, suggesting that this may be related to the general function of the CIA-targeting complex in the biogenesis of the bulk of cytosolic and nuclear Fe-S proteins.

### 21.6.2 XPD and the Rad3 family of DNA helicases

XPD is a Fe-S cluster-requiring DNA helicase that functions as a component of the general transcription complex TFIIH [147]. TFIIH is a highly conserved 10-subunit complex consisting of XPB, p63, p52, p44, p34, p8, XPD, CDK7, cyclin H, and MAT1 that is required for both transcription initiation and NER. XPD is thought to play at least two distinct roles in TFIIH function. First, it physically bridges the CDK-activating subcomplex (CDK7, cyclin H, and MAT1) to the rest of the TFIIH core complex. Second, XPD is a member of the superfamily 2 (SF2) class of helicases and possesses a 5′-3′ DNA helicase activity that unwinds the region surrounding the DNA lesion during NER. In a landmark finding in 2006, the White group used a combination of biochemical and spectroscopic approaches to identify a metal-binding site in archaeal XPD in which a [4Fe-4S] cluster was coordinated by four highly conserved cysteine ligands [148]. Multiple high-resolution structural studies subsequently verified the presence of a [4Fe-4S] domain that together with a novel “Arch” domain abuts the two helicase domains to form a channel through which they hypothesized ssDNA was extruded [149–151]. In this model, the [4Fe-4S] domain could potentially act as a “ploughshare” that is physically responsible for DNA strand separation. Considerable functional data have been reported that are consistent with this model including the observation that mutation of any of the conserved cysteine ligands leads to loss of [4Fe-4S] cluster binding and a concomitant loss of strand displacement activity without a detectable loss of ATPase activity [148, 152]. Together, these data established XPD as the first Fe-S cluster-dependent helicase to be characterized and provided a novel link between DNA repair and iron metabolism.

In addition to XPD, higher eukaryotes contain multiple SF2 XPD-like paralogues including FANCJ, RTEL1, and DDX11/CHLR1 [153, 154]. FANCJ is a downstream component of the Fanconi anemia pathway and is involved in responding to and repairing DNA interstrand cross-links. RTEL1 is involved in maintaining genome stability by regulating telomere length and suppressing inappropriate homologous recombination. CHLR1 is required for sister chromatid cohesion and heterochromatin organization. The cysteine residues required for coordinating the Fe-S cluster are conserved across these paralogues, suggesting that Fe-S cluster binding is a conserved feature for this family of helicases. Moreover, direct experimental evidence that recombinant FANCJ co-purifies with Fe and that clinically relevant mutations in its Fe-S cluster domain reduce both Fe binding and helicase activity has been reported, thus highlighting the central role of the Fe-S domain across the entire helicase family [155].

### 21.6.3 Fe-S proteins involved in DNA replication

DNA replication requires the coordinated activities of a large number of cellular enzymes [156]. These enzymes include (but are not limited to) (1) DNA primase, the enzyme responsible for synthesizing the RNA primers that nucleate DNA synthesis, (2) multiple DNA polymerases that catalyze DNA template-dependent DNA synthesis, and (3) DNA2, a nuclease involved in Okazaki fragment processing that is essential for lagging strand replication. Interestingly, all of these factors have been shown to be Fe-S cluster proteins and are dependent on Fe-S cofactor binding for either structure stabilization or function.

Eukaryotic primases are heterodimeric enzymes consisting of large (Pri2) and small (Pri1) subunits [157]. Although RNA synthesis is carried out specifically by the Pri1 subunit, Pri2 is equally essential and is required for initiation, elongation, and regulation of primer length. Biochemical studies of the CTD of Pri2 showed that the purified protein was brownish in color and contained four conserved cysteine residues, suggesting that it contains an Fe-S cluster [158, 159]. The presence of the cluster was confirmed by EPR spectroscopy. Subsequent mutational and functional analyses demonstrated that the C-terminal Fe-S cluster domain is required for primase activity and plays a role in recognizing and binding the ss/dsDNA junction [158, 159].

Eukaryotic DNA replication depends on the coordinated activity of three DNA polymerase complexes – Pol-α, Pol-δ, and Pol-ε [160]. Pol-α forms a protein complex with DNA primase and is required for the initiation of DNA replication, whereas Pol-δ and Pol-ε are required for processive DNA elongation. The CTDs of Pol1, Pol2, and Pol3, i.e. the catalytic subunits of Pol-α, Pol-ε, and Pol-δ, respectively, each contain two cysteine-rich motifs termed CysA and CysB, which were previously thought to be Zn-binding sites. Although CysA is still believed to be a Zn-binding site, definitive evidence has recently emerged demonstrating that CysB actually binds an essential [4Fe-4S] cluster [161]. This includes data showing that the CTDs of Pol1, Pol2, and Pol3 from budding yeast co-purify with iron *in vivo* using $^{55}$Fe-radiolabeling assays and spectroscopic data showing that recombinant versions of these CTDs contain [4Fe-4S] clusters after purification from *E. coli*. The exact molecular function of these Fe-S clusters in Pol1, Pol2, and Pol3 remains unclear, although, at least for Pol3, it appears that Pol3 Fe-S cluster integrity is required for the assembly of the accessory subunits (Pol31-Pol32) of the Pol-δ complexes [161].

Dna2 is a multifunctional enzyme with distinct nuclease and helicase domains and plays essential roles in Okazaki fragment maturation during DNA replication, double-stranded DNA break repair, and telomere maintenance [156]. Embedded within the nuclease domain is a conserved four-cysteine motif ($CX_{248}CX_2CX_5C$), which has been shown to bind an Fe-S cluster using spectroscopic and mutational approaches [162]. Interestingly, mutation of any of the conserved cysteine residues to alanine impairs both nuclease and helicase activities without inhibiting its DNA-binding activity or causing gross structural rearrangements in the protein as assayed by protease sensitivity. Based on these findings, a role for DNA2's Fe-S cluster in mediating the dynamic conformational changes that are required for coupling the nuclease and helicase activities has been proposed.

### 21.6.4 DNA glycosylases as Fe-S proteins

Endonuclease III/MutY DNA glycosylases are a highly conserved family of DNA repair enzymes that play a key role in base excision repair (BER) [163]. They function by excising specific damaged bases from intact DNA helices, leaving an apurinic site that can be repaired by downstream enzymes in the BER pathway. *E. coli* endonuclease III, which catalyzes the removal of oxidatively damaged bases, was the first DNA repair enzyme demonstrated to require a Fe-S cluster for function [164]. Structural and functional studies have indicated that this Fe-S cluster plays both a structural role in positioning the DNA-binding residues of endonuclease III to facilitate DNA binding and a redox-mediated role in localizing the enzyme to sites of DNA damage *via* charge transport [165–167]. This charge transport function will be discussed in greater detail in Section 21.7. Unlike their *E. coli* counterparts, the human homologues of this family have not been extensively studied, although evidence demonstrating that they are Fe-S proteins has been reported [168].

## 21.7 Biochemical functions of Fe-S clusters in DNA metabolic enzymes

As the number of Fe-S proteins with roles in DNA metabolism has increased, understanding the molecular roles of their metal centers has become a major priority. For the majority of Fe-S proteins involved in DNA metabolism, the Fe-S cluster has been proposed to play a noncatalytic role in stabilizing the structure of the enzyme and potentially facilitating nucleic acid binding [169]. The major exception to this trend is a model developed by the Barton group in which the Fe-S clusters play a role in a novel redox process known as DNA charge transport [170, 171]. The basic premise of this model is that DNA can effectively conduct an electric charge *via* overlapping π-orbitals of stacked aromatic nucleotide bases. This charge transport has two important features. First, it can efficiently occur over long molecular distances, making it a potential mechanism for long distance signaling across a large part of a chromosome. Second, it is extremely sensitive to perturbations in the integrity of the base pair stack such as those that occur as a consequence of exposure to DNA damaging agents. These characteristics led the Barton group to explore whether the DNA repair machinery can exploit these features to better facilitate the cellular response to DNA damage.

Initial studies from the Barton laboratory focused on the DNA glycosylases MutY and endonuclease III (EndoIII) from *E. coli*. These glycosylases both contain an evolutionarily conserved [4Fe-4S] cluster and are components of the BER pathway that functions in the repair of damaged or modified bases [165, 166, 172]. Through a series of elegant biochemical studies, they have established a model in which these enzymes monitor the integrity of the genome using their Fe-S clusters to signal to one another through DNA charge transport. Their proposed model is shown in Fig. 21.5 and is as follows. First, a reduced Fe-S cluster enzyme is weakly bound to DNA as it scans the genome. This enzyme can be activated by oxidation potentially as a result of oxidative stress or other DNA damaging agents. This increases its affinity for DNA by 1000-fold and "locks" it onto the DNA. Second, as the enzyme is oxidized, it releases an electron that is transported along "healthy" DNA until it reaches a second oxidized DNA-bound enzyme nearby. Third, reduction of the second enzyme signals the integrity of the intervening DNA sequence promoting its release from the undamaged DNA

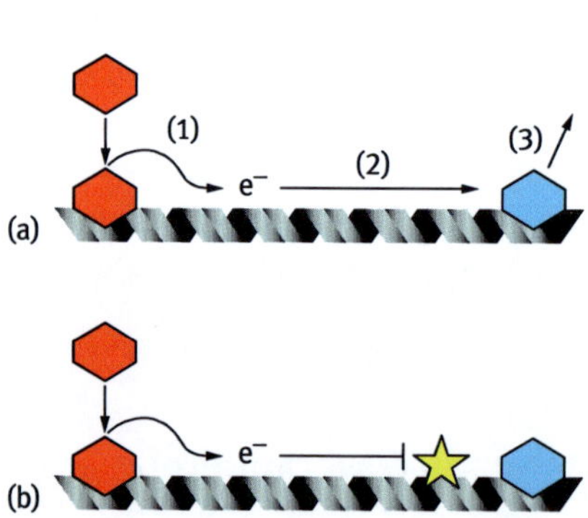


**Fig. 21.5:** DNA charge transport model for localization of Fe/S enzymes. (a) Step 1: Oxidation of a Fe/S DNA repair protein stabilizes its association with DNA. Step 2: If the surrounding DNA is undamaged, DNA charge transport occurs between the oxidized enzyme and the other nearby oxidized Fe/S DNA repair enzymes. Step 3: The reduction of DNA-bound Fe/S cluster proteins triggers their dissociation from the DNA and enables them to continue scanning the genome for lesions. (b) If the intervening DNA between two DNA-bound Fe/S enzymes contains damaged or mismatched bases that block DNA charge transport, then the Fe/S cluster enzymes remain bound in close proximity to the DNA lesion and facilitate their repair.

region. However, if these two DNA-bound enzymes are flanking a damaged base, then DNA charge transport is blocked and the oxidized DNA repair enzymes remain bound near the DNA lesion where they can initiate the repair process.

Although these charge transport studies were previously limited to these DNA glycosylases, more recent work has expanded the breadth of DNA repair enzymes that potentially function using DNA charge transport [173]. This includes work on the helicase XPD, which is essential for both transcription initiation and NER and contains a [4Fe-4S] cluster that possesses a physiologically relevant redox potential when bound to DNA. In atomic force microscopy experiments, it was shown that XPD redistributes to DNA lesions in a manner that is dependent on an intact Fe-S cluster and consistent with the DNA charge transport model for repair protein localization. It was also demonstrated that XPD could cooperate with EndoIII to localize to DNA lesions, raising the intriguing possibility that scanning the genome for different types of DNA damage could occur *via* coordinated DNA charge transport between different DNA repair pathways.

Despite strong *in vitro* evidence supporting this DNA charge transport model for DNA repair, *in vivo* data for this model remains sparse [165]. The strongest *in vivo* experiments supporting charge transport come from genetic experiments performed in *E. coli*. It was shown that MutY activity was partially lost in strains deficient for EndoIII. Importantly, this phenotype could be rescued by the introduction of a catalytically inactive EndoIII mutant, suggesting that EndoIII's ability to regulate MutY activity was independent of its own enzymatic activity. Because the mutant EndoIII retained its capacity for charge transport, the authors concluded that DNA-mediated signaling between MutY and EndoIII likely accounted for the cooperation found between the two enzymes.

## 21.8 Interplay among Fe-S proteins, genome stability, and tumorigenesis

During the multistep development of cancer, normal cells acquire new biological capabilities that enable their progression to a neoplastic state [174]. The acquisition of these capabilities is enabled by chance genomic alterations that give rise to heritable mutant phenotypes. To accelerate the accumulation of these mutations, cancer cells invariably disrupt critical cellular pathways that normally function to ensure proper maintenance of the genome. Tumor-associated genomic instability thus becomes an essential and enabling characteristic of tumor pathogenesis [175]. Although the contribution of iron metabolism and Fe-S protein assembly defects to promoting genome instability remains largely unclear, several lines of evidence suggest that they may play both positive and negative roles in this process.

The inability to assemble Fe-S clusters in proteins involved in DNA metabolism is likely to promote genomic instability in a multitude of ways. Fe-S proteins are essential components of a wide range of DNA repair pathways (NER, BER, telomere maintenance, DNA interstrand cross-link repair, DNA double-strand break repair) that are collectively required to respond to a broad spectrum of DNA lesions including base modifications, UV-induced thymidine dimers, DNA interstrand cross-links, and DNA strand breaks [169, 176]. From the tumorigenesis point of view, this strong dependency of DNA repair on proper Fe-S protein assembly makes it an attractive target whereby inactivation of a single pathway (Fe-S protein assembly) can broadly disrupt the majority of the cellular DNA repair pathways and strongly promote genomic instability. As described earlier, many DNA replication factors are Fe-S proteins including DNA polymerase, DNA primases, and nucleases involved in Okazaki fragment maturation such as DNA2. Thus, inhibition of Fe-S protein maturation would also lead to misregulation and/or inactivation of these replication factors, resulting in profound impacts on genomic integrity. Successful DNA replication requires that the entire nuclear genome be replicated once and only once. Previous studies have shown that misregulation of DNA replication factors is sufficient to cause either over-replication or under-replication of specific genomic regions and thus directly breach genomic integrity mechanisms [177].

Although intriguing, this proposed model of how inhibiting Fe-S protein biogenesis may induce genomic instability through the disruption of key DNA repair and replication pathways remains largely speculative. The strongest evidence for a physiological relevance of this connection comes from work by the Gottschling group exploring aging in budding yeast [178, 179]. They demonstrated that yeast undergo an age-dependent increase in genomic instability, specifically increased loss of heterozygosity (LOH), that particularly is induced by mitochondrial dysfunction [178, 180]. Specifically, their studies showed that an age-associated or chemically induced loss of mtDNA decreased the mitochondrial membrane potential and an increased frequency of nuclear LOH. This loss of mitochondrial function correlated with a transcriptional signature that was similar to iron starvation, suggesting the cellular iron metabolism was disrupted [179]. As Fe-S protein biogenesis is central for cellular iron regulation (see in Section 21.5), they hypothesized that the increased nuclear LOH resulted from a Fe-S protein assembly defect during the "crisis" stemming from the loss of mtDNA. Further credence to this model came from the observation that conditional inactivation of the CIA factor Nar1 led to increased nuclear LOH [179]. Interestingly, it also been shown that human cultured cell lines that have lost their mtDNA exhibit a decreased capacity for the repair of oxidatively damaged DNA and an increase in their nuclear mutation rates, suggesting that the role of mitochondria in maintaining genomic integrity is highly conserved [181].

If defects in Fe-S cluster biogenesis can drive genomic instability, then it would also be expected that reduced Fe-S protein biogenesis would promote tumorigenesis. Although evidence in this regard is scarce, several key observations have been reported. First, mice in which the Fe-S cluster biogenesis protein frataxin gene has been knocked out in hepatocytes show a reduced life span, resulting from a dramatic increase in hepatic tumor incidence [182]. Second, a gene expression signature that includes ISCU, ISCA1, and CIAO1 is effective at stratifying breast cancer patients based on metastasis-free long-term survival [183]. Third, mutations that specifically disrupt the assembly of Fe-S clusters in XPD and FANCJ give rise to xeroderma pigmentosum and Fanconi anemia, two syndromes that are characterized by a predisposition to cancer [148, 155].

Additional indirect evidence that may support a connection between Fe-S protein biogenesis and tumorigenesis comes from broader work looking at the role of excess iron in tumorigenesis. Both animal and population studies have clearly established a link between cancer susceptibility and excess iron resulting from either high dietary intake or genetic disorders such as hemochromatosis, which lead to iron overload and deposition in liver, heart, and endocrine organs [184, 185]. The mechanism by which iron overload promotes tumorigenesis results, at least in part, from the ability of iron to generate reactive oxygen species *via* the Fenton reaction, which in turn leads to oxidative damage of cellular macromolecules. These high levels of oxidative stress can also impact Fe-S cluster metabolism and the downstream genome stability pathways through multiple mechanisms. First, the Fe-S clusters of DNA repair holoenzymes can be directly damaged by oxidation, leading to their inactivation [6, 186]. Second, oxidative stress can also lead to mitochondrial dysfunction and reduced Fe-S cluster biogenesis [12, 187]. Together, these two mechanisms may offer a plausible route by which excess iron can disrupt Fe-S cluster-dependent genomic stability pathways and thereby promote tumor-associated genome stability.

## 21.9 Summary

It is essential to emphasize that although the connection between Fe-S protein biogenesis and genome stability and tumorigenesis may be reasonable based on existing data, this issue remains highly speculative. Genetic data indicating that components of the Fe-S cluster machinery pathway are either mutated or inactivated during the course of tumorigenesis are completely lacking. Similarly, a systematic analysis of the integrity of the Fe-S protein assembly pathways or the activity of Fe-S proteins in primary tumors has not been reported. Additional work in this area is clearly needed to establish the pathological relevance of these pathways with respect to cancer.

To better understand the role of Fe-S protein assembly in tumorigenesis, it will also be necessary to focus on how substrate specificity and prioritization is determined by the biogenesis machinery. Considering that different subsets of Fe-S proteins are essential for cell proliferation, viability, and genome maintenance, it will be interesting to see what cellular mechanisms might be available for the differential maturation of different sets of Fe-S substrates. For example, precancerous cells would benefit from maintaining those Fe-S proteins necessary for growth and proliferation, while blocking the biogenesis of Fe-S proteins required for genome stability. It is not clear, however, to what extent this type of differential regulation is possible and whether cellular mechanisms even exist for prioritizing the maturation of one set of Fe-S proteins over another. Bacterial systems contain an alternate Fe-S cluster assembly pathway known as the SUF system that is activated in cells undergoing physiological stress or iron deprivation. This system may provide a potential paradigm for how cells might regulate Fe-S cluster assembly of a distinct group of client apo-proteins under defined physiological conditions [8, 188]. Examples of specialized Fe-S protein assembly pathways in eukaryotes are provided by the dedicated maturation of IRP1 *via* CIA2A (instead of CIA2B; Fig. 21.4b) and the MMS19-independent maturation of GPAT [120, 122]. Understanding the global regulation of the Fe-S protein assembly pathways and how the integrity of the genome stability pathways can be influenced by altering the flux of Fe-S clusters to specific substrates will be a crucial step in providing the conceptual framework necessary for elucidating how these pathways might be co-opted in the context of tumorigenesis.

A major impetus for elucidating the complex relationships governing Fe-S cluster assembly and genome stability is to assess whether these pathways may offer a novel avenue for therapeutic intervention in the context of cancer. One therapeutic strategy could entail the development of small molecule activators of Fe-S protein assembly capable of stimulating the maturation of Fe-S cluster-dependent DNA repair enzymes and thereby promote genome stability pathways. Alternatively, pharmacological inhibition of Fe-S protein biogenesis could lead to inactivation of genome stability pathways and thus sensitize tumors to chemotherapy. Preliminary studies looking at how the inhibition of Fe-S protein assembly could interfere with cell proliferation in the context of tumorigenesis could be performed using iron-chelating agents such as deferoxamine, whose anti-tumor activity is already being examined clinically [189].

## Acknowledgments

We thank the members of our groups for excellent work and stimulating discussions. R.L. acknowledges generous support from Deutsche Forschungsgemeinschaft (SFB 593, SFB 987, and GRK 1216), von Behring-Röntgen Stiftung, LOEWE program of state Hessen, and Max-Planck Gesellschaft.

## References

[1] Kispal G, Csere P, Prohl C, Lill R. The mitochondrial proteins Atm1p and Nfs1p are required for biogenesis of cytosolic Fe/S proteins. EMBO J 1999;18:3981–9.
[2] Schilke B, Voisine C, Beinert H, Craig E. Evidence for a conserved system for iron metabolism in the mitochondria of Saccharomyces cerevisiae. Proc Natl Acad Sci USA 1999;96:10206–11.
[3] Lill R, Mühlenhoff U. Maturation of iron-sulfur proteins in eukaryotes: mechanisms, connected processes, and diseases. Annu Rev Biochem 2008;77:669–700.
[4] Lill R. Function and biogenesis iron-sulphur proteins. Nature 2009;460:831–8.

[5] Lill R, Hoffmann B, Molik S, et al. The role of mitochondria in cellular iron-sulfur protein biogenesis and iron metabolism. Biochim Biophys Acta 2012;1823:1491–508.
[6] Rouault TA. Biogenesis of iron-sulfur clusters in mammalian cells: new insights and relevance to human disease. Dis Model Mech 2012;5:155–64.
[7] Johnson DC, Dean DR, Smith AD, Johnson MK. Structure, function and formation of biological iron-sulfur clusters. Annu Rev Biochem 2005;74:247–81.
[8] Py B, Barras F. Building Fe-S proteins: bacterial strategies. Nat Rev Microbiol 2010;8:436–46.
[9] Biederbick A, Stehling O, Rösser R, et al. Role of human mitochondrial Nfs1 in cytosolic iron-sulfur protein biogenesis and iron regulation. Mol Cell Biol 2006;26:5675–87.
[10] Gerber J, Neumann K, Prohl C, Mühlenhoff U, Lill R. The yeast scaffold proteins Isu1p and Isu2p are required inside mitochondria for maturation of cytosolic Fe/S proteins. Mol Cell Biol 2004;24:4848–57.
[11] Fosset C, Chauveau MJ, Guillon B, Canal F, Drapier JC, Bouton C. RNA silencing of mitochondrial m-Nfs1 reduces Fe-S enzyme activity both in mitochondria and cytosol of mammalian cells. J Biol Chem 2006;281:25398–406.
[12] Stehling O, Lill R. The role of mitochondria in cellular iron-sulfur protein biogenesis: mechanisms, connected processes, and diseases. Cold Spring Harb Perspect Biol 2013;5:a011312.
[13] Lill R, Mühlenhoff U. Iron-sulfur protein biogenesis in eukaryotes: components and mechanisms. Annu Rev Cell Dev Biol 2006;22:457–86.
[14] Embley TM, Martin W. Eukaryotic evolution, changes and challenges. Nature 2006;440:623–30.
[15] Molik S, Lill R. Role of mitosomes in cellular iron-sulfur protein biogenesis. J Endocyt Cell Res 2012;23:77–85.
[16] Shiflett AM, Johnson PJ. Mitochondrion-related organelles in eukaryotic protists. Annu Rev Microbiol 2010;64:409–29.
[17] van der Giezen M, Tovar J. Degenerate mitochondria. EMBO Rep 2005;6:525–30.
[18] Hjort K, Goldberg AV, Tsaousis AD, Hirt RP, Embley TM. Diversity and reductive evolution of mitochondria among microbial eukaryotes. Philos Trans R Soc Lond B Biol Sci 2010;365: 713–27.
[19] Goldberg AV, Molik S, Tsaousis AD, et al. Localization and functionality of microsporidian iron-sulphur cluster assembly proteins. Nature 2008;452:624–8.
[20] Tovar J, Leon-Avila G, Sanchez LB, et al. Mitochondrial remnant organelles of Giardia function in iron-sulphur protein maturation. Nature 2003;426:172–6.
[21] Li H, Outten CE. Monothiol CGFS glutaredoxins and BolA-like proteins: [2Fe-2S] binding partners in iron homeostasis. Biochemistry 2012;51:4377–89.
[22] Philpott CC, Leidgens S, Frey AG. Metabolic remodeling in iron-deficient fungi. Biochim Biophys Acta 2012;1823:1509–20.
[23] Rawat S, Stemmler TL. Key players and their role during mitochondrial iron-sulfur cluster biosynthesis. Chemistry 2011;17:746–53.
[24] Neupert W, Herrmann JM. Translocation of proteins into mitochondria. Annu Rev Biochem 2007;76:723–49.
[25] Schmidt O, Pfanner N, Meisinger C. Mitochondrial protein import: from proteomics to functional mechanisms. Nat Rev Mol Cell Biol 2010;11:655–67.
[26] Garland SA, Hoff K, Vickery LE, Culotta VC. Saccharomyces cerevisiae ISU1 and ISU2: members of a well-conserved gene family for iron-sulfur cluster assembly. J Mol Biol 1999;294:897–907.
[27] Mühlenhoff U, Gerber J, Richhardt N, Lill R. Components involved in assembly and dislocation of iron-sulfur clusters on the scaffold protein Isu1p. EMBO J 2003;22:4815–25.
[28] Tong WH, Rouault TA. Functions of mitochondrial ISCU and cytosolic ISCU in mammalian iron-sulfur cluster biogenesis and iron homeostasis. Cell Metab 2006;3:199–210.

[29] Adam AC, Bornhövd C, Prokisch H, Neupert W, Hell K. The Nfs1 interacting protein Isd11 has an essential role in Fe/S cluster biogenesis in mitochondria. EMBO J 2006;25:174–83.
[30] Wiedemann N, Urzica E, Guiard B, et al. Essential role of Isd11 in iron-sulfur cluster synthesis on Isu scaffold proteins. EMBO J 2006;25:184–95.
[31] Shi Y, Ghosh MC, Tong WH, Rouault TA. Human ISD11 is essential for both iron-sulfur cluster assembly and maintenance of normal cellular iron homeostasis. Hum Mol Genet 2009;18:3014–25.
[32] Foury F, Roganti T. Deletion of the mitochondrial carrier genes MRS3 and MRS4 suppresses mitochondrial iron accumulation in a yeast frataxin-deficient strain. J Biol Chem 2002;277:24475–83.
[33] Mühlenhoff U, Stadler J, Richhardt N, et al. A specific role of the yeast mitochondrial carriers Mrs3/4p in mitochondrial iron acquisition under iron-limiting conditions. J Biol Chem 2003;278:40612–20.
[34] Zhang Y, Lyver ER, Knight SA, Pain D, Lesuisse E, Dancis A. Mrs3p, Mrs4p, and frataxin provide iron for Fe-S cluster synthesis in mitochondria. J Biol Chem 2006;281:22493–502.
[35] Shaw GC, Cope JJ, Li L, et al. Mitoferrin is essential for erythroid iron assimilation. Nature 2006;440:96–100.
[36] Yoon H, Zhang Y, Pain J, et al. Rim2, a pyrimidine nucleotide exchanger, is needed for iron utilization in mitochondria. Biochem J 2011;440:137–46.
[37] Froschauer EM, Rietzschel N, Hassler MR, et al. The mitochondrial carrier Rim2 co-imports pyrimidine nucleotides and iron. Biochem J 2013;455:57–65.
[38] Gerber J, Mühlenhoff U, Lill R. An interaction between frataxin and Isu1/Nfs1 that is crucial for Fe/S cluster synthesis on Isu1. EMBO Rep 2003;4:906–11.
[39] Schmucker S, Martelli A, Colin F, et al. Mammalian frataxin: an essential function for cellular viability through an interaction with a preformed ISCU/NFS1/ISD11 iron-sulfur assembly complex. PLoS One 2011;6:e16199.
[40] Stemmler TL, Lesuisse E, Pain D, Dancis A. Frataxin and mitochondrial FeS cluster biogenesis. J Biol Chem 2010;285:26737–43.
[41] Stehling O, Elsässer HP, Brückel B, Mühlenhoff U, Lill R. Iron-sulfur protein maturation in human cells: evidence for a function of frataxin. Hum Mol Genet 2004;13:3007–15.
[42] Tsai CL, Barondeau DP. Human frataxin is an allosteric switch that activates the Fe-S cluster biosynthetic complex. Biochemistry 2010;49:9132–9.
[43] Yoon H, Golla R, Lesuisse E, et al. Mutation in the Fe-S scaffold protein Isu bypasses frataxin deletion. Biochem J 2012;441:473–80.
[44] Lange H, Kaut A, Kispal G, Lill R. A mitochondrial ferredoxin is essential for biogenesis of cellular iron-sulfur proteins. Proc Natl Acad Sci USA 2000;97:1050–5.
[45] Li J, Saxena S, Pain D, Dancis A. Adrenodoxin reductase homolog (Arh1p) of yeast mitochondria required for iron homeostasis. J Biol Chem 2001;276:1503–9.
[46] Sheftel AD, Stehling O, Pierik AJ, et al. Humans possess two mitochondrial ferredoxins, Fdx1 and Fdx2, with distinct roles in steroidogenesis, heme, and Fe/S cluster biosynthesis. Proc Natl Acad Sci USA 2010;107:11775–80.
[47] Shi Y, Ghosh M, Kovtunovych G, Crooks DR, Rouault TA. Both human ferredoxins 1 and 2 and ferredoxin reductase are important for iron-sulfur cluster biogenesis. Biochim Biophys Acta 2012;1823:484–92.
[48] Chandramouli K, Unciuleac MC, Naik S, Dean DR, Huynh BH, Johnson MK. Formation and properties of [4Fe-4S] clusters on the IscU scaffold protein. Biochemistry 2007;46:6804–11.
[49] Unciuleac MC, Chandramouli K, Naik S, et al. In vitro activation of apo-aconitase using a [4Fe-4S] cluster-loaded form of the IscU [Fe-S] cluster scaffolding protein. Biochemistry 2007;46:6812–21.

[50] Barros MH, Nobrega FG, Tzagoloff A. Mitochondrial ferredoxin is required for heme A synthesis in Saccharomyces cerevisiae. J Biol Chem 2002;277:9997–10002.
[51] Pierrel F, Hamelin O, Douki T, et al. Involvement of mitochondrial ferredoxin and para-aminobenzoic acid in yeast coenzyme Q biosynthesis. Chem Biol 2010;17:449–59.
[52] Mühlenhoff U, Richter N, Pines O, Pierik AJ, Lill R. Specialized function of yeast Isa1 and Isa2 proteins in the maturation of mitochondrial [4Fe-4S] proteins. J Biol Chem 2011;286:41205–16.
[53] Dutkiewicz R, Marszalek J, Schilke B, Craig EA, Lill R, Mühlenhoff U. The Hsp70 chaperone Ssq1p is dispensable for iron-sulfur cluster formation on the scaffold protein Isu1p. J Biol Chem 2006;281:7801–8.
[54] Kampinga HH, Craig EA. The HSP70 chaperone machinery: J proteins as drivers of functional specificity. Nat Rev Mol Cell Biol 2010;11:579–92.
[55] Uhrigshardt H, Singh A, Kovtunovych G, Ghosh M, Rouault TA. Characterization of the human HSC20, an unusual DnaJ type III protein, involved in iron-sulfur cluster biogenesis. Hum Mol Genet 2010;19:3816–34.
[56] Vickery LE, Cupp-Vickery JR. Molecular chaperones HscA/Ssq1 and HscB/Jac1 and their roles in iron-sulfur protein maturation. Crit Rev Biochem Mol Biol 2007;42:95–111.
[57] Silberg JJ, Tapley TL, Hoff KG, Vickery LE. Regulation of the HscA ATPase reaction cycle by the co-chaperone HscB and the iron-sulfur cluster assembly protein IscU. J Biol Chem 2004;279:53924–31.
[58] Knieszner H, Schilke B, Dutkiewicz R, et al. Compensation for a defective interaction of the Hsp70 Ssq1 with the mitochondrial Fe-S cluster scaffold Isu. J Biol Chem 2005;280:28966–72.
[59] Ciesielski SJ, Schilke BA, Osipiuk J, et al. Interaction of J-protein co-chaperone Jac1 with Fe-S scaffold Isu is indispensable in vivo and conserved in evolution. J Mol Biol 2012;417:1–12.
[60] Dutkiewicz R, Schilke B, Cheng S, Knieszner H, Craig EA, Marszalek J. Sequence-specific interaction between mitochondrial Fe-S scaffold protein Isu and Hsp70 Ssq1 is essential for their in vivo function. J Biol Chem 2004;279:29167–74.
[61] Hoff KG, Cupp-Vickery JR, Vickery LE. Contributions of the LPPVK motif of the iron-sulfur template protein IscU to interactions with the Hsc66-Hsc20 chaperone system. J Biol Chem 2003;278:37582–9.
[62] Bonomi F, Iametti S, Morleo A, Ta D, Vickery LE. Studies on the mechanism of catalysis of iron-sulfur cluster transfer from IscU[2Fe2S] by HscA/HscB chaperones. Biochemistry 2008;47:12795–801.
[63] Bonomi F, Iametti S, Morleo A, Ta D, Vickery LE. Facilitated transfer of IscU-[2Fe2S] clusters by chaperone-mediated ligand exchange. Biochemistry 2011;50:9641–50.
[64] Dutkiewicz R, Schilke B, Kneiszner H, Walter W, Craig EA, Marszalek J. Ssq1, a mitochondrial Hsp70 involved in iron-sulfur (Fe/S) center biogenesis: similarities to and differences from its bacterial counterpart. J Biol Chem 2003;278:29719–27.
[65] Uzarska MA, Dutkiewicz R, Freibert SA, Lill R, Mühlenhoff U. The mitochondrial Hsp70 chaperone Ssq1 facilitates Fe/S cluster transfer from Isu1 to Grx5 by complex formation. Mol Biol Cell 2013;24:1830–41.
[66] Rodriguez-Manzaneque MT, Tamarit J, Belli G, Ros J, Herrero E. Grx5 is a mitochondrial glutaredoxin required for the activity of iron/sulfur enzymes. Mol Biol Cell 2002;13:1109–21.
[67] Wingert RA, Galloway JL, Barut B, et al. Deficiency of glutaredoxin 5 reveals Fe-S clusters are required for vertebrate haem synthesis. Nature 2005;436:1035–9.
[68] Camaschella C, Campanella A, De Falco L, et al. The human counterpart of zebrafish shiraz shows sideroblastic-like microcytic anemia and iron overload. Blood 2007;110:1353–8.
[69] Ye H, Jeong SY, Ghosh MC, et al. Glutaredoxin 5 deficiency causes sideroblastic anemia by specifically impairing heme biosynthesis and depleting cytosolic iron in human erythroblasts. J Clin Invest 2010;120:1749–61.

[70] Bandyopadhyay S, Gama F, Molina-Navarro MM, et al. Chloroplast monothiol glutaredoxins as scaffold proteins for the assembly and delivery of [2Fe-2S] clusters. EMBO J 2008;27:1122–33.
[71] Molina-Navarro MM, Casas C, Piedrafita L, Belli G, Herrero E. Prokaryotic and eukaryotic monothiol glutaredoxins are able to perform the functions of Grx5 in the biogenesis of Fe/S clusters in yeast mitochondria. FEBS Lett 2006;580:2273–80.
[72] Herrero E, de la Torre-Ruiz MA. Monothiol glutaredoxins: a common domain for multiple functions. Cell Mol Life Sci 2007;64:1518–30.
[73] Iwema T, Picciocchi A, Traore DA, Ferrer JL, Chauvat F, Jacquamet L. Structural basis for delivery of the intact [Fe2S2] cluster by monothiol glutaredoxin. Biochemistry 2009;48:6041–3.
[74] Rouhier N, Unno H, Bandyopadhyay S, et al. Functional, structural, and spectroscopic characterization of a glutathione-ligated [2Fe-2S] cluster in poplar glutaredoxin C1. Proc Natl Acad Sci USA 2007;104:7379–84.
[75] Rada P, Smid O, Sutak R, et al. The monothiol single-domain glutaredoxin is conserved in the highly reduced mitochondria of Giardia intestinalis. Eukaryot Cell 2009;8:1584–91.
[76] Picciocchi A, Saguez C, Boussac A, Cassier-Chauvat C, Chauvat F. CGFS-type monothiol glutaredoxins from the cyanobacterium Synechocystis PCC6803 and other evolutionary distant model organisms possess a glutathione-ligated [2Fe-2S] cluster. Biochemistry 2007;46: 15018–26.
[77] Belli G, Polaina J, Tamarit J, et al. Structure-function analysis of yeast Grx5 monothiol glutaredoxin defines essential amino acids for the function of the protein. J Biol Chem 2002;277:37590–6.
[78] Jensen LT, Culotta VC. Role of Saccharomyces cerevisiae ISA1 and ISA2 in iron homeostasis. Mol Cell Biol 2000;20:3918–27.
[79] Kaut A, Lange H, Diekert K, Kispal G, Lill R. Isa1p Is a Component of the mitochondrial machinery for maturation of cellular iron-sulfur proteins and requires conserved cysteine residues for function. J Biol Chem 2000;275:15955–61.
[80] Pelzer W, Mühlenhoff U, Diekert K, Siegmund K, Kispal G, Lill R. Mitochondrial Isa2p plays a crucial role in the maturation of cellular iron-sulfur proteins. FEBS Lett 2000;476:134–9.
[81] Mühlenhoff U, Gerl MJ, Flauger B, et al. The ISC proteins Isa1 and Isa2 are required for the function but not for the de novo synthesis of the Fe/S clusters of biotin synthase in Saccharomyces cerevisiae. Eukaryot Cell 2007;6:495–504.
[82] Sheftel AD, Wilbrecht C, Stehling O, et al. The human mitochondrial ISCA1, ISCA2, and IBA57 proteins are required for [4Fe-4S] protein maturation. Mol Biol Cell 2012;23:1157–66.
[83] Gelling C, Dawes IW, Richhardt N, Lill R, Mühlenhoff U. Mitochondrial Iba57p is required for Fe/S cluster formation on aconitase and activation of radical SAM enzymes. Mol Cell Biol 2008;28:1851–61.
[84] Bych K, Kerscher S, Netz DJ, et al. The iron-sulphur protein Ind1 is required for effective complex I assembly. EMBO J 2008;27:1736–46.
[85] Sheftel AD, Stehling O, Pierik AJ, et al. Human Ind1, an iron-sulfur cluster assembly factor for respiratory complex I. Mol Cell Biol 2009;29:6059–73.
[86] Tong WH, Jameson GN, Huynh BH, Rouault TA. Subcellular compartmentalization of human Nfu, an iron-sulfur cluster scaffold protein, and its ability to assemble a [4Fe-4S] cluster. Proc Natl Acad Sci USA 2003;100:9762–7.
[87] Cameron JM, Janer A, Levandovskiy V, et al. Mutations in iron-sulfur cluster scaffold genes NFU1 and BOLA3 cause a fatal deficiency of multiple respiratory chain and 2-oxoacid dehydrogenase enzymes. Am J Hum Genet 2011;89:486–95.
[88] Navarro-Sastre A, Tort F, Stehling O, et al. A fatal mitochondrial disease is associated with defective NFU1 function in the maturation of a subset of mitochondrial Fe-S proteins. Am J Hum Genet 2011;89:656–67.

[89] Hiltunen JK, Autio KJ, Schonauer MS, Kursu VA, Dieckmann CL, Kastaniotis AJ. Mitochondrial fatty acid synthesis and respiration. Biochim Biophys Acta 2010;1797:1195–202.
[90] Haack TB, Rolinski B, Haberberger B, et al. Homozygous missense mutation in BOLA3 causes multiple mitochondrial dysfunctions syndrome in two siblings. J Inherit Metab Dis 2013;36:55–62.
[91] Aldea M, Garrido T, Hernandez-Chico C, Vicente M, Kushner SR. Induction of a growth-phase-dependent promoter triggers transcription of bolA, an Escherichia coli morphogene. EMBO J 1989;8:3923–31.
[92] Hess DC, Myers CL, Huttenhower C, et al. Computationally driven, quantitative experiments discover genes required for mitochondrial biogenesis. PLoS Genet 2009;5:e1000407.
[93] Huynen MA, Spronk CA, Gabaldon T, Snel B. Combining data from genomes, Y2H and 3D structure indicates that BolA is a reductase interacting with a glutaredoxin. FEBS Lett 2005;579:591–6.
[94] Kumanovics A, Chen O, Li L, et al. Identification of FRA1 and FRA2 as genes involved in regulating the yeast iron regulon in response to decreased mitochondrial iron-sulfur cluster synthesis. J Biol Chem 2008;283:10276–86.
[95] Li H, Mapolelo DT, Dingra NN, et al. The yeast iron regulatory proteins Grx3/4 and Fra2 form heterodimeric complexes containing a [2Fe-2S] cluster with cysteinyl and histidyl ligation. Biochemistry 2009;48:9569–81.
[96] Li H, Mapolelo DT, Randeniya S, Johnson MK, Outten CE. Human glutaredoxin 3 forms [2Fe-2S]-bridged complexes with human BolA2. Biochemistry 2012;51:1687–96.
[97] Land T, Rouault TA. Targeting of a human iron-sulfur cluster assembly enzyme, nifs, to different subcellular compartments is regulated through alternative AUG utilization. Mol Cell 1998;2:807–15.
[98] Nakai Y, Nakai M, Hayashi H, Kagamiyama H. Nuclear Localization of Yeast Nfs1p Is Required for Cell Survival. J Biol Chem 2001;276:8314–20.
[99] Mühlenhoff U, Balk J, Richhardt N, et al. Functional characterization of the eukaryotic cysteine desulfurase Nfs1p from Saccharomyces cerevisiae. J Biol Chem 2004;279:36906–15.
[100] Crooks DR, Jeong SY, Tong WH, et al. Tissue specificity of a human mitochondrial disease: differentiation-enhanced mis-splicing of the Fe-S scaffold gene ISCU renders patient cells more sensitive to oxidative stress in ISCU myopathy. J Biol Chem 2012;287:40119–30.
[101] Nakai Y, Nakai M, Lill R, Suzuki T, Hayashi H. Thio modification of yeast cytosolic tRNA is an iron-sulfur protein-dependent pathway. Mol Cell Biol 2007;27:2841–7.
[102] Pondarre C, Antiochos BB, Campagna DR, et al. The mitochondrial ATP-binding cassette transporter Abcb7 is essential in mice and participates in cytosolic iron-sulphur cluster biogenesis. Hum Mol Genet 2006;15:953–64.
[103] Cavadini P, Biasiotto G, Poli M, et al. RNA silencing of the mitochondrial ABCB7 transporter in HeLa cells causes an iron-deficient phenotype with mitochondrial iron overload. Blood 2007;109:3552–9.
[104] Sipos K, Lange H, Fekete Z, Ullmann P, Lill R, Kispal G. Maturation of cytosolic iron-sulfur proteins requires glutathione. J Biol Chem 2002;277:26944–9.
[105] Kispal G, Csere P, Guiard B, Lill R. The ABC transporter Atm1p is required for mitochondrial iron homeostasis. FEBS Lett 1997;418:346–50.
[106] Mesecke N, Terziyska N, Kozany C, et al. A disulfide relay system in the intermembrane space of mitochondria that mediates protein import. Cell 2005;121:1059–70.
[107] Sharma AK, Pallesen LJ, Spang RJ, Walden WE. Cytosolic iron-sulfur cluster assembly (CIA) system: factors, mechanism, and relevance to cellular iron regulation. J Biol Chem 2010;285:26745–51.
[108] Netz DJ, Pierik AJ, Macarenhas J, Stehling O, Lill R. Maturation of cytosolic and nuclear iron-sulfur proteins. Trends Cell Biol 2014;in press.

[109] Netz DJ, Pierik AJ, Stümpfig M, Mühlenhoff U, Lill R. The Cfd1-Nbp35 complex acts as a scaffold for iron-sulfur protein assembly in the yeast cytosol. Nat Chem Biol 2007;3:278–86.
[110] Netz DJ, Pierik AJ, Stuempfig M, et al. A bridging [4Fe-4S] cluster and nucleotide binding are essential for the function of the Cfd1-Nbp35 complex as a scaffold in iron-sulfur protein maturation. J Biol Chem 2012;287:12365–78.
[111] Netz DJ, Stumpfig M, Dore C, Mühlenhoff U, Pierik AJ, Lill R. Tah18 transfers electrons to Dre2 in cytosolic iron-sulfur protein biogenesis. Nat Chem Biol 2010;6:758–65.
[112] Pallesen LJ, Solodovnikova N, Sharma AK, Walden WE. Interaction with Cfd1 Increases the Kinetic Lability of FeS on the Nbp35 Scaffold. J Biol Chem 2013.
[113] Zhang Y, Lyver ER, Nakamaru-Ogiso E, et al. Dre2, a conserved eukaryotic Fe/S cluster protein, functions in cytosolic Fe/S protein biogenesis. Mol Cell Biol 2008;28:5569–82.
[114] Banci L, Bertini I, Calderone V, et al. Molecular view of an electron transfer process essential for iron-sulfur protein biogenesis. Proc Natl Acad Sci USA 2013;110:7136–41.
[115] Balk J, Pierik AJ, Aguilar Netz D, Mühlenhoff U, Lill R. The hydrogenase-like Nar1p is essential for maturation of cytosolic and nuclear iron-sulphur proteins. EMBO J 2004;23:2105–15.
[116] Song D, Lee FS. A role for IOP1 in mammalian cytosolic iron-sulfur protein biogenesis. J Biol Chem 2008;283:9231–8.
[117] Urzica E, Pierik AJ, Muhlenhoff U, Lill R. Crucial role of conserved cysteine residues in the assembly of two iron-sulfur clusters on the CIA protein Nar1. Biochemistry 2009;48:4946–58.
[118] Srinivasan V, Netz DJA, Webert H, et al. Structure of the yeast WD40 domain protein Cia1, a component acting late in iron-sulfur protein biogenesis. Structure 2007;15:1246–57.
[119] Weerapana E, Wang C, Simon GM, et al. Quantitative reactivity profiling predicts functional cysteines in proteomes. Nature 2010;468:790–5.
[120] Stehling O, Vashisht AA, Mascarenhas J, et al. MMS19 assembles iron-sulfur proteins required for DNA metabolism and genomic integrity. Science 2012;337:195–9.
[121] Gari K, Leon Ortiz AM, Borel V, Flynn H, Skehel JM, Boulton SJ. MMS19 links cytoplasmic iron-sulfur cluster assembly to DNA metabolism. Science 2012;337:243–5.
[122] Stehling O, Mascarenhas J, Vashisht AA, et al. Human CIA2A-FAM96A and CIA2B-FAM96B integrate iron homeostasis and maturation of different subsets of cytosolic-nuclear iron-sulfur proteins. Cell Metab 2013;18:187–98.
[123] van Wietmarschen N, Moradian A, Morin GB, Lansdorp PM, Uringa EJ. The mammalian proteins MMS19, MIP18, and ANT2 are involved in cytoplasmic iron-sulfur cluster protein assembly. J Biol Chem 2012;287:43351–8.
[124] Seki M, Takeda Y, Iwai K, Tanaka K. IOP1 protein is an external component of the human cytosolic iron-sulfur cluster assembly (CIA) machinery and functions in the MMS19 protein-dependent CIA pathway. J Biol Chem 2013;288:16680–9.
[125] Mühlenhoff U, Molik S, Godoy JR, et al. Cytosolic monothiol glutaredoxins function in intracellular iron sensing and trafficking via their bound iron-sulfur cluster. Cell Metab 2010;12:373–85.
[126] Haunhorst P, Hanschmann EM, Brautigam L, et al. Crucial function of vertebrate glutaredoxin 3 (PICOT) in iron homeostasis and hemoglobin maturation. Mol Biol Cell 2013;24:1895–903.
[127] Zhang Y, Liu L, Wu X, An X, Stubbe J, Huang M. Investigation of in vivo diferric tyrosyl radical formation in Saccharomyces cerevisiae Rnr2 protein: requirement of Rnr4 and contribution of Grx3/4 AND Dre2 proteins. J Biol Chem 2011;286:41499–509.
[128] Stehling O, Netz DJ, Niggemeyer B, et al. Human Nbp35 is essential for both cytosolic iron-sulfur protein assembly and iron homeostasis. Mol Cell Biol 2008;28:5517–28.
[129] Krogan NJ, Cagney G, Yu H, et al. Global landscape of protein complexes in the yeast Saccharomyces cerevisiae. Nature 2006;440:637–43.

[130] Rual JF, Venkatesan K, Hao T, et al. Towards a proteome-scale map of the human protein-protein interaction network. Nature 2005;437:1173–8.
[131] Tarassov K, Messier V, Landry CR, et al. An in vivo map of the yeast protein interactome. Science 2008;320:1465–70.
[132] Collins SR, Kemmeren P, Zhao XC, et al. Toward a comprehensive atlas of the physical interactome of Saccharomyces cerevisiae. Mol Cell Proteomics 2007;6:439–50.
[133] Pace NJ, Weerapana E. Diverse functional roles of reactive cysteines. ACS Chem Biol 2013;8:283–96.
[134] Luo D, Bernard DG, Balk J, Hai H, Cui X. The DUF59 family gene AE7 acts in the cytosolic iron-sulfur cluster assembly pathway to maintain nuclear genome integrity in Arabidopsis. Plant Cell 2012;24:4135–48.
[135] Schwenkert S, Netz DJ, Frazzon J, et al. Chloroplast HCF101 is a scaffold protein for [4Fe-4S] cluster assembly. Biochem J 2010;425:207–14.
[136] Almeida MS, Herrmann T, Peti W, Wilson IA, Wüthrich K. NMR structure of the conserved hypothetical protein TM0487 from Thermotoga maritima: implications for 216 homologous DUF59 proteins. Protein Sci 2005;14:2880–6.
[137] Mas C, Chen KE, Brereton IM, Martin JL, Hill JM. Backbone resonance assignments of the monomeric DUF59 domain of human Fam96a. Biomol NMR Assign 2012.
[138] Anderson CP, Shen M, Eisenstein RS, Leibold EA. Mammalian iron metabolism and its control by iron regulatory proteins. Biochim Biophys Acta 2012;1823:1468–83.
[139] Thompson JW, Bruick RK. Protein degradation and iron homeostasis. Biochim Biophys Acta 2012;1823:1484–90.
[140] Prakash L, Prakash S. Three additional genes involved in pyrimidine dimer removal in Saccharomyces cerevisiae: RAD7, RAD14 and MMS19. Mol Gen Genet 1979;176:351–9.
[141] Thomas D, Barbey R, Henry D, Surdin-Kerjan Y. Physiological analysis of mutants of Saccharomyces cerevisiae impaired in sulphate assimilation. J Gen Microbiol 1992;138:2021–8.
[142] Lauder S, Bankmann M, Guzder SN, Sung P, Prakash L, Prakash S. Dual requirement for the yeast MMS19 gene in DNA repair and RNA polymerase II transcription. Mol Cell Biol 1996;16:6783–93.
[143] Kou H, Zhou Y, Gorospe RM, Wang Z. Mms19 protein functions in nucleotide excision repair by sustaining an adequate cellular concentration of the TFIIH component Rad3. Proc Natl Acad Sci USA 2008;105:15714–9.
[144] Queimado L, Rao M, Schultz RA, et al. Cloning the human and mouse MMS19 genes and functional complementation of a yeast mms19 deletion mutant. Nucleic Acids Res 2001;29:1884–91.
[145] Seroz T, Winkler GS, Auriol J, et al. Cloning of a human homolog of the yeast nucleotide excision repair gene MMS19 and interaction with transcription repair factor TFIIH via the XPB and XPD helicases. Nucleic Acids Res 2000;28:4506–13.
[146] Ito S, Tan LJ, Andoh D, et al. MMXD, a TFIIH-independent XPD-MMS19 protein complex involved in chromosome segregation. Mol Cell 2010;39:632–40.
[147] Compe E, Egly JM. TFIIH: when transcription met DNA repair. Nat Rev Mol Cell Biol 2012;13:343–54.
[148] Rudolf J, Makrantoni V, Ingledew WJ, Stark MJ, White MF. The DNA repair helicases XPD and FancJ have essential iron-sulfur domains. Mol Cell 2006;23:801–8.
[149] Fan L, Fuss JO, Cheng QJ, et al. XPD helicase structures and activities: insights into the cancer and aging phenotypes from XPD mutations. Cell 2008;133:789–800.
[150] Liu H, Rudolf J, Johnson KA, et al. Structure of the DNA repair helicase XPD. Cell 2008;133:801–12.
[151] Wolski SC, Kuper J, Hanzelmann P, et al. Crystal structure of the FeS cluster-containing nucleotide excision repair helicase XPD. PLoS Biol 2008;6:e149.

[152] Pugh RA, Honda M, Leesley H, et al. The Iron-containing Domain Is Essential in Rad3 Helicases for Coupling of ATP Hydrolysis to DNA Translocation and for Targeting the Helicase to the Single-stranded DNA-Double-stranded DNA Junction. J Biol Chem 2008;283:1732–43.
[153] White MF. Structure, function and evolution of the XPD family of iron-sulfur-containing 5′->3′ DNA helicases. Biochem Soc Trans 2009;37:547–51.
[154] Wu Y, Suhasini AN, Brosh RM Jr. Welcome the family of FANCJ-like helicases to the block of genome stability maintenance proteins. Cell Mol Life Sci 2009;66:1209–22.
[155] Wu Y, Sommers JA, Suhasini AN, et al. Fanconi anemia group J mutation abolishes its DNA repair function by uncoupling DNA translocation from helicase activity or disruption of protein-DNA complexes. Blood 2010;116:3780–91.
[156] Bell SP, Dutta A. DNA replication in eukaryotic cells. Annu Rev Biochem 2002;71:333–74.
[157] Arezi B, Kuchta RD. Eukaryotic DNA primase. Trends Biochem Sci 2000;25:572–6.
[158] Klinge S, Hirst J, Maman JD, Krude T, Pellegrini L. An iron-sulfur domain of the eukaryotic primase is essential for RNA primer synthesis. Nat Struct Mol Biol 2007;14:875–7.
[159] Weiner BE, Huang H, Dattilo BM, Nilges MJ, Fanning E, Chazin WJ. An iron-sulfur cluster in the C-terminal domain of the p58 subunit of human DNA primase. J Biol Chem 2007;282:33444–51.
[160] Kunkel TA, Burgers PM. Dividing the workload at a eukaryotic replication fork. Trends Cell Biol 2008;18:521–7.
[161] Netz DJ, Stith CM, Stumpfig M, et al. Eukaryotic DNA polymerases require an iron-sulfur cluster for the formation of active complexes. Nat Chem Biol 2012;8:125–32.
[162] Pokharel S, Campbell JL. Cross talk between the nuclease and helicase activities of Dna2: role of an essential iron-sulfur cluster domain. Nucleic Acids Res 2012;40:7821–30.
[163] Robertson AB, Klungland A, Rognes T, Leiros I. DNA repair in mammalian cells: Base excision repair: the long and short of it. Cell Mol Life Sci 2009;66:981–93.
[164] Cunningham RP, Asahara H, Bank JF, et al. Endonuclease III is an iron-sulfur protein. Biochemistry 1989;28:4450–5.
[165] Boal AK, Genereux JC, Sontz PA, Gralnick JA, Newman DK, Barton JK. Redox signaling between DNA repair proteins for efficient lesion detection. Proc Natl Acad Sci USA 2009;106:15237–42.
[166] Boal AK, Yavin E, Barton JK. DNA repair glycosylases with a [4Fe-4S] cluster: a redox cofactor for DNA-mediated charge transport? J Inorg Biochem 2007;101:1913–21.
[167] Kuo CF, McRee DE, Fisher CL, O'Handley SF, Cunningham RP, Tainer JA. Atomic structure of the DNA repair [4Fe-4S] enzyme endonuclease III. Science 1992;258:434–40.
[168] McGoldrick JP, Yeh YC, Solomon M, Essigmann JM, Lu AL. Characterization of a mammalian homolog of the Escherichia coli MutY mismatch repair protein. Mol Cell Biol 1995;15:989–96.
[169] White MF, Dillingham MS. Iron-sulphur clusters in nucleic acid processing enzymes. Current opinion in structural biology 2011.
[170] Genereux JC, Boal AK, Barton JK. DNA-mediated charge transport in redox sensing and signaling. J Am Chem Soc 2010;132:891–905.
[171] Sontz PA, Muren NB, Barton JK. DNA charge transport for sensing and signaling. Acc Chem Res 2012;45:1792–800.
[172] Romano CA, Sontz PA, Barton JK. Mutants of the base excision repair glycosylase, endonuclease III: DNA charge transport as a first step in lesion detection. Biochemistry 2011;50:6133–45.
[173] Sontz PA, Mui TP, Fuss JO, Tainer JA, Barton JK. DNA charge transport as a first step in coordinating the detection of lesions by repair proteins. Proc Natl Acad Sci USA 2012;109:1856–61.
[174] Hanahan D, Weinberg RA. Hallmarks of cancer: the next generation. Cell 2011;144:646–74.
[175] Cassidy LD, Venkitaraman AR. Genome instability mechanisms and the structure of cancer genomes. Curr Opin Genet Dev 2012;22:10–3.

[176] Wu Y, Brosh RM Jr. DNA helicase and helicase-nuclease enzymes with a conserved iron-sulfur cluster. Nucleic Acids Res 2012;40:4247–60.
[177] Hook SS, Lin JJ, Dutta A. Mechanisms to control rereplication and implications for cancer. Curr Opin Cell Biol 2007;19:663–71.
[178] McMurray MA, Gottschling DE. An age-induced switch to a hyper-recombinational state. Science 2003;301:1908–11.
[179] Veatch JR, McMurray MA, Nelson ZW, Gottschling DE. Mitochondrial dysfunction leads to nuclear genome instability via an iron-sulfur cluster defect. Cell 2009;137:1247–58.
[180] McMurray MA, Gottschling DE. Aging and genetic instability in yeast. Current opinion in microbiology 2004;7:673–9.
[181] Minocherhomji S, Tollefsbol TO, Singh KK. Mitochondrial regulation of epigenetics and its role in human diseases. Epigenetics 2012;7:326–34.
[182] Thierbach R, Schulz TJ, Isken F, et al. Targeted disruption of hepatic frataxin expression causes impaired mitochondrial function, decreased life span and tumor growth in mice. Hum Mol Genet 2005;14:3857–64.
[183] Miller LD, Coffman LG, Chou JW, et al. An iron regulatory gene signature predicts outcome in breast cancer. Cancer Res 2011;71:6728–37.
[184] Pra D, Franke SI, Henriques JA, Fenech M. Iron and genome stability: an update. Mutat Res 2012;733:92–9.
[185] Torti SV, Torti FM. Iron and cancer: more ore to be mined. Nat Rev Cancer 2013;13:342–55.
[186] Flint DH, Tuminello JF, Emptage MH. The inactivation of Fe-S cluster containing hydro-lyases by superoxide. J Biol Chem 1993;268:22369–76.
[187] Gao X, Campian JL, Qian M, Sun XF, Eaton JW. Mitochondrial DNA damage in iron overload. J Biol Chem 2009;284:4767–75.
[188] Ayala-Castro C, Saini A, Outten FW. Fe-S cluster assembly pathways in bacteria. Microbiol Mol Biol Rev 2008;72:110–25.
[189] Yu Y, Gutierrez E, Kovacevic Z, et al. Iron chelators for the treatment of cancer. Curr Med Chem 2012;19:2689–702.

# 22 Iron-sulfur cluster assembly in plants

Hong Ye

## 22.1 Introduction

Iron is an essential micromineral nutrient. To overcome the insolubility of ferric iron in soil, plants reduce ferric iron to increase solubility or secrete chelators and absorb iron chelates to increase acquisition of iron (strategies I and II). The translocation, distribution, and delivery of iron throughout the whole plant involve a number of specific transporters and chelators. Inside the plant cell, iron is utilized to assemble heme and iron-sulfur (Fe-S) clusters. Fe-S cluster biosynthesis takes place separately in major subcellular compartments. It is mediated by the SUF system in plastids, by the ISC system in mitochondria, and by the CIA system in the cytosolic and nuclear compartments. As an essential pathway, the Fe-S cluster assembly machinery is highly conserved across the green lineage. In contrast to bacteria, plants lack the machinery for nitrogen fixation, but the legume plants often have a symbiotic relationship with nitrogen-fixing bacteria that use nitrogenase to assimilate nitrogen in specialized structures that surround their roots. The potential impact of the research on Fe-S protein biogenesis on agriculture will likely be quite significant.

## 22.2 Iron uptake, translocation, and distribution

Iron is an essential mineral nutrient for plants. Plants take up iron from soil, in addition to water and all other minerals. Iron contents in plants are variable; for instance, iron concentrations in major staple foods are 4.31 mg/100 g for rice and 0.27 mg/100 g for cassava (www.usda.gov), whereas the iron content in some other plants is 10 mg/100 g [1]. Despite the fact that iron concentration in soil is high compared with other minerals, most of the soil iron is in the form of insoluble ferric hydroxides and hence is not bioavailable. The solubility of iron depends on the pH of soil [2]. Iron in low-pH soil is more soluble and hence more bioavailable, whereas iron in high-pH soil is more insoluble and hence less bioavailable. Therefore, plants grown in basic soil often suffer iron deficiency, demonstrating dwarf size and chlorotic (yellowing of the leaves) symptoms.

To take up iron from soil at the roots, plants generally employ one of two different strategies, either strategy I or strategy II [1–6]. Dicotyledenous plants, for example, *Arabidopsis thaliana*, use strategy I and express ferric reductases and ferrous iron transporters on the cell membrane of roots. These plants also express a proton pump to acidify the root surface. The ferric reductase FRO2 reduces ferric iron into much more soluble ferrous iron in the root rhizosphere, the soil rich in plant exudates that surrounds the root [7, 8], and subsequently, ferrous iron is imported into root cells by the ferrous transporter, IRT1 [9–11]. Because of their central roles in iron uptake, the gene expression levels of IRT1 and FRO2 are highly inducible by iron deficiency. However, inside cells, free iron is toxic because it catalyzes the formation of reactive oxygen species (ROS) in the Haber-Weiss cycle; therefore, most iron is likely contained within chelated complexes in plants. An important intracellular chelator is nicotianamine (NA), a compound derived from methionine. The Fe-NA chelate is transported symplastically (through pores that allow cytoplasmic contacts between neighboring cells) through plasmodesmata (pores in the plant cell wall) toward the vasculature. At the boundary between the endodermis and the vasculature, iron is transported across the plasma membrane into the xylem, likely by the transporter ferroportin [1, 2]. The Fe-citrate chelate is translocated in the xylem from the root to the shoot driven by the force of transpiration. In leaves, iron is unloaded from the vasculature and transported into mesophyll cells by members of the Yellow Stripe-Like (YSL) family of transporters.

In contrast, monocots, which include the graminaceous plants, for example, rice (*Oryza sativa*), use strategy II for iron uptake [1, 2]. Rice plants secrete phytosiderophores of the mugineic acid (MA) family into the rhizosphere, which release insoluble ferric iron from soil particles to form an iron chelator complex. The resulting Fe-MA chelate is directly taken up into root cells by members of the YSL family of transporters. However, a recent study has shown that in addition to strategy II, rice is also able to take up ferrous iron through the IRT1 transporter [12]. The subsequent steps of translocation from root cells to vasculature and from vasculature to leaves are similar between strategy I and II plants.

In plant cells, iron is delivered into major organelles including chloroplasts, mitochondria, and vacuoles, where it is either stored or utilized for the synthesis of Fe-S clusters, heme, and nonheme iron proteins. Iron is imported into mitochondria by a mitoferrin-like transporter on the inner membrane [13]. To maintain iron homeostasis in the compartment, iron should be exportable. However, no potential mitochondrial iron exporter has yet been identified. Iron is imported into chloroplasts by PIC1 [14, 15], whereas iron may be exported from chloroplasts by the YSL4 and YSL6 transporters [16]. Chloroplasts are the major sink of iron in plant leaf cells, accommodating up to 80%–90% of total cellular iron in leaf cells. However, it is unknown whether the photosynthetic green organelle is important for the regulation of overall cellular iron homeostasis. Iron is used for the Fe-S cluster and heme synthesis in chloroplasts, and some iron is stored in ferritin, the iron storage protein of plants [17–19]. Unlike ferritin of mammalian cells, which is in the cytosol, the ferritin in plant cells is in chloroplasts/plastids [19]. To protect from toxicity, iron should be associated with some iron chaperone or chelated by chelators in plant cytosol, such as citrate, NA, and phytosiderophores. Potential iron chaperones in the cytosol of plant cells have not been identified. Some have proposed that PCBP family proteins are iron chaperones in eukaryotic cells [20], whereas others propose that the complex of monothiol glutaredoxin and glutathione bridging an Fe-S cluster facilitates iron delivery in cytosol and nuclei [21].

The excess iron in cytosol is delivered to the vacuole for storage, imported by the VIT1 transporter, and stored as ferric citrate [22, 23]; the vacuolar iron can be exported by NRAMP3/4 transporters [1].

More than 60 Fe-S proteins have been confirmed or estimated in *Arabidopsis* [24]. These proteins have various physiological functions. Fe-S proteins in plastids are involved in chlorophyll synthesis, nitrite reduction, sulfite reduction, redox homeostasis, and photosynthesis [25]. Fe-S proteins in mitochondria are involved in molybdenum cofactor (Moco) biosynthesis, electron transfer in complexes I and II, glutamate synthesis, and biotin synthesis. Fe-S proteins in cytosol and nuclei are involved in abscisic acid biosynthesis, ribosome assembly, DNA replication, and DNA repair.

## 22.3 Fe-S cluster assembly

Despite the simplicity in the structure of Fe-S clusters, the biological assembly of an Fe-S cluster is extremely complex [26–30]. Research on plant Fe-S cluster biosynthesis has mainly focused on the model plant, *A. thaliana*. To date, 42 genes have been identified in *Arabidopsis* with a proposed function in Fe-S assembly [24, 31, 32]. As listed in Tab. 22.1, all these genes are encoded in the nuclear genome, and the expressed proteins are targeted to chloroplasts, mitochondria, and cytosol, respectively. Because these compartments are physically separated by membranes, it is thought that Fe-S biosynthesis takes place independently in each of the subcellular compartments [25, 26, 33–35]. Due to the sequence similarity to their orthologues in bacteria, the Fe-S

**Tab. 22.1** List of proteins that are involved in the Fe-S cluster assembly in plant cells.

| | Protein name | Locus | Function | Reference |
|---|---|---|---|---|
| Plastid | NFS2 | At1g08490 | Cysteine desulfurase | Pilon-Smits *et al.* (2003) |
| | SUFE1 | At4g26500 | Activator of NFS2 | Xu *et al.* (2006) |
| | SUFE2 | At1g67810 | Activator of NFS2 | Murthy *et al.* (2007) |
| | SUFE3 | At5g50210 | Activator of NFS2 | Murthy *et al.* (2007) |
| | SUFA | At1g10500 | Cluster transfer | Abdel-Ghany *et al.* (2005) [53] |
| | NFU1 | At4g01940 | Scaffold | Léon *et al.* (2003) [48] |
| | NFU2 | At5g49940 | Scaffold | Touraine *et al.* (2004) [49] |
| | NFU3 | At4g25910 | Scaffold | Léon *et al.* (2003) [48] |
| | SUFB | At4g04770 | Scaffold | Xu *et al.* (2005) [51] |
| | SUFC | At3g10670 | Scaffold | Xu *et al.* (2006) [43] |
| | SUFD | At1g32500 | Scaffold | Xu *et al.* (2005) [51] |
| | HCF101 | At3g24430 | 4Fe-4S insertion | Schwenkert *et al.* (2010) [56] |
| | GRXS14 | At3g54900 | 2Fe-2S transfer | Bandyopadhyay *et al.* (2008) [57] |
| | GRXS16 | At2g38270 | 2Fe-2S transfer | Cheng *et al.* (2006) [58] |

(*Continued*)

**Tab. 22.1:** List of proteins that are involved in the Fe-S cluster assembly in plant cells. (*Continued*)

| | Protein name | Locus | Function | Reference |
|---|---|---|---|---|
| Mitochondria | NFS1 | At5g65720 | Cysteine desulfurase | Frazzon *et al.* (2007) [60] |
| | ISD11 | At5g61220 | Interacting with NFS1 | Heis *et al.* (2011) [62] |
| | ISU1 | At4g22220 | Scaffold | Frazzon *et al.* (2007) [60] |
| | ISU2 | At3g01020 | Scaffold | Léon *et al.* (2005) [74] |
| | ISU3 | At4g04080 | Scaffold | Frazzon *et al.* (2007) [60] |
| | ISA1 | At2g16710 | Cluster transfer | Uncharacterized |
| | ISA2 | At2g36260 | Cluster transfer | Uncharacterized |
| | ISA3 | At5g03905 | Cluster transfer | Uncharacterized |
| | NFU4 | At3g20970 | Scaffold | Léon *et al.* (2003) [48] |
| | NFU5 | At1g51390 | Scaffold | Léon *et al.* (2003) [48] |
| | ADX1 | At4g21090 | Electron transfer | Takubo *et al.* (2003) [75] |
| | ADX2 | At4g05450 | Electron transfer | Picciocchi *et al.* (2003) [76] |
| | ADXR | At4g32360 | Electron transfer | Takubo *et al.* (2003) [75] |
| | FH | At4g03240 | Iron donor | Busi *et al.* (2006) [67] |
| | HSCA1 | At4g37910 | HSP70-type chaperone | Xu *et al.* (2009) [77] |
| | HSCA2 | At5g09590 | HSP70-type chaperone | Uncharacterized |
| | HSCB | At5g06410 | Co-chaperone | Xu *et al.* (2009) [77] |
| | INDL | At4g19540 | Cluster transfer/ insertion | Bych *et al.* (2008) [81] |
| | IBA57 | At4g12130 | Cluster transfer | Waller *et al.* (2010) [78] |
| | GRXS15 | At3g15660 | Glutaredoxin | Bandyopadhyay *et al.* (2008) [57] |
| | ATM3 | At5g58270 | ABC transporter | Bernard *et al.* (2009) [86] |
| | ERV1 | At1g49880 | Sulfhydryl oxidase | Levitan *et al.* (2004) [90] |
| Cytosol | TAH18 | At3g02280 | Electron transfer | Varadarajan *et al.* (2010) [94] |
| | DRE2 | At5g18400 | Electron transfer | Varadarajan *et al.* (2010) [94] |
| | NBP35 | At5g50960 | Scaffold | Bych *et al.* (2008) [81] |
| | NAR1 | At4g16440 | [FeFe]-hydrogenase-like | Cavazza C.*et al.* (2008) [95] |
| | CIA1 | At2g26060 | WD40 protein | Srinivasan *et al.* (2007) [97] |
| | CIA2 | At1g68310 | DUF59 domain | Luo *et al.* (2012) [89] |
| | MMS19 | At5g48120 | Interacting with CIA | Stehling *et al.* (2012) [99] |

The proposed function and subcellular localization for each protein is as indicated.

assembly machinery in chloroplasts/plastids is termed as SUF (mobilization of sulfur) system, whereas mitochondrial Fe-S assembly is termed as ISC (iron-sulfur cluster) system, and the Fe-S biosynthesis in cytosol is referred to as the CIA system (cytosolic iron-sulfur assembly).

### 22.3.1 SUF system in plastids

The chloroplast/plastid SUF machinery of *Arabidopsis* consists of at least 14 proteins (Fig. 22.1 and Tab. 22.1), which are NFS2, SUFE1, SUFE2, SUFE3, SUFA, NFU1, NFU2, NFU3, SUFB, SUFC, SUFD, HCF101, GRXS14, and GRXS16.

NFS2 (nitrogen fixation, S protein) is the cysteine desulfurase that provides sulfur for Fe-S cluster assembly [36–38]. In isolated form, NFS2 is a dual function enzyme,

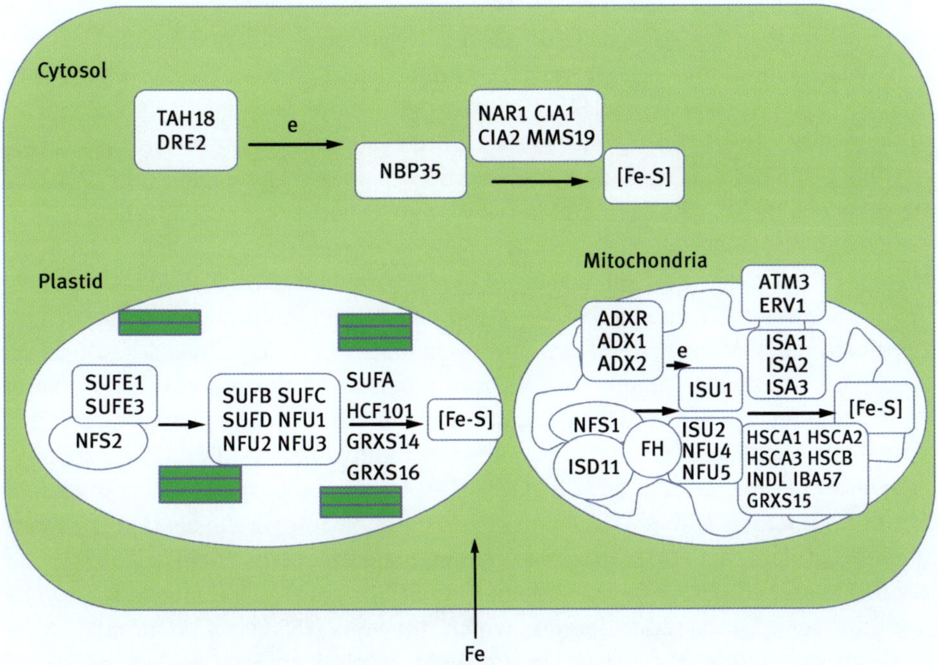


**Fig. 22.1:** Working model of the Fe-S cluster biosynthesis in a plant cell. About 40 genes have been confirmed or estimated to be involved in the Fe-S cluster assembly in plant cells. All these genes are nuclear genome encoded. The expressed proteins are translocated to plastids, mitochondria, and cytosol, where Fe-S clusters are assembled separately. The reaction is initiated by the donation of sulfur and iron, which are assembled into an Fe-S cluster on a major scaffold. Facilitated by multiple delivery or carrier proteins, the preassembled Fe-S cluster is delivered to target proteins, resulting in the maturation of Fe-S proteins. In each of the reactions, the arrows from left to right indicate the upstream to downstream steps in the Fe-S assembly pathway.

with both cysteine desulfurase (CysD) activity and selenocysteine lyase (SL) activity. The CysD activity catalyzes the removal of a sulfur from cysteine to produce sulfur and alanine, whereas its SL activity catalyzes the lysis of selenocysteine to produce selenide and alanine. The two enzymatic activities use the same core active site, but the CysD activity requires an additional cysteine residue that is not required for the SL activity. Homozygous, knockout deletions of NFS2 in plants by T-DNA insertion have not been reported, consistent with a crucial role of NFS2. RNAi knockdown plants of NFS2 exhibit severe phenotypes including chlorosis, chloroplast disorganization, leaf necrosis, and early death; concomitantly, various chloroplast Fe-S proteins are decreased in both abundance and activity. In contrast, mitochondrial Fe-S proteins and respiration remain unaffected, suggesting that Fe-S assembly is independent in mitochondria and chloroplasts [39]. Overexpression of NFS2 enhances selenium and sulfur accumulation in plants and increases tolerance to selenate exposure [36, 40], suggesting that NFS2 might be a part of a system to prevent selenium toxicity in plants. The physiological relevance of the SL enzyme activity may not be essential to plants because some plant species do not require any selenium. In contrast, the CysD activity is essential to plants. However, when measured *in vitro*, the CysD activity of NFS2 was much weaker than its SL activity [36, 38]. This observation suggested that an additional factor was required in planta for the CysD activity. The subsequent discovery that SUFE proteins strongly stimulated NFS2 provided a possible explanation for the assay results [41, 42]. The *Arabidopsis* genome encodes three SUFE genes, SUFE1, SUFE2, and SUFE3. As demonstrated by *in vitro* biochemical assays, each of the SUFE proteins can stimulate the CysD activity of the NFS2 protein 40- to 70-fold and increase Fe-S cluster assembly at least 20-fold. NFS2 and SUFE form a protein complex and together function as the cysteine desulfurase in plant chloroplasts/plastids [41, 42].

SUFE1 is a fusion protein with an N-terminal SUFE domain and a C-terminal BOLA domain [42, 43]. The SUFE domain is responsible for CysD-stimulating activity, whereas the function of the BOLA domain is currently unknown. Bioinformatics analysis and experimental results for BOLA-like proteins in other species have suggested that BOLA interacts with glutaredoxin [44–47]. Hence, it is very likely that the C-terminal BOLA domain of SUFE1 may interact with glutaredoxins in chloroplasts, such as GRXS14 and GRXS16. The SUFE domain of SUFE1 physically interacts with NFS2 to form a hetero-tetrameric complex, which increases NFS2 enzyme affinity for its substrate, cysteine, and increases Vmax 40-fold. T-DNA insertion knockout of SUFE1, a method for generating loss-of-function mutations in plants, is embryonic lethal, suggesting that the essential role of SUFE1 cannot be substituted by other SUFE paralogues. SUFE2 and SUFE3 share a SUFE domain with SUFE1. They also stimulate the CysD activity of NFS2 in a similar manner [41]. Because SUFE2, which consists solely of the SUFE region, is specifically expressed in flowers, especially in pollen, it is possible that SUFE2 may specifically enhance Fe-S cluster biosynthesis in pollen, the male gametophyte of flowering plants. SUFE3 is another fusion protein with an N-terminal SUFE domain and C-terminal NadA domain. NadA is a quinolinate synthase, required for NAD biosynthesis, and it is a [4Fe-4S] protein. SUFE3 stimulates CysD activity and enhances Fe-S cluster assembly and delivery to the NadA domain of SUFE3. The resulting holo-SUFE3 is the NadA enzyme of plants, involved in a critical step of NAD biosynthesis. Knockout of SUFE3 in *Arabidopsis* is embryonic lethal.

NFU1, NFU2, and NFU3 (nitrogen fixation, U protein) are scaffold proteins [48–50]. Their protein sequences consist of an N-terminal NFU domain and a C-terminal B domain, whereas mitochondrial NFU4 and NFU5 proteins consist of an N-terminal N domain and a C-terminal NFU domain. These NFU proteins share a conserved CXXC motif in the NFU domain. They are capable of assembling Fe-S clusters and delivering them to target proteins. Plants in which NFU2 is inactivated exhibit a dwarf phenotype with pale-green leaves, decreased activity of [4Fe-4S] sulfite reductase, and drastically impaired photosystem I (PSI) and ferredoxin accumulation in chloroplasts [49, 50]. However, glutamate synthase and Rieske Fe-S proteins are not affected in the mutants, indicating that the NFU2 scaffold is responsible for a subset of target proteins. SUFB, SUFC, and SUFD form a complex and together function as a multimolecular scaffold [51, 52]. The SUFBCD complex can assemble Fe-S clusters *in vitro* and transfer them to target proteins. SUFC is the subunit that possesses ABC/ATPase activity. SUFC knockout mutant plants are embryonic lethal, and mutant embryos contain abnormal plastids with disorganized thylakoid structures, which are membrane-bound compartments inside chloroplasts. SUFA is an alternative scaffold that probably functions in the delivery of Fe-S clusters [53, 54]. In an *in vitro* reconstitution experiment, SUFA was able to acquire a transient [2Fe-2S] cluster. The holo-SUFA was stable and could be purified by chromatography. When incubated with ferredoxin, a target protein, SUFA could donate its [2Fe-2S] cluster to generate holo-ferredoxin [53].

HCF101 (high chlorophyll fluorescence 101) is an alternative scaffold that functions in Fe-S cluster targeting [55, 56]. It is specifically required for the biogenesis of [4Fe-4S] proteins, such as PSI and ferredoxin-thioredoxin reductase (FTR) in chloroplasts, but not for [2Fe-2S] proteins such as ferredoxin. In the HCF101-deficient mutant plants, both the activity and protein level of [4Fe-4S] proteins are reduced. HCF101 belongs to the [4Fe-4S] cluster-containing P-loop NTPase superfamily. HCF101 protein binds a [4Fe-4S] cluster through ligation to three HCF101-specific cysteine residues. The reconstituted cluster can be transferred from HCF101 to a [4Fe-4S] apo-protein [56]. GRXS14 and GRXS16 possess the canonical CGFS active site characteristic of monothiol glutaredoxins. They function as alternative scaffold proteins [57, 58] that can bind [2Fe-2S] clusters transiently, before rapidly and quantitatively transferring them to the apoprotein form of ferredoxin in chloroplasts. A recent structural and biochemical study has identified that the *Arabidopsis* GRXS16 is a two-domain protein [59]. Its N-terminal domain has DNA endonuclease activity, whereas the C-terminal GRX domain has redox activity and Fe-S cluster biosynthetic activity. GRXS16 may link Fe-S cluster assembly, iron homeostasis, the ROS response, and DNA metabolism in chloroplasts of higher plants. Moreover, these two plastid glutaredoxins, GRXS14 and GRXS16, are potential candidate proteins that physically interact with the BOLA domain of SUFE1 (Fig. 22.1).

### 22.3.2 ISC system in mitochondria

The mitochondrial ISC machinery of *Arabidopsis* consists of 22 proteins (Fig. 22.1, Tab. 22.1). They are NFS1, ISD11, ISU1, ISU2, ISU3, ISA1, ISA2, ISA3, NFU4, NFU5, ADX1, ADX2, ADXR, FH, HSCA1, HSCA2, HSCB, INDL, IBA57, GRXS15, ATM3, and ERV1.

NFS1 is the cysteine desulfurase [60, 61], and an accessory protein, ISD11, is required to stimulate the cysteine desulfurase activity of NFS1 enzyme [62]. The NFS1-ISD11 complex forms persulfide. It has been suggested that ISD11 induces a conformational change in NFS1 and brings the substrate and the active site cysteine in proximity to form persulfide. Hence, ISD11 could regulate cysteine desulfurase activity of NFS1 enzyme [63]. Frataxin (FH) is a potential iron donor for Fe-S cluster assembly [64–67]. Numerous studies have suggested that frataxin may function as an iron-binding chaperone for Fe-S cluster assembly, particularly in humans [68–72]. However, the function of the plant frataxin orthologue, FH, remains elusive and reports on FH roles are quite variable. Meanwhile, FH is thought to be an iron chaperone that delivers iron for Fe-S clusters. FH insertion mutant plants exhibit decreased activity of mitochondrial Fe-S proteins, such as aconitase and succinate dehydrogenase. In addition, FH mutant plants also manifest alteration in transcripts from the heme biosynthesis pathway, diminished total heme content, and deficient catalase activity. Thus, FH also plays a role in heme biosynthesis of higher plants [66]. FH is involved in energy conversion and oxidative phosphorylation, iron handling, and response to oxidative stress. FH may have other functions in different tissues and in different organisms [73].

ISU1 is the central scaffold protein in mitochondria. As paralogues to ISU1, ISU2 and ISU3 are alternative scaffolds [74]. Among all three ISU family proteins, ISU1 has the highest physiological importance. In agreement with this, ISU1 is constitutively expressed at constant levels, whereas ISU2 and ISU3 expression is minimal or nondetectable. None of the ISU1–ISU3 deletion mutant plants seem to be lethal, probably suggesting that ISU1–ISU3 functions are redundant [60]. It is possible that ISU1 is the central housekeeping scaffold in plants and is constantly expressed at high levels. In normal plant cells, expression of the other ISU paralogues is repressed. However, in the absence of ISU1, expression of the other ISU paralogues is induced, and the increased ISU2 and ISU3 proteins could substitute for the function of ISU1. Adrenodoxins ADX1 and ADX2 and adrenodoxin reductase, ADXR, together form the reducing system that provides electrons for *de novo* Fe-S cluster assembly in the central scaffold ISU1. In addition to Fe-S cluster assembly, the ADX-ADXR low-potential electron-transfer chain transfers electrons from NADH to cytochrome *c* [75] and also reconstitutes a functional plant biotin synthase complex [76]. The transcript levels of ADX and ADXR are high in flowers, whereas the protein level of ADXR is high in the leaf, stem, and flower. HSCA1/HSCA2 and HSCB are the chaperones and co-chaperone, respectively, that work together in the delivery of Fe-S clusters from the ISU1 scaffold to target apo-proteins [21]. The preassembled Fe-S cluster on ISU1 needs to be labilized by the dedicated chaperone system through sequential protein-protein interactions.

A recent study has reported that HSCA1 and HSCB are localized to both mitochondria and cytosol in *A. thaliana* [77]. It would be interesting to see whether some fraction of ISU1 is also localized in cytosol because ISU1 is the physiologically relevant partner for the chaperone system. HSCB is highly expressed in anthers, the part of the stamen where pollen is produced, and trichomes, cells that project from the surface of the epidermis. The HSCB insertion mutant shows reduced seed production, reduced wax production, which is needed to protect the plant surface in a terrestrial environment, inappropriate trichome development, and dramatically reduced activities of Fe-S enzymes including aconitase and succinate dehydrogenase [77].

GRXS15, a monothiol glutaredoxin in mitochondria, is an alternative scaffold. GRXS15 helps to transfer the Fe-S cluster from ISU1 to apo-proteins, likely *via* transient binding of the Fe-S cluster in a glutathione-GRXS15 complex [57]. NFU4 and NFU5 are alternative scaffolds, and their mitochondrial localization is confirmed [48]. Their function is likely to deliver Fe-S clusters to a subset of target apo-proteins, which remain to be identified. ISA1–ISA3 are putative Fe-S cluster delivery proteins, which are not yet characterized in plants. IBA57 is a tetrahydrofolate-binding protein. *In vivo* evidence indicates that IBA57 is required for the activity of MiaB, a [4Fe-4S]-containing tRNA modification enzyme [78]. The expression of IBA57 is high in flowers, long narrow seed-pods known as siliques, and seeds. Insertional mutagenesis of IBA57 is embryonic lethal. ISA1, ISA2, ISA3, and IBA57 may form a protein complex that functions in the targeting of [4Fe-4S] clusters to specific mitochondrial apo-proteins, including aconitase and radical SAM enzymes [79, 80]. INDL is an assembly factor that may be specific to complex I of mitochondria. INDL may bind and transfer a [4Fe-4S] cluster to acceptor Fe-S proteins in complex I [81]. The molecular details of INDL function remain to be characterized in plants.

ATM3 is an ABC transporter located in the inner membrane of mitochondria [82–85]. It is thought to export a substance from the matrix of mitochondria to cytosol, but molecular details on functions of this transporter and its substrate(s) are unknown so far. However, several lines of evidence have strongly indicated that ATM3 is not required for the Fe-S biosynthesis in mitochondria but is rather needed for Fe-S assembly in cytosol [86–88]. In the absence of ATM3, cytosolic Fe-S biosynthesis is defective. The substrate(s) exported by ATM3 are thought to be a component used for the Fe-S cluster assembly in cytosol, which, however, remains to be characterized. In addition to a function in the maturation of extramitochondrial Fe-S proteins, ATM3 has a crucial role in the molybdenum cofactor (Moco) biosynthesis. It has been observed that cyclic pyranopterin monophosphate (cPMP), an intermediate of Moco biosynthesis, accumulates in the mitochondria of ATM3 mutant plants (Moco synthesis is discussed in greater detail in Chapter 19 by Leimkühler in this volume). It is therefore reasonable to presume that ATM3 has a function in the export of cPMP from mitochondria to cytosol [87]. An ATM3 knockout mutant (*Sta1) Arabidopsis* demonstrates dwarfism and chlorosis phenotype. It also shows altered morphology of leaf and cell nuclei [88]. In a comprehensive study using an allelic mutant series of ATM3, the activity of a cytosolic

Fe-S protein aconitase was strongly decreased, whereas mitochondrial and plastid Fe-S proteins were not affected. In contrast to ATM3, orthologous mutants in yeast and mammals accumulate iron in mitochondria, whereas the plant ATM3 mutants do not display a dramatic iron homeostasis defect and do not accumulate iron in mitochondria [86]. The difference between plants and other eukaryotes concerning the effects of ATM3 mutation on cellular iron homeostasis may be because of the presence of chloroplasts in plant cells, the organelle known to be more important in acquiring cellular iron. Recently, mutagenesis of ATM3 was found to result in defective genome integrity in the plant, *A. thaliana* [89]. These data support that ATM3 function is required for the maturation of Fe-S proteins in cytosol and nuclei, many of which are the genome DNA repair enzymes. ERV1 is a redox-active and FAD-containing sulfhydryl oxidase. It is located in the intermembrane space of mitochondria. Similar to orthologues in yeast and animal cells, ERV1 is required for the maturation of Fe-S proteins in cytosol. *In vitro* assays with purified protein and artificial substrates demonstrate a preference of ERV1 for dithiols that have a defined space between the thiol groups, suggesting that their substrate is thioredoxin-like [90].

### 22.3.3 CIA system in cytosol

The cytosolic CIA machinery of *Arabidopsis* consists of seven proteins (Fig. 22.1 and Tab. 22.1), involving TAH18, DRE2, NBP35, NAR1, CIA1, CIA2, and MMS19. The nuclear genome of *Arabidopsis* encodes only two cysteine desulfurases, NFS1 and NFS2. They are localized in the mitochondria and plastids, respectively, led by the targeting peptide in each protein [36, 37, 60]. There is not an exclusively cytosolic cysteine desulfurase present in plants, which is also true for other eukaryotes. The sulfur source for the Fe-S cluster assembly in cytosol is not clear. However, the sulfur for Fe-S assembly in cytosol might be provided by two mechanisms. One possibility is that ATM3 may export the sulfur to the cytosol for Fe-S assembly. The second possible mechanism is to generate a cytosolic cysteine desulfurase by alternative initiation site utilization of the mitochondrial NFS1 mRNA, which is believed the case in humans [34, 91] but has never been reported in plants.

NBP35 is the scaffold for Fe-S cluster assembly in cytosol of plants [92]. It is constitutively expressed in planta. NBP35 binds a [4Fe-4S] cluster in the C terminus and a stable [4Fe-4S] cluster in the N terminus. Holo-NBP35 is able to transfer an Fe-S cluster to an apo-protein *in vitro*. NBP35 mutant plants show an arrest of embryo development [92]. NBP35-deficient *Arabidopsis* mutants are seedling lethal, suggesting an essential function [93]. TAH18 is a diflavin reductase, which is able to reduce cytochrome *c* *in vitro*. It forms a protein complex with DRE2, and these two proteins together provide electrons for Fe-S assembly on the NBP35 scaffold [94]. Loss of TAH18 function results in early embryonic lethality in plants. TAH18 also has possible roles in the control of cell death and chromosomal segregation in mitosis [94]. Consistently, the promoter of

TAH18 is activated during cell cycle progression, probably because many DNA replication and DNA repair proteins are Fe-S proteins, and increased expression of TAH18 is required to help synthesize more Fe-S clusters for the DNA metabolism proteins during cell cycle progression.

NAR1 is an [FeFe]-hydrogenase-like gene. The recombinant NAR1 protein coordinates two Fe-S clusters. NAR1 knockdown results in a dwarf phenotype under normoxia, but the phenotype is indistinguishable from wild type under hypoxic conditions. This result suggests that the exposure to low oxygen might alleviate the severity of the phenotype caused by NAR1 knockdown [95]. Another study reports that the expression of some hypoxia-responsive genes, including ferredoxin (Fdx5) and two [FeFe]-hydrogenases, HydA1 and HydA2, could be upregulated by hypoxia treatment in both NAR1 knockdown and wild-type plants. This result indicates that NAR1 is not involved in regulating the hypoxia response of green algae [96]. The activity of cytosolic Fe-S enzymes, including aldehyde oxidase and xanthine dehydrogenase, is decreased in NAR1 knockdown lines, whereas Fe-S protein activities in the chloroplast and mitochondria are unaffected. These results indicate that NAR1 is required for the activity of cytosolic Fe-S enzymes but not for Fe-S proteins in chloroplasts or mitochondria [96].

CIA1 is a WD40-repeat protein, an essential and conserved member of the Fe-S assembly CIA machinery. The structure of CIA1 protein folds into a β propeller with seven blades arranged around a central axis. The conserved top surface residue R127 performs a critical function in cytosolic Fe-S protein assembly [97]. CIA2, also named AE7, is a DUF59 family gene that acts in the CIA Fe-S cluster assembly pathway to maintain nuclear genome integrity [89]. CIA2 mutations result in lower activities of the cytosolic [4Fe-4S] enzyme aconitase and the nuclear [4Fe-4S] enzyme DNA glycosylase, ROS1. The severe CIA2 mutant allele is embryonic lethal, whereas the weaker CIA2 mutant alleles are viable but exhibit highly increased DNA damage, which results in cell cycle arrest through the DNA damage response pathway [89].

MMS19, a conserved HEAT repeat protein, is a component of the CIA pathway for Fe-S cluster assembly and is implicated in DNA repair. MMS19 is encoded in the genome of plants, for instance, in *Arabidopsis*, but it has not yet been fully characterized in plants. Studies on its orthologues in mammalian cells indicate that MMS19 is a member of the cytosolic Fe-S protein assembly (CIA) machinery [98–100]. MMS19 functions as part of the CIA-targeting complex that specifically facilitates Fe-S cluster insertion into apo-proteins involved in methionine biosynthesis, DNA replication, DNA repair, and telomere maintenance. However, MMS19 is not required for targeting Fe-S clusters to other cytosolic apo-proteins, such as aconitase/IRP1 and glutamine amidotransferase (GPAT) [99]. By co-immunoprecipitation and mass spectrometry analysis, 12 known Fe-S proteins have been identified in the MMS19 complex, including XPD, FANCJ, DNA polymerase D, DNA primase, and DNA2. Consistent with an essential role of MMS19 in DNA metabolism, MMS19 knockout is embryonically lethal in mice [100]. As in other eukaryotic cells, NAR1, CIA1, CIA2, and MMS19 proteins form a stable complex in plant cells [89]. This protein complex may be termed as "CIA-targeting complex," which physically interacts with and transfers Fe-S clusters to various DNA metabolism Fe-S proteins.

## 22.4 Regulation of cellular iron homeostasis by Fe-S cluster biosynthesis

In mammals, cellular regulation of iron metabolism involves the cytosolic Fe-S protein IRP1, providing a tight link between Fe-S protein biogenesis and mitochondrial iron homeostasis [101, 102]. Similarly, cellular iron homeostasis in yeast cells is also tightly linked to mitochondrial ISC machinery [21, 28]. The mitochondrial inner membrane transporter ATM3 homologues export an unknown signal "X" into cytosol. The signal reflects iron status in mitochondria, and it affects CIA activity in cytosol. In the deficiency of signal "X," iron overaccumulates in mitochondria, and cellular iron homeostasis is impaired [33, 34].

Except for the plastid SUF system that is unique to plants, the mitochondrial ISC and cytosolic CIA machineries in plant cells are highly homologous to those in yeast and mammalian cells. However, to date, there is no strong evidence of an association between mitochondrial Fe-S assembly and cellular iron homeostasis in plants. Defects in mitochondrial ISC machinery or plant ATM3 do not cause iron accumulation in mitochondria of plant cells, although they decrease Fe-S protein activity in cytosol [86, 87]. The reason for nonaccumulation of iron in the mitochondria of ATM3 mutant plant cells may be because chloroplasts are also involved in cellular iron homeostasis or that the chloroplast rather than mitochondrion may develop iron homeostasis problems. Indeed, it has been reported in an ATM3 mutant (*Sta1*) plant that the mutant shows a chlorosis phenotype [88], a classical indication of iron deficiency in plants. Although chloroplasts have normal levels of iron, the mutant plant cells cannot properly utilize iron. The interplay among mitochondria ISC, plastids SUF, and cellular iron homeostasis in plants needs to be further elucidated.

## 22.5 Conservation of Fe-S cluster assembly genes across the green lineage

The green lineage is a diverse group of photosynthetic organisms, including unicellular green algae, mosses, ferns, and seed plants. The evolution has resulted in increasing levels of complexity, from the simplest green algae to the more complex higher plants. As an essential biological pathway, the Fe-S cluster biosynthetic pathway is conserved with the evolution of plants. Comparative genomic analysis of Fe-S cluster biosynthesis genes could reveal valuable information. To date, the genomes of over 40 plant and photosynthetic species have been sequenced, and gene sequences are available in the Phytozome and NCBI databases. In this chapter, five representative plant and photosynthetic species have been selected for a comparison of Fe-S biosynthesis genes, including green algae (*Chlamydomonas reinhardtii*), rice (*O. sativa*), tomato (*Solanum lycopersicum*), *Arabidopsis* (*A. thaliana*), and soybean (*Glycine max*).

Among all Fe-S synthesis genes (Tab. 22.1), the majority are conserved in the five species compared and probably in all plants. Only eight genes are not conserved in all five species (Tab. 22.2), including SUFE2, NFU3, ISU2, ISU3, ISA2, NFU5, ADX2, and MMS19. Lack of conservation suggests that these genes are not essential for all plants. It may be that the losses of some non-conserved genes are compensated by paralogous genes. Indeed, except for SUFE2 and MMS19, all other non-conserved genes belong to a gene family, in which all members have similar functions and can compensate for each other. SUFE1–SUFE3 share the SUFE domain, but they are not in the same gene family. The presence or absence of a BOLA or NadA domain makes them work differently. MMS19 is the only gene without a paralogue that is not conserved in all five species. MMS19 is encoded in the genome of all four multicellular plants including monocots and dicots, but it is not present in the unicellular algae, *Chlamydomonas* (Tab. 22.2). MMS19 is a central component of the NAR1-CIA1-CIA2-MMS19 complex in mammalian cytosol required for targeting Fe-S clusters to apoproteins involved in DNA replication and DNA repair [89, 99, 100]. Hence, MMS19 may be essential for the maintenance of genome integrity in multicellular plants. Unicellular green algae may employ a MMS19-independent mechanism for DNA repair. Most of the nonconserved genes are scaffolds, including NFU3, ISU2, ISU3, ISA2, and NFU5. An Fe-S biosynthesis machinery typically contains multiple scaffolds for delivering the preassembled [Fe-S] clusters to various target proteins. The

**Tab. 22.2:** The Fe-S biosynthesis genes that are not conserved across green lineage using five species as examples.

| | *Chlamydomonas* | Rice | Tomato | *Arabidopsis* | Soybean |
|---|---|---|---|---|---|
| Plastid | | | | | |
| SUFE2 | – | – | + | + | + |
| NFU3 | – | + | + | + | + |
| Mitochondria | | | | | |
| ISU2 | – | + | + | + | + |
| ISU3 | – | – | – | + | + |
| ISA2 | – | + | + | + | + |
| NFU5 | – | – | – | + | – |
| ADX2 | – | + | + | + | + |
| Cytosol | | | | | |
| MMS19 | – | + | + | + | + |

–, Gene is not encoded in the genome of the specific species; +, gene is encoded in the species.

least important nonconserved gene is NFU5, as it is encoded only in the genome of *Arabidopsis* among the five species compared here (Tab. 22.2). NFU5 and all other four NFU genes have been identified and characterized in *Arabidopsis*. NFU5 is an alternative scaffold in mitochondria of *Arabidopsis*.

The SUFE2 gene is present in all three dicot plants but not in the monocot rice (Tab. 22.2). Interestingly, more systematic phylogenetic analysis reveals that the SUFE2 gene is encoded in genomes of the majority of dicot plants (>90% of sequenced dicot species). In contrast, it is not present in all six monocot plant genomes, including *Zea mays, Sorghum bicolor, O. sativa, Brachypodium distachyon, Setaria italica, Panicum virgatum*. It seems that SUFE2 is perhaps a dicot-specific gene. As SUFE2 expression is flower-specific and high in pollen, SUFE2 has been thought to specifically function in pollen Fe-S cluster biosynthesis [41]. Hence, it is speculated that SUFE2 may be important for Fe-S assembly in pollens of dicots, but it is not required for Fe-S assembly in pollens of monocots.

## 22.6 Potential significance to agriculture

Gene mutations in Fe-S cluster biosynthesis are known to cause many severe diseases and even death to humans [26, 33, 34]. The research on mammalian Fe-S biosynthesis is of great medical importance. Gene mutations in Fe-S biosynthesis do not necessarily cause diseases in plants, but the resulting severe phenotypes, such as dwarf, chlorosis, and even embryonic lethality, highlight that Fe-S cluster assembly is an essential and basic biological pathway. Research on plant Fe-S biosynthesis is of potential significance to agriculture, based on the following reasons. First, iron is an important mineral nutrient to humans, and iron deficiency causes anemia and disorders. The majority of the world's population acquires iron nutrition from plant foods (crops, grains, vegetables, fruits), and the iron contents in plant foods therefore affect human health. Further research on plant Fe-S cluster biosynthesis could help us to identify central gene(s) in regulating cellular iron metabolism, which could be used in genetic engineering to produce iron-fortified crops. Second, Fe-S proteins are involved in the assimilation of nitrogen, sulfur, molybdenum, and iron nutrition [24, 25, 31]. These minerals are components of fertilizers widely used in agriculture. The Fe-S protein biogenesis in crops would affect the efficiency in the uptake and utilization of mineral nutrients from fertilizers. Excellent gene alleles in the Fe-S protein biogenesis could be used as selection markers in molecular breeding to achieve high efficiency of crops in the uptake and utilization of mineral nutrients, and reduction in the use of fertilizers, which would be good for both agriculture and environmental protection. Third, toxic heavy metal pollution in soil is increasingly a problem to agriculture. Some of the Fe-S cluster assembly genes are already known to have functions in improving plant tolerance to heavy metals. For example, the overexpression of NFS2 in plants results in higher tolerance to toxic levels of selenium [40], whereas increased expression of ATM3 increases the tolerance of plants to cadmium [103]. Better understanding of plant Fe-S cluster assembly could help us to develop strategies to reduce toxic heavy metal accumulation in our foods.

## Acknowledgments

The author thanks Dr Marinus Pilon (Colorado State University) for the critical reading of the manuscript, and thanks also to Lu Qin and Xuejiao Liang for making the figure and tables. The work in the author's laboratory is supported by grants from 100-Talent Program of Chinese Academy of Sciences, National Basic Research Program of China (973 Program 2013CB127102), Foundation of Key Laboratory of Plant Resources Conservation, and Sustainable Utilization, South China Botanical Garden, Chinese Academy of Sciences.

## References

[1] Palmer CM, Guerinot ML. Facing the challenges of Cu, Fe and Zn homeostasis in plants. Nat Chem Biol 2009;5:333–40.
[2] Kobayashi T, Nishizawa NK. Iron uptake, translocation, and regulation in higher plants. Annu Rev Plant Biol 2012;63:131–52.
[3] Morrissey J, Guerinot ML. Iron uptake and transport in plants: the good, the bad, and the ionome. Chem Rev 2009;109:4553–67.
[4] Jeong J, Guerinot ML. Homing in on iron homeostasis in plants. Trends Plant Sci 2009;14:280–5.
[5] Grotz N, Guerinot ML. Molecular aspects of Cu, Fe and Zn homeostasis in plants. Biochim Biophys Acta 2006;1763:595–608.
[6] Colangelo EP, Guerinot ML. Put the metal to the petal: metal uptake and transport throughout plants. Curr Opinion Plant Biol 2006;9:322–30.
[7] Connolly EL, Campbell NH, Grotz N, Prichard CL, Guerinot ML. Overexpression of the FRO2 ferric chelate reductase confers tolerance to growth on low iron and uncovers posttranscriptional control. Plant Physiol 2003;133:1102–10.
[8] Vasconcelos M, Eckert H, Arahana V, Graef G, Grusak MA, Clemente T. Molecular and phenotypic characterization of transgenic soybean expressing the Arabidopsis ferric chelate reductase gene, FRO2. Planta 2006;224:1116–28.
[9] Connolly EL, Fett JP, Guerinot ML. Expression of the IRT1 metal transporter is controlled by metals at the levels of transcript and protein accumulation. Plant Cell 2002;14:1347–57.
[10] Korshunova YO, Eide D, Clark WG, Guerinot ML, Pakrasi HB. The IRT1 protein from Arabidopsis thaliana is a metal transporter with a broad substrate range. Plant Mol Biol 1999;40:37–44.
[11] Vert G, Grotz N, Dedaldechamp F, et al. IRT1, an Arabidopsis transporter essential for iron uptake from the soil and for plant growth. Plant Cell 2002;14:1223–33.
[12] Ishimaru Y, Suzuki M, Tsukamoto T, et al. Rice plants take up iron as an Fe3+-phytosiderophore and as Fe2+. Plant J Cell Mol Biol 2006;45:335–46.
[13] Bashir K, Ishimaru Y, Shimo H, et al. The rice mitochondrial iron transporter is essential for plant growth. Nat Commun 2011;2:322.
[14] Duy D, Stube R, Wanner G, Philippar K. The chloroplast permease PIC1 regulates plant growth and development by directing homeostasis and transport of iron. Plant Physiol 2011;155:1709–22.

[15] Duy D, Wanner G, Meda AR, von Wiren N, Soll J, Philippar K. PIC1, an ancient permease in Arabidopsis chloroplasts, mediates iron transport. Plant Cell 2007;19:986–1006.
[16] Divol F, Couch D, Conejero G, Roschzttardtz H, Mari S, Curie C. The Arabidopsis yellow stripe like4 and 6 transporters control iron release from the chloroplast. Plant Cell 2013;25:1040–55.
[17] Briat JF. Plant ferritin and human iron deficiency. Nat Biotechnol 1999;17:621.
[18] Petit JM, van Wuytswinkel O, Briat JF, Lobreaux S. Characterization of an iron-dependent regulatory sequence involved in the transcriptional control of AtFer1 and ZmFer1 plant ferritin genes by iron. J Biol Chem 2001;276:5584–90.
[19] Briat JF, Duc C, Ravet K, Gaymard F. Ferritins and iron storage in plants. Biochim Biophys Acta 2010;1800:806–14.
[20] Leidgens S, Bullough KZ, Shi H, et al. Each member of the PCBP family exhibits iron chaperone activity toward ferritin. J Biol Chem 2013;288:17791–802.
[21] Lill R, Hoffmann B, Molik S, et al. The role of mitochondria in cellular iron-sulfur protein biogenesis and iron metabolism. Biochim Biophys Acta 2012;1823:1491–508.
[22] Kim SA, Punshon T, Lanzirotti A, et al. Localization of iron in Arabidopsis seed requires the vacuolar membrane transporter VIT1. Science 2006;314:1295–8.
[23] Zhang Y, Xu YH, Yi HY, Gong JM. Vacuolar membrane transporters OsVIT1 and OsVIT2 modulate iron translocation between flag leaves and seeds in rice. Plant J Cell Mol Biol 2012;72:400–10.
[24] Balk J, Pilon M. Ancient and essential: the assembly of iron-sulfur clusters in plants. Trends Plant Sci 2011;16:218–26.
[25] Ye H, Pilon M, Pilon-Smits EA. CpNifS-dependent iron-sulfur cluster biogenesis in chloroplasts. New Phytol 2006;171:285–92.
[26] Rouault TA, Tong WH. Iron-sulphur cluster biogenesis and mitochondrial iron homeostasis. Nat Rev Mol Cell Biol 2005;6:345–51.
[27] Johnson DC, Dean DR, Smith AD, Johnson MK. Structure, function, and formation of biological iron-sulfur clusters. Annu Rev Biochem 2005;74:247–81.
[28] Lill R, Muhlenhoff U. Maturation of iron-sulfur proteins in eukaryotes: mechanisms, connected processes, and diseases. Annu Rev Biochem 2008;77:669–700.
[29] Frazzon J, Dean DR. Formation of iron-sulfur clusters in bacteria: an emerging field in bioinorganic chemistry. Curr Opin Chem Biol 2003;7:166–73.
[30] Frazzon J, Fick JR, Dean DR. Biosynthesis of iron-sulphur clusters is a complex and highly conserved process. Biochem Soc Trans 2002;30:680–5.
[31] Balk J, Lobreaux S. Biogenesis of iron-sulfur proteins in plants. Trends Plant Sci 2005;10:324–31.
[32] Pilon M, Abdel-Ghany SE, Van Hoewyk D, Ye H, Pilon-Smits EA. Biogenesis of iron-sulfur cluster proteins in plastids. Genet Eng 2006;27:101–17.
[33] Ye H, Rouault TA. Human iron-sulfur cluster assembly, cellular iron homeostasis, and disease. Biochemistry 2010;49:4945–56.
[34] Rouault TA. Biogenesis of iron-sulfur clusters in mammalian cells: new insights and relevance to human disease. Dis Model Mech 2012;5:155–64.
[35] Xu XM, Moller SG. Iron-sulfur clusters: biogenesis, molecular mechanisms, and their functional significance. Antioxid Redox Signal 2011;15:271–307.
[36] Pilon-Smits EA, Garifullina GF, Abdel-Ghany S, et al. Characterization of a NifS-like chloroplast protein from Arabidopsis. Implications for its role in sulfur and selenium metabolism. Plant Physiol 2002;130:1309–18.
[37] Léon S, Touraine B, Briat JF, Lobreaux S. The AtNFS2 gene from Arabidopsis thaliana encodes a NifS-like plastidial cysteine desulphurase. Biochem J 2002;366:557–64.
[38] Ye H, Garifullina GF, Abdel-Ghany SE, Zhang L, Pilon-Smits EA, Pilon M. The chloroplast NifS-like protein of Arabidopsis thaliana is required for iron-sulfur cluster formation in ferredoxin. Planta 2005;220:602–8.

[39] Van Hoewyk D, Abdel-Ghany SE, Cohu CM, et al. Chloroplast iron-sulfur cluster protein maturation requires the essential cysteine desulfurase CpNifS. Proc Natl Acad Sci USA 2007;104:5686–91.
[40] Van Hoewyk D, Garifullina GF, Ackley AR, et al. Overexpression of AtCpNifS enhances selenium tolerance and accumulation in Arabidopsis. Plant Physiol 2005;139:1518–28.
[41] M NM, Ollagnier-de-Choudens S, Sanakis Y, et al. Characterization of Arabidopsis thaliana SufE2 and SufE3: functions in chloroplast iron-sulfur cluster assembly and Nad synthesis. J Biol Chem 2007;282:18254–64.
[42] Ye H, Abdel-Ghany SE, Anderson TD, Pilon-Smits EA, Pilon M. CpSufE activates the cysteine desulfurase CpNifS for chloroplastic Fe-S cluster formation. J Biol Chem 2006;281:8958–69.
[43] Xu XM, Moller SG. AtSufE is an essential activator of plastidic and mitochondrial desulfurases in Arabidopsis. EMBO J 2006;25:900–9.
[44] Willems P, Wanschers BF, Esseling J, et al. BOLA1 is an aerobic protein that prevents mitochondrial morphology changes induced by glutathione depletion. Antioxid Redox Signal 2013;18:129–38.
[45] Li H, Mapolelo DT, Randeniya S, Johnson MK, Outten CE. Human glutaredoxin 3 forms [2Fe-2S]-bridged complexes with human BolA2. Biochemistry 2012;51:1687–96.
[46] Couturier J, Jacquot JP, Rouhier N. Evolution and diversity of glutaredoxins in photosynthetic organisms. Cell Mol Life Sci 2009;66:2539–57.
[47] Huynen MA, Spronk CA, Gabaldon T, Snel B. Combining data from genomes, Y2H and 3D structure indicates that BolA is a reductase interacting with a glutaredoxin. FEBS Lett 2005;579:591–6.
[48] Léon S, Touraine B, Ribot C, Briat JF, Lobreaux S. Iron-sulphur cluster assembly in plants: distinct NFU proteins in mitochondria and plastids from Arabidopsis thaliana. Biochem J 2003;371:823–30.
[49] Touraine B, Boutin JP, Marion-Poll A, Briat JF, Peltier G, Lobreaux S. Nfu2: a scaffold protein required for [4Fe-4S] and ferredoxin iron-sulphur cluster assembly in Arabidopsis chloroplasts. Plant J Cell Mol Biol 2004;40:101–11.
[50] Yabe T, Morimoto K, Kikuchi S, Nishio K, Terashima I, Nakai M. The Arabidopsis chloroplastic NifU-like protein CnfU, which can act as an iron-sulfur cluster scaffold protein, is required for biogenesis of ferredoxin and photosystem I. Plant Cell 2004;16:993–1007.
[51] Xu XM, Adams S, Chua NH, Moller SG. AtNAP1 represents an atypical SufB protein in Arabidopsis plastids. J Biol Chem 2005;280:6648–54.
[52] Xu XM, Moller SG. AtNAP7 is a plastidic SufC-like ATP-binding cassette/ATPase essential for Arabidopsis embryogenesis. Proc Natl Acad Sci USA 2004;101:9143–8.
[53] Abdel-Ghany SE, Ye H, Garifullina GF, Zhang L, Pilon-Smits EA, Pilon M. Iron-sulfur cluster biogenesis in chloroplasts. Involvement of the scaffold protein CpIscA. Plant Physiol 2005;138:161–72.
[54] Yabe T, Nakai M. Arabidopsis AtIscA-I is affected by deficiency of Fe-S cluster biosynthetic scaffold AtCnfU-V. Biochem Biophys Res Commun 2006;340:1047–52.
[55] Lezhneva L, Amann K, Meurer J. The universally conserved HCF101 protein is involved in assembly of [4Fe-4S]-cluster-containing complexes in Arabidopsis thaliana chloroplasts. Plant J Cell Mol Biol 2004;37:174–85.
[56] Schwenkert S, Netz DJ, Frazzon J, et al. Chloroplast HCF101 is a scaffold protein for [4Fe-4S] cluster assembly. Biochem J 2010;425:207–14.
[57] Bandyopadhyay S, Gama F, Molina-Navarro MM, et al. Chloroplast monothiol glutaredoxins as scaffold proteins for the assembly and delivery of [2Fe-2S] clusters. EMBO J 2008;27:1122–33.
[58] Cheng NH, Liu JZ, Brock A, Nelson RS, Hirschi KD. AtGRXcp, an Arabidopsis chloroplastic glutaredoxin, is critical for protection against protein oxidative damage. J Biol Chem 2006;281:26280–8.

[59] Liu X, Liu S, Feng Y, et al. Structural insights into the N-terminal GIY-YIG endonuclease activity of Arabidopsis glutaredoxin AtGRXS16 in chloroplasts. Proc Natl Acad Sci USA 2013;110:9565–70.
[60] Frazzon AP, Ramirez MV, Warek U, et al. Functional analysis of Arabidopsis genes involved in mitochondrial iron-sulfur cluster assembly. Plant Mol Biol 2007;64:225–40.
[61] Turowski VR, Busi MV, Gomez-Casati DF. Structural and functional studies of the mitochondrial cysteine desulfurase from Arabidopsis thaliana. Mol Plant 2012;5:1001–10.
[62] Heis MD, Ditmer EM, de Oliveira LA, Frazzon AP, Margis R, Frazzon J. Differential expression of cysteine desulfurases in soybean. BMC Plant Biol 2011;11:166.
[63] Pandey A, Golla R, Yoon H, Dancis A, Pain D. Persulfide formation on mitochondrial cysteine desulfurase: enzyme activation by a eukaryote-specific interacting protein and Fe-S cluster synthesis. Biochem J 2012;448:171–87.
[64] Busi MV, Zabaleta EJ, Araya A, Gomez-Casati DF. Functional and molecular characterization of the frataxin homolog from Arabidopsis thaliana. FEBS Lett 2004;576:141–4.
[65] Maliandi MV, Busi MV, Clemente M, Zabaleta EJ, Araya A, Gomez-Casati DF. Expression and one-step purification of recombinant Arabidopsis thaliana frataxin homolog (AtFH). Protein Expr Purif 2007;51:157–61.
[66] Maliandi MV, Busi MV, Turowski VR, Leaden L, Araya A, Gomez-Casati DF. The mitochondrial protein frataxin is essential for heme biosynthesis in plants. FEBS J 2011;278:470–81.
[67] Busi MV, Maliandi MV, Valdez H, et al. Deficiency of Arabidopsis thaliana frataxin alters activity of mitochondrial Fe-S proteins and induces oxidative stress. Plant J Cell Mol Biol 2006;48:873–82.
[68] Stemmler TL, Lesuisse E, Pain D, Dancis A. Frataxin and mitochondrial FeS cluster biogenesis. J Biol Chem 2010;285:26737–43.
[69] Bencze KZ, Kondapalli KC, Cook JD, et al. The structure and function of frataxin. Crit Rev Biochem Mol Biol 2006;41:269–91.
[70] Wilson RB. Frataxin and frataxin deficiency in Friedreich's ataxia. J Neurol Sci 2003;207:103–5.
[71] Patel PI, Isaya G. Friedreich ataxia: from GAA triplet-repeat expansion to frataxin deficiency. Am J Hum Genet 2001;69:15–24.
[72] Palau F. Friedreich's ataxia and frataxin: molecular genetics, evolution and pathogenesis (Review). Int J Mol Med 2001;7:581–9.
[73] Busi MV, Gomez-Casati DF. Exploring frataxin function. IUBMB Life 2012;64:56–63.
[74] Léon S, Touraine B, Briat JF, Lobreaux S. Mitochondrial localization of Arabidopsis thaliana Isu Fe-S scaffold proteins. FEBS Lett 2005;579:1930–4.
[75] Takubo K, Morikawa T, Nonaka Y, et al. Identification and molecular characterization of mitochondrial ferredoxins and ferredoxin reductase from Arabidopsis. Plant Mol Biol 2003;52:817–30.
[76] Picciocchi A, Douce R, Alban C. The plant biotin synthase reaction. Identification and characterization of essential mitochondrial accessory protein components. J Biol Chem 2003;278:24966–75.
[77] Xu XM, Lin H, Latijnhouwers M, Moller SG. Dual localized AtHscB involved in iron sulfur protein biogenesis in Arabidopsis. PloS One 2009;4:e7662.
[78] Waller JC, Alvarez S, Naponelli V, et al. A role for tetrahydrofolates in the metabolism of iron-sulfur clusters in all domains of life. Proc Natl Acad Sci USA 2010;107:10412–7.
[79] Muhlenhoff U, Richter N, Pines O, Pierik AJ, Lill R. Specialized function of yeast Isa1 and Isa2 proteins in the maturation of mitochondrial [4Fe-4S] proteins. J Biol Chem 2011;286:41205–16.
[80] Gelling C, Dawes IW, Richhardt N, Lill R, Muhlenhoff U. Mitochondrial Iba57p is required for Fe/S cluster formation on aconitase and activation of radical SAM enzymes. Mol Cell Biol 2008;28:1851–61.

[81] Bych K, Kerscher S, Netz DJ, et al. The iron-sulphur protein Ind1 is required for effective complex I assembly. EMBO J 2008;27:1736–46.
[82] Chen S, Sanchez-Fernandez R, Lyver ER, Dancis A, Rea PA. Functional characterization of AtATM1, AtATM2, and AtATM3, a subfamily of Arabidopsis half-molecule ATP-binding cassette transporters implicated in iron homeostasis. J Biol Chem 2007;282:21561–71.
[83] Camaschella C. Hereditary sideroblastic anemias: pathophysiology, diagnosis, and treatment. Semin Hematol 2009;46:371–7.
[84] Lill R, Kispal G. Mitochondrial ABC transporters. Res Microbiol 2001;152:331–40.
[85] Shimada Y, Okuno S, Kawai A, et al. Cloning and chromosomal mapping of a novel ABC transporter gene (hABC7), a candidate for X-linked sideroblastic anemia with spinocerebellar ataxia. J Hum Genet 1998;43:115–22.
[86] Bernard DG, Cheng Y, Zhao Y, Balk J. An allelic mutant series of ATM3 reveals its key role in the biogenesis of cytosolic iron-sulfur proteins in Arabidopsis. Plant Physiol 2009;151:590–602.
[87] Teschner J, Lachmann N, Schulze J, et al. A novel role for Arabidopsis mitochondrial ABC transporter ATM3 in molybdenum cofactor biosynthesis. Plant Cell 2010;22:468–80.
[88] Kushnir S, Babiychuk E, Storozhenko S, et al. A mutation of the mitochondrial ABC transporter Sta1 leads to dwarfism and chlorosis in the Arabidopsis mutant starik. Plant Cell 2001;13:89–100.
[89] Luo D, Bernard DG, Balk J, Hai H, Cui X. The DUF59 family gene AE7 acts in the cytosolic iron-sulfur cluster assembly pathway to maintain nuclear genome integrity in Arabidopsis. Plant Cell 2012;24:4135–48.
[90] Levitan A, Danon A, Lisowsky T. Unique features of plant mitochondrial sulfhydryl oxidase. J Biol Chem 2004;279:20002–8.
[91] Land T, Rouault TA. Targeting of a human iron-sulfur cluster assembly enzyme, nifs, to different subcellular compartments is regulated through alternative AUG utilization. Mol Cell 1998;2:807–15.
[92] Bych K, Netz DJ, Vigani G, et al. The essential cytosolic iron-sulfur protein Nbp35 acts without Cfd1 partner in the green lineage. J Biol Chem 2008;283:35797–804.
[93] Kohbushi H, Nakai Y, Kikuchi S, Yabe T, Hori H, Nakai M. Arabidopsis cytosolic Nbp35 homodimer can assemble both [2Fe-2S] and [4Fe-4S] clusters in two distinct domains. Biochem Biophys Res Commun 2009;378:810–5.
[94] Varadarajan J, Guilleminot J, Saint-Jore-Dupas C, et al. ATR3 encodes a diflavin reductase essential for Arabidopsis embryo development. New Phytol 2010;187:67–82.
[95] Cavazza C, Martin L, Mondy S, Gaillard J, Ratet P, Fontecilla-Camps JC. The possible role of an [FeFe]-hydrogenase-like protein in the plant responses to changing atmospheric oxygen levels. J Inorg Biochem 2008;102:1359–65.
[96] Godman JE, Molnar A, Baulcombe DC, Balk J. RNA silencing of hydrogenase(-like) genes and investigation of their physiological roles in the green alga Chlamydomonas reinhardtii. Biochem J 2010;431:345–51.
[97] Srinivasan V, Netz DJ, Webert H, et al. Structure of the yeast WD40 domain protein Cia1, a component acting late in iron-sulfur protein biogenesis. Structure 2007;15:1246–57.
[98] Papatriantafyllou M. DNA Metabolism: MMS19: CIA agent for DNA-linked affairs. Nat Rev Mol Cell Biol 2012;13:538.
[99] Stehling O, Vashisht AA, Mascarenhas J, et al. MMS19 assembles iron-sulfur proteins required for DNA metabolism and genomic integrity. Science 2012;337:195–9.
[100] Gari K, Leon Ortiz AM, Borel V, Flynn H, Skehel JM, Boulton SJ. MMS19 links cytoplasmic iron-sulfur cluster assembly to DNA metabolism. Science 2012;337:243–5.
[101] Rouault TA. The role of iron regulatory proteins in mammalian iron homeostasis and disease. Nat Chem Biol 2006;2:406–14.

[102] Ghosh MC, Zhang DL, Jeong SY, et al. Deletion of iron regulatory protein 1 causes polycythemia and pulmonary hypertension in mice through translational derepression of HIF2alpha. Cell Metab 2013;17:271–81.
[103] Kim DY, Bovet L, Kushnir S, Noh EW, Martinoia E, Lee Y. AtATM3 is involved in heavy metal resistance in Arabidopsis. Plant Physiol 2006;140:922–32.

# 23 Origin and evolution of Fe-S proteins and enzymes

Eric S. Boyd, Gerrit J. Schut, Eric M. Shepard, Joan B. Broderick, Michael W. W. Adams and John W. Peters

## 23.1 Introduction

The strong relationship between the structure and reactivity of various Fe-S minerals and their derivatives and biological Fe-S clusters is beyond coincidental and provides insights into key aspects of the transition between the abiotic and the biotic earth. The near-universal occurrence and the functional diversity of Fe-S clusters in biology provide a strong basis for their essential role in biology and almost assuredly indicates that their use was a property of the Last Universal Common Ancestor (LUCA) ~3.5 to 3.8 billion years ago and of more primitive life forms. The parallels between biological Fe-S clusters and Fe-S mineral structure and reactivity is one of the main arguments in support of an "Fe-S world" theory for the origin of life. There is a great deal of merit in arguments supporting a central role for Fe-S moieties in the origin and evolution of life given their ubiquitous role and central importance in electron transfer reactions and energy conservation in all modes of extant life. However, in addition to Fe-S clusters, energy transformation reactions require nucleotide moieties, suggesting that the partnership between the "Fe-S world" and the "RNA world" was likely a very early protobiological event. This chapter brings together topics of Fe-S cluster structure, function, and biosynthesis, which have been emphasized in preceding chapters of this book, in the context of plausible pathways for their integration into biology as an important component of the origin and evolution of life.

## 23.2 Fe-S chemistry and the origin of life

A number of chemical properties inherent to Fe-S minerals and their derivatives compel us to think that they had a central role in the origin of life. Wächterhäuser's "Fe-S world" suggests that the first protocells could have existed where Fe-S compounds were not only at the heart of catalysis but could potentially function in heredity and compartmentalization [1, 2]. Others share the view that compartmentalization and even heredity may have been inorganic in nature, much in the same manner as the pioneers of the original "RNA world" theory suggested that the first generation of life involved RNA molecules with the combined roles of heredity and catalysis and energy metabolism [3, 4]. Many put the constraints for defining life as having the elements of a gene-encoded energy metabolism. Thus, both the "Fe-S world" and "RNA world" theories as they were originally drafted essentially suggest that the first life forms functioned with a single class of molecules. For biochemists working with various metalloenzymes, the idea that first-generation metabolism could function without mineral- or metal-mediated chemistry seems impractical and unlikely. Although it is easy to envision that nucleotides had a role in metabolism from the onset of life, there is no rational reason to suggest that they functioned in the absence of minerals or mineral-derived metals and metal clusters. Wächtershäuser championed the idea that life arose in and around hydrothermal systems *via* a metabolism-first scenario founded on the reductive citric acid cycle [5], an idea further explored by Russell and Hall [6]. A key feature of this theory is the notion that reactions are catalyzed by pyrite, or other metal sulfides, such as Ni-Fe-S phases. Russell and colleagues [7, 8] have more recently proposed a metabolism-first theory in which ensembles of molecules are contained in iron-sulfide "membranes" (proto-membranes). A common notion in metabolism-first theories is that mineral catalysts were gradually replaced with protein-based biocatalysts. To support this notion, it is often noted that the active metal centers in many proteins, cofactors, and pigments are in fact remnants of their mineral-based precursors [9–13].

Whatever the merits of the "Fe-S world," "RNA world," and "metabolism-first" theories for the origin of life, the strong relationship between the structure and the reactivity of Fe-S minerals and the biological Fe-S clusters associated with extant life is believed by many to be too strong to be purely coincidental. The implications of this compelling relationship are that one can envision a path from minerals to enzymes and that perhaps the earliest life forms lived vicariously through aspects of the reactivity of metals and minerals in their local environment. In this scenario, the selective pressure for the adaptation of metals in biology would be to refine their reactivity for specific functions and to release life from dependency on mineral surfaces, which would inherently lower the barrier for diversification into new ecological niches. The shared attributes of the abiotic and biotic world are invaluable for understanding potential mechanisms for the emergence of initial life and life's functional diversification over the past >3.8 billion years of evolution. A more detailed understanding of the commonalities and parallels between the structure and reactivity of Fe-S minerals and the active sites of Fe-S proteins and enzymes could contribute significantly to developing more advanced and rational theories for the origin of life that can be probed experimentally.

Fe-S clusters are well represented in both the abiotic [9] and biotic worlds [14, 15]. For example, Fe-S minerals, in particular pyrite ($FeS_2$) and pyrrhotite ($Fe_{1-x}S$), are common on earth, whereas troilite (FeS) is common in iron meteorites. Moreover, various proteins and enzymes that have well defined Fe-S clusters are found in nearly all life forms. In addition to the essential role of Fe-S clusters in biological electron transfer, complex Fe-S clusters are known to be at the catalytic sites of many enzymes, including carbon monoxide dehydrogenase, hydrogenase, and nitrogenase [16]. These enzymes catalyze the activation of gases that were important on the early earth and in present-day biology, namely, hydrogen, nitrogen, carbon monoxide, and carbon dioxide, reactions that are often characterized by high activation barriers [17–20].

Given the similarity of enzymatic Fe-S clusters and iron-sulfide minerals and their derivatives, it has been suggested that biocatalysis was preceded by mineral-based catalysis on the prebiotic earth, with emerging life recruiting and optimizing mineral-based catalysis over time [8, 13, 21, 22].

In some respects it is really not just the parallels that can be drawn between Fe-S minerals and biological Fe-S clusters that allow us to rationalize potential pathways connecting the two, but it is perhaps even more important to understand the differences. As the simplest example of an Fe-S mineral, pyrite consists of Fe and S arranged in a lattice of alternating disulfides ($S_2$) intermingled with bound octahedral Fe in what is best described as the $2^+$ oxidation state. In contrast, the simplest biological Fe-S clusters contain Fe typically in the $2^+$ or $3^+$ oxidation state bound in a tetrahedral geometry to sulfide ($S^{2-}$) and thiolate sulfurs (Fig. 23.1). When thinking about these two types of Fe-S structures, one can envision a continuum in which mineral surfaces are transformed through time by fracture and weathering. Subsequent extraction of

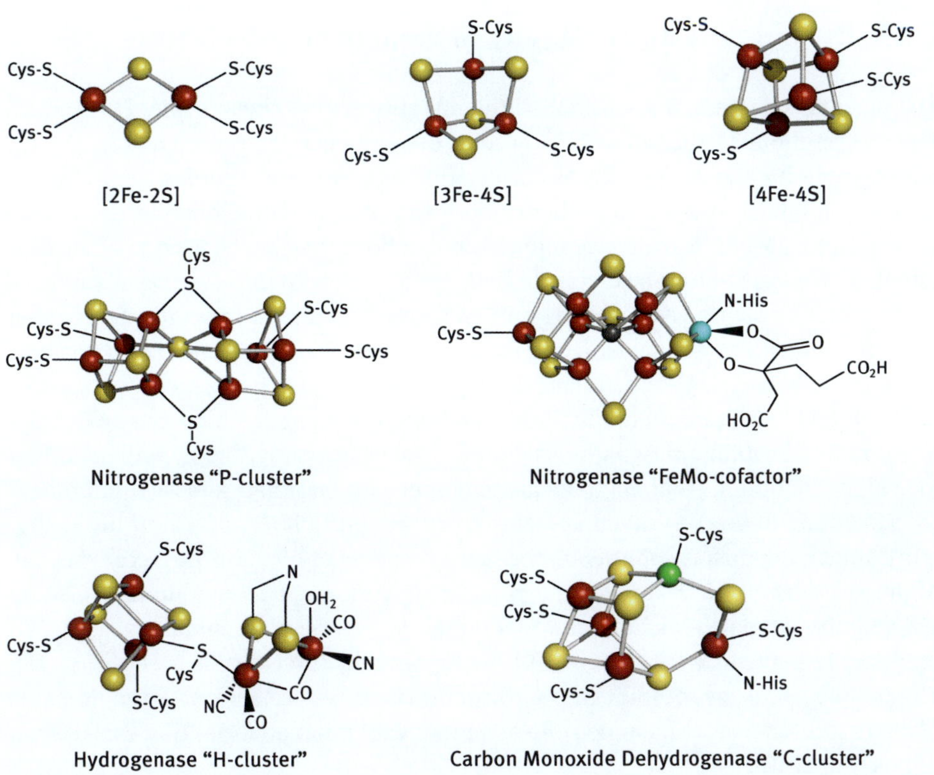


**Fig. 23.1:** Examples of simple (top) and more complex (middle and bottom) biological Fe-S clusters. Clusters are depicted in ball-and-stick fashion with ligands denoted by text. Color scheme: red, iron; yellow, sulfur; gray, carbon; cyan, molybdenum; green, nickel.

Fe ions and Fe-S fragments at these weathered surfaces by thiolate or carboxylate-based chelators would result in the liberation of protobiological Fe-S clusters. Fe-S clusters of this sort have been generated in mimetic inorganic synthesis and have been demonstrated to essentially self-assemble and remain stable under anaerobic conditions [23–25].

In biological settings, the reactions catalyzed by the complex Fe-S clusters associated with carbon monoxide dehydrogenase, acetyl-CoA synthase, hydrogenase, and nitrogenase, which all include the synthesis of small organic molecules and the redox dependent interconversion of small molecules, only act to provide additional enticement to invoke roles for Fe-S clusters in prebiotic chemistry and protobiology (Fig. 23.1). There is really no evidence of a step wise type progression that relates Fe-S mineral and biological moieties; however, the ability to envision such a hypothetical trajectory highlights the necessity for further examination of the properties and genesis of Fe-S clusters in the context of understanding the transition between abiotic and biotic earth and the origin of life.

## 23.3 The ubiquity and antiquity of biological Fe-S clusters

The previous chapters firmly establish the ubiquity of Fe-S clusters and their importance as essential components of living cells. In addition to the central role in general redox reactions and trafficking electrons, simple and more complex Fe-S clusters have key catalytic roles in a number of both catabolic and anabolic metabolic processes [17, 18, 26–28]. The aforementioned connections that can be used to relate Fe-S minerals in nature and Fe-S clusters in biology coupled with the selective advantages of reactions involving Fe-S proteins and enzymes lead one to generally accept that these functionalities are ancient, primordial and even likely predate coded proteins and the LUCA [29–31]. Although there are clear limitations of phylogenetic work, especially for addressing origin of life problems, there is a great deal that can be learned about rational evolutionary paths for proteins as well as assigning hypothetical biological functionalities that might be ancient or even primordial. A loose requirement for something to be considered ancient or predating the LUCA of extant life is ubiquity among organisms comprising the three domains of life. This property was one of the key motivating factors for the pioneering work of Carl Woese in establishing 16S/18S ribosomal RNA as a standard for defining taxonomic classifications and evolutionary relationships between organisms forming the three domains of life [32, 33]. The protein synthesis machinery, the ribosome, is conserved across all the domains of life in form and function and is one of a number of functionalities that was likely a property of LUCA.

In addition to the ribosome, other functionalities exist that are universal among organisms throughout all domains of life. Several Fe-S protein and enzyme families can be placed into this category and probably the most obvious examples are the ferredoxins [34, 35] and the radical *S*-adenosylmethionine (SAM) enzymes [36, 37]. Both are large families that are involved in a variety of cellular functions, occur in all domains of life, and are universally present in deeply rooted organisms within each domain. As touched on throughout this book, ferredoxins are key components in electron transfer in a broad range of metabolic processes. They are widely and integrally distributed among both simple and complex metabolic pathways where they can exist as independent electron transfer functionalities or as domains associated with more complex redox enzymes. Examples of the latter include the numerous ferredoxin binding domains present in the N- and C-terminal domains of [FeFe]-hydrogenase. Moreover, ferredoxins are involved in a number of widely distributed components of energy metabolism including respiration and photosynthesis. Ferredoxins and ferredoxin-like domains of enzymes, especially those related to the 2-[4Fe-4S] ferredoxins, which are the most abundant, exist with a significant level of primary sequence and structural conservation and can be readily identified by the relative location of Cys residues in the primary sequence, or what is commonly referred to as Cys motifs. Motifs for binding [4Fe-4S] clusters that have been clearly conserved through all manner and form of extant life, are strongly suggestive of their antiquity and imply an origin that predates the Last Universal Common Ancestor (LUCA) of extant life.

The radical SAM enzyme family is defined by a conserved motif of Cys residues that are involved in binding a [4Fe-4S] cluster [36]. The Fe-S cluster in radical SAM enzymes is bound differently than in ferredoxins where four cysteine thiolates coordinate the Fe ions of the cluster. In radical SAM proteins, the motif is defined by just three cysteines that coordinate only three of the Fe ions of the [4Fe-4S] cluster, leaving one unique Fe site free to bind the co-substrate SAM (Fig. 23.2) [38, 39]. The reductive cleavage of SAM that occurs at this "site-differentiated" Fe ion is common to all radical SAM enzymes and the 5′-deoxyadenosyl radical product is associated with a large variety of radical reactions involving hydrogen atom abstraction. Although the reactions in which enzymes of the radical SAM family are involved are diverse, there are a couple of observations concerning the nature of the reactions and their evolutionary characteristics that would indicate a potentially primordial origin. Similar to ferredoxins, radical SAM enzymes are present and widely distributed through all domains of life and their functionalities are associated with processes often ascribed to primitive metabolisms (e.g. hydrogenotrophic methanogenesis), and as such are present in numerous deeply rooted lineages. It has been suggested that the radical SAM enzyme anaerobic ribonucleotide reductase, responsible for the conversion of ribonucleotides to deoxyribonucleotides, is more ancient than other forms of ribonucleotide reductases [40]. If true, this implies that DNA synthesis is dependent on Fe-S cluster chemistry or biochemistry, consistent with an origin for Fe-S cluster chemistry/biochemistry prior to the divergence from LUCA. Moreover, this observation may suggest that SAM-mediated Fe-S cluster chemistry/biochemistry predates encoded proteins, at least DNA-encoded proteins. The ferredoxins and radical SAM

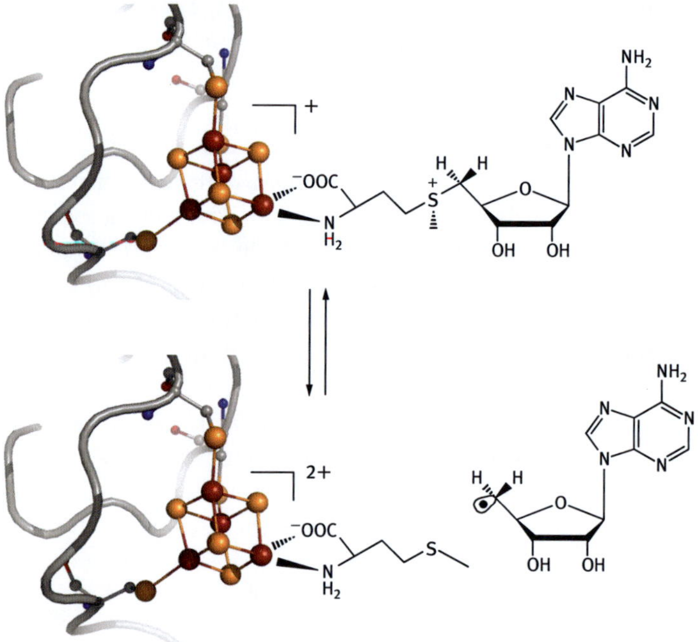


**Fig. 23.2:** Coordination of SAM to a site-differentiated [4Fe-4S] cluster followed by its reductive cleavage to L-methionine and the 5′-deoxyadenosyl radical species.

enzymes are just two specific examples of ancient Fe-S cluster proteins, but there are a number of other connections that can be drawn to suggest that Fe-S chemistry is as much of a hallmark of life as nucleotides and nucleic acids.

Increased application of phylogenetic and bioinformatic tools clearly point to the ubiquity of Fe-S clusters in early life and suggest a primitive origin for their use in biology that predates the proposed appearance of LUCA ~3.8 billion years ago. Although it is unclear what type of environment may have precipitated the emergence of life and which may have sustained early life, hydrothermal vents harbor a number of features that lead many to suggest these as potential originators of life [7, 41]. Most notable among these are the strong redox gradients present in these systems that are created by the merging of very reduced and hot (~350°C) hydrothermal fluids with lower temperature sea water, creating a suitable energetic driving force for the generation of mineral-based catalysts. As an example, the presence of elevated concentrations of iron and sulfide in these environments react spontaneously to form Fe-S clusters (nanoparticles) that can be further reduced by sulfide to generate Fe-$S_2$ and molecular $H_2$ [42]. Fe-S and Fe-$S_2$ produced through these anoxic abiotic reaction pathways under the sulfidic and ferruginous conditions present at hydrothermal vents may have catalyzed primitive reactions that were adapted and refined by biology as protein catalysts, as outlined in detail in previous sections. However, the critical questions become how life met its Fe-S cluster requirements for enzyme-based catalysis as it diversified away from sulfidic and ferruginous environments and how the mechanisms that life employs to generate simple Fe-S clusters may have been impacted by the advent of oxygenic photosynthesis and the gradual oxidation of early earth environments.

Life's requirements for Fe-S clusters has led to the emergence of redundant mechanisms to control their synthesis, namely the iron-sulfur cluster (ISC), sulfur formation (SUF), the nitrogenase system (NIF), and the APB systems [43–45]. It has been suggested previously that the various systems can be categorized as those used for general housekeeping (ISC), those used under oxidative stress (SUF), and those used for the synthesis of Fe-S clusters associated with specific processes, e.g. NIF, nitrogenase. Although the evolutionary trajectory of these systems is complex and remains unclear to date, the presence of multiple pathways in some organisms speaks to the central role of Fe-S clusters in biology. Unlike ISC, which exhibits a wide distribution among bacteria and eukaryotes, the SUF system of Fe-S cluster assembly is present in anaerobic methanogens and several archaea, as well as bacteria, which may point to its primitive nature and origin prior to the origin of oxygenic photosynthesis. The near-universal occurrence of SUF in oxygenic cyanobacteria and other aerobes suggests that SUF proteins evolved mechanisms to deal with $O_2$, and these adaptations likely underpin the utility of this system in repairing Fe-S clusters during the oxidative stress response [43]. Further bioinformatic work is needed to tease apart the phylogenetic distribution and complex evolutionary history of the major Fe-S cluster assembly systems and to provide insight into the role of major geological events (e.g. rise of oxygen, nitrogen crisis) in their evolutionary history.

Among the components of the major Fe-S cluster assembly systems are several proteins that have shared evolutionary history and that are common among the different assembly systems [43]. Similarities between the homologues present challenges in delineating the taxonomic distribution of the major Fe-S cluster assembly systems, but at the same time provide opportunities to place the pathways in evolutionary time, relative to each other. Homologous copies of cysteine desulfurase (encoded by *nifS, iscS, sufS/E*) [46], which functions to abstract sulfide from cysteine for use as a sulfur source in Fe-S cluster synthesis [47], are often encoded by each of the three major systems. As an interesting side note, methanogens do not encode for a homologue of cysteine desulfurase and instead presumably derive sulfur for Fe-S cluster biosynthesis from sulfide, a phenotype that has been attributed to the distribution of methanogens primarily within ecological niches that are anoxic and rich in sulfide [48]. However, the availability of sulfide in the environment is unlikely to explain the absence of a recognizable cysteine desulfurase, as other physiological groups that do not inhabit sulfidic environments also lack homologues of cysteine desulfurases. Like cysteine desulfurase, U-type scaffold (IscU, NifU, and SufU) and A-type scaffold (NifA, IscA, and SufA) proteins are encoded by each system where they function as platforms for Fe-S cluster synthesis [43]. These scaffolds harbor numerous cysteine-rich regions that coordinate cluster synthesis; the Fe-S clusters formed at these sites are often only bound transiently and are ultimately transferred to apoenzymes.

A critical challenge for all extant life is protecting redox-sensitive Fe-S clusters from oxidative damage due to molecular oxygen or oxygen radicals. This raises the central question as to what the response of Fe-S cluster assembly processes was to the origin of oxygenic photosynthesis and the gradual rise of oxygen. The rise of oxygen would have created at least two problems for biology. First, early oceans were reduced and replete with ferrous iron, which may have acted as the selective pressure leading to the emergence of and proliferation of Fe-dependent enzymes and proteins [49]. However, the gradual production of oxygen would eventually overcome the "buffering capacity" of the reduced ocean, due in part to the abundance of reduced iron [50]. Simultaneously, the once-abundant soluble $Fe^{2+}$ pools in oceans gradually diminished as iron was oxidized and became less soluble, placing strong selection on the development of mechanisms for acquiring and sequestering iron, which may have been achieved through the use of chelators and perhaps scaffold proteins or ferritins, respectively. Second, once the sinks for oxygen had been overcome and oxygen began to accumulate in natural systems, biology would need to develop mechanisms to shield intracellular Fe-S clusters from coming into direct contact with oxygen, or retreat to anoxic niches where oxygen would not be freely available to destabilize Fe-S clusters. Accordingly, the Fe-S cluster assembly process is likely a sentinel system that accurately reflects selection in the face of oxygen-induced oxidative stress, a feature that should be clearly depicted in the evolutionary history of constituent proteins. Alternatively, it is possible that the redundant Fe-S cluster assembly processes evolved in response to the requirement for a stress-resistant cluster assembly pathway or quick production of Fe-S clusters in actively $O_2$ respiring cells, which would be evident in a taxonomic profiling of the distribution of the various Fe-S assembly systems. Further phylogenetic and taxonomic analysis of the distribution of SUF and ISC on taxonomic trees, especially during the emergence of cyanobacteria, would go a long way toward defining the role of oxygen in the evolution of biological Fe-S cluster assembly.

## 23.4 Early energy conversion

Electron transfer-linked ion pumping across an ion-impenetrable barrier likely represents the most ancient form of energy conservation. We propose that Fe-S minerals provided the raw materials that enabled Fe-S clusters to both function in the capacity to act as electron wires and to catalyze redox-linked reactions. Moreover, Fe-S clusters have potentially a third function in being able to couple electron transfer and proton translocation, thereby affecting the first form of biological energy conservation. A good example is the class of energy-conserving, membrane-bound [NiFe]-hydrogenase (MBH) found in hyperthermophilic organisms. These grow at extreme temperatures and are thought by some to be the most ancient of extant microbes. These MBH enzymes catalyze the coupling of electron transfer from ferredoxin through a highly conserved Fe-S cluster to proton reduction (hydrogen gas production), and this is also coupled to thermodynamically unfavorable ion pumping [51]. Phylogenetic analyses of these enzymes demonstrated that this is a blueprint for ancestral energy-conserving mechanisms and that this mechanism of electron transfer through these Fe-S clusters is highly conserved and used throughout almost all present life forms [52]. In some cases the oxidation of simple one-carbon compounds, such as carbon monoxide, and formate, is coupled to $H_2$ production and energy conservation based on the MBH mechanism. This Fe-S-dependent MBH system also provided the framework for the evolution of modern-day oxygen-dependent respiratory systems that use NADH as the source of electrons and the Fe-S electron wire and ion-translocating mechanisms are without doubt evolutionarily related to those in the MBH-type systems. It is therefore clear that these types of Fe-S clusters are extremely well conserved throughout respiratory systems and that ancestral life performing some kind of respiration most likely contained homologues of these electron wire modules.

Recent new insights and the discovery of electron bifurcation mechanisms have reinforced the central role of Fe-S clusters in diverse and arguably ancient metabolic processes. Electron bifurcation refers to the mechanism by which endergonic and exergonic reactions are coupled in energy conservation transformations. The key component of bifurcating mechanisms is the essential and central role of a two-electron donor/acceptor group, which, for all characterized bifurcating mechanisms, is a flavin that allows the switching/splitting of electrons. Energy conservation by electron bifurcation was first proposed in 2008 [53], in which an exergonic reaction drives an endergonic reaction without the involvement of an ion gradient. In this case, the cytoplasmic enzyme complex butyryl-CoA dehydrogenase catalyzes the reduction of crotonyl-CoA ($E_0{}' = -10$ mV) to butyryl-CoA with NADH ($E_0{}' = -320$ mV) as the electron donor, and this is coupled to the endergonic reduction of ferredoxin ($E_m = -410$ mV). As a consequence of this pair of linked redox reactions, the oxidation of NADH can be coupled to $H_2$ production and consequently associates electron bifurcation and respiration processes. The flavin-containing systems characterized to date are butyryl-CoA dehydrogenase, bifurcating [FeFe]-hydrogenase, hydrogenase-heterodisulfide reductase, *trans*-hydrogenase (NfnAB), bifurcating NADP [FeFe]-hydrogenase/formate dehydrogenase complex, and caffeyl-CoA reductase [54]. By mixing or splitting high/low-potential electrons through the flavin, unfavorable redox reactions can be driven by simultaneous exergonic redox reactions. Electron bifurcation is also assumed to be an ancient energy-conserving mechanism. Indeed, flavins (including the methanogenic cofactor $F_{420}$, described in the following paragraph) can be considered remnants of an RNA world. In addition, flavins are often loosely associated with proteins and it can be argued that other, more simple redox active centers such as Fe-S clusters played their present-day role in an ancient bifurcation system. The importance of Fe-S clusters in bifurcation systems is striking in the case of the [FeFe]-hydrogenase. That some of these enzymes have a bifurcation mechanism was first demonstrated with the [FeFe]-hydrogenase of the hyperthermophilic bacterium *Thermotoga maritima* (Fig. 23.3) [55]. This organism contains a standard Embden-Meyerhof glycolytic pathway that produces both NADH and reduced ferredoxin as electron carriers. The exergonic electron transfer from ferredoxin to the [FeFe]-hydrogenase active site is used to pull the endergonic transfer of electrons from NADH, ultimately producing $H_2$ from both reduced ferredoxin and NADH simultaneously. Interestingly, some of the Fe-S clusters that act as electron wires in the [FeFe]-hydrogenase are highly conserved in the present-day versions of Complex I, again emphasizing the evolutionary importance and the catalytic power of Fe-S clusters [52].

The current examples of bifurcation systems are highly evolved, indicating a long evolutionary history, although they are mostly found within strict anaerobic organisms. In methanogens, hydrogenase-heterodisulfide reductase ($H_2$ to heterodisulfide and ferredoxin) does not involve nicotinamide nucleotides, although the central methanogenic pathway uses the flavin-like cofactor $F_{420}$, supporting the notion that methanogens are present-day representatives of an ancient organism. Meanwhile, some butyrate-producing clostridial species use a bifurcating butyryl-CoA dehydrogenase (NADH to crotonyl-CoA and ferredoxin) that does not rely on Fe-S clusters (other than the ferredoxin) but rather contains only FAD as a cofactor [56]. Although an "all Fe-S" bifurcating system lacking flavin has yet to be identified, the ability of Fe-S clusters to carry out two-electron transfer reactions, e.g. $[4Fe\text{-}4S]^{1+,2+,3+}$, and to carry out proton-coupled (hydride) transfer reactions, e.g. in [FeFe]-hydrogenase, argues that

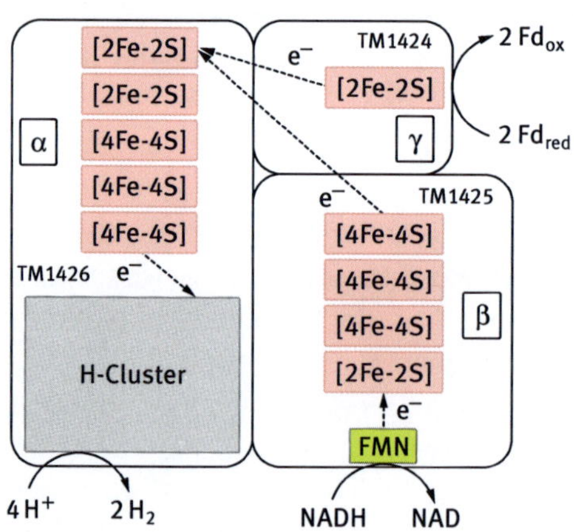


**Fig. 23.3:** The proposed cofactor content and pathway of electron flow in the bifurcating [FeFe]-hydrogenase of *T. maritima*. Reduced ferredoxin and NADH serve as electron sources, with the reducing equivalents shuttled among various Fe-S cluster centers until ultimately being delivered to the catalytic H-cluster site where $H_2$ is produced. The detailed structure of the H-cluster is shown in Fig. 23.1.

these moieties could have substituted for the role currently played by flavin in early bifurcating systems.

Electron bifurcation, together with substrate-level phosphorylation, and electron transport or chemiosmotic phosphorylation make up the three known mechanisms of biological energy conservation. Formally, substrate-level phosphorylation does not involve redox reactions *per se*, but the phosphorylating enzymes are typically part of oxidative pathways. In the case of the oxidation of glucose to acetate, the two oxidation steps [at the level of glyceraldehyde-3-phosphate (GAP) and pyruvate] were probably linked to ferredoxin reduction in the earliest forms of these pathways, such as those now present in several hyperthermophilic anaerobic microorganisms, which contain GAP and pyruvate oxidoreductases (GAPOR and POR). Note that GAPOR generates 3-phosphoglycerate, which cannot be used for substrate-level phosphorylation, rather than 1,3-bisphosphoglycerate, which can. Hence, energy is conserved in the GAPOR reaction in the form of a low-potential reductant (reduced ferredoxin), which uses the reducing equivalents to generate $H_2$ that is coupled to ion transport *via* the energy-conserving [NiFe]-MBH described in previous paragraphs. Most anaerobes have replaced GAPOR with GAP dehydrogenase (GAPDH) in which NAD is reduced (rather than ferredoxin) and energy is conserved in the form of 1,3-bisphosphoglycerate, which can be used for ATP synthesis. As discussed above, bifurcating [FeFe]-hydrogenase enables reductant from both reduced ferredoxin and NADH to be diposed of as $H_2$, as with the anaerobe *T. maritima*. Aerobes have similarly replaced POR with NAD-linked pyruvate dehydrogenase, but in this case, energy is conserved by aerobic respiration. In any event, electron transfer reactions involving Fe-S clusters dictate the mechanism in which energy can be conserved in pathways, for example, *via* GAPOR or GAPDH. Accordingly, a unifying theme among all three mechanisms of energy conservation in biological systems is oxidation-reduction chemistry mediated in part by Fe-S clusters that in many cases are ferredoxins.

A hallmark in much of this chemistry is the aforementioned pairing of Fe-S proteins and nucleotides, which reinforces the thought that the intimate relationship between Fe-S protein motifs and nucleotides as coenzymes and cofactors in metabolic pathways was forged very early. In the context of the origin of life, the Fe-S moieties in prebiotic energy-transducing reactions are attractive given their observed versatility in accessing a large range of oxidation-reduction potentials. This allows them to participate in a wide range of oxidation-reduction reactions, and as such, it is hard to imagine a nucleotide-based primordial energy metabolism that did not involve partnering with an Fe-S functionality. The aforementioned versatility of the simple Fe-S cluster cubanes in biology and their ubiquitous involvement in oxidative and reductive processes indeed place them at a seminal crossroads in the origin and evolution of early life on Earth.

## 23.5 Evolution of complex Fe-S cluster containing proteins

Our most recent work on Fe-S cluster biosynthesis has revealed that the parallels between the prebiotic Fe-S mineral reactivity and Fe-S enzymes go beyond common chemical reactivity. We have demonstrated that the stepwise modification of Fe-S clusters in biology occur by reactions of the type that can also be clearly rationalized in terms of the conditions on an early abiotic Earth [57]. These observations include radical-based reactions that transform Fe-S clusters by ligand modification and which tune their reactivity [13, 58–61]. Through the simple stepwise introduction of specific nonprotein ligands and other (non-iron) metals, simple Fe-S clusters are transformed into highly reactive metal sites that perform Hadean (representing the earliest eon in the Earth's history) chemistry. Such reactions are present today as the key fundamental biochemical transformations that support hydrogen, nitrogen, and carbon metabolism in microbial life. The results strongly suggest that the pathway from minerals to enzymes and the origin of simple metabolism is preserved in today's biochemistry.

Our work and the work of others suggest that there are two paradigms for generating the metal-containing enzyme active sites (Fig. 23.4) [57]. The first involves ancestral proteins with defined cavities serving as organic nests for the binding of reactive mineral clusters. In this model, these mineral clusters are refined stepwise through evolutionary time to improve reactivity through the synthesis of a well-defined cofactor with a combination of small molecule and chelating ligands. This paradigm appears to apply for nitrogenase and [FeFe]-hydrogenase. A second paradigm appears to exist for [NiFe]-hydrogenase and carbon monoxide dehydrogenase [62, 63]. These enzymes do not acquire an intact preformed metal cofactor, but rather each metal ion or a group of metal ions is modified on the pre-catalytic protein in a stepwise fashion. From our preliminary work, we have found that the apparently subtle differences in these paradigms have profound implications on the two types of biosynthetic pathways involved, the maturation proteins that these pathways contain, and consequently on how the two pathways involved in the synthesis of the

**Fig. 23.4:** (right) The path from mineral catalysis to biocatalysis represented to begin with Fe-S-based minerals (shown here as pyrite) and radiating out *via* hypothetical modified cluster intermediates to complex Fe-S enzymes. The top half depicts [FeFe]-hydrogenases (left) and [NiFe]-hydrogenases (right), which both catalyze reversible $H_2$ oxidation reactions. The bottom half shows nitrogenase catalyzing $N_2$ reduction (left) and carbon monoxide dehydrogenase catalyzing reversible CO oxidation (right). The representation is defined by the two evolutionary paradigms: paradigm 1, [FeFe]-hydrogenase and nitrogenase; paradigm 2: [NiFe]-hydrogenase and carbon monoxide dehydrogenase). Color scheme: red, iron; orange, sulfur; gray, carbon; cyan, molybdenum; green, nickel. (left) The maturation pathways of nitrogenase and [FeFe]-hydrogenase enzymes that represent paradigm 1. Importantly, the panel highlights the central role that radical SAM chemistry plays in the biosynthetic process. The term "sub" in the [FeFe]-hydrogenase pathway reflects synthesis of the [4Fe-4S] and 2Fe subclusters of the H-cluster, respectively.

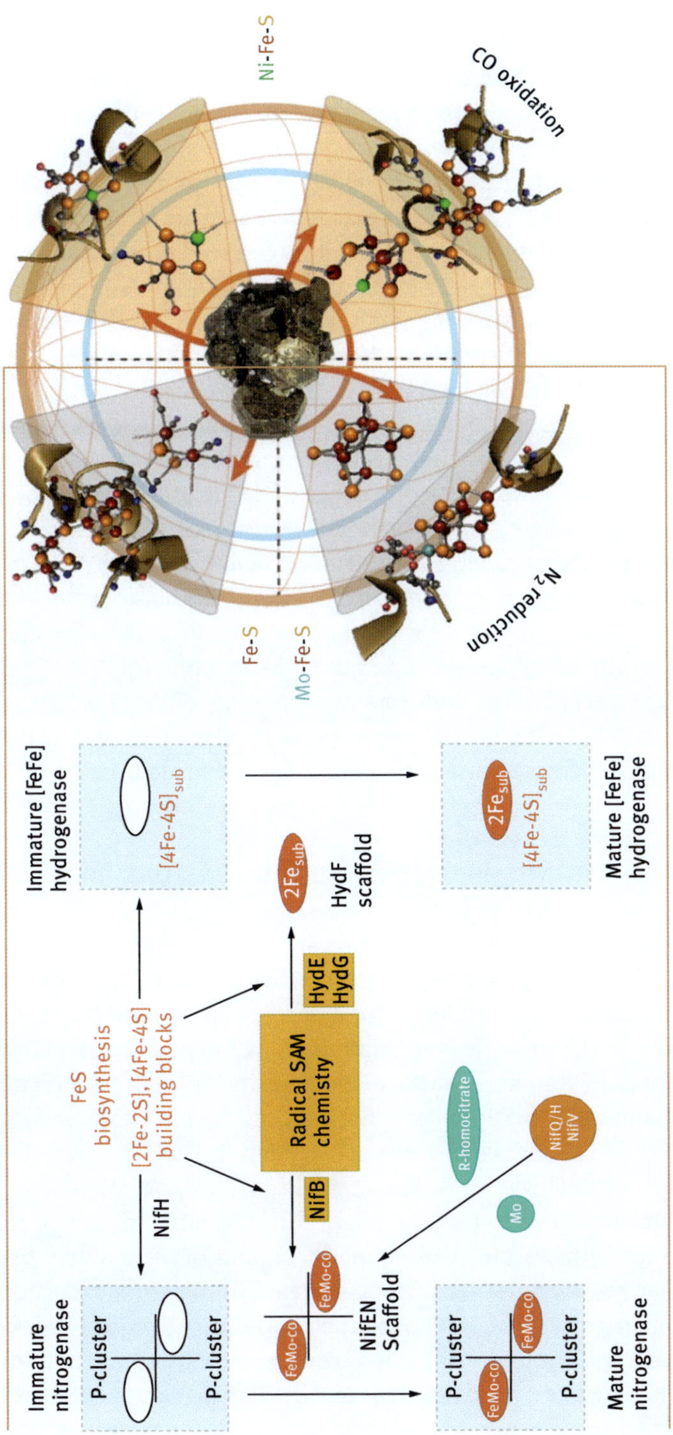
Reversible H$_2$ oxidation
Ni-Fe-S
CO oxidation
N$_2$ reduction
Fe-S
Mo-Fe-S
Paradigm1
FeS
biosynthesis
[2Fe-2S], [4Fe-4S]
building blocks
NifH
Immature
nitrogenase
P-cluster
P-cluster
Immature [FeFe]
hydrogenase
[4Fe-4S]$_{sub}$
Radical SAM
chemistry
NifB
HydE
HydG
2Fe$_{sub}$
HydF
scaffold
R-homocitrate
Mo
NifQ/H
NifV
FeMo-co
FeMo-co
NifEN
Scaffold
P-cluster
FeMo-co
FeMo-co
P-cluster
Mature
nitrogenase
2Fe$_{sub}$
[4Fe-4S]$_{sub}$
Mature [FeFe]
hydrogenase

Fe-S-based catalytic sites likely evolved from abiotic mineral precursors. The fact that the biosynthesis of active site clusters in these two types of enzyme follow different paradigms is somewhat surprising especially for the [NiFe]- and [FeFe]-hydrogenases, which catalyze the same chemical reaction despite being evolutionarily unrelated (i.e. convergent evolution) [62]. Our studies suggest that the nitrogenase and [FeFe]-hydrogenase structural proteins observed in extant biology were later innovations in biological evolution, likely emerging to refine chemistry already catalyzed by unique radical-based enzymes [57, 64–66]. The radical enzymes involved in nitrogenase and [FeFe]-hydrogenase cofactor biosynthesis catalyze the key Fe-S modifications that are responsible for the specialized reactivity of these enzymes and are the key determinants of their ancestry [59]. Such radical-based enzymes are an interesting example of the marriage between elements of the Fe-S world and the RNA world where the nucleotide SAM promotes unique chemical reactivity that can be firmly placed as the type of reactivity that could have promoted mineral modifications in the prebiotic Earth to facilitate prebiotic chemistry [13, 40]. The radical SAM centric Fe-S cofactor reactivity underpins the first paradigm for cluster assembly. In contrast to the first paradigm, preliminary phylogenetic analysis of [NiFe]-hydrogenase, as an example of the second paradigm, appears to indicate that the structural proteins are more ancient than the biosynthetic components, and perhaps are primordial. In contrast to the synthesis and insertion of intact cofactors, the assembly of these proteins involves modification of metal ions that are already inserted into the protein and that will eventually become catalytic once those metal ions are suitably modified.

## 23.6 The path from minerals to Fe-S proteins and enzymes

Although the relationship between Fe-S minerals and protein-based Fe-S clusters is too strong to be coincidental, visualizing the series of steps that might link the two is not as clear-cut. We can perhaps envision a form of early life or protolife that might have lived vicariously off the capacity for oxidation-reduction reactions and the catalytic potential provided by a mineral surface. Life constrained to a mineral surface would be under constant selective pressure to generate mechanisms to effect a release from the dependency on the surface and as such expand its ecological distribution. From this perspective, initial levels of Fe homeostasis, cluster biosynthesis, and first generation functional metal clusters were likely satisfied simultaneously *via* the weathering of mineral surfaces and subsequent chelation of metals and clusters. Organic nesting of metal ions not only released reactive metal clusters from the surface but also concurrently brought about functional diversification. There likely would be ongoing selective pressure that would continue to drive the formation and refinement of such organically nested iron ions and the fact that many cofactors have a combination of protein and nonprotein organic ligand architectures may illustrate this sort of evolutionary progression.

## References


[1] Wachtershauser G. Before enzymes and templates: theory of surface metabolism. Microbiol Rev 1988;52:452–84.
[2] Wachtershauser G. Groundworks for an evolutionary biochemistry: the iron-sulphur world. Prog Biophys Mol Biol 1992;58:85–201.
[3] Gilbert W. Origin of life: the RNA world. Nature 1986;319:618.
[4] Cech TR. The RNA worlds in context. Cold Spring Harbor Perspect Biol 2012;4:a006742.
[5] Wachtershauser G. Evolution of the first metabolic cycles. Proc Natl Acad Sci USA 1990;87:200–4.
[6] Russell MJ, Hall AJ. The emergence of life from iron monosulphide bubbles at a submarine hydrothermal redox and pH front. J Geol Soc 1997;154:377–402.
[7] Russell MJ, Daniel RM, Hall AJ, Sherringham JA. A hydrothermally precipitated catalytic iron sulphide membrane as a first step toward life. J Mol Evol 1994;39:231–43.
[8] Martin W, Russell MJ. On the origins of cells: a hypothesis for the evolutionary transitions from abiotic geochemistry to chemoautotrophic prokaryotes, and from prokaryotes to nucleated cells. Philos Trans R Soc London B Biol Sci 2003;358:59–83;discussion 83–55.
[9] Russell MJ, Martin W. The rocky roots of the acetyl-CoA pathway. Trends Biochem Sci 2004;29: 358–63.
[10] Nitschke W, McGlynn SE, Milner-White EJ, Russell MJ. On the antiquity of metalloenzymes and their substrates in bioenergetics. Biochim Biophys Acta 2013;1827:871–81.
[11] Hall DO, Cammack R, Rao KK. The iron-sulphur proteins: evolution of a ubiquitous protein from model systems to higher organisms. Orig Life 1974;5:363–86.
[12] Wachtershauser G. Origin of life in an iron-sulfur world. New York: Cambridge University Press; 1998.
[13] McGlynn SE, Mulder DM, Shepard EM, Broderick JB, Peters JW. Hydrogenase cluster biosynthesis: organometallic chemistry nature's way. Dalton Trans 2009:4274–85.
[14] Johnson MK, Smith AD. Iron-sulfur Proteins. In: King RB, ed. Encyclopedia of inorganic chemistry. 2nd ed. John Wiley & Sons; 2005;2589–619.
[15] Johnson DC, Dean DR, Smith AD, Johnson MK. Structure, function, and formation of biological iron-sulfur clusters. Annu Rev Biochem 2005;74:247–81.
[16] Drennan CL, Peters JW. Surprising cofactors in metalloenzymes. Curr Opin Struct Biol 2003;13:220–6.
[17] Bender G, Pierce E, Hill JA, Darty JE, Ragsdale SW. Metal centers in the anaerobic microbial metabolism of CO and $CO_2$. Metallomics 2011;3:797–815.
[18] Peters JW. Structure and mechanism of iron-only hydrogenases. Curr Opin Struct Biol 1999;9:670–6.
[19] Seefeldt LC, Hoffman BM, Dean DR. Electron transfer in nitrogenase catalysis. Curr Opin Chem Biol 2012;16:19–25.
[20] Igarashi RY, Seefeldt LC. Nitrogen fixation: the mechanism of the Mo-dependent nitrogenase. Crit Rev Biochem Mol Biol 2003;38:351–84.
[21] Russell MJ, Hall AJ, Cairns-Smith AG, Braterman PS. Submarine hot springs and the origin of life. Nature 1988;336:117.
[22] Wachtershauser G. Pyrite formation, the first energy source for life: a hypothesis. Microbiol Rev 1988;10:207–10.
[23] Daley CJ, Holm RH. Reactions of site-differentiated [Fe4S4]2+, 1+ clusters with sulfonium cations: reactivity analogues of biotin synthase and other members of the S-adenosyl-methionine enzyme family. J Inorg Biochem 2003;97:287–98.
[24] Daley CJ, Holm RH. Reactivity of [Fe4S4(SR)4]2-,3-clusters with sulfonium cations: analogue reaction systems for the initial step in biotin synthase catalysis. Inorg Chem 2001;40:2785–93.

[25] Hagen KS, Watson AD, Holm RH. Synthetic routes to Fe2S2, Fe3S4, Fe4S4, and Fe6S9 clusters from the common precursor [Fe(SC2H5)]4]2-: structures and properties. [Fe3S4(SR)4]3- and [Fe6S9(SC2H5)]2]4-, examples of the newest types of Fe-S-SR clusters. JACS 1983;105:3905–13.

[26] Shepard EM, Broderick JB. S-Adenosylmethionine and iron-sulfur clusters in biological radical reactions: the radical SAM superfamily. In: Comprehensive natural products chemistry. Elsevier; 2009.

[27] Horner DS, Heil B, Happe T, Embley TM. Iron hydrogenases – ancient enzymes in modern eukaryotes. Trends Biochem Sci 2002;27:148–53.

[28] Seefeldt LC, Yang ZY, Duval S, Dean DR. Nitrogenase reduction of carbon-containing compounds. Biochim Biophys Acta 2013;1827:1102–11.

[29] Doolittle WF, Brown JR. Tempo, mode, the progenote, and the universal root Proc Natl Acad Sci USA 1994;91:6721–8.

[30] Penny D, Poole A. The nature of the last universal common ancestor. Curr Opin Genet Dev 1999;9:672–7.

[31] Koonin EV. Comparative genomics, minimal gene-sets and the last universal common ancestor Nat Rev Microbiol 2003;1:127–36.

[32] Woese CR, Fox GE. Phylogenetic structure of the prokaryotic domain: the primary kingdoms. Proc Natl Acad Sci USA 1977;74:5088–90.

[33] Woese CR, Magrum LJ, Fox GE. Archaebacteria. J Mol Evol 1978;11:245–51.

[34] Fitch WM, Bruschi M. The evolution of prokaryotic ferredoxins – with a general method correcting for unobserved substitutions in less branched lineages. Mol Biol Evol 1987;4: 381–94.

[35] Meyer J. The evolution of ferredoxins. Trends Ecol Evol 1988;3:222–6.

[36] Sofia HJ, Chen G, Hetzler BG, Reyes-Spindola JF, Miller NE. Radical SAM, a novel protein superfamily linking unresolved steps in familiar biosynthetic pathways with radical mechanisms: functional characterization using new analysis and information visualization methods. Nucleic Acids Res 2001;29:1097–106.

[37] Frey PA, Hegeman AD, Ruzicka FJ. The radical SAM superfamily. Crit Rev Biochem Mol Biol 2008;43:63–88.

[38] Krebs C, Broderick WE, Henshaw TF, Broderick JB, Huynh BH. Coordination of adenosyl-methionine to a unique iron site of the [4Fe-4S] of pyruvate formate-lyase activating enzyme: a Mossbauer spectroscopic study. J Am Chem Soc 2002;124:912–3.

[39] Vey JL, Drennan CL. Structural insights into radical generation by the radical SAM superfamily. Chem Rev 2011;111:2487–506.

[40] Stubbe J. Ribonucleotide reductases: the link between an RNA and a DNA world? Curr Opin Struct Biol 2000;10:731–6.

[41] Russell MJ. The alkaline solution to the emergence of life: energy, entropy and early evolution Acta Biotheor 2007;55:133–79.

[42] Drobner E, Huber H, Wachtershauser G, Rose D, Stetter KO. Pyrite formation linked with hydrogen evolution under anaerobic conditions. Nature 1990;346:742–4.

[43] Ayala-Castro C, Saini A, Outten FW. Fe-S cluster assembly pathways in bacteria. Microbiol Mol Biol Rev 2008;72:110–25.

[44] Bandyopadhyay S, Chandramouli K, Johnson MK. Iron-sulfur cluster biosynthesis. Biochem Soc Trans 2008;36:1112–9.

[45] Fontecave M, Ollagnier-de-Choudens S. Iron-sulfur cluster biosynthesis in bacteria: mechanisms of cluster assembly and transfer. Arch Biochem Biophys 2008;474:226–37.

[46] Zheng L, White RH, Cash VL, Jack RF, Dean DR. Cysteine desulfurase activity indicates a role for NIFS in metallocluster biosynthesis. Proc Natl Acad Sci USA 1993;90:2754–8.

[47] Yuvaniyama P, Agar JN, Cash VL, Johnson MK, Dean DR. NifS-directed assembly of a transient [2Fe-2S] cluster within the NifU protein. Proc Natl Acad Sci USA 2000;97:599–604.

[48] Liu Y, Sieprawska-Lupa M, Whitman WB, White RH. Cysteine is not the sulfur source for iron-sulfur cluster and methionine biosynthesis in the methanogenic archaeon Methanococcus maripaludis. J Biol Chem 2010;285:31923–9.
[49] David LA, Alm EJ. Rapid evolutionary innovation during an Archaean genetic expansion. Nature 2011;469:93–6.
[50] Anbar AD. Oceans. Elements and evolution. Science 2008;322:1481–3.
[51] Sapra R, Bagramyan K, Adams MW. A simple energy-conserving system: proton reduction coupled to proton translocation. Proc Natl Acad Sci USA 2003;100:7545–50.
[52] Schut GJ, Boyd ES, Peters JW, Adams MW. The modular respiratory complexes involved in hydrogen and sulfur metabolism by heterotrophic hyperthermophilic archaea and their evolutionary implications. FEMS Microbiol Rev 2013;37:182–203.
[53] Herrmann G, Jayamani E, Mai G, Buckel W. Energy conservation via electron-transferring flavoprotein in anaerobic bacteria. J Bacteriol 2008;190:784–91.
[54] Buckel W, Thauer RK. Energy conservation via electron bifurcating ferredoxin reduction and proton/Na(+) translocating ferredoxin oxidation. Biochim Biophys Acta 2013;1827:94–113.
[55] Schut GJ, Adams MW. The iron-hydrogenase of Thermotoga maritima utilizes ferredoxin and NADH synergistically: a new perspective on anaerobic hydrogen production. J Bacteriol 2009;191:4451–7.
[56] Li F, Hinderberger J, Seedorf H, Zhang J, Buckel W, Thauer RK. Coupled ferredoxin and crotonyl coenzyme A (CoA) reduction with NADH catalyzed by the butyryl-CoA dehydrogenase/Etf complex from Clostridium kluyveri. J Bacteriol 2008;190:843–50.
[57] Peters JW, Broderick JB. Emerging paradigms for complex iron-sulfur cofactor assembly and insertion. Annu Rev Biochem 2012;81:429–50.
[58] Driesener RC, Challand MR, McGlynn SE, Shepard EM, Boyd ES, Broderick JB, Peters JW, Roach PL. [FeFe]-hydrogenase cyanide ligands derived from S-adenosylmethionine-dependent cleavage of tyrosine. Angew Chemie 2010;49:1687–90.
[59] Shepard EM, Boyd ES, Broderick JB, Peters JW. Biosynthesis of complex iron-sulfur enzymes. Curr Opin Chem Biol 2011;15:319–27.
[60] Shepard EM, Duffus BR, George SJ, McGlynn SE, Challand MR, Swanson KD, Roach PL, Cramer SP, Peters JW, Broderick JB. [FeFe]-hydrogenase maturation: HydG-catalyzed synthesis of carbon monoxide. J Am Chem Soc 2010;132:9247–9.
[61] Wiig JA, Hu Y, Lee CC, Ribbe MW. Radical SAM-dependent carbon insertion into the nitrogenase M-cluster. Science 2012;337:1672–5.
[62] Swanson KD, Duffus BR, Beard TE, Peters JW, Broderick JB. Cyanide and Carbon monoxide ligand formation in hydrogenase biosynthesis. Eur J Inorg Chem 2011:935–47.
[63] Kung Y, Drennan CL. A role for nickel-iron cofactors in biological carbon monoxide and carbon dioxide utilization. Curr Opin Chem Biol 2011;15:276–83.
[64] Boyd ES, Anbar AD, Miller S, Hamilton TL, Lavin M, Peters JW. A late methanogen origin for molybdenum-dependent nitrogenase. Geobiology 2011;9:221–32.
[65] Boyd ES, Peters JW. New insights into the evolutionary history of biological nitrogen fixation. Front Microbiol 2013;4:201.
[66] Mulder DW, Boyd ES, Sarma R, et al. Stepwise [FeFe]-hydrogenase H-cluster assembly revealed in the structure of HydA(DeltaEFG). Nature 2010;465:248–51.

# Index